Linear Algebra

Linear Algebra

Algorithms, Applications, and Techniques

Third Edition

Richard Bronson
Professor Emeritus of Mathematics
Fairleigh Dickinson University

Gabriel B. Costa
Professor of Mathematical Sciences
United States Military Academy
Associate Professor of Mathematics and Computer Science
Seton Hall University

John T. Saccoman
Professor of Mathematics and Computer Science
Seton Hall University

ELSEVIER

AMSTERDAM • BOSTON • HEIDELBERG • LONDON
NEW YORK • OXFORD • PARIS • SAN DIEGO
SAN FRANCISCO • SINGAPORE • SYDNEY • TOKYO
Academic Press is an imprint of Elsevier

Academic Press in an imprint of Elsevier

225, Wyman Street, Waltham, MA 02451, USA
The Boulevard, Langford Lane, Kidlington, Oxford OX5 1GB, UK
Radarweg 29, PO Box 211, 1000 AE Amsterdam, The Netherlands

First Edition 1995
Second Edition 2007

Library of Congress Cataloging-in Publication Data
Application submitted

British Library Cataloguing in Publication Data
A catalogue record for this book is available from the British Library

ISBN: 978-0-12-391420-0

For information on all Academic Press Publications
visit our web site at store.elsevier.com

Printed and bound in USA
13 14 15 16 17 10 9 8 7 6 5 4 3 2 1

Dedication

To Evy – R.B.

To my teaching colleagues at West Point and Seton Hall, especially to the Godfather, Dr. John J. Saccoman and to the Generals, Drs. Frank Giordano, Chris Arney, Gary Krahn and Mike Phillips – G.B.C.

To Ryan and Mary Erin – J.T.S.

Contents

Preface

As technology advances, so does our need to understand and characterize it. This is one of the traditional roles of mathematics, and now that we are more than a decade into the 21st century, no area of mathematics has been more versatile successful in this endeavor than that of linear algebra. The elements of linear algebra are the essential underpinnings of a wide range of modern applications, from mathematical modeling in economics to optimization procedures in airline scheduling and inventory control. Linear algebra furnishes today's analysts in business, engineering, and the social sciences with the tools they need to describe and define the theories that drive their disciplines. It also provides mathematicians with compact constructs for presenting central ideas in probability, differential equations, and operations research.

The third edition of this book presents the fundamental structures of linear algebra and develops the foundation for using those structures. Many of the concepts in linear algebra are abstract; indeed, linear algebra introduces students to formal deductive analysis. Formulating proofs and logical reasoning are skills that require nurturing, and it has been our aim to provide this.

We have streamlined our approach, in this third edition, while striving to have the material presented in a more logical and orderly manner. Regarding mathematical rigor, in the early sections, the proofs are relatively simple, not more than a few lines in length, and deal with concrete structures, such as matrices. Complexity builds as the book progresses.

We have also introduced some graph theoretical concepts in this edition of *Linear Algebra*. Matrices associated with graphs have been studied extensively, and we attempt to introduce the reader to some of these matrices and their applications.

A number of learning aids are included to assist readers. New concepts are carefully introduced and tied to the reader's experience. In the beginning, the basic concepts of matrix algebra are made concrete by relating them to a store's inventory. Linear transformations are tied to more familiar functions, and vector spaces are introduced in the context of column matrices. Illustrations give geometrical insight on the number of solutions to simultaneous linear equations, vector arithmetic, determinants, and projections to list just a few.

As in the previous edition, we have highlighted material to emphasize important ideas throughout the text. Computational methods—for calculating the inverse of a matrix, performing a Gram-Schmidt orthonormalization process, or the like—are presented as a sequence of operational steps. Theorems are clearly marked, and there is a summary of important terms and concepts at the end of each chapter. Each section ends with numerous exercises of progressive difficulty, allowing readers to gain proficiency in the techniques presented and expand their understanding of the underlying theory.

For about two-thirds of the text, the only prerequisite for understanding this material is a facility with high-school algebra. These topics can be covered in any course of 10 weeks or more in duration. Depending on the background of the readers, selected applications and numerical methods may also be considered in a quarter system.

We would like to thank the many people who helped shape the focus and content of this book; in particular, the administrative and educational leaders at Fairleigh Dickinson University, Seton Hall University and West Point.

I, Gabriel Costa, would particularly like to thank my Archbishop, the Most Reverend John J. Myers, J.C.D., D.D., for his continued support and blessing throughout the years. I would also like to acknowledge Dr. Bethany Kubik for her professional assistance. I, John T. Saccoman, would like to acknowledge the influence of the late Professor Frank Boesch of Stevens Institute of Technology, and the assistance of Dr. Sarah Bleiler and Dr. John J. Saccoman.

Lastly, our heartfelt gratitude is given to the bevy of professionals at Elsevier, whom we have been privileged to work. They have provided us with valuable suggestions and technical expertise throughout this endeavor.

About the Authors

Richard Bronson has written eleven books in mathematics, many translated into multiple languages. He has published children's poetry in magazines, including *Highlights for Children*. He was on the editorial staff of *Simulation Magazine* and *SIAM News* and the children's magazine, *Kids Club*. He is the author of the political thriller *Antispin*.

Gabriel B. Costa is a Catholic Priest and a professor of Mathematical Sciences at the United States Military Academy at West Point, where he also functions as an associate chaplain. He is on an extended Academic Leave from Seton Hall University. He has co-authored several books in mathematics and three books in the area of sabermetrics.

John T. Saccoman is Professor and Chair of the Department of Mathematics and Computer Science at Seton Hall University. He has published more than a dozen papers in the area of network reliability synthesis. He has also co-authored three books in sabermetrics.

All three authors received their doctorates in mathematics from Stevens Institute of Technology.

CHAPTER 1

Matrices

Chapter Outline

1.1 BASIC CONCEPTS

We live in a complex world of finite resources, competing demands, and information streams that must be analyzed before resources can be allocated fairly to the demands for those resources. Any mechanism that makes the processing of information more manageable is a mechanism to be valued.

Consider an inventory of T-shirts for one department of a large store. The T-shirt comes in three different sizes and five colors, and each evening, the department's supervisor prepares an inventory report for management. A paragraph from such a report dealing with the T-shirts is reproduced in Figure 1.1.

This report is not easy to analyze. In particular, one must read the entire paragraph to determine the number of sand-colored, small T-shirts in current stock. In contrast, the rectangular array of data presented in Figure 1.2 summarizes the same information better. Using Figure 1.2, we see at a glance that no small, sand-colored T-shirts are in stock.

A matrix is a rectangular array of elements arranged in horizontal rows and vertical columns.

A *matrix* is a rectangular array of elements arranged in horizontal rows and vertical columns. The array in Figure 1.1 is a matrix, as are

> T-shirts
>
> Nine teal small and five teal medium; eight plum small and six plum medium; large sizes are nearly depleted with only three sand, one rose, and two peach still available; we also have three medium rose, five medium sand, one peach medium, and seven peach small.

FIGURE 1.1

S =

	Rose	Teal	Plum	Sand	Peach
small	0	9	8	0	7
medium	3	5	6	5	1
large	1	0	0	3	2

FIGURE 1.2

$$\mathbf{L} = \begin{bmatrix} 1 & 3 \\ 5 & 2 \\ 0 & -1 \end{bmatrix}, \tag{1.1}$$

$$\mathbf{M} = \begin{bmatrix} 4 & 1 & 1 \\ 3 & 2 & 1 \\ 0 & 4 & 2 \end{bmatrix}, \tag{1.2}$$

and

$$\mathbf{N} = \begin{bmatrix} 19.5 \\ -\pi \\ \sqrt{2} \end{bmatrix}. \tag{1.3}$$

The rows and columns of a matrix may be labeled, as in Figure 1.1, or not labeled, as in matrices (1.1) through (1.3).

The matrix in Equation (1.1) has three rows and two columns; it is said to have *order* (or size) 3×2 (read three by two). By convention, the row index is always given before the column index. The matrix in Equation (1.2) has order 3×3, whereas that in Equation (1.3) has order 3×1. The order of the stock matrix in Figure 1.2 is 3×5.

The entries of a matrix are called *elements*. We use uppercase boldface letters to denote matrices and lowercase letters for elements. The letter identifier for an element is generally the same letter as its host matrix. Two subscripts are attached to element labels to identify their location in a matrix; the first subscript specifies the row position and the second subscript the column position. Thus, l_{12} denotes the element in the first row and second column of a matrix **L**; for the matrix **L** in

Equation (1.2), $l_{12}=3$. Similarly, m_{32} denotes the element in the third row and second column of a matrix **M**; for the matrix **M** in Equation (1.3), $m_{32}=4$. In general, a matrix **A** of order $p \times n$ has the form

$$\mathbf{A} = \begin{bmatrix} a_{11} & a_{12} & a_{13} & \dots & a_{1n} \\ a_{21} & a_{22} & a_{23} & \dots & a_{2n} \\ a_{31} & a_{32} & a_{33} & \dots & a_{3n} \\ \vdots & \vdots & \vdots & \ddots & \vdots \\ a_{p1} & a_{p2} & a_{p3} & \dots & a_{pn} \end{bmatrix} \tag{1.4}$$

which is often abbreviated to $[a_{ij}]_{p\times n}$ or just $[a_{ij}]$, where a_{ij} denotes an element in the ith row and jth column.

Any element having its row index equal to its column index is a *diagonal element*. Diagonal elements of a matrix are the elements in the 1-1 position, 2-2 position, 3-3 position, and so on, for as many elements of this type that exist in a particular matrix. Matrix (1.1) has 1 and 2 as its diagonal elements, whereas matrix (1.2) has 4, 2, and 2 as its diagonal elements. Matrix (1.3) has only 19.5 as a diagonal element.

A matrix is *square* if it has the same number of rows as columns. In general, a square matrix has the form

$$\begin{bmatrix} a_{11} & a_{12} & a_{13} & \dots & a_{1n} \\ a_{21} & a_{22} & a_{23} & \dots & a_{2n} \\ a_{31} & a_{32} & a_{33} & \dots & a_{3n} \\ \vdots & \vdots & \vdots & \ddots & \vdots \\ a_{n1} & a_{n2} & a_{n3} & \dots & a_{nn} \end{bmatrix}$$

with the elements $a_{11}, a_{22}, a_{33}, \dots, a_{nn}$ forming the *main* (or principal) *diagonal*.

The elements of a matrix need not be numbers; they can be functions or, as we shall see later, matrices themselves. Hence

$$\begin{bmatrix} \int_0^1 (t^2+1)dt & t^3 & \sqrt{3t} & 2 \end{bmatrix},$$

$$\begin{bmatrix} \sin\theta & \cos\theta \\ -\cos\theta & \sin\theta \end{bmatrix},$$

and

$$\begin{bmatrix} x^2 & x \\ e^x & \frac{d}{dx}\ln x \\ 5 & x+2 \end{bmatrix}$$

are all good examples of matrices.

A *row matrix* is a matrix having a single row; a *column matrix* is a matrix having a single column. The elements of such a matrix are commonly called its *components*, and the number of components its *dimension*. We use lowercase boldface letters to distinguish row matrices and column matrices from more general matrices. Thus,

$$\mathbf{x} = \begin{bmatrix} 1 \\ 2 \\ 3 \end{bmatrix}$$

is a 3-dimensional column vector, whereas

$$\mathbf{u} = \begin{bmatrix} t & 2t & -t & 0 \end{bmatrix}$$

An n-tuple is a row matrix or a column matrix having n-components.

is a 4-dimensional row vector. The term *n-tuple* refers to either a row matrix or a column matrix having dimension n. In particular, $\mathbf{x}$ is a 3-tuple because it has three components while $\mathbf{u}$ is a 4-tuple because it has four components.

Two matrices are equal if they have the same order and if their corresponding elements are equal.

Two matrices $\mathbf{A}=[a_{ij}]$ and $\mathbf{B}=[b_{ij}]$ are *equal* if they have the same order and if their corresponding elements are equal; that is, both $\mathbf{A}$ and $\mathbf{B}$ have order $p \times n$ and $a_{ij}=b_{ij}$ $(i=1, 2, 3, \ldots, p;\ j=1, 2, \ldots, n)$. Thus, the equality

$$\begin{bmatrix} 5x+2y \\ x-y \end{bmatrix} = \begin{bmatrix} 7 \\ 1 \end{bmatrix}$$

implies that $5x+2y=7$ and $x-3y=1$.

Figure 1.2 lists a stock matrix for T-shirts as

$$\mathbf{S} = \begin{bmatrix} \text{Rose} & \text{Teal} & \text{Plum} & \text{Sand} & \text{Peach} \\ 0 & 9 & 8 & 0 & 7 \\ 3 & 5 & 6 & 5 & 1 \\ 1 & 0 & 0 & 3 & 2 \end{bmatrix} \begin{matrix} \\ \text{small} \\ \text{medium} \\ \text{large} \end{matrix}$$

If the overnight arrival of new T-shirts is given by the delivery matrix

$$\mathbf{D} = \begin{bmatrix} \text{Rose} & \text{Teal} & \text{Plum} & \text{Sand} & \text{Peach} \\ 9 & 0 & 0 & 9 & 0 \\ 3 & 3 & 3 & 3 & 3 \\ 6 & 8 & 8 & 6 & 6 \end{bmatrix} \begin{matrix} \\ \text{small} \\ \text{medium} \\ \text{large} \end{matrix}$$

then the new inventory matrix is

$$\mathbf{S}+\mathbf{D} = \begin{bmatrix} \text{Rose} & \text{Teal} & \text{Plum} & \text{Sand} & \text{Peach} \\ 9 & 9 & 8 & 9 & 7 \\ 6 & 8 & 9 & 8 & 4 \\ 7 & 8 & 8 & 9 & 8 \end{bmatrix} \begin{matrix} \\ \text{small} \\ \text{medium} \\ \text{large} \end{matrix}$$

The sum of two matrices of the same order is the matrix obtained by adding together corresponding elements of the original two matrices.

The *sum of two matrices of the same order* is a matrix obtained by adding together corresponding elements of the original two matrices; that is, if both $\mathbf{A}=[a_{ij}]$ and $\mathbf{B}=[b_{ij}]$ have order $p \times n$, then $\mathbf{A}+\mathbf{B}=[a_{ij}+b_{ij}](i=1,2,3,\ldots,p; j=1,2,\ldots,n)$. Addition is not defined for matrices of different orders.

Example 1

$$\begin{bmatrix} 5 & 1 \\ 7 & 3 \\ -2 & -1 \end{bmatrix} + \begin{bmatrix} -6 & 3 \\ 2 & -1 \\ 4 & 1 \end{bmatrix} = \begin{bmatrix} 5+(-6) & 1+3 \\ 7+2 & 3+(-1) \\ -2+4 & -1+1 \end{bmatrix} = \begin{bmatrix} -1 & 4 \\ 9 & 2 \\ 2 & 0 \end{bmatrix},$$

and

$$\begin{bmatrix} t^2 & 5 \\ 3t & 0 \end{bmatrix} + \begin{bmatrix} 1 & -6 \\ t & -t \end{bmatrix} = \begin{bmatrix} t^2+1 & -1 \\ 4t & -t \end{bmatrix}.$$

The matrices

$$\begin{bmatrix} 5 & 0 \\ -1 & 2 \\ 2 & 1 \end{bmatrix} \quad \text{and} \quad \begin{bmatrix} -6 & 2 \\ 1 & 1 \end{bmatrix}$$

cannot be added because they are not of the same order.

▶THEOREM 1

*If matrices **A**, **B**, and **C** all have the same order, then*

(a) *the commutative law of addition holds; that is,*

$$\mathbf{A}+\mathbf{B}=\mathbf{B}+\mathbf{A},$$

(b) *the associative law of addition holds; that is,*

$$\mathbf{A}+(\mathbf{B}+\mathbf{C})=(\mathbf{A}+\mathbf{B})+\mathbf{C}.◀$$

Proof: We leave the proof of part *(a)* as an exercise (see Problem 38). To prove part *(b)*, we set $\mathbf{A}=[a_{ij}]$, $\mathbf{B}=[b_{ij}]$, and $\mathbf{C}=[c_{ij}]$. Then

$$\begin{aligned}
\mathbf{A}+(\mathbf{B}+\mathbf{C}) &= [a_{ij}] + ([b_{ij}] + [c_{ij}]) \\
&= [a_{ij}] + [b_{ij} + c_{ij}] && \text{definition of matrix addition} \\
&= [a_{ij} + (b_{ij} + c_{ij})] && \text{definition of matrix addition} \\
&= [(a_{ij} + b_{ij}) + c_{ij}] && \text{associative property of regular addition} \\
&= [(a_{ij} + b_{ij})] + [c_{ij}] && \text{definition of matrix addition} \\
&= ([a_{ij}] + [b_{ij}]) + [c_{ij}] && \text{definition of matrix addition} \\
&= (\mathbf{A}+\mathbf{B})+\mathbf{C}
\end{aligned}$$

The difference **A** − **B** of two matrices of the same order is the matrix obtained by subtracting from the elements of **A** the corresponding elements of **B**.

We define the zero matrix **0** to be a matrix consisting of only zero elements. When a zero matrix has the same order as another matrix **A**, we have the additional property

$$\mathbf{A}+\mathbf{0}=\mathbf{A} \tag{1.5}$$

Subtraction of matrices is defined analogously to addition; the orders of the matrices must be identical and the operation is performed elementwise on all entries in corresponding locations.

Example 2

$$\begin{bmatrix} 5 & 1 \\ 7 & 3 \\ -2 & -1 \end{bmatrix} - \begin{bmatrix} -6 & 3 \\ 2 & -1 \\ 4 & 1 \end{bmatrix} = \begin{bmatrix} 5-(-6) & 1-3 \\ 7-2 & 3-(-1) \\ -2-4 & -1-1 \end{bmatrix} = \begin{bmatrix} 11 & -2 \\ 5 & 4 \\ -6 & -2 \end{bmatrix}$$

Example 3 The inventory of T-shirts at the beginning of a business day is given by the stock matrix

$$\mathbf{S} = \begin{bmatrix} \text{Rose} & \text{Teal} & \text{Plum} & \text{Sand} & \text{Peach} \\ 9 & 9 & 8 & 9 & 7 \\ 6 & 8 & 9 & 8 & 4 \\ 7 & 8 & 8 & 9 & 8 \end{bmatrix} \begin{matrix} \\ \text{small} \\ \text{medium} \\ \text{large} \end{matrix}$$

What will the stock matrix be at the end of the day if sales for the day are five small rose, three medium rose, two large rose, five large teal, five large plum, four medium plum, and one each of large sand and large peach?

Solution: Purchases for the day can be tabulated as

$$\mathbf{P} = \begin{bmatrix} \text{Rose} & \text{Teal} & \text{Plum} & \text{Sand} & \text{Peach} \\ 5 & 0 & 0 & 0 & 0 \\ 3 & 0 & 4 & 0 & 0 \\ 2 & 5 & 5 & 1 & 1 \end{bmatrix} \begin{matrix} \\ \text{small} \\ \text{medium} \\ \text{large} \end{matrix}$$

The stock matrix at the end of the day is

$$\mathbf{S} - \mathbf{P} = \begin{bmatrix} \text{Rose} & \text{Teal} & \text{Plum} & \text{Sand} & \text{Peach} \\ 4 & 9 & 8 & 9 & 7 \\ 3 & 8 & 5 & 8 & 4 \\ 5 & 3 & 3 & 8 & 7 \end{bmatrix} \begin{matrix} \\ \text{small} \\ \text{medium} \\ \text{large} \end{matrix}$$

A matrix **A** can always be added to itself, forming the sum **A** + **A**. If **A** tabulates inventory, **A** + **A** represents a doubling of that inventory, and we would like to write

$$\mathbf{A} + \mathbf{A} = 2\mathbf{A} \tag{1.6}$$

The product of a scalar λ by a matrix **A** is the matrix obtained by multiplying every element of **A** by λ.

The right side of Equation (1.6) is a number times a matrix, a product known as *scalar multiplication*. If the equality in Equation (1.6) is to be true, we must define 2**A** as the matrix having each of its elements equal to twice the corresponding elements in **A**. This leads naturally to the following definition: If $\mathbf{A} = [a_{ij}]$ is a $p \times n$ matrix, and if λ is a real number, then

$$\lambda\mathbf{A} = [\lambda a_{ij}] (i = 1, 2, \ldots, p; \quad j = 1, 2, \ldots, n) \tag{1.7}$$

Equation (1.7) can also be extended to complex numbers λ, so we use the term *scalar* to stand for an arbitrary real number or an arbitrary complex number when we need to work in the complex plane. Because Equation (1.7) is true for all real numbers, it is also true when λ denotes a real-valued function.

Example 4

$$7\begin{bmatrix} 5 & 1 \\ 7 & 3 \\ -2 & -1 \end{bmatrix} = \begin{bmatrix} 35 & 7 \\ 49 & 21 \\ -14 & -7 \end{bmatrix} \quad \text{and} \quad t\begin{bmatrix} 1 & 0 \\ 3 & 2 \end{bmatrix} = \begin{bmatrix} t & 0 \\ 3t & 2t \end{bmatrix}$$

Example 5 Find $5\mathbf{A} - \frac{1}{2}\mathbf{B}$ if

$$\mathbf{A} = \begin{bmatrix} 4 & 1 \\ 0 & 3 \end{bmatrix} \quad \text{and} \quad \mathbf{B} = \begin{bmatrix} 6 & -20 \\ 18 & 8 \end{bmatrix}$$

Solution:

$$5\mathbf{A} - \frac{1}{2}\mathbf{B} = 5\begin{bmatrix} 4 & 1 \\ 0 & 3 \end{bmatrix} - \frac{1}{2}\begin{bmatrix} 6 & -20 \\ 18 & 8 \end{bmatrix}$$

$$= \begin{bmatrix} 20 & 5 \\ 0 & 15 \end{bmatrix} - \begin{bmatrix} 3 & -10 \\ 9 & 4 \end{bmatrix} = \begin{bmatrix} 17 & 15 \\ -9 & 11 \end{bmatrix}$$

▶THEOREM 2

If **A** *and* **B** *are matrices of the same order and if λ_1 and λ_2 denote scalars, then the following distributive laws hold:*

(a) $\lambda_1(\mathbf{A}+\mathbf{B}) = \lambda_1\mathbf{A} + \lambda_2\mathbf{B}$
(b) $(\lambda_1+\lambda_2)\mathbf{A} = \lambda_1\mathbf{A} + \lambda_2\mathbf{A}$
(c) $(\lambda_1\lambda_2)\mathbf{A} = \lambda_1(\lambda_2\mathbf{A})$ ◀

Proof: We leave the proofs of (*b*) and (*c*) as exercises (see Problems 40 and 41). To prove (*a*), we set $\mathbf{A} = [a_{ij}]$ and $\mathbf{B} = [b_{ij}]$. Then

$$\begin{aligned}
\lambda_1(\mathbf{A}+\mathbf{B}) &= \lambda_1\left(\left[a_{ij}\right] + \left[b_{ij}\right]\right) \\
&= \lambda_1\left[\left(a_{ij} + b_{ij}\right)\right] && \text{definition of matrix addition} \\
&= \left[\lambda_1\left(a_{ij} + b_{ij}\right)\right] && \text{definition of scalar multiplication} \\
&= \left[\left(\lambda_1 a_{ij} + \lambda_1 b_{ij}\right)\right] && \text{distributive property of scalars} \\
&= \left[\lambda_1 a_{ij}\right] + \left[\lambda_1 b_{ij}\right] && \text{definition of matrix addition} \\
&= \lambda_1\left[a_{ij}\right] + \lambda_1\left[b_{ij}\right] && \text{definition of scalar multiplication} \\
&= \lambda_1\,\mathbf{A} + \lambda_1\mathbf{B}
\end{aligned}$$

Problems 1.1

(1) Determine the orders of the following matrices:

$$A = \begin{bmatrix} 1 & 2 \\ 3 & 4 \end{bmatrix}, \quad B = \begin{bmatrix} 5 & 6 \\ 7 & 8 \end{bmatrix}, \quad C = \begin{bmatrix} -1 & 0 \\ 3 & -3 \end{bmatrix},$$

$$D = \begin{bmatrix} 3 & 1 \\ -1 & 2 \\ 3 & -2 \\ 2 & 6 \end{bmatrix}, \quad E = \begin{bmatrix} -2 & 2 \\ 0 & -2 \\ 5 & -3 \\ 5 & 1 \end{bmatrix}, \quad F = \begin{bmatrix} 0 & 1 \\ -1 & 0 \\ 0 & 0 \\ 2 & 2 \end{bmatrix},$$

$$G = \begin{bmatrix} 1/2 & 1/3 & 1/4 \\ 2/3 & 3/5 & -5/6 \end{bmatrix}, \quad H = \begin{bmatrix} \sqrt{2} & \sqrt{3} & \sqrt{5} \\ \sqrt{2} & \sqrt{5} & \sqrt{2} \\ \sqrt{5} & \sqrt{2} & \sqrt{3} \end{bmatrix},$$

$$J = \begin{bmatrix} 0 & 0 & 0 & 0 & 0 \end{bmatrix}.$$

(2) Find, if they exist, the elements in the 1-2 and 3-1 positions for each of the matrices defined in Problem 1.

(3) Find, if they exist, a_{11}, a_{21}, b_{32}, d_{32}, d_{23}, e_{22}, g_{23}, h_{33}, and j_{21} for the matrices defined in Problem 1.

(4) Determine which, if any, of the matrices defined in Problem 1 are square.

(5) Determine which, if any, of the matrices defined in Problem 1 are row matrices and which are column matrices.

(6) Construct a 4-dimensional column matrix having the value j as its jth component.

(7) Construct a 5-dimensional row matrix having the value i^2 as its ith component.

(8) Construct the 2×2 matrix **A** having $a_{ij} = (-1)^{i+j}$.

(9) Construct the 3×3 matrix **A** having $a_{ij} = i/j$.

(10) Construct the $n \times n$ matrix **B** having $b_{ij} = n - i - j$. What will this matrix be when specialized to the 3×3 case?

(11) Construct the 2×4 matrix **C** having

$$c_{ij} = \begin{cases} i & \text{when } i = 1 \\ j & \text{when } i = 2 \end{cases}$$

(12) Construct the 3×4 matrix **D** having

$$d_{ij} = \left\{ \begin{array}{ll} i + j & \text{when } i > j \\ 0 & \text{when } i = j \\ i - j & \text{when } i < j \end{array} \right\}$$

In Problems 13 through 30, perform the indicated operations on the matrices defined in Problem 1.

(13) 2A.	**(14)** −5A.	**(15)** 3D.	**(16)** 10E.
(17) −F.	**(18)** A+B.	**(19)** C+A.	**(20)** D+E.
(21) D+F.	**(22)** A+D.	**(23)** A−B.	**(24)** C−A.
(25) D−E.	**(26)** D−F.	**(27)** 2A+3B.	**(28)** 3A−2C.
(29) 0.1A+0.2C.	**(30)** −2E+F.		

The matrices **A** through **F** in Problems 31 through 36 are defined in Problem 1.

(31) Find **X** if **A**+**X**=**B**.

(32) Find **Y** if 2**B**+**Y**=**C**.

(33) Find **X** if 3**D**−**X**=**E**.

(34) Find **Y** if **E**−2**Y**=**F**.

(35) Find **R** if 4**A**+5**R**=10**C**.

(36) Find **S** if 3 **F**−2**S**=**D**.

(37) Find 6A−θB if

$$\mathbf{A} = \begin{bmatrix} \theta^2 & 2\theta - 1 \\ 4 & 1/\theta \end{bmatrix} \quad \text{and} \quad \mathbf{B} = \begin{bmatrix} \theta^2 - 1 & 6 \\ 3/\theta & \theta^2 + 2\theta + 1 \end{bmatrix}.$$

(38) Prove part (*a*) of Theorem 1.

(39) Prove that if **0** is a zero matrix having the same order as **A**, then **A**+**0**=**A**.

(40) Prove part (*b*) of Theorem 2.

(41) Prove part (*c*) of Theorem 2.

(42) Store 1 of a three-store chain has 3 refrigerators, 5 stoves, 3 washing machines, and 4 dryers in stock. Store 2 has in stock no refrigerators, 2 stoves, 9 washing machines, and 5 dryers; while store 3 has in stock 4 refrigerators, 2 stoves, and no washing machines or dryers. Present the inventory of the entire chain as a matrix.

(43) The number of damaged items delivered by the SleepTight Mattress Company from its various plants during the past year is given by the damage matrix

$$\begin{bmatrix} 80 & 12 & 16 \\ 50 & 40 & 16 \\ 90 & 10 & 50 \end{bmatrix}$$

The rows pertain to its three plants in Michigan, Texas, and Utah; the columns pertain to its regular model, its firm model, and its extra-firm model, respectively. The company's goal for next year is to reduce by 10% the number of damaged regular mattresses shipped by each plant, to reduce by 20% the number of damaged firm mattresses shipped by its Texas plant, to reduce by 30% the number of damaged extra-firm mattresses shipped by its Utah plant, and to keep all other entries the same as last year. What will next year's damage matrix be if all goals are realized?

(44) On January 1, Ms. Smith buys three certificates of deposit from different institutions, all maturing in one year. The first is for \$1000 at 7%, the second is for \$2000 at 7.5%, and the third is for \$3000 at 7.25%. All interest rates are effective on an annual basis. Represent in a matrix all the relevant information regarding Ms. Smith's investments.

(45) (a) Mr. Jones owns 200 shares of IBM and 150 shares of AT&T. Construct a 1×2 portfolio matrix that reflects Mr. Jones' holdings.

(b) Over the next year, Mr. Jones triples his holdings in each company. What is his new portfolio matrix?

(c) The following year, Mr. Jones sells shares of each company in his portfolio. The number of shares sold is given by the matrix $[50\ 100]$, where the first component refers to shares of IBM stock. What is his new portfolio matrix?

(46) The inventory of an appliance store can be given by a 1×4 matrix in which the first entry represents the number of television sets, the second entry the number of air conditioners, the third entry the number of refrigerators, and the fourth entry the number of dishwashers.

(a) Determine the inventory given on January 1 by $[15\ 2\ 8\ 6]$.

(b) January sales are given by $[4\ 0\ 2\ 3]$. What is the inventory matrix on February 1?

(c) February sales are given by $[5\ 0\ 3\ 3]$, and new stock added in February is given by $[3\ 2\ 7\ 8]$. What is the inventory matrix on March 1?

(47) The daily gasoline supply of a local service station is given by a 1×3 matrix in which the first entry represents gallons of regular, the second entry gallons of premium, and the third entry gallons of super.

(a) Determine the supply of gasoline at the close of business on Monday given by $[14{,}000\quad 8000\quad 6000]$.

(b) Tuesday's sales are given by $[3500\ 2000\ 1500]$. What is the inventory matrix at day's end?

(c) Wednesday's sales are given by $[5000\ 1500\ 1200]$. In addition, the station received a delivery of 30,000 gallons of regular, 10,000 gallons of premium, but no super. What is the inventory at day's end?

1.2 MATRIX MULTIPLICATION

Matrix multiplication is the first operation where our intuition fails. First, two matrices are *not* multiplied together elementwise. Second, it is not always possible to multiply matrices of the same order while often it is possible to multiply matrices of different orders. Our purpose in introducing a new construct, such as the matrix, is to use it to enhance our understanding of real-world phenomena and to solve problems that were previously difficult to solve. A matrix is just a table of values, and not really new. Operations on tables, such as matrix addition, are new, but all operations considered in Section 1.1 are natural extensions of the analogous operations on real numbers. If we expect to use matrices to analyze problems differently, we must change something, and that something is the way we multiply matrices.

The motivation for matrix multiplication comes from the desire to solve systems of linear equations with the same ease and in the same way as one linear equation in one variable. A linear equation in one variable has the general form

$$[\text{constant}]\cdot[\text{variable}] = \text{constant}$$

We solve for the variable by dividing the entire equation by the multiplicative constant on the left. We want to mimic this process for many equations in many variables. Ideally, we want a single master equation of the form

$$\begin{bmatrix} \text{package} \\ \text{of} \\ \text{constants} \end{bmatrix} \cdot \begin{bmatrix} \text{package} \\ \text{of} \\ \text{variables} \end{bmatrix} = \begin{bmatrix} \text{package} \\ \text{of} \\ \text{constants} \end{bmatrix}$$

which we can divide by the package of constants on the left to solve for all the variables at one time. To do this, we need an arithmetic of *"packages,"* first to define the multiplication of such *"packages"* and then to divide *"packages"* to solve for the unknowns. The *"packages"* are, of course, matrices.

A simple system of two linear equations in two unknowns is

$$\begin{aligned} 2x + 3y &= 10 \\ 4x + 5y &= 20 \end{aligned} \tag{1.8}$$

Combining all the coefficients of the variables on the left of each equation into a *coefficient matrix,* all the variables into column matrix of variables, and the constants on the right of each equation into another column matrix, we generate the matrix system

$$\begin{bmatrix} 2 & 3 \\ 4 & 5 \end{bmatrix} \cdot \begin{bmatrix} x \\ y \end{bmatrix} = \begin{bmatrix} 10 \\ 20 \end{bmatrix} \tag{1.9}$$

We want to define matrix multiplication so that system (1.9) is equivalent to system (1.8); that is, we want multiplication defined so that

$$\begin{bmatrix} 2 & 3 \\ 4 & 5 \end{bmatrix} \cdot \begin{bmatrix} x \\ y \end{bmatrix} = \begin{bmatrix} (2x + 3y) \\ (4x + 5y) \end{bmatrix} \tag{1.10}$$

Then system (1.9) becomes

$$\begin{bmatrix} (2x+3y) \\ (4x+5y) \end{bmatrix} = \begin{bmatrix} 10 \\ 20 \end{bmatrix}$$

which, from our definition of matrix equality, is equivalent to system (1.8).

The product of two matrices **AB** is defined if the number of columns of **A** equals the number of rows of **B**.

We shall define the *product* **AB** of two matrices **A** and **B** when the number of columns of **A** is equal to the number of rows of **B**, and the result will be a matrix having the same number of rows as **A** and the same number of columns as **B**. Thus, if **A** and **B** are

$$\mathbf{A} = \begin{bmatrix} 6 & 1 & 0 \\ -1 & 2 & 1 \end{bmatrix} \quad \text{and} \quad \mathbf{B} = \begin{bmatrix} -1 & 0 & 1 & 0 \\ 3 & 2 & -2 & 1 \\ 4 & 1 & 1 & 0 \end{bmatrix}$$

then the product **AB** is defined, because **A** has three columns and **B** has three rows. Furthermore, the product **AB** will be 2×4 matrix, because **A** has two rows and **B** has four columns. In contrast, the product **BA** is not defined, because the number of columns in **B** is a different number from the number of rows in **A**.

A simple schematic for matrix multiplication is to write the orders of the matrices to be multiplied next to each other in the sequence the multiplication is to be done and then check whether the abutting numbers match. If the numbers match, then the multiplication is defined *and* the order of the product matrix is found by deleting the matching numbers and collapsing the two "×" symbols into one. If the abutting numbers do not match, then the product is not defined.

In particular, if **AB** is to be found for **A** having order 2×3 and **B** having order 3×4, we write

$$(2 \times 3) \; (3 \times 4) \tag{1.11}$$

where the abutting numbers are distinguished by the curved arrow. These abutting numbers are equal, both are 3, hence the multiplication is defined. Furthermore, by deleting the abutting threes in Equation (1.11), we are left with 2×2, which is the order of the product **AB**. In contrast, the product **BA** yields the schematic

$$(3 \times 4) \; (2 \times 3)$$

where we write the order of **B** before the order of **A** because that is the order of the proposed multiplication. The abutting numbers are again distinguished by the curved arrow, but here the abutting numbers are not equal, one is 4 and the other is 2, so the product **BA** is not defined. In general, if **A** is an $n \times r$ matrix and **B** is an $r \times p$ matrix, then the product **AB** is defined as an $n \times p$ matrix. The schematic is

$$(n \times r) \; (r \times p) = (n \times p) \tag{1.12}$$

When the product **AB** is considered, **A** is said to *premultiply* **B** while **B** is said to *postmultiply* **A**.

To calculate the *i-j* element of **AB**, when the multiplication is defined, multiply the elements in the *i*th row of **A** by the corresponding elements in the *j*th column of **B** and sum the results.

Knowing the order of a product is helpful in calculating the product. If **A** and **B** have the orders indicated in Equation (1.12), so that the multiplication is defined, we take as our motivation the multiplication in Equation (1.10) and calculate the *i-j* element $(i=1,2,\ldots,n;\ j=1,2,\ldots,p)$ of the product $\mathbf{AB}=\mathbf{C}=[c_{ij}]$ by multiplying the elements in the *i*th row of **A** by the corresponding elements in the *j*th row column of **B** and summing the results. That is,

$$\begin{bmatrix} a_{11} & a_{12} & \ldots & a_{1r} \\ a_{21} & a_{22} & \ldots & a_{2r} \\ \vdots & \vdots & \vdots & \vdots \\ a_{n1} & a_{n2} & \ldots & a_{nr} \end{bmatrix} \begin{bmatrix} b_{11} & b_{12} & \ldots & b_{1p} \\ b_{21} & b_{22} & \ldots & b_{2p} \\ \vdots & \vdots & \vdots & \vdots \\ b_{r1} & b_{r2} & \ldots & b_{rp} \end{bmatrix} = \begin{bmatrix} c_{11} & c_{12} & \ldots & a_{1p} \\ c_{21} & c_{22} & \ldots & c_{2p} \\ \vdots & \vdots & \vdots & \vdots \\ c_{n1} & c_{n2} & \ldots & c_{np} \end{bmatrix}$$

where

$$c_{ij} = a_{i1}b_{1j} + a_{i2}b_{2j} + a_{i3}b_{3j} + \cdots + a_{ir}b_{rj} = \sum_{k=1}^{r} a_{ik}b_{kj}$$

In particular, c_{11} is obtained by multiplying the elements in the first row of **A** by the corresponding elements in the first column of **B** and adding; hence

$$c_{11} = a_{11}b_{11} + a_{12}b_{21} + a_{13}b_{31} + \cdots + a_{1r}b_{r1}$$

The element c_{12} is obtained by multiplying the elements in the first row of **A** by the corresponding elements in the second column of **B** and adding; hence

$$c_{12} = a_{11}b_{12} + a_{12}b_{22} + a_{13}\,b_{32} + \cdots + a_{1r}b_{r2}$$

The element c_{35}, if it exists, is obtained by multiplying the elements in the third row of **A** by the corresponding elements in the fifth column of **B** and adding; hence

$$c_{35} = a_{31}b_{15} + a_{32}b_{25} + a_{33}b_{35} + \cdots + a_{3r}b_{r5}$$

Example 1 Find **AB** and **BA** for

$$\mathbf{A} = \begin{bmatrix} 1 & 2 & 3 \\ 4 & 5 & 6 \end{bmatrix} \quad \text{and} \quad \mathbf{B} = \begin{bmatrix} -7 & -8 \\ 9 & 10 \\ 0 & -11 \end{bmatrix}$$

Solution: **A** has order 2×3 and **B** has order 3×2, so our schematic for the product **AB** is

$$(2 \times 3)\ (3 \times 2)$$

The abutting numbers are both 3; hence the product **AB** is defined. Deleting both abutting numbers, we have 2×2 as the order of the product.

$$\mathbf{AB} = \begin{bmatrix} 1 & 2 & 3 \\ 4 & 5 & 6 \end{bmatrix} \begin{bmatrix} -7 & -8 \\ 9 & 10 \\ 0 & -11 \end{bmatrix}$$

$$= \begin{bmatrix} 1(-7) + 2(9) + 3(0) & 1(-8) + 2(10) + 3(-11) \\ 4(-7) + 5(9) + 6(0) & 4(-8) + 5(10) + 6(-11) \end{bmatrix}$$

$$= \begin{bmatrix} 11 & -21 \\ 17 & -48 \end{bmatrix}$$

Our schematic for the product **BA** is

$$(3 \times 2) \; (2 \times 3)$$

The abutting numbers are now both 2; hence the product **BA** is defined. Deleting both abutting numbers, we have 3×3 as the order of the product **BA**.

$$\mathbf{BA} = \begin{bmatrix} -7 & -8 \\ 9 & 10 \\ 0 & -11 \end{bmatrix} \begin{bmatrix} 1 & 2 & 3 \\ 4 & 5 & 6 \end{bmatrix}$$

$$= \begin{bmatrix} (-7)1 + (-8)4 & (-7)2 + (-8)5 & (-7)3 + (-8)6 \\ 9(1) + 10(4) & 9(2) + 10(5) & 9(3) + 10(6) \\ 0(1) + (-11)4 & 0(2) + (-11)5 & 0(3) + (-11)6 \end{bmatrix}$$

$$= \begin{bmatrix} -39 & -54 & -69 \\ 48 & 68 & 87 \\ -44 & -55 & -66 \end{bmatrix}$$

Example 2 Find **AB** and **BA** for

$$\mathbf{A} = \begin{bmatrix} 2 & 1 \\ -1 & 0 \\ 3 & 1 \end{bmatrix} \quad \text{and} \quad \mathbf{B} = \begin{bmatrix} 3 & 1 & 5 & -1 \\ 4 & -2 & 1 & 0 \end{bmatrix}$$

Solution: **A** has two columns and **B** has two rows, so the product **AB** is defined.

$$\mathbf{AB} = \begin{bmatrix} 2 & 1 \\ -1 & 0 \\ 3 & 1 \end{bmatrix} \begin{bmatrix} 3 & 1 & 5 & -1 \\ 4 & -2 & 1 & 0 \end{bmatrix}$$

$$= \begin{bmatrix} 2(3) + 1(4) & 2(1) + 1(-2) & 2(5) + 1(1) & 2(-1) + 1(0) \\ -1(3) + 0(4) & -1(1) + 0(-2) & -1(5) + 0(1) & -1(-1) + 0(0) \\ 3(3) + 1(4) & 3(1) + 1(-2) & 3(5) + 1(1) & 3(-1) + 1(0) \end{bmatrix}$$

$$= \begin{bmatrix} 10 & 0 & 11 & -2 \\ -3 & -1 & -5 & 1 \\ 13 & 1 & 16 & -3 \end{bmatrix}$$

In contrast, **B** has four columns and **A** has three rows, so the product **BA** is *not* defined.

Observe from Examples 1 and 2 that $\mathbf{AB} \neq \mathbf{BA}$! In Example 1, **AB** is a 2×2 matrix, whereas **BA** is a 3×3 matrix. In Example 2, **AB** is a 3×4 matrix, whereas **BA** is not defined. In general, the product of two matrices is not commutative.

In general, $\mathbf{AB} \neq \mathbf{BA}$.

Example 3 Find **AB** and **BA** for

$$\mathbf{A} = \begin{bmatrix} 3 & 1 \\ 0 & 4 \end{bmatrix} \quad \text{and} \quad \mathbf{B} = \begin{bmatrix} 1 & 1 \\ 0 & 2 \end{bmatrix}$$

Solution:

$$\begin{aligned} \mathbf{AB} &= \begin{bmatrix} 3 & 1 \\ 0 & 4 \end{bmatrix} \begin{bmatrix} 1 & 1 \\ 0 & 2 \end{bmatrix} \\ &= \begin{bmatrix} 3(1)+1(0) & 3(1)+1(2) \\ 0(1)+4(0) & 0(1)+4(2) \end{bmatrix} \\ &= \begin{bmatrix} 3 & 5 \\ 0 & 8 \end{bmatrix} \end{aligned}$$

$$\begin{aligned} \mathbf{BA} &= \begin{bmatrix} 1 & 1 \\ 0 & 2 \end{bmatrix} \begin{bmatrix} 3 & 1 \\ 0 & 4 \end{bmatrix} \\ &= \begin{bmatrix} 1(3)+1(0) & 1(1)+1(4) \\ 0(3)+2(0) & 0(1)+2(4) \end{bmatrix} \\ &= \begin{bmatrix} 3 & 5 \\ 0 & 8 \end{bmatrix} \end{aligned}$$

In Example 3, the products **AB** and **BA** are defined and equal. Although matrix multiplication is not commutative, as a general rule, some matrix products are commutative. Matrix multiplication also lacks other familiar properties besides commutativity. We know from our experiences with real numbers that if the product $ab=0$, then either $a=0$ or $b=0$ or both are zero. This is not true, in general, for matrices. Matrices exist for which $\mathbf{AB}=\mathbf{0}$ *without* either **A** or **B** being zero (see Problems 20 and 21). The cancellation law also does not hold for matrix multiplication. In general, the equation $\mathbf{AB}=\mathbf{AC}$ does *not* imply that $\mathbf{B}=\mathbf{C}$ (see Problems 22 and 23). Matrix multiplication, however, does retain some important properties.

▶THEOREM 1

If **A**, **B**, *and* **C** *have appropriate orders so that the following additions and multiplications are defined, then*

(a) $\mathbf{A}(\mathbf{BC})=(\mathbf{AB})\mathbf{C}$ (associate law of multiplication)
(b) $\mathbf{A}(\mathbf{B}+\mathbf{C})=\mathbf{AB}+\mathbf{AC}$ (left distributive law)
(c) $(\mathbf{B}+\mathbf{C})\mathbf{A}=\mathbf{BA}+\mathbf{CA}$ (right distributive law) ◀

Proof: We leave the proofs of parts (a) and (c) as exercises (see Problems 37 and 38). To prove part (b), we assume that $\mathbf{A}=[a_{ij}]$ is an $m \times n$ matrix and both $\mathbf{B}=[b_{ij}]$ and $\mathbf{C}=[c_{ij}]$ are $n \times p$ matrices. Then

$$
\begin{aligned}
\mathbf{A}(\mathbf{B}+\mathbf{C}) &= [a_{ij}]\left([b_{ij}]+[c_{ij}]\right) \\
&= [a_{ij}]\left[(b_{ij}+c_{ij})\right] && \text{definition of matrix addition} \\
&= \left[\sum_{k=1}^{n} a_{ik}(b_{kj}+c_{kj})\right] && \text{definition of matrix multiplication} \\
&= \left[\sum_{k=1}^{n} a_{ik}b_{kj}+a_{ik}c_{kj}\right] \\
&= \left[\sum_{k=1}^{n} a_{ik}b_{kj}+\sum_{k=1}^{n} a_{ik}c_{kj}\right] \\
&= \left[\sum_{k=1}^{n} a_{ik}b_{kj}\right]+\left[\sum_{k=1}^{n} a_{ik}b_{kj}\right] && \text{definition of matrix addition} \\
&= [a_{ij}][b_{ij}]+[a_{ij}][c_{ij}] && \text{definition of matrix mulitiplication}
\end{aligned}
$$

With multiplication defined as it is, we can decouple a system of linear equations so that all of the variables in the system are packaged together. In particular, the set of simultaneous linear equations

$$
\begin{aligned}
5x - 3y + 2z &= 14 \\
x + y - 4z &= -7 \\
7x - 3z &= 1
\end{aligned}
\tag{1.13}
$$

can be written as the matrix equation $\mathbf{Ax}=\mathbf{b}$ where

$$
\mathbf{A}=\begin{bmatrix} 5 & -3 & 2 \\ 1 & 1 & -4 \\ 7 & 0 & -3 \end{bmatrix}, \quad \mathbf{x}=\begin{bmatrix} x \\ y \\ z \end{bmatrix}, \quad \text{and} \quad \mathbf{b}=\begin{bmatrix} 14 \\ -7 \\ 1 \end{bmatrix}.
$$

The column matrix $\mathbf{x}$ lists all the variables in Equation (1.13), the column matrix $\mathbf{b}$ enumerates the constants on the right sides of the equations in Equation (1.13), and the matrix $\mathbf{A}$ holds the coefficients of the variables. $\mathbf{A}$ is known as a *coefficient matrix* and care must taken in constructing $\mathbf{A}$ to place all the x coefficients in the first column, all the y coefficients in the second column, and all the z coefficients in the third column. The zero in 3-2 location in $\mathbf{A}$ appears because the coefficient of y in the third equation of Equation (1.13) is zero. By redefining the matrices **A**, **x**, and **b** appropriately, we can represent any system of simultaneous linear equations by the matrix equation

Any system of simultaneous linear equations can be written as the matrix equation **Ax** = **b**.

$$\mathbf{Ax} = \mathbf{b} \tag{1.14}$$

Example 4 The system of linear equations

$$\begin{aligned} 2x + y - z &= 4 \\ 3x + 2y + 2w &= 0 \\ x - 2y + 3z + 4w &= -1 \end{aligned}$$

has the matrix form **Ax** = **b** with

$$\mathbf{A} = \begin{bmatrix} 2 & 1 & -1 & 0 \\ 3 & 2 & 0 & 2 \\ 1 & -2 & 3 & 4 \end{bmatrix}, \quad \mathbf{x} = \begin{bmatrix} x \\ y \\ z \\ w \end{bmatrix}, \quad \text{and} \quad \mathbf{b} = \begin{bmatrix} 4 \\ 0 \\ -1 \end{bmatrix}.$$

We have accomplished part of the goal we set in the beginning of this section: to write a system of simultaneous linear equations in the matrix form **Ax** = **b**, where all the variables are segregated into the column matrix **x**. All that remains is to develop a matrix operation to solve the matrix equation **Ax** = **b** for **x**. To do so, at least for a large class of *square* coefficient matrices, we first introduce some additional matrix notation and review the traditional techniques for solving systems of equations, because those techniques form the basis for the missing matrix operation.

Problems 1.2

(1) Determine the orders of the following products if the order of **A** is 2×4, the order of **B** is 4×2, the order of **C** is 4×1, the order of **D** is 1×2, and the order of **E** is 4×4.

(a) **AB**, (b) **BA**, (c) **AC**, (d) **CA**, (e) **CD**, (f) **AE**,
(g) **EB**, (h) **EA**, (i) **ABC**, (j) **DAE**, (k) **EBA**, (l) **EECD**.

In Problems 2 through 19, find the indicated products for

$$\mathbf{A} = \begin{bmatrix} 1 & 2 \\ 3 & 4 \end{bmatrix}, \quad \mathbf{B} = \begin{bmatrix} 5 & 6 \\ 7 & 8 \end{bmatrix}, \quad \mathbf{C} = \begin{bmatrix} -1 & 0 & 1 \\ 3 & -2 & 1 \end{bmatrix}, \quad \mathbf{D} = \begin{bmatrix} 1 & 1 \\ -1 & 3 \\ 2 & -2 \end{bmatrix},$$

$$\mathbf{E} = \begin{bmatrix} -2 & 2 & 1 \\ 0 & -2 & -1 \\ 1 & 0 & 1 \end{bmatrix}, \quad \mathbf{F} = \begin{bmatrix} 0 & 1 & 2 \\ -1 & -1 & 0 \\ 1 & 2 & 3 \end{bmatrix},$$

$$\mathbf{x} = [1 \quad -2], \quad \mathbf{y} = [1 \quad 2 \quad 1].$$

(2) AB. **(3)** BA. **(4)** AC. **(5)** BC. **(6)** CB. **(7)** xA.

(8) xB. **(9)** xC. **(10)** Ax. **(11)** CD. **(12)** DC. **(13)** yD.

(14) yC. **(15)** Dx. **(16)** xD. **(17)** EF. **(18)** FE. **(19)** yF.

(20) Find **AB** for $\mathbf{A} = \begin{bmatrix} 2 & 6 \\ 3 & 9 \end{bmatrix}$ and $\mathbf{B} = \begin{bmatrix} 3 & -6 \\ -1 & 2 \end{bmatrix}$. Note that $\mathbf{AB} = \mathbf{0}$ but neither **A** nor **B** equals the zero matrix.

(21) Find **AB** for $\mathbf{A} = \begin{bmatrix} 4 & 2 \\ 2 & 1 \end{bmatrix}$ and $\mathbf{B} = \begin{bmatrix} 3 & -4 \\ -6 & 8 \end{bmatrix}$.

(22) Find **AB** and **AC** for $\mathbf{A} = \begin{bmatrix} 4 & 2 \\ 2 & 1 \end{bmatrix}$, $\mathbf{B} = \begin{bmatrix} 1 & 1 \\ 2 & 1 \end{bmatrix}$, and $\mathbf{C} = \begin{bmatrix} 2 & 2 \\ 0 & -1 \end{bmatrix}$. What does this result imply about the cancellation law for matrices?

(23) Find **AB** and **CB** for $\mathbf{A} = \begin{bmatrix} 3 & 2 \\ 1 & 0 \end{bmatrix}$, $\mathbf{B} = \begin{bmatrix} 2 & 4 \\ 1 & 2 \end{bmatrix}$, and $\mathbf{C} = \begin{bmatrix} 1 & 6 \\ 3 & -4 \end{bmatrix}$. Show that $\mathbf{AB} = \mathbf{CB}$ but $\mathbf{A} \neq \mathbf{C}$.

(24) Calculate the product $\begin{bmatrix} 1 & 2 \\ 3 & 4 \end{bmatrix} \begin{bmatrix} x \\ y \end{bmatrix}$.

(25) Calculate the product $\begin{bmatrix} 1 & 0 & -1 \\ 3 & 1 & 1 \\ 1 & 3 & 0 \end{bmatrix} \begin{bmatrix} x \\ y \\ z \end{bmatrix}$.

(26) Calculate the product $\begin{bmatrix} a_{11} & a_{12} \\ a_{21} & a_{22} \end{bmatrix} \begin{bmatrix} x \\ y \end{bmatrix}$.

(27) Calculate the product $\begin{bmatrix} b_{11} & b_{12} & b_{13} \\ b_{21} & b_{22} & b_{23} \end{bmatrix} \begin{bmatrix} x \\ y \\ z \end{bmatrix}$.

(28) Evaluate the expression $\mathbf{A}^2 - 4\mathbf{A} - 5\mathbf{I}$ for the matrix $\mathbf{A} = \begin{bmatrix} 1 & 2 \\ 4 & 3 \end{bmatrix}$.

(29) Evaluate the expression $(\mathbf{A} - \mathbf{I})(\mathbf{A} + 2\mathbf{I})$ for the matrix $\mathbf{A} = \begin{bmatrix} 3 & 5 \\ -2 & 4 \end{bmatrix}$.

(30) Evaluate the expression $(\mathbf{I} - \mathbf{A})(\mathbf{A}^2 - \mathbf{I})$ for the matrix $\mathbf{A} = \begin{bmatrix} 2 & -1 & 1 \\ 3 & -2 & 1 \\ 0 & 0 & 1 \end{bmatrix}$.

(31) Use the definition of matrix multiplication to show that

$$j\text{th column of } (\mathbf{AB}) = \mathbf{A} \times (j\text{th column of } \mathbf{B}).$$

(32) Use the definition of matrix multiplication to show that

$$i\text{th row of}(\mathbf{AB}) = (i\text{th row of }\mathbf{A}) \times \mathbf{B}.$$

(33) Prove that if **A** has a row of zeros and **B** is any matrix for which the product **AB** is defined, then **AB** also has a row of zeros.

(34) Show by example that if **B** has a row of zeros and **A** is any matrix for which the product **AB** is defined, then **AB** need not have a row of zeros.

(35) Prove that if **B** has a column of zeros and **A** is any matrix for which the product **AB** is defined, then **AB** also has a column of zeros.

(36) Show by example that if **A** has a column of zeros and **B** is any matrix for which the product **AB** is defined, then **AB** need not have a column of zeros.

(37) Prove part (a) of Theorem 1.

(38) Prove part (c) of Theorem 1.

In Problems 39 through 50, write each system in matrix form $\mathbf{Ax}=\mathbf{b}$.

(39) $2x + 3y = 10$
$4x - 5y = 11$

(40) $5x + 20y = 80$
$-x + 4y = -64$

(41) $3x + 3y = 100$
$3x - 8y = 300$
$-x + 2y = 500$

(42) $x + 3y = 4$
$2x - y = 1$
$-2x - 6y = -8$
$4x - 9y = -5$
$-6x + 3y = -3$

(43) $x + y - z = 0$
$3x + 2y + 4z = 0$

(44) $2x - y = 12$
$-4x - z = 15$

(45) $x + 2y - 2z = -1$
$2x + y + z = 5$
$-x + y - z = -2$

(46) $2x + y - z = 0$
$x + 2y + z = 0$
$3x - y + 2z = 0$

(47) $x + z + y = 2$
$3z + 2x + y = 5$
$3y + x = 1$

(48) $x + 2y - z = 5$
$2x - y + 2z = 1$
$2x + 2y - z = 7$
$x + 2y + z = 3$

(49) $5x + 3y + 2z + 4w = 5$
$x + y + w = 0$
$3x + 2y + 2z = -3$
$x + y + 2z + 3w = 4$

(50) $2x - y + z - w = 1$
$x + 2y - z + 2w = -1$
$x - 3y + 2z - 3w = 2$

(51) The price schedule for a Chicago to Los Angeles flight is given by

$$\mathbf{p} = [200 \quad 350 \quad 500]$$

where row matrix elements pertain, respectively, to coach ticket prices, business-class ticket prices and first-class ticket prices. The number of tickets purchased in each class for a particular flight is given by the column matrix

$$\mathbf{n} = \begin{bmatrix} 130 \\ 20 \\ 10 \end{bmatrix}$$

Calculate the products (a) **pn** and (b) **np**, and determine the significance of each.

(52) The closing prices of a person's portfolio during the past week are tabulated as

$$\mathbf{P} = \begin{bmatrix} 40 & 40\frac{1}{2} & 40\frac{7}{8} & 41 & 41 \\ 3\frac{1}{4} & 3\frac{5}{8} & 3\frac{1}{2} & 4 & 3\frac{7}{8} \\ 10 & 9\frac{3}{4} & 10\frac{1}{8} & 10 & 9\frac{5}{8} \end{bmatrix}$$

where the columns pertain to the days of the week, Monday through Friday, and the rows pertain to the prices of Orchard Fruits, Lion Airways, and Arrow Oil. The person's holdings in each of these companies are given by the row matrix

$$\mathbf{h} = [100 \quad 500 \quad 400]$$

Calculate the products (a) **hP** and (b) **Ph**, and determine the significance of each.

(53) The time requirements for a company to produce three products is tabulated in

$$\mathbf{T} = \begin{bmatrix} 0.2 & 0.5 & 0.4 \\ 1.2 & 2.3 & 1.7 \\ 0.8 & 3.1 & 1.2 \end{bmatrix}$$

where the rows pertain to lamp bases, cabinets, and tables, respectively. The columns pertain to the hours of labor required for cutting the wood, assembling, and painting, respectively. The hourly wages of a carpenter to cut wood, of a craftsperson to assemble a product, and of a decorator to paint are given, respectively, by the columns of the matrix

$$\mathbf{w} = \begin{bmatrix} 10.50 \\ 14.00 \\ 12.25 \end{bmatrix}$$

Calculate the product **Tw** and determine its significance.

(54) Continuing with the information provided in the previous problem, assume further that the number of items on order for lamp bases, cabinets, and tables, respectively, are given in the rows of

$$\mathbf{q} = [1000 \quad 100 \quad 200]$$

Calculate the product **qTw** and determine its significance.

(55) The results of a flue epidemic at a college campus are collected in the matrix

$$\mathbf{F} = \begin{bmatrix} 0.20 & 0.20 & 0.15 & 0.15 \\ 0.10 & 0.30 & 0.30 & 0.40 \\ 0.70 & 0.50 & 0.55 & 0.45 \end{bmatrix}$$

where each element is a percent converted to a decimal. The columns pertain to freshmen, sophomores, juniors, and seniors, respectively; whereas the rows represent bedridden students, students who are infected but ambulatory, and well students, respectively. The male-female composition of each class is given by the matrix

$$\mathbf{C} = \begin{bmatrix} 1050 & 950 \\ 1100 & 1050 \\ 360 & 500 \\ 860 & 1000 \end{bmatrix}.$$

Calculate the product **FC** and determine its significance.

1.3 SPECIAL MATRICES

Certain types of matrices appear so frequently that it is advisable to discuss them separately. The *transpose* of a matrix **A**, denoted by $\mathbf{A}^{\mathrm{T}}$, is obtained by converting all the rows of **A** into the columns of $\mathbf{A}^{\mathrm{T}}$ while preserving the ordering of the rows/columns. The first row of **A** becomes the first column of $\mathbf{A}^{\mathrm{T}}$, the second row of **A** becomes the second column of $\mathbf{A}^{\mathrm{T}}$, and the last row of **A** becomes the last column of $\mathbf{A}^{\mathrm{T}}$. More formally, if $\mathbf{A} = [a_{ij}]$ is an $n \times p$ matrix, then the transpose of **A**, denoted by $\mathbf{A}^{\mathrm{T}} = \lfloor \mathbf{b}_{ij}^{\mathrm{T}} \rfloor$, is a $p \times n$ matrix where $a_{ij}^{\mathrm{T}} = a_{ji}$.

The transpose **A** is obtained by converting all the rows of **A** into columns while preserving the ordering of the rows/columns.

Example 1 If $\mathbf{A} = \begin{bmatrix} 1 & 2 & 3 \\ 4 & 5 & 6 \\ 7 & 8 & 9 \end{bmatrix}$, then $\mathbf{A}^{\mathrm{T}} = \begin{bmatrix} 1 & 4 & 7 \\ 2 & 5 & 8 \\ 3 & 6 & 9 \end{bmatrix}$, while the transpose of

$$\mathbf{B} = \begin{bmatrix} 1 & 2 & 3 & 4 \\ 5 & 6 & 7 & 8 \end{bmatrix} \text{ is } \mathbf{B}^{\mathrm{T}} = \begin{bmatrix} 1 & 5 \\ 2 & 6 \\ 3 & 7 \\ 4 & 8 \end{bmatrix}.$$

▶THEOREM 1

The following properties are true for any scalar λ and any matrices for which the indicated additions and multiplications are defined:

(a) $(\mathbf{A}^{\mathrm{T}})^{\mathrm{T}} = \mathbf{A}$
(b) $(\lambda\mathbf{A})^{\mathrm{T}} = \lambda\mathbf{A}^{\mathrm{T}}$
(c) $(\mathbf{A}+\mathbf{B})^{\mathrm{T}} = \mathbf{A}^{\mathrm{T}} + \mathbf{B}^{\mathrm{T}}$
(d) $(\mathbf{AB})^{\mathrm{T}} = \mathbf{B}^{\mathrm{T}}\mathbf{A}^{\mathrm{T}}$ ◀

Proof: We prove part (d) and leave the others as exercises (see Problems 21 through 23). Let $\mathbf{A}=[a_{ij}]$ and $\mathbf{B}=[b_{ij}]$ have orders $n \times m$ and $m \times p$, so that the product $\mathbf{AB}$ is defined. Then

$$\begin{aligned}
(\mathbf{AB})^{\mathrm{T}} &= \left([a_{ij}]\,[b_{ij}]\right)^{\mathrm{T}} \\
&= \left[\sum_{k=1}^{m} a_{ik}b_{kj}\right]^{\mathrm{T}} && \text{definition of matrix multiplication} \\
&= \left[\sum_{k=1}^{m} a_{jk}b_{ki}\right] && \text{definition of the transpose} \\
&= \left[\sum_{k=1}^{m} a^{\mathrm{T}}_{kj}b^{\mathrm{T}}_{ik}\right] && \text{definition of the transpose} \\
&= \left[\sum_{k=1}^{m} b^{\mathrm{T}}_{ik}a^{\mathrm{T}}_{kj}\right] \\
&= \left[b^{\mathrm{T}}_{ij}\right]\left[a^{\mathrm{T}}_{ij}\right] && \text{definition of matrix multiplication} \\
&= \mathbf{B}^{\mathrm{T}}\mathbf{A}^{\mathrm{T}}
\end{aligned}$$

Observation: The transpose of a product of matrices is *not* the product of the transposes but rather the *commuted* product of the transposes.

A matrix $\mathbf{A}$ is *symmetric* if it equals its own transpose; that is, if $\mathbf{A}=\mathbf{A}^{\mathrm{T}}$. A matrix $\mathbf{A}$ is *skew-symmetric* if it equals the negative of its transpose; that is, if $\mathbf{A}=-\mathbf{A}^{\mathrm{T}}$.

Example 2 $\mathbf{A}=\begin{bmatrix} 1 & 2 & 3 \\ 2 & 4 & 5 \\ 3 & 5 & 6 \end{bmatrix}$ is symmetric while $\mathbf{B}=\begin{bmatrix} 0 & 2 & -3 \\ -2 & 0 & 1 \\ 3 & -1 & 0 \end{bmatrix}$ is skew-symmetric.

A submatrix of a matrix **A** is a matrix obtained from **A** by removing any number of rows or columns from **A**.

A *submatrix* of a matrix $\mathbf{A}$ is a matrix obtained from $\mathbf{A}$ by removing any number of rows or columns from $\mathbf{A}$. In particular, if

$$\mathbf{A}=\begin{bmatrix} 1 & 2 & 3 & 4 \\ 5 & 6 & 7 & 8 \\ 9 & 10 & 11 & 12 \\ 13 & 14 & 15 & 16 \end{bmatrix} \tag{1.16}$$

then both $\mathbf{B}=\begin{bmatrix} 10 & 12 \\ 14 & 16 \end{bmatrix}$ and $\mathbf{C}=[2 \quad 3 \quad 4]$ are submatrices of $\mathbf{A}$. Here $\mathbf{B}$ is obtained by removing the first and second rows together with the first and third columns from $\mathbf{A}$, while $\mathbf{C}$ is obtained by removing from $\mathbf{A}$ the second, third, and fourth rows together with the first column. By removing no rows and no columns from $\mathbf{A}$, it follows that $\mathbf{A}$ is a submatrix of itself.

A matrix is *partitioned* if it is divided into submatrices by horizontal and vertical lines between rows and columns. By varying the choices of where to place the horizontal and vertical lines, one can partition a matrix in different ways. Thus,

A matrix is partitioned if it is divided into submatrices by horizontal and vertical lines between rows and columns.

$$\mathbf{AB} = \left[\begin{array}{c|c} \mathbf{CG}+\mathbf{DJ} & \mathbf{CH}+\mathbf{DK} \\ \hline \mathbf{EG}+\mathbf{FJ} & \mathbf{EH}+\mathbf{FK} \end{array}\right]$$

provided the partitioning was such that the indicated multiplications are defined.

Example 3 Find **AB** if

$$\mathbf{A} = \left[\begin{array}{cc|c} 3 & 1 & 0 \\ 2 & 0 & 1 \\ \hline 0 & 0 & 3 \\ 0 & 0 & 1 \\ \hline 0 & 0 & 0 \end{array}\right] \quad \text{and} \quad \mathbf{B} = \left[\begin{array}{cc|ccc} 2 & 1 & 0 & 0 & 0 \\ -1 & 1 & 0 & 0 & 0 \\ \hline 0 & 1 & 0 & 0 & 1 \end{array}\right]$$

Solution: From the indicated partitions, we find that

$$\mathbf{AB} = \left[\begin{array}{c|c}
\begin{bmatrix} 3 & 1 \\ 2 & 0 \end{bmatrix}\begin{bmatrix} 2 & 1 \\ -1 & 1 \end{bmatrix} + \begin{bmatrix} 0 \\ 0 \end{bmatrix}[0 \;\; 1] & \begin{bmatrix} 3 & 1 \\ 2 & 0 \end{bmatrix}\begin{bmatrix} 0 & 0 & 0 \\ 0 & 0 & 0 \end{bmatrix} + \begin{bmatrix} 0 \\ 0 \end{bmatrix}[0 \;\; 0 \;\; 1] \\ \hline
\begin{bmatrix} 0 & 0 \\ 0 & 0 \end{bmatrix}\begin{bmatrix} 2 & 1 \\ -1 & 1 \end{bmatrix} + \begin{bmatrix} 3 \\ 1 \end{bmatrix}[0 \;\; 1] & \begin{bmatrix} 0 & 0 \\ 0 & 0 \end{bmatrix}\begin{bmatrix} 0 & 0 & 0 \\ 0 & 0 & 0 \end{bmatrix} + \begin{bmatrix} 3 \\ 1 \end{bmatrix}[0 \;\; 0 \;\; 1] \\ \hline
[0 \;\; 0]\begin{bmatrix} 2 & 1 \\ -1 & 1 \end{bmatrix} + [0]\,[0 \;\; 1] & [0 \;\; 0]\begin{bmatrix} 0 & 0 & 0 \\ 0 & 0 & 0 \end{bmatrix} + [0]\,[0 \;\; 0 \;\; 1]
\end{array}\right]$$

$$= \left[\begin{array}{c|c}
\begin{bmatrix} 5 & 4 \\ 4 & 2 \end{bmatrix} + \begin{bmatrix} 0 & 0 \\ 0 & 0 \end{bmatrix} & \begin{bmatrix} 0 & 0 & 0 \\ 0 & 0 & 0 \end{bmatrix} + \begin{bmatrix} 0 & 0 & 0 \\ 0 & 0 & 0 \end{bmatrix} \\ \hline
\begin{bmatrix} 5 & 4 \\ 4 & 2 \end{bmatrix} + \begin{bmatrix} 0 & 0 \\ 0 & 0 \end{bmatrix} & \begin{bmatrix} 0 & 0 & 0 \\ 0 & 0 & 0 \end{bmatrix} + \begin{bmatrix} 0 & 0 & 0 \\ 0 & 0 & 0 \end{bmatrix} \\ \hline
[0 \;\; 0] + [0 \;\; 0] & [0 \;\; 0 \;\; 0] + [0 \;\; 0 \;\; 0]
\end{array}\right]$$

$$= \left[\begin{array}{cc|ccc} 5 & 4 & 0 & 0 & 0 \\ 4 & 2 & 0 & 0 & 0 \\ \hline 0 & 3 & 0 & 0 & 3 \\ 0 & 1 & 0 & 0 & 1 \\ \hline 0 & 0 & 0 & 0 & 0 \end{array}\right] = \begin{bmatrix} 5 & 4 & 0 & 0 & 0 \\ 4 & 2 & 0 & 0 & 0 \\ 0 & 3 & 0 & 0 & 3 \\ 0 & 1 & 0 & 0 & 1 \\ 0 & 0 & 0 & 0 & 0 \end{bmatrix}$$

Note that we partitioned to make maximum use of the zero submatrices of both **A** and **B**.

A *zero row* in a matrix is a row containing only zero elements, whereas a nonzero row is a row that contains at least one nonzero element.

▶DEFINITION 1

A matrix is in *row-reduced form* if it satisfies the following four conditions:

(i) All zero rows appear below nonzero rows when both types are present in the matrix.
(ii) The first nonzero element in any nonzero row is 1.
(iii) All elements directly below (that is, in the same column but in succeeding rows from) the first nonzero element of a nonzero row are zero.
(iv) The first nonzero element of any nonzero row appears in a later column (further to the right) than the first nonzero element in any preceding row. ◀

Row-reduced matrices are invaluable for solving sets of simultaneous linear equations. We shall use these matrices extensively in succeeding sections, but at present we are interested only in determining whether a given matrix is or is not in row-reduced form.

Example 4

$$\mathbf{A} = \begin{bmatrix} 1 & 1 & -2 & 4 & 7 \\ 0 & 0 & -6 & 5 & 7 \\ 0 & 0 & 0 & 0 & 0 \\ 0 & 0 & 0 & 0 & 0 \end{bmatrix}$$

is not in row-reduced form because the first nonzero element in the second row is not 1. If a_{23} was 1 instead of -6, then the matrix would be in row-reduced form.

$$\mathbf{B} = \begin{bmatrix} 1 & 2 & 3 \\ 0 & 0 & 0 \\ 0 & 0 & 1 \end{bmatrix}$$

is not in row-reduced form because the second row is a zero row and it appears before the third row, which is a nonzero row. If the second and third rows had been interchanged, then the matrix would be in row-reduced form.

$$\mathbf{C} = \begin{bmatrix} 1 & 2 & 3 & 4 \\ 0 & 0 & 1 & 2 \\ 0 & 1 & 0 & 5 \end{bmatrix}$$

is not in row-reduced form because the first nonzero element in row two appears in a later column, column 3, than the first nonzero element in row three. If the second and third rows had been interchanged, then the matrix would be in row-reduced form.

$$\mathbf{D} = \begin{bmatrix} 1 & -2 & 3 & 3 \\ 0 & 0 & 1 & -3 \\ 0 & 0 & 1 & 0 \end{bmatrix}$$

is not in row-reduced form because the first nonzero element in row two appears in the third column and everything below this element is not zero. Had d_{33} been zero instead of 1, then the matrix would be in row-reduced form.

For the remainder of this section, we restrict ourselves to square matrices, matrices having the same number of rows as columns. Recall that the main diagonal of an $n \times n$ matrix $\mathbf{A} = [a_{ij}]$ consists of all the diagonal elements $a_{11}, a_{22}, \ldots, a_{nn}$. A *diagonal matrix* is a square matrix having only zeros as non-diagonal elements. Thus,

$$\begin{bmatrix} 5 & 0 \\ 0 & -1 \end{bmatrix} \quad \text{and} \quad \begin{bmatrix} 3 & 0 & 0 \\ 0 & 3 & 0 \\ 0 & 0 & 3 \end{bmatrix}$$

are both diagonal matrices or orders 2×2 and 3×3, respectively. A square zero matrix is a special diagonal matrix having all its elements equal to zero.

An *identity matrix*, denoted as **I**, is a diagonal matrix having all its diagonal elements equal to 1. The 2×2 and 4×4 identity matrices are, respectively,

An identity matrix **I** is a diagonal matrix having all its diagonal elements equal to 1.

$$\begin{bmatrix} 1 & 0 \\ 0 & -1 \end{bmatrix} \quad \text{and} \quad \begin{bmatrix} 1 & 0 & 0 & 0 \\ 0 & 1 & 0 & 0 \\ 0 & 0 & 1 & 0 \\ 0 & 0 & 0 & 1 \end{bmatrix}$$

If **A** and **I** are square matrices of the same order, then

$$\mathbf{AI} = \mathbf{IA} = \mathbf{A}. \tag{1.17}$$

A *block diagonal* matrix **A** is one that can be partitioned into the form

$$\mathbf{A} = \begin{bmatrix} \mathbf{A}_1 & & & & \mathbf{0} \\ & \mathbf{A}_2 & & & \\ & & \mathbf{A}_3 & & \\ & & & \ddots & \\ & \mathbf{0} & & & \mathbf{A}_k \end{bmatrix}$$

where $\mathbf{A}_1, \mathbf{A}_2, \ldots, \mathbf{A}_k$ are square submatrices. Block diagonal matrices are particularly easy to multiply because in partitioned form they act as diagonal matrices.

A matrix $\mathbf{A} = [a_{ij}]$ is *upper triangular* if $a_{ij} = 0$ for $i > j$; that is, if all elements below the main diagonal are zero. If $a_{ij} = 0$ for $i < j$, that is, if all elements above the main diagonal are zero, then **A** is *lower triangular*. Examples of upper and lower triangular matrices are, respectively,

$$\begin{bmatrix} -1 & 2 & 4 & 1 \\ 0 & 1 & 3 & -1 \\ 0 & 0 & 2 & 5 \\ 0 & 0 & 0 & 5 \end{bmatrix} \quad \text{and} \quad \begin{bmatrix} 5 & 0 & 0 & 0 \\ -1 & 2 & 0 & 0 \\ 0 & 1 & 3 & 0 \\ 2 & 1 & 4 & 1 \end{bmatrix}$$

▶THEOREM 2

The product of two lower (upper) triangular matrices of the same order is also lower (upper) triangular.◀

Proof: We prove this proposition for lower triangular matrices and leave the upper triangular case as an exercise (see Problem 35). Let $\mathbf{A}=[a_{ij}]$ and $\mathbf{B}=[b_{ij}]$ both be $n\times n$ lower triangular matrices, and set $\mathbf{AB}=\mathbf{C}=[c_{ij}]$. We need to show that $\mathbf{C}$ is lower triangular, or equivalently, that $c_{ij}=0$ when $i<j$. Now

$$c_{ij}=\sum_{k=1}^{n}a_{ik}b_{kj}=\sum_{k=1}^{j-1}a_{ik}b_{kj}+\sum_{k=1}^{n}a_{ik}b_{kj}$$

We are given that both $\mathbf{A}$ and $\mathbf{B}$ are lower triangular, hence $a_{ik}=0$ when $i<k$ and $b_{kj}=0$ when $k<j$. Thus,

$$\sum_{k=1}^{j-1}a_{ik}b_{kj}=\sum_{k=1}^{j-1}a_{ik}(0)=0$$

because in this summation k is always less than j. Furthermore, if we restrict $i<j$, then

$$\sum_{k=1}^{n}a_{ik}b_{kj}=\sum_{k=1}^{n}(0)b_{kj}=0$$

because $i<j\le k$. Thus, $c_{ij}=0$ when $i<j$.

Finally, we define positive integral powers of matrix in the obvious manner: $\mathbf{A}^2=\mathbf{AA}$, $\mathbf{A}^3=\mathbf{AAA}=\mathbf{AA}^2$ and, in general, for any positive integer n

$$\mathrm{A}^n=\underbrace{\mathrm{AA}\ldots\mathrm{A}}_{n-\text{times}} \tag{1.18}$$

For $n=0$, we define $\mathbf{A}^0=\mathbf{I}$.

Example 5 If $\mathbf{A}=\begin{bmatrix}1 & -2\\ 1 & 3\end{bmatrix}$, then $\mathbf{A}^2=\begin{bmatrix}1 & -2\\ 1 & 3\end{bmatrix}\begin{bmatrix}1 & -2\\ 1 & 3\end{bmatrix}=\begin{bmatrix}-1 & -8\\ 4 & 7\end{bmatrix}$

It follows directly from part (d) of Theorem 1 that

$$\left(\mathbf{A}^2\right)^{\mathrm{T}}=(\mathbf{AA})^{\mathrm{T}}=\mathbf{A}^{\mathrm{T}}\mathbf{A}^{\mathrm{T}}=\left(\mathbf{A}^{\mathrm{T}}\right)^2,$$

which may be generalized to

$$\left(\mathbf{A}^n\right)^{\mathrm{T}}=\left(\mathbf{A}^{\mathrm{T}}\right)^n \tag{1.19}$$

for any positive integer n.

Problems 1.3

(1) For each of the following pairs of matrices **A** and **B**, find the products $(\mathbf{AB})^T$, $\mathbf{A}^T\mathbf{B}^T$, and $\mathbf{B}^T\mathbf{A}^T$ and verify that $(\mathbf{AB})^T = \mathbf{B}^T\mathbf{A}^T$.

(a) $\mathbf{A} = \begin{bmatrix} 3 & 0 \\ 4 & 1 \end{bmatrix}, \quad \mathbf{B} = \begin{bmatrix} -1 & 2 & 1 \\ 3 & -1 & 0 \end{bmatrix}.$

(b) $\mathbf{A} = \begin{bmatrix} 2 & 2 & 2 \\ 3 & 4 & 5 \end{bmatrix}, \quad \mathbf{B} = \begin{bmatrix} 1 & 2 \\ 3 & 4 \\ 5 & 6 \end{bmatrix}.$

(c) $\mathbf{A} = \begin{bmatrix} 1 & 5 & -1 \\ 2 & 1 & 3 \\ 0 & 7 & -8 \end{bmatrix}, \quad \mathbf{B} = \begin{bmatrix} 6 & 1 & 3 \\ 2 & 0 & -1 \\ -1 & -7 & 2 \end{bmatrix}.$

(2) Verify that $(\mathbf{A}+\mathbf{B})^T = \mathbf{A}^T + \mathbf{B}^T$ for the matrices given in part (c) of Problem 1.

(3) Find $\mathbf{x}^T\mathbf{x}$ and $\mathbf{x}\mathbf{x}^T$ for $\mathbf{x} = \begin{bmatrix} 2 \\ 3 \\ 4 \end{bmatrix}$.

(4) Simplify the following expressions:

(a) $(\mathbf{AB}^T)^T$

(b) $(\mathbf{A}+\mathbf{B}^T)^T + \mathbf{A}^T$

(c) $[\mathbf{A}^T(\mathbf{B}+\mathbf{C}^T)]^T$

(d) $[(\mathbf{AB})^T + \mathbf{C}]^T$

(e) $[(\mathbf{A}+\mathbf{A}^T)\,(\mathbf{A}-\mathbf{A}^T)]^T$.

(5) Which of the following matrices are submatrices of $\mathbf{A} = \begin{bmatrix} 1 & 2 & 3 \\ 4 & 5 & 6 \\ 7 & 8 & 9 \end{bmatrix}$?

(a) $\begin{bmatrix} 1 & 3 \\ 7 & 9 \end{bmatrix}$, (b) $[1]$, (c) $\begin{bmatrix} 1 & 2 \\ 8 & 9 \end{bmatrix}$, (d) $\begin{bmatrix} 4 & 6 \\ 7 & 9 \end{bmatrix}$.

(6) Identify all of the nonempty submatrices of $\mathbf{A} = \begin{bmatrix} a & b \\ c & d \end{bmatrix}$.

(7) Partition $\mathbf{A} = \begin{bmatrix} 4 & 1 & 0 & 0 \\ 2 & 2 & 0 & 0 \\ 0 & 0 & 1 & 0 \\ 0 & 0 & 1 & 2 \end{bmatrix}$ into block diagonal form and then calculate $\mathbf{A}^2$.

(8) Partition $\mathbf{B} = \begin{bmatrix} 3 & 2 & 0 & 0 \\ -1 & 1 & 0 & 0 \\ 0 & 0 & 2 & 1 \\ 0 & 0 & 1 & -1 \end{bmatrix}$ into block diagonal form and then calculate $\mathbf{B}^2$.

(9) Use the matrices defined in Problems (7) and (8), partitioned into block diagonal form, to calculate **AB**.

(10) Use partitioning to calculate $\mathbf{A}^2$ and $\mathbf{A}^3$ for

$$\mathbf{A} = \begin{bmatrix} 1 & 0 & 0 & 0 & 0 & 0 \\ 0 & 2 & 0 & 0 & 0 & 0 \\ 0 & 0 & 0 & 1 & 0 & 0 \\ 0 & 0 & 0 & 0 & 1 & 0 \\ 0 & 0 & 0 & 0 & 0 & 1 \\ 0 & 0 & 0 & 0 & 0 & 0 \end{bmatrix}.$$

What is $\mathbf{A}^n$ for any positive integer $n > 3$?

(11) Determine which, if any, of the following matrices are in row-reduced form:

$$\mathbf{A} = \begin{bmatrix} 0 & 1 & 0 & 4 & -7 \\ 0 & 0 & 0 & 1 & 2 \\ 0 & 0 & 0 & 0 & 1 \\ 0 & 0 & 0 & 0 & 0 \end{bmatrix}, \quad \mathbf{B} = \begin{bmatrix} 1 & 1 & 0 & 4 & -7 \\ 0 & 1 & 0 & 1 & 2 \\ 0 & 0 & 1 & 0 & 1 \\ 0 & 0 & 0 & 1 & 5 \end{bmatrix},$$

$$\mathbf{C} = \begin{bmatrix} 1 & 1 & 0 & 4 & -7 \\ 0 & 1 & 0 & 1 & 2 \\ 0 & 0 & 0 & 0 & 1 \\ 0 & 0 & 0 & 1 & 5 \end{bmatrix}, \quad \mathbf{D} = \begin{bmatrix} 0 & 1 & 0 & 4 & -7 \\ 0 & 0 & 0 & 0 & 0 \\ 0 & 0 & 1 & 0 & 1 \\ 0 & 0 & 0 & 0 & 0 \end{bmatrix},$$

$$\mathbf{E} = \begin{bmatrix} 2 & 2 & 2 \\ 0 & 2 & 2 \\ 0 & 0 & 2 \end{bmatrix}, \quad \mathbf{F} = \begin{bmatrix} 0 & 0 & 0 \\ 0 & 0 & 0 \\ 0 & 0 & 0 \end{bmatrix},$$

$$\mathbf{G} = \begin{bmatrix} 1 & 2 & 3 \\ 0 & 0 & 1 \\ 1 & 0 & 0 \end{bmatrix}, \quad \mathbf{H} = \begin{bmatrix} 0 & 0 & 0 \\ 0 & 1 & 0 \\ 0 & 0 & 0 \end{bmatrix},$$

$$\mathbf{J} = \begin{bmatrix} 0 & 1 & 1 \\ 1 & 0 & 2 \\ 0 & 0 & 0 \end{bmatrix}, \quad \mathbf{K} = \begin{bmatrix} 1 & 0 & 2 \\ 0 & -1 & 1 \\ 0 & 0 & 0 \end{bmatrix},$$

$$\mathbf{L} = \begin{bmatrix} 2 & 0 & 0 \\ 0 & 2 & 0 \\ 0 & 0 & 0 \end{bmatrix}, \quad \mathbf{M} = \begin{bmatrix} 1 & 1/2 & 1/3 \\ 0 & 1 & 1/4 \\ 0 & 0 & 1 \end{bmatrix},$$

$$\mathbf{N} = \begin{bmatrix} 1 & 0 & 0 \\ 0 & 0 & 1 \\ 0 & 0 & 0 \end{bmatrix}, \quad \mathbf{Q} = \begin{bmatrix} 0 & 1 \\ 1 & 0 \end{bmatrix},$$

$$R = \begin{bmatrix} 1 & 1 \\ 0 & 0 \end{bmatrix}, \quad S = \begin{bmatrix} 1 & 0 \\ 1 & 0 \end{bmatrix},$$

$$T = \begin{bmatrix} 1 & 12 \\ 0 & 1 \end{bmatrix}.$$

(12) Determine which, if any, of the matrices in Problem 11 are upper triangular.

(13) Must a square matrix in row-reduced form necessarily be upper triangular?

(14) Must an upper triangular matrix be in row-reduced form?

(15) Can a matrix be both upper triangular and lower triangular simultaneously?

(16) Show that $\mathbf{AB} = \mathbf{BA}$ for

$$\mathbf{A} = \begin{bmatrix} -1 & 0 & 0 \\ 0 & 3 & 0 \\ 0 & 0 & 1 \end{bmatrix}, \quad \text{and} \quad \mathbf{B} = \begin{bmatrix} 5 & 0 & 0 \\ 0 & 3 & 0 \\ 0 & 0 & 2 \end{bmatrix}.$$

(17) Prove that if $\mathbf{A}$ and $\mathbf{B}$ are diagonal matrices of the same order, then $\mathbf{AB} = \mathbf{BA}$.

(18) Does a 2×2 diagonal matrix commute with every other 2×2 matrix?

(19) Calculate the products $\mathbf{AD}$ and $\mathbf{BD}$ for

$$\mathbf{A} = \begin{bmatrix} 1 & 1 & 1 \\ 1 & 1 & 1 \\ 1 & 1 & 1 \end{bmatrix}, \quad \mathbf{B} = \begin{bmatrix} 0 & 1 & 2 \\ 3 & 4 & 5 \\ 6 & 7 & 8 \end{bmatrix}, \quad \text{and} \quad \mathbf{D} = \begin{bmatrix} 2 & 0 & 0 \\ 0 & 3 & 0 \\ 0 & 0 & -5 \end{bmatrix}.$$

What conclusions can you make about postmultiplying a square matrix by a diagonal matrix?

(20) Calculate the products $\mathbf{DA}$ and $\mathbf{DB}$ for the matrices defined in Problem 19. What conclusions can you make about premultiplying a square matrix by a diagonal matrix?

(21) Prove that $(\mathbf{A}^T)^T = \mathbf{A}$ for any matrix $\mathbf{A}$.

(22) Prove that $(\lambda\mathbf{A})^T = \lambda\mathbf{A}^T$ for any matrix $\mathbf{A}$ and any scalar λ.

(23) Prove that if $\mathbf{A}$ and $\mathbf{B}$ are matrices of the same order then $(\mathbf{A}+\mathbf{B})^T = \mathbf{A}^T + \mathbf{B}^T$.

(24) Let $\mathbf{A}$, $\mathbf{B}$, and $\mathbf{C}$ be matrices of orders $m \times p$, $p \times r$, and $r \times s$, respectively. Prove that $(\mathbf{ABC})^T = \mathbf{C}^T\mathbf{B}^T\mathbf{A}^T$.

(25) Prove that if $\mathbf{A}$ is a square matrix, then $\mathbf{B} = (\mathbf{A} + \mathbf{A}^T)/2$ is a symmetric matrix.

(26) Prove that if $\mathbf{A}$ is a square matrix, then $\mathbf{C} = (\mathbf{A} - \mathbf{A}^T)/2$ is a skew-symmetric matrix.

(27) Use the results of the last two problems to prove that any square matrix can be written as the sum of a symmetric matrix and a skew-symmetric matrix.

(28) Write the matrix **A** in part (c) of Problem 1 as the sum of a symmetric matrix and a skew-symmetric matrix.

(29) Write the matrix **B** in part (c) of Problem 1 as the sum of a symmetric matrix and a skew-symmetric matrix.

(30) Prove that $\mathbf{AA}^{\mathrm{T}}$ is symmetric for any matrix **A**.

(31) Prove that the diagonal elements of a skew-symmetric matrix must be zero.

(32) Prove that if a 2×2 matrix **A** commutes with *every* 2×2 diagonal matrix, then **A** must be diagonal. *Hint:* Consider, in particular, the diagonal matrix $\mathbf{D} = \begin{bmatrix} 1 & 0 \\ 0 & 0 \end{bmatrix}$.

(33) Prove that if a $n \times n$ matrix **A** commutes with every $n \times n$ diagonal matrix, the **A** must be diagonal.

(34) Prove that if $\mathbf{D} = [d_{ij}]$ is a diagonal matrix, then $\mathbf{D}^2 = [d_{ij}^2]$.

(35) Prove that the product of two upper triangular matrices is upper triangular.

1.4 LINEAR SYSTEMS OF EQUATIONS

Systems of simultaneous linear equations appear frequently in engineering and scientific problems. The need for efficient methods that solve such systems was one of the historical forces behind the introduction of matrices, and that need continues today, especially for solution techniques that are applicable to large systems containing hundreds of equations and hundreds of variables.

A system of m-linear equations in n-variables $x_1, x_2, \ldots, x_n$ has the general form

$$\begin{aligned} a_{11}x_1 + a_{12}x_2 + \cdots + a_{1n}x_n &= b_1 \\ a_{21}x_1 + a_{22}x_2 + \cdots + a_{2n}x_n &= b_2 \\ &\vdots \\ a_{m1}x_1 + a_{m2}x_2 + \cdots + a_{mn}x_n &= b_m \end{aligned} \tag{1.20}$$

where the coefficients a_{ij} $(i = 1, 2, \ldots, m;\ j = 1, 2, \ldots, n)$ and the quantities b_i are all known scalars. The variables in a linear equation appear only to the first power and are multiplied only by known scalars. Linear equations do *not* involve products of variables, variables raised to powers other than one, or variables appearing as arguments of transcendental functions.

For systems containing a few variables, it is common to denote the variables by distinct letters such as x, y, and z. Such labeling is impractical for systems involving hundreds of variables; instead a single letter identifies all variables with different numerical subscripts used to distinguished different variables, such as $x_1, x_2, \ldots, x_n$.

Example 1 The system

$$\begin{aligned} 2x + 3y - z &= 12,000 \\ 4x - 5y + 6z &= 35,600 \end{aligned}$$

of two equations in the variables x, y, and z is linear, as is the system

$$\begin{aligned} 20x_1 + 80x_2 + 35x_3 + 40x_4 + 55x_5 &= -0.005 \\ 90x_1 - 15x_2 - 70x_3 + 25x_4 + 55x_5 &= 0.015 \\ 30x_1 + 35x_2 - 35x_3 + 10x_4 + 65x_5 &= -0.015 \end{aligned}$$

of three equations with five variables $x_1, x_2, \ldots, x_5$. In contrast, the system

$$\begin{aligned} 2x + 3xy &= 25 \\ 4\sqrt{x} + \sin y &= 50 \end{aligned}$$

is not linear for many reasons: it contains a product xy of variables; it contains the variable x raised to the one-half power; and it contains the variable y as the argument of the transcendental sine function.

As shown in Section 1.2, any linear system of form (1.20) can be rewritten in the matrix form

$$\mathbf{Ax} = \mathbf{b} \qquad (1.14 \text{ repeated})$$

with

$$\mathbf{A} = \begin{bmatrix} a_{11} & a_{12} & \cdots & a_{1n} \\ a_{21} & a_{22} & \cdots & a_{2n} \\ \vdots & \vdots & \ddots & \vdots \\ a_{m1} & a_{m2} & \cdots & a_{mn} \end{bmatrix}, \quad \mathbf{x} = \begin{bmatrix} x_1 \\ x_2 \\ \vdots \\ x_n \end{bmatrix}, \quad \text{and} \quad \mathbf{b} = \begin{bmatrix} b_1 \\ b_2 \\ \vdots \\ b_m \end{bmatrix}.$$

If $m \neq n$, then **A** is *not* square and the dimensions of **x** and **b** will be different.

A *solution* to linear system (1.20) is a set of scalar values for the variables $x_1, x_2, \ldots, x_n$ that when substituted into each equation of the system makes each equation true.

A solution to linear system of equations is a set of scalar values for the variables that when substituted into each equation of the system makes each equation true.

Example 2 The scalar values $x=2$ and $y=3$ are a solution to the system

$$\begin{aligned} 3x + 2y &= 12 \\ 6x + 4y &= 24 \end{aligned}$$

A second solution is $x=-4$ and $y=12$. In contrast, the scalar values $x=1$, $y=2$, and $z=3$ are *not* a solution to the system

$$\begin{aligned} 2x + 3y + 4z &= 20 \\ 4x + 5y + 6z &= 32 \\ 7x + 8y + 9z &= 40 \end{aligned}$$

because these values do not make the third equation true, even though they do satisfy the first two equations of the system.

▶THEOREM 1

If $\mathbf{x}_1$ and $\mathbf{x}_2$ are two different solutions of $\mathbf{Ax}=\mathbf{b}$, then $\mathbf{z}=\alpha\mathbf{x}_1+\beta\mathbf{x}_2$ is also a solution for any real numbers α and β with $\alpha+\beta=1$.◀

Proof: $\mathbf{x}_1$ and $\mathbf{x}_2$ are given as solutions of $\mathbf{Ax}=\mathbf{b}$, hence $\mathbf{Ax}_1=\mathbf{b}$, and $\mathbf{Ax}_2=\mathbf{b}$. Then

$$\mathbf{Az} = \mathbf{A}(\alpha\mathbf{x}_1 + \beta\mathbf{x}_2) = \alpha(\mathbf{Ax}_1) + \beta(\mathbf{Ax}_2) = \alpha\mathbf{b} + \beta\mathbf{b} = (\alpha + \beta)\mathbf{b} = \mathbf{b}.$$

so $\mathbf{z}$ is also a solution.

Because there are infinitely many ways to form $\alpha+\beta=1$ (let α be any real number and set $\beta=1-\alpha$), it follows from Theorem 1 that once we identify two solutions we can combine them into infinitely many other solutions. Consequently, the number of possible solutions to a system of linear equations is either none, one, or infinitely many.

The graph of a linear equation in two variables is a line in the plane; hence a system of linear equations in two variables is depicted graphically by a set of lines. A solution to such a system is a set of coordinates for a point in the plane that lies on *all* the lines defined by the equations. In particular, the graphs of the equations in the system

$$\begin{aligned} x + y &= 1 \\ x - y &= 0 \end{aligned} \tag{1.21}$$

are shown in Figure 1.3. There is only one point of intersection, and the coordinates of this point $x=y=1/2$ is the unique solution to system (1.21). In contrast, the graphs of the equations in the system

$$\begin{aligned} x + y &= 1 \\ x + y &= 2 \end{aligned} \tag{1.22}$$

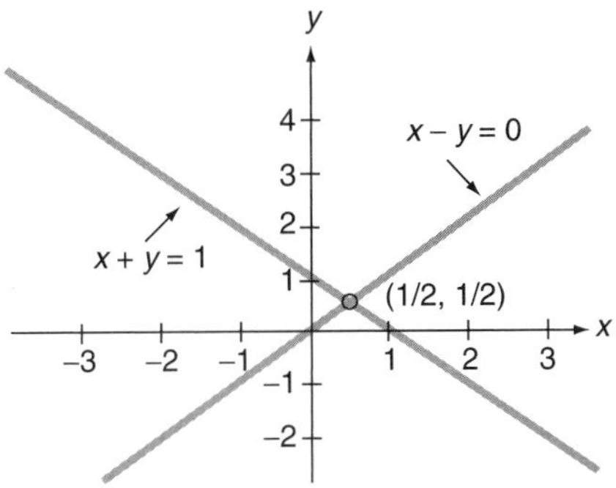

FIGURE 1.3

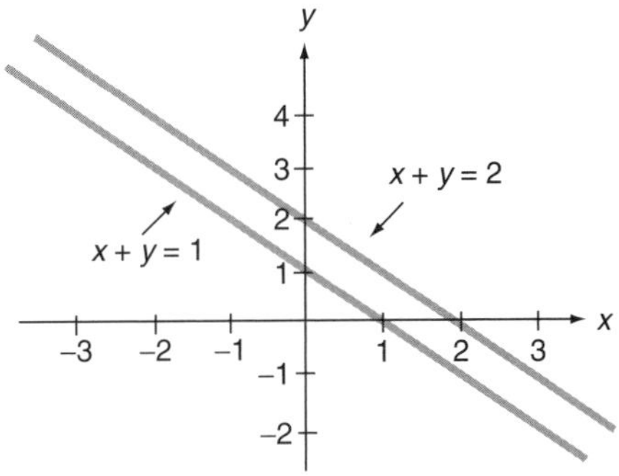

FIGURE 1.4

are shown in Figure 1.4. The lines are parallel and have no points of intersection, so system (1.22) has no solution. Finally, the graphs of the equations in the system

$$\begin{aligned} x + y &= 0 \\ 2x + 2y &= 0 \end{aligned} \tag{1.23}$$

are shown in Figure 1.5. The lines overlap, hence every point on either line is a point of intersection and system (1.23) has infinitely many solutions.

A system of simultaneous linear equations is *consistent* if it possesses at least one solution. If no solution exists, the system is *inconsistent.* Systems (1.21) and (1.23) are consistent; system (1.22) in inconsistent.

The graph of a linear equation in three variables is a plane in space; hence a system of linear equations in three variables is depicted graphically by a set of planes. A solution to such a system is the set of coordinates for a point in space that lies on *all* the planes defined by the equations. Such a system can have no solutions, one solution, or infinitely many solutions.

Figure 1.6 shows three planes that intersect at a single point, and it represents a system of three linear equations in three variables with a unique solution.

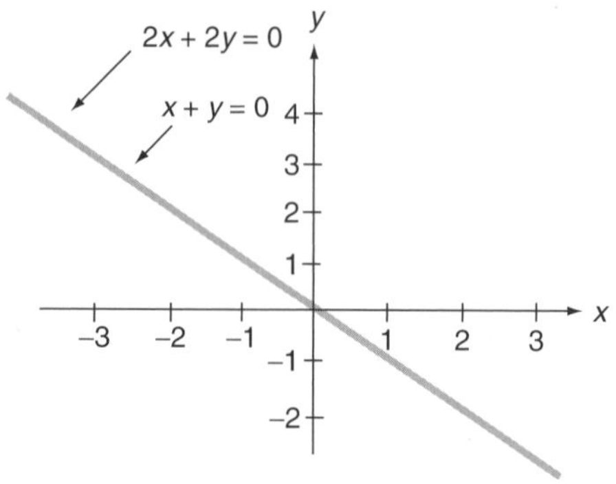

FIGURE 1.5

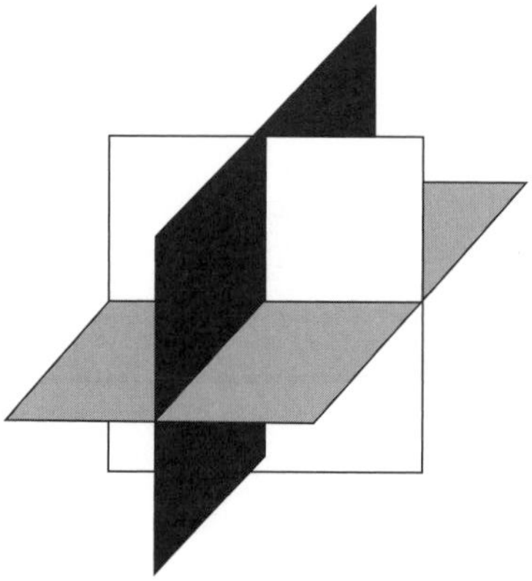

FIGURE 1.6

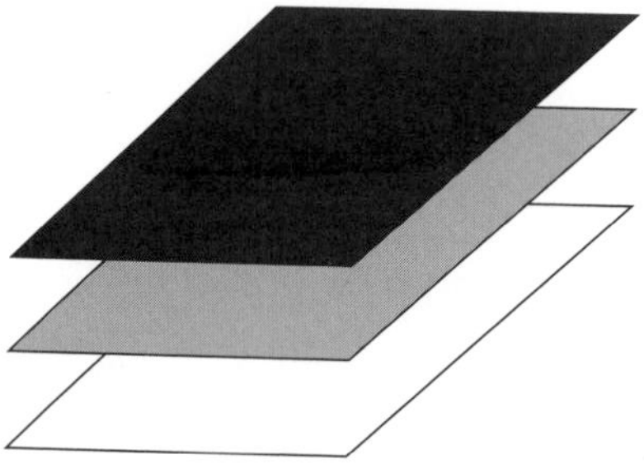

FIGURE 1.7

Figures 1.7 and 1.8 show systems of planes that have no points that lie on *all* three planes; each figure depicts a different system of three linear equations in three unknowns with no solutions. Figure 1.9 shows three planes intersecting at a line, and it represents a system of three equations in three variables with infinitely many solutions, one solution corresponding to each point on the line. A different example of infinitely many solutions is obtained by collapsing the

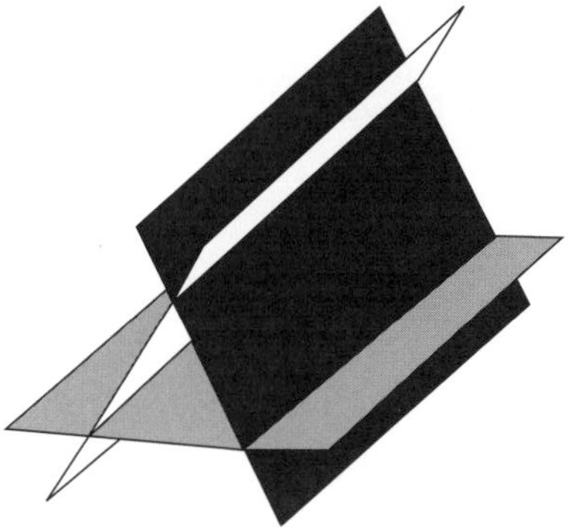

FIGURE 1.8

FIGURE 1.9

three planes in Figure 1.7 onto each other so that each plane is an exact copy of the others. Then every point on one plane is also on the other two.

System (1.20) is *homogeneous* if the right side of each equation is 0; that is, if $b_1 = b_2 = \ldots = b_m = 0$. In matrix form, we say that the system $\mathbf{Ax} = \mathbf{b}$ is homogeneous if $\mathbf{b} = \mathbf{0}$, a zero column matrix. If $\mathbf{b} \neq \mathbf{0}$, which implies that at least one component of $\mathbf{b}$ differs from 0, then the system of equations is *nonhomogeneous.* System (1.23) is homogeneous; systems (1.21) and (1.22) are nonhomogeneous. One solution to a homogeneous system of equations is obtained by setting all variables equal to 0. This solution is called the *trivial solution*. Thus, we have the following theorem.

A homogeneous system of linear equations has the matrix form $\mathbf{Ax} = \mathbf{0}$; one solution is the trivial solution $\mathbf{x} = \mathbf{0}$.

▶THEOREM 2

A homogeneous system of linear equations is consistent.◀

The augmented matrix for $\mathbf{Ax}=\mathbf{b}$ is the partitioned matrix $[\mathbf{A}|\mathbf{b}]$.

All the scalars contained in the system of equations $\mathbf{Ax}=\mathbf{b}$ appear in the coefficient matrix $\mathbf{A}$ and the column matrix $\mathbf{b}$. These scalars can be combined into the single partitioned matrix $[\mathbf{A}|\mathbf{b}]$, known as the *augmented matrix* for the system of equations.

Example 3 The system

$$\begin{aligned} x_1 + x_2 - 2x_3 &= -3 \\ 2x_1 + 5x_2 + 3x_3 &= 11 \\ -x_1 + 3x_2 + x_3 &= 5 \end{aligned}$$

can be written as the matrix equation

$$\begin{bmatrix} 1 & 1 & -2 \\ 2 & 5 & 3 \\ -1 & 3 & 1 \end{bmatrix} \begin{bmatrix} x_1 \\ x_2 \\ x_3 \end{bmatrix} = \begin{bmatrix} -3 \\ 11 \\ 5 \end{bmatrix}$$

which has as its augmented matrix

$$[\mathbf{A}|\mathbf{b}] = \left[\begin{array}{ccc|c} 1 & 1 & -2 & -3 \\ 2 & 5 & 3 & 11 \\ -1 & 3 & 1 & 5 \end{array}\right].$$

Example 4 Write the set of equation in x, y, and z associated with the augmented matrix

$$[\mathbf{A}|\mathbf{b}] = \left[\begin{array}{ccc|c} -2 & 1 & 3 & 8 \\ 0 & 4 & 5 & -3 \end{array}\right]$$

Solution:

$$\begin{aligned} -2x + y + 3z &= 8 \\ 4y + 5z &= -3 \end{aligned}$$

The traditional approach to solving a system of linear equations is to manipulate the equations so that the resulting equations are easy to solve and have the *same* solutions as the original equations. Three operations that alter equations but do *not* change their solutions are:

(i) Interchange the positions of any two equations.

(ii) Multiply an equation by a nonzero scalar.

(iii) Add to one equation a scalar times another equation.

If we restate these operations in words appropriate to an augmented matrix, we obtain the three *elementary row operations:*

(R_1) Interchange any two rows in a matrix.

(R_2) Multiply any row of a matrix by a nonzero scalar.

(R_3) Add to one row of a matrix a scalar times another row of that same matrix.

Gaussian elimination is a four-step matrix method, centered on the three elementary row operations, for solving simultaneous linear equations.

GAUSSIAN ELIMINATION

Step 1. Construct an augmented matrix for the given system of equations.
Step 2. Use elementary row operations to transform the augmented matrix into an augmented matrix in row-reduced form.
Step 3. Write the equations associated with the resulting augmented matrix.
Step 4. Solve the new set of equations by back substitution.

The new set of equations resulting from Step 3 is called the *derived set*, and it is solved easily by back-substitution. Each equation in the derived set is solved for the first unknown that appears in that equation with a nonzero coefficient, beginning with the last equation and sequentially moving through the system until we reach the first equation. By limiting Gaussian elimination to elementary row operations, we are assured that the derived set of equations has the same solutions as the original set.

Most of the work in Gaussian elimination occurs in the second step: the transformation of an augmented matrix to row-reduced form. In transforming a matrix to row-reduced form, it is advisable to adhere to three basic principles:

(i) Completely transform one column to the required form before considering another column.
(ii) Work on columns in order, from left to right.
(iii) Never use an operation that changes a zero in a previously transformed column.

Example 5 Use Gaussian elimination to solve the system

$$\begin{aligned} x + 3y &= 4, \\ 2x - y &= 1, \\ 3x + 2y &= 5, \\ 5x + 15y &= 20. \end{aligned}$$

Solution: The augmented matrix for this system is

$$\left[\begin{array}{rr|r} 1 & 3 & 4 \\ 2 & -1 & 1 \\ 3 & 2 & 5 \\ 5 & 15 & 20 \end{array}\right]$$

We transform this augmented matrix into row-reduced form using only the three elementary row operations. The first nonzero element in the first row appears in the 1-1 position, so use elementary row operation R_3 to transform all other elements in the first column to zero.

$$\begin{bmatrix} 1 & 3 & | & 4 \\ 2 & -1 & | & 1 \\ 3 & 2 & | & 5 \\ 5 & 15 & | & 20 \end{bmatrix} \rightarrow \begin{bmatrix} 1 & 3 & | & 4 \\ 0 & -7 & | & -7 \\ 3 & 2 & | & 5 \\ 5 & 15 & | & 20 \end{bmatrix}$$ by adding to the second row −2 times the first row

$$\rightarrow \begin{bmatrix} 1 & 3 & | & 4 \\ 0 & -7 & | & -7 \\ 0 & -7 & | & -7 \\ 5 & 15 & | & 20 \end{bmatrix}$$ by adding to the third row −3 times the first row

$$\rightarrow \begin{bmatrix} 1 & 3 & | & 4 \\ 0 & -7 & | & -7 \\ 0 & -7 & | & -7 \\ 0 & 0 & | & 0 \end{bmatrix}$$ by adding to the fourth row −5 times the first row

The first row and the first column are correctly formatted, so we turn our attention to the second row and second column. We use elementary row operations on the current augmented matrix to transform the first nonzero element in the second row to one and then all elements under it, in the second column, to zero. Thus,

$$\rightarrow \begin{bmatrix} 1 & 3 & | & 4 \\ 0 & -1 & | & -1 \\ 0 & -7 & | & -7 \\ 0 & 0 & | & 0 \end{bmatrix}$$ by multiplying the second row by −1/7

$$\rightarrow \begin{bmatrix} 1 & 3 & | & 4 \\ 0 & 1 & | & 1 \\ 0 & 0 & | & 0 \\ 0 & 0 & | & 0 \end{bmatrix}$$ by adding to the third row 7 times the second row

This augmented matrix is in row-reduced form, and the system of equations associated with it is the derived set

$$\begin{aligned} x + 3y &= 4 \\ y &= 1 \\ 0 &= 0 \\ 0 &= 0. \end{aligned}$$

Solving the second equation for y and then the first equation for x, we obtain $x = 1$ and $y = 1$ as the solution to both this last set of equations and also the original set of equations.

A pivot is transformed to unity prior to using it to cancel other elements to zero.

When one element in a matrix is used to convert another element to zero by elementary row operation R_3, the first element is called a, *pivot*. In Example 5, we used the element in the 1-1 position first to cancel the element in the 2-1 position and then to cancel the elements in the 3-1 and 4-1 positions. In each case, the unity element in the 1-1 position was the pivot. Later, we used the unity element in the 2-2 position to cancel the element −7 in the 3-2 position; here, the 2-2 element served as the pivot. We shall always use elementary row operation R_2 to transform a pivot to unity before using the pivot to transform other elements to zero.

Example 6 Use Gaussian elimination to solve the system

$$\begin{aligned} x + 2y + z &= 3, \\ 2x + 3y - z &= -6, \\ 3x - 2y - 4z &= -2. \end{aligned}$$

Solution: Transforming the augmented matrix for this system into row-reduced form using only elementary row operations, we obtain

$$\left[\begin{array}{ccc|c} 1 & 2 & 1 & 3 \\ 2 & 3 & -1 & -6 \\ 3 & -2 & -4 & -2 \end{array}\right] \to \left[\begin{array}{ccc|c} 1 & 2 & 1 & 3 \\ 0 & -1 & -3 & -12 \\ 3 & -2 & -4 & -2 \end{array}\right] \quad \begin{array}{l}\text{by adding to} \\ \text{the second row } -2 \\ \text{times the first row}\end{array}$$

$$\to \left[\begin{array}{ccc|c} 1 & 2 & 1 & 3 \\ 0 & -1 & -3 & -12 \\ 0 & -8 & -7 & -11 \end{array}\right] \quad \begin{array}{l}\text{by adding to} \\ \text{the third row } -3 \\ \text{times the first row}\end{array}$$

$$\to \left[\begin{array}{ccc|c} 1 & 2 & 1 & 3 \\ 0 & 1 & 3 & -12 \\ 0 & -8 & -7 & -11 \end{array}\right] \quad \begin{array}{l}\text{by multiplying the} \\ \text{second row by } -1\end{array}$$

$$\to \left[\begin{array}{ccc|c} 1 & 2 & 1 & 3 \\ 0 & 1 & 3 & 12 \\ 0 & 0 & 17 & 85 \end{array}\right] \quad \begin{array}{l}\text{by adding to the} \\ \text{third row 8 times} \\ \text{the second row}\end{array}$$

$$\to \left[\begin{array}{ccc|c} 1 & 2 & 1 & 3 \\ 0 & 1 & 3 & 12 \\ 0 & 0 & 1 & 5 \end{array}\right] \quad \begin{array}{l}\text{by multiplying the} \\ \text{third row by } 1/17\end{array}$$

This augmented matrix is in row-reduced form; the derived set is

$$\begin{aligned} x + 2y + z &= 3 \\ y + 3z &= 12 \\ z &= 5 \end{aligned}$$

Solving the third equation for z, then the second equation for y, and lastly, the first equation for x, we obtain $x=4$, $y=-3$, and $z=5$ as the solution to both this last system and the original system of equations.

Elementary row operation R_1 is used to move potential pivots into more useful locations by rearranging the positions of rows.

Example 7 Use Gaussian elimination to solve the system

$$\begin{aligned} 2x_3 + 3x_4 &= 0 \\ x_1 + 3x_3 + x_4 &= 0 \\ x_1 + x_2 + 2x_3 &= 0 \end{aligned}$$

Solution: The augmented matrix for this system is

$$\left[\begin{array}{cccc|c} 0 & 0 & 2 & 3 & 0 \\ 1 & 0 & 3 & 1 & 0 \\ 1 & 1 & 2 & 0 & 0 \end{array}\right]$$

Normally, we would use the element in the 1-1 position to transform to zero the two elements directly below it, but we cannot because the 1-1 element is itself zero. To move a nonzero element into the ideal pivot position, we interchange the first row with either of the other two rows. The choice is arbitrary.

$$\left[\begin{array}{cccc|c} 0 & 0 & 2 & 3 & 0 \\ 1 & 0 & 3 & 1 & 0 \\ 1 & 1 & 2 & 0 & 0 \end{array}\right] \rightarrow \left[\begin{array}{cccc|c} 1 & 0 & 3 & 1 & 0 \\ 0 & 0 & 2 & 3 & 0 \\ 1 & 1 & 2 & 0 & 0 \end{array}\right] \quad \text{by interchanging the first and second rows}$$

$$\rightarrow \left[\begin{array}{cccc|c} 1 & 0 & 3 & 1 & 0 \\ 0 & 0 & 2 & 3 & 0 \\ 0 & 1 & -1 & -1 & 0 \end{array}\right] \quad \text{by adding to the third row } -1 \text{ times the first row}$$

$$\rightarrow \left[\begin{array}{cccc|c} 1 & 0 & 3 & 1 & 0 \\ 0 & 1 & -1 & -1 & 0 \\ 0 & 0 & 2 & 3 & 0 \end{array}\right] \quad \text{by interchanging the second and third rows}$$

$$\rightarrow \left[\begin{array}{cccc|c} 1 & 0 & 3 & 1 & 0 \\ 0 & 1 & -1 & -1 & 0 \\ 0 & 0 & 1 & 3/2 & 0 \end{array}\right] \quad \text{by multiplying the third row by } 1/2$$

This augmented matrix is in row-reduced form; the derived set is

$$\begin{aligned} x_1 + 3x_3 + x_4 &= 0 \\ x_2 - x_3 - x_4 &= 0 \\ x_3 + \tfrac{3}{2}x_4 &= 0 \end{aligned}$$

We use the third equation to solve for x_3, then the second equation to solve for x_2, and lastly, the first equation to solve for x_1, because in each case those are the variables that appear first in the respective equations. There is *no* defining equation for x_4, so this variable remains arbitrary, and we solve for the other variables in terms of it. The solution to both this last set of equations and the original set of equations is $x_1=(7/2)x_4$, $x_2=(-1/2)x_4$ and $x_3=(-3/2)x_4$ with x_4 arbitrary. The solution can be written as the column matrix

If the solution to a derived set involves at least one arbitrary unknown, then the original system has infinitely many solutions.

$$\mathbf{x} = \begin{bmatrix} x_1 \\ x_2 \\ x_3 \\ x_4 \end{bmatrix} = \begin{bmatrix} (7/2)x_4 \\ (-1/2)x_4 \\ (-3/2)x_4 \\ x_4 \end{bmatrix} = \frac{x_4}{2}\begin{bmatrix} 7 \\ -1 \\ -3 \\ 2 \end{bmatrix}$$

Example 7 is a system of equations with infinitely many solutions, one for each real number assigned to the arbitrary variable x_4. Infinitely many solutions occur when the derived set of equations is consistent and has more unknowns than equations. If a derived set contains n variables and r equations, $n>r$, then each equation in the derived set is solved for the first variable in that equation with a nonzero coefficient; this defines r variables and leaves the remaining $n-r$ variables as arbitrary. These arbitrary variables may be chosen in infinitely many ways to produce solutions.

A homogeneous set of linear equations is always consistent. If such a system has more variables than equations, then its derived set will also have more variables than equations, resulting in infinitely many solutions. Thus, we have the following important result:

▸THEOREM 3

A homogeneous system of linear equations containing more variables than equations has infinitely many solutions. ◂

In contrast to homogeneous systems, a nonhomogeneous system may have no solutions. If a derived set of equations contains a false equation, such as $0=1$, that set is inconsistent because no values for the variables can make the false equation true. Because the derived set has the same solutions as the original set, it follows that the original set is also inconsistent.

If a derived set contains a false equation, then the original set of equations has no solution.

Example 8 Use Gaussian elimination to solve the system

$$x + 2y = 2,$$
$$3x + 6y = 7.$$

Solution: Transforming the augmented matrix for this system into row-reduced form, we obtain

$$\left[\begin{array}{cc|c} 1 & 2 & 2 \\ 3 & 6 & 7 \end{array}\right] \rightarrow \left[\begin{array}{cc|c} 1 & 2 & 2 \\ 0 & 0 & 1 \end{array}\right]$$ by adding the second row -3 times the first row

This augmented matrix is in row-reduced form; the derived set is

$$\begin{aligned} x + 2y &= 2 \\ 0 &= 1 \end{aligned}$$

No values of x and y can make this last equation true, so the derived set, as well as the original set of equations, has no solution.

Finally, we note that most augmented matrices can be transformed into a variety of row-reduced forms. If a row-reduced augmented matrix has two nonzero rows, then a different row-reduced augmented matrix is easily constructed by adding to the first row any nonzero constant times the second row. The equations associated with both augmented matrices, different as they may be, will have identical solutions.

Problems 1.4

(1) Determine whether the proposed values of x, y, and z are solutions to:

$$\begin{aligned} x + y + 2z &= 2, \\ x - y - 2z &= 0, \\ x + 2y + 2z &= 1. \end{aligned}$$

(a) $x=1, \quad y=-3, \quad z=2.$ (b) $x=1, \quad y=-1, \quad z=1.$

(2) Determine whether the proposed values of x_1, x_2, and x_3 are solutions to:

$$\begin{aligned} x_1 + 2x_2 + 3x_3 &= 6, \\ x_1 - 3x_2 + 2x_3 &= 0, \\ 3x_1 - 4x_2 + 7x_3 &= 6. \end{aligned}$$

(a) $x_1=1, \quad x_2=1, \ x_3=1.$
(b) $x_1=2, \quad x_2=2, \ x_3=0.$
(c) $x_1=14, \ x_2=2, \ x_3=-4.$

(3) Find a value for k such that $x=2$ and $y=k$ is a solution of the system

$$\begin{aligned} 3x + 5y &= 11, \\ 2x - 7y &= -3. \end{aligned}$$

(4) Find a value for k such that $x=2k$, $y=-k$, and $z=0$ is a solution of the system

$$\begin{aligned} x+2y+z&=0,\\ -2x-4y+2z&=0,\\ 3x-6y-4z&=1. \end{aligned}$$

(5) Find a value for k such that $x=2k$, $y=-k$, and $z=0$ is a solution of the system

$$\begin{aligned} x+2y+2z&=0,\\ 2x+4y+2z&=0,\\ -3x-6y-4z&=0. \end{aligned}$$

In Problems 6 through 11, write the set of equations associated with the given augmented matrix and the specified variables and then solve.

(6) $\left[\begin{array}{cc|c} 1 & 2 & 5 \\ 0 & 1 & 8 \end{array}\right]$ for x and y.

(7) $\left[\begin{array}{ccc|c} 1 & -2 & 3 & 10 \\ 0 & 1 & -5 & -3 \\ 0 & 0 & 1 & 4 \end{array}\right]$ for x, y, and z.

(8) $\left[\begin{array}{ccc|c} 1 & -3 & 12 & 40 \\ 0 & 1 & -6 & -200 \\ 0 & 0 & 1 & 25 \end{array}\right]$ for x_1, x_2, and x_3.

(9) $\left[\begin{array}{ccc|c} 1 & 3 & 0 & -8 \\ 0 & 1 & 4 & 2 \\ 0 & 0 & 0 & 0 \end{array}\right]$ for x, y, and z.

(10) $\left[\begin{array}{ccc|c} 1 & -7 & 2 & 0 \\ 0 & 1 & -1 & 0 \\ 0 & 0 & 0 & 0 \end{array}\right]$ for x_1, x_2, and x_3.

(11) $\left[\begin{array}{ccc|c} 1 & -1 & 0 & 1 \\ 0 & 1 & -2 & 2 \\ 0 & 0 & 1 & -3 \\ 0 & 0 & 0 & 1 \end{array}\right]$ for x_1, x_2, and x_3.

In Problems 12 through 29, use Gaussian elimination to solve the given system of equations.

(12) $\begin{aligned} x-2y&=5,\\ -3x+7y&=8. \end{aligned}$

(13) $\begin{aligned} 4x+24y&=20,\\ 2x+11y&=-8. \end{aligned}$

(14) $$-y = 6,\quad 2x + 7y = -5.$$

(15) $$-x + 3y = 0,\quad 3x + 5y = 0.$$

(16) $$-x + 3y = 0,\quad 3x - 9y = 0.$$

(17) $$x + 2y + 3z = 4,\quad -x - 7 + 2z = 3,\quad -2x + 3y = 0.$$

(18) $$y - 2z = 4,\quad x + 3y + 2z = 1,\quad -2x + 3y + z = 2.$$

(19) $$x + 3y + 2z = 0,\quad -x - 4y + 3z = -1,\quad 2x - z = 3,\quad 2x - y + 4z = 2.$$

(20) $$2x + 4y - z = 0,\quad -4x - 8y + 2z = 0,\quad -2x - 4y + z = -1.$$

(21) $$-3x + 6y - 3z = 0,\quad x - 2y + z = 0,\quad x - 2y + z = 0.$$

(22) $$-3x + 3y - 3z = 0,\quad x - y + 2z = 0,\quad 2x - 2y + z = 0,\quad x + y + z = 0.$$

(23) $$-3x_1 + 6x_2 - 3x_3 = 0,\quad x_1 - x_2 + x_3 = 0.$$

(24) $$x_1 - x_2 + 2x_3 = 0,\quad 2x_1 - 2x_2 + 4x_3 = 0.$$

(25) $$x_1 + 2x_2 = -3,\quad 3x_1 + x_2 = 1.$$

(26) $$x_1 + 2x_2 + x_3 = -1,\quad 2x_1 - 3x_2 + 2x_3 = 4.$$

(27) $$x_1 + 2x_2 = 5,\quad -3x_1 + x_2 = 13,\quad 4x_1 + 3x_2 = 0.$$

(28) $$2x_1 + 4x_2 = 2,\quad 3x_1 + 2x_2 + x_3 = 8,\quad 5x_1 - 3x_2 + 7x_3 = 15.$$

(29) $$2x_1 + 3x_2 - 4x_3 = 2,\quad 3x_1 - 2x_2 = -1,\quad 8x_1 - x_2 - 4x_3 = 10.$$

(30) Show graphically that the number of solutions to a linear system of two equations in three variables is either none or infinitely many.

(31) Let $\mathbf{y}$ be a solution to $\mathbf{Ax}=\mathbf{b}$ and let $\mathbf{z}$ be a solution to the associated homogeneous system $\mathbf{Ax}=\mathbf{0}$. Prove that $\mathbf{u}=\mathbf{y}+\mathbf{z}$ is also a solution to $\mathbf{Ax}=\mathbf{b}$.

(32) Let $\mathbf{y}$ and $\mathbf{z}$ be as defined in Problem 31.
(a) For what scalars α is $\mathbf{u}=\mathbf{y}+\alpha\mathbf{z}$ also a solution to $\mathbf{Ax}=\mathbf{b}$?
(b) For what scalars α is $\mathbf{u}=\alpha\mathbf{y}+\mathbf{z}$ also a solution to $\mathbf{Ax}=\mathbf{b}$?

In Problems 33 through 40, establish a set of equations that models each process and then solve.

(33) A manufacturer receives daily shipments of 70,000 springs and 45,000 pounds of stuffing for producing regular and support mattresses. Regular mattresses r require 50 springs and 30 pounds of stuffing; support mattresses s require 60 springs and 40 pounds of stuffing. How many mattresses of each type should be produced daily to utilize all available inventory?

(34) A manufacturer produces desks and bookcases. Desks d require 5 hours of cutting time and 10 hours of assembling time. Bookcases b require 15 minutes of cutting time and 1 hour of assembling time. Each day the manufacturer has available 200 hours for cutting and 500 hours for assembling. How many desks and bookcases should be scheduled for completion each day to utilize all available workpower?

(35) A mining company has a contract to supply 70,000 tons of low-grade ore, 181,000 tons of medium-grade ore, and 41,000 tons of high-grade ore to a supplier. The company has three mines that it can work. Mine A produces 8000 tons of low-grade ore, 5000 tons of medium-grade ore, and 1000 tons of high-grade ore during each day of operation. Mine B produces 3000 tons of low-grade ore, 12,000 tons of medium-grade ore, and 3000 tons of high-grade ore for each day it is in operation. The figures for mine C are 1000, 10,000, and 2000, respectively. How many days must each mine operate to meet contractual demands without producing a surplus?

(36) A small company computes its end-of-the- year bonus b as 5% of the net profit after city and state taxes have been paid. The city tax c is 2% of taxable income, while the state tax s is 3% of taxable income with credit allowed for the city tax as a pretax deduction. This year, taxable income was \$400,000. What is the bonus?

(37) A gasoline producer has \$800,000 in fixed annual costs and incurs an additional variable cost of \$30 per barrel B of gasoline. The total cost C is the sum of the fixed and variable costs. The net sales S is computed on a wholesale price of \$40 per barrel.

(a) Show that C, B, and S are related by two simultaneous equations.

(b) How many barrels must be produced to break even, that is, for net sales to equal cost?

(38) **(Leontief Closed Models)** A closed economic model involves a society in which all the goods and services produced by members of the society are consumed by those members. No goods and services are imported from without and none are exported. Such a system involves N members, each of whom produces goods or services and charges for their use. The problem is to determine the prices each member should charge for his or her labor so that everyone breaks even after one year. For simplicity, it is assumed that each member produces one unit per year.

Consider a simple closed system limited to a farmer, a carpenter, and a weaver. The farmer produces one unit of food each year, the carpenter produces one unit of finished wood products each year, and the weaver produces one unit of clothing each year. Let p_1 denote the farmer's annual income (that is, the price she charges for her unit of food), let p_2 denote the carpenter's annual income (that is, the price he charges for his unit of finished wood products), and let p_3 denote the weaver's annual income. Assume on an annual basis that the farmer and the carpenter consume 40% each of the available food, while the weaver eats the remaining 20%. Assume that the carpenter uses 25% of the wood products he makes, while the farmer uses 30% and the weaver uses 45%. Assume further that the farmer uses 50% of the weaver's clothing while the carpenter uses 35% and the weaver consumes the remaining 15%. Show that a break-even equation for the farmer is

$$0.40p_1 + 0.30p_2 + 0.50p_3 = p_1$$

while the break-even equation for the carpenter is

$$0.40p_1 + 0.25p_2 + 035p_3 = p_2$$

What is the break-even equation for the weaver? Rewrite all three equations as a homogeneous system and then find the annual incomes of each sector.

(39) Paul, Jim, and Mary decide to help each other build houses. Paul will spend half his time on his own house and a quarter of his time on each of the houses of Jim and Mary. Jim will spend one third of his time on each of the three houses under construction. Mary will spend one sixth of her time on Paul's house, one third on Jim's house, and one half of her time on her own house. For tax purposes, each must place a price on his or her labor, but they want to do so in a way that each will break-even. Show that the process of determining break-even wages is a Leontief closed model containing three homogeneous equations and then find the wages of each person.

(40) Four third-world countries each grow a different fruit for export and each uses the income from that fruit to pay for imports of the fruits from the other countries. Country A exports 20% of its fruit to country B, 30% to country C, 35% to country D, and uses the rest of its fruit for internal consumption. Country B exports 10% of its fruit to country A, 15% to country C, 35% to country D, and retains the rest for its own citizens. Country C does not export to country A; it divides its crop equally between countries B and D and its own people. Country D does not consume its own fruit; all is for export with 15% going to country A, 40% to country B, and 45% to country C. Show that the problem of determining prices on the annual harvests of fruit so that each country breaks even is equivalent to solving four homogeneous equations in four unknowns and then find the prices.

Gaussian elimination is often programmed for computer implementation, but because all computers store numbers as a finite string of digits, round-off error can be significant. A popular strategy for minimizing round-off errors is *partial pivoting,* which requires that a pivot always be larger than or equal in absolute value than any element below the pivot in the same column. This is accomplished by using elementary row operation R_1 to interchange rows whenever necessary. In Problems 41 through 46, determine the first pivot under a partial pivoting strategy for the given augmented matrix.

(41) $\left[\begin{array}{cc|c} 1 & 3 & 35 \\ 4 & 8 & 15 \end{array}\right]$

(42) $\left[\begin{array}{cc|c} 1 & -2 & -5 \\ 5 & 3 & 85 \end{array}\right]$

(43) $\left[\begin{array}{ccc|c} -2 & 8 & -3 & 100 \\ 4 & 5 & 4 & 75 \\ -3 & -1 & 2 & 250 \end{array}\right]$

(44) $\left[\begin{array}{ccc|c} 1 & 2 & 3 & 4 \\ 5 & 6 & 7 & 8 \\ 9 & 10 & 11 & 12 \end{array}\right]$

(45) $\left[\begin{array}{ccc|c} 1 & 8 & 8 & 400 \\ 0 & 1 & 7 & 800 \\ 0 & 3 & 9 & 600 \end{array}\right]$

(46) $\left[\begin{array}{cccc|c} 0 & 2 & 3 & 4 & 0 \\ 1 & 0.4 & 0.8 & 0.1 & 90 \\ 4 & 10 & 1 & 8 & 40 \end{array}\right]$

1.5 DETERMINANTS

Every linear transformation from one finite-dimensional vector space $\mathbb{V}$ back to itself can be represented by a square matrix. Each matrix representation is basis dependent, and, in general, a linear transformation will have a different matrix representation for each basis in $\mathbb{V}$. Some of these matrix representations may be simpler than others. In this chapter, we begin the process of identifying bases that generate simple matrix representations for linear transformations. This search will focus on a special class of vectors known as eigenvectors and will use some of the basic properties of determinants.

Every square matrix has associated with it a scalar called its *determinant*. Until very recently, determinants were central to the study of linear algebra, the hub around which much of the theory revolved. Determinants were used for calculating inverses, solving systems of simultaneous equations, and a host of other applications. No more. In their place are other techniques, often based on elementary row operations, which are more efficient and better adapted to computers. The applications of determinants have been reduced to less than a handful.

Determinants are defined in terms of permutations on positive integers. The theory is arduous and, once completed, gives way to simpler methods for calculating determinants. Because we make such limited use of determinants, we will not develop its theory here, restricting ourselves instead to the standard computational techniques.

Determinants are defined only for *square* matrices. Given a square matrix **A**, we use det(**A**) or |**A**| to designate the determinant of **A**. If a matrix can be exhibited, we designate its determinant by replacing the brackets with vertical straight lines. For example, if

$$\mathbf{A} = \begin{bmatrix} 1 & 2 & 3 \\ 4 & 5 & 6 \\ 7 & 8 & 9 \end{bmatrix} \tag{1.24}$$

then

$$\det(\mathbf{A}) = \begin{vmatrix} 1 & 2 & 3 \\ 4 & 5 & 6 \\ 7 & 8 & 9 \end{vmatrix} \tag{1.25}$$

We cannot overemphasize the fact that Equations (1.24) and (1.25) represent entirely different structures. Equation (1.24) is a matrix, a rectangular array of elements, whereas Equation (1.24) represents a scalar, a number associated with the matrix in Equation (1.25).

The determinant of a 1×1 matrix [a] is the scalar a; the determinant of a 2×2 matrix is the product of its diagonal terms less the product of its off-diagonal terms.

The determinant of a 1×1 matrix $[a]$ is defined as the scalar a. Thus, the determinant of the matrix [5] is 5 and the determinant of the matrix $[-3]$ is -3. The determinant of a 2×2 matrix $\begin{bmatrix} a & b \\ c & d \end{bmatrix}$ is defined as the scalar $ad - bc$.

Example 1 $\det\begin{bmatrix} 1 & 2 \\ 3 & 4 \end{bmatrix} = \begin{vmatrix} 1 & 2 \\ 4 & 3 \end{vmatrix} = 1(3) - 2(4) = 3 - 8 = -5$, while $\det\begin{bmatrix} 2 & -1 \\ 4 & 3 \end{bmatrix} = \begin{vmatrix} 2 & -1 \\ 4 & 3 \end{vmatrix} = 2(3) - (-1)(4) = 6 + 4 = 10$.

We could list separate rules for calculating determinants of 3×3, 4×4, and higher order matrices, but this is unnecessary. Instead we develop a method based on minors and cofactors that lets us reduce determinants of order $n > 2$ (if **A** has order $n \times n$, then det(**A**) is said to have order n) to a sum of determinants of order 2.

A minor of a matrix **A** is the determinant of any square submatrix of **A**.

A *minor* of a matrix **A** is the determinant of any square submatrix of **A**. A minor is formed from a square matrix **A** by removing an equal number of rows and columns from **A** and then taking the determinant of the resulting submatrix. In particular, if

$$\mathbf{A} = \begin{bmatrix} 1 & 2 & 3 \\ 4 & 5 & 6 \\ 7 & 8 & 9 \end{bmatrix}$$

then $\begin{bmatrix} 1 & 2 \\ 7 & 8 \end{bmatrix}$ and $\begin{bmatrix} 5 & 6 \\ 8 & 9 \end{bmatrix}$ are both minors because the matrices $\begin{bmatrix} 1 & 2 \\ 7 & 8 \end{bmatrix}$ and $\begin{bmatrix} 5 & 6 \\ 8 & 9 \end{bmatrix}$ are both submatrices of **A**. In contrast, $\begin{vmatrix} 1 & 2 \\ 8 & 9 \end{vmatrix}$ and $|1 \quad 2|$ are not minors because $\begin{bmatrix} 1 & 2 \\ 8 & 9 \end{bmatrix}$ is not a submatrix of **A** and $[1 \quad 2]$, although a submatrix of **A**, is not square.

If $\mathbf{A} = [a_{ij}]$ is a square matrix, then the *cofactor of the element* a_{ij} is the scalar obtained by multiplying $(-1)^{i+j}$ with the minor obtained from **A** by removing the *j*th row and *j*th column. In other words, to compute the cofactor of an element a_{ij} in a matrix **A**, first form a submatrix of **A** by deleting from **A** both the row and column in which the element a_{ij} appears, then calculate the determinant of that submatrix, and finally multiply the determinant by the number $(-1)^{i+j}$.

The cofactor of the element a_{ij} in a square matrix **A** is the product of $(-1)^{i+j}$ with the minor obtained from **A** by deleting its *i*th row and *j*th column.

Example 2 To find the cofactor of the element 4 in the matrix

$$\mathbf{A} = \begin{bmatrix} 1 & 2 & 3 \\ 4 & 5 & 6 \\ 7 & 8 & 9 \end{bmatrix}$$

we note that 4 appears in the second row and first column, hence $i=2$, $j=1$, and $(-1)^{i+j} = (-1)^{2+1} = (-1)^3 = -1$. The submatrix of **A** obtained by deleting the second row and first column is

$$\begin{bmatrix} 2 & 3 \\ 8 & 9 \end{bmatrix}$$

which has a determinant equal to $2(9) - 3(8) = -6$. The cofactor of 4 is $(-1)(-6) = 6$.

The element 9 appears in the third row and third column of **A**, hence $i=3$, $j=3$, and $(-1)^{i+j} = (-1)^{3+3} = (-1)^6 = 1$. The submatrix of **A** obtained by deleting the third row and third column is $\begin{bmatrix} 1 & 2 \\ 4 & 5 \end{bmatrix}$, which has a determinant equal to $1(5) - 2(4) = -3$. The cofactor of 9 is $(1)(-3) = -3$.

A cofactor is the product of a minor with the number $(-1)^{i+j}$. This number $(-1)^{i+j}$ is either $+1$ or -1, depending on the values of *i* and *j*, and its effect is to impart a plus or minus sign in front of the minor of interest. A useful schematic for quickly evaluating $(-1)^{i+j}$ is to use the sign in the *i-j*th position of the patterned matrix:

$$\begin{bmatrix} + & - & + & - & + \\ - & + & - & + & - \\ + & - & + & - & + \\ - & + & - & + & - \\ + & - & + & - & + \end{bmatrix}$$

EXPANSION BY COFACTORS (TO CALCULATE THE DETERMINANT OF A SQUARE MATRIX):

Step 1. Pick any one row or any one column of the matrix (dealer's choice).

Step 2. Calculate the cofactor of each element in the row or column selected.

Step 3. Multiply each element in the selected row or column by its cofactor and sum the results.

We now can find the determinant of any square matrix.

Example 3 Find det(A) for $\mathbf{A} = \begin{bmatrix} 3 & 5 & 0 \\ -1 & 2 & 1 \\ 3 & -6 & 4 \end{bmatrix}$.

Solution: We arbitrarily expand by the second column. Thus,

$$\begin{aligned} |\mathbf{A}| &= 5\,(\text{cofactor of } 5) + 2\,(\text{cofactor of } 2) + (-6)(\text{cofactor of } -6) \\ &= 5(-1)^{1+2}\begin{vmatrix} -1 & 1 \\ 3 & 4 \end{vmatrix} + 2(-1)^{2+2}\begin{vmatrix} 3 & 0 \\ 3 & 4 \end{vmatrix} + (-6)(-1)^{3+2}\begin{vmatrix} 3 & 0 \\ -1 & 1 \end{vmatrix} \\ &= 5(-1)(-4-3) + 2(1)(12-0) + (-6)(-1)(3-0) \\ &= (-5)(-7) + 2(12) + 6(3) = 77 \end{aligned}$$

Example 4 Redo Example 3 expanding by the first row.

Solution:

$$\begin{aligned} |\mathbf{A}| &= 3\,(\text{cofactor of } 3) + 5\,(\text{cofactor of} 2) + 0\,(\text{cofactor of } 0) \\ &= 3(-1)^{1+1}\begin{vmatrix} 2 & 1 \\ -6 & 4 \end{vmatrix} + 5(-1)^{1+2}\begin{vmatrix} -1 & 1 \\ 3 & 4 \end{vmatrix} + 0 \\ &= 3(1)(8+6) + 5(-1)(-4-3) + 0 \\ &= 3(14) + (-5)(-7) = 77 \end{aligned}$$

Expanding by a row or column containing the most zeros minimizes the number of computations needed to evaluate a determinant.

Examples 3 and 4 illustrate two important properties of expansion by cofactors. First, the value of a determinant is the same regardless of which row or column selected and second, expanding by a row or column containing zeros significantly reduces the number of computations.

Example 5 Find det(A) for $\mathbf{A} = \begin{bmatrix} 1 & 0 & 5 & 2 \\ -1 & 4 & 1 & 0 \\ 3 & 0 & 4 & 1 \\ -2 & 1 & 1 & 3 \end{bmatrix}$.

Solution: The row or column containing the most zeros is, for this matrix, the second column, so we expand by it.

$$|\mathbf{A}| = 0\,(\text{cofactor of } 0) + 4\,(\text{cofactor of } 4) + 0\,(\text{cofactor of zero}) + 1\,(\text{cofactor of } 1)$$

$$= 0 + 4(-1)^{2+2}\begin{vmatrix} 1 & 5 & 2 \\ 3 & 4 & 1 \\ -2 & 1 & 3 \end{vmatrix} + 0 + 1(-1)^{4+2}\begin{vmatrix} 1 & 5 & 2 \\ -1 & 1 & 0 \\ 3 & 4 & 1 \end{vmatrix}$$

$$= 4\begin{vmatrix} 1 & 5 & 2 \\ 3 & 4 & 1 \\ -2 & 1 & 3 \end{vmatrix} + \begin{vmatrix} 1 & 5 & 2 \\ -1 & 1 & 0 \\ 3 & 4 & 1 \end{vmatrix}$$

Using expansion of cofactors on each of these two determinants of order 3, we calculate

$$\begin{vmatrix} 1 & 5 & 2 \\ 3 & 4 & 1 \\ -2 & 1 & 3 \end{vmatrix} = 1(-1)^{1+1}\begin{vmatrix} 4 & 1 \\ 1 & 3 \end{vmatrix} + 5(-1)^{1+2}\begin{vmatrix} 3 & 1 \\ -2 & 3 \end{vmatrix} + 2(-1)^{1+3}\begin{vmatrix} 3 & 4 \\ -2 & 1 \end{vmatrix}$$

$$= 11 - 55 + 22 = -22 \text{(expanding by the first row)}$$

and

$$\begin{vmatrix} 1 & 5 & 2 \\ -1 & 1 & 0 \\ 3 & 4 & 1 \end{vmatrix} = 2(-1)^{1+3}\begin{vmatrix} -1 & 1 \\ 3 & 4 \end{vmatrix} + 0 + 1(-1)^{3+3}\begin{vmatrix} 1 & 5 \\ -1 & 1 \end{vmatrix}$$

$$= -14 + 6 = -8 \text{ (expanding by the third column)}$$

Consequently, $|\mathbf{A}| = 4(-22) + 1(-8) = -96$.

With no zero entries, the determinant of a 3×3 matrix requires $3 \cdot 2 = 3!$ multiplications, a 4×4 matrix requires $4 \cdot 3 \cdot 2 = 4!$ multiplications, and an $n \times n$ matrix requires $n!$ multiplications. Note that $10! = 3{,}628{,}000$ and $13!$ is over 1 billion, so the number of multiplications needed to evaluate a determinant becomes prohibitive as the order of a matrix increases. Clearly, calculating a determinant is a complicated and time-consuming process, one that is avoided whenever possible.

Another complicated operation is matrix multiplication, which is why the following result is so surprising. Its proof, however, is beyond the scope of this book.

> ▶**THEOREM 1**
>
> *If* **A** *and* **B** *are of the same order, then* det(**AB**) = det(**A**)det(**B**). ◀

Example 6 Verify Theorem 1 for $\mathbf{A} = \begin{bmatrix} 2 & 3 \\ 1 & 4 \end{bmatrix}$ and $\mathbf{B} = \begin{bmatrix} 6 & -1 \\ 7 & 4 \end{bmatrix}$.

Solution: $|\mathbf{A}| = 5$ and $|\mathbf{B}| = 31$. Also $\mathbf{AB} = \begin{bmatrix} 33 & 10 \\ 34 & 15 \end{bmatrix}$, hence $|\mathbf{AB}| = 155 = |\mathbf{A}||\mathbf{B}|$.

Any two column matrices in $\mathbb{R}^2$ that do not lie on the same straight line form the sides of a parallelogram, as illustrated in Figure 1.10. Here the column matrices

$$\mathbf{u} = \begin{bmatrix} a_1 \\ a_2 \end{bmatrix} \quad \text{and} \quad \mathbf{v} = \begin{bmatrix} b_1 \\ b_2 \end{bmatrix} \tag{1.26}$$

appear graphically as directed line segments with the tip of **u** falling on the point $A = (a_1, a_2)$ in the x-y plane and the tip of **v** falling on the point $B = (b_1, b^2)$. The parallelogram generated by these two vectors is $OACB$, where O denotes the origin and $C = (a_1 + b_1,\ a_2 + b_2)$. To calculate the area of this parallelogram, we note that

Area of parallelogram $OACB$

$$= \text{area of triangle } OPB + \text{area of trapezoid } PRCB - \text{area of triangle } OQA - \text{area of trapezoid } QRCA$$

$$= \frac{1}{2}b_1b_2 + \frac{1}{2}a_1(b_2 + a_2 + b_2) - \frac{1}{2}a_1a_2 + \frac{1}{2}b_1(a_2 + a_2 + b_2)$$

$$= a_1b_2 - a_2b_1 = \begin{vmatrix} a_1 & b_1 \\ a_2 & b_2 \end{vmatrix}$$

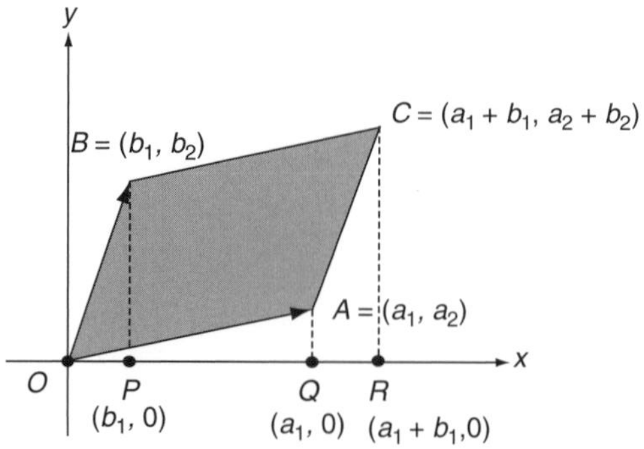

FIGURE 1.10

If we interchange the positions of the two columns in this last determinant, a quick computation shows that the resulting determinant is the *negative* of the area of the parallelogram. Because the area of the parallelogram is the same regardless which vector in Equation (A.3) is listed first and which second, we avoid any concern about ordering by simply placing absolute values around the determinant. Thus, we have proven:

▶THEOREM 2

If $\mathbf{u}=[a_1 \quad a_2]^T$ and $\mathbf{v}=[b_1 \quad b_2]$ are two column matrices in $\mathbb{R}^2$, then the area of the parallelogram generated by $\mathbf{u}$ and $\mathbf{v}$ is $|\det[\mathbf{u}\ \mathbf{v}]|$. ◀

Example 7 The area of the parallelogram defined by the column matrices $\mathbf{u}=\begin{bmatrix}6\\2\end{bmatrix}$ and $\mathbf{v}=\begin{bmatrix}6\\2\end{bmatrix}$ is $\left|\det\begin{bmatrix}-4 & 6\\ 4 & 2\end{bmatrix}\right| = |-32| = 32$ square units. These column matrices and the parallelogram they generate are illustrated in Figure 1.11.

Example 8 The area of the parallelogram defined by the column matrices $\mathbf{u}=\begin{bmatrix}-3\\-1\end{bmatrix}$ and $\mathbf{v}=\mathbf{v}=\begin{bmatrix}6\\2\end{bmatrix}$ is $\left|\det\begin{bmatrix}-3 & 6\\ -1 & 2\end{bmatrix}\right| = |0| = 0$ square units.

These vectors are illustrated in Figure 1.12. Because both vectors lie on the same straight line, the parallelogram generated by these vectors collapses into the line segment joining (−3, −1) and (6, 2), which has zero area.

Expansion by cofactors is often a tedious procedure for calculating determinants, especially for matrices of large order. Triangular matrices, however, contain many zeros and have determinants that are particularly easy to evaluate.

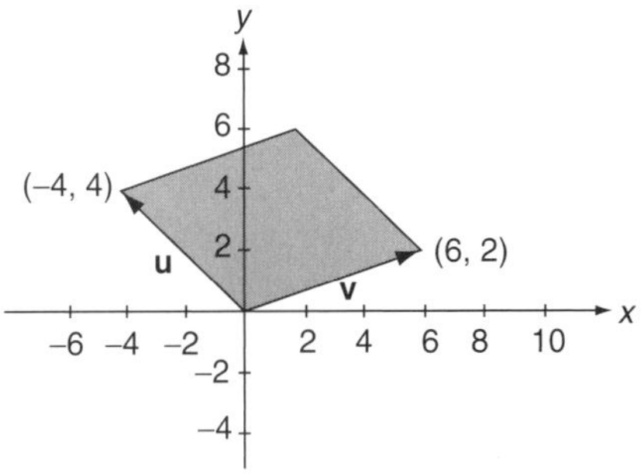

FIGURE 1.11

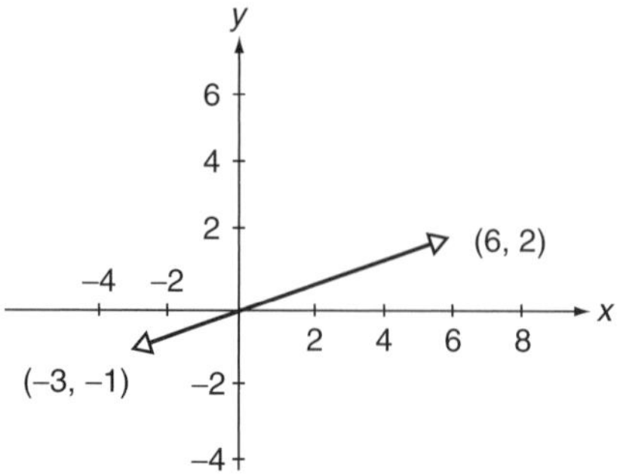

FIGURE 1.12

> **▶THEOREM 3**
>
> *The determinant of an upper or lower triangular matrix is the product of the elements on the main diagonal.◀*

Proof: We shall prove the proposition for upper triangular matrices by mathematical induction on the order of the determinant. (For an explanation of this proof technique, please refer to Appendix E.) The proof for lower triangular matrices is nearly identical and is left as an exercise for the reader. We first show that the proposition is true for all 1×1 upper triangular matrices and then we show that *if* the proposition is true for all $(k-1)\times(k-1)$ upper triangular matrices, then it must also be true for all $k \times k$ upper triangular matrices.

A 1×1 upper triangular matrix has the general form $\mathbf{A} = [a_{11}]$, containing a single diagonal element. Its determinant is a_{11}, which is the product of all diagonal elements in $\mathbf{A}$, thus the proposition is true for $n = 1$.

We now *assume* that the proposition is true for all $(k-1)\times(k-1)$ upper triangular matrices, and we use this assumption to prove the proposition for all $k \times k$ upper triangular matrices $\mathbf{A}$. Such a matrix has the general form

$$\mathbf{A} = \begin{bmatrix} a_{11} & a_{12} & a_{13} & \cdots & a_{1,\,k-1} & a_{1k} \\ 0 & a_{22} & a_{23} & \cdots & a_{2,\,k-1} & a_{2k} \\ 0 & 0 & a_{33} & \cdots & a_{3,\,k-1} & a_{3k} \\ \vdots & \vdots & \vdots & & \vdots & \vdots \\ 0 & 0 & 0 & \cdots & 0 & a_{kk} \end{bmatrix}$$

Evaluating det(**A**) by expansion by cofactors using the cofactors of the elements in the first column, the column containing the most zeros, we obtain

$$\det(\mathbf{A}) = a_{11} \cdot \det(\mathbf{B}) \tag{1.27}$$

where

$$\mathbf{B} = \begin{bmatrix} a_{22} & a_{23} & \cdots & a_{2,\,k-1} & a_{2k} \\ 0 & a_{33} & \cdots & a_{3,\,k-1} & a_{3k} \\ \vdots & \vdots & \vdots & & \vdots \\ 0 & 0 & \cdots & 0 & a_{kk} \end{bmatrix}$$

Matrix **B** is an upper triangular matrix of order $(k-1)\times(k-1)$ so by the induction hypothesis its determinant is the product of its diagonal elements. Consequently, $\det(\mathbf{B}) = a_{22}a_{33}\cdots a_{kk}$, and Equation (1.27) becomes $\det(\mathbf{A}) = a_{11}a_{22}a_{33}\cdots a_{kk}$, which is the product of the diagonal elements of **A**. Thus, Theorem 1 is proved by mathematical induction.

Example 10

$$\det\begin{bmatrix} 2 & 6 & -4 & 1 \\ 0 & 5 & 7 & -4 \\ 0 & 0 & -5 & 8 \\ 0 & 0 & 0 & 3 \end{bmatrix} = 2(5)(-5)(3) = -150$$

Because diagonal matrices are both upper and lower triangular, the following corollary is immediate.

▶COROLLARY 1

The determinant of a diagonal matrix is the product of the elements on its main diagonal. ◀

Expansion by a row or column having many zeros simplifies the calculation of a determinant; expansion by a zero row or zero column, when it exists, makes the process trivial. Multiplying each zero element by its cofactor yields zero products that when summed are still 0. We have, therefore, Theorem 2.

▶THEOREM 4

If a square matrix has a zero row or a zero column, then its determinant is 0. ◀

A useful property of determinants involves a square matrix and its transpose.

▶THEOREM 5

For any square matrix **A**, *det(***A***) = det(***A***^{T}).* ◀

Proof: (by mathematical induction on the order of the determinant): A 1×1 matrix has the general form $A = [a_{11}]$. Here $A = A^T$, hence $|A| = a_{11} = |A^T|$, and the proposition is true for $n = 1$.

We now *assume* that the proposition is true for all $(k-1) \times (k-1)$ matrices, and we use this assumption to prove the proposition for all $k \times k$ matrices A. Such a matrix has the general form

$$A = \begin{bmatrix} a_{11} & a_{12} & a_{13} & \cdots & a_{1k} \\ a_{21} & a_{22} & a_{23} & \cdots & a_{2k} \\ a_{31} & a_{32} & a_{33} & \cdots & a_{3k} \\ \vdots & \vdots & \vdots & \ddots & \vdots \\ a_{k1} & a_{k2} & a_{k3} & \cdots & a_{kk} \end{bmatrix}$$

Evaluating det(A) by expansion by cofactors using the first column, we obtain

$$\det(A) = a_{11}(-1)^{1+1}\det\begin{bmatrix} a_{22} & a_{23} & \cdots & a_{2k} \\ a_{32} & a_{33} & \cdots & a_{3k} \\ \vdots & \vdots & \ddots & \vdots \\ a_{k2} & a_{k3} & \cdots & a_{kk} \end{bmatrix}$$

$$+a_{21}(-1)^{2+1}\det\begin{bmatrix} a_{12} & a_{13} & \cdots & a_{1k} \\ a_{32} & a_{33} & \cdots & a_{3k} \\ \vdots & \vdots & \ddots & \vdots \\ a_{k2} & a_{k3} & \cdots & a_{kk} \end{bmatrix}$$

$$+\cdots + a_{k1}(-1)^{k+1}\det\begin{bmatrix} a_{12} & a_{13} & \cdots & a_{1k} \\ a_{22} & a_{23} & \cdots & a_{2k} \\ \vdots & \vdots & \ddots & \vdots \\ a_{k-1,\,2} & a_{k-1,\,3} & \cdots & a_{k-1,\,k} \end{bmatrix}$$

Each of the matrices on the right side of this last equality has order $(k-1) \times (k-1)$ so by the induction hypothesis each of their determinants equals, respectively, the determinants of their transposes. Consequently,

$$\det(\mathbf{A}) = a_{11}(-1)^{1+1}\det\begin{bmatrix} a_{22} & a_{23} & \cdots & a_{k2} \\ a_{23} & a_{33} & \cdots & a_{k3} \\ \vdots & \vdots & \ddots & \vdots \\ a_{2k} & a_{3k} & \cdots & a_{kk} \end{bmatrix}$$

$$+a_{21}(-1)^{1+2}\det\begin{bmatrix} a_{12} & a_{32} & \cdots & a_{k2} \\ a_{13} & a_{33} & \cdots & a_{k3} \\ \vdots & \vdots & \ddots & \vdots \\ a_{1k} & a_{3k} & \cdots & a_{kk} \end{bmatrix}$$

$$+\cdots + a_{k1}(-1)^{1+k}\det\begin{bmatrix} a_{12} & a_{22} & \cdots & a_{k-1,\,2} \\ a_{13} & a_{23} & \cdots & a_{k-1,\,3} \\ \vdots & \vdots & \ddots & \vdots \\ a_{1k} & a_{2k} & \cdots & a_{k-1,\,k} \end{bmatrix}$$

$$= \det\begin{bmatrix} a_{11} & a_{21} & a_{31} & \cdots & a_{k1} \\ a_{12} & a_{22} & a_{32} & \cdots & a_{k2} \\ a_{13} & a_{23} & a_{33} & \cdots & a_{k3} \\ \vdots & \vdots & \vdots & \ddots & \vdots \\ a_{1k} & a_{2k} & a_{3k} & \cdots & a_{kk} \end{bmatrix}$$

where this last determinant is evaluated by expansion by cofactors using its first row. Since this last matrix is $\mathbf{A}^T$, we have $\det(\mathbf{A}) = \det(\mathbf{A}^T)$, and Theorem 5 is proven by mathematical induction.

An elegant method for substantially reducing the number of arithmetic operations needed to evaluate determinants of matrices whose elements are all constants is based on elementary row operations. For the sake of expediency, we state the relevant properties and then demonstrate their validity for 3×3 matrices.

▶THEOREM 6

*If matrix **B** is obtained from a square matrix **A** by interchanging the position of two rows in **A** (the first elementary row operation), then* $|\mathbf{B}| = -|\mathbf{A}|$. ◀

Demonstration of Validity: Consider

$$\mathbf{A} = \begin{bmatrix} a_{11} & a_{12} & a_{13} \\ a_{21} & a_{22} & a_{23} \\ a_{31} & a_{32} & a_{33} \end{bmatrix} \tag{1.28}$$

Expanding $|\mathbf{A}|$ by cofactors using its third row, we obtain

$$|\mathbf{A}| = a_{31}(a_{12}a_{23} - a_{13}a_{22}) - a_{32}(a_{11}a_{23} - a_{13}a_{21}) + a_{33}(a_{11}a_{22} - a_{12}a_{21})$$

Now consider the matrix **B** obtained from **A** by interchanging the positions of the second and third rows of **A**:

$$\mathbf{B} = \begin{bmatrix} a_{11} & a_{12} & a_{13} \\ a_{31} & a_{32} & a_{33} \\ a_{21} & a_{22} & a_{23} \end{bmatrix}$$

Expanding $|\mathbf{B}|$ by cofactors using its second row, we obtain

$$|\mathbf{B}| = -a_{31}(a_{12}a_{23} - a_{13}a_{22}) - a_{32}(a_{11}a_{23} - a_{13}a_{21}) - a_{33}(a_{11}a_{22} - a_{12}a_{21})$$

Thus, $|\mathbf{B}| = -|\mathbf{A}|$. Through similar reasoning, we can show that the result is valid regardless of which two rows of **A** are interchanged.

As an immediate consequence of Theorem 6, we have the following corollary:

►COROLLARY 2

If two rows of a square matrix are identical, then its determinant is 0.◄

Proof: The matrix remains unaltered if the two identical rows are interchanged, hence its determinant must remain constant. It follows from Theorem 6, however, that interchanging two rows of a matrix changes the sign of its determinant. Thus, the determinant must, on the one hand, remain the same and, on the other hand, change sign. The only way both conditions are met simultaneously is for the determinant to be 0.

►THEOREM 7

*If matrix **B** is obtained from a square matrix **A** by multiplying every element in one row of **A** by the scalar λ (the second elementary row operation), then $|\mathbf{B}| = \lambda|\mathbf{A}|$.*◄

Demonstration of Validity: Consider the matrix **A** given in Equation (1.28) and construct **B** from **A** by multiplying the first row of **A** by λ. Then expanding $|\mathbf{B}|$ by cofactors using its first row, we obtain

$$|\mathbf{B}| = \begin{vmatrix} \lambda a_{11} & \lambda a_{12} & \lambda a_{13} \\ a_{21} & a_{22} & a_{23} \\ a_{31} & a_{32} & a_{33} \end{vmatrix} = \lambda a_{11}\begin{vmatrix} a_{22} & a_{23} \\ a_{32} & a_{33} \end{vmatrix} - \lambda a_{12}\begin{vmatrix} a_{21} & a_{23} \\ a_{31} & a_{33} \end{vmatrix} + \lambda a_{13}\begin{vmatrix} a_{21} & a_{23} \\ a_{31} & a_{32} \end{vmatrix}$$

$$= \lambda \left(a_{11} \begin{vmatrix} a_{22} & a_{23} \\ a_{32} & a_{33} \end{vmatrix} - a_{12} \begin{vmatrix} a_{21} & a_{23} \\ a_{31} & a_{33} \end{vmatrix} + a_{13} \begin{vmatrix} a_{21} & a_{22} \\ a_{31} & a_{32} \end{vmatrix} \right)$$

$$= \lambda \begin{vmatrix} a_{11} & a_{12} & a_{13} \\ a_{21} & a_{22} & a_{23} \\ a_{31} & a_{32} & a_{33} \end{vmatrix} = \lambda|\mathbf{A}|$$

Through similar reasoning, we can show that the result is valid regardless of which row of **A** is multiplied by λ.

Multiplying a scalar times a matrix multiplies every element of the matrix by that scalar. In contrast, it follows from Theorem 7 that a scalar times a determinant is equivalent to multiplying *one* row of the associated matrix by the scalar and then evaluating the determinant of the resulting matrix. Thus,

$$\det\begin{bmatrix} 8 & 16 \\ 3 & 4 \end{bmatrix} = 8\det\begin{bmatrix} 1 & 2 \\ 3 & 4 \end{bmatrix} = \det\begin{bmatrix} 1 & 2 \\ 24 & 32 \end{bmatrix}$$

while

$$\det\left\{8\begin{bmatrix} 1 & 2 \\ 3 & 4 \end{bmatrix}\right\} = \det\begin{bmatrix} 8 & 16 \\ 24 & 32 \end{bmatrix} = 8\det\begin{bmatrix} 1 & 2 \\ 24 & 32 \end{bmatrix} = 8(8)\det\begin{bmatrix} 1 & 2 \\ 3 & 4 \end{bmatrix}$$

Therefore, as an immediate extension of Theorem 7, we have the next corollary.

►COROLLARY 3

If **A** *is an* $n \times n$ *matrix and* λ *a scalar, then* $\det(\lambda\mathbf{A}) = \lambda^n \det(\mathbf{A})$. ◄

Applying the first two elementary row operations to a matrix changes the determinant of the matrix. Surprisingly, the third elementary row operation has no effect on the determinant of a matrix.

►THEOREM 8

If matrix **B** *is obtained from a square matrix* **A** *by adding to one row of* **A** *a scalar times another row of* **A** *(the third elementary row operation), then* $|\mathbf{B}| = |\mathbf{A}|$. ◄

Demonstration of Validity: Consider the matrix **A** given in Equation (1.28) and construct **B** from **A** by adding to the third row of **A** the scalar λ times the first row of **A**. Thus,

$$\mathbf{B} = \begin{bmatrix} a_{11} & a_{12} & a_{13} \\ a_{21} & a_{22} & a_{23} \\ a_{31} + \lambda a_{11} & a_{32} + \lambda a_{12} & a_{33} + \lambda a_{13} \end{bmatrix}$$

Expanding $|\mathbf{B}|$ by cofactors using its third row, we obtain

$$|\mathbf{B}| = (a_{31} + \lambda a_{11}) \begin{vmatrix} a_{12} & a_{13} \\ a_{22} & a_{23} \end{vmatrix} - (a_{32} + \lambda a_{12}) \begin{vmatrix} a_{11} & a_{13} \\ a_{21} & a_{23} \end{vmatrix}$$

$$= a_{31} \begin{vmatrix} a_{12} & a_{13} \\ a_{22} & a_{23} \end{vmatrix} - a_{32} \begin{vmatrix} a_{11} & a_{13} \\ a_{21} & a_{23} \end{vmatrix} - a_{33} \begin{vmatrix} a_{11} & a_{13} \\ a_{21} & a_{22} \end{vmatrix}$$

$$+ \lambda \left\{ a_{11} \begin{vmatrix} a_{12} & a_{13} \\ a_{22} & a_{23} \end{vmatrix} - a_{12} \begin{vmatrix} a_{11} & a_{13} \\ a_{21} & a_{23} \end{vmatrix} + a_{13} \begin{vmatrix} a_{11} & a_{13} \\ a_{21} & a_{22} \end{vmatrix} \right\}$$

The first three terms of this sum are exactly $|\mathbf{A}|$ (expand det(**A**) by its third row) while the last three terms of the sum are

$$\lambda \begin{vmatrix} a_{11} & a_{12} & a_{13} \\ a_{21} & a_{22} & a_{23} \\ a_{11} & a_{12} & a_{13} \end{vmatrix}$$

(expand this determinant by its third row). Thus,

$$|\mathbf{B}| = |\mathbf{A}| + \lambda \begin{vmatrix} a_{11} & a_{12} & a_{13} \\ a_{21} & a_{22} & a_{23} \\ a_{11} & a_{12} & a_{13} \end{vmatrix}$$

It follows from Corollary 2 that this last determinant is 0 because its first and third rows are identical, hence $|\mathbf{B}| = |\mathbf{A}|$.

Example 11 Without expanding, use the properties of determinants to show that

$$\begin{vmatrix} a & b & c \\ r & s & t \\ x & y & z \end{vmatrix} = \begin{vmatrix} a - r & b - s & c - t \\ r + 2x & s + 2y & t + 2z \\ x & y & z \end{vmatrix}$$

Solution:

$$\begin{vmatrix} a & b & c \\ r & s & t \\ x & y & z \end{vmatrix} = \begin{vmatrix} a-r & b-s & c-t \\ r & s & t \\ x & y & z \end{vmatrix} \quad \begin{array}{l}\text{Theorem 8 : adding} \\ \text{to the first row } -1 \\ \text{times the second row}\end{array}$$

$$= \begin{vmatrix} a-r & b-s & c-t \\ r+2x & s+2y & t+2z \\ x & y & z \end{vmatrix} \quad \begin{array}{l}\text{Theorem 8 : adding} \\ \text{to the second row 2} \\ \text{times the third row}\end{array}$$

Pivotal condensation is an efficient algorithm for calculating the determinant of a matrix whose elements are all constants. Elementary row operations are used to transform a matrix to row-reduced form, because such a matrix is upper triangular and its determinant is easy to evaluate using Theorem 3. A record is kept of all the changes made to the determinant of a matrix while reducing the matrix to row-reduced form. The product of these changes with the determinant of the row-reduced matrix is the determinant of the original matrix.

Pivotal Condensation Transform a matrix into row-reduced form using elementary row operations, keeping a record of the changes made. Evaluate the determinant by using Theorems 3, 6, 7, and 8.

Example 12 Use pivotal condensation to evaluate $\det\begin{bmatrix} 1 & 2 & 3 \\ -2 & 3 & 2 \\ 3 & -1 & 1 \end{bmatrix}$.

Solution:

$$\begin{vmatrix} 1 & 2 & 3 \\ -2 & 3 & 2 \\ 3 & -1 & 1 \end{vmatrix} = \begin{vmatrix} 1 & 2 & 3 \\ 0 & 7 & 8 \\ 3 & -1 & 1 \end{vmatrix} \quad \begin{array}{l}\text{Theorem 8 : adding} \\ \text{to the second row 2} \\ \text{times the first row}\end{array}$$

$$= \begin{vmatrix} 1 & 2 & 3 \\ 0 & 7 & 8 \\ 3 & -7 & -8 \end{vmatrix} \quad \begin{array}{l}\text{Theorem 8 : adding} \\ \text{to the third row } -3 \\ \text{times the first row}\end{array}$$

$$= 7\begin{vmatrix} 1 & 2 & 3 \\ 0 & 1 & 8/7 \\ 3 & -7 & -8 \end{vmatrix} \quad \begin{array}{l}\text{Theorem 8 : applied} \\ \text{to the second row}\end{array}$$

$$= 7\begin{vmatrix} 1 & 2 & 3 \\ 0 & 1 & 8/7 \\ 0 & 0 & 0 \end{vmatrix} \quad \begin{array}{l}\text{Theorem 8 : adding} \\ \text{to the third row 7} \\ \text{times the second row}\end{array}$$

$$= 7(0) = 0 \qquad \text{Theorem 3}$$

Example 13 Use pivotal condensation to evaluate $\det\begin{bmatrix} 0 & -1 & 4 \\ 1 & -5 & 1 \\ -6 & 2 & -3 \end{bmatrix}$.

Solution:

$$\begin{vmatrix} 0 & -1 & 4 \\ 1 & -5 & 1 \\ -6 & 2 & -3 \end{vmatrix} = (-1)\begin{vmatrix} 0 & -5 & 1 \\ 0 & -1 & 4 \\ -6 & 2 & -3 \end{vmatrix} \quad \begin{matrix}\text{Theorem 6 : interchanging} \\ \text{the first and second rows}\end{matrix}$$

$$= (-1)\begin{vmatrix} 1 & -5 & 1 \\ 0 & -1 & 4 \\ 0 & -28 & 3 \end{vmatrix} \quad \begin{matrix}\text{Theorem 8 : adding} \\ \text{to the third row 6} \\ \text{times the first row}\end{matrix}$$

$$= (-1)(-1)\begin{vmatrix} 1 & 5 & 1 \\ 0 & 1 & -4 \\ 0 & -28 & 3 \end{vmatrix} \quad \begin{matrix}\text{Theorem 7 : applied} \\ \text{to the second row}\end{matrix}$$

$$= \begin{vmatrix} 1 & -5 & 1 \\ 0 & 1 & -4 \\ 0 & 0 & -109 \end{vmatrix} \quad \begin{matrix}\text{Theorem 8 : adding} \\ \text{to the third row 28} \\ \text{times the second row}\end{matrix}$$

$$= -(109)\begin{vmatrix} 1 & -5 & 1 \\ 0 & 1 & -4 \\ 0 & 0 & 1 \end{vmatrix} \quad \begin{matrix}\text{Theorem 7 : applied} \\ \text{to the third row}\end{matrix}$$

$$= (-109)(1) = -109 \quad \text{Theorem 3}$$

It follows from Theorem 6 that any property about determinants dealing with row operations is equally true for the analogous operations on columns, because a row operation on the transpose of a matrix is the same as a column operation on the matrix itself. Therefore, if two columns of a matrix are interchanged, its determinant changes sign; if two columns of a matrix are identical, its determinant is 0; multiplying a determinant by a scalar is equivalent to multiplying one column of the matrix by that scalar and then evaluating the new determinant; and the third elementary column operation when applied to a matrix does not change the determinant of the matrix.

Any property about determinants dealing with row operations is equally true for the analogous operations on columns.

We have from Theorem 6 of Section 2.7 that a square matrix has an inverse if and only if the matrix can be transformed by elementary row operations to row-reduced form with all ones on its main diagonal. Using pivotal condensation, we also have that a matrix can be transformed by elementary row operations to row-reduced form with all ones on its main diagonal if and only if its determinant is nonzero. Thus, we have Theorem 9.

> **▶THEOREM 9**
>
> *A square matrix has an inverse if and only if its determinant is nonzero.*◀

The matrix given in Example 12 does not have an inverse because its determinant is 0, while the matrix given in Example 4 is invertible because its determinant is nonzero. Inverses, when they exist, are obtained by the method developed in Section 2.4. Techniques also exist for finding inverses using determinants, but they are far less efficient and rarely used in practice.

If a determinant of a matrix is nonzero, then its determinant and that of its inverse are related.

> **▶THEOREM 10**
>
> *If a matrix $\mathbf{A}$ is invertible, then $\det(\mathbf{A}^{-1}) = 1/\det(\mathbf{A})$.*◀

Proof: If $\mathbf{A}$ is invertible, then $\det(\mathbf{A}) \neq 0$ and $\mathbf{A}\mathbf{A}^{-1} = \mathbf{I}$. Therefore,

$$\begin{aligned} \det\left(\mathbf{A}\mathbf{A}^{-1}\right) &= \det(\mathbf{I}) \\ \det\left(\mathbf{A}\mathbf{A}^{-1}\right) &= 1 \\ \det(\mathbf{A})\cdot\det\left(\mathbf{A}^{-1}\right) &= 1 \\ \det\left(\mathbf{A}^{-1}\right) &= 1/\det(\mathbf{A}) \end{aligned}$$

> **▶THEOREM 11**
>
> *Similar matrices have the same determinant.*◀

Proof: If $\mathbf{A}$ and $\mathbf{B}$ are similar matrices, then there exists an invertible matrix $\mathbf{P}$ such that $\mathbf{A} = \mathbf{P}^{-1}\,\mathbf{B}\mathbf{P}$. It follows from Theorem 1 and Theorem 10 that

$$\begin{aligned} \det(\mathbf{A}) = \det\left(\mathbf{P}^{-1}\mathbf{B}\mathbf{P}\right) &= \det\left(\mathbf{P}^{-1}\right)\det(\mathbf{B})\det(\mathbf{P}) \\ &= [1/\det(\mathbf{P})]\det(\mathbf{B})\det(\mathbf{P}) = \det(\mathbf{B}) \end{aligned}$$

Problems 1.5

In Problems 1 through 31, find the determinants of the given matrices.

(1) $\begin{bmatrix} 3 & 4 \\ 5 & 6 \end{bmatrix}$. **(2)** $\begin{bmatrix} 3 & -4 \\ 5 & 6 \end{bmatrix}$. **(3)** $\begin{bmatrix} 3 & 4 \\ -5 & 6 \end{bmatrix}$.

(4) $\begin{bmatrix} 5 & 6 \\ 7 & 8 \end{bmatrix}$. **(5)** $\begin{bmatrix} 5 & 6 \\ -7 & 8 \end{bmatrix}$. **(6)** $\begin{bmatrix} 5 & 6 \\ 7 & -8 \end{bmatrix}$.

(7) $\begin{bmatrix} 1 & -1 \\ 2 & 7 \end{bmatrix}$.

(8) $\begin{bmatrix} -2 & -3 \\ -4 & 4 \end{bmatrix}$.

(9) $\begin{bmatrix} 3 & -1 \\ -3 & 8 \end{bmatrix}$.

(10) $\begin{bmatrix} 1 & 2 & -2 \\ 0 & 2 & 3 \\ 0 & 0 & -3 \end{bmatrix}$.

(11) $\begin{bmatrix} 3 & 2 & -2 \\ 1 & 0 & 4 \\ 2 & 0 & -3 \end{bmatrix}$.

(12) $\begin{bmatrix} 1 & -2 & -2 \\ 7 & 3 & -3 \\ 0 & 0 & -3 \end{bmatrix}$.

(13) $\begin{bmatrix} 2 & 0 & -1 \\ 1 & 1 & 1 \\ 3 & 2 & -3 \end{bmatrix}$.

(14) $\begin{bmatrix} 3 & 5 & 2 \\ -1 & 0 & 4 \\ -2 & 2 & 7 \end{bmatrix}$.

(15) $\begin{bmatrix} 1 & -3 & -3 \\ 2 & 8 & 3 \\ 4 & 5 & 0 \end{bmatrix}$.

(16) $\begin{bmatrix} 2 & 1 & -9 \\ 3 & -1 & 1 \\ 3 & -1 & 2 \end{bmatrix}$.

(17) $\begin{bmatrix} -1 & 3 & 3 \\ 1 & 1 & 4 \\ -1 & 1 & 2 \end{bmatrix}$.

(18) $\begin{bmatrix} 1 & -3 & -3 \\ 2 & 8 & 3 \\ 3 & 5 & 1 \end{bmatrix}$.

(19) $\begin{bmatrix} 2 & 1 & 3 \\ 3 & -1 & 2 \\ 2 & 3 & 5 \end{bmatrix}$.

(20) $\begin{bmatrix} -1 & 3 & 3 \\ 4 & 5 & 6 \\ -1 & 3 & 3 \end{bmatrix}$.

(21) $\begin{bmatrix} 1 & 2 & -3 \\ 5 & 5 & 1 \\ 2 & -5 & -1 \end{bmatrix}$.

(22) $\begin{bmatrix} -4 & 0 & 0 \\ 2 & -1 & 0 \\ 3 & 1 & -2 \end{bmatrix}$.

(23) $\begin{bmatrix} 1 & 3 & 2 \\ -1 & 4 & 1 \\ 5 & 3 & 8 \end{bmatrix}$.

(24) $\begin{bmatrix} 3 & -2 & 0 \\ 1 & 1 & 2 \\ -3 & 4 & 1 \end{bmatrix}$.

(25) $\begin{bmatrix} -4 & 0 & 0 & 0 \\ 1 & -5 & 0 & 0 \\ 2 & 1 & -2 & 0 \\ 3 & 1 & -2 & 1 \end{bmatrix}$.

(26) $\begin{bmatrix} -1 & 2 & 1 & 2 \\ 1 & 0 & 3 & -1 \\ 2 & 2 & -1 & 1 \\ 2 & 0 & -3 & 2 \end{bmatrix}$.

(27) $\begin{bmatrix} 1 & 1 & 2 & -2 \\ 1 & 5 & 2 & -1 \\ -2 & -2 & 1 & 3 \\ -3 & 4 & -1 & 8 \end{bmatrix}$.

(28) $\begin{bmatrix} -1 & 3 & 2 & -2 \\ 1 & -5 & -4 & 6 \\ 3 & -6 & 1 & 1 \\ 3 & -4 & 3 & -3 \end{bmatrix}$.

(29) $\begin{bmatrix} 1 & 1 & 0 & -2 \\ 1 & 5 & 0 & -1 \\ -2 & -2 & 0 & 3 \\ -3 & 4 & 0 & 8 \end{bmatrix}$.

(30) $\begin{bmatrix} 1 & 2 & 1 & -1 \\ 4 & 0 & 3 & 0 \\ 1 & 1 & 0 & 5 \\ 2 & -2 & 1 & 1 \end{bmatrix}$.

(31) $\begin{bmatrix} 11 & 1 & 0 & 9 & 0 \\ 2 & 1 & 1 & 0 & 0 \\ 4 & -1 & 1 & 0 & 0 \\ 3 & 2 & 2 & 1 & 0 \\ 0 & 0 & 1 & 2 & 0 \end{bmatrix}$.

(32) Find t so that $\begin{vmatrix} t & 2t \\ 1 & t \end{vmatrix} = 0$

(33) Find t so that $\begin{vmatrix} t-2 & t \\ 3 & t+2 \end{vmatrix} = 0$.

(34) Find λ so that $\begin{vmatrix} 4-\lambda & 2 \\ -1 & 1-\lambda \end{vmatrix} = 0.$

(35) Find λ so that $\begin{vmatrix} 1-\lambda & 5 \\ 1 & -1-\lambda \end{vmatrix} = 0.$

In Problems 36 through 43, find $\det(\mathbf{A} - \lambda\mathbf{I})$ when $\mathbf{A}$ is:

(36) The matrix defined in Problem 1.

(37) The matrix defined in Problem 2.

(38) The matrix defined in Problem 4.

(39) The matrix defined in Problem 7.

(40) The matrix defined in Problem 11.

(41) The matrix defined in Problem 12.

(42) The matrix defined in Problem 13.

(43) The matrix defined in Problem 14.

(44) Verify Theorem 1 for $\mathbf{A} = \begin{bmatrix} 6 & 1 \\ 1 & 2 \end{bmatrix}$ and $\mathbf{B} = \begin{bmatrix} 3 & -1 \\ 2 & 1 \end{bmatrix}$.

(45) Find the area of the parallelogram generated by the vectors $[-1 \quad 3]^{\mathrm{T}}$ and $[2 \quad -3]^{\mathrm{T}}$.

(46) Find the area of the parallelogram generated by the vectors $[1 \quad -5]^{\mathrm{T}}$ and $[-4 \quad -4]^{\mathrm{T}}$.

(47) Find the area of the parallelogram generated by the vectors $[2 \quad 4]^{\mathrm{T}}$ and $[3 \quad -8]^{\mathrm{T}}$.

In Problems 48 through 65, find the determinants of the given matrices using pivotal condensation.

(48) $\begin{bmatrix} 1 & 2 & -2 \\ 1 & 3 & 3 \\ 2 & 5 & 0 \end{bmatrix}$. **(49)** $\begin{bmatrix} 1 & 2 & 3 \\ 4 & 5 & 6 \\ 7 & 8 & 9 \end{bmatrix}$. **(50)** $\begin{bmatrix} 3 & -4 & 2 \\ -1 & 5 & 7 \\ 1 & 9 & -6 \end{bmatrix}$.

(51) $\begin{bmatrix} -1 & 3 & 3 \\ 1 & 1 & 4 \\ -1 & 1 & 2 \end{bmatrix}$. **(52)** $\begin{bmatrix} 1 & -3 & -3 \\ 2 & 8 & 4 \\ 3 & 5 & 1 \end{bmatrix}$. **(53)** $\begin{bmatrix} 2 & 1 & -9 \\ 3 & -1 & 1 \\ 3 & -1 & 2 \end{bmatrix}$.

(54) $\begin{bmatrix} 2 & 1 & 3 \\ 3 & -1 & 2 \\ 2 & 3 & 5 \end{bmatrix}$. **(55)** $\begin{bmatrix} -1 & 3 & 3 \\ 4 & 5 & 6 \\ -1 & 3 & 3 \end{bmatrix}$. **(56)** $\begin{bmatrix} 1 & 2 & -3 \\ 5 & 5 & 1 \\ 2 & -5 & -1 \end{bmatrix}$.

(57) $\begin{bmatrix} 2 & 0 & -1 \\ 1 & 1 & 1 \\ 3 & 2 & -3 \end{bmatrix}$. **(58)** $\begin{bmatrix} 3 & 5 & 2 \\ -1 & 0 & 4 \\ -2 & 2 & 7 \end{bmatrix}$. **(59)** $\begin{bmatrix} 1 & -3 & -3 \\ 2 & 8 & 3 \\ 4 & 5 & 0 \end{bmatrix}$.

(60) $\begin{bmatrix} 3 & 5 & 4 & 6 \\ -2 & 1 & 0 & 7 \\ -5 & 4 & 7 & 2 \\ 8 & -3 & 1 & 1 \end{bmatrix}$.

(61) $\begin{bmatrix} -1 & 2 & 1 & 2 \\ 1 & 0 & 3 & -1 \\ 2 & 2 & -1 & 1 \\ 2 & 0 & -3 & 2 \end{bmatrix}$.

(62) $\begin{bmatrix} 1 & 1 & 2 & -2 \\ 1 & 5 & 2 & -1 \\ -2 & -2 & 1 & 3 \\ -3 & 4 & -1 & 8 \end{bmatrix}$.

(63) $\begin{bmatrix} -1 & 3 & 2 & -2 \\ 1 & -5 & -4 & 6 \\ 3 & -6 & 1 & 1 \\ 3 & -4 & 3 & -3 \end{bmatrix}$.

(64) $\begin{bmatrix} 1 & 1 & 0 & -2 \\ 1 & 5 & 0 & -1 \\ -2 & -2 & 0 & 3 \\ -3 & 4 & 0 & 8 \end{bmatrix}$.

(65) $\begin{bmatrix} -2 & 0 & 1 & 3 \\ 4 & 0 & 2 & -2 \\ -3 & 1 & 0 & 1 \\ 5 & 4 & 1 & 7 \end{bmatrix}$.

In Problems 66 through 72, use the properties of determinants to prove the stated identities.

(66) $\begin{vmatrix} a & b & c \\ r & s & t \\ x & y & z \end{vmatrix} = -\frac{1}{4}\begin{vmatrix} 2a & 4b & 2c \\ -r & -2s & -t \\ x & 2y & z \end{vmatrix}$.

(67) $\begin{vmatrix} a-3x & b-3y & c-3z \\ a+5x & b+5y & c+5z \\ x & y & z \end{vmatrix} = 0.$

(68) $\begin{vmatrix} 2a & 3a & c \\ 2r & 3t & t \\ 2x & 3x & z \end{vmatrix} = 0.$

(69) $\begin{vmatrix} a & b & c \\ r & s & t \\ x & y & z \end{vmatrix} = \begin{vmatrix} a & x & r \\ b & y & s \\ c & z & t \end{vmatrix}$.

(70) $\begin{vmatrix} a & r & x \\ b & s & y \\ c & t & z \end{vmatrix} = \begin{vmatrix} a+x & r-x & x \\ b+y & s-y & y \\ c+z & t-z & z \end{vmatrix}$.

(71) $-12\begin{vmatrix} a & r & x \\ b & s & y \\ c & t & z \end{vmatrix} = \begin{vmatrix} 2a & 3r & x \\ 4b & 6s & 2y \\ -2c & -3t & -z \end{vmatrix}$.

(72) $5\begin{vmatrix} a & r & x \\ b & s & y \\ c & t & z \end{vmatrix} = \begin{vmatrix} a-3b & r-3s & x-3y \\ b-2c & s-2t & y-2z \\ 5c & 5t & 5z \end{vmatrix}$.

(73) Verify Theorem 5 directly for the matrices in Problems 48 through 51.

(74) Verify Corollary 3 directly for $\lambda = 3$ and $\mathbf{A} = \begin{bmatrix} 1 & 3 \\ -3 & 4 \end{bmatrix}$.

(75) Verify Corollary 3 directly for $\lambda = -2$ and $\mathbf{A} = \begin{bmatrix} 2 & 3 \\ -3 & -2 \end{bmatrix}$.

(76) Verify Corollary 3 directly for $\lambda = -1$ and $\mathbf{A}$ given by the matrix in Problem 1.

1.6 THE INVERSE

In Section 1.2, we defined matrix multiplication so that any system of linear equations can be written in the matrix form

$$\mathbf{Ax} = \mathbf{b} \qquad (1.14 \text{ repeated})$$

with the intent of solving this equation for $\mathbf{x}$ and obtaining all the variables in the original system *at the same time*. Unfortunately, we cannot divide (Equation 1.14) by the coefficient matrix $\mathbf{A}$ because matrix division is an undefined operation. An equally good operation is, however, available to us.

Division of real numbers is equivalent to multiplication by reciprocals. We can solve the linear equation $5x = 20$ for the variable x either by dividing the equation by 5 or by multiplying the equation by 0.2, the reciprocal of 5. A real number b is the reciprocal of a if and only if $ab = 1$, in which case we write $b = a^{-1}$. The concept of reciprocals can be extended to matrices. The matrix counterpart of the number 1 is an identity matrix $\mathbf{I}$, and the word inverse is used for a matrix $\mathbf{A}$ instead of reciprocal even though the notation $\mathbf{A}^{-1}$ is retained. Thus, a matrix $\mathbf{B}$ is an *inverse* of a matrix $\mathbf{A}$ if

An $n \times n$ matrix $\mathbf{A}^{-1}$ is the inverse of an $n \times n$ matrix $\mathbf{A}$ if $\mathbf{AA}^{-1} = \mathbf{A}^{-1}\mathbf{A} = \mathbf{I}$.

$$\mathbf{AB} = \mathbf{BA} = \mathbf{I} \qquad (1.29)$$

in which case we write $\mathbf{B} = \mathbf{A}^{-1}$.

The requirement that a matrix commute with its inverse implies that both matrices are square and of the same order. Thus, inverses are only defined for square matrices. If a square matrix $\mathbf{A}$ has an inverse, then $\mathbf{A}$ is said to be *invertible* or *nonsingular*; if $\mathbf{A}$ does not have an inverse, then $\mathbf{A}$ is said to be *singular*.

Example 1 The matrix $\mathbf{B} = \begin{bmatrix} -2 & 1 \\ 1.5 & -0.5 \end{bmatrix}$ is an inverse of $\mathbf{A} = \begin{bmatrix} 1 & 2 \\ 3 & 4 \end{bmatrix}$ because

$$\mathbf{AB} = \begin{bmatrix} 1 & 2 \\ 3 & 4 \end{bmatrix}\begin{bmatrix} -2 & 1 \\ 1.5 & -0.5 \end{bmatrix} = \begin{bmatrix} 1 & 0 \\ 0 & 1 \end{bmatrix} = \begin{bmatrix} -2 & 1 \\ 1.5 & -0.5 \end{bmatrix}\begin{bmatrix} 1 & 2 \\ 3 & 4 \end{bmatrix} = \mathbf{BA}$$

and we write

$$\mathbf{A}^{-1} = \begin{bmatrix} 1 & 2 \\ 3 & 4 \end{bmatrix}^{-1} = \begin{bmatrix} -2 & 1 \\ 1.5 & -0.5 \end{bmatrix}$$

In contrast, $\mathbf{C} = \begin{bmatrix} 1 & 1/2 \\ 1/3 & 1/4 \end{bmatrix}$ is not an inverse of $\mathbf{A}$ because

$$\mathbf{AC} = \begin{bmatrix} 1 & 2 \\ 3 & 4 \end{bmatrix} \begin{bmatrix} 1 & 1/2 \\ 1/3 & 1/4 \end{bmatrix} = \begin{bmatrix} 5/3 & 1 \\ 13/3 & 5/2 \end{bmatrix} \neq \mathbf{I}$$

Equation (1.24) is a test for checking whether one matrix is an inverse of another matrix. In Section 2.6, we prove that if $\mathbf{AB} = \mathbf{I}$ for two square matrices of the same order, then **A** and **B** commute under multiplication and $\mathbf{BA} = \mathbf{I}$. If we borrow this result, we reduce the checking procedure by half. A square matrix **B** is an inverse of a square matrix **A** if either $\mathbf{AB} = \mathbf{I}$ or $\mathbf{BA} = \mathbf{I}$; each equality guarantees the other. We also show later in this section that an inverse is unique; that is, if a square matrix has an inverse, it has only one.

We can write the inverses of some simple matrices by inspections. The inverse of a diagonal matrix **D** having all nonzero elements on its main diagonal is a diagonal matrix whose diagonal elements are the reciprocals of the corresponding diagonal elements of **D**. The inverse of

$$\mathbf{D} = \begin{bmatrix} \lambda_1 & 0 & 0 & \cdots & 0 \\ 0 & \lambda_2 & 0 & \cdots & 0 \\ 0 & 0 & \lambda_3 & \cdots & 0 \\ \vdots & \vdots & \vdots & \ddots & \vdots \\ 0 & 0 & 0 & \cdots & \lambda_k \end{bmatrix} \text{ is } \mathbf{D}^{-1} = \begin{bmatrix} 1/\lambda_1 & 0 & 0 & \cdots & 0 \\ 0 & 1/\lambda_2 & 0 & \cdots & 0 \\ 0 & 0 & 1/\lambda_3 & \cdots & 0 \\ \vdots & \vdots & \vdots & \ddots & \vdots \\ 0 & 0 & 0 & \cdots & 1/\lambda_k \end{bmatrix}$$

if none of the diagonal elements is zero. It is easy to show that if any diagonal element in a diagonal matrix is zero, then that matrix is singular (see Problem 56).

An elementary matrix **E** is a square matrix that generates an elementary row operation on a matrix **A** (which need not be square) under the multiplication **EA**. Elementary matrices are constructed by applying the desired elementary row operation to an identity matrix of appropriate order. That order is a square matrix having as many columns as there are rows in **A** so that the multiplication **EA** is defined. Identity matrices contain many zeros, and because nothing is accomplished by interchanging the positions of zeros, or multiplying zeros by constants, or adding zeros together, the construction of an elementary matrix can be simplified.

An elementary matrix **E** is a square matrix that generates an elementary row operation on a matrix **A** under the multiplication **EA**.

CREATING ELEMENTARY MATRICES:

(i) To construct an elementary matrix that interchanges the ith row with the jth row, begin with an identity matrix **I**. First interchange the 1 in the i-i position with the 0 in the j-i position and then interchange the 1 in the j-j position with the 0 in the i-j position.

(ii) To construct an elementary matrix that multiplies the ith row of a matrix by the nonzero scalar k, begin with an identity matrix **I** and replace the 1 in the i-i position with k.

(iii) To construct an elementary matrix that adds to the jth row of a matrix the scalar k times the ith row of that matrix, begin with an identity matrix and replace the 0 in the j-i position with k.

Example 2 Find elementary matrices that when multiplied on the right by any 3×5 matrix **A** will (a) interchange the first and second rows of **A**, (b) multiply the third row of **A** by -0.5, and (c) add to the third row of **A** 4 times its second row.

Solution:

$$\text{(a)} \begin{bmatrix} 0 & 1 & 0 \\ 1 & 0 & 0 \\ 0 & 0 & 1 \end{bmatrix}, \quad \text{(b)} \begin{bmatrix} 1 & 0 & 0 \\ 0 & 1 & 0 \\ 0 & 0 & -0.5 \end{bmatrix}, \quad \text{(c)} \begin{bmatrix} 1 & 0 & 0 \\ 0 & 1 & 0 \\ 0 & 4 & 1 \end{bmatrix}.$$

Example 3 Find elementary matrices that when multiplied on the right by any 4×3 matrix **A** will (a) interchange the second and fourth rows of **A**, (b) multiply the third row of **A** by 3, and (c) add to the fourth row of **A** -5 times its second row.

Solution:

$$\text{(a)} \begin{bmatrix} 1 & 0 & 0 & 0 \\ 0 & 0 & 0 & 1 \\ 0 & 0 & 1 & 0 \\ 0 & 1 & 0 & 0 \end{bmatrix}, \quad \text{(b)} \begin{bmatrix} 1 & 0 & 0 & 0 \\ 0 & 1 & 0 & 0 \\ 0 & 0 & 3 & 0 \\ 0 & 0 & 0 & 1 \end{bmatrix}, \quad \text{(c)} \begin{bmatrix} 1 & 0 & 0 & 0 \\ 0 & 0 & 0 & 0 \\ 0 & 0 & 1 & 0 \\ 0 & -5 & 0 & 1 \end{bmatrix}.$$

▶THEOREM 1

(a) *The inverse of an elementary matrix that interchanges two rows is the elementary matrix itself.*

(b) *The inverse of an elementary matrix that multiplies one row by a nonzero scalar k is a matrix obtained by replacing the scalar k in the elementary matrix by 1/k.*

(c) *The inverse of an elementary matrix that adds to one row a constant k times another row is a matrix obtained by replacing the scalar k in the elementary matrix by −k.*◀

Proof:

(a) Let **E** be an elementary matrix that has the effect interchanging the ith and ith rows of a matrix. **E** comes from interchanging the ith and jth rows of the identity matrix having the same order as **E**. Then $\mathbf{EE}=\mathbf{I}$, because interchanging the positions of the ith row of an identity matrix with jth row twice in succession does not alter the original matrix. With $\mathbf{EE}=\mathbf{I}$, it follows that $\mathbf{E}^{-1}=\mathbf{E}$.

(b) Let **E** be an elementary matrix that has the effect of multiplying the ith row of a matrix by a nonzero scalar k, and let **F** be an elementary matrix that has the effect of multiplying the ith row of a matrix by a nonzero scalar $1/k$. **E** comes from multiplying the ith of the identity matrix having the same order as **E** by k. Then $\mathbf{FE}=\mathbf{I}$, because multiplying the ith row of an identity matrix first by k and then by $1/k$ does not alter the original matrix. With $\mathbf{FE}=\mathbf{I}$, it follows that $\mathbf{F}=\mathbf{E}^{-1}$.

(c) The proof is similar to the part (b) and is left as an exercise for the reader (see Problem 63).

Example 4 The inverses of the elementary matrices found in Example 2 are, respectively,

$$\text{(a)}\quad \begin{bmatrix} 0 & 1 & 0 \\ 1 & 0 & 0 \\ 0 & 0 & 1 \end{bmatrix}, \quad \text{(b)}\quad \begin{bmatrix} 1 & 0 & 0 \\ 0 & 1 & 0 \\ 0 & 0 & -2 \end{bmatrix}, \quad \text{(c)}\quad \begin{bmatrix} 1 & 0 & 0 \\ 0 & 1 & 0 \\ 0 & -4 & 1 \end{bmatrix}.$$

The inverses of the elementary matrices found in Example 3 are, respectively,

$$\text{(a)}\quad \begin{bmatrix} 1 & 0 & 0 & 0 \\ 0 & 0 & 0 & 1 \\ 0 & 0 & 1 & 0 \\ 0 & 1 & 0 & 0 \end{bmatrix}, \quad \text{(b)}\quad \begin{bmatrix} 1 & 0 & 0 & 0 \\ 0 & 1 & 0 & 0 \\ 0 & 0 & 1/3 & 0 \\ 0 & 0 & 0 & 1 \end{bmatrix}, \quad \text{(c)}\quad \begin{bmatrix} 1 & 0 & 0 & 0 \\ 0 & 0 & 0 & 0 \\ 0 & 0 & 1 & 0 \\ 0 & 5 & 0 & 1 \end{bmatrix}.$$

Elementary row operations are the backbone of a popular method for calculating inverses. We shall show in Section 2.6 that a square matrix is invertible if and only if it can be transformed into a row-reduced matrix having all ones on the main diagonal. If such a transformation is possible, then the original matrix can be reduced still further, all the way to an identity matrix. This is done by applying elementary row operation R_3—adding to one row of a matrix a scalar times another row of the same matrix—to each column, *beginning with the last column and moving sequentially towards the first column*, placing zeros in all positions above the diagonal elements.

Example 5 Use elementary row operations to transform the row-reduced matrix

$$\mathbf{A} = \begin{bmatrix} 1 & 2 & 1 \\ 0 & 1 & 3 \\ 0 & 0 & 1 \end{bmatrix}$$

to the identity matrix.

Solution:

$$\begin{bmatrix} 1 & 2 & 1 \\ 0 & 1 & 3 \\ 0 & 0 & 1 \end{bmatrix} \rightarrow \begin{bmatrix} 1 & 2 & 1 \\ 0 & 1 & 0 \\ 0 & 0 & 1 \end{bmatrix} \quad \text{by adding to the second row } -3 \text{ times the third row}$$

$$\rightarrow \begin{bmatrix} 1 & 2 & 0 \\ 0 & 1 & 0 \\ 0 & 0 & 1 \end{bmatrix} \quad \text{by adding to the first row } -1 \text{ times the third row}$$

$$\rightarrow \begin{bmatrix} 1 & 0 & 1 \\ 0 & 1 & 0 \\ 0 & 0 & 1 \end{bmatrix} \quad \text{by adding to the first row } -2 \text{ times the second row}$$

Thus, a square **A** has an inverse if and only if **A** can be transformed into an identity matrix with elementary row operations. Because each elementary row operation can be represented by an elementary matrix, we conclude that a matrix **A** has an inverse if and only if there exists a sequence of elementary matrices $\mathbf{E}_1, \mathbf{E}_2, \ldots, \mathbf{E}_k$ such that

$$\mathbf{E}_k, \mathbf{E}_{k-1} \ldots \mathbf{E}_2\mathbf{E}_1\mathbf{A} = \mathbf{I}$$

Denoting the product of these elementary matrices by **B**, we have $\mathbf{BA}=\mathbf{I}$, which implies that $\mathbf{B}=\mathbf{A}^{-1}$. To calculate the inverse of a matrix **A**, we need only record the product of the elementary row operations used to transform **A** to **I**. This is accomplished by applying the same elementary row operations to both **A** and **I** simultaneously.

CALCULATING INVERSES

Step 1. Create an augmented matrix [**A** | **I**], where **A** is the $n \times n$ matrix to be inverted and **I** is the $n \times n$ identity matrix.

Step 2. Use elementary row operations on [**A** | **I**] to transform the left partition **A** to row-reduced form, applying each operation to the full augmented matrix.

Step 3. If the left partition of the row-reduced matrix has zero elements on its main diagonal, stop: **A** does not have inverse. Otherwise, continue.

Step 4. Use elementary row operations on the row-reduced augmented matrix to transform the left partition to the $n \times n$ identity matrix, applying each operation to the full augmented matrix.

Step 5. The right partition of the final augmented matrix is the inverse of **A**.

Example 6 Find the inverse of $\mathbf{A} = \begin{bmatrix} 1 & 2 \\ 3 & 4 \end{bmatrix}$.

Solution:

$$\left[\begin{array}{cc|cc} 1 & 2 & 1 & 0 \\ 3 & 4 & 0 & 1 \end{array}\right] \rightarrow \left[\begin{array}{cc|cc} 1 & 2 & 1 & 0 \\ 0 & -2 & -3 & 1 \end{array}\right]$$ by adding to the second row -3 times the first row

$$\rightarrow \left[\begin{array}{cc|cc} 0 & 0 & -2 & 0 \\ 0 & 1 & 3/2 & -1/2 \end{array}\right]$$ by multiplying the second row by $-1/2$

A has been transformed into row-reduced form with a main diagonal of only ones; **A** has an inverse. Continuing with the transformation process, we get

$$\rightarrow \left[\begin{array}{cc|cc} 0 & 0 & -2 & 0 \\ 0 & 1 & 3/2 & -1/2 \end{array}\right]$$ by adding to the first row –2 times the second row

Thus,

$$\mathbf{A}^{-1} = \begin{bmatrix} -2 & 1 \\ 3/2 & -1/2 \end{bmatrix}$$

Example 7 Find the inverse of $\mathbf{A} = \begin{bmatrix} 5 & 8 & 1 \\ 0 & 2 & 1 \\ 4 & 3 & -1 \end{bmatrix}$

Solution:

$$\left[\begin{array}{ccc|ccc} 5 & 8 & 1 & 1 & 0 & 0 \\ 0 & 2 & 1 & 0 & 1 & 0 \\ 4 & 3 & -1 & 0 & 0 & 1 \end{array}\right]$$

$$\rightarrow \left[\begin{array}{ccc|ccc} 1 & 1.6 & 0.2 & 0.2 & 0 & 0 \\ 0 & 2 & 1 & 0 & 1 & 0 \\ 4 & 3 & -1 & 0 & 0 & 1 \end{array}\right]$$ by multiplying the first row by 0.2

$$\rightarrow \left[\begin{array}{ccc|ccc} 1 & 1.6 & 0.2 & 0.2 & 0 & 0 \\ 0 & 2 & 1 & 0 & 1 & 0 \\ 0 & -3.4 & -1.8 & -0.8 & 0 & 1 \end{array}\right]$$ by adding to the third row -4 times the first row

$$\rightarrow \left[\begin{array}{ccc|ccc} 1 & 1.6 & 0.2 & 0.2 & 0 & 0 \\ 0 & 1 & 0.5 & 0 & 0.5 & 0 \\ 0 & -3.4 & -1.8 & -0.8 & 0 & 1 \end{array}\right]$$ by multiplying the second row by 1/2

$$\rightarrow \left[\begin{array}{ccc|ccc} 1 & 1.6 & 0.2 & 0.2 & 0 & 0 \\ 0 & 1 & 0.5 & 0 & 0.5 & 0 \\ 0 & 0 & -0.1 & -0.8 & 1.7 & 1 \end{array}\right]$$ by adding to the third row 3.4 times the second row

$$\rightarrow \left[\begin{array}{ccc|ccc} 1 & 1.6 & 0.2 & 0.2 & 0 & 0 \\ 0 & 1 & 0.5 & 0 & 0.5 & 0 \\ 0 & 0 & 1 & 8 & -17 & -10 \end{array}\right]$$ by multiplying the third row by -10

A has been transformed into row-reduced form with a main diagonal of only ones; **A** has an inverse. Continuing with the transformation process, we get

$$\rightarrow \left[\begin{array}{ccc|ccc} 1 & 1.6 & 0.2 & 0.2 & 0 & 0 \\ 0 & 1 & 0 & -4 & 9 & 5 \\ 0 & 0 & 1 & 8 & -17 & -10 \end{array}\right]$$ by adding to the second row -0.5 times the third row

$$\rightarrow \left[\begin{array}{ccc|ccc} 1 & 1.6 & 0 & -1.4 & 3.4 & 2 \\ 0 & 1 & 0 & -4 & 9 & 5 \\ 0 & 0 & 1 & 8 & -17 & -10 \end{array}\right]$$ by adding to the first row -0.2 times the third row

$$\rightarrow \left[\begin{array}{ccc|ccc} 1 & 0 & 0 & 5 & -11 & 6 \\ 0 & 1 & 0 & -4 & 9 & 5 \\ 0 & 0 & 1 & 8 & -17 & -10 \end{array}\right]$$ by adding to the first row −1.6 times the second row

Thus,

$$\mathbf{A}^{-1} = \begin{bmatrix} 5 & -11 & -6 \\ -4 & 9 & 5 \\ 8 & -17 & -10 \end{bmatrix}$$

Example 8 Find the inverse of $\mathbf{A} = \begin{bmatrix} 0 & 1 & 1 \\ 1 & 1 & 1 \\ 1 & 1 & 3 \end{bmatrix}$.

Solution:

$$\left[\begin{array}{ccc|ccc} 0 & 1 & 1 & 1 & 0 & 0 \\ 1 & 1 & 1 & 0 & 1 & 0 \\ 1 & 1 & 3 & 0 & 0 & 1 \end{array}\right]$$

$$\rightarrow \left[\begin{array}{ccc|ccc} 1 & 1 & 1 & 0 & 1 & 0 \\ 0 & 1 & 1 & 1 & 0 & 0 \\ 1 & 1 & 3 & 0 & 0 & 1 \end{array}\right]$$ by interchanging the first and second rows

$$\rightarrow \left[\begin{array}{ccc|ccc} 1 & 1 & 1 & 0 & 1 & 0 \\ 0 & 1 & 1 & 1 & 0 & 0 \\ 0 & 0 & 2 & 0 & -1 & 1 \end{array}\right]$$ by adding to the third row −1 times the first row

$$\rightarrow \left[\begin{array}{ccc|ccc} 1 & 1 & 1 & 0 & 1 & 0 \\ 0 & 1 & 1 & 1 & 0 & 0 \\ 0 & 0 & 1 & 0 & -1/2 & 1/2 \end{array}\right]$$ by multiplying the third row by 1/2

$$\rightarrow \left[\begin{array}{ccc|ccc} 1 & 1 & 1 & 0 & 1 & 0 \\ 0 & 1 & 0 & 1 & 1/2 & -1/2 \\ 0 & 0 & 1 & 0 & -1/2 & 1/2 \end{array}\right]$$ by adding to the second row −1 times the third row

$$\rightarrow \left[\begin{array}{ccc|ccc} 1 & 1 & 0 & 0 & 3/2 & -1/2 \\ 0 & 1 & 0 & 1 & 1/2 & -1/2 \\ 0 & 0 & 1 & 0 & -1/2 & 1/2 \end{array}\right]$$ by adding to the first row −1 times the third row

$$\rightarrow \left[\begin{array}{ccc|ccc} 1 & 0 & 0 & -1 & 1 & 0 \\ 0 & 1 & 0 & 1 & 1/2 & -1/2 \\ 0 & 0 & 1 & 0 & -1/2 & 1/2 \end{array}\right]$$ by adding to the first row −1 times the second row

Thus,

$$\mathbf{A}^{-1} = \begin{bmatrix} -1 & 1 & 0 \\ 1 & 1/2 & -1/2 \\ 0 & -1/2 & 1/2 \end{bmatrix}.$$

Example 9 Find the inverse of $\mathbf{A} = \begin{bmatrix} 1 & 2 \\ 2 & 4 \end{bmatrix}$.

Solution:

$$\left[\begin{array}{cc|cc} 1 & 2 & 1 & 0 \\ 2 & 4 & 0 & 0 \end{array}\right] \to \left[\begin{array}{cc|cc} 1 & 2 & 1 & 0 \\ 0 & 0 & -2 & 1 \end{array}\right]$$ by adding to the second row −2 times the first row

A has been transformed into row-reduced form. Because the main diagonal contains a zero entry, **A** does not have an inverse; **A** is singular.

▶THEOREM 2

The inverse of a matrix is unique.◀

Proof: If **B** and **C** are both inverses of the matrix **A**, then

$$\mathbf{AB} = \mathbf{I}, \quad \mathbf{BA} = \mathbf{I}, \quad \mathbf{AC} = \mathbf{I}, \quad \text{and} \quad \mathbf{CA} = \mathbf{I}.$$

It now follows that

$$\mathbf{C} = \mathbf{CI} = \mathbf{C}(\mathbf{AB}) = (\mathbf{CA})\mathbf{B} = \mathbf{IB} = \mathbf{B}.$$

Thus, if **B** and **C** are both inverses of **A**, they must be equal; hence, the inverse is unique.

Using Theorem 2, we can prove some useful properties of inverses.

▶THEOREM 3

If **A** *and* **B** *are* $n \times n$ *nonsingular matrices, then*

(a) $(\mathbf{A}^{-1})^{-1} = \mathbf{A}$,
(b) $(\mathbf{AB})^{-1} = \mathbf{B}^{-1}\mathbf{A}^{-1}$,
(c) $(\mathbf{A}^{\mathsf{T}})^{-1} = (\mathbf{A}^{-1})^{\mathsf{T}}$,
(d) $(\lambda\mathbf{A})^{-1} = (1/\lambda)\mathbf{A}^{-1}$, *if* λ *is a nonzero scalar.*◀

Proof: We prove parts (b) and (c) and leave parts (a) and (d) as exercises (see Problems 59 and 60). To prove (b), we note that

$$\left(B^{-1}A^{-1}\right)(AB) = B^{-1}\left(A^{-1}A\right)B = B^{-1}IB = B^{-1}B = I.$$

Thus, $B^{-1}A^{-1}$ is *an* inverse of **AB**. Because the inverse is unique, it follows that $(AB)^{-1} = B^{-1}A^{-1}$.

To prove (c), we note that

$$(A^T)\,(A^{-1})^T = (A^{-1}A)^T = I^T = I.$$

Thus, $(A^{-1})^T$ is *an* inverse of A^T. Because the inverse is unique, it follows that $(A^T)^{-1} = (A^{-1})^T$.

The process of finding an inverse is known as *inversion*, and, interestingly, some matrix forms are preserved under this process.

▶THEOREM 4

(a) *The inverse of a nonsingular symmetric matrix is symmetric.*
(b) *The inverse of a nonsingular upper or lower triangular matrix is again an upper or lower triangular matrix, respectively.*◀

Proof: If A is symmetric, then $A^T = A$. Combining this observation with part (c) of Theorem 2, we find that

$$(A^{-1})^T = (A^T)^{-1} = (A)^{-1}$$

so A^{-1} also equals its transpose and is symmetric. This proves part (a). Part (b) is immediate from Theorem 2 and the constructive procedure used for calculating inverses. The details are left as an exercise (see Problem 62).

A system of simultaneously linear equations has the matrix form

$$Ax = b \qquad (1.14 \text{ repeated})$$

If the coefficient matrix **A** is invertible, we can premultiply both sides of Equation (1.14) by A^{-1} to obtain

$$\begin{aligned} A^{-1}(Ax) &= A^{-1}b \\ \left(A^{-1}A\right)x &= A^{-1}b \\ Ix &= A^{-1}b \end{aligned}$$

or

$$x = A^{-1}b \qquad (1.30)$$

The matrix equation **Ax** = **b** has **x** = $\mathbf{A}^{-1}\mathbf{b}$ as its solution if the coefficient matrix **A** is invertible.

This is precisely the form we sought in Section 1.2. With this formula, we can solve for all the variables in a system of linear equations *at the same time.*

Example 10 The system of equations

$$\begin{aligned} x + 2y &= 150 \\ 3x + 4y &= 250 \end{aligned}$$

can be written as $\mathbf{Ax}=\mathbf{b}$ with

$$\mathbf{A} = \begin{bmatrix} 1 & 2 \\ 3 & 4 \end{bmatrix}, \quad \mathbf{x} = \begin{bmatrix} x \\ y \end{bmatrix}, \quad \text{and} \quad \mathbf{b} = \begin{bmatrix} 150 \\ 250 \end{bmatrix}$$

Using the results of Example 6, we have that the coefficient matrix **A** is invertible and

$$\begin{bmatrix} x \\ y \end{bmatrix} = \mathbf{x} = \mathbf{A}^{-1}\mathbf{b} = \begin{bmatrix} -2 & 1 \\ 3/2 & -1/2 \end{bmatrix} \begin{bmatrix} 150 \\ 250 \end{bmatrix} = \begin{bmatrix} -50 \\ 100 \end{bmatrix}.$$

Hence, $x=-50$ and $y=100$.

Example 11 The system of equations

$$\begin{aligned} 5x + 8y + z &= 2 \\ 2y + z &= -1 \\ 4x + 3y - z &= 3 \end{aligned}$$

can be written as $\mathbf{Ax}=\mathbf{b}$ with

$$\mathbf{A} = \begin{bmatrix} 5 & 8 & 1 \\ 0 & 2 & 1 \\ 4 & 3 & -1 \end{bmatrix}, \quad \mathbf{x} = \begin{bmatrix} x \\ y \\ z \end{bmatrix}, \quad \text{and} \quad \mathbf{b} = \begin{bmatrix} 2 \\ -1 \\ 3 \end{bmatrix}.$$

Using the results of Example 7, we have that the coefficient matrix **A** is invertible and

$$\begin{bmatrix} x \\ y \end{bmatrix} = \mathbf{x} = \mathbf{A}^{-1}\mathbf{b} = \begin{bmatrix} 5 & -11 & -6 \\ -4 & 9 & 5 \\ 8 & -17 & -10 \end{bmatrix}, = \begin{bmatrix} 2 \\ -1 \\ 3 \end{bmatrix} \begin{bmatrix} 3 \\ -2 \\ 3 \end{bmatrix}$$

Hence, $x=3$, $y=-2$, and $z=3$.

Not only does the invertibility of the coefficient matrix **A** provide us with a solution to the system $\mathbf{Ax}=\mathbf{b}$, it also provides us with a means to show that this solution is the *only* solution to the system.

►THEOREM 5

If **A** *is invertible, then the system of simultaneous linear equations defined by* $\mathbf{Ax}=\mathbf{b}$ *has a unique (one and only one) solution.*◄

Proof: Define $\mathbf{w} = \mathbf{A}^{-1}\mathbf{b}$. Then

$$\mathbf{Aw} = \mathbf{AA}^{-1}\mathbf{b} = \mathbf{Ib} = \mathbf{b} \tag{1.31}$$

and $\mathbf{w}$ is one solution to the system $\mathbf{Ax} = \mathbf{b}$. Let $\mathbf{y}$ be another solution to this system. Then necessarily

$$\mathbf{Ay} = \mathbf{b} \tag{1.32}$$

Equations (1.26) and (1.27) imply that

$$\mathbf{Aw} = \mathbf{Ay}$$

Premultiplying both sides of this last equation by $\mathbf{A}^{-1}$, we find

$$\begin{aligned} \mathbf{A}^{-1}(\mathbf{Aw}) &= \mathbf{A}^{-1}(\mathbf{Ay}) \\ (\mathbf{A}^{-1}\mathbf{A})\mathbf{w} &= (\mathbf{A}^{-1}\mathbf{A})\mathbf{y} \\ \mathbf{Iw} &= \mathbf{Iy} \end{aligned}$$

or

$$\mathbf{w} = \mathbf{y}$$

Thus, if $\mathbf{y}$ is a solution of $\mathbf{Ax} = \mathbf{b}$, then it must equal $\mathbf{w}$. Therefore, $\mathbf{w} = \mathbf{A}^{-1}\mathbf{b}$ is the only solution to this system.

If $\mathbf{A}$ is singular, so that $\mathbf{A}^{-1}$ does not exist, then Equation (1.25) is *not* valid and other methods, such as Gaussian elimination, must be used to solve the given system of simultaneous equations.

Problems 1.6

(1) Determine if any of the following matrices are inverses for $\mathbf{A} = \begin{bmatrix} 1 & 3 \\ 2 & 9 \end{bmatrix}$:

(a) $\begin{bmatrix} 1 & 1/3 \\ 1/2 & 1/9 \end{bmatrix}$, (b) $\begin{bmatrix} -1 & -3 \\ -2 & -9 \end{bmatrix}$,

(c) $\begin{bmatrix} 3 & -1 \\ -2/3 & 1/3 \end{bmatrix}$, (d) $\begin{bmatrix} 9 & -3 \\ -2 & 1 \end{bmatrix}$.

(2) Determine if any of the following matrices are inverses for $\mathbf{A} = \begin{bmatrix} 1 & 1 \\ 1 & 1 \end{bmatrix}$:

(a) $\begin{bmatrix} 1 & 1 \\ 1 & 1 \end{bmatrix}$, (b) $\begin{bmatrix} -1 & 1 \\ 1 & -1 \end{bmatrix}$,

(c) $\begin{bmatrix} 1 & 1 \\ -1 & -1 \end{bmatrix}$, (d) $\begin{bmatrix} 2 & -1 \\ -1 & 2 \end{bmatrix}$.

In Problems 3 through 12, find elementary matrices that when multiplied on the right by the given matrix $\mathbf{A}$ will generate the specified result.

(3) Interchange the order of the first and second rows of a 2×2 matrix **A**.

(4) Multiply the first row of a 2×2 matrix **A** by 3.

(5) Multiply the second row of a 2×2 matrix **A** by -5.

(6) Multiply the second row of a 3×3 matrix **A** by -5.

(7) Add to the second row of a 2×2 matrix **A** three times its first row.

(8) Add to the first row of a 2×2 matrix **A** three times its second row.

(9) Add to the second row of a 3×3 matrix **A** three times its third row.

(10) Add to the third row of a 3×4 matrix **A** five times its first row.

(11) Interchange the order of the second and fourth rows of a 6×6 matrix **A**.

(12) Multiply the second row of a 2×5 matrix **A** by 7.

In Problems 13 through 22, find the inverses of the given elementary matrices.

(13) $\begin{bmatrix} 2 & 0 \\ 0 & 1 \end{bmatrix}$ **(14)** $\begin{bmatrix} 1 & 2 \\ 0 & 1 \end{bmatrix}$ **(15)** $\begin{bmatrix} 1 & 0 \\ -3 & 1 \end{bmatrix}$

(16) $\begin{bmatrix} 1 & 0 \\ 1 & 1 \end{bmatrix}$ **(17)** $\begin{bmatrix} 1 & 0 & 0 \\ 0 & 2 & 0 \\ 0 & 0 & 1 \end{bmatrix}$ **(18)** $\begin{bmatrix} 0 & 1 & 0 \\ 1 & 0 & 0 \\ 0 & 0 & 1 \end{bmatrix}$

(19) $\begin{bmatrix} 1 & 0 & 3 \\ 0 & 1 & 0 \\ 0 & 0 & 1 \end{bmatrix}$ **(20)** $\begin{bmatrix} 1 & 0 & 0 \\ 0 & 1 & -2 \\ 0 & 0 & 1 \end{bmatrix}$ **(21)** $\begin{bmatrix} 1 & 0 & 0 & 0 \\ 0 & 1 & 0 & 0 \\ 0 & 0 & 0 & 1 \\ 0 & 0 & 1 & 0 \end{bmatrix}$

(22) $\begin{bmatrix} 1 & 0 & 0 & 0 \\ 0 & 1 & 0 & 0 \\ -3 & 0 & 1 & 0 \\ 0 & 0 & 0 & 1 \end{bmatrix}$

In Problems 23 through 39, find the inverses of the given matrices, if they exist.

(23) $\begin{bmatrix} 1 & 1 \\ 3 & 4 \end{bmatrix}$ **(24)** $\begin{bmatrix} 2 & 1 \\ 1 & 2 \end{bmatrix}$ **(25)** $\begin{bmatrix} 4 & 4 \\ 4 & 4 \end{bmatrix}$

(26) $\begin{bmatrix} 1 & 1 & 0 \\ 1 & 0 & 1 \\ 0 & 1 & 1 \end{bmatrix}$ **(27)** $\begin{bmatrix} 0 & 0 & 1 \\ 1 & 0 & 0 \\ 0 & 1 & 0 \end{bmatrix}$ **(28)** $\begin{bmatrix} 2 & 0 & -1 \\ 0 & 1 & 2 \\ 3 & 1 & 1 \end{bmatrix}$

(29) $\begin{bmatrix} 1 & 2 & 3 \\ 4 & 5 & 6 \\ 7 & 8 & 9 \end{bmatrix}$ **(30)** $\begin{bmatrix} 2 & 0 & 0 \\ 5 & 1 & 0 \\ 4 & 1 & 1 \end{bmatrix}$ **(31)** $\begin{bmatrix} 2 & 1 & 5 \\ 0 & 3 & -1 \\ 0 & 0 & 2 \end{bmatrix}$

(32) $\begin{bmatrix} 3 & 2 & 1 \\ 4 & 0 & 1 \\ 3 & 9 & 2 \end{bmatrix}$

(33) $\begin{bmatrix} 1 & 2 & -1 \\ 2 & 0 & 1 \\ -1 & 1 & 3 \end{bmatrix}$

(34) $\begin{bmatrix} 1 & 2 & 1 \\ 3 & -2 & -4 \\ 2 & 3 & -1 \end{bmatrix}$

(35) $\begin{bmatrix} 2 & 4 & 1 \\ 3 & -4 & -4 \\ 5 & 0 & -1 \end{bmatrix}$

(36) $\begin{bmatrix} 5 & 0 & -1 \\ 2 & -1 & 2 \\ 2 & 3 & -1 \end{bmatrix}$

(37) $\begin{bmatrix} 3 & 1 & 1 \\ 1 & 3 & -1 \\ 2 & 3 & -1 \end{bmatrix}$

(38) $\begin{bmatrix} 1 & 1 & 1 & 2 \\ 0 & 1 & -1 & 1 \\ 0 & 0 & 2 & 3 \\ 0 & 0 & 0 & -2 \end{bmatrix}$

(39) $\begin{bmatrix} 1 & 0 & 0 & 0 \\ 2 & -1 & 0 & 0 \\ 4 & 6 & 2 & 0 \\ 3 & 2 & 4 & -1 \end{bmatrix}$

(40) Show directly that the inverse of $\mathbf{A} = \begin{bmatrix} a & b \\ c & d \end{bmatrix}$, when $ad - bc \neq 0$ is

$$\mathbf{A}^{-1} = \frac{1}{ad - bc} \begin{bmatrix} d & -b \\ -c & a \end{bmatrix},$$

(41) Use the result of Problem (40) to calculate the inverses of

(a) $\begin{bmatrix} 1 & 1 \\ 3 & 4 \end{bmatrix}$ and (b) $\begin{bmatrix} 1 & 1/2 \\ 1/2 & 1/3 \end{bmatrix}$

In Problems 42 through 51, use matrix inversion, if possible, to solve the given systems of equations:

(42) $x + 2y = -3$
$3x + y = 1$

(43) $a + 2b = 5$
$-3a + b = 13$

(44) $4x + 2y = 6$
$2x - 3y = 1$

(45) $4l - p = 1$
$5l - 2p = -1$

(46) $2x + 3y = 8$
$6x + 9y = 24$

(47) $x + 2y - z = -1$
$2x + 3y + 2z = 5$
$y - z = 2$

(48) $2x + 3y - z = 4$
$-x - 2y + z = -2$
$3x - y = 2$

(49) $60l + 30m + 20n = 0$
$30l + 20m + 15n = -10$
$20l + 15m + 12n = -10$

(50) $2r + 3s - 4t = 12$
$3r - 2s = -1$
$8r - s - 4t = 10$

(51) $x + 2y - 2z = -1$
$2x + y + z = 5$
$-x + y - z = -2$

(52) Solve each of the following systems using the *same* inverse:

(a) $3x + 5y = 10$
$2x + 3y = 20$

(b) $3x + 5y = -8$
$2x + 3y = 22$

(c) $3x + 5y = 0.2$
$2x + 3y = 0.5$

(d) $3x + 5y = 0$
$2x + 3y = 5$

(53) Solve each of the following systems using the *same* inverse:

(a) $2x + 4y = 2$
$3x + 2y + z = 8$
$5x - 3y + 7z = 15$

(b) $2x + 4y = 3$
$3x + 2y + z = 8$
$5x - 3y + 7z = 15$

(c) $2x + 4y = 2$
$3x + 2y + z = 9$
$5x - 3y + 7z = 15$

(d) $2x + 4y = 1$
$3x + 2y + z = 7$
$5x - 3y + 7z = 14$

(54) If **A** is nonsingular matrix, we may define $\mathbf{A}^{-n} = (\mathbf{A}^{-1})^n$, for any positive integer n. Use this definition to find $\mathbf{A}^{-2}$ and $\mathbf{A}^{-3}$ for the following matrices:

(a) $\begin{bmatrix} 1 & 1 \\ 2 & 3 \end{bmatrix}$, (b) $\begin{bmatrix} 2 & 5 \\ 1 & 2 \end{bmatrix}$, (c) $\begin{bmatrix} 1 & 1 \\ 3 & 4 \end{bmatrix}$,

(d) $\begin{bmatrix} 1 & 1 & 1 \\ 0 & 1 & 1 \\ 0 & 0 & 1 \end{bmatrix}$, (e) $\begin{bmatrix} 1 & 2 & -1 \\ 0 & 1 & -1 \\ 0 & 0 & 1 \end{bmatrix}$.

(55) Prove that a square zero matrix does not have an inverse.

(56) Prove that if a diagonal matrix has at least one zero on its main diagonal, then that matrix does not have an inverse.

(57) Prove that if $\mathbf{A}^2 = \mathbf{I}$, then $\mathbf{A}^{-1} = \mathbf{A}$.

(58) If **A** is symmetric, prove the identity $(\mathbf{BA}^{-1})^{\mathrm{T}}(\mathbf{A}^{-1}\mathbf{B}^{T})^{-1} = \mathbf{I}$.

(59) Prove that if **A** is invertible, then $(\mathbf{A}^{-1})^{-1} = \mathbf{A}$.

(60) Prove that if **A** is invertible and if λ is a nonzero scalar, then $(\lambda\mathbf{A})^{-1} = (1/\lambda)\mathbf{A}^{-1}$.

(61) Prove that if **A**, **B**, and **C** are $n \times n$ nonsingular matrices, then $(\mathbf{ABC})^{-1} = \mathbf{C}^{-1}\mathbf{B}^{-1}\mathbf{A}^{-1}$.

(62) Prove that the inverse of a nonsingular upper (lower) triangular matrix is itself upper (lower) triangular.

(63) Prove part (c) of Theorem 1.

(64) Show that if **A** can be partitioned into the block diagonal form

$$\mathbf{A} = \begin{bmatrix} \mathbf{A}_1 & & & & 0 \\ & \mathbf{A}_2 & & & \\ & & A_3 & & \\ & & & \ddots & \\ & 0 & & & \mathbf{A}_k \end{bmatrix}$$

with $\mathbf{A}_1, \mathbf{A}_2, \ldots, \mathbf{A}_n$ all invertible, then

$$\mathbf{A}^{-1} = \begin{bmatrix} \mathbf{A}_1^{-1} & & & & 0 \\ & \mathbf{A}_2^{-1} & & & \\ & & \mathbf{A}_3^{-1} & & \\ & & & \ddots & \\ & 0 & & & \mathbf{A}_k^{-1} \end{bmatrix}$$

1.7 LU DECOMPOSITION

Matrix inversion of elementary matrices is at the core of still another popular method, known as **LU** decomposition, for solving simultaneous equations in the matrix form $\mathbf{Ax}=\mathbf{b}$. The method rests on factoring a nonsingular coefficient matrix **A** into the product of a lower triangular matrix **L** with an upper triangular matrix **U**. Generally, there are many such factorizations. If **L** is required to have all diagonal elements equal to 1, then the decomposition, when it exists, is unique and we may write

$$\mathbf{A} = \mathbf{LU} \tag{1.33}$$

with

$$\mathbf{L} = \begin{bmatrix} 1 & 0 & 0 & \cdots & 0 \\ l_{21} & 1 & 0 & \cdots & 0 \\ l_{31} & l_{32} & 1 & \cdots & 0 \\ \vdots & \vdots & \vdots & \ddots & \vdots \\ l_{n1} & l_{n2} & l_{n3} & \cdots & 1 \end{bmatrix}$$

$$\mathbf{U} = \begin{bmatrix} u_{11} & u_{12} & u_{13} & \cdots & u_{1n} \\ 0 & u_{22} & u_{23} & \cdots & u_{2n} \\ 0 & 0 & u_{23} & \cdots & u_{3n} \\ \vdots & \vdots & \vdots & \ddots & \vdots \\ 0 & 0 & 0 & \cdots & u_{nn} \end{bmatrix}$$

To decompose **A** into form (1.33), we first transform **A** to upper triangular form using just the third elementary row operation R_3. This is similar to transforming a matrix to row-reduced form, except we no longer use the first two elementary row operations. We do not interchange rows, and we do not multiply rows by nonzero constants. Consequently, we no longer require that the first nonzero element of each nonzero row be 1, and if any of the pivots are 0—which would indicate a row interchange in the transformation to row-reduced form—then the decomposition scheme we seek cannot be done.

Example 1 Use the third elementary row operation to transform the matrix

$$\mathbf{A} = \begin{bmatrix} 2 & -1 & 3 \\ 4 & 2 & 1 \\ -6 & -1 & 2 \end{bmatrix}$$

into upper triangular form.

Solution:

$$\mathbf{A} = \begin{bmatrix} 2 & -1 & 3 \\ 4 & 2 & 1 \\ -6 & -1 & 2 \end{bmatrix} \to \begin{bmatrix} 2 & -1 & 3 \\ 0 & 4 & -5 \\ -6 & -1 & 2 \end{bmatrix}$$ by adding to the second row − 2 times the first row

$$\to \begin{bmatrix} 2 & -1 & 3 \\ 0 & 4 & -5 \\ 0 & -4 & 11 \end{bmatrix}$$ by adding to the third row 3 times the first row

$$\to \begin{bmatrix} 2 & -1 & 3 \\ 0 & 4 & -5 \\ 0 & 0 & 6 \end{bmatrix}$$ by adding to the third row 1 times the second row

If a square matrix **A** can be reduced to upper triangular form **U** by a sequence of elementary row operations of the third type, then there exists a sequence of elementary matrices $\mathbf{E}_{21}$, $\mathbf{E}_{31}$, $\mathbf{E}_{41}$, ..., $\mathbf{E}_{n,n-1}$ such that

$$(\mathbf{E}_{n,n-1} \ldots \mathbf{E}_{41}\mathbf{E}_{31}\mathbf{E}_{21})\mathbf{A} = \mathbf{U} \tag{1.34}$$

where $\mathbf{E}_{21}$ denotes the elementary matrix that places a 0 in the 2-1 position, $\mathbf{E}_{31}$ denotes the elementary matrix that places a 0 in the 3-1 position, $\mathbf{E}_{41}$ denotes the elementary matrix that places a 0 in the 4-1 position, and so on. Since elementary matrices have inverses, we can write Equation (1.29) as

$$\mathbf{A} = \left(\mathbf{E}_{21}^{-1}\mathbf{E}_{31}^{-1}\mathbf{E}_{41}^{-1} \ldots \mathbf{E}_{n,n-1}^{-1}\right)\mathbf{U} \tag{1.35}$$

Each elementary matrix in Equation (1.34) is lower triangular. It follows from Theorem 4 of Section 1.5 that each of the inverses in Equation (1.35) are lower triangular and then from Theorem 2 of Section 1.3 that the product of these lower triangular inverses is itself lower triangular. If we set

$$\mathbf{L} = \left(\mathbf{E}_{21}^{-1}\mathbf{E}_{31}^{-1}\mathbf{E}_{41}^{-1} \ldots \mathbf{E}_{n,\,n-1}^{-1}\right)$$

then **L** is lower triangular and Equation (1.35) may be rewritten as **A** = **LU**, which is the decomposition we seek.

A square matrix **A** has an **LU** decomposition if **A** can be transformed to upper triangular form using only the third elementary row operation.

Example 2 Construct an **LU** decomposition for the matrix given in Example 1.

Solution: The elementary matrices associated with the elementary row operations described in Example 1 are

$$\mathbf{E}_{21} = \begin{bmatrix} 1 & 0 & 0 \\ -2 & 1 & 0 \\ 0 & 0 & 1 \end{bmatrix}, \quad \mathbf{E}_{31} = \begin{bmatrix} 1 & 0 & 0 \\ 0 & 1 & 0 \\ -3 & 0 & 1 \end{bmatrix}, \quad \text{and} \quad \mathbf{E}_{32} = \begin{bmatrix} 1 & 0 & 0 \\ 0 & 1 & 0 \\ 0 & -1 & 1 \end{bmatrix}$$

with inverses given respectively by

$$\mathbf{E}_{21}^{-1} = \begin{bmatrix} 1 & 0 & 0 \\ 2 & 1 & 0 \\ 0 & 0 & 1 \end{bmatrix}, \quad \mathbf{E}_{31}^{-1} = \begin{bmatrix} 1 & 0 & 0 \\ 0 & 1 & 0 \\ -3 & 0 & 1 \end{bmatrix}, \quad \text{and} \quad \mathbf{E}_{32}^{-1} = \begin{bmatrix} 1 & 0 & 0 \\ 0 & 1 & 0 \\ 0 & -1 & 1 \end{bmatrix}.$$

Then,

$$\begin{bmatrix} 2 & -1 & 3 \\ 4 & 2 & 1 \\ -6 & -1 & 2 \end{bmatrix} = \begin{bmatrix} 1 & 0 & 0 \\ 2 & 1 & 0 \\ 0 & 0 & 1 \end{bmatrix} \begin{bmatrix} 1 & 0 & 0 \\ 0 & 1 & 0 \\ -3 & 0 & 1 \end{bmatrix} \begin{bmatrix} 1 & 0 & 0 \\ 0 & 1 & 0 \\ 0 & -1 & 1 \end{bmatrix} \begin{bmatrix} 2 & -1 & 3 \\ 0 & 4 & -5 \\ 0 & 0 & 6 \end{bmatrix}$$

or, upon multiplying together the inverses of the elementary matrices,

$$\begin{bmatrix} 2 & -1 & 3 \\ 4 & 2 & 1 \\ -6 & -1 & 2 \end{bmatrix} = \begin{bmatrix} 1 & 0 & 0 \\ 2 & 1 & 0 \\ -3 & -1 & 1 \end{bmatrix} \begin{bmatrix} 2 & -1 & 3 \\ 0 & 4 & -5 \\ 0 & 0 & 6 \end{bmatrix}.$$

Example 2 suggests an important simplification of the decomposition process. Note that the elements in **L** located below the main diagonal are *the negatives of the scalars* used in the elementary row operations in Example 1 to reduce **A** to upper triangular form! This is no coincidence.

▶OBSERVATION 1

If, in transforming a square matrix **A** to upper triangular form, a zero is placed in the *i-j* position by adding to row *i* a scalar *k* times row *j*, then the *i-j* element of **L** in the **LU** decomposition of **A** is $-k$. ◀

We summarize the decomposition process as follows: Use only the third elementary row operation to transform a square matrix **A** to upper triangular form. If this is not possible, because of a zero pivot, then stop. Otherwise, the **LU** decomposition is found by defining the resulting upper triangular matrix as **U** and constructing the lower triangular matrix **L** according to Observation 1.

Example 3 Construct an **LU** decomposition for the matrix

$$A = \begin{bmatrix} 2 & 1 & 2 & 3 \\ 6 & 2 & 4 & 8 \\ 1 & -1 & 0 & 4 \\ 0 & 1 & -3 & -4 \end{bmatrix}$$

Solution: Transforming A to upper triangular form, we get

$$\begin{bmatrix} 2 & 1 & 2 & 3 \\ 6 & 2 & 4 & 8 \\ 1 & -1 & 0 & 4 \\ 0 & 1 & -3 & -4 \end{bmatrix} \rightarrow \begin{bmatrix} 2 & 1 & 2 & 3 \\ 0 & -1 & -2 & -1 \\ 1 & -1 & 0 & 4 \\ 0 & 1 & -3 & -4 \end{bmatrix}$$ by adding to the second row -3 times the first row

$$\rightarrow \begin{bmatrix} 2 & 1 & 2 & 3 \\ 0 & -1 & -2 & -1 \\ 0 & -\frac{3}{2} & -1 & \frac{5}{2} \\ 0 & 1 & -3 & -4 \end{bmatrix}$$ by adding to the third row $-1/2$ times the first row

$$\rightarrow \begin{bmatrix} 2 & 1 & 2 & 3 \\ 0 & -1 & -2 & -1 \\ 0 & 0 & 2 & 4 \\ 0 & 1 & -3 & -4 \end{bmatrix}$$ by adding to the third row $-3/2$ times the second row

$$\rightarrow \begin{bmatrix} 2 & 1 & 2 & 3 \\ 0 & -1 & -2 & -1 \\ 0 & 0 & 2 & 4 \\ 0 & 0 & -5 & -5 \end{bmatrix}$$ by adding to the fourth row 1 times the second row

$$\rightarrow \begin{bmatrix} 2 & 1 & 2 & 3 \\ 0 & -1 & -2 & -1 \\ 0 & 0 & 2 & 4 \\ 0 & 0 & 0 & 5 \end{bmatrix}$$ by adding to the fourth row 5/2 times the third row

We now have an upper triangular matrix **U**. To get the lower triangular matrix **L** in the decomposition, we note that we used the scalar -3 to place a 0 in the 2-1 position, so its negative $-(-3)=3$ goes into the 2-1 position of **L**. We used the scalar $-1/2$ to place a 0 in the 3-1 position in the second step of the preceding triangularization process, so its negative, $1/2$, becomes the 3-1 element in **L**; we used the scalar $5/2$ to place a 0 in the 4-3 position during the last step of the triangularization process, so its negative, $-5/2$, becomes the 4-3 element in **L**. Continuing in this manner, we generate the decomposition

$$\begin{bmatrix} 2 & 1 & 2 & 3 \\ 6 & 2 & 4 & 8 \\ 1 & -1 & 0 & 4 \\ 0 & 1 & -3 & -4 \end{bmatrix} = \begin{bmatrix} 1 & 0 & 0 & 0 \\ 3 & 1 & 0 & 0 \\ \frac{1}{2} & \frac{3}{2} & 1 & 0 \\ 0 & -1 & -\frac{5}{2} & 1 \end{bmatrix} \begin{bmatrix} 2 & 1 & 2 & 3 \\ 0 & -1 & -2 & -1 \\ 0 & 0 & 2 & 4 \\ 0 & 0 & 0 & 5 \end{bmatrix}$$

LU decompositions, when they exist, are used to solve systems of simultaneous linear equations. If a square matrix **A** can be factored into **A** = **LU**, then the system of equations **Ax** = **b** can be written as **L(Ux)** = **b**. To find **x**, we first solve the system

$$\mathbf{Ly} = \mathbf{b} \tag{1.36}$$

for **y**, and then once **y** is determined, we solve the system

$$\mathbf{Ux} = \mathbf{y} \tag{1.37}$$

for **x**. Both systems (1.36) and (1.37) are easy to solve, the first by forward substitution and the second by backward substitution.

If **A** = **LU** for a square matrix **A**, then the equation **Ax** = **b** is solved by first solving the equation **Ly** = **b** for **y** and then solving the equation **Ux** = **y** for **x**.

Example 4 Solve the system of equations:

$$\begin{aligned} 2x - y + 3z &= 9 \\ 4x + 2y + z &= 9 \\ -6x - y + 2z &= 12 \end{aligned}$$

Solution: This system has the matrix form

$$\begin{bmatrix} 2 & -1 & 3 \\ 4 & 2 & 1 \\ -6 & -1 & 2 \end{bmatrix} \begin{bmatrix} x \\ y \\ z \end{bmatrix} = \begin{bmatrix} 9 \\ 9 \\ 12 \end{bmatrix}$$

The **LU** decomposition for the coefficient matrix **A** is given in Example 2. If we define the components of **y** by α, β, and γ, respectively, the matrix system **Ly** = **b** is

$$\begin{bmatrix} 1 & 0 & 0 \\ 2 & 1 & 0 \\ -3 & -1 & 1 \end{bmatrix} \begin{bmatrix} \alpha \\ \beta \\ \gamma \end{bmatrix} = \begin{bmatrix} 9 \\ 9 \\ 12 \end{bmatrix}$$

which is equivalent to the system of equations

$$\begin{aligned} \alpha &= 9 \\ 2\alpha + \beta &= 9 \\ -3\alpha - \beta + \gamma &= 12 \end{aligned}$$

Solving this system from top to bottom, we get $\alpha = 9$, $\beta = -9$, and $\gamma = 30$. Consequently, the matrix system **Ux** = **y** is

$$\begin{bmatrix} 2 & -1 & 3 \\ 0 & 4 & -5 \\ 0 & 0 & 6 \end{bmatrix} \begin{bmatrix} x \\ y \\ z \end{bmatrix} = \begin{bmatrix} 9 \\ -9 \\ 30 \end{bmatrix}$$

which is equivalent to the system of equations

$$\begin{aligned} 2x - y + 3z &= 9 \\ 4y - 5z &= -9 \\ 6z &= 30 \end{aligned}$$

Solving this system from bottom to top, we obtain the final solution $x = -1$, $y = 4$, and $z = 5$.

Example 5 Solve the system

$$\begin{aligned} 2a + b + 2c + 3d &= 5 \\ 6a + 2b + 4c + 8d &= 8 \\ a - b + 4d &= -4 \\ b - 3c - 4d &= -3 \end{aligned}$$

Solution: The matrix representation for this system has as its coefficient matrix the matrix A of Example 3. Define

$$\mathbf{y} = [\alpha, \beta, \gamma, \delta]^T$$

Then, using the decomposition determined in Example 3, we can write the matrix system $\mathbf{Ly} = \mathbf{b}$ as the system of equations

$$\begin{aligned} \alpha &= 5 \\ 3\alpha + \beta &= 8 \\ \frac{1}{2}\alpha + \frac{3}{2}\beta + \gamma &= -4 \\ -\beta - \frac{5}{2}\gamma + \delta &= -3 \end{aligned}$$

which has as its solution $\alpha = 5$, $\beta = -7$, $\gamma = 4$, and $\delta = 0$. Thus, the matrix system $\mathbf{Ux} = \mathbf{y}$ is equivalent to the system of equations

$$\begin{aligned} 2a + b + 2c + 3d &= 5 \\ -b - 2c - d &= -7 \\ 2c + 4d &= 4 \\ 5d &= 0 \end{aligned}$$

Solving this set from bottom to top, we calculate the final solution as $a = -1$, $b = 3$, $c = 2$, and $d = 0$.

Problems 1.7

In Problems 1 through 14, **A** and **b** are given. Construct an **LU** decomposition for the matrix **A** and then use it to solve the system **Ax** = **b** for **x**.

(1) $\mathbf{A} = \begin{bmatrix} 1 & 1 \\ 3 & 4 \end{bmatrix}, \quad \mathbf{b} = \begin{bmatrix} 1 \\ -6 \end{bmatrix}.$ **(2)** $\mathbf{A} = \begin{bmatrix} 2 & 1 \\ 1 & 2 \end{bmatrix}, \quad \mathbf{b} = \begin{bmatrix} 11 \\ -2 \end{bmatrix}.$

(3) $\mathbf{A} = \begin{bmatrix} 8 & 3 \\ 5 & 2 \end{bmatrix}, \quad \mathbf{b} = \begin{bmatrix} 625 \\ 550 \end{bmatrix}.$ **(4)** $\mathbf{A} = \begin{bmatrix} 1 & 1 & 0 \\ 1 & 0 & 1 \\ 0 & 1 & 1 \end{bmatrix}, \quad \mathbf{b} = \begin{bmatrix} 4 \\ 1 \\ -1 \end{bmatrix}.$

(5) $\mathbf{A} = \begin{bmatrix} -1 & 2 & 0 \\ 1 & -3 & 1 \\ 2 & -2 & 3 \end{bmatrix}, \quad \mathbf{b} = \begin{bmatrix} -1 \\ -2 \\ 3 \end{bmatrix}.$

(6) $\mathbf{A} = \begin{bmatrix} 2 & 1 & 3 \\ 4 & 1 & 0 \\ -2 & -1 & -2 \end{bmatrix}, \quad \mathbf{b} = \begin{bmatrix} 10 \\ -40 \\ 0 \end{bmatrix}.$

(7) $\mathbf{A} = \begin{bmatrix} 3 & 2 & 1 \\ 4 & 0 & 1 \\ 3 & 9 & 2 \end{bmatrix}, \quad \mathbf{b} = \begin{bmatrix} 50 \\ 80 \\ 20 \end{bmatrix}.$

(8) $\mathbf{A} = \begin{bmatrix} 1 & 2 & -1 \\ 2 & 0 & 1 \\ -1 & 1 & 3 \end{bmatrix}, \quad \mathbf{b} = \begin{bmatrix} 80 \\ 159 \\ -75 \end{bmatrix}.$

(9) $\mathbf{A} = \begin{bmatrix} 1 & 2 & -1 \\ 0 & 2 & 1 \\ 0 & 0 & 1 \end{bmatrix}, \quad \mathbf{b} = \begin{bmatrix} 8 \\ -1 \\ 5 \end{bmatrix}.$

(10) $\mathbf{A} = \begin{bmatrix} 1 & 0 & 0 \\ 3 & 2 & 0 \\ 1 & 1 & 2 \end{bmatrix}, \quad \mathbf{b} = \begin{bmatrix} 2 \\ 4 \\ 2 \end{bmatrix}.$

(11) $\mathbf{A} = \begin{bmatrix} 1 & 0 & 1 & 1 \\ 1 & 1 & 0 & 1 \\ 1 & 1 & 1 & 0 \\ 0 & 1 & 1 & 1 \end{bmatrix}, \quad \mathbf{b} = \begin{bmatrix} 4 \\ -3 \\ -2 \\ -2 \end{bmatrix}.$

(12) $\mathbf{A} = \begin{bmatrix} 2 & 1 & -1 & 3 \\ 1 & 4 & 2 & 1 \\ 0 & 0 & -1 & 1 \\ 0 & 1 & 1 & 1 \end{bmatrix}, \quad \mathbf{b} = \begin{bmatrix} 1000 \\ 200 \\ 100 \\ 100 \end{bmatrix}.$

(13) $\mathbf{A} = \begin{bmatrix} 1 & 2 & 1 & 1 \\ 1 & 1 & 2 & 1 \\ 1 & 1 & 1 & 2 \\ 0 & 1 & 1 & 1 \end{bmatrix}, \quad \mathbf{b} = \begin{bmatrix} 30 \\ 30 \\ 10 \\ 10 \end{bmatrix}.$

(14) $\mathbf{A} = \begin{bmatrix} 2 & 0 & 2 & 0 \\ 2 & 2 & 0 & 6 \\ -4 & 3 & 1 & 1 \\ 1 & 0 & 3 & 1 \end{bmatrix}, \quad \mathbf{b} = \begin{bmatrix} -2 \\ 4 \\ 9 \\ 4 \end{bmatrix}.$

(15) (a) Use **LU** decomposition to solve the system

$$\begin{aligned} -x + 2y &= -9 \\ 2x + 3y &= 4 \end{aligned}$$

(b) Use the decomposition to solve the preceding system when the right sides of the equations are replaced by 1 and -1, respectively.

(16) (a) Use **LU** decomposition to solve the system

$$\begin{aligned} x + 3y - z &= -1 \\ 2x + 5y + z &= 4 \\ 2x + 7y - 4z &= -6 \end{aligned}$$

(b) Use the decomposition to solve the preceding system when the right side of each equation is replaced by 10, 10, and 10, respectively.

(17) Solve the system $\mathbf{Ax}=\mathbf{b}$ for the following vectors $\mathbf{b}$ when $\mathbf{A}$ is given as in Problem 4:

(a) $\begin{bmatrix} 5 \\ 7 \\ -4 \end{bmatrix}$, (b) $\begin{bmatrix} 2 \\ 2 \\ 0 \end{bmatrix}$, (c) $\begin{bmatrix} 40 \\ 50 \\ 20 \end{bmatrix}$, (d) $\begin{bmatrix} 1 \\ 1 \\ 3 \end{bmatrix}$.

(18) Solve the system $\mathbf{Ax}=\mathbf{b}$ for the following vectors $\mathbf{b}$ when $\mathbf{A}$ is given as in Problem 13:

(a) $\begin{bmatrix} -1 \\ 1 \\ 1 \\ 1 \end{bmatrix}$, (b) $\begin{bmatrix} 0 \\ 0 \\ 0 \\ 0 \end{bmatrix}$, (c) $\begin{bmatrix} 190 \\ 130 \\ 160 \\ 60 \end{bmatrix}$, (d) $\begin{bmatrix} 1 \\ 1 \\ 1 \\ 1 \end{bmatrix}$.

(19) Show that **LU** decomposition cannot be used to solve the system

$$\begin{aligned} 2y + z &= -1 \\ x + y + 3z &= 8 \\ 2x - y - z &= 1 \end{aligned}$$

but that the decomposition can be used if the first two equations are interchanged.

(20) Show that **LU** decomposition cannot be used to solve the system

$$\begin{aligned} x + 2y + z &= 2 \\ 2x + 4y - z &= 7 \\ x + y + 2z &= 2 \end{aligned}$$

but that the decomposition can be used if the first and third equations are interchanged.

(21) (a) Show that the **LU** decomposition procedure given in this section cannot be applied to

$$\mathbf{A} = \begin{bmatrix} 0 & 2 \\ 0 & 9 \end{bmatrix}$$

(b) Verify that **A** = **LU**, when

$$\mathbf{L} = \begin{bmatrix} 1 & 0 \\ 1 & 1 \end{bmatrix} \quad \text{and} \quad \mathbf{U} = \begin{bmatrix} 0 & 2 \\ 0 & 7 \end{bmatrix}$$

(c) Verify that **A** = **LU**, when

$$\mathbf{L} = \begin{bmatrix} 1 & 0 \\ 3 & 1 \end{bmatrix} \quad \text{and} \quad \mathbf{U} = \begin{bmatrix} 0 & 2 \\ 0 & 3 \end{bmatrix}$$

(d) Why do you think the **LU** decomposition procedure fails for this **A**? What might explain the fact that **A** has more than one **LU** decomposition?

CHAPTER 1 REVIEW

Important Terms

augmented matrix
block diagonal matrix
coefficient matrix
cofactor
column matrix
component
consistent equations
derived set
determinant
diagonal element
diagonal matrix
dimension
directed line segment
element
elementary matrix
elementary row operations
equivalent directed line segments
expansion by cofactor
Gaussian elimination
homogeneous equations
identity matrix
inconsistent equations
inverse
invertible matrix
linear equation
lower triangular matrix
LU decomposition
main diagonal
mathematical induction
matrix

nonhomogeneous equations
nonsingular matrix
n-tuple
order
partitioned matrix
pivot
pivotal condersation
power of a matrix
row matrix
row-reduced form
scalar
singular matrix
skew-symmetric matrix
square
submatrix
symmetric matrix
transpose
trivial solution
upper triangular matrix
zero matrix

Important Concepts

Section 1.1

- Two matrices are equal if they have the same order and if their corresponding elements are equal.
- The sum of two matrices of the same order is a matrix obtained by adding together corresponding elements of the original two matrices. Matrix addition is commutative and associative.
- The difference of two matrices of the same order is a matrix obtained by subtracting corresponding elements of the original two matrices.
- The product of a scalar by a matrix is the matrix obtained by multiplying every element of the matrix by the scalar.

Section 1.2

- The product **AB** of two matrices is defined only if the number of columns of **A** equals the number of rows of **B**. Then the *i-j* element of the product is obtained by multiplying the elements in the *i*th row of **A** by the corresponding elements in they *j*th column of **B** and summing the results.
- Matrix multiplication is *not* commutative. The associative law of multiplication as well as the left and right distributive laws for multiplication are valid.
- A system of linear equations may be written as the single matrix equation $\mathbf{Ax}=\mathbf{b}$.

Section 1.3

- The transpose of a matrix **A** is obtained by converting all the rows of **A** into columns while preserving the ordering of the rows/columns.
- The product of two lower (upper) triangular matrices of the same order is also a lower (upper) triangular matrix.

Section 1.4

- A system of simultaneous linear equations has either no solutions, one solution, or infinitely many solutions.
- A homogeneous system of linear equations is always consistent and admits the trivial solution as one solution.

- A linear equation in two variables graphs as a straight line. The coordinates of a point in the plane is a solution to a system of equations in two variables if and only if the point lies simultaneously on the straight line graph of every equation in the system.
- A linear equation in three variables graphs as a plane. The coordinates of a point in space is a solution to a system of equations in three variables if and only if the point lies simultaneously on the planes that represent every equation in the system.
- The heart of Gaussian elimination is the transformation of an augmented matrix to row-reduced form using only elementary row operations.
- If the solution to a derived set involves at least one arbitrary unknown, then the original set of equations has infinitely many solutions.
- A homogeneous system of linear equations having more variables than equations has infinitely many solutions.
- If a derived set contains a false equation, then the original set of equations has no solution.

Section 1.5

- Every square matrix has a number associated with it.
- Minors and cofactors are used to evaluate determinants.
- The determinant of the product of square matrices of the same size is the product of the determinants of the matrices.

Section 1.6

- An inverse, if it exists, is unique.
- The inverse of a diagonal matrix $\mathbf{D}$ with no zero elements on its main diagonal is another diagonal matrix having diagonal elements that are the reciprocals of the diagonal elements of $\mathbf{D}$.
- The inverse of an elementary matrix is again an elementary matrix.
- The inverse of a nonsingular upper (lower) triangular matrix is again an upper (lower) triangular matrix.
- A square matrix has an inverse if it can be transformed by elementary row operations to an upper triangular matrix with no zero elements on the main diagonal.
- The matrix equation $\mathbf{Ax}=\mathbf{b}$ has as its solution $\mathbf{x}=\mathbf{A}^{-1}\mathbf{b}$ if the $\mathbf{A}$ is invertible.

Section 1.7

- A square matrix $\mathbf{A}$ has an $\mathbf{LU}$ decomposition if $\mathbf{A}$ can be transformed to upper triangular form using only the third elementary row operation.
- If $\mathbf{A}=\mathbf{LU}$ for a square matrix $\mathbf{A}$, then the equation $\mathbf{Ax}=\mathbf{b}$ is solved by first solving the equation $\mathbf{Ly}=\mathbf{b}$ for $\mathbf{y}$ and then solving the equation $\mathbf{Ux}=\mathbf{y}$ for $\mathbf{x}$.

CHAPTER 2
Vector Spaces

Chapter Outline

2.1 PROPERTIES OF $\mathbb{R}^n$

At the core of mathematical analysis is the process of identifying fundamental structures that appear with some regularity in different situations, developing them in the abstract, and then applying the resulting knowledge base back to the individual situations. In this way, one can understand simultaneously many different situations by investigating the properties that govern all of them. Matrices would seem to have little in common with polynomials, which in turn appear to have little in common with directed line segments, yet they share fundamental characteristics that, when fully developed, provide a richer understanding of them all.

In order to motivate the ensuing discussion of these fundamental characteristics, we first present some of the properties of a common mathematical structure that should be familiar to readers of this text—the real number system. Points on the plane in an x-y coordinate system are identified by an ordered pair of real numbers; points in space are located by an ordered triplet of real numbers. These are just two examples of the more general concept of an ordered array of n-real numbers known as an *n-tuple*. We write an n-tuple as a $1 \times n$ row matrix. The elements in the row matrix are real numbers and the number of elements (columns) n is

$\mathbb{R}^n$ is the set of ordered arrays of n real numbers. This set is represented either by the set of all n-dimensional row matrices or by the set of all n-dimensional column matrices.

the *dimension* of the row matrix. The set of all n-tuples is often referred to as *n-space* and denoted by $\mathbb{R}^n$. In particular, the ordered pair [1 2] is a member of $\mathbb{R}^2$; it is a 2-tuple of dimension two. The ordered triplet [10 20 30] is a member of $\mathbb{R}^3$; it is a 3-tuple of dimension three. The p-tuple $\mathbf{a} = [a_1\ a_2\ a_3 \ldots a_p]$, where a_j ($j = 1, 2, \ldots, p$) is a real number, is a member of $\mathbb{R}^p$, and has dimension p.

An ordered array of real numbers also can be written as a column matrix, and often is. Here we work exclusively with row matrix representations, but only as a matter of convenience. We could work equally well with column matrices.

Row matrices are special types of matrices, those matrices having only one row, so the basic matrix operations defined in Section 1.1 remain valid for n-tuples represented as row matrices. This means we know how to add and subtract n-tuples of the same dimension and how to multiple a real number times an n-tuple (scalar multiplication). If we restrict ourselves to $\mathbb{R}^2$ and $\mathbb{R}^3$, we can describe these operations geometrically.

A two-dimensional row matrix $\mathbf{v} = [a\,b]$ is identified with the point (a, b) on x-y plane, measured a units along the horizontal x-axis from the origin and then b units parallel to the vertical y-axis. If we draw a directed line segment, or arrow, beginning at the origin and ending at the point (a, b), then this arrow, as shown in Figure 2.1, is a geometrical representation of the row matrix $[a\,b]$. It follows immediately from Pythagoras's theorem that the length or *magnitude* of $\mathbf{v}$, denoted by $\|\mathbf{v}\|$, is

$$\|\mathbf{v}\| = \|[a\ \ b]\| = \sqrt{a^2 + b^2}$$

and from elementary trigonometry that the angle θ satisfies the equation

$$\tan\theta = \frac{b}{a}$$

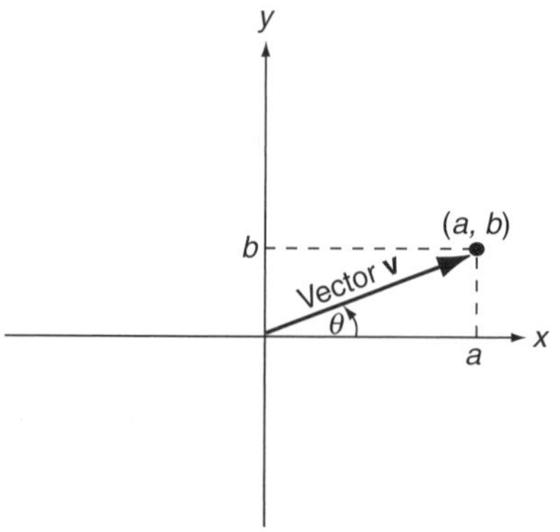

FIGURE 2.1

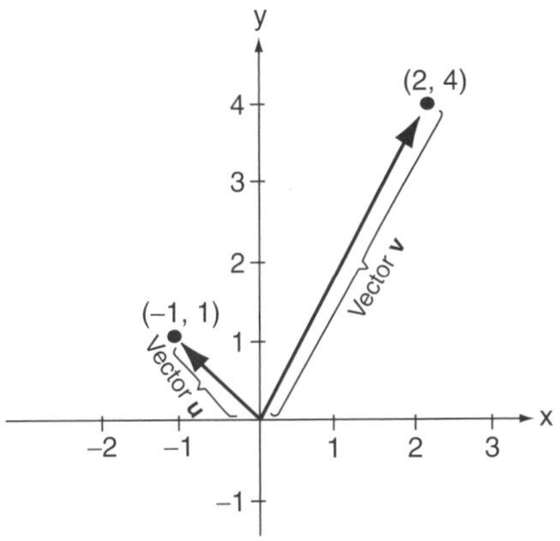

FIGURE 2.2

Example 1 Represent the row matrices $\mathbf{v}=[2\,4]$ and $\mathbf{u}=[-1\,1]$ geometrically and then determine the magnitude of each and the angle each makes with the horizontal x-axis.

Solution: The row matrices are graphed in Figure 2.2. For $\mathbf{v}$, we have

$$\|\mathbf{v}\| = \sqrt{(2)^2 + (4)^2} \approx 4.47, \tan\theta = \frac{4}{2} = 2, \text{and}\,\theta \approx 63.4^\circ$$

For $\mathbf{u}$, similar computations yield

$$\|\mathbf{u}\| = \sqrt{(-1)^2 + (1)^2} \approx 1.14, \tan\theta = \frac{1}{-1} = -1, \text{and}\,\theta \approx 135^\circ$$

To geometrically construct the sum of two row matrices $\mathbf{u}$ and $\mathbf{v}$ in $\mathbb{R}^2$, graph $\mathbf{u}$ and $\mathbf{v}$ on the same coordinate system, translate $\mathbf{v}$ so its initial point coincides with the terminal point of $\mathbf{u}$, *being careful to preserve both the magnitude and direction of* $\mathbf{v}$, and then draw an arrow from the origin to the terminal point of $\mathbf{v}$ after translation. This arrow geometrically represents the sum $\mathbf{u}+\mathbf{v}$. The process is illustrated in Figure 2.3 for the row matrices $\mathbf{u}=[-1\,1]$ and $\mathbf{v}=[2\,4]$.

To graph $\mathbf{u}+\mathbf{v}$ in $\mathbb{R}^2$, graph $\mathbf{u}$ and $\mathbf{v}$ on the same coordinate system, translate $\mathbf{v}$ so its initial point coincides with the terminal point of $\mathbf{u}$, and then draw an arrow from the origin to the terminal point of $\mathbf{v}$ after translation. To graph $\mathbf{u}-\mathbf{v}$ in $\mathbb{R}^2$, graph $\mathbf{u}$ and $\mathbf{v}$ on the same coordinate system and then draw an arrow from the terminal point of $\mathbf{v}$ to the terminal point of $\mathbf{u}$.

To construct the difference of two row matrices $\mathbf{u}-\mathbf{v}$ geometrically, graph both $\mathbf{u}$ and $\mathbf{v}$ normally and construct an arrow from the terminal point of $\mathbf{v}$ to the terminal point of $\mathbf{u}$. This arrow geometrically represents the difference $\mathbf{u}-\mathbf{v}$. The process is depicted in Figure 2.4 for $\mathbf{u}=[-1\,1]$ and $\mathbf{v}=[2\,4]$.

Translating an arrow (directed line segment) that represents a two-dimensional row matrix from one location in the plane to another does not affect the representation, providing both the magnitude and direction as defined by the angle the arrow makes with the positive x-axis are preserved. Many physical phenomena such as velocity and force are completely described by their magnitudes and directions. A wind velocity of 60 miles per hour in the northwest direction is a complete description of that velocity, and *it is independent of where that wind*

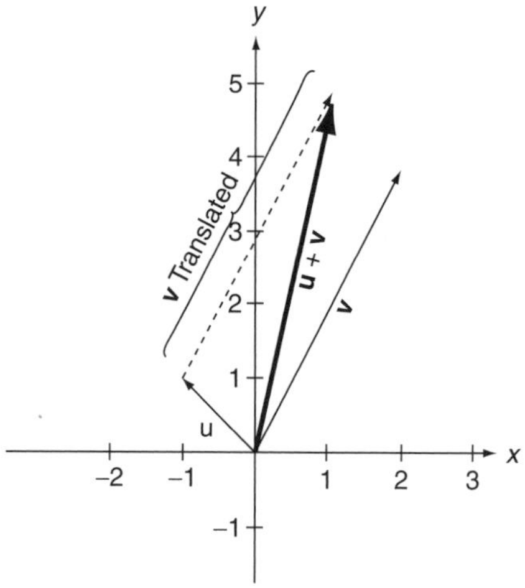

FIGURE 2.3

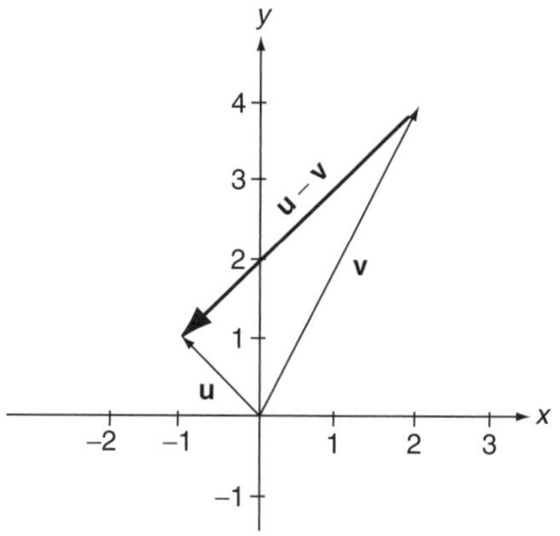

FIGURE 2.4

occurs, be it Lawrence, Kansas, or Portland, Oregon. This independence is the rationale behind translating row matrices geometrically. Geometrically, two-dimensional row matrices having the same magnitude and direction are call *equivalent*, and they are regarded as being equal even though they may be located at different positions in the plane. The four arrows drawn in Figure 2.5 are all geometrical representations of the same row matrix $[1 \; -3]$.

To recapture a row matrix from the directed line segment that represents it, we translate the directed line segment so that its tail lies on the origin and then read the coordinates of its tip. Alternatively, we note that if a directed line segment **w** does not originate at the origin, then it can be expressed as the difference between a directed line segment **u** that begins at the origin and ends at the

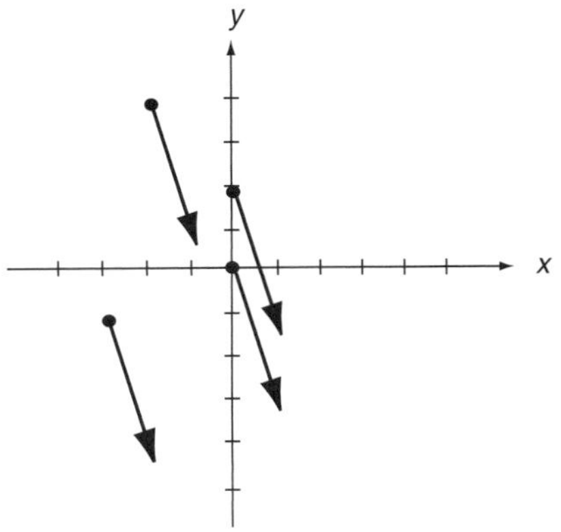

FIGURE 2.5

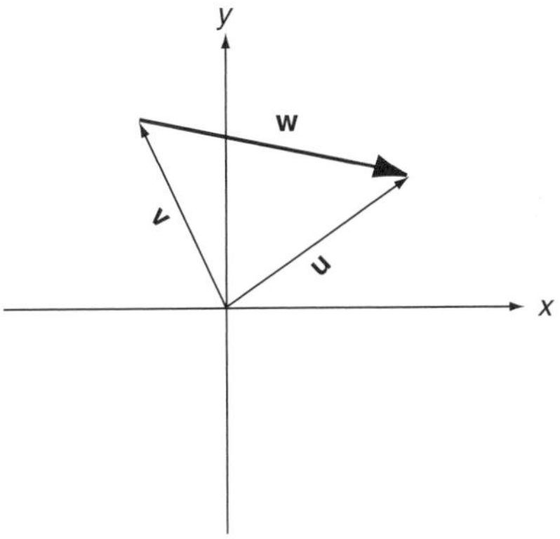

FIGURE 2.6

tip of **w** and a directed line segment **v** that originates at the origin and ends at the tail of **w** as shown in Figure 2.6. Therefore, if the tip of **w** is at the point (x_2, y_2) and the tail at the point (x_1, y_1), then **u** represents the row matrix $[x_2 y_2]$, **v** represents the row matrix $[x_1 y_1]$, and **w** is the difference $\mathbf{w} = \mathbf{u} - \mathbf{v} = [x_2 - x_1 \; y_2 - y_1]$.

Example 2 Determine the two-dimensional row matrix associated with the directed line segments **w** and **z** shown in Figure 2.7.

Solution: The tip of the directed line segment **w** is at the point (40, 30) while its tail lies on the point (10, −20), so

$$\mathbf{w} = [40 - 10 \quad 30 - (-20)] = [30 \quad 50]$$

The tip of the directed line segment **z** is at the point (−10, 30) while its tail lies on the point (−50, 50), so

$$\mathbf{z} = [-10 - (-50) \quad 30 - 50] = [40 \quad -20]$$

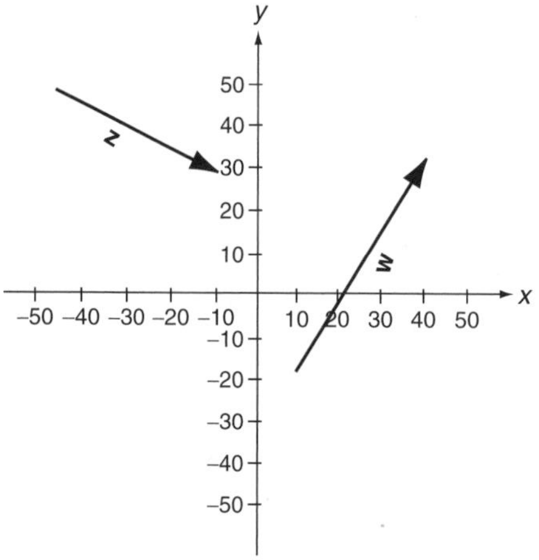

FIGURE 2.7

The graph of $k_{\mathbf{u}}$ in $\mathbb{R}^2$ is a directed line segment having length $|k|$ times the length of **u** with the same direction as **u** when the scalar k is positive and the opposite direction to **u** when k is negative.

A scalar multiplication $k\mathbf{u}$ is defined geometrically in $\mathbb{R}^2$ to be a directed line segment having length $|k|$ times the length of **u**, in the same direction as **u** when k is positive and in the opposite direction to **u** when k is negative. Effectively, $k\mathbf{u}$ is an elongation of the directed line segment representing **u** when $|k|$ is greater than 1, or a contraction of **u** by a factor of $|k|$ when $|k|$ is less than 1, followed by no rotation when k is positive or a rotation of 180° when k is negative.

Example 3 Find $-2\mathbf{u}$ and $1/2\ \mathbf{v}$ geometrically for the row matrices $\mathbf{u} = [-1\ 1]$ and $\mathbf{v} = [2\ 4]$.

Solution: To construct $-2\mathbf{u}$, we double the length of **u** and then rotate the resulting arrow by 180°. To construct $1/2\ \mathbf{v}$, we halve the length of **v** and effect no rotation. These constructions are illustrated in Figure 2.8.

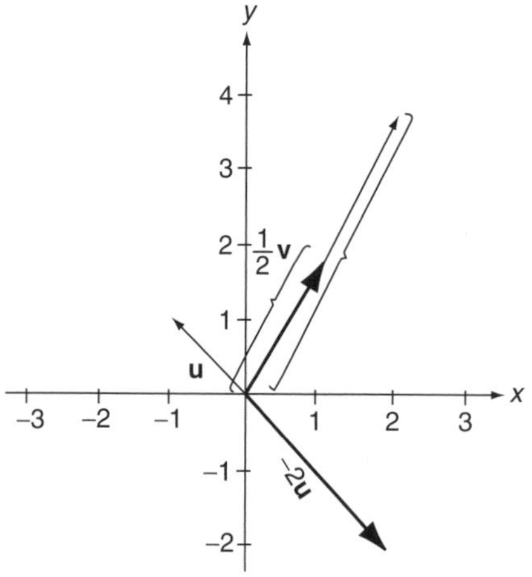

FIGURE 2.8

To graphically depict a three-dimensional row matrix, we first construct a rectangular coordinate system defined by three mutually perpendicular lines, representing the axes, that intersect at their respective origins. For convenience, we denote these axes as the x-axis, the y-axis, and the z-axis, and their point of intersection as the origin.

Rectangular coordinate systems are of two types: *right-handed systems* and *left-handed systems.* An xyz system is right-handed if the thumb of the right hand points in the direction of the positive z-axis when the fingers of the right hand are curled naturally—in a way that does not break the finger bones—from the positive x-axis towards the positive y-axis. In a left-handed system, the thumb of the left hand points in the positive z-axis when the fingers of the left hand are curled naturally from the positive x-axis towards the positive y-axis. Both types of systems are illustrated in Figure 2.9. In this book, we shall only use right-handed coordinate systems when graphing in space.

A three-dimensional row matrix $\mathbf{v} = [a\,b\,c]$ is identified with the point (a, b, c) in an xyz-coordinate system, measured a units along the x-axis from the origin, then b units parallel to the y-axis, and then finally c units parallel to the z-axis. An arrow or directed line segment having its tail at the origin and its tip at the point (a, b, c) represents the row matrix $\mathbf{v}$ geometrically. The geometrical representations of the row matrices $\mathbf{u} = [2\,4\,6]$ and $\mathbf{v} = [5\,2\,-3]$ are illustrated in Figures 2.10 and 2.11, respectively.

All of the geometrical processes developed for the addition, subtraction, and scalar multiplication of 2-tuples extend directly to 3-tuples. In particular, to graph $\mathbf{u} - \mathbf{v}$, first graph both directed line segments normally and then construct an arrow from the tip of $\mathbf{v}$ to the tip of $\mathbf{u}$. Multiplication of a directed line segment $\mathbf{u}$ by the scalar k is again an elongation of $\mathbf{u}$ by $|k|$ when $|k|$ is greater than unity and a contraction of $\mathbf{u}$ by $|k|$ when $|k|$ is less than unity, followed by no rotation when k is positive or a rotation of 180° when k is negative. If a directed line segment has its tip at the point (x_2, y_2, z_2) and its tail at the point (x_1, y_1, z_1), then the row matrix associated with it is $[(x_2 - x_1)\,(y_2 - y_1)\,(z_2 - z_1)]$.

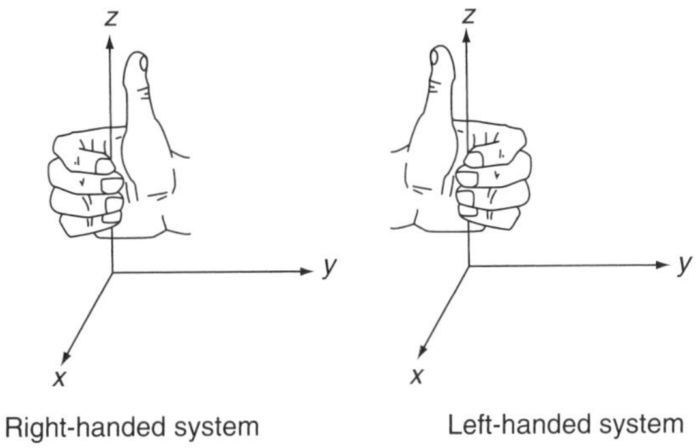

FIGURE 2.9

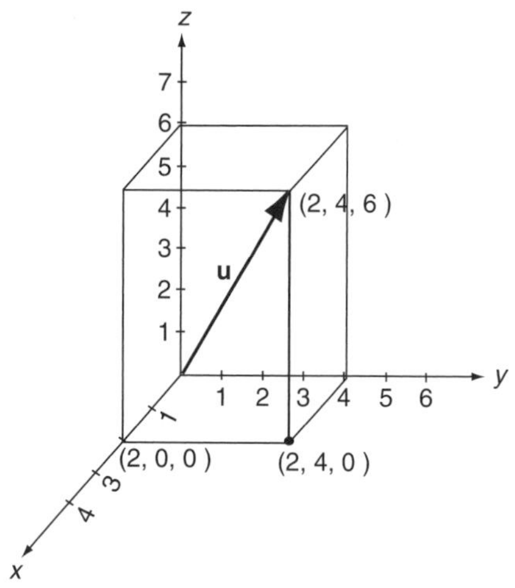

FIGURE 2.10

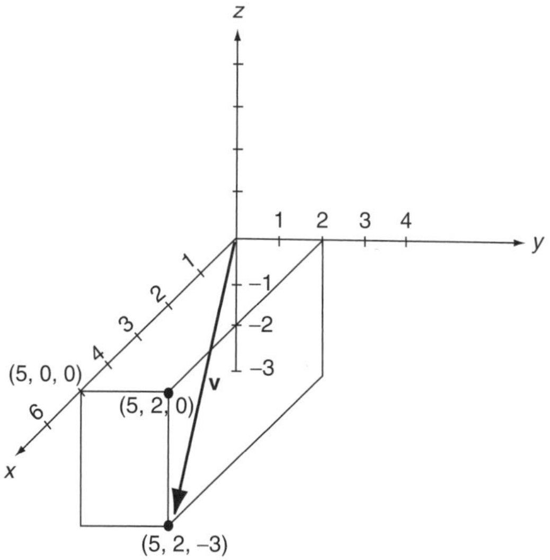

FIGURE 2.11

Although geometrical representations for $\mathbb{R}^n$ are limited to $n \leq 3$, the concept of magnitude can be extended to all n-tuples. We define the magnitude of the n-dimensional row matrix $\mathbf{a} = [a_1\ a_2\ a_3 \ldots a_n]$ as

$$\|\mathbf{a}\| = \sqrt{a_1^2 + a_2^2 + a_3^2 + \cdots + a_n^2} \tag{2.1}$$

Example 4 The magnitude of the 4-tuple $\mathbf{a} = [1\,2\,3\,4]$ is

$$\|\mathbf{a}\| = \sqrt{(1)^2 + (2)^2 + (3)^2 + (4)^2} = \sqrt{30}$$

while the magnitude of the 5-tuple $\mathbf{u} = [-4\,-5\,0\,5\,4]$ is

$$\|\mathbf{u}\| = \sqrt{(-4)^2 + (-5)^2 + (0)^2 + (5)^2 + (4)^2} = \sqrt{82}$$

An n-tuple is *normalized* if it has a magnitude equal to one. Any n-tuple (row matrix) is normalized by multiplying the n-tuple by the reciprocal of its magnitude.

An n-tuple is normalized if it has a magnitude equal to one.

Example 5 As shown in Example 4, $\mathbf{a} = [1\,2\,3\,4]$ has magnitude $\|\mathbf{a}\| = \sqrt{30}$, so

$$\frac{1}{\sqrt{30}}[1 \quad 2 \quad 3 \quad 4] = \left[\frac{1}{\sqrt{30}} \quad \frac{2}{\sqrt{30}} \quad \frac{3}{\sqrt{30}} \quad \frac{4}{\sqrt{30}}\right]$$

is normalized. Similarly, $\mathbf{u} = [-4\,-5\,0\,5\,4]$ has magnitude $\|\mathbf{u}\| = \sqrt{82}$, so

$$\frac{1}{\sqrt{82}}[-4 \quad -5 \quad 0 \quad 5 \quad 4] = \left[\frac{-4}{\sqrt{82}} \quad \frac{-5}{\sqrt{82}} \quad 0 \quad \frac{5}{\sqrt{82}} \quad \frac{4}{\sqrt{82}}\right]$$

is normalized.

Two row matrices of the same dimension can be added and subtracted but they cannot be multiplied. Multiplication of a $1 \times n$ matrix by another $1 \times n$ matrix is undefined. Scalar multiplication of row matrices is defined but inversion is not defined for row matrices of dimension greater than 1, because such row matrices are not square. Thus, row matrices, and therefore n-tuples, do not possess all the properties of real numbers. Listing the properties that n-tuples do share with real numbers and then developing an algebra around those properties is the focus of the next chapter.

In preparation for our work in vectors and vector spaces later in this chapter, we list some of the important properties shared by all n-tuples. If $\mathbf{a}$, $\mathbf{b}$, and $\mathbf{c}$ denote row matrices of the same dimension n, then it follows from Theorem 1 of Section 1.1 that

$$\mathbf{a} + \mathbf{b} = \mathbf{b} + \mathbf{c} \tag{2.2}$$

and

$$\mathbf{a} + (\mathbf{b} + \mathbf{c}) = (\mathbf{a} + \mathbf{b}) + \mathbf{c} \tag{2.3}$$

If we define the zero row matrix of dimension n as $\mathbf{0} = [0\,0\,0 \ldots 0]$, the row matrix having entries of zero in each of its n-columns, then it follows from Equation (1.5) that

$$\mathbf{a} + \mathbf{0} = \mathbf{a} \tag{2.4}$$

Setting $\mathbf{a}=[a_1 a_2 a_3 \ldots a_n]$ and $-\mathbf{a}=(-1)\mathbf{a}=[-a_1 - a_2 - a_3 \ldots - a_n]$, we also have

$$\mathbf{a}+(-\mathbf{a})=0 \tag{2.5}$$

It follows from Theorem 2 of Section 1.1 that if λ_1 and λ_2 denote arbitrary real numbers, then

$$\lambda_1(\mathbf{a}+\mathbf{b})=\lambda_1\mathbf{a}+\lambda_1\mathbf{b} \tag{2.6}$$

$$(\lambda_1+\lambda_2)\mathbf{a}=\lambda_1\mathbf{a}+\lambda_1\mathbf{a} \tag{2.7}$$

and

$$(\lambda_1\lambda_2)\mathbf{a}=\lambda_1(\lambda_2\mathbf{a}) \tag{2.8}$$

In addition,

$$1(\mathbf{a})=\mathbf{a} \tag{2.9}$$

Problems 2.1

In Problems 1 through 16, geometrically construct the indicated 2-tuple operations for

$$\mathbf{u}=[3\ -1],\ \mathbf{v}=[-2\ 5],\ \mathbf{w}=[-4\ -4],\ \mathbf{x}=[3\ 5],\text{ and }\mathbf{y}=[0\ -2].$$

(1) $\mathbf{u}+\mathbf{v}$. **(2)** $\mathbf{u}+\mathbf{w}$. **(3)** $\mathbf{v}+\mathbf{w}$. **(4)** $\mathbf{x}+\mathbf{y}$.

(5) $\mathbf{x}-\mathbf{y}$. **(6)** $\mathbf{y}-\mathbf{x}$. **(7)** $\mathbf{u}-\mathbf{v}$. **(8)** $\mathbf{w}-\mathbf{u}$.

(9) $\mathbf{u}-\mathbf{w}$. **(10)** $2\mathbf{x}$. **(11)** $3\mathbf{x}$. **(12)** $-2\mathbf{x}$.

(13) $\frac{1}{2}\mathbf{u}$. **(14)** $-\frac{1}{2}\mathbf{u}$. **(15)** $\frac{1}{3}\mathbf{v}$. **(16)** $-\frac{1}{4}\mathbf{w}$.

(17) Determine the angle that each directed line segment representation for the following row matrices makes with the positive horizontal x-axis:

(a) $\mathbf{u}=[3\ -1]$, (c) $\mathbf{w}=[-4\ -4]$, (e) $\mathbf{y}=[0\ -2]$.

(b) $\mathbf{v}=[-2\ 5]$, (d) $\mathbf{x}=[3\ 5]$,

(18) For arbitrary two-dimensional row matrices $\mathbf{u}$ and $\mathbf{v}$, construct on the same graph the sums $\mathbf{u}+\mathbf{v}$ and $\mathbf{v}+\mathbf{u}$. Show that $\mathbf{u}+\mathbf{v}=\mathbf{v}+\mathbf{u}$, and show for each that the sum is the diagonal of a parallelogram having as two of its sides the directed line segments that represent $\mathbf{u}$ and $\mathbf{v}$.

In Problems 19 through 29, determine the magnitudes of the given n-tuples.

(19) $[1\ -1]$. **(20)** $[3\ 4]$. **(21)** $[1\ 2]$.

(22) $[-1\ -1\ 1]$. **(23)** $[1/2\ 1/2\ 1/2]$. **(24)** $[1\ 1\ 1]$.

(25) $[2\ 1\ -1\ 3]$. **(26)** $[1\ -1\ 1\ -1]$. **(27)** $[1\ 0\ 1\ 0]$.

(28) $[0\ -1\ 5\ 3\ 2]$. **(29)** $[1\ 1\ 1\ 1\ 1]$.

In Problems 30 through 39, graph the indicated *n*-tuples.

(30) $[3\ 1\ 2]$. **(31)** $[1\ 2\ 3]$. **(32)** $[-1\ 2\ 3]$.

(33) $[-1\ 2\ -3]$. **(34)** $[20\ -50\ 10]$. **(35)** $[1000\ 100]$.

(36) $[2\ 2\ 2]$. **(37)** $[-2\ -1\ 2]$. **(38)** $[1000\ -500\ 200]$.

(39) $[-400\ -50\ -300]$.

In Problems 40 through 48, determine which, if any, of the given row matrices are normalized.

(40) $[1\ 1]$. **(41)** $[1/2\ 1/2]$. **(42)** $\left[\frac{1}{\sqrt{2}}\ \frac{-1}{\sqrt{2}}\right]$

(43) $[0\ 1\ 0]$. **(44)** $[1/2\ 1/3\ 1/6]$.

(45) $\left[\frac{1}{\sqrt{3}}\ \frac{1}{\sqrt{3}}\ \frac{1}{\sqrt{3}}\right]$. **(46)** $[1/2\ 1/2\ 1/2\ 1/2]$.

(47) $[1/6\ 5/6\ 3/6\ 1/6]$. **(48)** $\left[\frac{-1}{\sqrt{3}}\ 0\ \frac{1}{\sqrt{3}}\ \frac{-1}{\sqrt{3}}\right]$.

2.2 VECTORS

As stated earlier, matrices, polynomials and directed line segments would seem, on the surface, to have little in common. However, they share fundamental characteristics that provide a richer understanding of them all. What are some of these fundamental properties? First, they can be added. A matrix can be added to a matrix of the same order and the result is another matrix of that order. A directed line segment in the plane can be added to another directed line segment in the plane and the result is again a directed line segment of the same type. Thus, we have the concept of *closure under addition:* objects in a particular set are defined and an operation of addition is established on those objects so that the operation is doable and the result is again another object in the same set. Second, we also have the concept of *closure under scalar multiplication.* We know how to multiply a matrix or a directed line segment or a polynomial by a scalar, and the result is always another object of the same type. Also, we know that the commutative and associate laws hold for addition (see, for example, Theorem 1 in Section 1.1). Other properties are so obvious we take them for granted. If we multiply a matrix, directed line segment, or polynomial by the number 1 we always get back the original object. If we add to any matrix, polynomial, or directed line segment, respectively, the zero matrix of appropriate order, the zero polynomial, or the zero directed line segment, we always get back the original object.

Thus, we have very quickly identified a series of common characteristics. Are there others? More interesting, what is the smallest number of characteristics that we need to identify so that all the other characteristics immediately follow? To begin, we create a new label to apply to any set of objects that have these characteristics, *vector space*, and we refer to the objects in this set as *vectors*. We then show that matrices, directed line segments, n-tuples, polynomials, and even continuous functions are just individual examples of vector spaces. Just as cake, ice cream, pie, and JELL-O are all examples of the more general term *dessert*, so too will matrices, directed line segments, and polynomials be examples of the more general term *vectors*.

▶DEFINITION 1

A set of objects $\mathbb{V}=\{\mathbf{u}, \mathbf{v}, \mathbf{w}, \ldots\}$ and scalars $\{\alpha, \beta, \gamma, \ldots\}$ along with a binary operation of vector addition $\oplus$ on the objects and a scalar multiplication $\odot$ is a *vector space* if it possesses the following 10 properties:

Addition

(A1) Closure under addition: If $\mathbf{u}$ and $\mathbf{v}$ belong to $\mathbb{V}$, then so too does $\mathbf{u}\oplus\mathbf{v}$.
(A2) Commutative law for addition: If $\mathbf{u}$ and $\mathbf{v}$ belong to $\mathbb{V}$, then $\mathbf{u}\oplus\mathbf{v}=\mathbf{v}\oplus\mathbf{u}$.
(A3) Associative law for addition: If $\mathbf{u}$, $\mathbf{v}$, and $\mathbf{w}$ belong to $\mathbb{V}$, then $\mathbf{u}\oplus(\mathbf{v}\oplus\mathbf{w})=(\mathbf{u}\oplus\mathbf{v})\oplus\mathbf{w}$.
(A4) There exists a zero vector in $\mathbb{V}$ denoted by $\mathbf{0}$ such that for every vector $\mathbf{u}$ in $\mathbb{V}$, $\mathbf{u}\oplus\mathbf{0}=\mathbf{u}$.
(A5) For every vector $\mathbf{u}$ in $\mathbb{V}$ there exists a vector $-\mathbf{u}$, called the *additive inverse* of $\mathbf{u}$, such that $\mathbf{u}\oplus-\mathbf{u}=\mathbf{0}$.

Scalar Multiplication

(S1) Closure under scalar multiplication: If $\mathbf{u}$ belongs to $\mathbb{V}$, then so too does $\alpha\odot\mathbf{u}$ for any scalar α.
(S2) For any two scalars α and β and any vector $\mathbf{u}$ in $\mathbb{V}$, $\alpha\odot(\beta\odot\mathbf{u})=(\alpha\beta)\odot\mathbf{u}$.
(S3) For any vector $\mathbf{u}$ in $\mathbb{V}$, $\mathbf{1}\otimes\mathbf{u}=\mathbf{u}$.
(S4) For any two scalars α and β and any vector $\mathbf{u}$ in $\mathbb{V}$, $(\alpha+\beta)\odot\mathbf{u}=\alpha\odot\mathbf{u}\oplus\beta\odot\mathbf{u}$.
(S5) For any scalar α and any two vectors $\mathbf{u}$ and $\mathbf{v}$ in $\mathbb{V}$, $\alpha\odot(\mathbf{u}\oplus\mathbf{v})=\alpha\odot\mathbf{u}\oplus\alpha\odot\mathbf{u}$.◀

If the scalars are restricted to be real numbers, then $\mathbb{V}$ is called a *real vector space*; if the scalars are allowed to be complex numbers, then $\mathbb{V}$ is called a *complex vector space*. Throughout this book we shall assume that all scalars are real and that we are dealing with real vector spaces, unless an exception is noted. When we need to deal with complex scalars, we shall say so explicitly.

In set notation, $\in$ is read "belongs to" and the vertical line segment | is read "such that."

Since vector spaces are sets, it is convenient to use set notation. We denote sets by upper case letters in an outline font, such as $\mathbb{V}$ and $\mathbb{R}$. The format for a subset $\mathbb{S}$ of a set $\mathbb{W}$ is $\mathbb{S}=\{w\in\mathbb{W}|\ \textit{property } A\}$. The $\in$ is read "belongs to" or "is a member of" and the vertical line segment | is read "such that." An element w belongs to $\mathbb{S}$ only if w is a member of $\mathbb{W}$ and if w satisfies property A. In particular, the set

$$\mathbb{S}=\left\{[x \quad y \quad z]\in\mathbb{R}^3 \middle| y=0\right\}$$

is the set of all real 3-tuples, represented as row matrices, with a second component of zero.

Example 1 Determine whether $\mathbb{S}=\left\{[x\,y\,z]\in\mathbb{R}^3 \middle| y=0\right\}$ is a vector space under regular addition and scalar multiplication.

Solution: Following our convention, it is assumed that the scalars are real. Arbitrary vectors **u** and **v** in $\mathbb{S}$ have the form $\mathbf{u}=[a\ 0\ b]$ and $\mathbf{v}=[c\ 0\ d]$ with a, b, c, and d all real. Now,

$$\mathbf{u}\oplus\mathbf{v}=[a\quad 0\quad b]+[c\quad 0\quad d]=[a+c\quad 0\quad b+d]$$

and, for any real scalar α,

$$\alpha\odot\mathbf{u}=\alpha[a\quad 0\quad b]=[\alpha a\quad 0\quad \alpha b]$$

which are again three-dimensional row matrices having real components, of which the second one is 0. Thus, $\mathbb{S}$ is closed under vector addition and scalar multiplication and both properties A1 and S1 are satisfied.

To prove property A2, we observe that

$$\begin{aligned}\mathbf{u}\oplus\mathbf{v}&=[a\quad 0\quad b]+[c\quad 0\quad d]=[a+c\quad 0\quad b+d]\\&=[c+a\quad 0\quad d+b]=[c\quad 0\quad d]+[a\quad 0\quad b]\\&=\mathbf{v}\oplus\mathbf{u}\end{aligned}$$

To prove property A3, we set $\mathbf{w}=[e\ 0 f]$, with e and f representing real numbers, and note that

$$\begin{aligned}(\mathbf{u}\oplus\mathbf{v})\oplus\mathbf{w}&=([a\quad 0\quad b]+[c\quad 0\quad d]+[e\quad 0\quad f])\\&=[a+c\quad 0\quad b+d]+[e\quad 0\quad f]\\&=[(a+c)+e\quad 0\quad (b+d)+f]\\&=[a+(c+e)\quad 0\quad b+(d+f)]\\&=[a\quad 0\quad b]+[c+d\quad 0\quad d+f]\\&=[a\quad 0\quad b]+([c\quad 0\quad d]+[e\quad 0\quad f])\\&=\mathbf{u}\oplus(\mathbf{v}\oplus\mathbf{w})\end{aligned}$$

The row matrix $[0\,0\,0]$ is an element of $\mathbb{S}$. If we denote it as the zero vector $\mathbf{0}$, then

$$\begin{aligned}\mathbf{u}\oplus\mathbf{0}&=[a\quad 0\quad b]+[0\quad 0\quad 0]=[a+0\quad 0+0\quad b+0]\\&=[a\quad 0\quad b]=\mathbf{u}\end{aligned}$$

so property A4 is satisfied. Furthermore, if we define, $-\mathbf{u}=[-a\ 0 -b]$, then

$$\begin{aligned}\mathbf{u}\oplus-\mathbf{u}&=[a\quad 0\quad b]+[-a\quad 0\quad -b]=[a+-a\quad 0+0\quad b+-b]\\&=[0\quad 0\quad 0]=\mathbf{0}\end{aligned}$$

and property A5 is valid.

For any two real numbers α and β, we have that

$$\begin{aligned}&\alpha\odot(\beta\odot\mathbf{u})=\alpha\odot(\beta[a\quad 0\quad b])=\alpha\odot[\beta a\quad 0\quad \beta b]=\alpha[\beta a\quad 0\quad \beta b])\\&=(\alpha\beta)[a\quad 0\quad b]=(\alpha\beta)\odot\mathbf{u}\end{aligned}$$

so property S2 holds. In addition,

$$1 \odot \mathbf{u} = 1[a \quad 0 \quad b] = [1a \quad 0 \quad 1b] = [a \quad 0 \quad b] = \mathbf{u}$$

so property S3 is valid. To verify properties S4 and S5, we note that

$$\begin{aligned}
(\alpha + \beta) \odot \mathbf{u} &= (\alpha + \beta)[a \quad 0 \quad b] \\
&= [(\alpha + \beta)a \quad (\alpha + \beta)0 \quad (\alpha + \beta)b] \\
&= [\alpha a + \beta a \quad 0 \quad \alpha b + \beta b] \\
&= [\alpha a \quad 0 \quad \alpha b] + [\beta a \quad 0 \quad \beta b] \\
&= \alpha[a \quad 0 \quad b] + \beta[a \quad 0 \quad b] \\
&= \alpha \odot [a \quad 0 \quad b] + \beta \odot [a \quad 0 \quad b] \\
&= \alpha \odot \mathbf{u} \oplus \beta \mathbf{u}
\end{aligned}$$

and

$$\begin{aligned}
\alpha \odot (\mathbf{u} \oplus \mathbf{v}) &= \alpha \odot ([a \quad 0 \quad b] + [c \quad 0 \quad d]) \\
&= \alpha \odot [a + c \quad 0 \quad b + d] \\
&= [\alpha(a + c) \quad \alpha(0) \quad \alpha(b + d)] \\
&= [\alpha a + \alpha c \quad 0 \quad \alpha b + \alpha d] \\
&= [\alpha a \quad 0 \quad \alpha b] + [\alpha c \quad 0 \quad \alpha d] \\
&= \alpha[a \quad 0 \quad b] + \alpha[c \quad 0 \quad d] \\
&= \alpha \odot \mathbf{u} \oplus \alpha \odot \mathbf{v}
\end{aligned}$$

Therefore, all 10 properties are valid, and $\mathbb{S}$ is a vector space.

Example 2 Determine whether the set $\mathbb{M}_{p \times n}$ of all $p \times n$ real matrices under matrix addition and scalar multiplication is a vector space.

The set $\mathbb{M}_{p \times n}$ of all $p \times n$ real matrices under matrix addition and scalar multiplication is a vector space.

Solution: This is a vector space for any fixed values of p and n because all 10 properties follow immediately from our work in Chapter 1. The sum of two real $p \times n$ matrices is again a matrix of the same order, as is the product of a real number with a real matrix of this order. Thus, properties A1 and S1 are satisfied. Properties A2 through A4 are precisely Theorem 1 in Section 1.1 and Equation (1.5). If $\mathbf{A} = [a_{ij}]$, then $-\mathbf{A} = [-a_{ij}]$ is another element in the set and

$$\mathbf{A} \oplus -\mathbf{A} = [a_{ij}] + [-a_{ij}] = [(a_{ij} + -a_{ij})] = 0$$

which verifies property A5. Properties S2, S4, and S5 are Theorem 2 in Section 1.1. Property S3 is immediate from the definition of scalar multiplication.

The set $\mathbb{R}^n$ of n-tuples under standard addition and scalar multiplication for n-tuples is a vector space.

It follows from Example 2 that the set of all real 3×3 matrices ($p = n = 3$) is a vector space, as is $\mathbb{M}_{2 \times 6}$ the set of all real 2×6 matrices ($p = 2$ and $n = 6$). Also, $\mathbb{R}^n$ is a vector space, for any positive integer n, because $\mathbb{R}^n$ is $\mathbb{M}_{1 \times n}$ when we take

$\mathbb{R}^n$ to be the set of all n-dimensional real row matrices, and $\mathbb{R}^n$ is $\mathbb{M}_{n\times 1}$ when we take $\mathbb{R}^n$ to be the set of all n-dimensional real column matrices.

Example 3 Determine whether the set of all 2×2 real matrices is a vector space under regular scalar multiplication but with vector addition defined to be matrix multiplication. That is,

$$\mathbf{u}\oplus\mathbf{v}=\mathbf{uv}$$

Solution: This is not a vector space because it does not satisfy property A2. In particular,

$$\begin{bmatrix}1&2\\3&4\end{bmatrix}\oplus\begin{bmatrix}5&6\\7&8\end{bmatrix}=\begin{bmatrix}1&2\\3&4\end{bmatrix}\begin{bmatrix}5&6\\7&8\end{bmatrix}=\begin{bmatrix}19&22\\43&50\end{bmatrix}$$
$$\neq\begin{bmatrix}23&34\\31&46\end{bmatrix}=\begin{bmatrix}5&6\\7&8\end{bmatrix}\begin{bmatrix}1&2\\3&4\end{bmatrix}=\begin{bmatrix}5&6\\7&8\end{bmatrix}\oplus\begin{bmatrix}1&2\\3&4\end{bmatrix}$$

We use the $\oplus$ symbol to emphasize that vector addition may be nonstandard, as it is in Example 3. The notation denotes a well-defined process for combining two vectors together, regardless of how unconventional that process may be. Generally, vector addition is standard, and many writers discard the $\oplus$ notation in favor of the more conventional $+$ symbol whenever a standard addition is in effect. We shall, too, in later sections. For now, however, we want to stress that a vector space does not require a standard vector addition, only a well-defined operation for combining two vectors that satisfies the properties listed in Definition 1, so we shall retain the $\oplus$ notation a while longer.

The symbol $\oplus$ emphasizes that vector addition may be nonstandard.

Example 4 Redo Example 3 with the matrices restricted to being diagonal.

Solution: Diagonal matrices do commute under matrix multiplication, hence property A2 is now satisfied. The set is closed under vector addition, because the product of 2×2 diagonal matrices is again a diagonal matrix. Property A3 also holds, because matrix multiplication is associative. With vector addition defined to be matrix multiplication, the zero vector becomes the 2×2 identity matrix; for any matrix $\mathbf{A}$ in the set, $\mathbf{A}\oplus\mathbf{0}=\mathbf{AI}=\mathbf{A}$. To verify property A5, we must show that every real diagonal matrix $\mathbf{A}$ has an additive inverse $-\mathbf{A}$ with the property $\mathbf{A}\oplus-\mathbf{A}=\mathbf{0}$. Given that we have just identified the zero vector to be the identity matrix and vector addition to be matrix multiplication, the statement $\mathbf{A}\oplus-\mathbf{A}=\mathbf{0}$ is equivalent to the statement $\mathbf{A}(-\mathbf{A})=\mathbf{I}$. Property A5 is valid if and only if every matrix in the set has an inverse, in which case we take $-\mathbf{A}=\mathbf{A}^{-1}$. But, a diagonal matrix with at least one 0 on its main diagonal does not have an inverse. In particular the matrix,

$$\mathbf{A}=\begin{bmatrix}1&0\\0&0\end{bmatrix}$$

has no inverse. Thus, property A5 does not hold in general, and the given set is *not* a vector space.

Example 5 Redo Example 3 with the matrices restricted to being diagonal and all elements on the main diagonal restricted to being nonzero.

Solution: Repeating the reasoning used in Example 4, we find that properties A1-A5 are satisfied for this set. This set, however, is not closed under scalar multiplication. Whenever we multiply a matrix in the set by the zero scalar, we get

$$0 \odot \mathbf{A} = 0\mathbf{A} = \begin{bmatrix} 0 & 0 \\ 0 & 0 \end{bmatrix}$$

which is no longer a diagonal matrix with nonzero elements on the main diagonal and, therefore, not an element of the original set. Thus, the given set is *not* a vector space.

Example 6 Determine whether the set of nth degree polynomials in the variable t with real coefficients is a vector space under standard addition and scalar multiplication for polynomials if the scalars are restricted also to being real.

Solution: Arbitrary vectors $\mathbf{u}$ and $\mathbf{v}$ in this set are polynomials of the form

$$\mathbf{u} = a_n t^n + a_{n-1} t^{n-1} + \cdots + a_1 t + a_0$$

$$\mathbf{v} = b_n t^n + b_{n-1} t^{n-1} + \cdots + b_1 t + b_0$$

with a_j and b_j $(j=0, 1, \ldots, n)$ all real, and both a_n and b_n nonzero. Here,

$$\begin{aligned} \mathbf{u} \oplus \mathbf{v} &= \left(a_n t^n + a_{n-1} t^{n-1} + \cdots + a_1 t + a_0\right) \\ &\quad + \left(b_n t^n + b_{n-1} t^{n-1} + \cdots + b_1 t + b_0\right) \\ &= (a_n + b_n)t^n + (a_{n-1} + b_{n-1})t^{n-1} + \cdots + (a_1 + b_1)t + (a_0 + b_0) \end{aligned}$$

Note that when $a_n = -b_n$, $\mathbf{u} \oplus \mathbf{v}$ is no longer an nth degree polynomial, but rather a polynomial of degree less than n, which is not an element of the given set. Thus, the set is not closed under vector addition and is *not* a vector space.

Example 7 Determine whether the set $\mathbb{P}^n$ containing the identically zero polynomial and all polynomials of degree *n or less* in the variable t with real coefficients is a vector space under standard addition and scalar multiplication for polynomials, if the scalars also are restricted to being real.

Solution: If $\mathbf{u} \in \mathbb{P}^n$ and $\mathbf{v} \in \mathbb{P}^n$, then $\mathbf{u}$ and $\mathbf{v}$ have the form

$$\mathbf{u} = a_n t^n + a_{n-1} t^{n-1} + \cdots + a_1 t + a_0$$

$$\mathbf{v} = b_n t^n + b_{n-1} t^{n-1} + \cdots + b_1 t + b_0$$

with a_j and b_j $(j=0, 1, \ldots, n)$ real and possibly 0. Using the results of Example 6, we see that the sum of two polynomials of degree n or less is either another polynomial of the same type or the zero polynomial when $\mathbf{u}$ and $\mathbf{v}$ have their corresponding coefficients equal in absolute value but opposite in sign. Thus, property A1 is satisfied. If we define the zero vector to be the zero polynomial, then

$$\begin{aligned}\mathbf{u} \oplus \mathbf{0} &= \left(a_nt^n + a_{n-1}t^{n-1} + \cdots + a_1t + a_0\right) \\ &\quad + \left(0t^n + 0t^{n-1} + \cdots + 0t + 0\right) \\ &= (a_n + 0)t^n + (a_{n-1} + 0)t^{n-1} + \cdots + (a_1 + 0)t + (a_0 + 0) \\ &= \mathbf{u}\end{aligned}$$

Thus, property A4 is satisfied. Setting

$$\mathbf{u} = -a_nt^n - a_{n-1}t^{n-1} - \cdots - a_1t - a_0$$

we note that property A5 is also satisfied. Now,

$$\begin{aligned}\mathbf{u} \oplus \mathbf{v} &= \left(a_nt^n + a_{n-1}t^{n-1} + \cdots + a_1t + a_0\right) \\ &\quad + \left(b_nt^n + b_{n-1}t^{n-1} + \cdots + b_1t + b_0\right) \\ &= (a_n + b_n)t^n + (a_{n-1} + b_{n-1})t^{n-1} + \cdots + (a_1 + b_1)t + (a_0 + b_0) \\ &= (b_n + a_n)t^n + (b_{n-1} + a_{n-1})t^{n-1} + \cdots + (b_1 + a_1)t + (b_0 + a_0) \\ &= \left(b_nt^n + b_{n-1}t^{n-1} + \cdots + b_1t + b_0\right) \\ &\quad + \left(a_nt^n + a_{n-1}t^{n-1} + \cdots + a_1t + a_0\right) \\ &= \mathbf{v} \oplus \mathbf{u}\end{aligned}$$

so property A2 is satisfied. Property A3 is verified in a similar manner. For any real number α, we have

$$\begin{aligned}\alpha \odot \mathbf{u} &= \alpha\left(a_nt + a_{n-1}t^{n-1} + \cdots + a_1t + a_0\right) \\ &= (\alpha a_n)t + (\alpha a_{n-1})t^{n-1} + (\alpha a_1)t + (\alpha a_0)\end{aligned}$$

which is again an element in the original set, so the set is closed under scalar multiplication. Setting $\alpha = 1$ in the preceding equation also verifies property S3. The remaining three properties follow in a straightforward manner, so $\mathbb{P}^n$ is a vector space.

The set $\mathbb{P}^n$ of all polynomials of degree less than or equal to n, including the identically zero polynomial, under normal addition and scalar multiplication for polynomials is a vector space.

Example 8 Determine whether the set of two-dimensional column matrices with all real components is a vector space under regular addition but with scalar multiplication defined as

$$\alpha \odot \begin{bmatrix} a \\ b \end{bmatrix} = \begin{bmatrix} -\alpha a \\ -\alpha b \end{bmatrix}$$

Solution: Following convention, the scalars are assumed to be real numbers. Since column matrices are matrices, it follows from our work in Chapter 1 that properties A1 through A5 hold. It is clear from the definition of scalar multiplication that the set is closed under this operation; the result of multiplying a real two-dimensional column matrix by a real number is again a real two-dimensional column matrix. To check property S2, we note that for any two real numbers α and β and for any vector

$$\mathbf{u} = \begin{bmatrix} a \\ b \end{bmatrix}$$

we have

$$(\alpha\beta) \odot \mathbf{u} = (\alpha\beta) \odot \begin{bmatrix} a \\ b \end{bmatrix} = \begin{bmatrix} -(\alpha\beta)a \\ -(\alpha\beta)b \end{bmatrix} = \begin{bmatrix} -\alpha\beta a \\ -\alpha\beta b \end{bmatrix}$$

while

$$\alpha \odot (\beta \odot \mathbf{u}) = \alpha \odot \left(\beta \odot \begin{bmatrix} a \\ b \end{bmatrix}\right) = \alpha \odot \begin{bmatrix} -\beta a \\ -\beta b \end{bmatrix} = \begin{bmatrix} (-\alpha)(-\beta a) \\ (-\alpha)(-\beta b) \end{bmatrix} = \begin{bmatrix} \alpha\beta a \\ \alpha\beta b \end{bmatrix}$$

These two expressions are not equal whenever α and β are nonzero, so property S2 does not hold and the given set is *not* a vector space.

Property S3 is also violated with this scalar multiplication. For any vector

$$\mathbf{u} = \begin{bmatrix} a \\ b \end{bmatrix}$$

we have

$$1 \odot \mathbf{u} = 1 \odot \begin{bmatrix} a \\ b \end{bmatrix} = \begin{bmatrix} -a \\ -b \end{bmatrix} \neq \mathbf{u}$$

The symbol $\odot$ emphasizes that scalar multiplication maybe nonstandard.

Thus, we conclude again that the given set is not a vector space.

We use the $\odot$ symbol to emphasize that scalar multiplication may be nonstandard, as it was in Example 8. The $\odot$ symbol denotes a well-defined process for combining a scalar with a vector, regardless of how unconventional the process may be. In truth, scalar multiplication is generally quite standard, and many writers discard the $\odot$ notation whenever it is in effect. We shall, too, in later sections. For now, however, we want to retain this notation to stress that a vector space does not require a standard scalar multiplication, only a well-defined process for combining scalars and vectors that satisfies properties S1 through S5.

Example 9 Determine whether the set of three-dimensional row matrices with all components real and equal is a vector space under regular addition and scalar multiplication if *the scalars are complex numbers.*

Solution: An arbitrary vector in this set has the form $\mathbf{u} = [a\ a\ a]$, where a is real. This is *not* a vector space, because the set violates property S1. In particular, if α is any complex number with a nonzero imaginary part, then $\alpha \odot \mathbf{u}$ does not have real components. For instance, with $\alpha = 3i$ and $\mathbf{u} = [1\ 1\ 1]$, we have

$$\alpha \odot \mathbf{u} = (3i)[1 \quad 1 \quad 1] = [3i \quad 3i \quad 3i]$$

which is not a real-valued vector; the components of the row matrix are complex, not real. Thus, the original set is not closed under scalar multiplication. The reader can verify that all the other properties given in Definition 1 are applicable. However, as soon as we find one property that is not satisfied, we can immediately conclude the given set is not a vector space.

The purpose of defining a vector space in the abstract is to create a single mathematical structure that embodies the characteristics of many different well-known sets, and then to develop facts about each of those sets simultaneously by studying the abstract structure. If a fact is true for vector spaces in general, then that fact is true for $\mathbb{M}_{p\times n}$, the set of all $p\times n$ real matrices under regular matrix addition and scalar multiplication, as well as $\mathbb{R}^n$ and $\mathbb{P}^n$, the set of all polynomials of degree less than or equal to n including the zero polynomial, and any other set we may subsequently show is a vector set.

We first inquire about the zero vector. Does it have properties normally associated with the word *zero*? If we multiply the zero vector by a nonzero scalar, must the result be the zero vector again? If we multiply any vector by the number 0, is the result the zero vector? The answer in both cases is affirmative, but both results must be proven. We cannot just take them for granted! The zero vector is not the number 0, and there is no reason to expect (although one might hope) that facts about the number 0 are transferable to other structures that just happen to have the word *zero* as part of their name.

> **►THEOREM 1**
>
> *For any vector* $\mathbf{u}$ *in a vector space* $\mathbb{V}$, $0\odot\mathbf{u}=\mathbf{0}$.◄

Proof: Because a vector space is closed under scalar multiplication, we know that $0\odot\mathbf{u}$ is a vector in $\mathbb{V}$ (whether it is the zero vector is still to be determined). As a consequence of property A5, $0\odot\mathbf{u}$ must possess an additive inverse, denoted by $-0\odot\mathbf{u}$, such that

$$(0\odot\mathbf{u})\oplus(-0\odot\mathbf{u})=\mathbf{0} \tag{2.10}$$

Furthermore,

$$\begin{aligned} 0\odot\mathbf{u} &= (0+0)\odot\mathbf{u} && \text{A property of the } \textit{number } 0 \\ &= 0\odot\mathbf{u}\oplus 0\odot\mathbf{u} && \text{Property S4 of vector spaces} \end{aligned}$$

If we add the vector $-0\odot\mathbf{u}$ to each side of this last equation, we get

$$\begin{aligned} 0\odot\mathbf{u}\oplus -0\odot\mathbf{u} &= (0\odot\mathbf{u}\oplus 0\odot\mathbf{u})\oplus -0\odot\mathbf{u} \\ \mathbf{0} &= (0\odot\mathbf{u}\oplus 0\odot\mathbf{u})\oplus -0\odot\mathbf{u} && \text{From Eq.(2.10)} \\ \mathbf{0} &= 0\odot\mathbf{u}\oplus(0\odot\mathbf{u}\oplus -0\odot\mathbf{u}) && \text{Property A3} \\ \mathbf{0} &= (0\odot\mathbf{u})\oplus\mathbf{0} && \text{From Eq.(2.10)} \\ \mathbf{0} &= 0\odot\mathbf{u} && \text{Property A4} \end{aligned}$$

which proves Theorem 1 using just the properties of a vector space.

> **►THEOREM 2**
>
> *In any vector space* $\mathbb{V}$, $\alpha\odot\mathbf{0}=\mathbf{0}$, *for every scalar* α.◄

Proof: $0 \in \mathbb{V}$, hence $\alpha \odot 0 \in \mathbb{V}$, because a vector space is closed under scalar multiplication. It follows from property A5 that $\alpha \odot \mathbf{0}$ has an additive inverse, denoted by $-\alpha \odot \mathbf{0}$, such that

$$(a \odot 0) \oplus (-a \odot 0) = 0 \tag{2.11}$$

Furthermore,

$$\begin{aligned} a \odot 0 &= a \odot (0 \oplus 0) && \text{Property A4} \\ &= a \odot 0 \oplus a \odot 0 && \text{Property S5} \end{aligned}$$

Adding $-\alpha \odot \mathbf{0}$ to both sides of this last equation, we get

$$\begin{aligned} \alpha \odot 0 \oplus -\alpha \odot 0 &= (\alpha \odot 0 \oplus \alpha \odot 0) \oplus -\alpha \odot 0 && \\ 0 &= (\alpha \odot 0 \oplus \alpha \odot 0) \oplus -\alpha \odot 0 && \text{From Eq.(2.11)} \\ 0 &= \alpha \odot 0 \oplus (\alpha \odot 0 \oplus -\alpha \odot 0) && \text{Property A3} \\ 0 &= (\alpha \odot 0) \oplus 0 && \text{From Eq.(2.11)} \\ 0 &= \alpha \odot 0 && \text{Property A4} \end{aligned}$$

Thus, Theorem 2 follows directly from the properties of a vector space.

Property A4 asserts that every vector space has a zero vector, and property A5 assures us that every vector in a vector space $\mathbb{V}$ has an additive inverse. Neither property indicates whether there is only one zero element or many or whether a vector can have more than one additive inverse. The next two theorems do.

▶THEOREM 3

The additive inverse of any vector $\mathbf{v}$ *in a vector space* $\mathbb{V}$ *is unique.*◀

Proof: Let $\mathbf{v}_1$ and $\mathbf{v}_2$ denote additive inverses of the same vector $\mathbf{v}$. Then,

$$\mathbf{v} \oplus \mathbf{v}_1 = 0 \tag{2.12}$$
$$\mathbf{v} \oplus \mathbf{v}_2 = 0 \tag{2.13}$$

It now follows that

$$\begin{aligned} \mathbf{v}_1 &= \mathbf{v}_1 \oplus 0 && \text{Property A4} \\ &= \mathbf{v}_1 \oplus (\mathbf{v} \oplus \mathbf{v}_2) && \text{From Eq.(2.13)} \\ &= (\mathbf{v}_1 \oplus \mathbf{v}) \oplus \mathbf{v}_2 && \text{Property A3} \\ &= (\mathbf{v} \oplus \mathbf{v}_1) \oplus \mathbf{v}_2 && \text{Property A2} \\ &= 0 \oplus \mathbf{v}_2 && \text{From Eq.(2.12)} \\ &= \mathbf{v}_2 \oplus 0 && \text{Property A2} \\ &= \mathbf{v}_2 && \text{Property A4} \end{aligned}$$

▸THEOREM 4

The zero vector in a vector space $\mathbb{V}$ is unique.◂

Proof: This proof is similar to the previous one and is left as an exercise for the reader. (See Problem 34.)

▸THEOREM 5

For any vector $\mathbf{w}$ *in a vector space* $\mathbb{V}$, $-1 \odot \mathbf{w} = -\mathbf{w}$.◂

Proof: We need to show that $-1 \odot \mathbf{w}$ is the additive inverse of $\mathbf{w}$. First,

$$\begin{aligned}(-1 \odot \mathbf{w}) \oplus \mathbf{w} &= (-1 \odot \mathbf{w}) \oplus (1 \odot \mathbf{w}) && \text{Property S3}\\ &= (-1 + 1) \odot \mathbf{w} && \text{Property S5}\\ &= 0 \odot \mathbf{w} && \text{Property of real numbers}\\ &= \mathbf{0} && \text{Theorem 1}\end{aligned}$$

Therefore, $-1 \odot \mathbf{w}$ is an additive inverse of $\mathbf{w}$. By definition, $-\mathbf{w}$ is an additive inverse of $\mathbf{w}$, and because additive inverses are unique (Theorem 3), it follows that $-1 \odot \mathbf{w} = -\mathbf{w}$.

▸THEOREM 6

For any vector $\mathbf{w}$ *in a vector space* $\mathbb{V}$, $-(-\mathbf{w}) = \mathbf{w}$.◂

Proof: By definition, $-\mathbf{w}$ is the additive inverse of $\mathbf{w}$. It then follows that $\mathbf{w}$ is the additive inverse of $-\mathbf{w}$ (see Problem 33). Furthermore,

$$\begin{aligned}-\mathbf{w} \oplus -(-\mathbf{w}) &= -1 \odot \mathbf{w} \oplus -(-\mathbf{w}) && \text{Theorem 5}\\ &= -1 \odot \mathbf{w} \oplus -1 \odot (-\mathbf{w}) && \text{Theorem 5}\\ &= -1 \odot (\mathbf{w} \oplus -\mathbf{w}) && \text{Property S5}\\ &= -1 \odot \mathbf{0} && \text{Property A5}\\ &= \mathbf{0} && \text{Theorem 2}\end{aligned}$$

Therefore, $-(-\mathbf{w})$ is an additive inverse of $-\mathbf{w}$. Since $\mathbf{w}$ is also an additive inverse of $-\mathbf{w}$, it follows from Theorem 3 that the two are equal.

▸THEOREM 7

Let α *be a scalar and* $\mathbf{u}$ *a vector in a vector space* $\mathbb{V}$. *If* $\alpha \odot \mathbf{u} = \mathbf{0}$, *then either* $\alpha = 0$ *or* $\mathbf{u} = \mathbf{0}$.◂

Proof: We are given

$$\alpha \odot \mathbf{u} = \mathbf{0} \tag{2.14}$$

Now either α is 0 or it is not. If α is 0, the theorem is proven. If α is not 0, we form the scalar $1/\alpha$ and then multiply Equation (2.14) by $1/\alpha$, obtaining

$$(1/\alpha) \odot (\alpha \odot \mathbf{u}) = (1/\alpha) \odot \mathbf{0}$$

$$(1/\alpha) \odot (\alpha \odot \mathbf{u}) = \mathbf{0} \qquad \text{Theorem 2}$$

$$\left(\frac{1}{\alpha}\alpha\right) \odot \mathbf{u} = \mathbf{0} \qquad \text{Property S2}$$

$$1 \odot \mathbf{u} = \mathbf{0} \qquad \text{Property of numbers}$$

$$\mathbf{u} = \mathbf{0} \qquad \text{Property S3}$$

Problems 2.2

In Problems 1 through 32 a set of objects is given together with a definition for vector addition and scalar multiplication. Determine which are vector spaces, and for those that are not, identify at least one property that fails to hold.

(1) $\left\{ \begin{bmatrix} a & b \\ c & d \end{bmatrix} \right\} \in \mathbb{M}_{2\times 2} \middle| b = 0$ under standard matrix addition and scalar multiplication.

(2) $\left\{ \begin{bmatrix} a & b \\ c & d \end{bmatrix} \right\} \in \mathbb{M}_{2\times 2} \middle| c = 1$ under standard matrix addition and scalar multiplication.

(3) The set of all 2×2 real matrices $\mathbf{A} = [a_{ij}]$ with $a_{11} = -a_{22}$ under standard matrix addition and scalar multiplication.

(4) The set of all 3×3 real upper triangular matrices under standard matrix addition and scalar multiplication.

(5) The set of all 3×3 real lower triangular matrices of the form

$$\begin{bmatrix} 1 & 0 & 0 \\ a & 1 & 0 \\ b & c & 1 \end{bmatrix}$$

under standard matrix addition and scalar multiplication.

(6) $\{[a \quad b] \in \mathbb{R}^2 | a + b = 2\}$ under standard matrix addition and scalar multiplication.

(7) $\{[a \quad b] \in \mathbb{R}^2 | a = b\}$ under standard matrix addition and scalar multiplication.

(8) The set consisting of the single element $\mathbf{0}$ with vector addition and scalar multiplication defined as $\mathbf{0} \oplus \mathbf{0} = \mathbf{0}$ and $\alpha \odot \mathbf{0} = \mathbf{0}$ for any real number α.

(9) The set of all real two-dimensional row matrices $\{[a\ b]\}$ with standard matrix addition but scalar multiplication defined as $\alpha \odot [a\ b] = [0\ 0]$.

(10) The set of all real two-dimensional row matrices $\{[a\ b]\}$ with standard matrix addition but scalar multiplication defined as $\alpha \odot [a\ b] = [0\ \alpha b]$.

(11) The set of all real two-dimensional row matrices $\{[a\ b]\}$ with standard matrix addition but scalar multiplication defined as $\alpha \odot [a\ b] - [2\alpha a\ 2\alpha b]$.

(12) The set of all real two-dimensional row matrices $\{[a\ b]\}$ with standard matrix addition but scalar multiplication defined as $\alpha \odot [a\ b] = [5a\ 5b]$.

(13) The set of all real three-dimensional row matrices $\{[a\ b\ c]\}$ with standard scalar multiplication but vector addition defined as

$$[a \quad b \quad c] \oplus [x \quad y \quad z] = [a + x \quad b + y + 1 \quad c + z]$$

(14) The set of all real three-dimensional row matrices $\{[a\ b\ c]\}$ with standard scalar multiplication but vector addition defined as

$$[a \quad b \quad c] \oplus [x \quad y \quad z] = [a \quad b + y \quad c]$$

(15) The set of all real three-dimensional row matrices $\{[a\ b\ c]\}$ with standard matrix addition but scalar multiplication defined as $\alpha \odot [a\ b\ c] - [\alpha a\ \alpha b\ 1]$.

(16) The set of all real three-dimensional row matrices $\{[a\ b\ c]\}$ with positive components under standard matrix addition but scalar multiplication defined as

$$\alpha \odot [a \quad b \quad c] = [a^\alpha \quad b^\alpha \quad c^\alpha]$$

(17) The set of all real numbers (by convention, the scalars are also real numbers) with $a \oplus b = a \odot b = ab$, the standard multiplication of numbers.

(18) The set of all positive real numbers with $a \oplus b = ab$, the standard multiplication of numbers, and $a \odot b = ab$.

(19) The set of all solutions of the homogeneous set of linear equations $\mathbf{Ax} = \mathbf{0}$, under standard matrix addition and scalar multiplication.

(20) The set of all solutions of the set of linear equations $\mathbf{Ax} = \mathbf{b}$, $\mathbf{b} \neq \mathbf{0}$, under standard matrix addition and scalar multiplication.

(21) $\{p(t) \in \mathbb{P}^3 | p(0) = 0\}$ under standard addition and scalar multiplication of polynomials.

(22) The set of all ordered pairs of real numbers such that $(a, b) \oplus (c, d) = (a + c + 1, b + d + 1)$ and $k \odot (a, b) = (ka, kb)$.

(23) The set of all ordered pairs of real numbers such that $(a, b, c) \oplus (d, e, f) = (a+d, b+e, c+f)$ and $k \odot (a, b, c) = (0, 0, 0)$.

(24) The set of all ordered triples of real numbers of the form $(1, a, b)$ such that $(1, a, b) \oplus (1, c, d) = (1, a+b, c+d)$ and $k \odot (1, a, b) = (1, ka, kb)$.

(25) Let $\mathbf{w}$ be a vector in a vector space $\mathbb{V}$. Prove that if $-\mathbf{w}$ is the additive inverse of $\mathbf{w}$ then the reverse is also true: $\mathbf{w}$ is the additive inverse of $-\mathbf{w}$.

(26) Prove Theorem 4.

(27) Prove that $\mathbf{v} \oplus (\mathbf{u} - \mathbf{v}) - \mathbf{u}$ if $\mathbf{u} - \mathbf{v}$ is shorthand for $\mathbf{u} \oplus -\mathbf{v}$.

(28) Prove that if $\mathbf{u} \oplus \mathbf{v} = \mathbf{u} \oplus \mathbf{w}$, then $\mathbf{v} - \mathbf{w}$.

(29) Prove that $\mathbf{u} \oplus \mathbf{u} - 2\mathbf{u}$ if $2\mathbf{u}$ is shorthand for $2 \odot \mathbf{u}$.

(30) Prove that the only solution to the equation $\mathbf{u} \oplus \mathbf{u} - 2\mathbf{v}$ is $\mathbf{u} = \mathbf{v}$.

(31) Prove that if $\mathbf{u} \neq \mathbf{0}$ and $\alpha \odot \mathbf{u} = \beta \oplus \mathbf{u}$, then $\alpha = \beta$.

(32) Prove that the additive inverse of the zero vector is the zero vector.

2.3 SUBSPACES

To show that a set of objects $\mathbb{S}$ is a vector space, we must verify that all 10 properties of a vector space are satisfied, the 5 properties involving vector addition and the 5 properties involving scalar multiplication. This process, however, can be shortened considerably if the set of objects is a subset of a *known* vector space $\mathbb{V}$. Then, instead of 10 properties, we need only verify the 2 closure properties, because the other 8 properties follow immediately from these 2 and the fact that $\mathbb{S}$ is a subset of a known vector space.

A subspace of a vector space $\mathbb{V}$ is a subset of $\mathbb{V}$ that is a vector space in its own right.

We define a nonempty subset $\mathbb{S}$ of a vector space $\mathbb{V}$ as a *subspace* of $\mathbb{V}$ if $\mathbb{S}$ is itself a vector space under the *same* operations of vector addition and scalar multiplication defined on $\mathbb{V}$.

> **▸THEOREM 1**
>
> *Let $\mathbb{S}$ be a nonempty subset of a vector space $\mathbb{V}$ with operations $\oplus$ and $\odot$. $\mathbb{S}$ is a subspace of $\mathbb{V}$ if and only if the following two closure conditions hold:*
>
> (i) *Closure under addition: If $\mathbf{u} \in \mathbb{S}$ and $\mathbf{v} \in \mathbb{S}$, then $\mathbf{u} \oplus \mathbf{v} \in \mathbb{S}$.*
> (ii) *Closure under scalar multiplication: If $\mathbf{u} \in \mathbb{S}$ and α is any scalar, then $\alpha \odot \mathbf{u} \in \mathbb{S}$.* ◀

Proof: If $\mathbb{S}$ is a vector space, then it must satisfy all 10 properties of a vector space, in particular the closure properties defined by conditions (i) and (ii). Thus, if $\mathbb{S}$ is a vector space, then (i) and (ii) are satisfied.

We now show the converse: If conditions (i) and (ii) are satisfied, then $\mathbb{S}$ is a vector space; that is, all 10 properties of a vector space specified in Definition 1 of Section 2.2 follow from the closure properties *and* the fact that $\mathbb{S}$ is a subset of a known vector space $\mathbb{V}$. Conditions (i) and (ii) are precisely Properties A1 and S1. Properties A2, A3, and S2 through S5 follow for elements in $\mathbb{S}$ because these elements are also in $\mathbb{V}$ and $\mathbb{V}$ is known to be a vector space whose elements satisfy *all* the properties of a vector space. In particular, to verify Property A2, we let **u** and **v** denote arbitrary elements in $\mathbb{S}$. Because $\mathbb{S}$ is a subset of $\mathbb{V}$, it follows that **u** and **v** are in $\mathbb{V}$. Because $\mathbb{V}$ *is* a vector space, we have $\mathbf{u} \oplus \mathbf{v} = \mathbf{v} \oplus \mathbf{u}$. To verify S3, we let **u** again denote an arbitrary element in $\mathbb{S}$. Because $\mathbb{S}$ is a subset of $\mathbb{V}$, it follows that **u** is an element of $\mathbb{V}$. Because $\mathbb{V}$ *is* a vector space, we have $1 \odot \mathbf{u} = \mathbf{u}$.

All that remains is to verify that the zero vector and additive inverses of elements in $\mathbb{S}$ are themselves members of $\mathbb{S}$. Because $\mathbb{S}$ is nonempty, it must contain at least one element, which we denote as **u**. Then, for the zero scalar, 0, we know that $0 \odot \mathbf{u}$ is in $\mathbb{S}$, as a result of condition (ii), and this vector is the zero vector as a result of Theorem 1 of the previous section. Thus, Property A4 is satisfied. If **u** is an element of $\mathbb{S}$, then the product $-1 \odot \mathbf{u}$ is also an element of $\mathbb{S}$, as a result of condition (ii); it follows from Theorem 5 of the previous section that $-1 \odot \mathbf{u}$ is the additive inverse of **u**, so Property A5 is also satisfied.

►CONVENTION

For the remainder of this book, we drop the $\oplus$ and $\odot$ symbols in favor of the traditional sum symbol (+) and scalar multiplication denoted by juxtaposition. All vector spaces will involve standard vector addition and scalar multiplication, unless noted otherwise.◄

We use Theorem 1 to significantly shorten the work required to show that some sets are vector spaces!

Example 1 Determine whether $\left\{ \begin{bmatrix} a & b \\ c & d \end{bmatrix} \in \mathbb{M}_{2\times2} \middle| b = c = 0 \right\}$ is a vector space under standard matrix addition and scalar multiplication.

Solution: The set $\mathbb{S}$ of 2×2 real matrices with zeros in the 1-2 and 2-1 positions is a subset of $\mathbb{M}_{2\times2}$, and $\mathbb{M}_{2\times2}$ is a vector space (see Example 4 in Section 2.2 with $p = n = 2$). Thus, Theorem 1 is applicable, and instead of verifying all 10 properties of a vector space, we need only verify closure in $\mathbb{S}$ under matrix addition and scalar multiplication.

Arbitrary elements **u** and **v** in $\mathbb{S}$ have the form

$$\mathbf{u} = \begin{bmatrix} a & 0 \\ 0 & b \end{bmatrix} \quad \text{and} \quad \mathbf{v} = \begin{bmatrix} c & 0 \\ 0 & d \end{bmatrix}$$

for any real numbers a, b, c, and d. Here

$$\mathbf{u} + \mathbf{v} = \begin{bmatrix} a+c & 0 \\ 0 & b+d \end{bmatrix}$$

and for any real scalar α,

$$\alpha \mathbf{u} = \begin{bmatrix} \alpha a & 0 \\ 0 & \alpha b \end{bmatrix}$$

Because these matrices are again elements in $\mathbb{S}$, each having zeros in their 1-2 and 2-1 positions, it follows from Theorem 1 that $\mathbb{S}$ is a subspace of $\mathbb{M}_{2\times 2}$. The set $\mathbb{S}$ is therefore a vector space.

Example 2 Determine whether the set $\mathbb{S} = \{[x\, y\, z] \in \mathbb{R}^3 | y = 0\}$ is a vector space under standard matrix addition and scalar multiplication.

Solution: We first observe that $\mathbb{S}$ is a subset of $\mathbb{R}^3$, considered as row matrices, which we know is a vector space from our work in Section 2.2. Thus, Theorem 1 is applicable. Arbitrary elements **u** and **v** in $\mathbb{S}$ have the form

$$\mathbf{u} = [a \quad 0 \quad b] \quad \text{and} \quad \mathbf{v} = [c \quad 0 \quad d]$$

It follows that

$$\mathbf{u} + \mathbf{v} = [a + c \quad 0 \quad b + d] \in \mathbb{S}$$

and for any real scalar α,

$$\alpha \mathbf{u} = [\alpha a \quad 0 \quad \alpha B] \in \mathbb{S}$$

If a set is a subset of a known vector space, then the simplest way to show the set is a vector space is to show the set is a subspace.

Thus, $\mathbb{S}$ is closed under addition and scalar multiplication, and it follows from Theorem 1 that $\mathbb{S}$ is a subspace of $\mathbb{R}^3$. The set $\mathbb{S}$ is therefore a vector space.

Compare Example 2 to Example 1 of Section 2.2. In both, we were asked to prove that the same set is a vector space. In Section 2.2, we did this by verifying all 10 properties of a vector space; in Example 2, we verified the 2 properties of a subspace. Clearly it is simpler to verify 2 properties than 10; thus, it is simpler to show that a set is vector space by showing it is a subspace rather than demonstrating directly that the set is a vector space. To do so, however, we must *recognize* that the given set is a subset of *known* vector space, in this case $\mathbb{R}^3$.

The subspace in Example 2 has an interesting graphical representation. $\mathbb{R}^3$, the set of all 3-tuples, is represented geometrically by all points in three-space. The set $\mathbb{S}$ in Example 2 is the set of all points in $\mathbb{R}^3$ having a second component of 0.

In an x, y, z coordinate system, these points fill the entire x-z plane, which is illustrated graphically by the shaded plane in Figure 2.12.

Example 3 Determine whether the set $\mathbb{S}$, illustrated graphically by the shaded plane in Figure 2.13, is a subspace of $\mathbb{R}^3$.

Solution: The shaded plane is parallel to the y-z plane, intersecting the x-axis at $x = 3$. The x-coordinate of any point on this plane is fixed at $x = 3$, and the plane is defined as

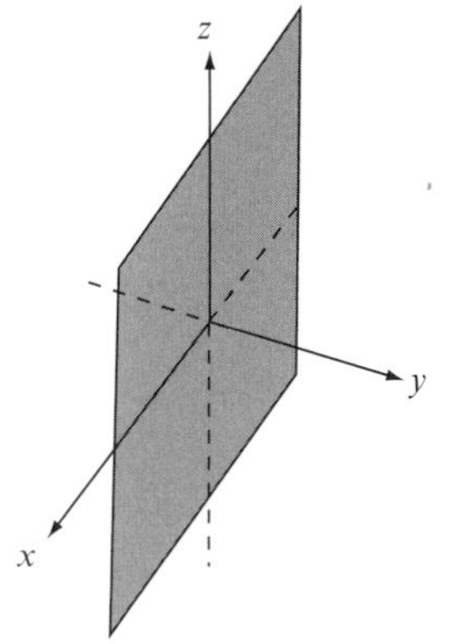

FIGURE 2.12

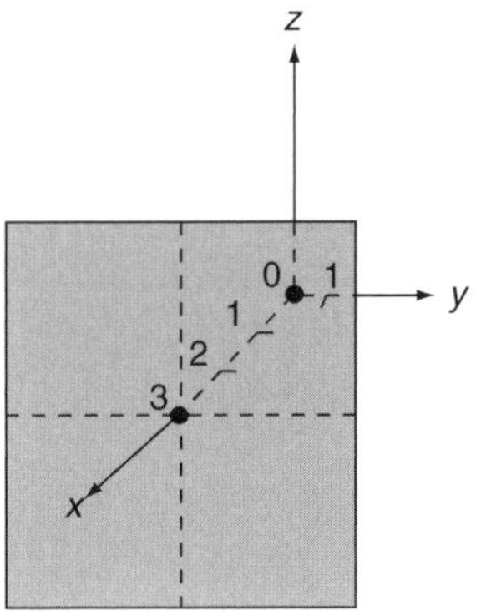

FIGURE 2.13

$$\mathbb{S} = \{[x \quad y \quad z] \in \mathbb{R}^3 | x = 3\}$$

Elements **u** and **v** in $\mathbb{S}$ have the form

$$\mathbf{u} = [3 \quad a \quad b] \quad \text{and} \quad \mathbf{v} = [3 \quad c \quad d]$$

for some choice of the scalars a, b, c, and d. Here

$$\mathbf{u} + \mathbf{v} = [6 \quad a + c \quad b + d]$$

which is *not* an element of $\mathbb{S}$ because its first component is not 3. Condition (i) of Theorem 1 is violated. The set $\mathbb{S}$ is not closed under addition and, therefore, is *not* a subspace.

As an alternative solution to Example 3, we note that the set $\mathbb{S}$ does not contain the zero vector, and therefore cannot be a vector space. The zero vector in $\mathbb{R}^3$ is $\mathbf{0} = [0\,0\,0]$, and this vector is clearly not in $\mathbb{S}$ because all elements in $\mathbb{S}$ have a first

If a subset of a vector space does not include the zero vector, that subset cannot be a subspace.

component of 3. Often we can determine by inspection whether the zero vector of a vector space is included in a given subset. If the zero vector is *not* included, we may conclude immediately that the subset is not a vector space and, therefore, not a subspace. If the zero vector is part of the set, then the two closure properties must be verified before one can determine whether the given set is a subspace.

One simple subspace associated with any vector space is the following:

▸THEOREM 2

For any vector space $\mathbb{V}$, the subset containing only the zero vector is a subspace.◂

Proof: It follows from the definition of a zero vector that $\mathbf{0}+\mathbf{0}=\mathbf{0}$. It also follows from Theorem 2 of Section 2.2 that $\alpha\mathbf{0}=\mathbf{0}$ for any scalar α. Both closure conditions of Theorem 1 are satisfied, and the set $\mathbb{S}$ containing just the single element $\mathbf{0}$ is a subspace.

Example 4 Determine whether the set $\mathbb{S}=\{[\alpha\ 2\alpha\ 4\alpha]\,|\,\alpha \text{ is a real number}\}$ is a subspace of $\mathbb{R}^3$.

Solution: Setting $\alpha=0$, we see that the zero vector, $\mathbf{0}=[0\,0\ 0]$, of $\mathbb{R}^3$ is an element of $\mathbb{S}$, so we can make *no* conclusion a priori about $\mathbb{S}$ as a subspace. We must apply Theorem 1 directly. Elements $\mathbf{u}$ and $\mathbf{v}$ in $\mathbb{S}$ have the form

$$\mathbf{u}=[t\quad 2t\quad 4t]\quad \text{and}\quad \mathbf{v}=[s\quad 2s\quad 4s]$$

for some choice of the scalars s and t. Therefore,

$$\begin{aligned}\mathbf{u}+\mathbf{v}&=[t+s\quad 2t+2s\quad 4t+4s]\\&=[(t+s)\quad 2(t+s)\quad 4(t+s)]\in\mathbb{S}\end{aligned}$$

and for any real scalar α,

$$\begin{aligned}\alpha\mathbf{u}&=[\alpha t\quad \alpha(2t)\quad \alpha(4t)]\\&=[(\alpha t)\quad 2(\alpha t)\quad 4(\alpha t)]\in\mathbb{S}\end{aligned}$$

Because $\mathbb{S}$ is closed under vector addition and scalar multiplication, it follows from Theorem 1 that $\mathbb{S}$ is a subspace of $\mathbb{R}^3$.

The subspace in Example 4 also has an interesting graphical representation. If we rewrite an arbitrary vector $\mathbf{u}$ as

$$\mathbf{u}=[t\quad 2t\quad 4t]=t[1\quad 2\quad 4]$$

we see that every vector is a scalar multiple of the directed line segment having its tail at the origin and its tip at the point $(1, 2, 4)$. Because t can be any real number, zero, positive or negative, we can reach any point on the line that contains this directed line segment. Thus, the subspace $\mathbb{S}$ is represented graphically by the straight line in $\mathbb{R}^3$ illustrated in Figure 2.14.

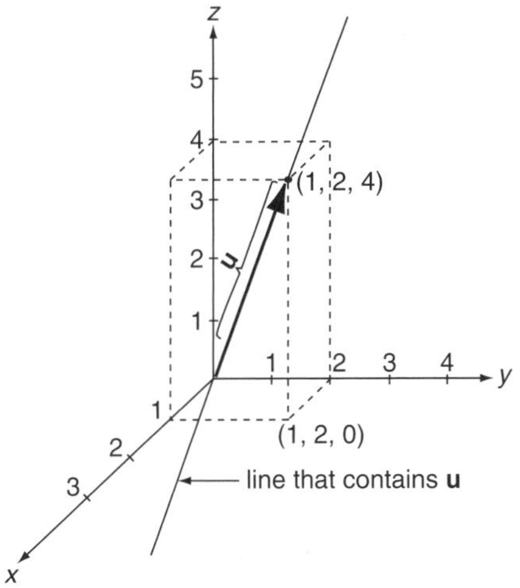

FIGURE 2.14

As a result of Examples 2 through 4 and Theorem 2, one might suspect that a proper subset $\mathbb{S}$ of $\mathbb{R}^3$ is a subspace if and only if $\mathbb{S}$ is the zero vector or else the graph of $\mathbb{S}$ is either a straight line through the origin or a plane that contains the origin. This is indeed the case as we shall prove in Section 2.5.

Lines through the origin and planes that contain the origin are subspaces of $\mathbb{R}^3$.

The two conditions specified in Theorem 1 can be collapsed into a single condition.

▶THEOREM 3

A nonempty subset $\mathbb{S}$ of a vector space $\mathbb{V}$ is a subspace of $\mathbb{V}$ if and only if whenever **u** *and* **v** *are any two elements in $\mathbb{S}$ and α and β are any two scalars, then*

$$\alpha\mathbf{u} + \beta\mathbf{v} \tag{2.15}$$

is also in $\mathbb{S}$.◀

Proof: If $\mathbb{S}$ is a subspace, then it must satisfy the two conditions of Theorem 1. In particular, if **u** is an element of $\mathbb{S}$ and α a scalar, then $\alpha\mathbf{u}$ is in $\mathbb{S}$ as a consequence of condition (ii). Similarly, $\beta\mathbf{v}$ must be an element of $\mathbb{S}$ whenever **v** is an element and β is a scalar. Knowing that $\alpha\mathbf{u}$ and $\beta\mathbf{v}$ are two elements in $\mathbb{S}$, we may conclude that their sum, given by Equation (2.15), is also in $\mathbb{S}$ as a consequence of condition (i).

Conversely, if Equation (2.15) is an element in $\mathbb{S}$ for all values of the scalars α and β, then condition (i) of Theorem 1 follows by setting $\alpha = \beta = 1$. Condition (ii) follows by setting $\beta = 0$ and leaving α arbitrary.

Example 5 Determine whether $\mathbb{S} = \{p(t) \in \mathbb{P}^2 | p(2) = 1\}$ is a subspace of $\mathbb{P}^2$.

Solution: $\mathbb{P}^2$ is a vector space (see Example 7 of Section 2.2 with $n=2$). The zero vector $\mathbf{0}$ in $\mathbb{P}^2$ has the property $\mathbf{0}(t)=0$ for all real values of t. Thus, $\mathbf{0}(2)=0\neq 1$, the zero vector is not in $\mathbb{S}$, and $\mathbb{S}$ is not a subspace.

Example 6 Determine whether $\mathbb{S} = \{p(t) \in \mathbb{P}^2 | p(2) = 0\}$ is a subspace of $\mathbb{P}^2$.

Solution: Let $\mathbf{u}=p$ and $\mathbf{v}=q$ be any two polynomials in $\mathbb{S}$. Then $p(2)=0$ and $q(2)=0$. Set $\mathbf{w}=\alpha\mathbf{u}+\beta\mathbf{v}$, for arbitrary values of the scalars α and β. Then $\mathbf{w}$ is also a polynomial of degree two or less or the zero polynomial. Furthermore,

$$\mathbf{w}(2) = (\alpha p + \beta q)(2) = \alpha p(2) + \beta q(2) = \alpha 0 + \beta 0 = 0,$$

so $\mathbf{w}$ is also an element of $\mathbb{S}$. It follows from Theorem 3 that $\mathbb{S}$ is a subspace of $\mathbb{P}^n$.

A vector **u** is a linear combination of a finite number of other vectors if **u** can be written as a sum of scalar multiples of those vectors.

Expression (2.6) in Theorem 3 is a special case of a linear combination. We say that a vector $\mathbf{u}$ in a vector space $\mathbb{V}$ is a *linear combination* of the vectors $\mathbf{v}_1, \mathbf{v}_2, \ldots \mathbf{v}_n$ in $\mathbb{V}$ if there exists scalars $d_1, d_2, \ldots, d_n$ such that

$$\mathbf{u} = d_1\mathbf{v}_1 + d_2\mathbf{v}_2 + \cdots + d_n\mathbf{v}_n \tag{2.16}$$

Example 7 Determine whether $\mathbf{u}=[1\ 2\ 3]$ is a linear combination of

$$\mathbf{v}_1 = [1 \quad 1 \quad 1], \quad \mathbf{v}_2 = [2 \quad 4 \quad 0], \quad \text{and} \quad \mathbf{v}_3 = [0 \quad 0 \quad 1]$$

Solution: These vectors are all in the vector space $\mathbb{R}^3$, considered as row matrices. We seek scalars d_1, d_2, and d_3 that satisfy the equation

$$[1 \quad 2 \quad 3] = d_1[1 \quad 1 \quad 1] + d_2[2 \quad 4 \quad 0] + d_3[0 \quad 0 \quad 1]$$

or

$$[1 \quad 2 \quad 3] = [d_1 + 2d_2 \quad d_1 + 4d_2 \quad d_1 + d_3]$$

This last matrix equation is equivalent to the system of equations

$$\begin{aligned} 1 &= d_1 + 2d_2 \\ 2 &= d_1 + 4d_2 \\ 3 &= d_1 + d_3 \end{aligned}$$

Using Gaussian elimination, we find that the only solution to this system is $d_1=0$, $d_2=1/2$, and $d_3=3$. Thus,

$$[1 \quad 2 \quad 3] = 0[1 \quad 1 \quad 1] + \tfrac{1}{2}[2 \quad 4 \quad 0] + 3[0 \quad 0 \quad 1]$$

and the vector $\mathbf{u}=[1\ 2\ 3]$ is a linear combination of the other three.

Example 8 Determine whether $\mathbf{u} = \begin{bmatrix} -1 & 0 \\ 2 & 4 \end{bmatrix}$ is a linear combination of

$$\mathbf{v}_1 = \begin{bmatrix} 1 & 1 \\ 2 & 2 \end{bmatrix} \quad \text{and} \quad \mathbf{v}_2 = \begin{bmatrix} 3 & 2 \\ 3 & 5 \end{bmatrix}$$

Solution: These vectors are in the vector space $\mathbb{M}_{2\times 2}$. We seek scalars d_1 and d_2 that satisfy the equation

$$\begin{bmatrix} -1 & 0 \\ 2 & 4 \end{bmatrix} = d_1 \begin{bmatrix} 1 & 1 \\ 2 & 2 \end{bmatrix} + d_2 \begin{bmatrix} 3 & 2 \\ 3 & 5 \end{bmatrix} \tag{2.17}$$

or

$$\begin{bmatrix} -1 & 0 \\ 2 & 4 \end{bmatrix} = \begin{bmatrix} d_1 + 3d_2 & d_1 + 2d_2 \\ 2d_1 + 3d_2 & 2d_1 + 5d_2 \end{bmatrix}$$

which is equivalent to the system of equations

$$\begin{aligned} -1 &= d_1 + 3d_2 \\ 0 &= d_1 + 2d_2 \\ 2 &= 2d_1 + 3d_2 \\ 4 &= 2d_1 + 5d_2 \end{aligned}$$

Using Gaussian elimination, we find that this system has no solution. There are *no* values of d_1 and d_2 that satisfy Equation (2.8), and, therefore, $\mathbf{u}$ is *not* a linear combination of $\mathbf{v}_1$ and $\mathbf{v}_2$.

The set of *all* linear combinations of a finite set of vectors, $\mathbb{S} = \{\mathbf{v}_1, \mathbf{v}_2, \ldots, \mathbf{v}_n\}$, is called the *span* of $\mathbb{S}$, denoted as *span* $\{\mathbf{v}_1, \mathbf{v}_2, \ldots, \mathbf{v}_n\}$ or simply *span*$(\mathbb{S})$. Thus, the span of the polynomial set $\{t^2, t, 1\}$ is $\mathbb{P}^2$ because every polynomial $p(t)$ in $\mathbb{P}^2$ can be written as

The span of a finite number of vectors is the set of all linear combinations of those vectors.

$$p(t) = d_1 t^2 + d_2 t + d_3(1)$$

for some choice of the scalars d_1, d_2, and d_3. The span of the set $\{[1\,0\,0\,0], [0\,1\,0\,0]\}$ are all row-vectors of the form $[d_1\, d_2\, 0\, 0]$ for any choice of the real numbers d_2 and d_3.

The span of a finite set of vectors is useful because it is a subspace! Thus, we create subspaces conveniently by forming all linear combinations of just a few vectors.

►THEOREM 4

The span of a set of vectors $\mathbb{S} = \{\mathbf{v}_1, \mathbf{v}_2, \ldots, \mathbf{v}_n\}$ in a vector space $\mathbb{V}$ is a subspace of $\mathbb{V}$. ◄

Proof: Let $\mathbf{u}$ and $\mathbf{w}$ be elements of *span* $(\mathbb{S})$. Then

$$\mathbf{u} = d_1\mathbf{v}_1 + d_2\mathbf{v}_2 + \cdots + d_n\mathbf{v}_n \quad \text{and} \quad \mathbf{w} = c_1\mathbf{v}_1 + c_2\mathbf{v}_2 + \cdots + c_n\mathbf{v}_n$$

for some choice of the scalars d_1 through d_n and c_1 through c_n. It follows that

$$\begin{aligned}\alpha\mathbf{u} + \beta\mathbf{w} &= \alpha(d_1\mathbf{v}_1 + d_2\mathbf{v}_2 + \cdots + d_n\mathbf{v}_n) + \beta(c_1\mathbf{v}_1 + c_2\mathbf{v}_2 + \cdots + c_n\mathbf{v}_n)\\ &= (\alpha d_1)\mathbf{v}_1 + (\alpha d_2)\mathbf{v}_2 + \cdots + (\alpha d_n)\mathbf{v}_n + (\beta c_1)\mathbf{v}_1 + (\beta c_2)\mathbf{v}_2 + \cdots + (\beta c_n)\mathbf{v}_n\\ &= (\alpha d_1 + \beta c_1)\mathbf{v}_1 + (\alpha d_2 + \beta c_2)\mathbf{v}_2 + \cdots + (\alpha d_n + \beta c_n)\mathbf{v}_n\end{aligned}$$

Each quantity in parentheses on the right side of this last equation is a combination of scalars of the form $\alpha d_j + \beta c_j$ (for $j = 1, 2, \ldots, n$) and is, therefore, itself a scalar. Thus, $\alpha\mathbf{u} + \beta\mathbf{w}$ is a linear combination of the vectors in $\mathbb{S}$ and a member of *span*($\mathbb{S}$). It follows from Theorem 3 that *span*($\mathbb{S}$) is a subspace of $\mathbb{V}$.

Not only is the *span*($\mathbb{S}$) a subspace that includes the vectors in $\mathbb{S}$, but it is the smallest such subspace. We formalize this statement in the following theorem, the proof of which is left as an exercise for the reader (see Problem 50).

►THEOREM 5

If $\mathbb{S} = \{\mathbf{v}_1\mathbf{v}_2, \ldots, \mathbf{v}_n\}$ is a set of vectors in a vector space $\mathbb{V}$ and if $\mathbb{W}$ is a subspace of $\mathbb{V}$ that contains all the vectors in $\mathbb{S}$, then $\mathbb{W}$ contains all the vectors in span($\mathbb{S}$). ◄

Problems 2.3

In Problems 1 through 23, determine whether each set is a subspace of the indicated vector space.

(1) $\mathbb{S} = \{[a \quad b] \in \mathbb{R}^2 | a = 0\}$.

(2) $\mathbb{S} = \{[a \quad b] \in \mathbb{R}^2 | a = -b\}$.

(3) $\mathbb{S} = \{[a \quad b] \in \mathbb{R}^2 | b = -5a\}$.

(4) $\mathbb{S} = \{[a \quad b] \in \mathbb{R}^2 | b = a + 3\}$.

(5) $\mathbb{S} = \{[a \quad b] \in \mathbb{R}^2 | b \geq a\}$.

(6) $\mathbb{S} = \{[a \quad b] \in \mathbb{R}^2 | a = b = 0\}$.

(7) $\mathbb{S} = \{[a \quad b \quad c] \in \mathbb{R}^3 | a = b\}$.

(8) $\mathbb{S} = \{[a \quad b \quad c] \in \mathbb{R}^3 | b = 0\}$.

(9) $\mathbb{S} = \{[a \quad b \quad c] \in \mathbb{R}^3 | a = b + 1\}$.

(10) $\mathbb{S} = \{[a \quad b \quad c] \in \mathbb{R}^3 | c = a - b\}$.

(11) $\mathbb{S} = \{[a \quad b \quad c] \in \mathbb{R}^3 | c = ab\}$.

(12) $\mathbb{S} = \{[a \quad b \quad c] \in \mathbb{R}^3 | a = b \text{ and } c = 0\}$.

(13) $\mathbb{S} = \{[a \quad b \quad c] \in \mathbb{R}^3 | b = 3a \text{ and } c = a + 3\}$.

(14) $\mathbb{S} = \left\{ \begin{bmatrix} a & 2a & 0 \\ 0 & a & 2a \end{bmatrix} \middle| a \text{ is real} \right\}$ as a subset of $\mathbb{M}_{2\times 3}$.

(15) $\left\{ \begin{bmatrix} a & b & c \\ d & e & f \end{bmatrix} . \in \mathbb{M}_{2\times 3} \middle| c = e = f = 0 \right\}$.

(16) $\left\{ \begin{bmatrix} a & b & c \\ d & e & f \end{bmatrix} . \in \mathbb{M}_{2\times 3} \middle| c = e = f = 1 \right\}$.

(17) $\mathbb{S} = \left\{ \begin{bmatrix} 0 & 0 & 0 \\ 0 & 0 & 0 \end{bmatrix} \right\}$ as a subset of $\mathbb{M}_{2\times 3}$.

(18) $\mathbb{S} = \{\mathbf{A} \in \mathbb{M}_{3\times 3} | \mathbf{A} \text{ is lower triangular}\}$.

(19) $\mathbb{S} = \{\mathbf{A} \in \mathbb{M}_{3\times 3} | \mathbf{A} \text{ is a diagonal matrix}\}$.

(20) $\mathbb{S} = \left\{ \begin{bmatrix} a & a^2 & a^3 \\ a^2 & a & a^2 \\ a^3 & a^2 & a \end{bmatrix} \middle| a \textit{ is real} \right\}$ as a subset of $\mathbb{M}_{3\times 3}$.

(21) $\mathbb{S} = \{\mathbf{A} \in \mathbb{M}_{2\times 2} | \mathbf{A} \text{ is invertible}\}$.

(22) $\mathbb{S} = \{\mathbf{A} \in \mathbb{M}_{2\times 2} | \mathbf{A} \text{ is singular}\}$.

(23) $\mathbb{S} = \{f(t) \in \mathbb{C}[-1, 1] | f(-t) = -f(t)\}$.

(24) Determine whether $\mathbf{u}$ is a linear combination of $\mathbf{v}_1 = [1\ 2]$ and $\mathbf{v}_2 = [3\ 6]$.

(a) $\mathbf{u} = [2\ 4]$, (b) $\mathbf{u} = [2\ -4]$,

(c) $\mathbf{u} = [-3\ -6]$, (d) $\mathbf{u} = [2\ 2]$.

(25) Determine which, if any, of the vectors $\mathbf{u}$ defined in the previous problem are in span $\{\mathbf{v}_1, \mathbf{v}_2\}$.

(26) Determine whether $\mathbf{u}$ is a linear combination of $\mathbf{v}_1 = [1\ 0\ 1]$ and $\mathbf{v}_1 = [1\ 1\ 1]$.

(a) $\mathbf{u} = [3\ 2\ 3]$, (b) $\mathbf{u} = [3\ 3\ 2]$,

(c) $\mathbf{u} = [0\ 0\ 0]$, (d) $\mathbf{u} = [0\ 1\ 1]$.

(27) Determine which, if any, of the vectors $\mathbf{u}$ defined in the previous problem are in *span* $\{\mathbf{v}_1, \mathbf{v}_2\}$.

(28) Determine whether the following vectors are linear combinations of

$$\mathbf{v}_1 = \begin{bmatrix} 1 \\ 0 \\ 0 \end{bmatrix}, \quad \mathbf{v}_2 = \begin{bmatrix} 1 \\ 1 \\ 0 \end{bmatrix}, \quad \mathbf{v}_3 = \begin{bmatrix} 1 \\ 1 \\ 1 \end{bmatrix}.$$

(a) $\begin{bmatrix} 1 \\ 0 \\ 1 \end{bmatrix}$, (b) $\begin{bmatrix} 1 \\ 0 \\ -1 \end{bmatrix}$, (c) $\begin{bmatrix} 2 \\ 2 \\ 0 \end{bmatrix}$,

(d) $\begin{bmatrix} 2 \\ 2 \\ 4 \end{bmatrix}$, (e) $\begin{bmatrix} 1 \\ 2 \\ 3 \end{bmatrix}$, (f) $\begin{bmatrix} 2 \\ -5 \end{bmatrix}$.

(29) Determine whether the following matrices are linear combinations of

$$\mathbf{A}_1 = \begin{bmatrix} 1 & 0 \\ 0 & 0 \end{bmatrix}, \quad \mathbf{A}_2 = \begin{bmatrix} 0 & 1 \\ 0 & 0 \end{bmatrix}, \quad \mathbf{A}_3 = \begin{bmatrix} 1 & 1 \\ 1 & 0 \end{bmatrix}.$$

(a) $\begin{bmatrix} 0 & 1 \\ 1 & 1 \end{bmatrix}$, (b) $\begin{bmatrix} 1 & 2 \\ 3 & 0 \end{bmatrix}$, (c) $\begin{bmatrix} 1 & 1 \\ 0 & 0 \end{bmatrix}$,

(d) $\begin{bmatrix} 0 & 0 \\ 0 & 0 \end{bmatrix}$, (e) $\begin{bmatrix} 2 & 0 \\ -2 & 0 \end{bmatrix}$, (f) $\begin{bmatrix} 0 & 0 \\ 1 & -1 \end{bmatrix}$.

(30) Determine which, if any, of the matrices given in parts (a) through (f) of the previous problem are in *span* $\{\mathbf{A}_1, \mathbf{A}_2, \mathbf{A}_3\}$.

(31) Determine whether the following polynomials are linear combinations of

$$\{t^3 + t^2, t^3 + t, t^2 + t\}.$$

(a) t^3+t^2+t, (b) $2t^3-t$, (c) $5t$, (d) $2t^2+1$.

(32) Find *span* $\{\mathbf{v}_1, \mathbf{v}_2\}$ for the vectors given in Problem 24.

(33) Find *span* $\{\mathbf{A}_1, \mathbf{A}_2, \mathbf{A}_3\}$ for the matrices given in Problem 29.

(34) Find *span* $\{p_1(t), p_2(t), p_3(t)\}$ for the polynomial given in Problem 31.

(35) Describe the graph of all points in the set $\mathbb{S}$ described in Problem 3.

For problems 36 and 37, let S represent a set of vectors of the form $[x\ y\ z]$ in $\mathbb{R}^3$ that satisfy the given equations. Determine if S forms a subspace in $\mathbb{R}^3$.

(36) $\mathbb{S} = \{[x\ y\ z] | x - 2y + z = 0\}$

(37) $\mathbb{S} = \{[x\ y\ z] | x + y - z = 1\}$

(38) Determine if {[1,1,1], [2,2,0] [3,0,0]} spans $\mathbb{R}^3$.

(39) Determine if {[1,1,2], [1,0,1] [2,1,3]} spans $\mathbb{R}^3$.

(40) Show that $\mathbb{P}^2$ is a subspace of $\mathbb{P}^3$. Generalize to $\mathbb{P}^m$ and $\mathbb{P}^n$ when $m<n$.

(41) Show that if $\mathbf{u}$ is a linear combination of the vectors $\mathbf{v}_1, \mathbf{v}_2, \ldots \mathbf{v}_n$ and if each $\mathbf{v}_i$ is a linear combination of the vectors $(i=1, 2, \ldots, n)$, then $\mathbf{u}$ can also be expressed as a linear combination of $\mathbf{w}_1, \mathbf{w}_2, \ldots \mathbf{w}_m$.

(42) Let $\mathbf{A}$ be an $n \times n$ matrix and both $\mathbf{x}$ and $\mathbf{y}$ $n \times 1$ column matrices. Prove that if $\mathbf{y} - \mathbf{Ax}$, then $\mathbf{y}$ is a linear combination of the columns of $\mathbf{A}$.

(43) Show that the set of solutions of the matrix equation $\mathbf{Ax} = \mathbf{0}$, where $\mathbf{A}$ is a $p \times n$ matrix, is a subspace of $\mathbb{R}^n$.

(44) Show that the set of solutions of the matrix equation $\mathbf{Ax} = \mathbf{b}$, where $\mathbf{A}$ is a $p \times n$ matrix, is not a subspace of $\mathbb{R}^n$ when $\mathbf{b} \neq \mathbf{0}$.

(45) Prove that $span\{\mathbf{u}, \mathbf{v}\} = span\{\mathbf{u}+\mathbf{v}, \mathbf{u}-\mathbf{v}\}$.

(46) Prove that $span\{\mathbf{u}, \mathbf{v}, \mathbf{w}\} = span\{\mathbf{u}+\mathbf{v}, \mathbf{v}+\mathbf{w}, \mathbf{u}+\mathbf{w}\}$.

(47) Prove that $span\{\mathbf{u}, \mathbf{v}, \mathbf{0}\} = span\{\mathbf{u}, \mathbf{v}\}$.

(48) Prove Theorem 5.

2.4 LINEAR INDEPENDENCE

Most vector spaces contain infinitely many vectors. In particular, if $\mathbf{u}$ is a nonzero vector of a vector space $\mathbb{V}$ and if the scalars are real numbers, then it follows from the closure property of scalar multiplication that $\alpha\mathbf{u} \in \mathbb{V}$ for *every* real number α. It is useful, therefore, to determine whether a vector space can be completely characterized by just a few representatives. If so, we can describe a vector space by its representatives. Instead of listing all the vectors in a vector space, which are often infinitely many in number, we simplify the identification of a vector space by listing only its representatives. We then use those representatives to study the entire vector space.

Efficiently characterizing a vector space by its representatives is one of the major goals in linear algebra, where by *efficiently* we mean listing as few representatives as possible. We devote this section and the next to determining properties that such a set of representatives must possess.

A set of vectors $\{\mathbf{v}_1, \mathbf{v}_2, \ldots, \mathbf{v}_n\}$ in a vector space $\mathbb{V}$ is *linearly dependent* if there exist scalars, $c_1, c_2, \ldots, c_n$, not all zero, such that

$$c_1\mathbf{v}_1 + c_2\mathbf{v}_2 + \cdots + c_n\mathbf{v}_n = 0 \tag{2.18}$$

The vectors are *linearly independent* if the only set of scalars that satisfies Equation (2.18) is the set $c_1 = c_2 = \ldots = c_n = 0$.

The set of vectors $\{\mathbf{v}_1, \mathbf{v}_2, \ldots, \mathbf{v}_n\}$ is linearly independent if the only set of scalars that satisfy $c_1\mathbf{v}_1 + c_2\mathbf{v}_2 + \cdots + c_n\mathbf{v}_n = 0$ is $c_1 = c_2 = \ldots = c_n = 0$.

To test whether a given set of vectors is linearly independent, we first form vector Equation (2.18) and ask, "What values for the c's satisfy this equation?" Clearly, $c_1 = c_2 = \ldots = c_n = 0$ is a suitable set. If this is the only set of values that satisfies Equation (2.18), then the vectors are linearly independent. If there exists a set of values that is not all zero, then the vectors are linearly dependent.

It is not necessary for all the c's to be different from zero for a set of vectors to be linearly dependent. Consider the vectors $\mathbf{v}_1 = [1\,2]$, $\mathbf{v}_2 = [1\,4]$, and $\mathbf{v}_3 = [2\,4]$. The constants $c_1 = 2$, $c_2 = 0$, and $c_3 = -1$ is a set of scalars, *not all zero*, such that $c_1\mathbf{v}_1 + c_2\mathbf{v}_2 + c_3\mathbf{v}_3 = \mathbf{0}$. Thus, this set is linearly dependent.

Example 1 Is the set $\{[1\,2], [3\,4]\}$ in $\mathbb{R}^2$ linearly independent?

Solution: Here $\mathbf{v}_1=[1\ 2]$, $\mathbf{v}_2=[3\ 4]$, and Equation (2.18) becomes

$$c_1[1\quad 2]+c_2[3\quad 4]=[0\quad 0]$$

This vector equation can be rewritten as

$$[c_1\quad 2c_1]+[3c_2\quad 4c_2]=[0\quad 0]$$

or as

$$[c_1+3c_2\quad 2c_1+4c_2]=[0\quad 0]$$

Equating components, we generate the system

$$\begin{aligned} c_1+3c_2&=0\\ 2c_1+4c_2&=0 \end{aligned}$$

which has as its only $c_1=c_2=0$. Consequently, the original set of vectors is linearly independent.

Example 2 Determine whether the set of column matrices in $\mathbb{R}^3$

$$\left\{\begin{bmatrix}2\\6\\-2\end{bmatrix},\begin{bmatrix}3\\1\\2\end{bmatrix},\begin{bmatrix}8\\16\\-3\end{bmatrix}\right\}$$

is linearly independent.

Solution: Equation (2.18) becomes

$$c_1\begin{bmatrix}2\\6\\-2\end{bmatrix}+c_2\begin{bmatrix}3\\1\\2\end{bmatrix}+c_3\begin{bmatrix}8\\16\\-3\end{bmatrix}=\begin{bmatrix}0\\0\\0\end{bmatrix} \tag{2.19}$$

which can be rewritten as

$$\begin{bmatrix}2c_1\\6c_1\\-2c_1\end{bmatrix}+\begin{bmatrix}3c_2\\c_2\\2c_2\end{bmatrix}+\begin{bmatrix}8c_3\\16c_3\\-3c_3\end{bmatrix}=\begin{bmatrix}0\\0\\0\end{bmatrix}$$

or

$$\begin{bmatrix}2c_1+3c_2+8c_3\\6c_1+c_2+16c_3\\-2c_1+2c_2-3c_3\end{bmatrix}=\begin{bmatrix}0\\0\\0\end{bmatrix}$$

This matrix equation is equivalent to the homogeneous system of equations

$$\begin{aligned} 2c_1+3c_2+8c_3&=0\\ 6c_1+c_2+16c_3&=0\\ -2c_1+2c_2-3c_3&=0 \end{aligned}$$

Using Gaussian elimination, we find the solution to this system is $c_1 = -2.5, c_3$, $c_2 = -c_3$, c_3 arbitrary. Setting $c_3 = 2$, we obtain $c_1 = -5$, $c_2 = -2$, $c_3 = 2$ as a particular nonzero set of constants that satisfies Equation (2.18). The original set of vectors is linearly dependent.

Example 3 Determine whether the set of matrices

$$\left\{ \begin{bmatrix} 1 & 1 \\ 0 & 0 \end{bmatrix}, \begin{bmatrix} 0 & 1 \\ 0 & 1 \end{bmatrix}, \begin{bmatrix} 0 & 0 \\ 1 & 1 \end{bmatrix}, \begin{bmatrix} 1 & 0 \\ 1 & 1 \end{bmatrix}, \begin{bmatrix} 1 & 1 \\ 0 & 1 \end{bmatrix} \right\}$$

in $\mathbb{M}_{2\times 2}$ is linearly independent.

Solution: Equation (2.17) becomes

$$c_1 \begin{bmatrix} 1 & 1 \\ 0 & 0 \end{bmatrix} + c_2 \begin{bmatrix} 0 & 1 \\ 0 & 1 \end{bmatrix} + c_3 \begin{bmatrix} 0 & 0 \\ 1 & 1 \end{bmatrix} + c_4 \begin{bmatrix} 1 & 0 \\ 1 & 1 \end{bmatrix} + c_5 \begin{bmatrix} 1 & 1 \\ 0 & 1 \end{bmatrix} = \begin{bmatrix} 0 & 0 \\ 0 & 0 \end{bmatrix}$$

or

$$\begin{bmatrix} c_1 + c_4 + c_5 & c_1 + c_2 + c_5 \\ c_3 + c_4 & c_2 + c_3 + c_4 + c_5 \end{bmatrix} = \begin{bmatrix} 0 & 0 \\ 0 & 0 \end{bmatrix}$$

which is equivalent to the homogeneous system of equations

$$\begin{aligned} c_1 + c_4 + c_5 &= 0 \\ c_1 c_2 + c_5 &= 0 \\ c_3 + c_4 &= 0 \\ c_2 + c_3 + c_4 + c_5 &= 0 \end{aligned}$$

This system has more unknowns than equations, so it follows from Theorem 3 of Section 1.4 that there are infinitely many solutions, all but one of which are nontrivial. Because nontrivial solutions exist to Equation (2.18), the set of vectors is linearly dependent.

Example 4 Determine whether the set $\{t^2 + 2t - 3,\ t^2 + 5t,\ 2t^2 - 4\}$ of vectors in $\mathbb{P}^2$ is linearly independent.

Solution: Equation (2.18) becomes

$$c_1\left(t^2 + 2t - 3\right) + c_2\left(t^2 + 5t\right) + c_3\left(2t^2 - 4\right) = 0$$

or

$$(c_1 + c_2 + 2c_3)t^3 + (2c_1 + 5c_2)t + (-3c_1 - 4c_3) = 0t^2 + 0t + 0$$

Equating coefficients of like powers of t, we generate the system of equations

$$\begin{aligned} c_1 + c_2 + 2c_3 &= 0 \\ 2c_1 + 5c_2 &= 0 \\ -3c_1 - 4c_3 &= 0 \end{aligned}$$

Using Gaussian elimination, we find that this system admits only the trivial solution $c_1 = c_2 = c_3 = 0$. The given set of vectors is linearly independent.

The defining equations for linear combinations and linear dependence, Equations (2.16) and (2.18), are similar, so we should not be surprised to find that the concepts are related.

►THEOREM 1

A finite set of vectors is linearly dependent if and only if one of the vectors is a linear combination of the vectors that precede it, in the ordering established by the listing of vectors in the set. ◄

Proof: First, we must prove that if a set of vectors is linearly dependent, then one of the vectors is a linear combination of other vectors that are listed before it in the set. Second, we must show the converse: if one of the vectors of a given set is a linear combination of the vectors that precede it, then the set is linearly dependent.

Let $\{\mathbf{v}_1, \mathbf{v}_2, \ldots, \mathbf{v}_n\}$ be a linearly dependent set. Then there exists scalars $c_1, c_2, \ldots, c_n$, not all zero, such that Equation (2.18) is satisfied. Let c_i be the last nonzero scalar. At the very worst $i = n$ when $c_n \neq 0$, but if $c_n = 0$, then $i < n$. Equation (2.18) becomes

$$c_1\mathbf{v}_1 + c_2\mathbf{v}_2 + \cdots + c_{i-1}\mathbf{v}_{i-1} + c_i\mathbf{v}_i + 0\mathbf{v}_{i+1} + 0\mathbf{v}_{i+2} + \cdots + 0\mathbf{v}_n = \mathbf{0}$$

which can be rewritten as

$$\mathbf{v}_i = -\frac{c_1}{c_i}\mathbf{v}_1 - \frac{c_2}{c_i}\mathbf{v}_2 - \cdots - \frac{c_{i-1}}{c_i}\mathbf{v}_{i-1} \tag{2.20}$$

Consequently, $\mathbf{v}_i$, is a linear combination of $\mathbf{v}_1, \mathbf{v}_2, \ldots, \mathbf{v}_{i-1}$, with coefficients $d_1 = -c_1/c_i$, $d_2 = -c_2/c_i$, $\ldots$, $d_{i-1} = -c_{i-1}/c_i$.

Now let one vector of the set $\{\mathbf{v}_1, \mathbf{v}_2, \ldots, \mathbf{v}_n\}$, say $\mathbf{v}_i$, be a linear combination of the vectors in the set that precede it, namely, $\mathbf{v}_1, \mathbf{v}_2, \ldots, \mathbf{v}_{i-1}$. Then there exist scalars $d_1, d_2, \ldots, d_{i-1}$ such that

$$\mathbf{v}_i = d_1\mathbf{v}_1 + d_2\mathbf{v}_2 + \cdots + d_{i-1}\mathbf{v}_{i-1}$$

which can be rewritten as

$$d_1\mathbf{v}_1 + d_2\mathbf{v}_2 + \cdots + d_{i-1}\mathbf{v}_{i-1} + (-1)\mathbf{v}_i + 0\mathbf{v}_{i+1} + 0\mathbf{v}_{i+2} + \cdots + 0\mathbf{v}_n = \mathbf{0}$$

This is Equation (2.18) with $c_j = d_j$ $(j = 1, 2, \ldots, i-1)$, $c_i = -1$, and $c_j = 0$ $(j = i+1, i+2, \ldots, n)$. Because this is a set of scalars not all zero, in particular $c_i = -1$, it follows that the original set of vectors is linearly dependent.

It is not necessary for every vector in a given set to be a linear combination of preceding vectors if that set is linearly dependent, but only that at least *one* vector in the set have this property. For example, the set $\{[1\,0], [2\,0], [0\,1]\}$ is linearly dependent because

$$-2[1 \quad 0] + 1[2 \quad 0] + 0[0 \quad 1] = [0 \quad 0]$$

Here $[0\,1]$ cannot be written as a linear combination of the preceding two vectors; however, $[2\,0]$ can be written as a linear combination of the vector that precedes it, namely, $[2\ 0] = 2[1\,0]$.

►THEOREM 2

A subset of a vector space $\mathbb{V}$ consisting of the single vector $\mathbf{u}$ is linearly dependent if and only if $\mathbf{u} = 0$.◄

Proof: If the set $\{\mathbf{u}\}$ is linearly dependent, then there exists a nonzero scalar c that satisfies the vector equation

$$c\mathbf{u} = 0 \tag{2.21}$$

It then follows from Theorem 7 of Section 2.2 that $\mathbf{u} = 0$. Conversely, if $\mathbf{u} = 0$, then it follows from Theorem 1 of Section 2.2 that Equation (2.21) is valid for any scalar c. Thus nonzero scalars exist that satisfy Equation (2.21) and the set $\{\mathbf{u}\}$ is linearly dependent.

►THEOREM 3

A subset of a vector space $\mathbb{V}$ consisting of two distinct vectors is linearly dependent if and only if one vector is a scalar multiple of the other.◄

Proof: If the set $\{\mathbf{v}_1, \mathbf{v}_2\}$ is linearly dependent, then it follows from Theorem 1 that $\mathbf{v}_2$ can be written as a linear combination of $\mathbf{v}_1$. That is, $\mathbf{v}_2 = d_1\mathbf{v}_1$, which means that $\mathbf{v}_2$ is a scalar multiple of $\mathbf{v}_1$.

Conversely, if one of the two vectors can be written as a scalar multiple of the other, then either $\mathbf{v}_2 = \alpha\mathbf{v}_1$ or $\mathbf{v}_1 = \alpha\mathbf{v}_2$ for some scalar α. This implies, respectively, that either

$$\alpha\mathbf{v}_1 + (-1)\mathbf{v}_2 = 0 \quad \text{or} \quad (1)\mathbf{v}_1 - \alpha\mathbf{v}_2 = 0$$

Both equations are in the form of Equation (2.18), the first with $c_1 = \alpha$, $c_2 = -1$ and the second with $c_1 = 1$, $c_2 = -\alpha$. Either way, we have a set of scalars, not all zero, that satisfy Equation (2.18), whereupon the set $\{\mathbf{v}_1, \mathbf{v}_2\}$ is linearly dependent.

Theorem 3 has an interesting geometrical representation in both $\mathbb{R}^2$ and $\mathbb{R}^3$. We know from our work in Section 1.7 that a scalar multiple of a nonzero vector in $\mathbb{R}^2$ or $\mathbb{R}^3$ is an elongation of the nonzero vector (when the scalar in absolute value is greater than unity) or a contraction of that nonzero vector (when the

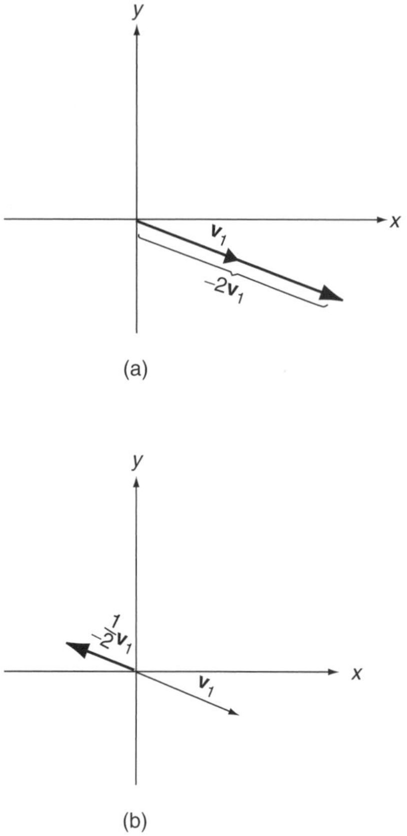

FIGURE 2.15

scalar in absolute value is less than unity), followed by a rotation of 180° if the scalar is negative. Figure 2.15 illustrates two possibilities in $\mathbb{R}^2$ for a particular nonzero vector $\mathbf{v}_2$. If $\mathbf{v}_2 = 2\mathbf{v}_1$, we have the situation depicted in Figure 2.15a; if, however, $\mathbf{v}_2 = -1/2\mathbf{v}_1$, we have the situation depicted in Figure 2.15b. Either way, both vectors lie on the same straight line. The same situation prevails in $\mathbb{R}^3$. We conclude that two vectors are linearly dependent in either $\mathbb{R}^2$ or $\mathbb{R}^3$ if and only if both vectors lie on the same line. Alternatively, two vectors are linearly independent in either $\mathbb{R}^2$ or $\mathbb{R}^3$ if and only if they do *not* lie on the same line.

Two vectors are linearly dependent in $\mathbb{R}^2$, or $\mathbb{R}^3$ if and only if they lie on the same line.

A set of three vectors in $\mathbb{R}^3$, $\{\mathbf{v}_1, \mathbf{v}_2, \mathbf{v}_3\}$, is linearly dependent if any two of the vectors lie on the same straight line (see Problem 31). If no two vectors lie on the same straight line but the set is linearly dependent, then it follows from Theorem 1 that $\mathbf{v}_3$ must be a linear combination of $\mathbf{v}_1$ and $\mathbf{v}_2$ (see Problem 32). In such a case, there exist scalars d_1 and d_2 such that $\mathbf{v}_3 = d_1\mathbf{v}_1 + d_2\mathbf{v}_2$. This situation is illustrated graphically in Figure 2.16 for the particular case where both vectors $\mathbf{v}_1$ and $\mathbf{v}_2$ are in the x-y plane, d_1 is a positive real number that is less than unity, and d_2 is a positive real number that is slightly greater than unity. It follows from our work in Section 1.7 that $\mathbf{v}_3 = d_1\mathbf{v}_1 + d_2\mathbf{v}_2$ is another vector in the x-y plane. The situation

A set of three vectors in $\mathbb{R}^3$ is linearly dependent if and only if all three vectors lie on the same line or all lie in the same plane.

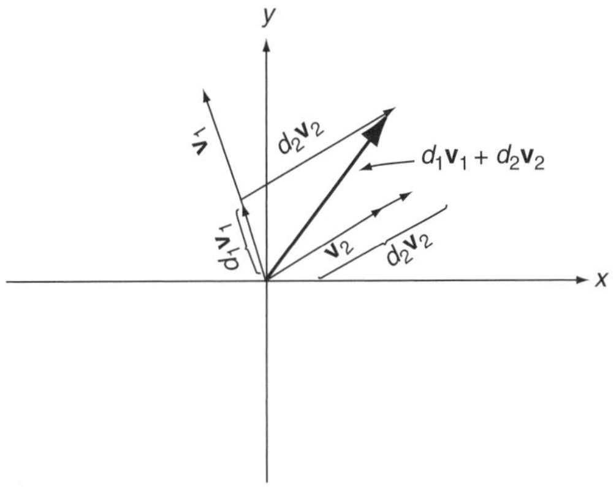

FIGURE 2.16

is analogous for any two vectors in $\mathbb{R}^3$ that do not lie on the same line: any linear combination of the two vectors will lie in the plane formed by those two vectors. We see, therefore, that if a set of three vectors in $\mathbb{R}^3$ is linearly dependent, then either all three vectors lie on the same line or all three lie in the same plane.

►THEOREM 4

A set of vectors in a vector space $\mathbb{V}$ that contains the zero vector is linearly dependent.◄

Proof: Consider the set $\{\mathbf{v}_1, \mathbf{v}_2, \ldots, \mathbf{v}_n, \mathbf{0}\}$. Pick $c_1 = c_2 = \ldots = c_n = 0$ and $c_{n+1} = 5$ (any other nonzero number will do equally well). This is a set of scalars, not all zero, such that

$$c_1\mathbf{v}_1 + c_2\mathbf{v}_2 + \cdots + c_n\mathbf{v}_n + c_{n+1}\mathbf{0} = \mathbf{0}$$

Hence, the set of vectors is linearly dependent.

►THEOREM 5

If a set of vectors $\mathbb{S}$ in a vector space $\mathbb{V}$ is linearly independent, then any subset of $\mathbb{S}$ is also linearly independent.◄

Proof: See Problem 42.

►THEOREM 6

If a set of vectors $\mathbb{S}$ in a vector space $\mathbb{V}$ is linearly dependent, then any larger set containing $\mathbb{S}$ is also linearly dependent.◄

Proof: See Problem 43.

Problems 2.4

In Problems 1 through 30, determine whether each set is linearly independent.

(1) $\{[1 \quad 0], [0 \quad 1]\}$.

(2) $\{[1 \quad 1], [1 \quad -1]\}$.

(3) $\{[2 \quad -4], [-3 \quad 6]\}$.

(4) $\{[1 \quad 3], [2 \quad -1], [1 \quad 1]\}$.

(5) $\left\{\begin{bmatrix} 1 \\ 2 \end{bmatrix}, \begin{bmatrix} 3 \\ 4 \end{bmatrix}\right\}$.

(6) $\left\{\begin{bmatrix} 1 \\ -1 \end{bmatrix}, \begin{bmatrix} 1 \\ 1 \end{bmatrix}, \begin{bmatrix} 1 \\ 2 \end{bmatrix}\right\}$.

(7) $\left\{\begin{bmatrix} 1 \\ 0 \\ 1 \end{bmatrix}, \begin{bmatrix} 1 \\ 1 \\ 0 \end{bmatrix}, \begin{bmatrix} 0 \\ 1 \\ 1 \end{bmatrix}\right\}$.

(8) $\left\{\begin{bmatrix} 1 \\ 0 \\ 1 \end{bmatrix}, \begin{bmatrix} 1 \\ 0 \\ 2 \end{bmatrix}, \begin{bmatrix} 2 \\ 0 \\ 1 \end{bmatrix}\right\}$.

(9) $\left\{\begin{bmatrix} 1 \\ 0 \\ 1 \end{bmatrix}, \begin{bmatrix} 1 \\ 1 \\ 1 \end{bmatrix}, \begin{bmatrix} 0 \\ -1 \\ 1 \end{bmatrix}\right\}$.

(10) $\left\{\begin{bmatrix} 0 \\ 0 \\ 0 \end{bmatrix}, \begin{bmatrix} 3 \\ 2 \\ 1 \end{bmatrix}, \begin{bmatrix} 2 \\ 1 \\ 3 \end{bmatrix}\right\}$.

(11) $\left\{\begin{bmatrix} 1 \\ 2 \\ 3 \end{bmatrix}, \begin{bmatrix} 3 \\ 2 \\ 1 \end{bmatrix}, \begin{bmatrix} 2 \\ 1 \\ 3 \end{bmatrix}\right\}$.

(12) $\left\{\begin{bmatrix} 1 \\ 2 \\ 3 \end{bmatrix}, \begin{bmatrix} 3 \\ 2 \\ 1 \end{bmatrix}, \begin{bmatrix} 2 \\ 1 \\ 3 \end{bmatrix}, \begin{bmatrix} -1 \\ 2 \\ -3 \end{bmatrix}\right\}$.

(13) $\left\{\begin{bmatrix} 4 \\ 5 \\ 1 \end{bmatrix}, \begin{bmatrix} 3 \\ 0 \\ 2 \end{bmatrix}, \begin{bmatrix} 1 \\ 1 \\ 1 \end{bmatrix}\right\}$.

(14) $\{[1 \quad 1 \quad 0], [1 \quad -1 \quad 0]\}$.

(15) $\{[1 \quad 2 \quad 3], [-3 \quad -6 \quad -9]\}$.

(16) $\{[10 \quad 20 \quad 20], [10 \quad -10 \quad 10], [10 \quad 20 \quad 10]\}$.

(17) $\{[10 \quad 20 \quad 20], [10 \quad -10 \quad 10], [10 \quad 20 \quad 10], [20 \quad 10 \quad 20]\}$.

(18) $\{[2\ \ 1\ \ 1], [3\ \ -1\ \ 4], [1\ \ 3\ \ -2]\}$.

(19) $\left\{\begin{bmatrix}2\\1\\1\\3\end{bmatrix}, \begin{bmatrix}4\\-1\\2\\-1\end{bmatrix}, \begin{bmatrix}8\\1\\4\\5\end{bmatrix}\right\}$.

(20) $\left\{\begin{bmatrix}1 & 0\\0 & 0\end{bmatrix}, \begin{bmatrix}0 & 1\\0 & 0\end{bmatrix}, \begin{bmatrix}0 & 0\\1 & 0\end{bmatrix}, \begin{bmatrix}0 & 0\\0 & 1\end{bmatrix}\right\}$.

(21) $\left\{\begin{bmatrix}1 & 1\\0 & 0\end{bmatrix}, \begin{bmatrix}1 & 1\\1 & 1\end{bmatrix}, \begin{bmatrix}0 & 0\\1 & 1\end{bmatrix}\right\}$.

(22) $\left\{\begin{bmatrix}1 & 1\\0 & 0\end{bmatrix}, \begin{bmatrix}1 & 0\\1 & 1\end{bmatrix}, \begin{bmatrix}0 & 0\\1 & 1\end{bmatrix}\right\}$.

(23) $\left\{\begin{bmatrix}1 & 0\\1 & 1\end{bmatrix}, \begin{bmatrix}1 & 1\\1 & 0\end{bmatrix}, \begin{bmatrix}1 & 1\\0 & 1\end{bmatrix}, \begin{bmatrix}0 & 1\\1 & 1\end{bmatrix}\right\}$.

(24) $\left\{\begin{bmatrix}1 & 0\\1 & 1\end{bmatrix}, \begin{bmatrix}1 & 1\\1 & 0\end{bmatrix}, \begin{bmatrix}2 & 2\\0 & 2\end{bmatrix}, \begin{bmatrix}1 & 0\\2 & 0\end{bmatrix}\right\}$.

(25) $\{t, 2\}$.

(26) $\{t^3+t^2, t^3+t, t^2+t\}$.

(27) $\{t^3+t^2, t^2-t^2, t^3-3t^2\}$.

(28) $\{t^3+t^2, t^3-t^2, t^3-t, t^3+1\}$.

(29) $\{t^2+t, t^2+t-1, t^2+1, t\}$.

(30) $\{t^2+t, t^2+t-2, 1\}$.

(31) Consider a set of three vectors in $\mathbb{R}^3$. Prove that if two of the vectors lie on the same straight line, then the set must be linearly dependent.

(32) Consider a linearly dependent set of three vectors $\{\mathbf{v}_1, \mathbf{v}_2, \mathbf{v}_3\}$ in $\mathbb{R}^3$. Prove that if no two vectors lie on the same straight line, $\mathbf{v}_3$ must be a linear combination of $\mathbf{v}_1$ and $\mathbf{v}_2$.

(33) Prove that a set of vectors is linearly dependent if and only if one of the vectors is a linear combination of the vectors that follow it.

(34) Prove that if $\{\mathbf{u}, \mathbf{v}\}$ is linearly independent, then so too is $\{\mathbf{u}+\mathbf{v}, \mathbf{u}-\mathbf{v}\}$.

(35) Prove that if $\{\mathbf{v}_1, \mathbf{v}_2, \mathbf{v}_3\}$ is linearly independent, then so too is the set $\{\mathbf{u}_1, \mathbf{u}_2, \mathbf{u}_3\}$ where $\mathbf{u}_1 = \mathbf{v}_1 + \mathbf{v}_2 + \mathbf{v}_3$, $\mathbf{u}_2 = \mathbf{v}_2 + \mathbf{v}_3$, and $\mathbf{u}_3 = \mathbf{v}_3$.

(36) Prove that if $\{\mathbf{v}_1, \mathbf{v}_2, \mathbf{v}_3\}$ is linearly independent, then so too is the set $\{\mathbf{u}_1, \mathbf{u}_2, \mathbf{u}_3\}$ where $\mathbf{u}_1 = \mathbf{v}_1 + \mathbf{v}_2 + \mathbf{v}_3$, $\mathbf{u}_2 = \mathbf{v}_2 + \mathbf{v}_3$, and $\mathbf{u}_3 = \mathbf{v}_3$.

(37) Prove that if $\{\mathbf{v}_1, \mathbf{v}_2, \mathbf{v}_3\}$ is linearly independent, then so too is the set $\{\alpha_1\mathbf{v}_1, \alpha_2\mathbf{v}_2, \alpha_3\mathbf{v}_3\}$ for any choice of the nonzero scalars α_1, α_2, and α_3.

(38) Prove that the nonzero rows, considered as row matrices, of a row-reduced matrix is a linearly independent set.

(39) Let **A** be an $n \times n$ matrix and let $\{\mathbf{x}_1, \mathbf{x}_2, \ldots \mathbf{v}_k\}$ and $\{\mathbf{y}_1, \mathbf{y}_2, \ldots \mathbf{y}_k\}$ be two sets of n-dimensional column vectors having the property that $\mathbf{A}\mathbf{x}_i = \mathbf{y}_i (i = 1, 2, \ldots, k)$. Show that the set $\{\mathbf{x}_1, \mathbf{x}_2, \ldots \mathbf{v}_k\}$ is linearly independent if the set $\{\mathbf{y}_1, \mathbf{y}_2, \ldots \mathbf{y}_k\}$ is.

(40) What can be said about a set of vectors that contains as a proper subset a set of linearly independent vectors?

(41) What can be said about a subset of a linearly dependent set of vectors?

(42) Prove Theorem 5.

(43) Prove Theorem 6.

(44) An extension of Theorem 2 in Section 1.5 to $\mathbb{R}^3$ states that the volume of parallelpiped generated by three column matrices $\mathbf{u}_1$, $\mathbf{u}_2$, and $\mathbf{u}_3$ in $\mathbb{R}^3$ is $|\det[\mathbf{u}_1\, \mathbf{u}_2\, \mathbf{u}_3]|$. Find the volumes of the parallelepipeds defined by the vectors:

(a) $[1\,2\,1]^T$, $[2\,-1\,0]^T$, $[2\,1\,1]^T$.

(b) $[1\,2\,3]^T$, $[3\,2\,1]^T$, $[1\,1\,1]^T$.

(c) $[1\,0\,1]^T$, $[2\,1\,1]^T$, $[4\,3\,1]^T$.

(45) Use Problem 44 to show that the determinant of a 3×3 matrix with linearly dependent columns must be 0.

(46) What can be said about the determinant of an upper triangular matrix? A lower triangular matrix?

(47) What can be said about the determinant of a matrix containing a zero row? A zero column?

2.5 BASIS AND DIMENSION

We began the previous section with a quest for completely characterizing vector spaces by just a few of its representatives and determining the properties representatives must have if the characterization is to be an efficient one. One property we want is the ability to recreate *every* vector in a given vector space from its representatives; that is, we want the ability to combine representatives to generate all other vectors in a vector space. The only means we have for combining vectors is vector addition and scalar multiplication, so the only combinations available to us are linear combinations (see Section 2.3). We define a set of vectors $\mathbb{S}$ in a vector space $\mathbb{V}$ as a *spanning set* for $\mathbb{V}$ if every vector in $\mathbb{V}$ can be written as a linear combination of the vectors in $\mathbb{S}$; that is, if $\mathbb{V} = span\{\mathbb{S}\}$.

The set of vectors $\mathbb{S}$ is a spanning set for a vector space $\mathbb{V}$ if every vector in $\mathbb{V}$ can be written as a linear combination of vectors in $\mathbb{S}$.

Example 1 Determine whether any of the following sets are spanning sets for $\mathbb{R}^2$, considered as column matrices:

(a) $\mathbb{S}_1 = \left\{ e_1 = \begin{bmatrix} 1 \\ 0 \end{bmatrix}, \quad e_2 = \begin{bmatrix} 0 \\ 1 \end{bmatrix} \right\}$

(b) $\mathbb{S}_2 = \left\{ e_1 = \begin{bmatrix} 1 \\ 0 \end{bmatrix}, e_2 = \begin{bmatrix} 0 \\ 1 \end{bmatrix}, f_1 = \begin{bmatrix} 1 \\ 1 \end{bmatrix} \right\}$

(c) $\mathbb{S}_3 = \left\{ f_1 = \begin{bmatrix} 1 \\ 1 \end{bmatrix}, f_2 = \begin{bmatrix} 2 \\ 2 \end{bmatrix} \right\}$

Solution: An arbitrary column matrix $\mathbf{u} \in \mathbb{R}^2$ has the form

$$\mathrm{u} = \begin{bmatrix} a \\ b \end{bmatrix}$$

for some choice of the scalars a and b.

(a) Since

$$\begin{bmatrix} a \\ b \end{bmatrix} = a\begin{bmatrix} 1 \\ 0 \end{bmatrix} + b\begin{bmatrix} 0 \\ 1 \end{bmatrix}$$

it follows that every vector in $\mathbb{R}^2$ is a linear combination of $\mathbf{e}_1$ and $\mathbf{e}_2$. Thus, $\mathbb{S}_1$ is a spanning set for $\mathbb{R}^2$.

(b) Since

$$\begin{bmatrix} a \\ b \end{bmatrix} = a\begin{bmatrix} 1 \\ 0 \end{bmatrix} + b\begin{bmatrix} 0 \\ 1 \end{bmatrix} + 0\begin{bmatrix} 1 \\ 1 \end{bmatrix}$$

it follows that $\mathbb{S}_2$ is also a spanning set for $\mathbb{R}^2$.

(c) $\mathbb{S}_3$ is not a spanning set for $\mathbb{R}^2$. Every linear combination of vectors in $\mathbb{S}_3$ has identical first and second components. The vector $[1\ 2]^T$ does not have identical components and, therefore, cannot be written as a linear combination of $\mathbf{f}_1$ and $\mathbf{f}_2$.

If $\mathbb{S}$ is a spanning set for a vector space $\mathbb{V}$, then $\mathbb{S}$ is said to *span* $\mathbb{V}$. As a spanning set, $\mathbb{S}$ represents $\mathbb{V}$ completely because every vector in $\mathbb{V}$ can be gotten from the vectors in $\mathbb{S}$. If we also require that $\mathbb{S}$ be a linearly independent set, then we are guaranteed that no vector in $\mathbb{S}$ can be written as a linear combination of other vectors in $\mathbb{S}$ (Theorem 1 of Section 2.4). Linear independence ensures that the set $\mathbb{S}$ does not contain any superfluous vectors. A spanning set of vectors, that is, also a linearly independent set meets all our criteria for efficiently representing a given vector space. We call such a set a *basis.*

A basis for a vector space $\mathbb{V}$ is a set of vectors that is linearly independent and also spans $\mathbb{V}$.

Example 2 Determine whether the set $\mathbb{C} = \{t^2 + 2t - 3, t^2 + 5t, 2t^2 - 4\}$ is a basis for $\mathbb{P}^3$.

Solution: $\mathbb{C}$ is not a spanning set for $\mathbb{P}^3$, because t^3 is a third-degree polynomial in $\mathbb{P}^3$ and no linear combination of the vectors in $\mathbb{C}$ can equal it. Because $\mathbb{C}$ does not span $\mathbb{P}^3$, $\mathbb{C}$ *cannot* be a basis. We could show that $\mathbb{C}$ is linearly independent (see Example 4 of Section 2.4), but that is now irrelevant.

Example 3 Determine whether the set

$$\mathbb{D} = \left\{ \begin{bmatrix} 1 & 1 \\ 0 & 0 \end{bmatrix}, \begin{bmatrix} 0 & 1 \\ 0 & 1 \end{bmatrix}, \begin{bmatrix} 0 & 0 \\ 1 & 1 \end{bmatrix}, \begin{bmatrix} 1 & 0 \\ 1 & 1 \end{bmatrix}, \begin{bmatrix} 1 & 1 \\ 0 & 1 \end{bmatrix} \right\}$$

is a basis for $\mathbb{M}_{2\times 2}$.

Solution: It follows from Example 3 of Section 2.4 that $\mathbb{D}$ is linearly dependent, not independent, so $\mathbb{D}$ *cannot* be a basis. We could show that $\mathbb{D}$ does indeed span $\mathbb{M}_{2\times 2}$, but that no longer matters.

Example 4 Determine whether the set $\mathbb{S} = \left\{ \mathbf{e}_1 = \begin{bmatrix} 1 \\ 0 \end{bmatrix}, \mathbf{e}_2 = \begin{bmatrix} 0 \\ 1 \end{bmatrix} \right\}$ is a basis for $\mathbb{R}^2$, considered as column matrices.

Solution: We need to show that span $(\mathbb{S}) = \mathbb{R}^2$ and also that $\mathbb{S}$ is linearly independent. We showed in part (a) of Example 1 that $\mathbb{S}$ is a spanning set for $\mathbb{R}^2$. To demonstrate linear independence, we form the vector equation

$$c_1 \begin{bmatrix} 1 \\ 0 \end{bmatrix} + c_2 \begin{bmatrix} 0 \\ 1 \end{bmatrix} = \begin{bmatrix} 0 \\ 0 \end{bmatrix}$$

or

$$\begin{bmatrix} c_1 \\ c_2 \end{bmatrix} = \begin{bmatrix} 0 \\ 0 \end{bmatrix}$$

The only solution to this vector equation is $c_1 = c_2 = 0$, so the two vectors are linearly independent. It follows that $\mathbb{S}$ is a basis for $\mathbb{R}^2$.

A straightforward extension of Example 4 shows that a basis for $\mathbb{R}^n$, considered as column vectors, is the set of the n-tuples

$$\left\{ \mathbf{e}_1 = \begin{bmatrix} 1 \\ 0 \\ 0 \\ \vdots \\ 0 \\ 0 \end{bmatrix}, \mathbf{e}_2 = \begin{bmatrix} 0 \\ 1 \\ 0 \\ \vdots \\ 0 \\ 0 \end{bmatrix}, \mathbf{e}_3 = \begin{bmatrix} 0 \\ 0 \\ 1 \\ \vdots \\ 0 \\ 0 \end{bmatrix}, \ldots, \mathbf{e}_{n-1} = \begin{bmatrix} 0 \\ 0 \\ 0 \\ \vdots \\ 1 \\ 0 \end{bmatrix}, \mathbf{e}_n = \begin{bmatrix} 0 \\ 0 \\ 0 \\ \vdots \\ 1 \\ 0 \end{bmatrix} \right\} \tag{2.22}$$

where $\mathbf{e}_j$ $(j = 1, 2, 3, \ldots, n)$ has its jth component equal to unity and all other components equal to zero. This set is known as the *standard basis* for $\mathbb{R}^n$.

Example 5 Determine whether the set $\mathbb{B} = \left\{ f_1 = \begin{bmatrix} 1 \\ 1 \end{bmatrix}, f_2 = \begin{bmatrix} 1 \\ -1 \end{bmatrix} \right\}$ is a basis for $\mathbb{R}^2$, considered as column matrices.

Solution: An arbitrary vector $\mathbf{u}$ in $\mathbb{R}^2$ has the form

$$u = \begin{bmatrix} a \\ b \end{bmatrix}$$

for some choice of the scalars a and b. $\mathbb{B}$ is a spanning set for $\mathbb{R}^2$ if there exist scalars d_1 and d_2 such that

$$d_1\begin{bmatrix}1\\1\end{bmatrix}+d_2\begin{bmatrix}1\\-1\end{bmatrix}=\begin{bmatrix}a\\b\end{bmatrix} \tag{2.23}$$

Note that we do not actually have to find the scalars d_1 and d_2, we only need to show that they exist. System (2.14) is equivalent to the set of simultaneous equations

$$\begin{aligned} d_1 + d_2 &= a \\ d_1 - b_2 &= b \end{aligned}$$

which we solve by Gaussian elimination for the variables d_1 and d_2. The augmented matrix for this system is

$$\left[\begin{array}{cc|c}1 & 1 & a\\1 & -1 & b\end{array}\right] \rightarrow \left[\begin{array}{cc|c}1 & 1 & a\\0 & -2 & b-a\end{array}\right]$$ by adding to the second row -1 times the first row

$$\rightarrow \left[\begin{array}{cc|c}1 & 1 & a\\0 & 1 & \frac{1}{2}a-\frac{1}{2}b\end{array}\right]$$ by multiplying the second row by $-1/2$

The system of equations associated with this row-reduced augmented matrix is

$$\begin{aligned} d_1 + d_2 &= a \\ d_2 &= \frac{1}{2}a - \frac{1}{2}b \end{aligned} \tag{2.24}$$

System (2.24) has a solution for d_1 and d_2 for every choice of the scalars a and b. Therefore, there exist scalars d_1 and d_2 that satisfy Equation (2.23) and $\mathbb{B}$ is a spanning set for $\mathbb{R}^2$.

We next show that $\mathbb{B}$ is linearly independent, which is tantamount to showing that the only solution to the vector equation $d_1\mathbf{f}_1+d_2\mathbf{f}_2=\mathbf{0}$ is the trivial solution $d_1=d_2=0$. This vector equation is precisely Equation (2.23) with $a=b=0$, and it reduces to Equation (2.24) with $a=b=0$. Under these special conditions, the second equation of Equation (2.24) is $d_2=0$, and when it is substituted into the first equation we find $d_1=0$. Thus, $\mathbb{B}$ is also a linearly independent set, and a basis for $\mathbb{R}^2$.

►OBSERVATION

To show that a set of vectors is a basis for a vector space $\mathbb{V}$, first verify that the set spans $\mathbb{V}$. Much of the work can be reused to determine whether the set is also linearly independent.◄

A vector space is finite-dimensional if it has a basis containing a finite number of vectors.

A vector space $\mathbb{V}$ is *finite-dimensional* if it has a basis containing a finite number of vectors. In particular, $\mathbb{R}^2$ is finite-dimensional because, as shown in Example 4, it has a basis with two (a finite number) of the vectors. A vector space that is not finite-dimensional is called *infinite dimensional*, but we shall not consider such vector spaces in this book. It follows from Examples 4 and 5 that a finite-dimensional vector space can have different bases. The fact that different bases of a vector space must contain the same number of vectors is a consequence of the next two theorems.

▶THEOREM 1

If $\mathbb{S} = \{\mathbf{v}_1, \mathbf{v}_2, \ldots, \mathbf{v}_n\}$ is a basis for a vector space $\mathbb{V}$, then any set containing more than n vectors is linearly dependent.◀

Proof: Let $\mathbb{T} = \{\mathbf{u}_1, \mathbf{u}_2, \ldots, \mathbf{u}_p\}$ be a set of p vectors in $\mathbb{V}$ with $p > n$. We need to show that there exist scalars $c_1, c_2, \ldots, c_p$, not all zero, that satisfy the vector equation

$$c_1\mathbf{u}_1 + c_2\mathbf{u}_2 + \cdots + c_p\mathbf{u}_p = 0 \tag{2.25}$$

Because $\mathbb{S}$ is a spanning set for $\mathbb{V}$, it follows that every vector in $\mathbb{V}$, in particular those vectors in $\mathbb{T}$, can be written as a linear combination of the vectors in $\mathbb{S}$. Therefore,

$$\begin{aligned} \mathbf{u}_1 &= a_{11}\mathbf{v}_1 + a_{21}\mathbf{v}_2 + \cdots + a_{n1}\mathbf{v}_n \\ \mathbf{u}_2 &= a_{12}\mathbf{v}_2 + a_{22}\mathbf{v}_2 + \cdots + a_{n2}\mathbf{v}_n \\ &\vdots \\ \mathbf{u}_p &= a_{1p}\mathbf{v}_1 + a_{2p}\mathbf{v}_2 + \cdots + a_{np}\mathbf{v}_n \end{aligned} \tag{2.26}$$

for some values of the scalars a_{ij} $(i = 1, 2, \ldots, n; j = 1, 2, \ldots, p)$. Substituting the equations of system (2.26) into the left side of Equation (2.25) and rearranging, we obtain

$$\begin{aligned} &(c_1a_{11} + c_2a_{12} + \cdots + c_pa_{1p})\mathbf{v}_1 \\ &\quad + (c_1a_{21} + c_2a_{22} + \cdots + c_pa_{2p})\mathbf{v}_2 \\ &\qquad + \cdots + (c_1a_{n1} + c_2a_{n2} + \cdots c_pa_{np})\mathbf{v}_n = 0 \end{aligned}$$

Because $\mathbb{S}$ is a basis, it is a linearly independent set, and the only way the above equation can be satisfied is for each coefficient of $\mathbf{v}j$ $(j = 1, 2, \ldots, n)$ to be zero. Thus,

$$\begin{aligned} a_{11}c_1 + a_{12}c_2 + \cdots + a_{1p}c_p &= 0 \\ a_{21}c_1 + a_{22}c_2 + \cdots + a_{2p}c_p &= 0 \\ &\vdots \\ a_{n1}c_1 + a_{n2}c_2 + \cdots + a_{np}c_p &= 0 \end{aligned}$$

But this is a set of n-equations in p-unknowns, $c_1, c_2, \ldots, c_p$, with $p > n$, so it follows from Theorem 3 of Section 1.4 that this set has infinitely many solutions. Most of these solutions will be nontrivial, so there exist scalars, not all zero, that satisfy Equation (2.16).

As an immediate consequence of Theorem 1, we have

▶COROLLARY 1

If $\mathbb{S} = \{\mathbf{v}_1, \mathbf{v}_2, \ldots, \mathbf{v}_n\}$ is a basis for a vector space $\mathbb{V}$, then every linearly independent set of vectors in $\mathbb{V}$ must contain n or fewer vectors. ◀

We are now in the position to state and prove one of the fundamental principles of linear algebra.

▶THEOREM 2

Every basis for a finite-dimensional vector space must contain the same number of vectors. ◀

Proof: Let $\mathbb{S} = \{\mathbf{v}_1, \mathbf{v}_2, \ldots, \mathbf{v}_n\}$ and $\mathbb{T} = \{\mathbf{u}_1, \mathbf{u}_2, \ldots, \mathbf{u}_n\}$ be two bases for a finite-dimensional vector space $\mathbb{V}$. Because $\mathbb{S}$ is a basis and $\mathbb{T}$ is a linearly independent set, it follows from Corollary 1 that $p < n$. Reversing roles, $\mathbb{T}$ is a basis and $\mathbb{S}$ is a linearly independent set, so it follows from Corollary 1 that $n < p$. Together, both inequalities imply that $p = n$.

Because the number of vectors in a basis for a finite-dimensional vector space $\mathbb{V}$ is always the same, we can give that number a name. We call it the *dimension* of the $\mathbb{V}$ and denote it as $dim(\mathbb{V})$.

The dimension of a vector space is the number of vectors in a basis for that vector space.

The vector space containing just the zero vector is an anomaly. The only *nonempty* subset of this vector space is the vector space itself. But the subset $\{0\}$ is linearly dependent, as a consequence of Theorem 2 of Section 2.4 and, therefore, cannot be a basis. We define the dimension of the vector space containing just the zero vector to be zero, which is equivalent to saying that the empty set is the basis for this vector space.

Example 6 Determine the dimension of $\mathbb{P}^n$.

Solution: A basis for this vector space is $\mathbb{S} = \{t^n, t^{n-1}, \ldots, t, 1\}$. First, $\mathbb{S}$ is a spanning set, because if $p(t)$ is a vector in $\mathbb{P}^n$, then

$$p(t) = a_n t^n + a_{n-1} t^{n-1} + \cdots + a_1 t + a_0(1)$$

for some choice of the scalars a_j $(j = 0, 1, \ldots, n)$. Second, $\mathbb{S}$ is a linearly independent set, because the only solution to

$$c_n t^n + c_{n-1} t^{n-1} + \cdots + c_1 t + c_0(1) = 0 = 0t^n + 0t^{n-1} + \cdots + 0t + 0$$

is $c_0 = c_i = \ldots, = c_n = 0$. The basis $\mathbb{S}$ contains $n+1$ elements, and it follows that $dim(\mathbb{P}^n) = n + 1$. $\mathbb{S}$ is often called the *standard basis for* $\mathbb{P}^n$.

Example 7 The standard basis for $\mathbb{M}_{2\times 2}$ is

$$\mathbb{S} = \left\{ \begin{bmatrix} 1 & 0 \\ 0 & 0 \end{bmatrix}, \begin{bmatrix} 0 & 1 \\ 0 & 0 \end{bmatrix}, \begin{bmatrix} 0 & 0 \\ 1 & 0 \end{bmatrix}, \begin{bmatrix} 0 & 0 \\ 0 & 1 \end{bmatrix} \right\}$$

(See Problem 5.) Thus, $dim(\mathbb{M}_{2\times 2})=4$. More generally, the *standard basis for* $\mathbb{M}_{p\times n}$ is the set of pn matrices, each having a single 1 in a different position with all other entries equal to zero. Consequently, $dim(\mathbb{M}_{p\times n}) = pn$.

$dim(\mathbb{R}^n) = n$-
$dim(\mathbb{P}^n) = n+1$
$dim(\mathbb{M}_{p\times n}) = pn$

Example 8 The dimension of $\mathbb{R}^n$ is n. $\mathbb{R}^n = \mathbb{M}_{1\times n}$ when we represent n-tuples as row matrices, whereas $\mathbb{R}^n = \mathbb{M}_{n\times 1}$ when we represent n-tuples as column matrices. Either way, it follows from Example 7 that $dim(\mathbb{R}^n) = dim(\mathbb{M}_{1\times n}) = dim(\mathbb{M}_{n\times 1}) = n$. The standard basis for $\mathbb{R}^n$, considered as column matrices, is depicted in Equation (2.22).

As an immediate consequence of Theorem 1, we obtain one of the more important results in linear algebra.

▶THEOREM 3

In an n-dimensional vector space, every set of $n+1$ *or more vectors is linearly dependent.* ◀

Example 9 The set $\mathbb{A} = \{[1\ 5], [2\ -4], [-3\ -4]\}$ is a set of three vectors in the two-dimensional vector space $\mathbb{R}^2$, considered as row matrices. Therefore, $\mathbb{A}$ is linearly dependent. The set $\mathbb{R} = \{t^2+t,\ t^2-t,\ t+1,\ t-1\}$ is a set of four vectors in the three-dimensional vector space $\mathbb{P}^2$. Therefore, $\mathbb{R}$ is linearly dependent.

In Section 2.3, we surmised that lines through the origin and planes that include the origin are subspaces of $\mathbb{R}^3$. The following theorem formalizes this conjecture and provides a complete geometric interpretation of subspaces in $\mathbb{R}^3$.

▶THEOREM 4

Let $\mathbb{U}$ *be a subspace of* $\mathbb{R}^3$.

(i) *If* $dim(\mathbb{U}) = 0$, *then* $\mathbb{U}$ *contains just the origin.*
(ii) *If* $dim(\mathbb{U}) = 1$, *then the graph of* $\mathbb{U}$ *is a straight line through the origin.*
(iii) *If* $dim(\mathbb{U}) = 2$, *then the graph of* $\mathbb{U}$ *is a plane that includes the origin.* ◀

Proof: By definition, a vector space has dimension zero if and only if the vector space contains just the zero vector, which for $\mathbb{R}^3$ is the origin $[0\,0\,0]$. This proves part (i).

If $\mathbb{U}$ is a one-dimensional subspace, then it has a basis consisting of a single nonzero vector, which we denote as $\mathbf{u}$. Every vector in $\mathbb{U}$ can be written as a linear combination of vectors in a basis for $\mathbb{U}$, which here implies that every vector $\mathbf{v}$ in $\mathbb{U}$ is a scalar multiple of $\mathbf{u}$; that is, $\mathbf{v}=\alpha\mathbf{u}$ for some scalar α. The set of all such vectors graph as a line through the origin that contains $\mathbf{u}$ (see Figure 2.14 for the special case $\mathbf{u}=[1\ 2\ 4]$). In Figure 2.14, $\alpha>1$ generates a point on the line that is further from the origin than $\mathbf{u}$ but in the same direction as $\mathbf{u}$; $\alpha<1$ but still positive generates a point on the line that is closer to the origin than $\mathbf{u}$ but still in the same direction as $\mathbf{u}$; $\alpha < 0$ generates a point in the opposite direction of $\mathbf{u}$. Finally, if QJ is a two-dimensional subspace, then it has a basis consisting of two nonzero vectors, which we denote as $\mathbf{v}_1$ and $\mathbf{v}_2$. The vectors in such a basis must be linearly independent, so $\mathbf{v}_2$ cannot be a scalar multiple of $\mathbf{v}_1$. Therefore, $\mathbf{v}_2$ does not lie on the line through the origin containing $\mathbf{v}_i$. Any vector $\mathbf{v}$ in $\mathbb{U}$ can be written as a linear combination of $\mathbf{v}_1$ and $\mathbf{v}_2$, so

$$\mathbf{v} = \alpha\mathbf{v}_1 + \beta\mathbf{v}_2$$

for particular values of the scalars α and β. Consider the plane that contains the two basis vectors. From the geometric representation of vector addition and scalar multiplication in $\mathbb{R}^3$ developed in Section 1.5, it follows that every point in the plane containing the two basis vectors can be reached as a linear combination of $\mathbf{v}_1$ and $\mathbf{v}_2$ and that every linear combination of these two vectors is in the plane defined by those two vectors. (See Figure 2.17 where $\mathbf{v}$ denotes a point in the plane defined by $\mathbf{v}_1$ and $\mathbf{v}_2$; here $0 < \alpha < 1$ and β is negative.)

The standard basis in $\mathbb{R}^2$, considered as column vectors, consists of the two vectors

$$\mathbf{e}_1 = \begin{bmatrix} 1 \\ 0 \end{bmatrix} \quad \text{and} \quad \mathbf{e}_2 = \begin{bmatrix} 0 \\ 1 \end{bmatrix}$$

which in many engineering texts are denoted by $\mathbf{i}$ and $\mathbf{j}$, respectively. Both are graphed in Figure 2.18. For an arbitrary vector $\mathbf{v}$ in $\mathbb{R}^2$, we have

$$\mathbf{v} = \begin{bmatrix} a \\ b \end{bmatrix} = a\mathbf{e}_1 + b\mathbf{e}_2 = a\mathrm{i} + b\mathrm{j}$$

The standard basis in $\mathbb{R}^3$, considered as column vectors, consists of the three vectors

$$\mathbf{e}_1 = \begin{bmatrix} 1 \\ 0 \\ 0 \end{bmatrix}, \mathbf{e}_2 = \begin{bmatrix} 0 \\ 1 \\ 0 \end{bmatrix}, \text{and } \mathbf{e}_3 = \begin{bmatrix} 0 \\ 0 \\ 1 \end{bmatrix}$$

which in many engineering texts are denoted by $\mathbf{i}$, $\mathbf{j}$, and $\mathbf{k}$, respectively. These are graphed in Figure 2.19. For an arbitrary vector $\mathbf{v}$ in $\mathbb{R}^3$, we have

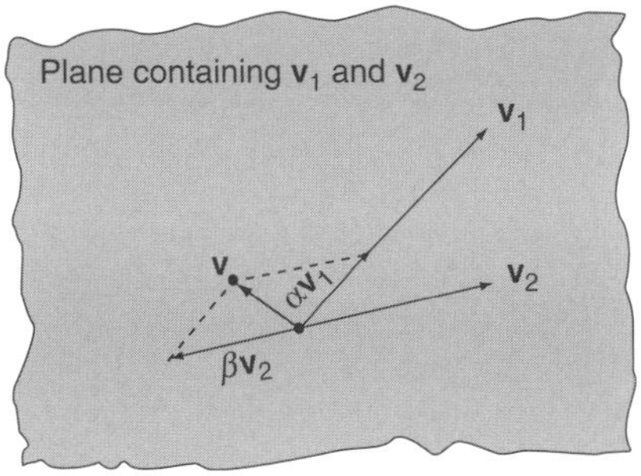

FIGURE 2.17

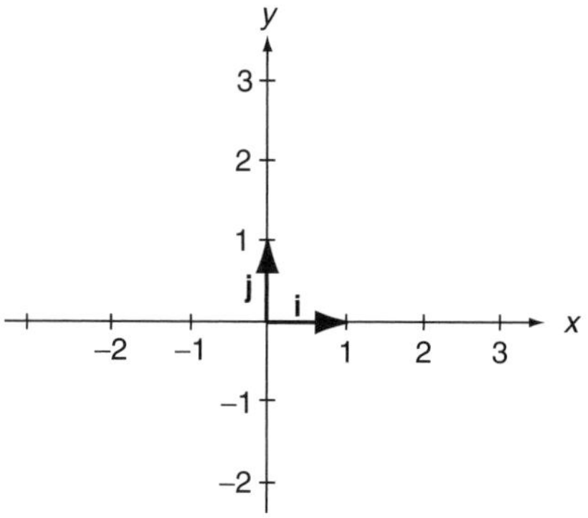

FIGURE 2.18

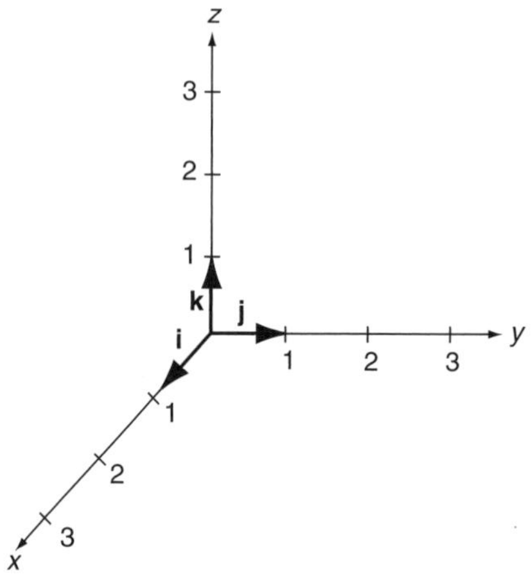

FIGURE 2.19

$$\mathbf{v} = \begin{bmatrix} a \\ b \\ c \end{bmatrix} = a\mathbf{e}_1 + b\mathbf{e}_2 + c\mathbf{e}_3 = a\mathbf{i} + b\mathbf{j} + c\mathbf{k}$$

More generally, if $\mathbb{S} = \{\mathbf{v}_1, \mathbf{v}_2, \ldots, \mathbf{v}_n\}$ is a basis for a vector space $\mathbb{V}$, then $\mathbb{S}$ is a spanning set $\mathbb{V}$. Consequently, if $\mathbf{v} \in \mathbb{V}$, then there exist scalars $d_1, d_2, \ldots, d_n$ such that

$$\mathbf{v} = d_1\mathbf{v}_1 + d_2\mathbf{v}_2 + \cdots + d_n\mathbf{v}_n \tag{2.27}$$

We shall prove shortly that this set of scalars is unique for each $\mathbf{v}$; that is, for each $\mathbf{v}$ there is one and only one set of scalars $d_1, d_2, \ldots, d_n$ that satisfies Equation (2.27). These scalars are called the *coordinates* of $\mathbf{v}$ with respect to $\mathbb{S}$ and are represented by the n-tuple

$$\mathbf{v} \leftrightarrow \begin{bmatrix} a_1 \\ b_2 \\ \vdots \\ d_n \end{bmatrix}_S$$

Example 10 Find the coordinate representations of the vector $\mathbf{v} = [7\ 2]^T$, first with respect to the standard basis $\mathbb{C} = \left\{[1\ 0]^T, [0\ 1]^T\right\}$ and then with respect to the basis $\mathbb{D}_1 = \left\{[1\ 1]^T, [1\ -1]^T\right\}$.

Solution: With respect to the standard basis, we have

$$\begin{bmatrix} 7 \\ 2 \end{bmatrix} = 7\begin{bmatrix} 1 \\ 0 \end{bmatrix} + 2\begin{bmatrix} 0 \\ 1 \end{bmatrix}$$

so the coordinates are 7 and 2 and the 2-tuple representation is

$$\begin{bmatrix} 7 \\ 2 \end{bmatrix} \leftrightarrow \begin{bmatrix} 7 \\ 2 \end{bmatrix}_{\mathbb{C}}$$

To determine the representation with respect to $\mathbb{S}_1$, we need to first write the given vector as a linear combination of the vectors in $\mathbb{S}_1$. We need values of the scalars d_1 and d_2 that satisfy the equation

$$\begin{bmatrix} 7 \\ 2 \end{bmatrix} = d_1\begin{bmatrix} 1 \\ 1 \end{bmatrix} + d_2\begin{bmatrix} 1 \\ -1 \end{bmatrix}$$

This is equivalent to the system of equations

$$\begin{aligned} d_1 + d_2 &= 7 \\ d_1 - d_2 &= 2 \end{aligned}$$

which admits the solution $d_1 = 9/2$ and $d_2 = 5/2$. These are the coordinates of $\mathbf{v}$ with respect to $\mathbb{S}_1$, and we may write

$$\begin{bmatrix} 7 \\ 2 \end{bmatrix} = \frac{9}{2}\begin{bmatrix} 1 \\ 1 \end{bmatrix} + \frac{5}{2}\begin{bmatrix} 1 \\ -1 \end{bmatrix} \leftrightarrow \begin{bmatrix} 9/2 \\ 5/2 \end{bmatrix}_{\mathbb{D}}$$

It was no accident in the previous example that the n-tuple representation of the vector $\mathbf{v}$ with respect to the standard basis was the vector itself. This is always the case for vectors in $\mathbb{R}^n$ with respect to the standard basis. Consequently, we drop the subscript notation on the n-tuple representation of the coordinates of a vector whenever we deal with the standard basis.

Example 11 Determine the coordinate representation of the matrix $\begin{bmatrix} 4 & 3 \\ 6 & 2 \end{bmatrix}$ with respect to the basis

$$\mathbb{S} = \left\{\begin{bmatrix} 0 & 1 \\ 1 & 1 \end{bmatrix}, \begin{bmatrix} 1 & 0 \\ 1 & 1 \end{bmatrix}, \begin{bmatrix} 1 & 1 \\ 0 & 1 \end{bmatrix}, \begin{bmatrix} 1 & 1 \\ 1 & 0 \end{bmatrix}\right\}$$

Solution: We first determine scalars d_1, d_2, d_3, and d_4 that satisfy the matrix equation

$$\begin{bmatrix} 4 & 3 \\ 6 & 2 \end{bmatrix} = d_1 \begin{bmatrix} 0 & 1 \\ 1 & 1 \end{bmatrix} + d_2 \begin{bmatrix} 1 & 0 \\ 1 & 1 \end{bmatrix} + d_3 \begin{bmatrix} 1 & 1 \\ 0 & 1 \end{bmatrix} + d_4 \begin{bmatrix} 1 & 1 \\ 1 & 0 \end{bmatrix}$$

This is equivalent to the system of equations

$$\begin{aligned} d_2 + d_3 + d_4 &= 4 \\ d_1 + d_3 + d_4 &= 3 \\ d_1 + d_2 + d_4 &= 6 \\ d_1 + d_2 + d_3 &= 2 \end{aligned}$$

which admits the solution $d_1 = 1$, $d_2 = 2$, $d_3 = -1$, and $d_4 = 3$. These are the coordinates of the given matrix with respect to $\mathbb{S}$, and we may write

$$\begin{bmatrix} 4 & 3 \\ 6 & 2 \end{bmatrix} \leftrightarrow \begin{bmatrix} 1 \\ 2 \\ -1 \\ 3 \end{bmatrix}_{\mathbb{S}}$$

The notation $\leftrightarrow$ signifies that the n-tuple on the right side equals the sum of the products of each coordinate times its corresponding vector in the basis. The subscript on the n-tuple denotes the basis under consideration. In Example 10, the notation

$$\begin{bmatrix} 9/2 \\ 5/2 \end{bmatrix}_{\mathbb{D}} \text{ denotes the sum } \frac{9}{2}\begin{bmatrix} 1 \\ 1 \end{bmatrix} + \frac{5}{2}\begin{bmatrix} 1 \\ -1 \end{bmatrix}$$

while in Example 11, the notation

$$\begin{bmatrix} 1 \\ 2 \\ -1 \\ 3 \end{bmatrix}_{\mathbb{S}} \text{ denotes the sum}$$

$$(1)\begin{bmatrix} 0 & 1 \\ 1 & 1 \end{bmatrix} + (2)\begin{bmatrix} 1 & 0 \\ 1 & 1 \end{bmatrix} + (-1)\begin{bmatrix} 1 & 1 \\ 0 & 1 \end{bmatrix} + (3)\begin{bmatrix} 1 & 1 \\ 1 & 0 \end{bmatrix}$$

Although a vector generally has different coordinate representations for different bases, a vector's coordinate representation with respect to any one basis is unique! In Example 10, we produced two coordinate representations for the vector $[7\,2]^{\mathrm{T}}$, one for each of two bases. Within each basis, however, there is one and only one coordinate representation for a vector. We formalize this fact in the following theorem.

▶THEOREM 5

Let $\{\mathbf{v}_1, \mathbf{v}_2, \ldots, \mathbf{v}_n\}$ be a basis for a vector space $\mathbb{V}$ and let $\mathbf{v} \in \mathbb{V}$. If

$$\mathbf{v} = c_1\mathbf{v}_1 + c_2\mathbf{v}_2 + \cdots + c_n \quad \text{and} \quad \mathbf{v} = d_1\mathbf{v}_1 + d_2\mathbf{v}_2 + \cdots + d_n\mathbf{v}_n$$

are two ways of expressing $\mathbf{v}$ as linear combinations of the basis vectors, then $c_i = d_i$ for each i ($i = 1, 2, \ldots, n$). ◀

Proof:

$$\begin{aligned} 0 &= \mathbf{v} - \mathbf{v} \\ &= (c_1\mathbf{v}_1 + c_2\mathbf{v}_2 + \cdots + c_n\mathbf{v}_n) - (d_1\mathbf{v}_1 + d_2\mathbf{v}_2 + \cdots + d_n\mathbf{v}_n) \\ &= (c_1 - d_1)\mathbf{v}_1 + (c_2 - d_2\mathbf{v}_2) + \cdots + (c_n - d_n)\mathbf{v}_n \end{aligned}$$

Vectors in a basis are linearly independent, so the only solution to the last equation is for each of the coefficients within the parentheses to be 0. Therefore, $(c_i - d_i) = 0$ for each value of i ($i = 1, 2, \ldots, n$), which implies that $c_i = d_i$.

We conclude this section with a two-part theorem, the proofs of which we leave as exercises for the reader (see Problems 18 and 22).

▶THEOREM 6

Let $\mathbb{V}$ be an n-dimensional vector space.

(i) *If $\mathbb{S}$ is a spanning set for $\mathbb{V}$, then some subset of $\mathbb{S}$ forms a basis for $\mathbb{V}$; that is, $\mathbb{S}$ can be reduced to a basis by deleting from $\mathbb{S}$ a suitable number (perhaps 0) of vectors.*

(ii) *If $\mathbb{S}$ is a linearly independent set of vectors in $\mathbb{V}$, then there exists a basis for $\mathbb{V}$ that includes in it all the vectors of $\mathbb{S}$; that is, $\mathbb{S}$ can be extended to a basis by augmenting onto it a suitable number (perhaps 0) of vectors.* ◀

Problems 2.5

(1) Determine which of the following sets are bases for $\mathbb{R}^2$, considered as row matrices.

(a) $\{[1 \quad 0], [1 \quad 1]\}$.

(b) $\{[1 \quad 0], [1 \quad 1]\}$.

(c) $\{[1 \quad 1], [1 \quad 2]\}$.

(d) $\{[1 \quad 2], [1 \quad 3]\}$.

(e) $\{[1 \quad 2], [2 \quad 4]\}$.

(f) $\{[10 \quad 20], [10 \quad -20]\}$.

(g) $\{[10 \quad 20], [-10 \quad -20]\}$.

(h) $\{[1 \quad 1], [1 \quad 2], [2 \quad 1]\}$

(2) Determine which of the following sets are bases for $\mathbb{R}^2$, considered as column vectors.

(a) $\left\{\begin{bmatrix}2\\3\end{bmatrix},\begin{bmatrix}2\\-3\end{bmatrix}\right\}$.

(b) $\left\{\begin{bmatrix}1\\2\end{bmatrix},\begin{bmatrix}0\\0\end{bmatrix}\right\}$.

(c) $\left\{\begin{bmatrix}1\\2\end{bmatrix},\begin{bmatrix}1\\-2\end{bmatrix}\right\}$.

(d) $\left\{\begin{bmatrix}1\\2\end{bmatrix},\begin{bmatrix}-1\\-2\end{bmatrix}\right\}$.

(e) $\left\{\begin{bmatrix}10\\20\end{bmatrix},\begin{bmatrix}20\\30\end{bmatrix}\right\}$.

(f) $\left\{\begin{bmatrix}50\\100\end{bmatrix},\begin{bmatrix}100\\150\end{bmatrix}\right\}$.

(g) $\left\{\begin{bmatrix}1\\2\end{bmatrix}\right\}$.

(h) $\left\{\begin{bmatrix}1\\2\end{bmatrix},\begin{bmatrix}1\\3\end{bmatrix},\begin{bmatrix}1\\4\end{bmatrix}\right\}$.

(3) Determine which of the following sets are bases for $\mathbb{R}^3$, considered as row vectors.

(a) $\{[1\ \ 0\ \ 0],[0\ \ 1\ \ 0],[0\ \ 0\ \ 1]\}$.

(b) $\{[1\ \ 1\ \ 0],[0\ \ 1\ \ 1],[1\ \ 0\ \ 1]\}$.

(c) $\{[1\ \ 0\ \ 0],[1\ \ 1\ \ 0],[1\ \ 1\ \ 1]\}$.

(d) $\{[1\ \ 1\ \ 0],[0\ \ 1\ \ 1],[1\ \ 2\ \ 1]\}$.

(e) $\{[1\ \ 1\ \ 0],[0\ \ 1\ \ 1],[1\ \ 3\ \ 1]\}$.

(f) $\{[1\ \ 1\ \ 0],[0\ \ 1\ \ 1],[1\ \ 4\ \ 1]\}$.

(g) $\{[1\ \ 2\ \ 3],[4\ \ 5\ \ 6],[0\ \ 0\ \ 0]\}$.

(h) $\{[1\ \ 2\ \ 3],[4\ \ 5\ \ 6],[7\ \ 8\ \ 9]\}$.

(4) Determine which of the following sets are bases for $\mathbb{R}^3$, considered as column vectors.

(a) $\left\{[1\ \ 2\ \ 1]^T,[1\ \ 2\ \ 0]^T\right\}$.

(b) $\left\{[1\ \ 2\ \ 0]^T,[1\ \ 2\ \ 1]^T,[1\ \ 2\ \ 2]^T\right\}$.

(c) $\left\{[1\ \ 2\ \ 0]^T,[1\ \ 2\ \ 1]^T,[2\ \ 4\ \ 1]^T\right\}$.

(d) $\left\{[1\ \ 2\ \ 0]^T,[2\ \ 4\ \ 0]^T,[2\ \ 4\ \ 1]^T\right\}$.

(e) $\left\{[1\ \ 2\ \ -3]^T,[1\ \ 2\ \ 0]^T,[1\ \ 0\ \ -3]^T\right\}$.

(f) $\left\{[1\ \ 1\ \ 1]^T,[2\ \ 1\ \ 1]^T,[2\ \ 2\ \ 1]^T\right\}$.

(g) $\left\{[2\ \ 1\ \ 1]^T,[2\ \ 2\ \ 1]^T,[2\ \ 2\ \ -1]^T\right\}$.

(h) $\left\{[1\ \ 2\ \ 1]^T,[1\ \ 3\ \ 1]^T,[1\ \ 4\ \ 1]^T,[1\ \ 5\ \ 1]^T\right\}$.

(5) Determine which of the following sets are bases for $\mathbb{M}_{2\times 2}$.

(a) $\left\{\begin{bmatrix}1 & 0\\0 & 0\end{bmatrix}, \begin{bmatrix}0 & 1\\0 & 0\end{bmatrix}, \begin{bmatrix}0 & 0\\1 & 0\end{bmatrix}, \begin{bmatrix}0 & 0\\0 & 1\end{bmatrix}\right\}$.

(b) $\left\{\begin{bmatrix}1 & 1\\0 & 0\end{bmatrix}, \begin{bmatrix}-1 & 1\\0 & 0\end{bmatrix}, \begin{bmatrix}0 & 0\\1 & 1\end{bmatrix}, \begin{bmatrix}0 & 0\\0 & -1\end{bmatrix}\right\}$.

(c) $\left\{\begin{bmatrix}1 & 0\\0 & 0\end{bmatrix}, \begin{bmatrix}1 & 1\\0 & 0\end{bmatrix}, \begin{bmatrix}1 & 1\\1 & 0\end{bmatrix}, \begin{bmatrix}1 & 1\\1 & 1\end{bmatrix}\right\}$.

(d) $\left\{\begin{bmatrix}1 & 1\\1 & 0\end{bmatrix}, \begin{bmatrix}1 & 1\\0 & 1\end{bmatrix}, \begin{bmatrix}1 & 0\\1 & 1\end{bmatrix}, \begin{bmatrix}0 & 1\\1 & 1\end{bmatrix}\right\}$.

(6) Determine which of the following sets are bases for $\mathbb{P}^1$.

(a) $\{t+1, t\}$.

(b) $\{t+1, 1\}$.

(c) $\{t+1, t, 1\}$.

(d) $\{t+1, t-1\}$.

(7) Determine which of the following sets are bases for $\mathbb{P}^2$.

(a) $\{t^2+t+1, t\}$.

(b) $\{t^2+t, t+1, t^2+1, 1\}$.

(c) $\{t^2+t+1, t+1, 1\}$.

(d) $\{t^2+t+1, t+1, t-1\}$.

(e) $\{t^2+t, t+1, t^2+1\}$.

(f) $\{t^2+t+1, t+1, t^2\}$.

(8) Determine which of the following sets are bases for $\mathbb{P}^3$.

(a) $\{t^3+t^2+t, t^2+t+1, t+1\}$.

(b) $\{t^3, t^2, t, 1\}$.

(c) $\{t^3+t^2+t, t^2+t+1, t+1, 1\}$.

(d) $\{t^3+t^2, t^2+t, t+1, 1\}$.

(e) $\{t^3+t^2+t, t^3+t^2, t^2+t, t, t+1, 1\}$.

(f) $\{t^3+t^2, t^3-t^2, t+1, t-1\}$.

(g) $\{t^3+t^2+1, t^3+t^2, t+1, t-1\}$.

(h) $\{t^3+t^2+t, t^3+t^2, t^2+t, t^3+t\}$.

(9) Find an n-tuple representation for the coordinates of $[1\ 3]$ with respect to the sets given in (a) Problem 1(a) and (b) Problem 1(d).

(10) Find an n-tuple representation for the coordinates of $[2\ 2]$ with respect to the sets given in (a) Problem 1(a) and (b) Problem 1(d).

(11) Find an n-tuple representation for the coordinates of $[1\ -1]$ with respect to the sets given in (a) Problem 1(a) and (b) Problem 1(b).

(12) Find an n-tuple representation for the coordinates of $[1\ -2]^{\mathrm{T}}$ with respect to the sets given in (a) Problem 2(c) and (b) Problem 2(e).

(13) Find an n-tuple representation for the coordinates of $[100\ -100]^{\mathrm{T}}$ with respect to the sets given in (a) Problem 2(e) and (b) Problem 2(f).

(14) Find an n-tuple representation for the coordinates of $[1\ 1\ 0]$ with respect to the sets given in (a) Problem 3(a), (b) Problem 3(b), and (c) Problem 3(c).

(15) Find an n-tuple representation for the coordinates of $t+2$ with respect to the sets given in (a) Problem 6(a) and (b) Problem 6(b).

(16) Find an n-tuple representation for the coordinates of t^2 with respect to the sets given in (a) Problem 8(c) and (b) Problem 8(d).

(17) Let $\mathbb{S}$ be a spanning set for a vector space $\mathbb{V}$, and let $\mathrm{v} \in \mathbb{S}$. Prove that if $\mathbf{v}$ is a linear combination of other vectors in the set, then the set that remains by deleting $\mathbf{v}$ from $\mathbb{S}$ is also a spanning set for $\mathbb{V}$.

(18) Show that any spanning set for a vector space $\mathbb{V}$ can be reduced to a basis by deleting from $\mathbb{S}$ a suitable number of vectors.

(19) Reduce the set displayed in Example 3 to a basis for $\mathbb{M}_{2\times2}$.

(20) Show that the set displayed in Problem 1(h) is a spanning set for $\mathbb{R}^2$ and reduce it to a basis.

(21) Show that the set displayed in Problem 7(b) is a spanning set for $\mathbb{P}^2$ and reduce it to a basis.

(22) Prove that any linearly independent set of vectors in a vector space $\mathbb{V}$ can be extended to a basis for $\mathbb{V}$. Hint: Append to the set a known basis and then use Problem 18.

(23) Extend the set displayed in Example 2 into a basis for $\mathbb{P}^3$.

(24) Show that the set displayed in Problem 4(a) is linearly independent and extend it into a basis for $\mathbb{R}^3$.

(25) Show that the set displayed in Problem 8(a) is linearly independent and extend it into a basis for $\mathbb{P}^3$.

(26) Prove that a spanning set for a vector space $\mathbb{V}$ cannot contain less elements then the dimension of $\mathbb{V}$.

(27) Prove that any set of two vectors in $\mathbb{R}^2$ is a basis if one vector is not a scalar multiple of the other.

(28) Let $\mathbb{W}$ be a subspace of a vector space $\mathbb{V}$ and let $\mathbb{S}$ be a basis for $\mathbb{W}$. Prove that $\mathbb{S}$ can be extended to a basis for $\mathbb{V}$.

(29) Let $\mathbb{W}$ be a subspace of a vector space $\mathbb{V}$. Prove that $\dim(\mathbb{W}) \leq \dim(\mathbb{V})$.

(30) Let $\mathbb{W}$ be a subspace of a vector space $\mathbb{V}$. Prove that if $\dim(\mathbb{W}) - \dim(\mathbb{V})$, then $\mathbb{W} = \mathbb{V}$.

(31) Prove that in an n-dimensional vector space $\mathbb{V}$ no set of $n-1$ vectors can span $\mathbb{V}$.

(32) Prove that if $\{\mathbf{v}_1, \mathbf{v}_2\}$ is a basis for a vector space, then so too is $\{\mathbf{u}_1, \mathbf{u}_2\}$, where $\mathbf{u}_1 = \mathbf{v}_1 + \mathbf{v}_2$, and $\mathbf{u}_2 = \mathbf{v}_1 - \mathbf{v}_2$.

(33) Prove that if $\{\mathbf{v}_1, \mathbf{v}_2, \mathbf{v}_3\}$ is a basis for a vector space, then so too is $\{\mathbf{u}_1, \mathbf{u}_2, \mathbf{u}_3\}$, where $\mathbf{u}_1 = \mathbf{v}_1 + \mathbf{v}_2 + \mathbf{v}_3$, $\mathbf{u}_2 = \mathbf{v}_2 - \mathbf{v}_3$, and $\mathbf{u}_3 = \mathbf{v}_3$.

(34) Prove that if $\{\mathbf{v}_1, \mathbf{v}_2, \ldots, \mathbf{v}_n\}$ is a basis for a vector space, then so too is $\{k_1\mathbf{v}_1, k_2\mathbf{v}_2, \ldots, k_n\mathbf{v}_n\}$ $\{\mathbf{u}_1, \mathbf{u}_2, \mathbf{u}_3\}$, where $k_1, k_2, \ldots, k_n$ is any set of nonzero scalars.

2.6 ROW SPACE OF A MATRIX

An $m \times n$ matrix $\mathbf{A}$ contains m-rows and n-columns. Each row, considered as a row matrix in its own right, is an element of $\mathbb{R}^n$, so it follows from Theorem 4 of Section 2.4 that the span of the rows, considered as row matrices, is a subspace. We call this subspace the *row space* of the matrix $\mathbf{A}$. The dimension of the row space is known as the *row rank* of A.

The row space of a matrix is the subspace spanned by the rows of the matrix; the dimension of the row space is the row rank.

Example 1 The matrix $\mathbf{A} = \begin{bmatrix} 1 & 2 & 3 \\ 4 & 5 & 6 \end{bmatrix}$ 1 has two rows, $[1\,2\,3]$ and $[4\,5\,6]$, both of which are elements of $\mathbb{R}^3$. The row space of $\mathbf{A}$ consists of all linear combinations of these two vectors; that is, if we set $\mathbb{S} = \{[1\ 2\ 3][4\ 5\ 6]\}$, then the row space of $\mathbf{A}$ is *span*$(\mathbb{S})$. The dimension of *span*$(\mathbb{S})$ is the row rank of A.

To determine the row rank of a matrix, we must identify a basis for its row space and then count the number of vectors in that bases. This sounds formidable, but as we shall see that it is really quite simple. For a row-reduced matrix, the procedure is trivial.

▶THEOREM 1

The nonzero rows of a row-reduced matrix form a basis for the row space of that matrix, and the row rank is the number of nonzero rows. ◀

Proof: Let $\mathbf{v}_1$ designate the first nonzero row, $\mathbf{v}_2$ the second nonzero row, and so on through $\mathbf{v}_r$, which designates the last nonzero row of the row-reduced matrix. This matrix may still have additional rows, but if so they are all zero. The row space of this matrix is *span* $\{\mathbf{v}_1, \mathbf{v}_2, \ldots, \mathbf{v}_r\}$. The zero rows, if any, will add nothing to the span.

We want to show the nonzero rows form a basis for the row space. Thus, we must show that these rows, considered as row matrices, span the subspace and are linearly independent. They clearly span the subspace, because that is precisely how the row space is formed. To determine linear independence, we consider the vector equation

$$c_1\mathbf{v}_2 + c_2\mathbf{v}_2 + \cdots + c_r\mathbf{v}_r = 0 \tag{2.28}$$

The first nonzero element in the first nonzero row of a row-reduced matrix must be one. Assume it appears in column j. Then, no other row has a nonzero element in column j. Consequently, when the left side of Equation (2.28) is computed, it will have c_1 as its jth component. Because the right side of Equation (2.28) is the zero vector, it follows that $c_1=0$. With $c_1=0$, Equation (2.28) reduces to

$$c_2\mathbf{v}_2 + c_3\mathbf{v}_3 + \cdots + c_r\mathbf{v}_r = 0$$

A similar argument then shows that $c_2=0$. With both $c_1=c_2=0$, Equation (2.28) becomes

$$c_3\mathbf{v}_3 + c_4\mathbf{v}_4 + \cdots + c_r\mathbf{v}_r = 0$$

A repetition of the same argument shows iteratively that $c_1, c_2, \ldots, c_r$ are all zero. Thus, the nonzero rows are linearly independent.

Example 2 Determine the row rank of the matrix

$$\mathbf{A} = \begin{bmatrix} 1 & 0 & -2 & 5 & 3 \\ 0 & 0 & 1 & -4 & 1 \\ 0 & 0 & 0 & 1 & 0 \\ 0 & 0 & 0 & 0 & 0 \end{bmatrix}$$

Solution: **A** is in row-reduced form. Because **A** contains three nonzero rows, the row rank of **A** is 3.

Most matrices are not in row-reduced form. All matrices, however, can be transformed to row-reduced form by elementary row operations, and such transformations do *not* alter the underlying row space.

▶THEOREM 2

If B is obtained from **A** *by an elementary row operation, then the row space of* **A** *is the same as the row space of* **B**.◀

Proof: We shall consider only the third elementary row operation and leave the proofs of the other two as exercises (see Problems 46 and 47). Let **B** be obtained from **A** by adding λ times the jth row of **A** to the kth row of **A**. Consequently, if we denote the rows of **A** by the set of row matrices

$\mathbb{A} = \{\mathbf{A}_1, \mathbf{A}_2, \ldots, \mathbf{A}_j, \ldots, \mathbf{A}_k, \ldots, \mathbf{A}_n\}$ and the rows of **B** by $\mathbb{B} = \{\mathbf{B}_1, \mathbf{B}_2, \ldots, \mathbf{B}_j, \ldots, \mathbf{B}_k, \ldots, \mathbf{B}_n\}$, then $\mathbf{B}_i, = \mathbf{A}_i$ for all $i = 1, 2, \ldots, n$ except $i = k$, and $\mathbf{B}_k = \mathbf{A}_k + \lambda\mathbf{A}_j$. We need to show that if $\mathbf{v}$ is any vector in the span of $\mathbb{A}$, then it is also in the span of $\mathbb{B}$ and vice versa.

If $\mathbf{v}$ is in the span of $\mathbb{A}$, then there exists constants $c_1, c_2, \ldots, c_n$ such that

$$\mathbf{v} = c_1\mathbf{A}_1 + c_2\mathbf{A}_2 + \cdots + c_j\mathbf{A}_j + \cdots + c_k\mathbf{A}_k + \cdots + c_n\mathbf{A}_n.$$

We may rearrange the right side of this equation to show that

$$\begin{aligned}\mathbf{v} &= c_1\mathbf{A}_1 + c_2\mathbf{A}_2 + \cdots + (c_j + \lambda c_k - \lambda c_k)\mathbf{A}_j + \cdots + c_k\mathbf{A}_k + \cdots + c_n\mathbf{A}_n \\ &= c_1\mathbf{A}_1 + c_2A_2 + \cdots + (c_j - \lambda c_k)\mathbf{A}_j + \cdots + c_k(\mathbf{A}_k + \lambda\mathbf{A}_j) + \cdots + c_n\mathbf{A}_n \\ &= c_1\mathbf{B}_1 + c_2\mathbf{B}_2 + \cdots + (c_j - \lambda c_k)\mathbf{B}_j + \cdots + c_k\mathbf{B}_k + \cdots + c_n\mathbf{B}_n\end{aligned}$$

Thus, $\mathbf{v}$ is also in the span of $\mathbb{B}$.

Conversely, if $\mathbf{v}$ is in the span of $\mathbb{B}$, then there exists constants $d_1, d_2, \ldots, d_n$ such that

$$\mathbf{v} = \alpha_1\mathbf{B}_1 + \alpha_2\mathbf{B}_2 + \cdots + d_j\mathbf{B}_j + \cdots + d_k\mathbf{B}_k + \cdots + d_n\mathbf{B}_n$$

We may rearrange the right side of this equation to show that

$$\begin{aligned}\mathbf{v} &= d_1\mathbf{A}_1 + d_2\mathbf{A}_2 + \cdots + d_j\mathbf{A}_j + \cdots + d_k(\mathbf{A}_k + \lambda\mathbf{A}_j) + \cdots + d_n\mathbf{A}_n \\ &= d_1\mathbf{A}_1 + d_2\mathbf{A}_2 + \cdots + (d_j + d_k\lambda)\mathbf{A}_j + \cdots + d_k\mathbf{A}_k + \cdots + d_n\mathbf{A}_n\end{aligned}$$

Thus, $\mathbf{v}$ is also in the span of $\mathbb{A}$.

As an immediate extension of Theorem 2, it follows that if **B** is obtained from **A** by a *series* of elementary row operations, then both **A** and **B** have the same row space. Together Theorems 1 and 2 suggest a powerful method for determining the row rank of any matrix. Simply use elementary row operations to transform a given matrix to row-reduced form and then count the number of nonzero rows.

To find the row rank of a matrix, use elementary row operations to transform the matrix to row-reduced form and then count the number of nonzero rows.

Example 3 Determine the row rank of

$$\mathbf{A} = \begin{bmatrix} 1 & 3 & 4 \\ 2 & -1 & 1 \\ 3 & 2 & 5 \\ 5 & 15 & 20 \end{bmatrix}$$

Solution: In Example 5 of Section 1.4, we transformed this matrix into the row-reduced form

$$\mathbf{B} = \begin{bmatrix} 1 & 3 & 4 \\ 0 & 1 & 1 \\ 0 & 0 & 0 \\ 0 & 0 & 0 \end{bmatrix}$$

Because **B** is obtained from **A** by elementary row operations, both matrices have the same row space and row rank. **B** has two nonzero rows, so its row rank, as well as the row rank of **A**, is 2.

Example 4 Determine the row rank of

$$\mathbf{A} = \begin{bmatrix} 1 & 2 & 1 & 3 \\ 2 & 3 & -1 & -6 \\ 3 & -2 & -4 & -2 \end{bmatrix}$$

Solution: In Example 6 of Section 1.4, we transformed this matrix into the row-reduced form

$$\mathbf{B} = \begin{bmatrix} 1 & 2 & 1 & 3 \\ 0 & 1 & 3 & 12 \\ 0 & 0 & 1 & 5 \end{bmatrix}$$

B has three nonzero rows, so its row rank, as well as the row rank of **A**, is 3.

A basis for the row space of a matrix is the set of nonzero rows of that matrix, after it has been transformed to row-reduced form by elementary row operations.

A basis for the row space of a matrix is equally obvious: namely, the set of nonzero rows in the row-reduced matrix. These vectors are linearly independent and, because they are linear combinations of the original rows, they span the same space.

Example 5 Find a basis for the row space of the matrix **A** given in Example 3.

Solution: The associated row-reduced matrix **B** (see Example 3) has as nonzero rows the row matrices $[1\ 3\ 4]$ and $[0\ 1\ 1]$. Together these two vectors are a basis for the row space of A.

Example 6 Find a basis for the row space of the matrix **A** given in Example 4.

Solution: The associated row-reduced matrix **B** (see Example 4) has as nonzero rows the row matrices $[1\ 2\ 1\ 3]$, $[0\ 1\ 3\ 12]$, and $[0\ 0\ 1\ 5]$. These three vectors form a basis for the row space of A.

To find a basis for a set of n-tuples, create a matrix having as its rows those n-tuples and then find a basis for the row space of that matrix.

A basis of the row space of a matrix **A** is a basis for the span of the rows of A. Thus, we can determine a basis for any set of n-tuples simply by creating a matrix **A** having as its rows those n-tuples and then finding a basis for the row space of **A**. This is an elegant procedure for describing the span of any finite set of vectors $\mathbb{S}$ in $\mathbb{R}^n$.

Example 7 Find a basis for the span of $\mathbb{S} = \left\{ \begin{bmatrix} 2 \\ 6 \\ -2 \end{bmatrix} \begin{bmatrix} 3 \\ 1 \\ 2 \end{bmatrix} \begin{bmatrix} 8 \\ 16 \\ -3 \end{bmatrix} \right\}$.

Solution: We create a matrix **A** having as its *rows* the vectors in $\mathbb{S}$. Note that the elements of $\mathbb{S}$ are column matrices, so we use their transposes as the rows of **A**. Thus,

$$\mathbf{A} = \begin{bmatrix} 2 & 6 & -2 \\ 3 & 1 & 2 \\ 8 & 16 & -3 \end{bmatrix}$$

Reducing this matrix to row-reduced form, we obtain

$$\begin{bmatrix} 1 & 3 & -1 \\ 0 & 1 & -5/8 \\ 0 & 0 & 0 \end{bmatrix}$$

The nonzero rows of this matrix, $[1\,3-1]$ and $[0\,1-5/8]$, form a basis for the row space of **A**. The set of transposes of these vectors

$$\mathbb{B} = \left\{ \begin{bmatrix} 1 \\ 3 \\ -1 \end{bmatrix} \begin{bmatrix} 0 \\ 1 \\ -5/8 \end{bmatrix} \right\}$$

is a basis for the span of $\mathbb{S}$, therefore, *span*$(\mathbb{S})$ is the set of all linear combinations of the vectors in $\mathbb{B}$.

We can extend this procedure to all finite-dimensional vector spaces, not just n-tuples. We know from Section 2.5 that every vector in a finite-dimensional vector space can be represented by an n-tuple. Therefore, to find a basis for the span of a set of vectors $\mathbb{S}$ that are *not* n-tuples, we first write coordinate representations for each vector in $\mathbb{S}$, generally with respect to a standard basis when one exists. We then create a matrix **A** having as its rows the coordinate representations of the vectors in $\mathbb{S}$. We use elementary row operations to identify a basis for the row space of **A**. This basis will consist of n-tuples. Transforming each n-tuple in this basis vector back to the original vector space provides a basis for the span of $\mathbb{S}$.

Example 8 Find a basis for the span of the vectors in

$$\mathbb{C} = \left\{ t^3 + 3t^2,\, 2t^3 + 2t - 2,\, t^3 - 6t^2 + 3t - 3,\, 3t^2 - t + 1 \right\}$$

Solution: The vectors in $\mathbb{C}$ are elements of the vector space $\mathbb{P}^3$, which has as its standard basis $\{t^3, t^2, t, 1\}$. With respect to this basis, the coordinate representations of the polynomials in $\mathbb{C}$ are

$$t^3 + 3t^2 \leftrightarrow \begin{bmatrix} 1 \\ 3 \\ 0 \\ 0 \end{bmatrix},\ 2t^3 + 2t - 2 \leftrightarrow \begin{bmatrix} 2 \\ 0 \\ 2 \\ -2 \end{bmatrix},$$

$$t^3 - 6t^2 + 3t - 3 \leftrightarrow \begin{bmatrix} 1 \\ -6 \\ 3 \\ -3 \end{bmatrix},\ \text{and } 3t^2 - t + 1 \leftrightarrow \begin{bmatrix} 0 \\ 3 \\ -1 \\ 1 \end{bmatrix}$$

We create a matrix **A** having as its *rows* these 4-tuples. Thus,

$$\mathbf{A} = \begin{bmatrix} 1 & 3 & 0 & 0 \\ 2 & 0 & 2 & -2 \\ 1 & -6 & 3 & -3 \\ 0 & 3 & -1 & 1 \end{bmatrix}$$

Reducing this matrix to row-reduced form, we obtain

$$\mathbf{B} = \begin{bmatrix} 1 & 3 & 0 & 0 \\ 0 & 1 & -1/3 & 1/3 \\ 0 & 0 & 0 & 0 \\ 0 & 0 & 0 & 0 \end{bmatrix}$$

The nonzero rows of **B**, namely, $[1\,3\,0\,0]$ and $[0\,1 - 1/3\,1/3]$, forma basis for the row space of **A**. The set of transposes of these vectors are coordinate representatives for the polynomials

$$\begin{bmatrix} 1 \\ 3 \\ 0 \\ 0 \end{bmatrix} \leftrightarrow t^3 + 3t^2, \quad \text{and} \quad \begin{bmatrix} 0 \\ 1 \\ -1/3 \\ 1/3 \end{bmatrix} \leftrightarrow t^2 - \frac{1}{3}t + \frac{1}{3}.$$

These two polynomials are a basis for *span*($\mathbb{C}$).

Example 9 Describe the span of the vectors in set

$$\mathbb{R} = \left\{ \begin{bmatrix} 1 & 1 \\ 0 & 0 \end{bmatrix}, \begin{bmatrix} 0 & 1 \\ 0 & 1 \end{bmatrix}, \begin{bmatrix} 1 & 0 \\ 0 & -1 \end{bmatrix}, \begin{bmatrix} 0 & 0 \\ 1 & -1 \end{bmatrix}, \begin{bmatrix} 0 & 1 \\ 1 & 0 \end{bmatrix} \right\}$$

Solution: The vectors in $\mathbb{R}$ are elements of the vector space $\mathbb{M}_{2\times 2}$, which has as its standard basis

$$\left\{ \begin{bmatrix} 1 & 0 \\ 0 & 0 \end{bmatrix}, \begin{bmatrix} 0 & 1 \\ 0 & 0 \end{bmatrix}, \begin{bmatrix} 0 & 0 \\ 1 & 0 \end{bmatrix}, \begin{bmatrix} 0 & 0 \\ 0 & 1 \end{bmatrix} \right\}$$

Coordinate representations of the matrices in $\mathbb{R}$ with respect to the standard basis are

$$\begin{bmatrix} 1 & 1 \\ 0 & 0 \end{bmatrix} = (1)\begin{bmatrix} 1 & 0 \\ 0 & 0 \end{bmatrix} + (1)\begin{bmatrix} 0 & 1 \\ 0 & 0 \end{bmatrix} + (0)\begin{bmatrix} 0 & 0 \\ 1 & 0 \end{bmatrix} + (0)\begin{bmatrix} 0 & 0 \\ 0 & 1 \end{bmatrix} \leftrightarrow \begin{bmatrix} 1 \\ 1 \\ 0 \\ 0 \end{bmatrix}$$

$$\begin{bmatrix} 0 & 1 \\ 0 & 1 \end{bmatrix} = (0)\begin{bmatrix} 1 & 0 \\ 0 & 0 \end{bmatrix} + (1)\begin{bmatrix} 0 & 1 \\ 0 & 0 \end{bmatrix} + (0)\begin{bmatrix} 0 & 0 \\ 1 & 0 \end{bmatrix} + (1)\begin{bmatrix} 0 & 0 \\ 0 & 1 \end{bmatrix} \leftrightarrow \begin{bmatrix} 0 \\ 1 \\ 0 \\ 1 \end{bmatrix}$$

$$\begin{bmatrix} 1 & 0 \\ 0 & -1 \end{bmatrix} = (1)\begin{bmatrix} 1 & 0 \\ 0 & 0 \end{bmatrix} + (0)\begin{bmatrix} 0 & 1 \\ 0 & 0 \end{bmatrix} + (0)\begin{bmatrix} 0 & 0 \\ 1 & 0 \end{bmatrix} + (-1)\begin{bmatrix} 0 & 0 \\ 0 & 1 \end{bmatrix} \leftrightarrow \begin{bmatrix} 1 \\ 0 \\ 0 \\ -1 \end{bmatrix}$$

$$\begin{bmatrix} 0 & 0 \\ 1 & -1 \end{bmatrix} = (0)\begin{bmatrix} 1 & 0 \\ 0 & 0 \end{bmatrix} + (0)\begin{bmatrix} 0 & 1 \\ 0 & 0 \end{bmatrix} + (1)\begin{bmatrix} 0 & 0 \\ 1 & 0 \end{bmatrix} + (-1)\begin{bmatrix} 0 & 0 \\ 0 & 1 \end{bmatrix} \leftrightarrow \begin{bmatrix} 0 \\ 0 \\ 1 \\ -1 \end{bmatrix}$$

$$\begin{bmatrix} 0 & 1 \\ 1 & 0 \end{bmatrix} = (0)\begin{bmatrix} 1 & 0 \\ 0 & 0 \end{bmatrix} + (1)\begin{bmatrix} 0 & 1 \\ 0 & 0 \end{bmatrix} + (1)\begin{bmatrix} 0 & 0 \\ 1 & 0 \end{bmatrix} + (0)\begin{bmatrix} 0 & 0 \\ 0 & 1 \end{bmatrix} \leftrightarrow \begin{bmatrix} 0 \\ 1 \\ 1 \\ 0 \end{bmatrix}$$

We create a matrix **A** having as its *rows* these 4-tuples. Thus,

$$\mathbf{A} = \begin{bmatrix} 1 & 1 & 0 & 0 \\ 0 & 1 & 0 & 1 \\ 1 & 0 & 0 & -1 \\ 0 & 0 & 1 & -1 \\ 0 & 1 & 1 & 0 \end{bmatrix}$$

Reducing this matrix to row-reduced form, we obtain

$$\mathbf{B} = \begin{bmatrix} 1 & 1 & 0 & 0 \\ 0 & 1 & 0 & 1 \\ 0 & 0 & 1 & -1 \\ 0 & 0 & 0 & 0 \\ 0 & 0 & 0 & 0 \end{bmatrix}$$

The nonzero rows of **B**, $[1\ 1\ 0\ 0]$, $[0\ 1\ 0\ 1]$, and $[0\ 0\ 1\ -1]$, form a basis for the row space of **B**. The set of transposes of these vectors are coordinate representatives for the matrices

$$\begin{bmatrix} 1 \\ 1 \\ 0 \\ 0 \end{bmatrix} \leftrightarrow (1)\begin{bmatrix} 1 & 0 \\ 0 & 0 \end{bmatrix} + (1)\begin{bmatrix} 0 & 1 \\ 0 & 0 \end{bmatrix} + (0)\begin{bmatrix} 0 & 0 \\ 1 & 0 \end{bmatrix} + (0)\begin{bmatrix} 0 & 0 \\ 0 & 1 \end{bmatrix} = \begin{bmatrix} 1 & 1 \\ 0 & 0 \end{bmatrix}$$

$$\begin{bmatrix} 0 \\ 1 \\ 0 \\ 1 \end{bmatrix} \leftrightarrow (0)\begin{bmatrix} 1 & 0 \\ 0 & 0 \end{bmatrix} + (1)\begin{bmatrix} 0 & 1 \\ 0 & 0 \end{bmatrix} + (0)\begin{bmatrix} 0 & 0 \\ 1 & 0 \end{bmatrix} + (1)\begin{bmatrix} 0 & 0 \\ 0 & 1 \end{bmatrix} = \begin{bmatrix} 0 & 1 \\ 0 & 0 \end{bmatrix}$$

$$\begin{bmatrix} 0 \\ 0 \\ 1 \\ -1 \end{bmatrix} \leftrightarrow (0)\begin{bmatrix} 1 & 0 \\ 0 & 0 \end{bmatrix} + (0)\begin{bmatrix} 0 & 1 \\ 0 & 0 \end{bmatrix} + (1)\begin{bmatrix} 0 & 0 \\ 1 & 0 \end{bmatrix} + (-1)\begin{bmatrix} 0 & 0 \\ 0 & 1 \end{bmatrix} = \begin{bmatrix} 0 & 0 \\ 1 & -1 \end{bmatrix}$$

These three matrices form a basis for *span*($\mathbb{R}$). Consequently, every matrix in the span of $\mathbb{R}$ must be a linear combination of these three matrices; that is, every matrix in *span*($\mathbb{S}$) must have the form

$$\alpha\begin{bmatrix} 1 & 1 \\ 0 & 0 \end{bmatrix} + \beta\begin{bmatrix} 0 & 1 \\ 0 & 1 \end{bmatrix} + \gamma\begin{bmatrix} 0 & 0 \\ 1 & -1 \end{bmatrix} = \begin{bmatrix} \alpha & \alpha+\beta \\ \gamma & \beta+\gamma \end{bmatrix}$$

for any choice of the scalars α, β, and γ.

Row rank is also useful for determining if a set of n-tuples is linearly independent.

▶THEOREM 3

Let $\mathbb{S}$ be a set of k n-tuples and let **A** *be the $k \times n$ matrix having as its rows the n-tuples in $\mathbb{S}$. $\mathbb{S}$ is linearly independent if and only if the row rank of* **A** *is k, the number of elements in $\mathbb{S}$.* ◀

Proof: Assume that the k n-tuples of $\mathbb{S}$ are linearly independent. Then these k n-tuples are a basis for *span*($\mathbb{S}$), which means that the dimension of *span*($\mathbb{S}$) is k. But the row rank of **A** is the dimension of the row space of **A**, and the row space of **A** is also *span*($\mathbb{S}$). Because every basis for the same vector space must contain the same number of elements (Theorem 2 of Section 2.5), it follows that the row rank of **A** equals k.

Conversely, if the row rank of **A** equals k, then a basis for *span*($\mathbb{S}$) must contain k n-tuples. The vectors in $\mathbb{S}$ are a spanning set for *span*($\mathbb{S}$), by definition. Now, either $\mathbb{S}$ is linearly independent or linearly dependent. If it is linearly dependent, then one vector must be a linear combination of vectors that precede it. Delete this vector from $\mathbb{S}$. The resulting set still spans $\mathbb{S}$. Keep deleting vectors until no vector is a linear combination of preceding vectors. At that point we have a linearly independent set that spans $\mathbb{S}$ that is a basis for *span*($\mathbb{S}$), which contains fewer than k vectors. This contradicts the fact that the dimension *of span*($\mathbb{S}$) equals k. Thus, $\mathbb{S}$ cannot be linearly dependent, which implies it must linearly independent.

Example 10 Determine whether the set

$$\mathbb{D} = \{[0 \;\; 1 \;\; 2 \;\; 3 \;\; 0], [1 \;\; 3 \;\; -1 \;\; 2 \;\; 1], \\ [2 \;\; 6 \;\; -1 \;\; -3 \;\; 1], [4 \;\; 0 \;\; 1 \;\; 0 \;\; 2]\}$$

is linearly independent.

Solution: We consider the matrix

$$\mathbf{A} = \begin{bmatrix} 0 & 1 & 2 & 0 \\ 1 & 3 & -1 & 1 \\ 2 & 6 & -1 & 1 \\ 4 & 0 & 1 & 2 \end{bmatrix}$$

which can be transformed (after the first two rows are interchanged) to the row-reduced form

$$\mathbf{B} = \begin{bmatrix} 1 & 3 & -1 & 2 & 1 \\ 0 & 1 & 2 & 3 & 0 \\ 0 & 0 & 1 & -7 & -1 \\ 0 & 0 & 0 & 1 & 27/231 \end{bmatrix}$$

Matrix **B** has four nonzero rows, hence the row rank of **B**, as well as the row rank of **A**, is four. There are four 5-tuples in $\mathbb{D}$, so it follows from Theorem 3 that $\mathbb{S}$ is linearly independent.

We can extend Theorem 3 to all finite-dimensional vector spaces, not just n-tuples. We represent every vector in a given set $\mathbb{S}$ by an n-tuple with respect to a basis and then apply Theorem 3 directly to the coordinate representations.

Example 11 Determine whether the set of four polynomials in Example 8 is linearly independent.

Solution: Coordinate representations for each of the given polynomials with respect to the standard basis in $\mathbb{P}^3$ were determined in Example 8. The matrix **A** in Example 8 has as its rows each coordinate representation. **A** can be transformed into the row-reduced form of the matrix **B** in Example 8. It follows that the row rank of **B** is two, which is also the row rank of **A**. This number is less than the number of elements in $\mathbb{S}$, hence $\mathbb{S}$ is linearly dependent.

Problems 2.6

In Problems 1 through 21, find a basis for *span*($\mathbb{S}$).

(1) $\mathbb{S} = \left\{ \begin{bmatrix} 1 \\ 1 \\ 2 \end{bmatrix} \begin{bmatrix} 2 \\ -1 \\ 0 \end{bmatrix} \begin{bmatrix} 4 \\ 1 \\ 4 \end{bmatrix} \right\}.$

(2) $\mathbb{S} = \left\{ \begin{bmatrix} 1 \\ 1 \\ 2 \end{bmatrix} \begin{bmatrix} 2 \\ 1 \\ 0 \end{bmatrix} \begin{bmatrix} 4 \\ 1 \\ 4 \end{bmatrix} \right\}.$

(3) $\mathbb{S} = \left\{ \begin{bmatrix} 2 \\ 1 \\ 2 \end{bmatrix}, \begin{bmatrix} -2 \\ -1 \\ -2 \end{bmatrix}, \begin{bmatrix} 4 \\ 2 \\ 4 \end{bmatrix}, \begin{bmatrix} -4 \\ -2 \\ -4 \end{bmatrix} \right\}.$

(4) $\mathbb{S} = \left\{ \begin{bmatrix} 1 \\ 0 \\ 2 \end{bmatrix}, \begin{bmatrix} -1 \\ 1 \\ -1 \end{bmatrix}, \begin{bmatrix} 0 \\ 1 \\ 1 \end{bmatrix}, \begin{bmatrix} -1 \\ 2 \\ 0 \end{bmatrix} \right\}.$

(5) $\mathbb{S} = \{[1 \ \ 2 \ \ -1 \ \ 1], [0 \ \ 1 \ \ 2 \ \ 1], [2 \ \ 3 \ \ -4 \ \ 1], [2 \ \ 4 \ \ -2 \ \ 2]\}.$

(6) $\mathbb{S} = \{[0 \ \ 1 \ \ 1 \ \ 1], [1 \ \ 0 \ \ 0 \ \ 1], [-1 \ \ 1 \ \ 1 \ \ 0], [1 \ \ 1 \ \ 0 \ \ 1]\}.$

(7) $\mathbb{S} = \{[1 \ \ 0 \ \ -1 \ \ 1], [3 \ \ 1 \ \ 0 \ \ 1], [1 \ \ 1 \ \ 2 \ \ -1], [3 \ \ 2 \ \ 3 \ \ -1],$
$[2 \ \ 1 \ \ 0 \ \ 0]\}.$

(8) $\mathbb{S}=\{[2\ \ 2\ \ 1\ \ 2],[1\ \ -1\ \ 0\ \ 1],[0\ \ -4\ \ -1\ \ 1],[1\ \ 0\ \ 2\ \ 1]\},$ $[0\ \ -1\ \ 2\ \ 2]\}$.

(9) $\mathbb{S}=\{[1\ \ 2\ \ 4\ \ 0],[2\ \ 4\ \ 8\ \ 0],[1\ \ -1\ \ 0\ \ 1],[4\ \ 2\ \ 8\ \ 2]\},$ $[4\ \ -1\ \ 4\ \ 3]\}$.

(10) $\mathbb{S}=\{t^2+t, t+1, t^2+1, 1\}$.

(11) $\mathbb{S}=\{t^2+t+1, 2t^2-2t+1, t^2-3t\}$.

(12) $\mathbb{S}=\{t, t+1, t-1, 1\}$.

(13) $\mathbb{S}=\{t^2+t, t-1, t^2+1\}$.

(14) $\mathbb{S}=\{t^2+t+1, t+1, t^2\}$.

(15) $\mathbb{S}=\{t^3+t^2-t, t^3+2t^2+1, 2t^3+3t^2-t+1, 3t^3+5t^2-t+2\}$.

(16) $\mathbb{S}=\{2t^3+t^2+1, t^2+t, 2t^3-t+1, t+1, 2t^3+2\}$.

(17) $\mathbb{S}=\{t^3+3t^2, t^2+1, t+1, t^3+4t^2+t+2, t^2+t+2\}$.

(18) $\mathbb{S}=\left\{\begin{bmatrix}1 & 0\\ 0 & 0\end{bmatrix},\begin{bmatrix}0 & 1\\ 0 & 0\end{bmatrix},\begin{bmatrix}1 & 2\\ 0 & 0\end{bmatrix},\begin{bmatrix}1 & 3\\ 0 & 0\end{bmatrix}\right\}$.

(19) $\mathbb{S}=\left\{\begin{bmatrix}1 & 1\\ 1 & 0\end{bmatrix},\begin{bmatrix}-1 & 1\\ 1 & 0\end{bmatrix},\begin{bmatrix}1 & -1\\ 1 & 0\end{bmatrix},\begin{bmatrix}1 & 1\\ -1 & 0\end{bmatrix}\right\}$.

(20) $\mathbb{S}=\left\{\begin{bmatrix}1 & 0\\ 0 & 1\end{bmatrix},\begin{bmatrix}0 & 1\\ 1 & 0\end{bmatrix},\begin{bmatrix}1 & 1\\ 1 & 1\end{bmatrix},\begin{bmatrix}1 & -1\\ -1 & 1\end{bmatrix}\right\}$.

(21) $\mathbb{S}=\left\{\begin{bmatrix}1 & 3\\ 1 & 2\end{bmatrix},\begin{bmatrix}1 & 2\\ 1 & 1\end{bmatrix},\begin{bmatrix}0 & 1\\ 0 & 1\end{bmatrix},\begin{bmatrix}2 & 7\\ 2 & 5\end{bmatrix}\right\}$.

In Problems 22 through 43, use row rank to determine whether the given sets are linearly independent.

(22) $\{[1\ \ 0][0\ \ 1]\}$.

(23) $\{[1\ \ 1][1\ \ -1]\}$.

(24) $\{[2\ \ -4][-3\ \ 6]\}$.

(25) $\left\{\begin{bmatrix}1\\ 0\\ 1\end{bmatrix},\begin{bmatrix}1\\ 1\\ 0\end{bmatrix},\begin{bmatrix}0\\ 1\\ 1\end{bmatrix}\right\}$.

(26) $\left\{\begin{bmatrix}1\\ 0\\ 1\end{bmatrix},\begin{bmatrix}1\\ 0\\ 2\end{bmatrix},\begin{bmatrix}2\\ 0\\ 1\end{bmatrix}\right\}$.

(27) $\left\{\begin{bmatrix}1\\ 0\\ 1\end{bmatrix},\begin{bmatrix}1\\ 1\\ 1\end{bmatrix},\begin{bmatrix}1\\ -1\\ 1\end{bmatrix}\right\}$.

(28) $\left\{\begin{bmatrix}0\\0\\0\end{bmatrix},\begin{bmatrix}3\\2\\1\end{bmatrix},\begin{bmatrix}2\\1\\3\end{bmatrix}\right\}.$

(29) $\left\{\begin{bmatrix}1\\2\\3\end{bmatrix},\begin{bmatrix}3\\2\\1\end{bmatrix},\begin{bmatrix}2\\1\\3\end{bmatrix}\right\}.$

(30) $\{[1 \quad 1 \quad 0][1 \quad -1 \quad 0]\}.$

(31) $\{[1 \quad 2 \quad 3][-3 \quad -6 \quad -9]\}.$

(32) $\{[10 \quad 20 \quad 20],[10 \quad -10 \quad 10][10 \quad 20 \quad 10]\}.$

(33) $\{[2 \quad 1 \quad 1],[3 \quad -1 \quad 4],[1 \quad 3 \quad -2]\}.$

(34) $\left\{\begin{bmatrix}1 & 1\\0 & 0\end{bmatrix},\begin{bmatrix}1 & 1\\1 & 1\end{bmatrix},\begin{bmatrix}0 & 0\\1 & 1\end{bmatrix}\right\}.$

(35) $\left\{\begin{bmatrix}1 & 1\\0 & 0\end{bmatrix},\begin{bmatrix}1 & 0\\1 & 1\end{bmatrix},\begin{bmatrix}0 & 0\\1 & 1\end{bmatrix}\right\}.$

(36) $\left\{\begin{bmatrix}1 & 0\\1 & 1\end{bmatrix},\begin{bmatrix}1 & 1\\1 & 0\end{bmatrix},\begin{bmatrix}1 & 1\\0 & 1\end{bmatrix},\begin{bmatrix}0 & 1\\1 & 1\end{bmatrix}\right\}.$

(37) $\left\{\begin{bmatrix}1 & 0\\1 & 1\end{bmatrix},\begin{bmatrix}1 & 1\\1 & 0\end{bmatrix},\begin{bmatrix}2 & 2\\0 & 2\end{bmatrix},\begin{bmatrix}1 & 0\\2 & 0\end{bmatrix}\right\}.$

(38) $\{t,2\}.$

(39) $\{t^3+t^2, t^3+t, t^2+t\}.$

(40) $\{t^3+t^2, t^3-t^2, t^3-3t^2\}.$

(41) $\{t^3+t^2, t^3-t^2, t^3-t, t^3+1\}.$

(42) $\{t^2+t, t^2+t-1, t^2+1, t\}.$

(43) $\{t^2+t, t^2+t-2, 1\}.$

(44) Can a 4×3 matrix have linearly independent rows?

(45) Prove that if the row rank of an $m \times n$ matrix is k, then $k \leq \text{minimum } \{m, n\}$.

(46) Prove that if a matrix **B** is obtained from a matrix **A** by interchanging the positions of any two rows of **A**, then both **A** and **B** have the same row space.

(47) Prove that if a matrix **B** is obtained from a matrix **A** by multiplying one row of **A** by a nonzero scalar, then both **A** and **B** have the same row space.

2.7 RANK OF A MATRIX

We began this chapter noting that much of mathematical analysis is identifying fundamental structures that appear with regularity in different situations, developing those structures in the abstract, and then applying the resulting knowledge base back to the individual situations to further our understanding of those situations. The fundamental structure we developed was that of a vector space. We now use our knowledge of this structure to further our understanding of sets of simultaneous linear equations and matrix inversion.

The column space of a matrix is the subspace spanned by the columns of the matrix; the dimension of the column space is the column rank.

In the last section we defined the row space of a matrix **A** to be the subspace spanned by the rows of **A**, considered as row matrices. We now define the *column space* of a matrix **A** to be the subspace spanned by the columns of **A**, considered as column matrices. The dimension of the column space is called the *column rank* of **A**.

Example 1 The matrix $\mathbf{A} = \begin{bmatrix} 1 & 2 & 3 \\ 4 & 5 & 6 \end{bmatrix}$ has three columns, all belonging to $\mathbb{R}^2$. The column space of **A** consists of all linear combinations of the columns of **A**; that is, if we set

$$\mathbb{T} = \left\{ \begin{bmatrix} 1 \\ 4 \end{bmatrix}, \begin{bmatrix} 2 \\ 5 \end{bmatrix}, \begin{bmatrix} 3 \\ 6 \end{bmatrix} \right\}$$

then the column space of **A** is $span(\mathbb{T})$. The dimension of $span(\mathbb{T})$ is the column rank of **A**.

The row space of a $p \times n$ matrix **A** is a subspace of $\mathbb{R}^n$ while its column space is a subspace of $\mathbb{R}^p$, and these are very different vector spaces when p and n are unequal. Surprisingly, both have the same dimension. The proof of this statement is a bit lengthy, so we separate it into two parts.

▶LEMMA 1

The column rank of a matrix is less than or equal to its row rank.◀

Proof: Let $\mathbf{A}_1, \mathbf{A}_2, \ldots, \mathbf{A}_p$ be the rows, considered as row matrices, of a $p \times n$ matrix $\mathbf{A} = [a_{ij}]$. Then

$$\mathbf{A}_i = [a_{i1} \quad a_{i2} \quad \ldots \quad a_{in}]; \quad (i = 1, 2, \ldots, p)$$

Let k denote the row rank of **A**. Thus, k is the dimension of the subspace spanned by the rows of **A**, and this subspace has a basis containing exactly k vectors. Designate one such basis as the set $\mathbb{B} = \{\mathbf{u}_1, \mathbf{u}_2, \ldots, \mathbf{u}_k\}$. Each vector in the basis is an n-tuple of the form

$$\mathbf{u}_i = [u_{i1} \quad u_{i2} \quad \ldots \quad u_{in}]; (i = 1, 2, \ldots, k)$$

Since $\mathbb{B}$ is a basis, every vector in the subspace spanned by the rows of **A** can be written as a linear combination of the vectors in $\mathbb{B}$, including the rows of **A** themselves. Thus,

$$\begin{aligned} \mathbf{A}_1 &= d_{11}\mathbf{u}_1 + d_{12}\mathbf{u}_2 + \cdots + d_{1k}\mathbf{u}_k \\ \mathbf{A}_2 &= d_{21}\mathbf{u}_1 + d_{22}\mathbf{u}_2 + \cdots + d_{2k}\mathbf{u}_k \\ &\vdots \\ \mathbf{A}_p &= d_{p1}\mathbf{u}_1 + d_{p2}\mathbf{u}_2 + \cdots + d_{pk}\mathbf{u}_k \end{aligned}$$

for some set of uniquely determined scalars d_{ij} ($i=1, 2, \ldots; j=1, 2, \ldots, k$). In each of the preceding individual equalities, both the left and right sides are n-tuples. If we consider just the jth component of each n-tuple ($j=1, 2, \ldots, n$), first the jth component of $\mathbf{A}_1$, then the jth component of $\mathbf{A}_2$, sequentially through the jth component of $\mathbf{A}_p$, we obtain the equalities

$$\begin{aligned} a_{1j} &= d_{11}u_{1j} + d_{12}u_{2j} + \cdots + d_{1k}u_{kj} \\ a_{2j} &= d_{21}u_{1j} + d_{22}u_{2j} + \cdots + d_{2k}u_{kj} \\ &\vdots \\ a_{pj} &= d_{p1}u_{1j} + d_{p2}u_{2j} + \cdots + d_{pk}u_{kj} \end{aligned}$$

which can be rewritten as the vector equation

$$\begin{bmatrix} a_{1j} \\ a_{2j} \\ \vdots \\ a_{pj} \end{bmatrix} = u_{1j}\begin{bmatrix} d_{11} \\ d_{21} \\ \vdots \\ d_{p1} \end{bmatrix} + u_{2j}\begin{bmatrix} d_{12} \\ d_{22} \\ \vdots \\ d_{p2} \end{bmatrix} + \cdots + u_{kj}\begin{bmatrix} d_{1k} \\ d_{2k} \\ \vdots \\ d_{pk} \end{bmatrix}$$

Thus, the jth column of $\mathbf{A}$ can be expressed as a linear combination of k vectors. Since this is true for each j, it follows that each column of $\mathbf{A}$ can be expressed as a linear combination of the same k vectors, which implies that the dimension of the column space of $\mathbf{A}$ is at most k. That is, the column rank of $\mathbf{A} \leq k =$ the row rank of $\mathbf{A}$.

▶THEOREM 1

The row rank of a matrix equals its column rank.◀

Proof: For any matrix $\mathbf{A}$, we may apply Lemma 1 to its transpose and conclude that the column rank of $\mathbf{A}^T$ is less than or equal to its row rank. But since the columns of $\mathbf{A}^T$ are the rows of $\mathbf{A}$ and vice versa, it follows that the row rank of $\mathbf{A}$ is less than or equal to its column rank. Combining this result with Lemma 1, we have Theorem 1.

Since the row rank and column rank of a matrix $\mathbf{A}$ are equal, we refer to them both simply as the *rank* of $\mathbf{A}$, denoted as $r(\mathbf{A})$.

With the concepts of vector space, basis, and rank in hand, we can give explicit criteria for determining when solutions to sets of simultaneous linear equations exist. In other words, we can develop a theory of solutions to complement our work in Chapter 1.

The rank of a matrix **A**, denoted as $r(\mathbf{A})$, is the row rank of **A**, which is also the column rank of **A**.

A system of m simultaneous linear equations in n unknowns has the form

$$\begin{aligned} a_{11}x_1 + a_{12}x_2 + \cdots + a_{1n}x_n &= b_1 \\ a_{21}x_1 + a_{22}x_2 + \cdots + a_{2n}x_n &= b_2 \\ &\vdots \\ a_{m1}x_1 + a_{m2}x_2 + \cdots + a_{mn}x_n &= b_m \end{aligned} \tag{2.29}$$

or the matrix form

$$\mathbf{Ax} = \mathbf{b} \tag{2.30}$$

If we denote the columns of **A** by the w-dimensional column matrices

$$\mathbf{A}_1 = \begin{bmatrix} a_{11} \\ a_{21} \\ \vdots \\ a_{m1} \end{bmatrix}, \quad \mathbf{A}_2 = \begin{bmatrix} a_{12} \\ a_{22} \\ \vdots \\ a_{m2} \end{bmatrix}, \quad \ldots, \quad \mathbf{A}_n = \begin{bmatrix} a_{1n} \\ a_{2n} \\ \vdots \\ a_{mn} \end{bmatrix}$$

then we can rewrite Equation (2.20) in the vector form

$$x_1\mathbf{A}_1 + x_2\mathbf{A}_2 + \cdots + x_n\mathbf{A}_n = \mathbf{b} \tag{2.31}$$

Example 2 The system of equations

$$\begin{aligned} x - 2y + 3z &= 7 \\ 4x + 5y - 6z &= 8 \end{aligned}$$

has the vector form

$$x\begin{bmatrix} 1 \\ 4 \end{bmatrix} + y\begin{bmatrix} -2 \\ 5 \end{bmatrix} + z\begin{bmatrix} 3 \\ -6 \end{bmatrix} = \begin{bmatrix} 7 \\ 8 \end{bmatrix}$$

Solving (2.29) or (2.30) is equivalent to finding scalars $x_12 \ldots, x_n$ that satisfy Equation (2.31). If such scalars exist, then the vector **b** is a linear combination of the vectors $\mathbf{A}_1, \mathbf{A}_2, \ldots, \mathbf{A}_n$. That is, **b** is in the span of $\{\mathbf{A}_1, \mathbf{A}_2, \ldots, \mathbf{A}_n\}$ or, equivalently, in the column space of **A**. Consequently, adjoining **b** to the set of vectors defined by the columns of **A** will not change the column rank of **A**. Therefore, the column rank of **A** must equal the column rank of $[\mathbf{A}|\mathbf{b}]$. On the other hand, if no scalars $x_1, x_2, \ldots, x_n$ satisfy Equation (2.31), then **b** is not a linear combination of $\mathbf{A}_1, \mathbf{A}_2, \ldots, \mathbf{A}_n$. That is, **b** is *not* in the span of $\{\mathbf{A}_1, \mathbf{A}_2, \ldots, \mathbf{A}_n\}$, in which case, the column rank of $[\mathbf{A}|\mathbf{b}]$ must be greater by 1 than the column rank of **A**. Since column rank equals row rank equals rank, we have proven Theorem 2.

▶THEOREM 2

The system $\mathbf{Ax} = \mathbf{b}$ *is consistent if and only if* $r(\mathbf{A}) = r[\mathbf{A}|\mathbf{b}]$. ◀

Example 3 Determine whether the following system of equations is consistent:

$$x + y - z = 1$$
$$x + y - z = 0$$

Solution:

$$A = \begin{bmatrix} 1 & 1 & -1 \\ 1 & 1 & -1 \end{bmatrix}, \; b = \begin{bmatrix} 1 \\ 0 \end{bmatrix}, \; [A|b] = \left[\begin{array}{ccc|c} 1 & 1 & -1 & 1 \\ 1 & 1 & -1 & 0 \end{array}\right]$$

[**A**|**b**] is transformed to row-reduced form

$$\left[\begin{array}{ccc|c} 1 & 1 & -1 & 1 \\ 1 & 1 & -1 & 0 \end{array}\right] \to \left[\begin{array}{ccc|c} 1 & 1 & -1 & 1 \\ 0 & 0 & 0 & -1 \end{array}\right] \quad \text{by adding to the second row } -1 \text{ times the first row}$$

$$\to \left[\begin{array}{ccc|c} 1 & 1 & -1 & 1 \\ 0 & 0 & 0 & 1 \end{array}\right] \quad \text{by multiplying the second row by } -1 \tag{2.32}$$

This matrix has two nonzero rows, hence $r[\mathbf{A}|\mathbf{b}]=2$. If we delete the last column from the matrix in Equation (2.32), we have **A** in the row-reduced form

$$\begin{bmatrix} 1 & 1 & -1 \\ 0 & 0 & 0 \end{bmatrix}$$

This matrix has one nonzero row, so $r(\mathbf{A})=1$. Since $r(\mathbf{A}) \neq r[\mathbf{A}|\mathbf{b}]$, it follows from Theorem 2 that the given set of equations has no solution and is not consistent.

Example 4 Determine whether the following system of equations is consistent:

$$x + y + w = 3$$
$$2x + 2y + 2w = 6$$
$$-x - y - w = -3$$

Solution:

$$\mathbf{A} = \begin{bmatrix} 1 & 1 & 1 \\ 2 & 2 & 2 \\ -1 & -1 & -1 \end{bmatrix}, \quad \mathbf{b} = \begin{bmatrix} 3 \\ 6 \\ -3 \end{bmatrix}, \quad [\mathbf{A}|\mathbf{b}] = \left[\begin{array}{ccc|c} 1 & 1 & 1 & 3 \\ 2 & 2 & 2 & 6 \\ -1 & -1 & -1 & -3 \end{array}\right]$$

By transforming both **A** and [**A**|**b**] to row-reduced form, we can show that $r(\mathbf{A})=r[\mathbf{A}|\mathbf{b}]=1$. Therefore, the original system is consistent.

Once a system is determined to be consistent, the following theorem specifies the number of solutions.

►THEOREM 3

If the system $\mathbf{Ax}=\mathbf{b}$ is consistent and if $r(\mathbf{A})=k$, then solutions to the system are expressible in terms of $n-k$ arbitrary unknowns, where n denotes the total number of unknowns in the system.◄

Proof: To determine the rank of the augmented matrix $[\mathbf{A}|\mathbf{b}]$, reduce the augmented matrix to row-reduced form and count the number of nonzero rows. With Gaussian elimination, we can solve the resulting row-reduced matrix for the variables associated with the first nonzero entry in each nonzero row. Thus, each nonzero row defines one variable and all other variables remain arbitrary.

Example 5 Determine the number of solutions to the system described in Example 4.

Solution: The system has three unknowns, x, y, and w, hence $n=3$. Here $r(\mathbf{A})=r[\mathbf{A}|\mathbf{b}]=1$, so $k=1$. The solutions are expressible in terms of $3-1=2$ arbitrary unknowns. Using Gaussian elimination, we find the solution as $x=3-y-w$ with both y and w arbitrary.

Example 6 Determine the number of solutions to the system

$$2x-3y+z=-1$$
$$x-y+2z=2$$
$$2x+y-3z=3$$

Solution:

$$\mathbf{A}=\begin{bmatrix}2 & -3 & 1\\ 1 & -1 & 2\\ 2 & 1 & -3\end{bmatrix},\quad \mathbf{b}=\begin{bmatrix}-1\\ 2\\ 3\end{bmatrix},\quad [\mathbf{A}|\mathbf{b}]=\left[\begin{array}{ccc|c}2 & -3 & 1 & -1\\ 1 & -1 & 2 & 2\\ 2 & 1 & -3 & 3\end{array}\right]$$

By transforming both $\mathbf{A}$ and $[\mathbf{A}|\mathbf{b}]$ to row-reduced form, we can show that $r(\mathbf{A})=r[\mathbf{A}|\mathbf{b}]=3$; hence, the given system is consistent. In this case, $n=3$ (three variables) and (rank) $k=3$; the solutions are expressible in terms of $3-3=0$ arbitrary unknowns. Thus, the solution is unique (none of the unknowns is arbitrary). Using Gaussian elimination, we find the solution as $x=y=2$, $z=1$.

A homogeneous system of simultaneous linear equations has the form

$$\begin{aligned}a_{11}x_1+a_{12}x_2+\cdots+a_{1n}x_n&=0\\ a_{21}x_1+a_{22}x_2+\cdots+a_{2n}x_n&=0\\ &\vdots\\ a_{m1}x_1+a_{m2}x_2+\cdots+a_{mn}x_n&=0\end{aligned}\tag{2.33}$$

or the matrix form

$$\mathbf{Ax}=0\tag{2.34}$$

A homogeneous system of equations is always consistent, and one solution is always the trivial solution.

Since Equation (2.34) is a special case of Equation (2.30) with $\mathbf{b}=\mathbf{0}$, Theorems 2 and 3 remain valid. Because of the simplified structure of a homogeneous system, however, we can draw conclusions about it that are not valid for nonhomogeneous systems. In particular, a homogeneous system is consistent, because the trivial solution $\mathbf{x}=\mathbf{0}$ is always a solution to $\mathbf{Ax}=\mathbf{0}$. Furthermore, if the rank of $\mathbf{A}$

equals the number of unknowns, then the solution is unique and the trivial solution is the only solution. On the other hand, it follows from Theorem 3 that if the rank of A is less than the number of unknowns, then the solution will be in terms of arbitrary unknowns. Since these arbitrary unknowns can be assigned nonzero values, nontrivial solutions exist. Thus, we have Theorem 4.

▶THEOREM 4

A homogeneous system of equations $\mathbf{Ax}=\mathbf{0}$ *in n unknowns will admit nontrivial solutions if and only if* $r(\mathbf{A})\neq n$.◀

The concept of rank also provides the tools to prove two results we simply stated in the previous chapter. We can now determine a criterion for the existence of an inverse and also show that, for square matrices, the equality $\mathbf{AB}=\mathbf{I}$ implies the equality $\mathbf{BA}=\mathbf{I}$. For convenience, we separate the analysis into segments.

▶LEMMA 2

Let **A** *and* **B** *be* $n\times n$ *matrices. If* $\mathbf{AB}=\mathbf{I}$, *then the system of equations* $\mathbf{Ax}=\mathbf{y}$ *has a solution for every choice of the vector* **y**.◀

Proof: Once **y** is specified, set $\mathbf{x}=\mathbf{By}$. Then

$$\mathbf{Ax} = \mathbf{A}(\mathbf{By}) = (\mathbf{AB})\mathbf{y} = \mathbf{Iy} = \mathbf{y}$$

hence $\mathbf{x}=\mathbf{By}$ is a solution of $\mathbf{Ax}=\mathbf{y}$.

▶LEMMA 3

If **A** *and* **B** *are* $n\times n$ *matrices with* $\mathbf{AB}=\mathbf{I}$, *then the rows of* **A**, *considered as n-dimensional row matrices, are linearly independent.*◀

Proof: Designate the rows of **A** by $\mathbf{A}_1, \mathbf{A}_2, \ldots, \mathbf{A}_n$, respectively, and the columns of **I** as the vectors $\mathbf{e}_1, \mathbf{e}_2, \ldots, \mathbf{e}_n$, respectively. It follows from Lemma 2 that the set of equations $\mathbf{Ax}=\mathbf{e}_j$ $(j=1, 2, \ldots, n)$ has a solution for each j. Denote these solutions by $\mathbf{x}_1, \mathbf{x}_2, \ldots, \mathbf{x}_n$ respectively. Therefore,

$$\mathbf{Ax}_j = \mathbf{e}_j \tag{2.35}$$

Since $\mathbf{e}_j$ is an n-dimensional column matrix having a unity element in row j and zeros elsewhere, it follows from Equation (2.26) that, for $i=1, 2, \ldots, n$,

$$i\text{th component of } \mathbf{Ax}_j = \begin{cases} 1 & \text{when } i=j \\ 0 & \text{when } i\neq j \end{cases}$$

This equation can be simplified if we make use of the *Kronecker delta* δ_{ij} defined as

$$i\text{th component of } \mathbf{A}\mathbf{x}_j = \begin{cases} 1 & \text{when } i = j \\ 0 & \text{when } i \neq j \end{cases} \tag{2.36}$$

Thus, Equation (2.35) may be written as

$$i\text{th component of } \mathbf{A}\mathbf{x}_j = \delta_{ij}$$

or, more simply, as

$$\mathbf{A}_i\mathbf{x}_j = \delta_{ij} \tag{2.37}$$

Now consider the vector equation

$$\sum_{i=1}^{n} c_i\mathbf{A}_i = 0 \tag{2.38}$$

We want to show that each constant c_i ($i=1, 2, \ldots, n$) must be 0. Multiplying both sides of Equation (2.38) on the right by the vector $\mathbf{x}_j$, and using Equations (2.36) and (2.37), we have

$$0 = 0\mathbf{x}_j = \left(\sum_{i=1}^{n} c_i\mathbf{A}_i\right)\mathbf{x}_j\sum_{i=1}^{n}(c_i\mathbf{A}_i)\mathbf{x}_j = \sum_{i=1}^{n} c_i\left(\mathbf{A}_i\mathbf{x}_j\right) = \sum_{i=1}^{n} c_i\delta_{ij} = c_j$$

Thus, for each $\mathbf{x}_j$ ($j=1, 2, \ldots, n$) we have $c_j=0$, which implies that $c_1=c_2=\cdots=c_n=0$ and that the rows of $\mathbf{A}$, namely, $\mathbf{A}_1, \mathbf{A}_2, \ldots, \mathbf{A}_n$, are linearly independent.

It follows directly from Lemma 3 and the definition of an inverse that if an $n \times n$ matrix $\mathbf{A}$ has an inverse, then $\mathbf{A}$ must have rank n. This in turn implies that if $\mathbf{A}$ does not have rank n, then $\mathbf{A}$ does not have an inverse. We also want the converse: that is, if $\mathbf{A}$ has rank n, then $\mathbf{A}$ has an inverse.

►LEMMA 4

If an $n \times n$ matrix $\mathbf{A}$ *has rank n, then there exists a square matrix* $\mathbf{C}$ *such that* $\mathbf{CA}=\mathbf{I}$.◄

Proof: If an $n \times n$ matrix $\mathbf{A}$ has rank n, then its row-reduced form is an upper triangular matrix with all elements on the main diagonal equal to 1. Using these diagonal elements as pivots, we can use elementary row operations to further transform $\mathbf{A}$ to an identity matrix. Corresponding to each elementary row operation is an elementary matrix. Therefore, if $\mathbf{A}$ has rank n, then there is a sequence of elementary matrices $\mathbf{E}_1, \mathbf{E}_2, \ldots, \mathbf{E}_{k-1}, \mathbf{E}_k$ such that

$$\mathbf{E}_k\mathbf{E}_{k-1}\ldots\mathbf{E}_2\mathbf{E}_1\mathbf{A} = \mathbf{I} \tag{2.39}$$

Setting

$$\mathbf{C} = \mathbf{E}_k\mathbf{E}_{k-1}\ldots\mathbf{E}_2\mathbf{E}_1$$

we have

$$\mathbf{CA} = \mathbf{I} \tag{2.40}$$

▶LEMMA 5

If **A** *and* **B** *are $n \times n$ matrices such that* **AB**=**I**, *then* **BA**=**I**.◀

Proof: If **AB**=**I**, then it follows from Lemma 3 that **A** has rank n. It then follows from Lemma 4 that there exists a matrix **C** such that **CA**=**I**. Consequently,

$$\mathbf{C} = \mathbf{CI} = \mathbf{C}(\mathbf{AB}) = (\mathbf{CA})\mathbf{B} = \mathbf{IB} = \mathbf{B}$$

so the equality **CA**=**I** implies that **BA**=**I**.

If we replace **A** by **C** and **B** by **A** in Lemma 5, we have that, if **C** and **A** are $n \times n$ matrices such that **CA**=**I**, then it is also true that

$$\mathbf{AC} = \mathbf{I} \tag{2.41}$$

Therefore, if **A** is an $n \times n$ matrix with rank n, then We want to show that each constant Equation (2.40) holds, whereupon Equation (2.41) also holds. Together Equations (2.40) and (2.41) imply that **C** is the inverse of **A**. Thus, we have proven Theorem 5.

▶THEOREM 5

An $n \times n$ matrix **A** *has an inverse if and only if* **A** *has rank n.*◀

In addition, we also have Theorem 6.

▶THEOREM 6

A square matrix has an inverse if and only if it can be transformed by elementary row operations to an upper triangular matrix with all elements on the main diagonal equal to 1.◀

Proof: An $n \times n$ matrix **A** has an inverse if and only if it has rank n (Theorem 5). It has rank n if and only if it can be transformed by elementary row operations into a row-reduced matrix **B** having rank n (Theorem 2 of Section 2.6). **B** has rank n if and only if it contains n nonzero rows (Theorem 1 of Section 2.6). **A** row-reduced, $n \times n$ matrix **B** has n nonzero rows if and only if it is upper triangular with just ones on its main diagonal.

Problems 2.7

In Problems 1 through 7, find the ranks of the given matrices.

(1) $\begin{bmatrix} 1 & 2 & 0 \\ 3 & 1 & -5 \end{bmatrix}$.

(2) $\begin{bmatrix} 2 & 8 & -6 \\ -1 & -4 & 3 \end{bmatrix}$.

(3) $\begin{bmatrix} 4 & 1 \\ 2 & 3 \\ 2 & 2 \end{bmatrix}$.

(4) $\begin{bmatrix} 4 & 8 \\ 6 & 12 \\ 9 & 18 \end{bmatrix}$.

(5) $\begin{bmatrix} 1 & 4 & -2 \\ 2 & 8 & -4 \\ -1 & -4 & 2 \end{bmatrix}$.

(6) $\begin{bmatrix} 1 & 2 & 4 & 2 \\ 1 & 1 & 3 & 2 \\ 1 & 4 & 6 & 2 \end{bmatrix}$.

(7) $\begin{bmatrix} 1 & 7 & 0 \\ 0 & 1 & 1 \\ 1 & 1 & 0 \end{bmatrix}$.

(8) What is the largest possible value for the rank of a 2×5 matrix?

(9) What is the largest possible value for the rank of a 4×3 matrix?

(10) What is the largest possible value for the rank of a 4×6 matrix?

(11) Show that the rows of a 5×3 matrix are linearly dependent.

(12) Show that the columns of a 2×4 matrix are linearly dependent.

(13) What is the rank of a zero matrix?

(14) Use the concept of rank to determine whether $[3\ 7]$ can be written as a linear combination of the following sets of vectors.

(a) $\{[1\ 2], [4\ 8]\}$ (b) $\{[1\ 2], [3\ 2]\}$.

(15) Use the concept of rank to determine whether $[2\ 3]$ can be written as a linear combination of the following sets of vectors.

(a) $\{[10\ 15], [4\ 6]\}$, (b) $\{[1\ 1], [1\ -1]\}$,

(c) $\{[2\ -4], [-3\ 6]\}$.

(16) Use the concept of rank to determine whether $[1\ 1\ 1]^{\mathrm{T}}$ can be written as a linear combination of the following sets of vectors.

(a) $\left\{ \begin{bmatrix} 1 \\ 0 \\ 1 \end{bmatrix}, \begin{bmatrix} 1 \\ 1 \\ 0 \end{bmatrix}, \begin{bmatrix} 0 \\ 1 \\ 1 \end{bmatrix} \right\}$, (b) $\left\{ \begin{bmatrix} 1 \\ 0 \\ 1 \end{bmatrix}, \begin{bmatrix} 1 \\ 0 \\ 2 \end{bmatrix}, \begin{bmatrix} 2 \\ 0 \\ 1 \end{bmatrix} \right\}$,

(c) $\left\{ \begin{bmatrix} 1 \\ 0 \\ 1 \end{bmatrix}, \begin{bmatrix} 1 \\ 1 \\ 1 \end{bmatrix}, \begin{bmatrix} 1 \\ -1 \\ 1 \end{bmatrix} \right\}$.

In Problems 17 through 25, discuss the solutions of the given systems of equations in terms of consistency and number of solutions. Check your answers by solving the systems wherever possible.

(17) $x - 2y = 0$
$x + y = 1$
$2x - y = 1$

(18) $x + y = 0$
$2x - 2y = 1$
$x - y = 0$

(19) $x + y + z = 1$
$x - y + z = 2$
$3x + y + 3z = 4$

(20) $x + 3y + 2z - w = 2$
$2x - y + z + w = 3$

(21) $2x - y + z = 0$
$x + 2y - z = 4$
$x + y + z = 1$

(22) $2x + 3y = 0$
$x - 4y = 0$

(23) $x - y + 2z = 0$
$2x + 3y - z = 0$
$-2x + 7y - 7z = 0$

(24) $x - y + 2z = 0$
$2x - 3y + 5z = 0$
$-2x + 7y - 9z = 0$

(25) $x - 2y + 3z + 3w = 0$
$y - 2z + 2w = 0$
$x + y - 3z + 9w = 0$

(26) Prove that if one row of a square matrix is a linear combination of another row, then the determinant of the matrix must be 0.

(27) Prove that if the determinant of an $n \times n$ matrix is 0, then the rank of that matrix must be less than n.

(28) Prove that if **A** and **B** are square matrices of the same order, then **AB** is nonsingular if and only if both **A** and **B** are nonsingular.

CHAPTER 2 REVIEW

Important Terms

additive inverse
basis
column rank
column space
coordinates
dimension
equivalent directed line segments
finite-dimensional vector space
linear combinations
linearly dependent vectors
linearly independent vectors
$\mathbb{M}_{p \times n}$
normalized n-tuple
$\mathbb{P}_n$
$\mathbb{R}^n$
rank
right-handed coordinate system
row rank
row space
span of vectors
spanning set

subspace	vector space
vector	zero vector

Important Concepts

Section 2.1

- Addition, subtraction, and scalar multiplication of 2-tuples can be done graphically in the plane.

Section 2.2

- The zero vector in a vector space is unique.
- The additive inverse of any vector **v** in a vector space is unique and is equal to $-1 \cdot \mathbf{v}$.

Section 2.3

- A nonempty subset $\mathbb{S}$ of a vector space $\mathbb{V}$ is a subspace of $\mathbb{V}$ if and only if $\mathbb{S}$ is closed under addition and scalar multiplication.
- If a subset of a vector space does not include the zero vector, then that subset cannot be a subspace.
- Lines through the origin and planes that contain the origin are subspaces of $\mathbb{R}^3$.
- The span of a set of vectors $\mathbb{S}$ in a vector space $\mathbb{V}$ is the smallest subspace of $\mathbb{V}$ that contains $\mathbb{S}$.

Section 2.4

- A set of vectors is linearly dependent if and only if one of the vectors is a linear combination of the vectors that precede it.
- Two vectors are linearly dependent in $\mathbb{R}^2$ or $\mathbb{R}^3$ if and only if they lie on the same line.
- A set of three vectors in $\mathbb{R}^3$ is linearly dependent if and only if all three vectors lie on the same line or all lie on the same plane.

Section 2.5

- $dim(\mathbb{R}^n) = n$; $dim(\mathbb{P}^n) = n + 1$; $dim(\mathbb{M}_{p \times n}) = pn$.
- Every basis for a finite-dimensional vector space contains the same number of vectors.
- In an n-dimensional vector space, every set of $n+1$ or more vectors is linearly dependent.
- A spanning set of vectors for a finite-dimensional vector space $\mathbb{V}$ can be reduced to a basis for $\mathbb{V}$; a linearly independent set of vectors in $\mathbb{V}$ can be expanded into a basis.

Section 2.6

- If matrix **B** is obtained from matrix **A** by an elementary row operation, then the row space of **A** is the same as the row space of **B**.

- To find the row rank of a matrix, use elementary row operations to transform the matrix to row-reduced form and then count the number of nonzero rows. The nonzero rows are a basis for the row space of the original matrix.

Section 2.7

- The row rank of a matrix equals its column rank.
- The system of equation $\mathbf{Ax}=\mathbf{b}$ is consistent if and only if the rank of $\mathbf{A}$ equals the rank of the augmented matrix $[\mathbf{A}|\mathbf{b}]$.
- If the system $\mathbf{Ax}=\mathbf{b}$ is consistent and if $r(\mathbf{A})=k$, then the solutions to the system are expressible in terms of $n-k$ arbitrary unknowns, where n denotes the total number of unknowns in the system.
- A homogeneous system of equations is always consistent, and one solution is always the trivial solution.
- An $n\times n$ matrix $\mathbf{A}$ has an inverse if and only if $\mathbf{A}$ has rank n.
- A square matrix has an inverse if and only if it can be transformed by elementary row operations to an upper triangular matrix with all unity elements on its main diagonal.

CHAPTER 3

Linear Transformations

Chapter Outline

3.1 FUNCTIONS

Relationships between items are at the heart of everyday interactions, and if mathematics is to successfully model or explain such interactions, then mathematics must account for relationships. In commerce, there are relationships between labor and production, between production and profit, and between profit and investment. In physics, there are relationships between force and acceleration, and between mass and energy. In sociology, there is a relationship between control and evasions. We need, therefore, mathematical structures to represent relationships. One such structure is a function.

A *function* is a rule of correspondence between two sets, generally called the *domain* and *range*, that assigns to each element in the domain exactly one element (but not necessarily a different one) in the range.

A function is a rule of correspondence between two sets, a domain and range, that assigns to each element in the domain exactly one element (but not necessarily a different one) in the range.

Example 1 The rules of correspondence described by the arrows in Figures 3.1 and 3.2 between the domain $\{A, B, C\}$ and the range $\{1, 2, 3, 4, 5\}$ are functions. In both cases, each element in the domain is assigned exactly one element in the range. In Figure 3.1, A is assigned 1, B is assigned 3, and C is assigned 5. Although some elements in the range are not paired with elements in the domain, this is of no consequence. A function must pair every element in the domain with an element in the range, but not vice versa. In Figure 3.2, each element in the domain is

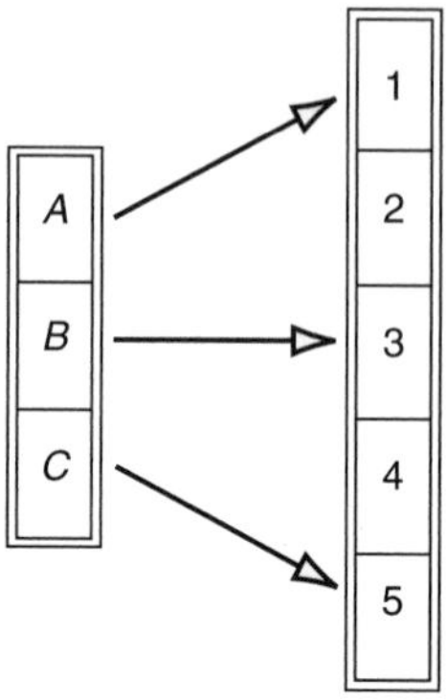

FIGURE 3.1

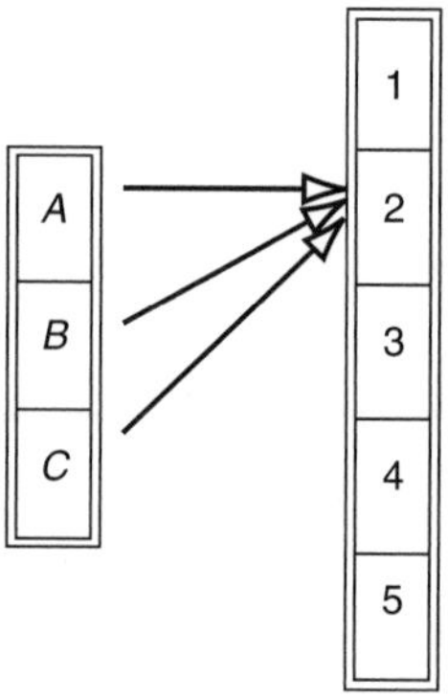

FIGURE 3.2

assigned the same element in the range, namely, 2. This too is of no consequence. A function must pair every element in the domain with an element in the range, but not necessarily with a different element.

Example 2 The rule of correspondence described by the arrows in Figure 3.3 between the domain and range, which are both the set of words {*dog, cat, bird*}, is *not* a function. The word *cat,* in the domain, is not matched with any element in the range. A function must match every element in the domain with an element in the range.

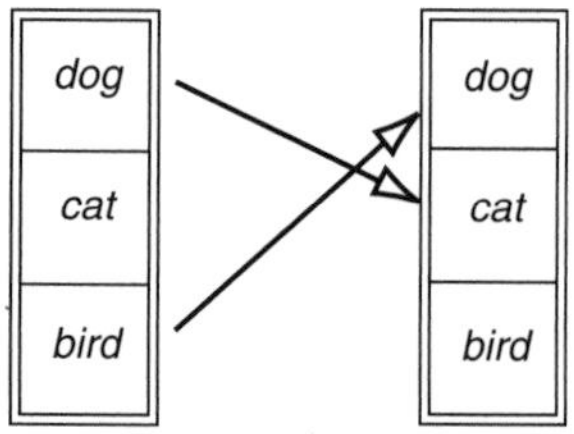

FIGURE 3.3

The *image* of a function consists of those elements in the range that are matched with elements in the domain. An element y in the range is in the image only if there is an element x in the domain such that x is assigned the value y by the rule of correspondence. In Figure 3.1, the image is the set $\{1, 3, 5\}$ because 1, 3, and 5 are the only elements in the range actually assigned to elements in the domain. In Figure 3.2, the image is the set $\{2\}$ because the number 2 is the only number in the range matched with elements in the domain.

The image of a function is the set of all elements in the range that are matched with elements in the domain by the rule of correspondence.

The domain and range of a function can be any type of set, ranging from sets of letters to sets of colors to sets of animals, while the rule of correspondence can be specified by arrows, tables, graphs, formulas, or words. If we restrict ourselves to sets of real numbers and rules of correspondence given by equations, then we have the functions studied most often in algebra and calculus.

Whenever we have two sets of numbers and a function f relating the arbitrary element x in the domain to the element y in the range through an equation, we say that y is a function of x and write $y=f(x)$. Letters other than x and y may be equally appropriate. The equation $R=f(N)$ is shorthand notation for the statement that we have a function consisting of two sets of numbers and an equation, where N and R denote elements in the domain and range, respectively. If the domain is not specified, it is assumed to be all real numbers for which the rule of correspondence makes sense; if the range is not specified, it is taken to be the set of all real numbers.

If we have a rule of correspondence defined by the formula $f(x)$, then we find the element in the range associated with a particular value of x by replacing x with that particular value in the formula. Thus, $f(2)$ is the effect of applying the rule of correspondence to the domain element 2, while $f(5)$ is the effect of applying the rule of correspondence to the domain element 5.

Example 3 Find $f(2)$, $f(5)$, and $f(-5)$ for $f(x)=1/x^2$.

Solution: The domain and range are not specified, so they assume their default values. The formula $1/x^2$ is computable for all real numbers except 0, so this becomes the domain. The range is the set of all real numbers. The image is all positive real numbers because those are the only numbers actually matched to elements in the domain by the formula. Now

$$\begin{aligned} f(2) &= 1/(2)^2 = 1/4 \quad = 0.25 \\ f(5) &= 1/(5)^2 = 1/25 \quad = 0.04 \\ f(-5) &= 1/(-5)^2 = 1/25 = 0.04 \end{aligned}$$

Problems 3.1

In Problems 1 through 16, the rules of correspondence are described by arrows. Determine whether the given relationships are functions and, for those that are, identify their images.

(1)

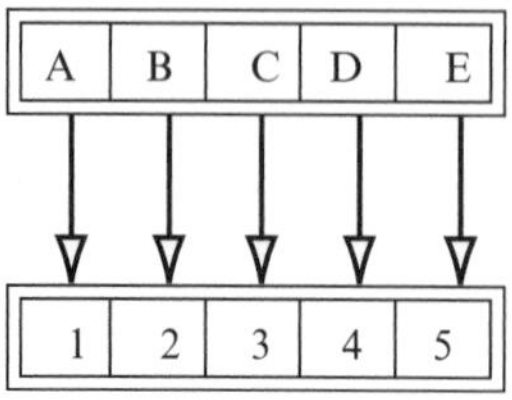

(2)

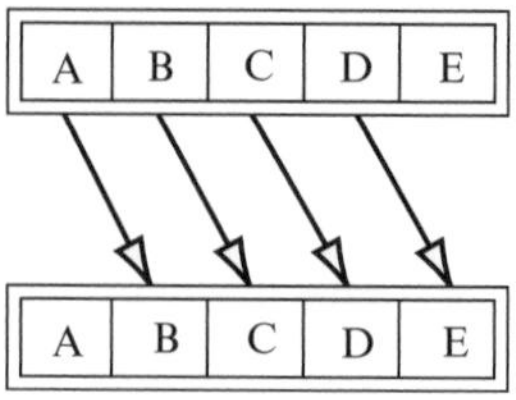

(3)

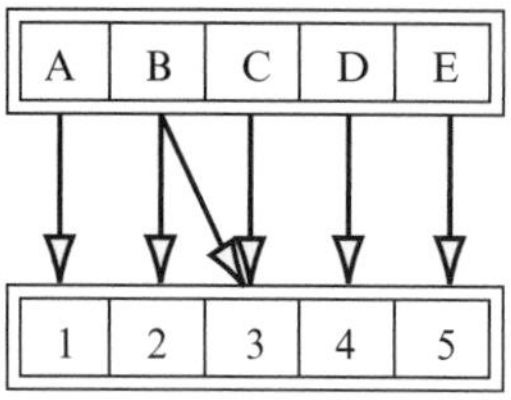

(4)

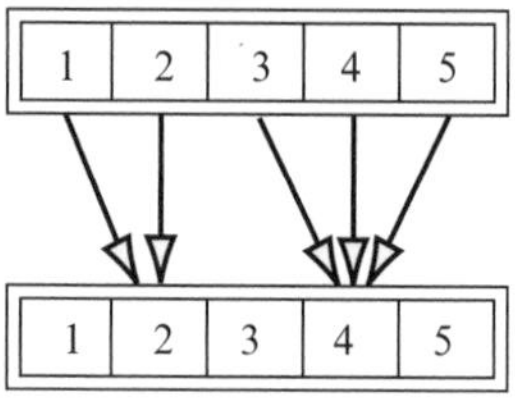

(5)

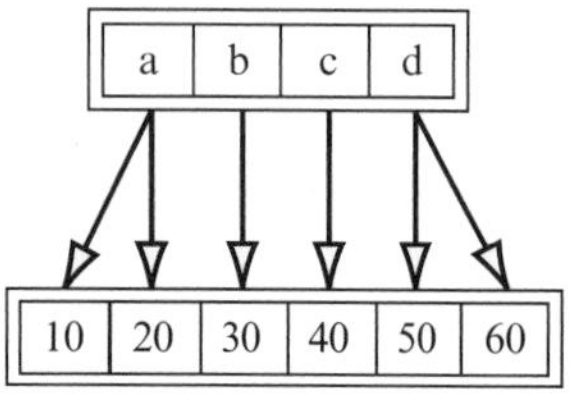

(6)

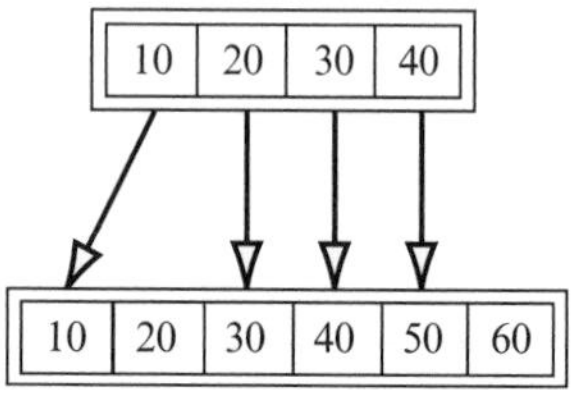

(7)

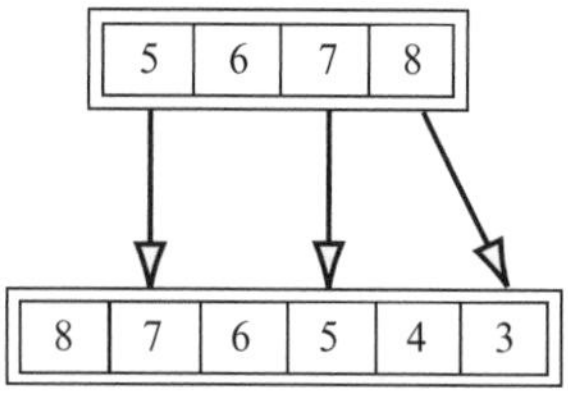

(8)

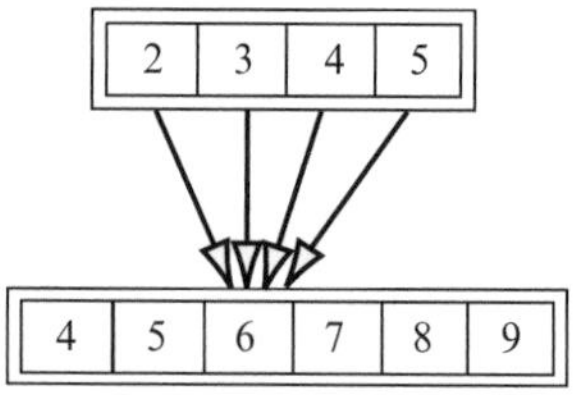

(9)

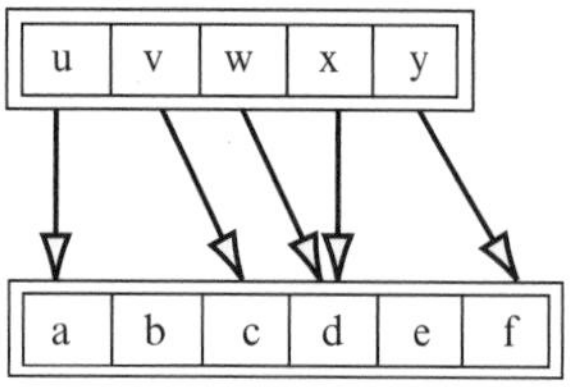

(10)

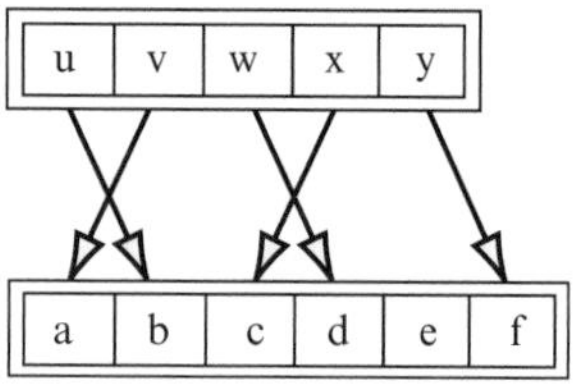

(11)

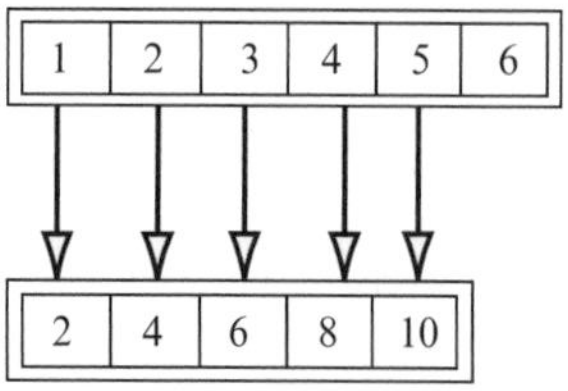

(12)

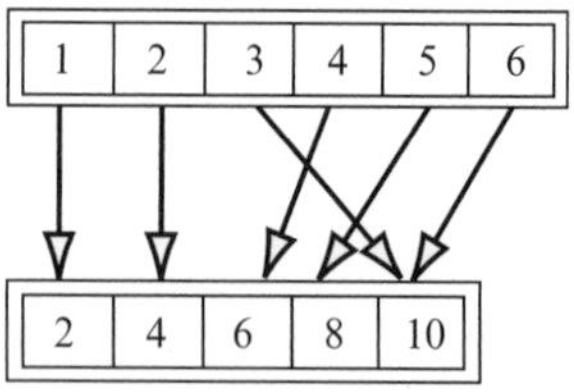

(13)

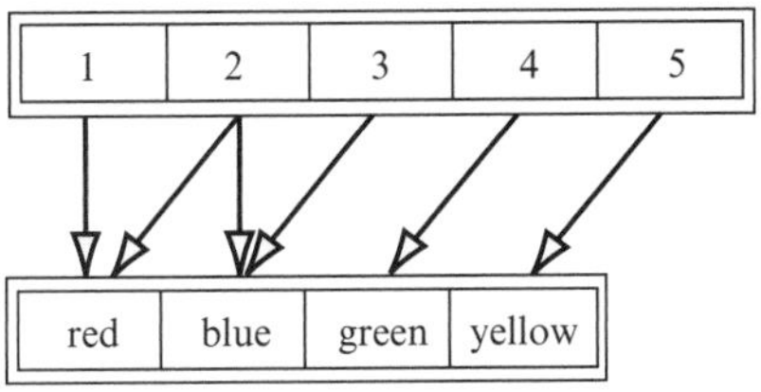

(14)

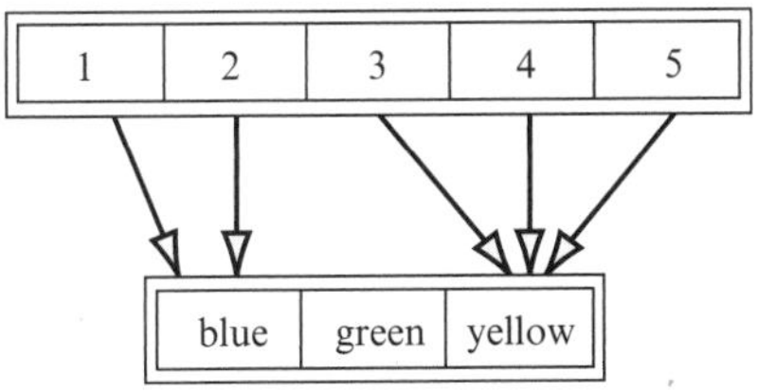

(15)

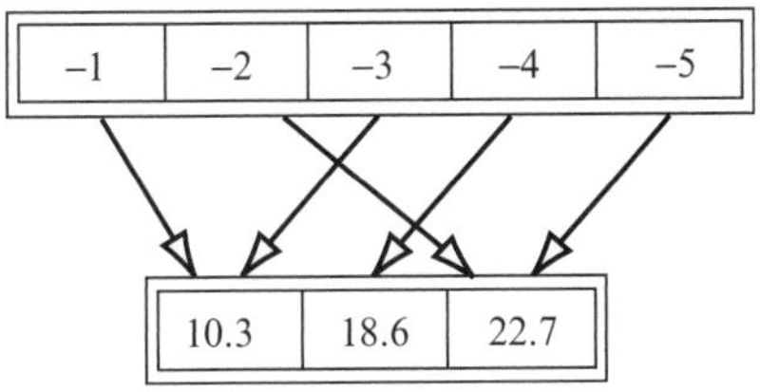

In Problems 16 through 18, determine whether the given tables represent functions where the rule of correspondence is to assign to each element in the top row the element directly below it in the bottom row.

(16)

x	1	2	3	4	5
y	10	18	23	18	10

(17)

x	1	2	3	4	5
y	10	10	20	20	20

(18)

x	1	2	3	4	5
y	10	18	23	29	34

In Problems 19 through 22, determine whether the specified correspondences constitute functions.

(19) The correspondence between people and their weights.

(20) The correspondence between people and their social security numbers.

(21) The correspondence between cars and the colors they are painted.

(22) The correspondence between stocks listed on the New York Stock Exchange and their closing prices on a given day.

In Problems 23 through 29, determine whether a domain exists on the horizontal axis so that the given graphs represent functions. The rule of correspondence assigns to each x value in the domain all y values on the vertical axis (the range) for which the points (x, y) lie on the graph.

(23)

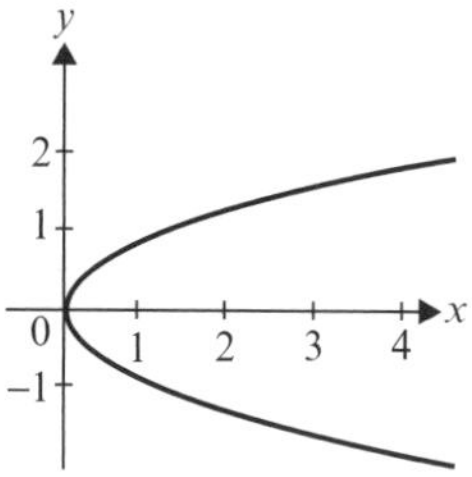

(24)

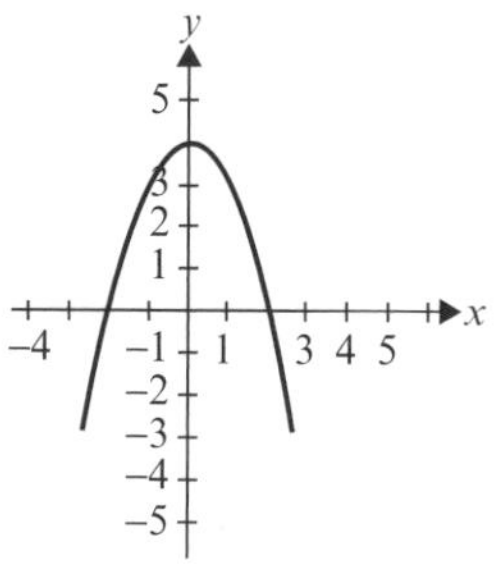

(25)

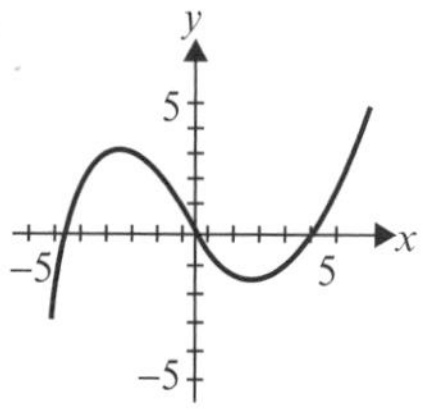

(26)

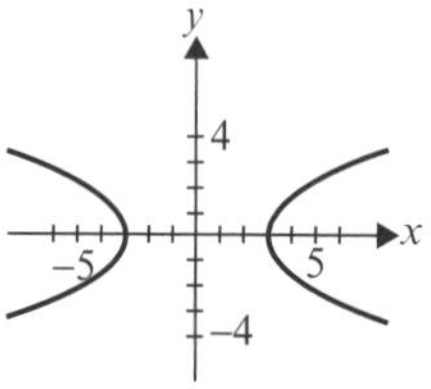

(27)

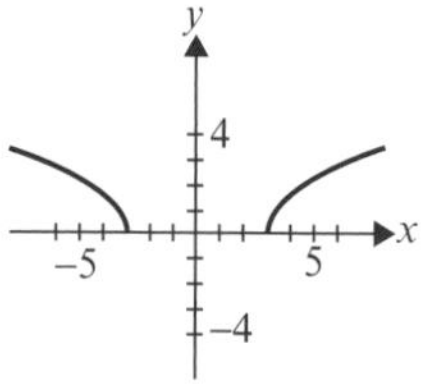

(28)

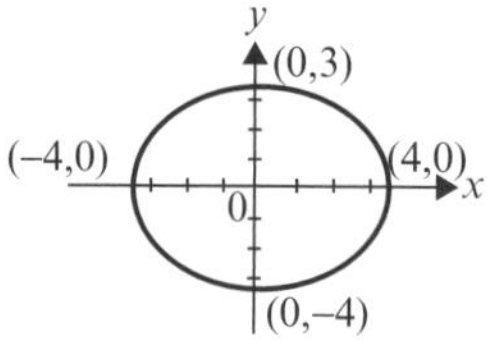

(29)

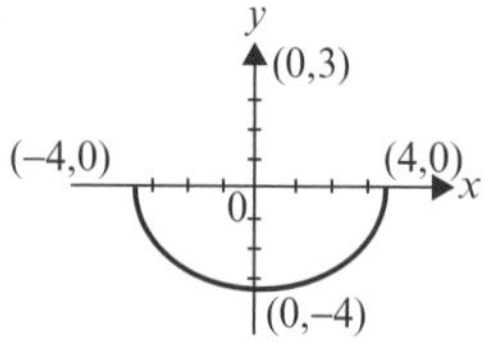

(30) Determine whether the following equations represent functions on the specified domains:

(a) $y = +\sqrt{x}$ for $-\infty < x < \infty$.

(b) $y = +\sqrt{x}$ for $0 < x < \infty$.

(c) $y = \pm\sqrt{x}$ for $0 < x < \infty$.

(d) $y = \sqrt[3]{x}$ for $-\infty < x < \infty$.

(31) Given the function $y = f(x) = x^2 - 3x + 2$ defined on all real numbers, find (a) $f(0)$, (b) $f(1)$, (c) $f(-1)$, (d) $f(2x)$.

(32) Given the function $y = f(x) = 2x^2 - x$ defined on all real numbers, find (a) $f(1)$, (b) $f(-1)$, (c) $f(2x)$, (d) $f(a+b)$.

(33) Given the function $y = f(x) = x^3 - 1$ defined on all real numbers, find (a) $f(-2)$, (b) $f(0)$, (c) $f(2z)$, (d) $f(a+b)$.

(34) A function is *onto* if its image equals its range. Determine whether either of the functions defined in Example 1 are onto.

(35) Determine which of the functions defined in Problems 1 through 15 are onto.

(36) A function is *one to one* if the equality $f(x) = f(z)$ implies that $x = z$; that is, if each element in the image is matched with one and only one element in the domain. Determine whether either of the functions defined in Example 1 are one to one.

(37) Determine which of the functions defined in Problems 1 through 15 are one to one.

3.2 LINEAR TRANSFORMATIONS

Two frequently used synonyms for the word *function* are *mapping* and *transformation*. In high-school algebra and calculus, the domain and range are restricted to subsets of the real numbers and the word *function* is used almost exclusively. In linear algebra, the domain and range are vector spaces and the word *transformation* is preferred.

A *transformation* T is a rule of correspondence between two vector spaces, a domain $\mathbb{V}$ and a range $\mathbb{W}$, that assigns to each element in $\mathbb{V}$ exactly one element (but not necessarily a different one) in $\mathbb{W}$. Such a transformation is denoted by the shorthand notation $T:\mathbb{V} \rightarrow \mathbb{W}$. We write $\mathbf{w} = T(\mathbf{v})$ whenever the vector $\mathbf{w}$ in $\mathbb{W}$ is matched with the vector $\mathbf{v}$ in $\mathbb{V}$ by the rule of correspondence associated with T. We will, on occasion, discard the parentheses and write $\mathbf{w} = T\mathbf{v}$ when there is no confusion as to what this notation signifies.

A transformation is a function with vector spaces for its domain and range.

The image of T is the set of all vectors in $\mathbb{W}$ that are matched with vectors in $\mathbb{V}$ under the rule of correspondence. Thus, $\mathbf{w}$ is in the image of T if and only if there exists a vector $\mathbf{v}$ in $\mathbb{V}$ such that $\mathbf{w} = T(\mathbf{v})$.

A transformation $T: \mathbb{V} \rightarrow \mathbb{W}$ is *linear* if for any two scalars, α and β, and any two vectors, $\mathbf{u}$ and $\mathbf{v}$, in $\mathbb{V}$ the following equality holds:

$$T(\alpha\mathbf{u} + \beta\mathbf{v}) = \alpha T(\mathbf{u}) + \beta T(\mathbf{v}) \tag{3.1}$$

For the special case $\alpha = \beta = 1$, (3.1) reduces to

$$T(\mathbf{u} + \mathbf{v}) = T(\mathbf{u}) + T(\mathbf{v}) \tag{3.2}$$

while for the special case $\beta=0$, (3.1) becomes

$$T(\alpha\mathbf{u}) = \alpha T(\mathbf{u}) \tag{3.3}$$

Verifying (3.1) is equivalent to verifying (3.2) and (3.3) separately (see Problem 47).

The left side of (3.1) is the mapping of the linear combination $\alpha\mathbf{u}+\beta\mathbf{v}$ from the vector space $\mathbb{V}$ into the vector space $\mathbb{W}$. If T is linear, then the result of mapping $\alpha\mathbf{u}+\beta\mathbf{v}$ into $\mathbb{W}$ is the same as separately mapping $\mathbf{u}$ and $\mathbf{v}$ into $\mathbb{W}$, designated as $T(\mathbf{u})$ and $T(\mathbf{v})$, and then forming the *identical* linear combination with $T(\mathbf{u})$ and $T(\mathbf{v})$ in $\mathbb{W}$ as was formed in $\mathbb{V}$ with $\mathbf{u}$ and $\mathbf{v}$; namely, α times the first vector plus β times the second vector. Linear combinations are fundamental to vector spaces because they involve the only operations, addition and scalar multiplication, guaranteed to exist in a vector space. Of all possible transformations, *linear* transformations are those special ones that preserve linear combinations.

A transformation is linear if it preserves linear combinations.

Example 1 Determine whether the transformation $T : \mathbb{V} \to \mathbb{V}$ defined by $T(\mathbf{v}) = k\mathbf{v}$ for all vectors $\mathbf{v}$ in $\mathbb{V}$ and any scalar k is linear.

Solution: In this example, $\mathbb{V} = \mathbb{W}$; that is, both the domain and the range are the same vector space. For any two vectors $\mathbf{u}$ and $\mathbf{v}$ in $\mathbb{V}$, we have

$$T(\alpha\mathbf{u} + \beta\mathbf{v}) = k(\alpha\mathbf{u} + \beta\mathbf{v}) = \alpha(k\mathbf{u}) + \beta(k\mathbf{v}) = \alpha T(\mathbf{u}) + \beta T(\mathbf{v})$$

Thus, (3.1) is valid, and the transformation is linear.

The linear transformation in Example 1 is called a *dilation.* In $\mathbb{R}^2$, a dilation reduces to a scalar multiple of a 2-tuple, having the geometrical effect of elongating $\mathbf{v}$ by a factor of $|k|$ when $|k|>1$ or contracting $\mathbf{v}$ by a factor of $|k|$ when $|k|<1$ followed by a rotation of 180° when k is negative and no rotation when k is positive. These dilations are illustrated in Figure 3.4. When $\mathbb{V} = \mathbb{R}^2$ and $k=-1$, the transformation $\mathbb{T}$ is sometimes called a *rotation through the origin.* It is illustrated in Figure 3.5.

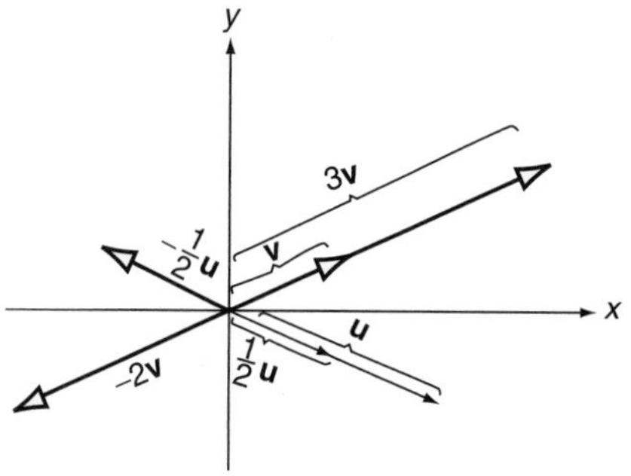

FIGURE 3.4

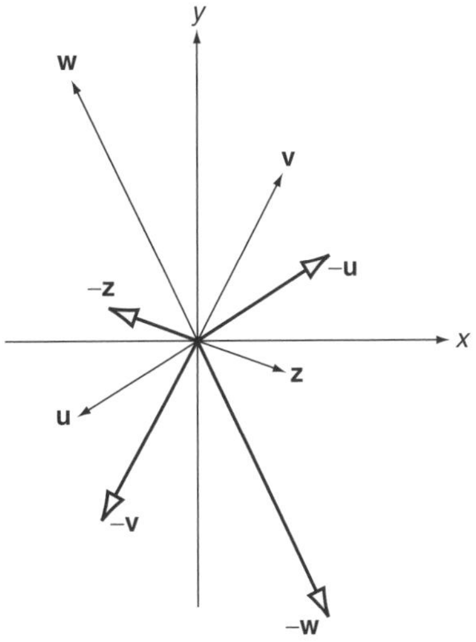

FIGURE 3.5

Example 2 Determine whether the transformation $T: \mathbb{V} \rightarrow \mathbb{W}$ defined by $T(\mathbf{v}) = 0$ for all vectors $\mathbf{v}$ in $\mathbb{V}$ is linear.

Solution: For any two scalars α and β and for any two vectors $\mathbf{u}$ and $\mathbf{v}$ in $\mathbb{V}$, we have

$$T(\alpha\mathbf{u} + \beta\mathbf{v}) = 0 = 0 + 0 = \alpha 0 + \beta 0 = \alpha T(\mathbf{u}) + \beta T(\mathbf{v})$$

Thus, (3.1) is valid, and T is linear. Transformations of this type are called *zero transformations* because they map all vectors in the domain into the zero vector in $\mathbb{W}$.

Example 3 Determine whether the transformation L is linear if $L: \mathbb{P}^3 \rightarrow \mathbb{P}^2$ is defined by

$$L\left(a_3t^3 + a_2t^2 + a_1t + a_0\right) = 3a_3t^3 + 2a_2t + a_1$$

where a_i $(i = 0, 1, 2, 3)$ denotes a real number.

Solution: A transformation is linear if it satisfies (3.1) or, equivalently, both (3.2) and (3.3). For practice, we try to validate (3.2) and (3.3). Setting

$$\mathbf{u} = a_3t^3 + a_2t^2 + a_1t + a_0 \quad \text{and} \quad \mathbf{v} = b_3t^3 + b_2t^2 + b_1t + b_0$$

we have $L(\mathbf{u}) = 3a_3t^2 + 2a_2t + a_1$, $L(\mathbf{v}) = 3b_3t^2 + 2b_2t + b_1$, and

$$\begin{aligned}
L(\mathbf{u} + \mathbf{v}) &= L\left(\left(a_3t^3 + a_2t^2 + a_1t + a_0\right) + \left(b_3t^3 + b_2t^2 + b_1t + b_0\right)\right) \\
&= L\left((a_3 + b_3)t^3 + (a_2 + b_2)t^2 + (a_1 + b_1)t + (a_0 + b_0)\right) \\
&= 3(a_3 + b_3)t^2 + 2(a_2 + b_2)t + (a_1 + b_1) \\
&= \left(3a_3t^2 + 2a_2t + a_1\right) + \left(3b_3t^2 + 2b_2t + b_1\right) \\
&= L(\mathbf{u}) + L(\mathbf{v})
\end{aligned}$$

For any scalar α, we have

$$\begin{aligned} L(\alpha\mathbf{u}) &= L\big(\alpha\big(a_3t^3 + a_2t^2 + a_1t + a_0\big)\big) \\ &= L\big((\alpha a_3)t^3 + (\alpha a_2)t^2 + (\alpha a_1)t + (\alpha a_0)\big) \\ &= 3(\alpha a_3)t^2 + 2(\alpha a_2)t + (\alpha a_1) \\ &= \alpha\big(3a_3t^2 + 2a_2t + a_1\big) \\ &= \alpha L(\mathbf{u}) \end{aligned}$$

Therefore, both (3.2) and (3.3) are satisfied, and L is linear. Readers familiar with elementary calculus will recognize this transformation as the derivative.

Example 4 Determine whether the transformation T is linear if $T\colon \mathbb{R}^2 \to \mathbb{R}^1$ is defined by $T[a\ b] = ab$ for all real numbers a and b.

Solution: This transformation maps 2-tuples into the product of its components. In particular, $T[2\ {-3}] = 2(-3) = -6$ and $T[1\ 0] = 1(0) = 0$. In general, setting $\mathbf{u} = [a\ b]$ and $\mathbf{v} = [c\ d]$, we have $T(\mathbf{u}) = ab$, $T(\mathbf{v}) = cd$, and

$$T(\mathbf{u}) + T(\mathbf{v}) = ab + cd \tag{3.4}$$

while

$$\begin{aligned} T(\mathbf{u}+\mathbf{v}) &= T([a\ b] + [c\ d]) \\ &= T[a + c\ b + d] \\ &= (a + c)(b + d) = ab + cd + cd + ad \end{aligned} \tag{3.5}$$

Equations (3.4) and (3.5) are generally *not* equal, hence (3.2) is not satisfied, and the transformation is not linear. In particular, for $\mathbf{u} = [2\ {-3}]$ and $\mathbf{v} = [1\ 0]$,

$$\begin{aligned} T(\mathbf{u}+\mathbf{v}) = T([2 - 3] + [1\ 0]) = T[3\ {-3}] = 3(-3) = -9 \\ \neq -6 + 0 = T[2 - 3] + T[1\ 0] = T\mathbf{u} + T\mathbf{v} \end{aligned}$$

We can also show that (3.3) does not hold, but this is redundant. *If either* (3.2) or (3.3) is violated, the transformation is not linear.

Example 5 Determine whether the transformation T is linear if $T\colon \mathbb{R}^2 \to \mathbb{R}^2$ is defined by $T[a\ b] = [a\ {-b}]$ for all real numbers a and b.

Solution: This transformation maps 2-tuples into 2-tuples by changing the sign of the second component. Here, $T[2\ 3] = [2\ {-3}]$, $T[0\ {-5}] = [0\ 5]$, and $T[-1\ 0] = [-1\ 0]$. In general, setting $\mathbf{u} = [a\ b]$ and $\mathbf{v} = [c\ d]$, we have $T(\mathbf{u}) = [a\ {-b}]$, $T(\mathbf{v}) = [c\ {-d}]$, and

$$\begin{aligned} T(\mathbf{u}+\mathbf{v}) &= T([a\ b] + [c\ d]) \\ &= T[a + c\ \ b + d] \\ &= [a + c - (b + d)] \\ &= [a + c - b\ {-d}] \\ &= [a\ {-b}] + [c\ {-d}] \\ &= T(\mathbf{u}) + T(\mathbf{v}) \end{aligned}$$

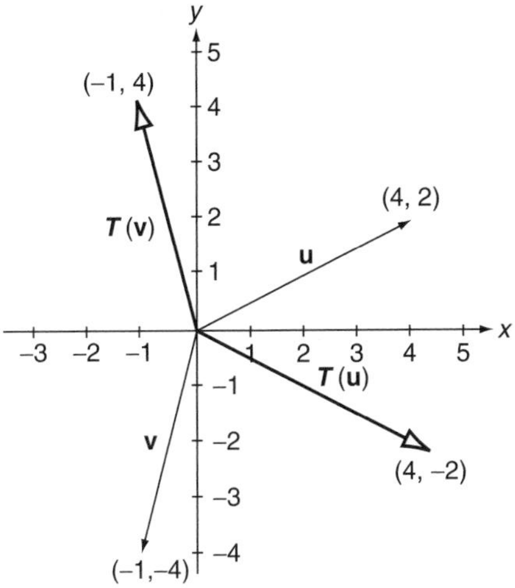

FIGURE 3.6

For any scalar α, we have

$$T(\alpha\mathbf{u}) = T(\alpha[a\ b]) = T[\alpha a\ \alpha b] = [\alpha a\ -\alpha b] = \alpha[a\ -b] = \alpha T(\mathbf{u})$$

Thus, (3.2) and (3.3) are satisfied, and the transformation is linear.

The linear transformation T defined in Example 5 is called a *reflection across the x-axis.* For vectors graphed on an *x-y* coordinate system, the transformation maps each vector into its mirror image across the horizontal axis. Some illustrations are given in Figure 3.6. The counterpart to T is the linear transformation $S : \mathbb{R}^2 \rightarrow \mathbb{R}^2$ defined by $S[a\ b] = [-a\ b]$, which is called a *reflection across the y-axis.* For vectors graphed on an *x-y* coordinate system, the transformation S maps each vector into its mirror image across the vertical axis. Some illustrations are given in Figure 3.7.

Example 6 Determine whether the transformation L is linear if $L : \mathbb{R}^2 \rightarrow \mathbb{R}^2$ is defined by $L[a\ b] = [a\ 0]$ for all real numbers a and b.

Solution: Here $L[-2\ 5] = [-2\ 0]$, $L[0\ 4] = [0\ 0]$, and $L[4\ 0] = [4\ 0]$. In general, setting $\mathbf{u} = [a\ b]$ and $\mathbf{v} = [c\ d]$, we have $L(\mathbf{u}) = [a\ 0]$, $L(\mathbf{v}) = [c\ 0]$, and for any scalars α and β,

$$\begin{aligned} L(\alpha\mathbf{u} + \beta\mathbf{v}) &= L(\alpha[a\ b] + \beta[c\ d]) \\ &= L[\alpha a + \beta c\ \ \alpha b + \beta d] \\ &= [\alpha a + \beta c\ \ 0] \\ &= \alpha[a\ 0] + \beta[c\ 0] \\ &= \alpha L(\mathbf{u}) + \beta L(\mathbf{v}) \end{aligned}$$

Equation (3.1) is satisfied, hence L is linear.

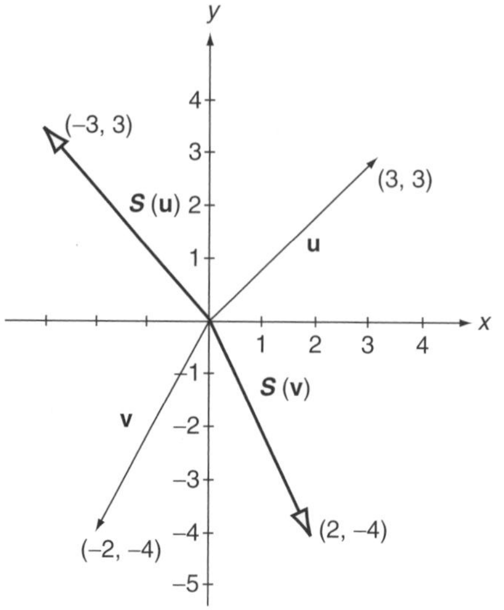

FIGURE 3.7

The linear transformation defined in Example 6 is called a *projection onto the x-axis*. Its counterpart, the transformation $M : \mathbb{R}^2 \rightarrow \mathbb{R}^2$ defined by $M[a\ b] = [0\ b]$ for all real numbers a and b, is also linear and is called a *projection onto the y-axis*. Some illustrations are given in Figure 3.8. Note that for any vector $\mathbf{v}$ in $\mathbb{R}^2$, $\mathbf{v} = L(\mathbf{v}) + M(\mathbf{v})$.

Example 7 Determine whether the transformation R is linear, if R is defined by

$$R\begin{bmatrix} a \\ b \end{bmatrix} = \begin{bmatrix} \cos\theta & -\sin\theta \\ \sin\theta & \cos\theta \end{bmatrix} \begin{bmatrix} a \\ b \end{bmatrix} = \begin{bmatrix} a\cos\theta - b\sin\theta \\ a\sin\theta + b\cos\theta \end{bmatrix}$$

where a and b denote arbitrary real numbers and θ is a constant.

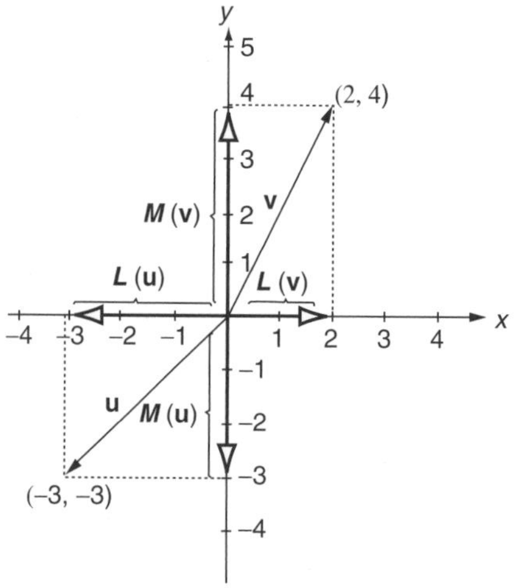

FIGURE 3.8

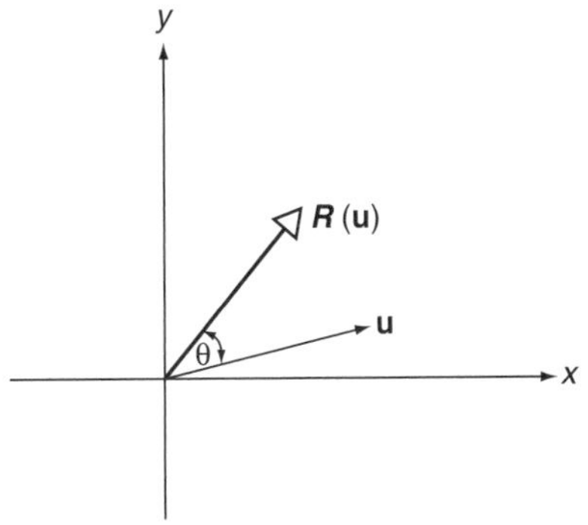

FIGURE 3.9

Solution: R is a transformation from $\mathbb{R}^2$ to $\mathbb{R}^2$ defined by a matrix multiplication. Setting

$$\mathbf{u}\begin{bmatrix} a \\ b \end{bmatrix}, \quad \mathbf{v} = \begin{bmatrix} c \\ b \end{bmatrix}, \quad \text{and} \quad \mathbf{A} = \begin{bmatrix} \cos\,\theta & -\sin\,\theta \\ \sin\,\theta & \cos\,\theta \end{bmatrix}$$

it follows directly from the properties of matrix multiplication that

$$R(\alpha\mathbf{u} + \beta\mathbf{v}) = \mathbf{A}(\alpha\mathbf{u} + \beta\mathbf{v}) = \alpha\mathbf{A}\mathbf{u} + \beta\mathbf{A}\mathbf{v} = \alpha R(\mathbf{u}) + \beta R(\mathbf{v})$$

for any choice of the scalars α and β. Equation (3.1) is valid, hence R is linear.

The linear transformation defined in Example 7 is called a *rotation*, because it has the geometric effect of rotating around the origin each vector $\mathbf{v}$ by the angle θ in the counterclockwise direction. This is illustrated in Figure 3.9.

The solution to Example 7 is extended easily to *any* linear transformation defined by matrix multiplication on n-tuples. Consequently, every matrix defines a linear transformation.

▶THEOREM 1

If $\mathbf{L}:\mathbb{R}^n \to \mathbb{R}^m$ is defined as $\mathbf{L}(\mathbf{u}) = \mathbf{A}\mathbf{u}$ for an $m \times n$ matrix $\mathbf{A}$, then $\mathbf{L}$ is linear.◀

Proof: It follows from the properties of matrices that for any two vectors $\mathbf{u}$ and $\mathbf{v}$ in $\mathbb{R}^n$, and any two scalars α and β, that

$$\begin{aligned} L(\alpha\mathbf{u} + \beta\mathbf{v}) &= \mathbf{A}(\alpha\mathbf{u} + \beta\mathbf{v}) = \mathbf{A}(\alpha\mathbf{u}) + \mathbf{A}(\beta\mathbf{v}) \\ &= \alpha(\mathbf{A}\mathbf{u}) + \beta(\mathbf{A}\mathbf{v}) = \alpha L(\mathbf{u}) + \beta L(\mathbf{v}) \end{aligned}$$

Problems 3.2

(1) Define $T : \mathbb{R}^2 \to \mathbb{R}^2$ by $T[a\ b] = [2a\ 3b]$. Find

(a) $T[2\ 3]$,

(b) $T[-1\ 5]$,

(c) $T[-8\ 200]$,

(d) $T[0\ -7]$.

(2) Redo Problem 1 with $T[a\ b] = [a+2\ b-2]$.

(3) Define $S: \mathbb{R}^3 \to \mathbb{R}^2$ by $S[a\ b\ c] = [a+b\ c]$. Find

(a) $S[1\ 2\ 3]$, (b) $S[-2\ 3\ -3]$,

(c) $S[2\ -2\ 0]$, (d) $S[1\ 4\ 3]$.

(4) Redo Problem 3 with $S[a\ b\ c] = [a-c\ c-b]$.

(5) Redo Problem 3 with $S[a\ b\ c] = [a+2b-3c\ 0]$.

(6) Define $N: \mathbb{R}^2 \to \mathbb{R}^3$ by $N[a\ b] = [a+b\ 2a+b\ b+2]$. Find

(a) $N[1\ 1]$, (b) $N[2\ -3]$,

(c) $N[3\ 0]$, (d) $N[0\ 0]$.

(7) Redo Problem 6 with $N[a\ b] = [a+b\ ab\ a-b]$.

(8) Define $\mathbf{P}: \mathbb{M}_{2\times 2} \to \mathbb{M}_{2\times 2}$ as $\mathbf{P}\begin{bmatrix} a & b \\ c & d \end{bmatrix} = \begin{bmatrix} c & a \\ d & b \end{bmatrix}$. Find

(a) $\mathbf{P}\begin{bmatrix} 1 & 2 \\ 3 & 4 \end{bmatrix}$, (b) $\mathbf{P}\begin{bmatrix} 1 & -1 \\ 3 & 3 \end{bmatrix}$,

(c) $\mathbf{P}\begin{bmatrix} 10 & 20 \\ -5 & 0 \end{bmatrix}$, (d) $\mathbf{P}\begin{bmatrix} 28 & -32 \\ 13 & 44 \end{bmatrix}$.

(9) Redo Problem 8 with $\mathbf{P}\begin{bmatrix} a & b \\ c & d \end{bmatrix} = \begin{bmatrix} a+b & 0 \\ 0 & c-d \end{bmatrix}$

(10) Define $T: \mathbb{P}^2 \to \mathbb{P}^2$ by $T(a_2t^2 + a_1t + a_0) = (a_2 - a_1)t^2 + (a_1 - a_0)t$. Find

(a) $T(2t^2 - 3t + 4)$, (b) $T(t^2 + 2t)$,

(c) $T(3t)$, (d) $T(-t^2 + 2t - 1)$.

In Problems 11 through 40, determine whether the given transformations are linear.

(11) $T: \mathbb{R}^2 \to \mathbb{R}^2$, $T[a\ b] = [2a\ 3b]$.

(12) $T: \mathbb{R}^2 \to \mathbb{R}^2$, $T[a\ b] = [a+2\ b-2]$.

(13) $T: \mathbb{R}^2 \to \mathbb{R}^2$, $T[a\ b] = [a\ 1]$.

(14) $S: \mathbb{R}^2 \to \mathbb{R}^2$, $S[a\ b] = [a^2\ b^2]$.

(15) $S: \mathbb{R}^3 \to \mathbb{R}^2$, $S[a\ b\ c] = [a+b\ c]$.

(16) $S: \mathbb{R}^3 \to \mathbb{R}^2$, $S[a\ b\ c] = [a-c\ c-b]$.

(17) $S: \mathbb{R}^3 \to \mathbb{R}^2$, $S[a\ b\ c] = [a+2b-3c\ 0]$.

(18) $S: \mathbb{R}^2 \to \mathbb{R}^3$, $S[a\ b] = [a+b\ 2a+b\ b+2]$.

(19) $S: \mathbb{R}^2 \to \mathbb{R}^3$, $S[a\ b] = [a\ 0\ b]$.

(20) $N: \mathbb{R}^2 \rightarrow \mathbb{R}^3$, $N[a\ b] = [0\ 0\ 0]$.

(21) $N: \mathbb{R}^2 \rightarrow \mathbb{R}^3$, $N[a\ b] = [a+b\ ab\ a-b]$.

(22) $N: \mathbb{R}^2 \rightarrow \mathbb{R}^3$, $N[a\ b] = [0\ 0\ 2a-5b]$.

(23) $T: \mathbb{R}^2 \rightarrow \mathbb{R}^3$, $T[a\ b] = [a+\ -a\ -8a]$.

(24) $T: \mathbb{R}^3 \rightarrow \mathbb{R}^1$, $T[a\ b\ c] = a - c$.

(25) $S: \mathbb{R}^3 \rightarrow \mathbb{R}^1$, $S[a\ b\ c] = abc$.

(26) $L: \mathbb{R}^3 \rightarrow \mathbb{R}^1$, $L[a\ b\ c] = 0$.

(27) $P: \mathbb{R}^3 \rightarrow \mathbb{R}^1$, $P[a\ b\ c] = 1$.

(28) $P: \mathbb{M}_{2\times 2} \rightarrow \mathbb{M}_{2\times 2}$, $P\begin{bmatrix} a & b \\ c & d \end{bmatrix} = \begin{bmatrix} c & a \\ d & b \end{bmatrix}$.

(29) $P: \mathbb{M}_{2\times 2} \rightarrow \mathbb{M}_{2\times 2}$, $P\begin{bmatrix} a & b \\ c & d \end{bmatrix} = \begin{bmatrix} a+b & 0 \\ 0 & c-d \end{bmatrix}$.

(30) $T: \mathbb{M}_{2\times 2} \rightarrow \mathbb{M}_{2\times 2}$, $T\begin{bmatrix} a & b \\ c & d \end{bmatrix} = \begin{bmatrix} 2d & 0 \\ 0 & 0 \end{bmatrix}$.

(31) $T: \mathbb{M}_{2\times 2} \rightarrow \mathbb{M}_{2\times 2}$, $T\begin{bmatrix} a & b \\ c & d \end{bmatrix} = \begin{bmatrix} ad & 0 \\ cd & 0 \end{bmatrix}$.

(32) $T: \mathbb{M}_{2\times 2} \rightarrow \mathbb{R}^1$, $T\begin{bmatrix} a & b \\ c & d \end{bmatrix} = ad - bc$.

(33) $R: \mathbb{M}_{2\times 2} \rightarrow \mathbb{R}^1$, $R\begin{bmatrix} a & b \\ c & d \end{bmatrix} = b + 2c - 3d$.

(34) $S: \mathbb{M}_{p\times n} \rightarrow \mathbb{M}_{n\times p}$, $S(\mathbf{A}) = \mathbf{A}^{\mathrm{T}}$.

(35) $S: \mathbb{M}_{p\times n} \rightarrow \mathbb{M}_{p\times n}$, $S(\mathbf{A}) = -\mathbf{A}$.

(36) $L: \mathbb{M}_{n\times n} \rightarrow \mathbb{M}_{n\times n}$, $L(\mathbf{A}) = \mathbf{A} - \mathbf{A}^{\mathrm{T}}$.

(37) $L: \mathbb{P}^2 \rightarrow \mathbb{P}^2$, $L(a_2t^2 + a_1t + a_0) = a_0t$.

(38) $T: \mathbb{P}^2 \rightarrow \mathbb{P}^2$, $T(a_2t^2 + a_1t + a_0) = a_2(t-1)^2 + a_1(t-1) + a_0$.

(39) $T: \mathbb{P}^2 \rightarrow \mathbb{P}^2$, $T(a_2t^2 + a_1t + a_0) = (a_2 - a1)(t^2 + (a_1 - a_0)t$.

(40) $S: \mathbb{P}^2 \rightarrow \mathbb{P}^2$, $S(a_2t^2 + a_1t + a_0) = (a_2 - 1)t^2$.

(41) Let $S: \mathbb{M}_{n\times n} \rightarrow \mathbb{R}^1$ map an $n \times n$ matrix into the sum of its diagonal elements. Such a transformation is known as the *trace*. Is it linear?

(42) Let $T: \mathbb{M}_{n\times n} \rightarrow \mathbb{M}_{n\times n}$ be defined as $T(\mathbf{A}) = \begin{cases} \mathbf{A}^{-1} & \text{if } \mathbf{A} \text{ is nonsingular} \\ 0 & \text{if } \mathbf{A} \text{ is singular} \end{cases}$. Is T linear?

(43) Let $I: \mathbb{V} \to \mathbb{V}$ denote the *identity transformation* defined by $\mathbf{I}(\mathbf{v}) = \mathbf{v}$ for all vectors $\mathbf{v}$ in $\mathbb{V}$. Show that I is linear.

(44) Let $L: \mathbb{V} \to \mathbb{V}$ denote a linear transformation and let $\{\mathbf{v}_1, \mathbf{v}_2, \ldots, \mathbf{v}_n\}$ be a basis for $\mathbb{V}$. Prove that if $L(\mathbf{v}_i) = \mathbf{v}_i$, for all i $(i = 1, 2, \ldots, n)$, then L must be the identity transformation.

(45) Let $0: \mathbb{V} \to \mathbb{W}$ denote the zero *transformation* defined by $\mathbf{0}(\mathbf{v}) = \mathbf{0}$ for all vectors $\mathbf{v}$ in $\mathbb{V}$. Show that 0 is linear.

(46) Let $L: \mathbb{V} \to \mathbb{W}$ denote a linear transformation and let $\{\mathbf{v}_1, \mathbf{v}_2, \ldots, \mathbf{v}_n\}$ be a basis for $\mathbb{V}$. Prove that if $L(\mathbf{v}_i) = \mathbf{0}$ for all $(i = 1, 2, \ldots, n)$, then L must be the zero transformation.

(47) Prove that Equations (3.2) and (3.3) imply (3.1).

(48) Determine whether $T: \mathbb{M}_{n\times n} \to \mathbb{M}_{n\times n}$ defined by $T(\mathbf{A}) = \mathbf{A}\mathbf{A}^{\mathrm{T}}$ is linear.

(49) Find $T(\mathbf{u} + 3\mathbf{v})$ for a linear transformation T if it is known that $T(\mathbf{u}) = 22$ and $T(\mathbf{v}) = -8$.

(50) Find $T(\mathbf{u})$ for a linear transformation T if it is known that $T(\mathbf{u} + \mathbf{v}) = 2\mathbf{u} + 3\mathbf{v}$ and $T(\mathbf{u} - \mathbf{v}) = 4\mathbf{u} + 5\mathbf{v}$.

(51) Find $T(\mathbf{v})$ for a linear transformation T if it is known that $T(\mathbf{u} + \mathbf{v}) = \mathbf{u}$ and $T(\mathbf{u}) = \mathbf{u} - 2\mathbf{v}$.

(52) Let $L: \mathbb{V} \to \mathbb{W}$ denote a linear transformation. Prove that $L(\mathbf{v}_1 + \mathbf{v}_2 + \mathbf{v}_3) = L(\mathbf{v}_1) + L(\mathbf{v}_2) + L(\mathbf{v}_3)$ for any three vectors $\mathbf{v}_1$, $\mathbf{v}_2$, and $\mathbf{v}_3$ in $\mathbb{V}$. Generalize this result to the sum of more than three vectors.

(53) Let $S: \mathbb{V} \to \mathbb{W}$ and $T: \mathbb{V} \to \mathbb{W}$ be two linear transformations. Their sum is another transformation from $\mathbb{V}$ into $\mathbb{W}$ defined by $(S + T)\mathbf{v} = S(\mathbf{v}) + T(\mathbf{v})$ for all $\mathbf{v}$ in $\mathbb{V}$. Prove that the transformation $S + T$ is linear.

(54) Let $T: \mathbb{V} \to \mathbb{W}$ be a linear transformation and k a given scalar. Define a new transformation $kT: \mathbb{V} \to \mathbb{W}$ by $(kT)\mathbf{v} = k(T\mathbf{v})$ for all $\mathbf{v}$ in $\mathbb{V}$. Prove that the transformation kT is linear.

(55) Let $S: \mathbb{V} \to \mathbb{W}$ and $T: \mathbb{V} \to \mathbb{V}$ be two linear transformations and define their product as another transformation from $\mathbb{V}$ into $\mathbb{V}$ defined by $(ST)\mathbf{v} - S(T\mathbf{v})$ for all $\mathbf{v}$ in $\mathbb{V}$. This product first applies T to a vector and then S to that result. Prove that the transformation ST is linear.

(56) Let $S: \mathbb{R}^2 \to \mathbb{R}^2$ be defined by $S[a\ b] = [2a + b\ 3a]$ and $T: \mathbb{R}^2 \to \mathbb{R}^2$ be defined by $T[a\ b] = [b - a]$. Find $ST(\mathbf{v})$ for the following vectors $\mathbf{v}$:

(a) $[1\ 2]$, (b) $[2\ 0]$, (c) $[-1\ 3]$,

(d) $[-1\ 1]$, (e) $[-2\ -2]$, (f) $[2\ -3]$.

(57) Find $TS(\mathbf{v})$ for the vectors and transformations given in the previous problem.

(58) Let $S: \mathbb{R}^2 \to \mathbb{R}^2$ be defined by $S[a\ b]=[a+b\ \ a-b]$ and $T: \mathbb{R}^2 \to \mathbb{R}^2$ be defined by $T[a\ b]=[2b\ 3b]$. Find $ST(\mathbf{v})$ for the following vectors $\mathbf{v}$:

(a) $[1\ 2]$, (b) $[2\ 0]$, (c) $[-1\ 3]$,
(d) $[-1\ 1]$, (e) $[-2\ -2]$, (f) $[2\ -3]$.

(59) Find $TS(\mathbf{v})$ for the vectors and transformations given in the previous problem.

(60) Let $S:\mathbb{R}^2 \to \mathbb{R}^2$ be defined by $S[a\ b]=[a\ \ a+2b]$ and $T:\mathbb{R}^2 \to \mathbb{R}^2$ be defined by $T[a\ b]=[a+2b\ \ a-2b]$. Find $ST(\mathbf{v})$ for the following vectors $\mathbf{v}$:

(a) $[1\ 2]$, (b) $[2\ 0]$, (c) $[-1\ 3]$,
(d) $[-1\ 1]$, (e) $[-2\ -2]$, (f) $[2\ -3]$.

(61) Let L be defined as in Example 6. Show that $L^2=L$.

(62) Let L and M be transformations from $\mathbb{R}^2$ into $\mathbb{R}^2$, the first a projection onto the x-axis and the second a projection onto the y-axis (see Example 6). Show that their product is the zero transformation.

3.3 MATRIX REPRESENTATIONS

We showed in Chapter 2 that any vector in a finite-dimensional vector space can be represented as an n-tuple with respect to a given basis. Consequently, we can study finite-dimensional vector spaces by analyzing n-tuples. We now show that every linear transformation from an n-dimensional vector space into an m-dimensional vector space can be represented by an $m \times n$ matrix. Thus, we can reduce the study of linear transformations on finite-dimensional vector space to the study of matrices!

Recall from Section 2.4 that there is only one way to express $\mathbf{v}$ as a linear combination of a given set of basis vectors. If $\mathbf{v}$ is any vector in a finite-dimensional vector space $\mathbb{V}$, and if $\mathbb{B} = \{\mathbf{v}_1, \mathbf{v}_2, \ldots, \mathbf{v}_n\}$ is a basis for $\mathbb{V}$, then there exists a unique set of scalars $c_1, c_2, \ldots, c_n$ such that

$$\mathbf{v} = c_1\mathbf{v}_1 + c_2\mathbf{v}_2 + \cdots + c_n\mathbf{v}_n \tag{3.6}$$

We write

$$v \leftrightarrow \begin{bmatrix} c_1 \\ c_2 \\ \vdots \\ c_n \end{bmatrix}_{\mathbb{B}} \tag{3.7}$$

to indicate that the n-tuple is a coordinate representation for the sum on the right side of (3.6). The subscript on the n-tuple denotes the underlying basis and emphasizes that the coordinate representation is basis dependent.

Example 1 Find a coordinate representation for the vector $\mathbf{v} = 4t^2 + 3t + 2$ in $\mathbb{P}^2$ with respect to the basis $\mathbb{C} = \{t^2 + t, t + 1, t - 1\}$.

Solution: To write $\mathbf{v}$ as a linear combination of the basis vectors, we must determine scalars c_1, c_2, and c_3 that satisfy the equation

$$\begin{aligned} 4t^2 + 3t + 2 &= c_1(t^2 + 1) + c_2(t + 1) + c_3(t - 1) \\ &= c_1 t^2 + (c_1 + c_2 + c_3)t + (c_2 - c_3) \end{aligned}$$

Equating coefficients of like powers of t, we generate the system of equations

$$\begin{aligned} c_1 &= 4 \\ c_1 + c_2 + c_3 &= 3 \\ c_3 - c_3 &= 2 \end{aligned}$$

which has as its solution $c_1 = 4$, $c_2 = 1/2$, and $c_3 = -3/2$. Accordingly (3.6) becomes

$$4t^2 + 3t + 2 = 4(t^2 + t) + (1/2)(t + 1) + (-3/2)(t - 1)$$

and (3.7) takes the form

$$4t^2 + 3t + 2 \leftrightarrow \begin{bmatrix} 4 \\ 1/2 \\ -3/2 \end{bmatrix}_{\mathbb{D}}$$

A linear transformation is described completely by its actions on a basis for the domain.

If $T: \mathbb{V} \to \mathbb{W}$ is a linear transformation and $\mathbf{v}$ is any vector in $\mathbb{V}$ expressed in form (3.6), then

$$\begin{aligned} T(\mathbf{v}) &= T(c_1\mathbf{v}_1 + c_2\mathbf{v}_2 + \cdots + c_n\mathbf{v}_n) \\ &= c_1 T(\mathbf{v}_1) + c_2 T(\mathbf{v}_2) + \cdots + c_n T(\mathbf{v}_n) \end{aligned} \tag{3.8}$$

Consequently, T is described completely by its actions on a basis. Once we know how T transforms the basis vectors, we can substitute those results into the right side of (3.8) and determine how T affects any vector $\mathbf{v}$ in $\mathbb{V}$.

Example 2 A linear transformation $T: \mathbb{R}^2 \to \mathbb{R}^3$ has the property that

$$T\begin{bmatrix} 1 \\ 0 \end{bmatrix} = \begin{bmatrix} 1 \\ 2 \\ 0 \end{bmatrix} \quad \text{and} \quad T\begin{bmatrix} 0 \\ 1 \end{bmatrix} = \begin{bmatrix} 0 \\ 3 \\ 4 \end{bmatrix}$$

Determine $T(\mathbf{v})$ for any vector $\mathbf{v} \in \mathbb{R}^2$.

Solution: If $\mathbf{v} \in \mathbb{R}^2$, then $\mathbf{v} = [a\ b]^{\mathrm{T}}$ for some choice of the real numbers a and b. The set $\{[1\ 0]^{\mathrm{T}}, [0\ 1]^{\mathrm{T}}\}$ is the standard basis for $\mathbb{R}^2$, and with respect to this basis

$$\begin{bmatrix} a \\ b \end{bmatrix} = a\begin{bmatrix} 1 \\ 0 \end{bmatrix} + b\begin{bmatrix} 0 \\ 1 \end{bmatrix}$$

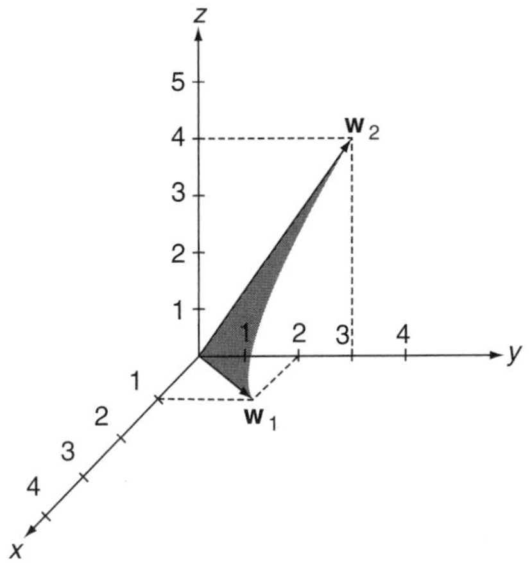

FIGURE 3.10

Consequently,

$$T\begin{bmatrix} a \\ b \end{bmatrix} = aT\begin{bmatrix} 1 \\ 0 \end{bmatrix} + bT\begin{bmatrix} 0 \\ 1 \end{bmatrix} = a\begin{bmatrix} 1 \\ 2 \\ 0 \end{bmatrix} + b\begin{bmatrix} 0 \\ 3 \\ 4 \end{bmatrix} = \begin{bmatrix} a \\ 2a+3b \\ 4b \end{bmatrix}$$

Example 2 has an interesting geometrical interpretation. We see from the solution that

$$T\left(a\begin{bmatrix} 1 \\ 0 \end{bmatrix} + b\begin{bmatrix} 0 \\ 1 \end{bmatrix}\right) = a\begin{bmatrix} 1 \\ 2 \\ 0 \end{bmatrix} + b\begin{bmatrix} 0 \\ 3 \\ 4 \end{bmatrix}$$

Thus, linear combinations of the vectors in the standard basis for $\mathbb{R}^2$ are mapped into linear combinations of the vectors $\mathbf{w}_1 = [1\ 2\ 0]^T$ and $\mathbf{w}_2 = [0\ 3\ 4]^T$. All linear combinations of the vectors in the standard basis for $\mathbb{R}^2$ generate the *x-y* plane. All linear combinations of $\mathbf{w}_1$ and $\mathbf{w}_2$ is the span of $\{\mathbf{w}_1, \mathbf{w}_2\}$, a plane in $\mathbb{R}^3$, which is *partially* illustrated by the shaded region in Figure 3.10. Thus, the linear transformation defined in Example 2 maps the *x-y* plane onto the plane spanned by $\{\mathbf{w}_1, \mathbf{w}_2\}$.

Example 3 A linear transformation $T:\mathbb{R}^2 \to \mathbb{R}^2$ has the property that

$$T\begin{bmatrix} 1 \\ 1 \end{bmatrix} = \begin{bmatrix} 5 \\ 6 \end{bmatrix} \quad \text{and} \quad T\begin{bmatrix} 1 \\ -1 \end{bmatrix} = \begin{bmatrix} 7 \\ 8 \end{bmatrix}$$

Determine $T\mathbf{v}$ for any vector $\mathbf{v} \in \mathbb{R}^2$.

Solution: The set of vectors $\{[1\ 1]^T, [1\ -1]^T\}$ is a basis for $\mathbb{R}^2$. If $\mathbf{v} = [a\ b]^T$ for some choice of the real numbers a and b, then

$$\begin{bmatrix} a \\ b \end{bmatrix} = \frac{a+b}{2}\begin{bmatrix} 1 \\ 1 \end{bmatrix} + \frac{a-b}{2}\begin{bmatrix} 1 \\ -1 \end{bmatrix}$$

and

$$T\begin{bmatrix} a \\ b \end{bmatrix} = \frac{a+b}{2}T\begin{bmatrix} 1 \\ 1 \end{bmatrix} + \frac{a-b}{2}T\begin{bmatrix} 1 \\ -1 \end{bmatrix}$$

$$= \frac{a+b}{2}\begin{bmatrix} 5 \\ 6 \end{bmatrix} + \frac{a-b}{2}\begin{bmatrix} 7 \\ 8 \end{bmatrix} = \begin{bmatrix} 6a-b \\ 7a-b \end{bmatrix}$$

Every linear transformation from one finite-dimensional vector space into another can be represented by a matrix.

With these two concepts—first, that any finite-dimensional vector can be represented as a basis dependent n-tuple, and second, that a linear transformation is completely described by its actions on a basis—we have the necessary tools to show that every linear transformation from one finite-dimensional vector space into another can be represented by a matrix. Let T designate a linear transformation from an n-dimensional vector space $\mathbb{V}$ into an m-dimensional vector space $\mathbb{W}$, and let $\mathbb{B} = \{\mathbf{v}_1, \mathbf{v}_2, \ldots, \mathbf{v}_n\}$ be a basis for $\mathbb{V}$ and $\mathbb{C} = \{\mathbf{w}_1, \mathbf{w}_2, \ldots, \mathbf{w}_m\}$ be a basis for $\mathbb{W}$. Then $T(\mathbf{v}_1)$, $T(\mathbf{v}_2)$, ..., $T(\mathbf{v}_n)$ are all vectors in $\mathbb{W}$ and each can be expressed as a linear combination of the basis vectors in $\mathbb{C}$.

In particular,

$$T(\mathbf{v}_1) = a_{11}\mathbf{w}_1 + a_{21}\mathbf{w}_2 + \cdots + a_{m1}\mathbf{w}_m$$

for some choice of the scalars a_{11}, a_{21}, ..., a_{m1},

$$T(\mathbf{v}_2) = a_{12}\mathbf{w}_1 + a_{22}\mathbf{w}_2 + \cdots + a_{m2}\mathbf{w}_m$$

for some choice of the scalars a_{12}, a_{22}, ..., a_{m1}, and, in general,

$$T(\mathbf{v}_j) = a_{1j}\mathbf{w}_1 + a_{2j}\mathbf{w}_2 + \cdots + a_{mj}\mathbf{w}_m \tag{3.9}$$

for some choice of the scalars $a_{1j}, a_{2j}, \ldots, a_{mj} (j = 1, 2, \ldots, m)$. The coordinate representations of these vectors are

$$T(\mathbf{v}_1) \leftrightarrow \begin{bmatrix} a_{11} \\ a_{21} \\ \vdots \\ a_{m1} \end{bmatrix}_{\mathbb{C}}, \quad T(\mathbf{v}_2) \leftrightarrow \begin{bmatrix} a_{12} \\ a_{22} \\ \vdots \\ a_{m2} \end{bmatrix}_{\mathbb{C}}, \ldots,$$

$$T(\mathbf{v}_j) \leftrightarrow \begin{bmatrix} a_{1j} \\ a_{2j} \\ \vdots \\ a_{mj} \end{bmatrix}_{\mathbb{C}}, \ldots, \quad T(\mathbf{v}_n) \leftrightarrow \begin{bmatrix} a_{1n} \\ a_{2n} \\ \vdots \\ a_{nm} \end{bmatrix}_{\mathbb{C}}$$

$\mathbf{A}_{\mathbb{B}}^{\mathbb{C}}$ denotes a matrix representation of a linear transformation with respect to the $\mathbb{B}$ basis in the domain and the $\mathbb{C}$ basis in the range.

If we use these n-tuples as the columns of a matrix $\mathbf{A}$, then, as we shall show shortly, $\mathbf{A}$ is the matrix representation of the linear transformation T. Because this matrix is basis dependent, in fact dependent on both the basis $\mathbb{B}$ in $\mathbb{V}$ and the basis $\mathbb{C}$ in $\mathbb{W}$, we write $\mathbf{A}_{\mathbb{B}}^{\mathbb{C}}$ to emphasize these dependencies. The notation $\mathbf{A}_{\mathbb{B}}^{\mathbb{C}}$ denotes the matrix representation of T with respect to the $\mathbb{B}$ basis in $\mathbb{V}$ and the $\mathbb{C}$ basis in $\mathbb{W}$. Often, the subscript $\mathbb{B}$ or the superscript $\mathbb{C}$ is deleted when either is the standard basis in $\mathbb{R}^n$ and $\mathbb{R}^m$, respectively.

Example 4 Find the matrix representation with respect to the standard basis in $\mathbb{R}^2$ and the standard basis $\mathbb{C} = \{t^2, t, 1\}$ in $\mathbb{P}^2$ for the linear transformation $T\colon \mathbb{R}^2 \to \mathbb{P}^2$ defined by

$$T\begin{bmatrix} a \\ b \end{bmatrix} = 2at^2 + (a+b)t + 3b$$

Solution:

$$T\begin{bmatrix} 1 \\ 0 \end{bmatrix} = (2)t^2 + (1)t + (0)1 \leftrightarrow \begin{bmatrix} 2 \\ 1 \\ 0 \end{bmatrix}_{\mathbb{C}}$$

and

$$T\begin{bmatrix} 0 \\ 1 \end{bmatrix} = (0)t^2 + (1)t + (3)1 \leftrightarrow \begin{bmatrix} 0 \\ 1 \\ 3 \end{bmatrix}_{\mathbb{C}}$$

so

$$\mathbf{A}^{\mathbb{C}} = \begin{bmatrix} 2 & 0 \\ 1 & 1 \\ 0 & 3 \end{bmatrix}^{\mathbb{C}}$$

We suppressed the subscript notation for the basis in the domain because it is the standard basis in $\mathbb{R}^2$.

Example 5 Redo Example 4 with the basis for the domain changed to $\mathbb{B} = \{[1\ 1]^{\mathrm{T}}, [1\ -1]^{\mathrm{T}}\}$.

Solution:

$$T\begin{bmatrix} 1 \\ 1 \end{bmatrix} = (2)t^2 + (2)t + (3)1 \leftrightarrow \begin{bmatrix} 2 \\ 2 \\ 3 \end{bmatrix}_{\mathbb{C}}$$

and

$$T\begin{bmatrix} 1 \\ -1 \end{bmatrix} = (2)t^2 + (0)t + (-3)1 \leftrightarrow \begin{bmatrix} 2 \\ 0 \\ -3 \end{bmatrix}_{\mathbb{C}}$$

hence,

$$\mathbf{A}_{\mathbb{B}}^{\mathbb{C}} = \begin{bmatrix} 2 & 2 \\ 2 & 0 \\ 3 & -3 \end{bmatrix}_{\mathbb{B}}^{\mathbb{C}}$$

Example 6 Find the matrix representation with respect to the standard basis in $\mathbb{R}^2$ and the basis $\mathbb{D}^2 = \{t^2 + t, t + 1, t - 1\}$ in $\mathbb{P}$ for the linear transformation $T\colon \mathbb{R}^2 \to \mathbb{P}^2$ defined by

$$T\begin{bmatrix} a \\ b \end{bmatrix} = (4a+b)t^2 + (3a)t + (2a-b)$$

Solution: Using the results of Example 1, we have

$$T\begin{bmatrix} 1 \\ 0 \end{bmatrix} = 4t^2 + 3t + 2 \leftrightarrow \begin{bmatrix} 4 \\ 1/2 \\ -3/2 \end{bmatrix}_{\mathbb{D}}$$

Similar reasoning yields

$$T\begin{bmatrix} 0 \\ 1 \end{bmatrix} = t^2 - 1 = (1)(t^2+t) + (-1)(t+1) + (0)(t-1) \leftrightarrow \begin{bmatrix} 1 \\ -1 \\ 0 \end{bmatrix}_{\mathbb{D}}$$

Thus,

$$\mathbf{A}^{\mathbb{D}} = \begin{bmatrix} 4 & 1 \\ 1/2 & -1 \\ -3/2 & 0 \end{bmatrix}^{\mathbb{D}}$$

Example 7 Find the matrix representation for the linear transformation $T: \mathbb{M}_{2\times 2} \to \mathbb{M}_{2\times 2}$ defined by

$$T\begin{bmatrix} a & b \\ c & d \end{bmatrix} = \begin{bmatrix} a+2b+3c & 2b-3c+4d \\ 3a-4b-5d & 0 \end{bmatrix}$$

with respect to the standard basis

$$\mathbb{B} = \left\{ \begin{bmatrix} 1 & 0 \\ 0 & 0 \end{bmatrix}, \begin{bmatrix} 0 & 1 \\ 0 & 0 \end{bmatrix}, \begin{bmatrix} 0 & 0 \\ 1 & 0 \end{bmatrix}, \begin{bmatrix} 0 & 0 \\ 0 & 1 \end{bmatrix} \right\}$$

Solution:

$$T\begin{bmatrix} 1 & 0 \\ 0 & 0 \end{bmatrix} = \begin{bmatrix} 1 & 0 \\ 3 & 0 \end{bmatrix} = (1)\begin{bmatrix} 1 & 0 \\ 0 & 0 \end{bmatrix} + (0)\begin{bmatrix} 0 & 1 \\ 0 & 0 \end{bmatrix} + (3)\begin{bmatrix} 0 & 0 \\ 1 & 0 \end{bmatrix}$$

$$+(0)\begin{bmatrix} 0 & 0 \\ 0 & 1 \end{bmatrix} \leftrightarrow \begin{bmatrix} 1 \\ 0 \\ 3 \\ 0 \end{bmatrix}_{\mathbb{B}}$$

$$T\begin{bmatrix} 0 & 1 \\ 0 & 0 \end{bmatrix} = \begin{bmatrix} 2 & 2 \\ -4 & 0 \end{bmatrix} = (2)\begin{bmatrix} 1 & 0 \\ 0 & 0 \end{bmatrix} + (2)\begin{bmatrix} 0 & 1 \\ 0 & 0 \end{bmatrix} + (-4)\begin{bmatrix} 0 & 0 \\ 1 & 0 \end{bmatrix}$$

$$+(0)\begin{bmatrix} 0 & 0 \\ 0 & 1 \end{bmatrix} \leftrightarrow \begin{bmatrix} 2 \\ 2 \\ -4 \\ 0 \end{bmatrix}_{\mathbb{B}}$$

$$T\begin{bmatrix}0 & 0\\1 & 0\end{bmatrix} = \begin{bmatrix}3 & -3\\0 & 0\end{bmatrix} = (3)\begin{bmatrix}1 & 0\\0 & 0\end{bmatrix} + (-3)\begin{bmatrix}0 & 1\\0 & 0\end{bmatrix} + (0)\begin{bmatrix}0 & 0\\1 & 0\end{bmatrix}$$
$$+(0)\begin{bmatrix}0 & 0\\0 & 1\end{bmatrix} \leftrightarrow \begin{bmatrix}3\\-3\\0\\0\end{bmatrix}_{\mathbb{B}}$$

$$T\begin{bmatrix}0 & 0\\0 & 1\end{bmatrix} = \begin{bmatrix}0 & 4\\-5 & 0\end{bmatrix} = (0)\begin{bmatrix}1 & 0\\0 & 0\end{bmatrix} + (4)\begin{bmatrix}0 & 1\\0 & 0\end{bmatrix} + (-5)\begin{bmatrix}0 & 0\\1 & 0\end{bmatrix}$$
$$+(0)\begin{bmatrix}0 & 0\\0 & 1\end{bmatrix} \leftrightarrow \begin{bmatrix}3\\4\\-5\\0\end{bmatrix}_{\mathbb{B}}$$

Therefore,

$$\mathbf{A}_{\mathbb{B}}^{\mathbb{B}} = \begin{bmatrix}1 & 2 & 3 & 0\\0 & 2 & -3 & 4\\3 & -4 & 0 & -5\\0 & 0 & 0 & 0\end{bmatrix}_{\mathbb{B}}^{\mathbb{B}}$$

To prove that $\mathbf{A}_{\mathbb{B}}^{\mathbb{C}}$, as we defined it, is a matrix representation for a linear transformation T, we begin with a vector $\mathbf{v}$ in the domain $\mathbb{V}$. If $\mathbb{B} = \{\mathbf{v}_1, \mathbf{v}_2, \ldots, \mathbf{v}_n\}$ is a basis for $\mathbb{V}$, then there exists a unique set of scalars $c_1, c_2, \ldots, c_n$ such that

$$\mathbf{v} = c_1\mathbf{v}_1 + c_2\mathbf{v}_2 + \cdots + c_n\mathbf{v}_n = \sum_{j=1}^{n} c_j\mathbf{v}_j$$

The coordinate representation of $\mathbf{v}$ with respect to the $\mathbb{B}$ basis is

$$\mathbf{v} \leftrightarrow \begin{bmatrix}c_1\\c_2\\\vdots\\c_n\end{bmatrix}_{\mathbb{B}}$$

Setting $\mathbf{w} = T(\mathbf{v})$, it follows from (3.8) and (3.9) that

$$\begin{aligned}\mathbf{w} &= T(\mathbf{v}) = T\left(\sum_{j=1}^{n} c_j\mathbf{v}_j\right)\\ &= \sum_{j=1}^{n} c_j\mathbf{T}(\mathbf{v}_j)\end{aligned}$$

$$= \sum_{j=1}^{n} c_j (a_{1j}\mathbf{w}_1 + a_{2j}\mathbf{w}_2 + \cdots + a_{mj}\mathbf{w}_m)$$

$$= \sum_{j=1}^{n} c_j \left(\sum_{j=1}^{n} a_{ij}\mathbf{w}_i \right)$$

$$= \sum_{i=1}^{m} \left(\sum_{j=1}^{n} a_{ij}c_j \right) \mathbf{w}_i$$

We now have $\mathbf{w}$ in terms of the basis vectors in $\mathbb{C} = \{\mathbf{w}_1, \mathbf{w}_2, \ldots, \mathbf{w}_m\}$. Since the summation in the last parentheses is the coefficient of each basis vector, we see that the coordinate representation for $\mathbf{w}$ with respect to the $\mathbb{C}$ basis is

$$T(\mathbf{v}) = \mathbf{w} \leftrightarrow \begin{bmatrix} \sum_{j=1}^{n} a_{1j}c_j \\ \sum_{j=1}^{n} a_{2j}c_j \\ \vdots \\ \sum_{j=1}^{n} a_{mj}c_j \end{bmatrix}_{\mathbb{C}}$$

This vector is the matrix product

$$\begin{bmatrix} a_{11} & a_{12} & \cdots & a_{1n} \\ a_{21} & a_{22} & \cdots & a_{2n} \\ \vdots & \vdots & \ddots & \vdots \\ a_{m1} & a_{m2} & \cdots & a_{mn} \end{bmatrix} \begin{bmatrix} c_1 \\ c_2 \\ \vdots \\ c_n \end{bmatrix}$$

Thus,

$$T(\mathbf{v}) = \mathbf{w} \leftrightarrow \mathbf{A}_{\mathbb{B}}^{\mathbb{C}}\mathbf{v}_{\mathbb{B}} \tag{3.10}$$

We can calculate $T(\mathbf{v})$ in two ways: first, *the direct approach* using the left side of (3.10), by evaluating directly how T affects $\mathbf{v}$; or second, *the indirect approach* using the right side of (3.10), by multiplying the matrix representation of T by the coordinate representation of $\mathbf{v}$ to obtain $\mathbf{A}_{\mathbb{B}}^{\mathbb{C}}\mathbf{v}_{\mathbb{B}}$, the m-tuple representation of $\mathbf{w}$, from which $\mathbf{w}$ itself is easily calculated. These two processes are shown schematically in Figure 3.11, the direct approach by the single solid arrow and the indirect approach by the path of three dashed arrows.

Example 8 Calculate $T\begin{bmatrix}1\\3\end{bmatrix}$ using both the direct and indirect approaches illustrated in Figure 3.11 for the linear transformation T: $\mathbb{R}^2 \to \mathbb{P}^2$ defined by $T\begin{bmatrix}a\\b\end{bmatrix}$ $= 2\text{ at}^{\,2}\ (a+b)t + 3b$. With the indirect approach, use $\mathbb{B} = \left\{[1\ \ 1]^{\mathrm{T}}, [1 - 1]^{\mathrm{T}}\right\}$ as the basis for $\mathbb{R}^2$ and $\mathbb{C} = \{t^2, t, 1\}$ as the basis for $\mathbb{P}^2$.

Solution: Using the direct approach, we have

$$T\begin{bmatrix}1\\3\end{bmatrix} = 2(1)t^2 + (1+3)t + 3(3) = 2t^2 + 4t + 9$$

Using the indirect approach, we first determine the coordinate representation for $[1\ 3]^{\mathrm{T}}$ with the respect to the $\mathbb{B}$ basis. It is

$$\begin{bmatrix}1\\3\end{bmatrix} = 2\begin{bmatrix}1\\1\end{bmatrix} + (-1)\begin{bmatrix}1\\-1\end{bmatrix} \leftrightarrow \begin{bmatrix}2\\-1\end{bmatrix}_{\mathbb{B}} = \mathbf{v}_{\mathbb{B}}$$

Then, using the results of Example 5, we have

$$\mathbf{A}_{\mathbb{B}}^{\mathbb{C}}\mathbf{v}_{\mathbb{B}} = \begin{bmatrix}2 & 2\\2 & 0\\3 & -3\end{bmatrix}_{\mathbb{B}}^{\mathbb{C}} \begin{bmatrix}2\\-1\end{bmatrix}_{\mathbb{B}} = \begin{bmatrix}2\\4\\9\end{bmatrix}_{\mathbb{C}} \leftrightarrow 2t^2 + 4t = 9$$

which is the same result obtained by the direct approach.

Example 9 Calculate $T\begin{bmatrix}2\\-3\end{bmatrix}$ using both the direct and indirect approaches illustrated in Figure 3.11 for the linear transformation and bases described in Example 6.

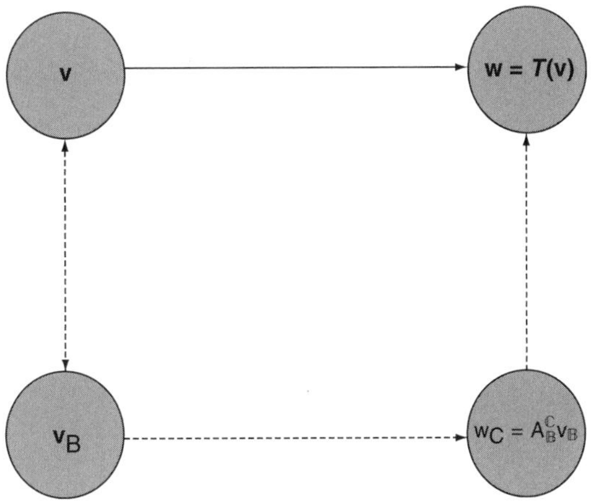

FIGURE 3.11

Solution: Using the direct approach, we have

$$T\begin{bmatrix} 2 \\ -3 \end{bmatrix} = [4(2) + (-3)]t^2 + 3(2)t + [2(2) - (-3)] = 5t^2 + 6t + 7$$

Using the indirect approach, we note that

$$\begin{bmatrix} 2 \\ -3 \end{bmatrix} = 2\begin{bmatrix} 1 \\ 0 \end{bmatrix} + (-3)\begin{bmatrix} 0 \\ 1 \end{bmatrix} \leftrightarrow \begin{bmatrix} 2 \\ -3 \end{bmatrix}_{\mathbb{B}} = \mathbf{v}_{\mathbb{B}}$$

Then, using the results of Example 6, we have

$$\mathbf{A}_{\mathbb{B}}^{\mathbb{C}}\mathbf{v}_{\mathbb{B}} \begin{bmatrix} 4 & 1 \\ 1/2 & -1 \\ -3/2 & 0 \end{bmatrix}_{\mathbb{B}}^{\mathbb{C}} \begin{bmatrix} 2 \\ -3 \end{bmatrix}_{\mathbb{B}} = \begin{bmatrix} 5 \\ 4 \\ -3 \end{bmatrix}_{\mathbb{C}}$$

$$\leftrightarrow 5\left(t^2 + t\right) + 4(t + 1) + (-3)(t - 1) = 5t^2 + 6t + 7$$

which is the same result obtained by the direct approach.

The direct approach illustrated in Figure 3.11 is clearly quicker. The indirect approach, however, is almost entirely in terms of matrices and matrix operations, which are conceptually easier to understand and more tangible. Theorem 1 of Section 2.2 states that every matrix represents a linear transformation. We just showed that every linear transformation can be represented by a matrix. Thus, matrices and linear transformations are equivalent concepts dressed somewhat differently. We can analyze one by studying the other.

The subscript-superscript notation we introduced on matrices and coordinate representations is actually helpful in tracking a linear transformation $T:\mathbb{V} \rightarrow \mathbb{W}$, where $\mathbb{V}$ and $\mathbb{W}$ are vector spaces of dimensions n and m, respectively. Suppose $\mathbf{w} = T(\mathbf{v})$. We let $\mathbf{v}_{\mathbb{B}}$ denote the coordinate representation of $\mathbf{v}$ with respect to a $\mathbb{B}$ basis and $\mathbf{w}_{\mathbb{C}}$ denote the coordinate representation of $\mathbf{w}$ with respect to a $\mathbb{C}$ basis. The indirect approach yields the matrix equation

$$\mathbf{w}_{\mathbb{C}} = \mathbf{A}_{\mathbb{B}}^{\mathbb{C}}\mathbf{v}_{\mathbb{B}}$$

The matrix $\mathbf{A}$ maps an n-tuple with respect to the $\mathbb{B}$ basis into an m-tuple with respect to the $\mathbb{C}$ basis. The subscript on $\mathbf{A}$ must match the subscript on $\mathbf{v}$. The superscript on $\mathbf{A}$ matches the subscript on $\mathbf{w}$. Figure 3.12 demonstrates the directional flow with arrows.

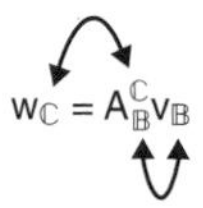

FIGURE 3.12

Problems 3.3

In Problems 1 through 25, find the matrix representation for $T : \mathbb{V} \to \mathbb{W}$ with respect to the given bases, $\mathbb{B}$ for a vector space $\mathbb{V}$ and $\mathbb{C}$ for a vector space $\mathbb{W}$.

(1) $T : \mathbb{R}^2 \to \mathbb{R}^3$ defined by $T\begin{bmatrix} a \\ b \end{bmatrix} = \begin{bmatrix} a+b \\ a-b \\ 2b \end{bmatrix}$, $\mathbb{B} = \left\{ \begin{bmatrix} 1 \\ 0 \end{bmatrix}, \begin{bmatrix} 1 \\ 1 \end{bmatrix} \right\}$, and

$\mathbb{C} = \left\{ \begin{bmatrix} 1 \\ 0 \\ 0 \end{bmatrix}, \begin{bmatrix} 0 \\ 1 \\ 0 \end{bmatrix}, \begin{bmatrix} 0 \\ 0 \\ 1 \end{bmatrix} \right\}$.

(2) Problem 1 with $\mathbb{B} = \left\{ \begin{bmatrix} 1 \\ 1 \end{bmatrix}, \begin{bmatrix} 1 \\ 2 \end{bmatrix} \right\}$.

(3) Problem 1 with $\mathbb{C} = \left\{ \begin{bmatrix} 1 \\ 1 \\ 0 \end{bmatrix}, \begin{bmatrix} 1 \\ 0 \\ 1 \end{bmatrix}, \begin{bmatrix} 0 \\ 1 \\ 0 \end{bmatrix} \right\}$.

(4) Problem 1 with $\mathbb{B} = \left\{ \begin{bmatrix} 1 \\ 1 \end{bmatrix}, \begin{bmatrix} 1 \\ 2 \end{bmatrix} \right\}$ and $\mathbb{C} = \left\{ \begin{bmatrix} 1 \\ 1 \\ 0 \end{bmatrix}, \begin{bmatrix} 1 \\ 0 \\ 1 \end{bmatrix}, \begin{bmatrix} 0 \\ 1 \\ 0 \end{bmatrix} \right\}$.

(5) $T : \mathbb{R}^3 \to \mathbb{R}^2$ defined by $T\begin{bmatrix} a \\ b \\ c \end{bmatrix} = \begin{bmatrix} 2a+3b-c \\ 4b+5c \end{bmatrix}$, $\mathbb{B} = \left\{ \begin{bmatrix} 1 \\ 0 \\ 0 \end{bmatrix}, \begin{bmatrix} 1 \\ 1 \\ 0 \end{bmatrix}, \begin{bmatrix} 1 \\ 1 \\ 1 \end{bmatrix} \right\}$,

and $\mathbb{C} = \left\{ \begin{bmatrix} 1 \\ 1 \end{bmatrix}, \begin{bmatrix} 0 \\ 1 \end{bmatrix} \right\}$.

(6) Problem 5 with $\mathbb{B} = \left\{ \begin{bmatrix} 1 \\ -1 \\ 0 \end{bmatrix}, \begin{bmatrix} 1 \\ 0 \\ -1 \end{bmatrix}, \begin{bmatrix} -1 \\ 1 \\ 1 \end{bmatrix} \right\}$.

(7) Problem 5 with $\mathbb{C} = \left\{ \begin{bmatrix} 1 \\ 1 \end{bmatrix}, \begin{bmatrix} 1 \\ -1 \end{bmatrix} \right\}$.

(8) Problem 2 with $\mathbb{B} = \left\{ \begin{bmatrix} 1 \\ -1 \\ 0 \end{bmatrix}, \begin{bmatrix} 1 \\ 0 \\ -1 \end{bmatrix}, \begin{bmatrix} -1 \\ 1 \\ 1 \end{bmatrix} \right\}$, and $\mathbb{C} = \left\{ \begin{bmatrix} 1 \\ 1 \end{bmatrix}, \begin{bmatrix} 1 \\ -1 \end{bmatrix} \right\}$.

(9) Problem 5 With $T\begin{bmatrix} a \\ b \\ c \end{bmatrix} = \begin{bmatrix} a+2b-3c \\ 9a-8b-7c \end{bmatrix}$.

(10) $T : \mathbb{R}^2 \to \mathbb{R}^2$ defined by $T\begin{bmatrix} a \\ b \end{bmatrix} = \begin{bmatrix} 25a+30b \\ -45a+50b \end{bmatrix}$,
$\mathbb{B} = \mathbb{C} =$ standard basis in $\mathbb{R}^2$.

(11) Problem 10 with $\mathbb{B} = \left\{ \begin{bmatrix} 10 \\ 10 \end{bmatrix}, \begin{bmatrix} 0 \\ 5 \end{bmatrix} \right\}$ and $\mathbb{C}$ again the standard basis.

(12) Problem 10 with $\mathbb{C} = \left\{ \begin{bmatrix} 10 \\ 10 \end{bmatrix}, \begin{bmatrix} 0 \\ 5 \end{bmatrix} \right\}$ and $\mathbb{B}$ again the standard basis.

(13) Problem 10 with $\mathbb{B} = \mathbb{C} = \left\{ \begin{bmatrix} 10 \\ 10 \end{bmatrix}, \begin{bmatrix} 0 \\ 5 \end{bmatrix} \right\}$

(14) Problem 10 with $\mathbb{B} = \left\{ \begin{bmatrix} 1 \\ -1 \end{bmatrix}, \begin{bmatrix} 1 \\ 2 \end{bmatrix} \right\}$ and $\mathbb{C} = \left\{ \begin{bmatrix} 1 \\ 2 \end{bmatrix}, \begin{bmatrix} 2 \\ 1 \end{bmatrix} \right\}$

(15) Problem 10 with $T\begin{bmatrix} a \\ b \end{bmatrix} = \begin{bmatrix} 2a \\ 3b - a \end{bmatrix}$.

(16) The transformation in Problem 15 with the bases of Problem 14.

(17) $T : \mathbb{P}^2 \to \mathbb{P}^3$ defined by $T(at^2+bt+c)=t(at^2+bt+c)$, $\mathbb{B} = \{t^2, t, 1\}$, and $\mathbb{C} = \{t^3, t^2, t, 1\}$.

(18) Problem 17 with $\mathbb{B} = \{t^2 + t, t^2 + 1, t + 1\}$ with $\mathbb{C} = \{t^2, t^2 + 1, t^2 - 1, t\}$.

(19) $T : \mathbb{P}^3 \to \mathbb{P}^2$ defined by $T(at^3 + bt^2 + ct + d) = 3at^2 + 2bt + c$,
$\mathbb{B} = \{t^3, t^2 + 1, t^2 - 1, t\}$ and $\mathbb{C} = \{t^2 + t, t^2 + 1, t + 1\}$

(20) $T : \mathbb{P}^2 \to \mathbb{R}^2$ defined by $T(at^2 + bt + c) = \begin{bmatrix} 2a + b \\ 3a - 4b + c \end{bmatrix}$,
$\mathbb{B} = \{t^2, t^2 - 1, t\}$ and $\mathbb{C} = \left\{ \begin{bmatrix} 1 \\ 1 \end{bmatrix}, \begin{bmatrix} 1 \\ -1 \end{bmatrix} \right\}$.

(21) $T : \mathbb{P}^2 \to \mathbb{R}^2$ defined by $T(at^2 + bt + c) = \begin{bmatrix} 2a + 3b \\ 4a - 5c \\ 6b + 7c \end{bmatrix}$, $\mathbb{B} = \{t^2, t^2 - 1, t\}$
and $\mathbb{C} = \left\{ \begin{bmatrix} 1 \\ 0 \\ 0 \end{bmatrix}, \begin{bmatrix} 1 \\ 1 \\ 0 \end{bmatrix}, \begin{bmatrix} 1 \\ 1 \\ 1 \end{bmatrix} \right\}$.

(22) $T : \mathbb{P}^2 \to \mathbb{M}_{2\times 2}$ defined by $T(at^2 + bt + c) = \begin{bmatrix} 2a + b & c - 3a \\ 4a - 5c & 6b + 7c \end{bmatrix}$
$\mathbb{B} = \{t^2, t^2 - 1, t\}$ and $\mathbb{C} = \left\{ \begin{bmatrix} 1 & 0 \\ 0 & 0 \end{bmatrix}, \begin{bmatrix} 1 & 1 \\ 0 & 0 \end{bmatrix}, \begin{bmatrix} 0 & 0 \\ 1 & 1 \end{bmatrix}, \begin{bmatrix} 0 & 0 \\ 1 & -1 \end{bmatrix} \right\}$.

(23) Problem 22 with $\mathbb{C} = \left\{ \begin{bmatrix} 1 & 0 \\ 0 & 0 \end{bmatrix}, \begin{bmatrix} 1 & 1 \\ 0 & 0 \end{bmatrix}, \begin{bmatrix} 1 & 1 \\ 0 & 1 \end{bmatrix}, \begin{bmatrix} 1 & 1 \\ 1 & 1 \end{bmatrix} \right\}$.

(24) T: $\mathbb{M}_{2\times 2} \to \mathbb{P}^3$ defined by $T\begin{bmatrix} a & b \\ c & d \end{bmatrix} = (a+b)t^3 + (a - 2b)t^2 + (2a - 3b + 4c)t + (a - d)$,

$\mathbb{B} = \left\{ \begin{bmatrix} 1 & 0 \\ 0 & 0 \end{bmatrix}, \begin{bmatrix} 1 & 1 \\ 0 & 0 \end{bmatrix}, \begin{bmatrix} 0 & 0 \\ 1 & 1 \end{bmatrix}, \begin{bmatrix} 0 & 0 \\ 1 & -1 \end{bmatrix} \right\}$, and $\mathbb{C} = \{t^3, t^2 - 1, t - 1, 1\}$.

(25) $T: \mathbb{M}_{2\times2} \rightarrow \mathbb{R}^2$ defined by $T\begin{bmatrix} a & b \\ c & d \end{bmatrix} = \begin{bmatrix} a+b+3c \\ b+c-5d \end{bmatrix}$,
$\mathbb{B} = \left\{ \begin{bmatrix} 1 & 0 \\ 0 & 0 \end{bmatrix}, \begin{bmatrix} 1 & 1 \\ 0 & 0 \end{bmatrix}, \begin{bmatrix} 0 & 0 \\ 1 & 1 \end{bmatrix}, \begin{bmatrix} 0 & 0 \\ 1 & -1 \end{bmatrix} \right\}$, and $\mathbb{C} = \left\{ \begin{bmatrix} 1 \\ 2 \end{bmatrix}, \begin{bmatrix} 2 \\ 1 \end{bmatrix} \right\}$.

In Problems 26 through 37, find the indicated mapping directly and by the indirect approach illustrated in Figure 3.11.

(26) $T\begin{bmatrix} 1 \\ 3 \end{bmatrix}$ with the information provided in Problem 1.

(27) $T\begin{bmatrix} 2 \\ -1 \end{bmatrix}$ with the information provided in Problem 1.

(28) $T\begin{bmatrix} -5 \\ 3 \end{bmatrix}$ with the information provided in Problem 2.

(29) $T\begin{bmatrix} 1 \\ 2 \\ 3 \end{bmatrix}$ with the information provided in Problem 5.

(30) $T\begin{bmatrix} 2 \\ 2 \\ 2 \end{bmatrix}$ with the information provided in Problem 5.

(31) $T\begin{bmatrix} 2 \\ -1 \\ -1 \end{bmatrix}$ with the information provided in Problem 5.

(32) $T\begin{bmatrix} 2 \\ -3 \end{bmatrix}$ with the information provided in Problem 10.

(33) $T(3t^2 - 2t)$ with the information provided in Problem 19.

(34) $T(3t^2 - 2t + 5)$ with the information provided in Problem 19.

(35) $T(t^2 - 2t - 1)$ with the information provided in Problem 20.

(36) $T(t^2 - 2t - 1)$ with the information provided in Problem 21.

(37) T(4) with the information provided in Problem 21.

(38) A matrix representation for $T : \mathbb{P}^1 \rightarrow \mathbb{P}^1$ is $\begin{bmatrix} 1 & 2 \\ 3 & 4 \end{bmatrix}_{\mathbb{B}}^{\mathbb{B}}$ with respect to $\mathbb{B} = \{t+1, t-1\}$. Find $T(at+b)$ for scalars a and b.

(39) A matrix representation for with respect to $T : \mathbb{P}^1 \rightarrow \mathbb{P}^1$ is $\begin{bmatrix} 1 & 2 \\ 3 & 4 \end{bmatrix}_{\mathbb{C}}^{\mathbb{C}}$ with respect to $\mathbb{C} = \{t+1, t+2\}$. Find $T(at+b)$ for scalars a and b.

(40) A matrix representation for $T: \mathbb{P}^2 \rightarrow \mathbb{P}^2$ is $\begin{bmatrix} 1 & 2 & 3 \\ 1 & 1 & 2 \\ 2 & 0 & 1 \end{bmatrix}_{\mathbb{B}}^{\mathbb{B}}$ with respect to $\mathbb{B} = \{t^2, t^2 + t, t^2 + t + 1\}$. Find $T(at^2 + bt + c)$ for scalars a, b, and c.

(41) A matrix representation for $T: \mathbb{M}_{2\times2} \rightarrow \mathbb{M}_{2\times2}$ is $\begin{bmatrix} 1 & 1 & 0 & 2 \\ 0 & 1 & 1 & 0 \\ 1 & 0 & 2 & 1 \\ 1 & 1 & 1 & 1 \end{bmatrix}_{\mathbb{B}}^{\mathbb{B}}$ with respect to the basis $\mathbb{B} = \left\{ \begin{bmatrix} 1 & 0 \\ 0 & 0 \end{bmatrix}, \begin{bmatrix} 1 & 1 \\ 0 & 0 \end{bmatrix}, \begin{bmatrix} 1 & 1 \\ 1 & 0 \end{bmatrix}, \begin{bmatrix} 1 & 1 \\ 1 & 1 \end{bmatrix} \right\}$. Find $T\begin{bmatrix} a & b \\ c & d \end{bmatrix}$ for scalars a, b, c, and d.

3.4 CHANGE OF BASIS

In general, a vector has many coordinate representations, a different one for each basis.

Coordinate representations for vectors in an n-dimensional vector space are basis dependent, and different bases generally result in different n-tuple representations for the same vector. In particular, we saw from Example 10 of Section 2.4 that the 2-tuple representation for $\mathbf{v} = [7\ 2]^{\mathrm{T}}$ is

$$\mathbf{v}_{\mathbb{S}} = \begin{bmatrix} 7 \\ 2 \end{bmatrix}_{\mathbb{C}} \tag{3.11}$$

with respect to the standard basis $= \{[1\ 0]^{\mathrm{T}}, [0\ 1]^{\mathrm{T}}\}$ for $\mathbb{R}^2$, but

$$\mathbf{v}_{\mathbb{D}} = \begin{bmatrix} 9/2 \\ 5/2 \end{bmatrix}_{\mathbb{D}} \tag{3.12}$$

with respect to the basis $\mathbb{D} = \left\{[1\ 1]^{\mathrm{T}},, [1 - 1]^{\mathrm{T}}\right\}$. It is natural to ask, therefore, whether *different* coordinate representations for *same* vector are related.

Let $\mathbb{C} = \{\mathbf{u}_1, \mathbf{u}_2, \ldots, \mathbf{u}_n\}$ and $\mathbb{D} = \{\mathbf{v}_1, \mathbf{v}_2, \ldots, \mathbf{v}_n\}$ be two bases for a vector space $\mathbb{V}$. If $\mathbf{v} \in \mathbb{V}$, the $\mathbf{v}$ can be expressed as a unique linear combination of the basis vectors in $\mathbb{C}$; that is, there exists a unique set of scalars $c_1, c_2, \ldots, c_n$ such that

$$\mathbf{v} = c_1\mathbf{u}_1 + c_2\mathbf{u}_2 + \cdots + c_n\mathbf{u}_n = \sum_{j=1}^{n} c_j\mathbf{u}_j \tag{3.13}$$

Similarly, if we consider the $\mathbb{D}$ basis instead, there exists a unique set of scalars $d_1, d_2, \ldots, d_n$ such that

$$\mathbf{v} = d_1\mathbf{v}_1 + d_2\mathbf{v}_2 + \cdots + d_n\mathbf{v}_n = \sum_{i=1}^{n} d_i\mathbf{v}_i \tag{3.14}$$

The coordinate representations of $\mathbf{v}$ with respect to $\mathbb{C}$ and $\mathbb{D}$, respectively, are

$$\mathbf{v}_{\mathbb{C}} = \begin{bmatrix} c_1 \\ c_2 \\ \vdots \\ c_n \end{bmatrix}_{\mathbb{C}} \quad \text{and} \quad \mathbf{v}_{\mathbb{D}} = \begin{bmatrix} d_1 \\ d_2 \\ \vdots \\ d_n \end{bmatrix}_{\mathbb{D}}$$

Now since each basis vector in $\mathbb{C}$ is also a vector in $\mathbb{V}$, it too can be expressed as a unique linear combination of the basis vectors in $\mathbb{D}$. In particular,

$$\mathbf{u}_1 = p_{11}\mathbf{v}_1 + p_{21}\mathbf{v}_2 + \cdots + p_{n1}\mathbf{v}_n$$

for some choice of the scalars $p_{11}, p_{21}, \ldots, p_{n1}$;

$$\mathbf{u}_2 = p_{12}\mathbf{v}_1 + p_{22}\mathbf{v}_2 + \cdots + p_{n2}\mathbf{v}_n$$

for some choice of the scalars $p_{12}, p_{22}, \ldots, p_{n2}$; and, in general,

$$\mathbf{u}_j = p_{1j}\mathbf{v}_1 + p_{2j}\mathbf{v}_2 + \cdots + p_{nj}\mathbf{v}_n = \sum_{i=1}^{n} p_{ij}\mathbf{v}_i \tag{3.15}$$

for some choice of the scalars $p_{1j}, p_{2j}, \ldots, p_{nj}$ $(j = 1, 2, \ldots, n)$. The n-tuple representations of these vectors with respect to the $\mathbb{D}$ basis are

$$\mathbf{u}_1 \leftrightarrow \begin{bmatrix} p_{11} \\ p_{21} \\ \vdots \\ p_{n1} \end{bmatrix}_{\mathbb{D}}, \ldots, \mathbf{u}_2 \leftrightarrow \begin{bmatrix} p_{12} \\ p_{22} \\ \vdots \\ p_{n2} \end{bmatrix}_{\mathbb{D}}, \ldots, \mathbf{u}_j \leftrightarrow \begin{bmatrix} p_{1j} \\ p_{2j} \\ \vdots \\ p_{nj} \end{bmatrix}_{\mathbb{D}}, \ldots, \mathbf{u}_n \leftrightarrow \begin{bmatrix} p_{1n} \\ p_{2n} \\ \vdots \\ p_{nn} \end{bmatrix}_{\mathbb{D}}$$

If we use these n-tuples as the columns of a matrix $\mathbf{P}$, then

$$\mathbf{P}_{\mathbb{C}}^{\mathbb{D}} = \begin{bmatrix} p_{11} & p_{12} & \cdots & p_{1j} & \cdots & p_{1n} \\ p_{21} & p_{22} & \cdots & p_{2j} & \cdots & p_{2n} \\ \vdots & \vdots & & \vdots & & \vdots \\ p_{n1} & p_{n2} & \cdots & p_{nj} & \cdots & p_{nn} \end{bmatrix}$$

where the subscript-superscript notation on $\mathbf{P}$ indicates that we are mapping from the $\mathbb{C}$ basis to the $\mathbb{D}$ basis. The matrix $\mathrm{P}_{\mathbb{C}}^{\mathbb{D}}$ is called the *transition matrix from the* $\mathbb{C}$ *basis to the* $\mathbb{D}$ *basis*. It follows from (3.13) and (3.15) that

$$\mathbf{v} = \sum_{j=1}^{n} c_j\mathbf{u}_j = \sum_{j=1}^{n} c_j\left(\sum_{j=1}^{n} p_{ij}\mathbf{v}_i\right) = \sum_{i=1}^{n}\left(\sum_{j=1}^{n} p_{ij}c_i\right)\mathbf{v}_i$$

But we also have from (3.14) that

$$\mathbf{v} = \sum_{i=1}^{n} d_i\mathbf{v}_j$$

and because this representation is unique (see Theorem 5 of Section 2.4), we may infer that

$$d_i = \sum_{j=1}^{n} p_{ij} c_j$$

Therefore,

$$\begin{bmatrix} d_1 \\ d_2 \\ \vdots \\ d_n \end{bmatrix}_{\mathbb{D}} = \begin{bmatrix} \sum_{j=1}^{n} p_{1j} cj \\ \sum_{j=1}^{n} p_{2j} cj \\ \vdots \\ \sum_{j=1}^{n} p_{nj} cj \end{bmatrix}$$

which can be written as the matrix product

$$\begin{bmatrix} d_1 \\ d_2 \\ \vdots \\ d_n \end{bmatrix}_{\mathbb{D}} = \begin{bmatrix} p_{11} & p_{12} & \cdots & p_{1j} & \cdots & p_{1n} \\ p_{21} & p_{22} & \cdots & p_{2j} & \cdots & p_{2n} \\ \vdots & \vdots & & \vdots & & \vdots \\ p_{n1} & p_{n2} & \cdots & p_{nj} & \cdots & p_{nn} \end{bmatrix} \begin{bmatrix} c_1 \\ c_2 \\ \vdots \\ c_n \end{bmatrix}_{\mathbb{C}}$$

or

$$\mathbf{v}_{\mathbb{D}} = \mathbf{P}_{\mathbb{C}}^{\mathbb{D}} \mathbf{v}_{\mathbb{C}} \tag{3.16}$$

We have proven:

▶THEOREM 1

If $\mathbf{v}_{\mathbb{C}}$ and $\mathbf{v}_{\mathbb{D}}$ are the n-tuple (coordinate) representations of a vector $\mathbf{v}$ with respect to the bases $\mathbb{C}$ and $\mathbb{D}$, respectively, and if $\mathbf{P}_j$ is the n-tuple representation of the jth basis vector in $\mathbb{C}$ ($j=1, 2, \ldots, n$) with respect to the $\mathbb{D}$ basis, then $\mathbf{v}_{\mathbb{D}} = \mathbf{P}_{\mathbb{C}}^{\mathbb{D}} \mathbf{v}_{\mathbb{C}}$ where the jth column of $\mathbf{P}_{\mathbb{C}}^{\mathbb{D}}$ is $\mathbf{P}_j$. ◀

Example 1 Find the transition matrix between the bases $\mathbb{C} = \left\{[1\ 0]^{\mathrm{T}},, [0\ 1]^{\mathrm{T}}\right\}$ and for $\mathbb{P}^1$ and $\mathbb{D} = \left\{[1\ 1]^{\mathrm{T}},, [1 - 1]^{\mathrm{T}}\right\}$ in $\mathbb{R}^2$, and verify Theorem 1 for the coordinate representations of $\mathbf{v} = [7\ 2]^{\mathrm{T}}$ with respect to each basis.

Solution: We have

$$\begin{bmatrix} 1 \\ 0 \end{bmatrix} = \frac{1}{2}\begin{bmatrix} 1 \\ 1 \end{bmatrix} + \frac{1}{2}\begin{bmatrix} 1 \\ -1 \end{bmatrix} \leftrightarrow \begin{bmatrix} 1/2 \\ 1/2 \end{bmatrix}_{\mathbb{D}}$$

and

$$\begin{bmatrix} 0 \\ 1 \end{bmatrix} = \frac{1}{2}\begin{bmatrix} 1 \\ 1 \end{bmatrix} - \frac{1}{2}\begin{bmatrix} 1 \\ -1 \end{bmatrix} \leftrightarrow \begin{bmatrix} 1/2 \\ -1/2 \end{bmatrix}_{\mathbb{D}}$$

and the transition matrix from $\mathbb{C}$ to $\mathbb{D}$ as

$$\mathbf{P}_{\mathbb{C}}^{\mathbb{D}} = \begin{bmatrix} 1/2 & 1/2 \\ 1/2 & -1/2 \end{bmatrix}$$

The coordinate representation of $[7\ 2]^{\mathrm{T}}$ with respect to the $\mathbb{C}$ and $\mathbb{D}$ bases were found in Example 10 of Section 2.4 to be, respectively,

$$\mathbf{v}_{\mathbb{C}} = \begin{bmatrix} 7 \\ 2 \end{bmatrix} \quad \text{and} \quad \mathbf{v}_{\mathbb{D}} = \begin{bmatrix} 9/2 \\ 5/2 \end{bmatrix}$$

Here

$$\mathbf{P}_{\mathbb{S}}^{\mathbb{B}}\mathbf{v}_{\mathbb{S}} = \begin{bmatrix} 1/2 & 1/2 \\ 1/2 & -1/2 \end{bmatrix}\begin{bmatrix} 7 \\ 2 \end{bmatrix} = \begin{bmatrix} 9/2 \\ 5/2 \end{bmatrix} = \mathbf{v}_{\mathbb{B}}$$

Although Theorem 1 involves the transition matrix from $\mathbb{C}$ to $\mathbb{D}$, it is equally valid in the reverse direction for the transition matrix from $\mathbb{D}$ to $\mathbb{C}$. If $\mathbf{P}_{\mathbb{D}}^{\mathbb{C}}$ represents this matrix, then

$$\mathbf{v}_{\mathbb{C}} = \mathbf{P}_{\mathbb{D}}^{\mathbb{C}}\mathbf{v}_{\mathbb{D}} \tag{3.17}$$

Example 2 Verify (3.17) for the bases and vector $\mathbf{v}$ described in Example 1.

Solution: As in Example 1, $\mathbb{C} = \{[1\ 0]^{\mathrm{T}}, [0\ 1]^{\mathrm{T}}\}$ and $\mathbb{D} = \left\{[1\ 1]^{\mathrm{T}},, [1-1]^{\mathrm{T}}\right\}$. Now, however,

$$\begin{bmatrix} 1 \\ 0 \end{bmatrix} = 1\begin{bmatrix} 1 \\ 0 \end{bmatrix} + 1\begin{bmatrix} 0 \\ 1 \end{bmatrix} \leftrightarrow \begin{bmatrix} 1 \\ 1 \end{bmatrix}_{\mathbb{C}}$$

and

$$\begin{bmatrix} 1 \\ -1 \end{bmatrix} = 1\begin{bmatrix} 1 \\ 0 \end{bmatrix} - 1\begin{bmatrix} 0 \\ 1 \end{bmatrix} \leftrightarrow \begin{bmatrix} 1 \\ -1 \end{bmatrix}_{\mathbb{C}}$$

and the transition matrix from $\mathbb{D}$ to $\mathbb{C}$ is

$$\mathbf{P}_{\mathbb{D}}^{\mathbb{C}} = \begin{bmatrix} 1 & 1 \\ 1 & -1 \end{bmatrix}$$

Here

$$\mathbf{P}_{\mathbb{D}}^{\mathbb{C}}\mathbf{v}_{\mathbb{D}} = \begin{bmatrix} 1 & 1 \\ 1 & -1 \end{bmatrix}\begin{bmatrix} 9/2 \\ 5/2 \end{bmatrix} = \begin{bmatrix} 7 \\ 2 \end{bmatrix} = \mathbf{v}_{\mathbb{C}}$$

Note that the subscript-superscript notation is helpful in tracking which transition matrix can multiply which coordinate representation. The subscript on the matrix must match the subscript on the vector being multiplied! The superscript on the transition matrix must match the subscript on the vector that results from the multiplication. Equation (3.16) is

$$\mathbf{v}_{\mathbf{D}} = \mathbf{P}_{\mathbf{C}}^{\mathbf{D}} \mathbf{v}_{\mathbf{C}}$$

while equation (3.17) is

$$\mathbf{v}_{\mathbf{C}} = \mathbf{P}_{\mathbf{D}}^{\mathbf{C}} \mathbf{v}_{\mathbf{D}}$$

The arrows show the matches that must occur if the multiplication is to be meaningful and if the equality is to be valid.

An observant reader will note that the transition matrix $\mathbf{P}_{\mathbb{D}}^{\mathbb{C}}$ found in Example 2 is the inverse of the transition matrix $\mathbf{P}_{\mathbb{C}}^{\mathbb{D}}$ found in Example 1. This is not a coincidence.

> **▸THEOREM 2**
>
> *The transition matrix from $\mathbb{C}$ to $\mathbb{D}$, where both $\mathbb{C}$ and $\mathbb{D}$ are bases for the same finite dimensional vector space, is invertible and its inverse is the transition matrix from $\mathbb{D}$ to $\mathbb{C}$.* ◂

Proof: Let $\mathbf{P}_{\mathbb{C}}^{\mathbb{D}}$ denote the transition matrix from basis $\mathbb{C}$ to basis $\mathbb{D}$ and let $\mathbf{P}_{\mathbb{D}}^{\mathbb{C}}$ be the transition matrix from $\mathbb{D}$ to $\mathbb{C}$. If the underlying vector space is n-dimensional, then both of these transition matrices have order $n \times n$, and their product is well defined. Denote this product as $\mathbf{A} = [a_{ij}]$. Then

$$\mathbf{P}_{\mathbb{C}}^{\mathbb{D}} \mathbf{P}_{\mathbb{D}}^{\mathbb{C}} = \mathbf{A} = \begin{bmatrix} a_{11} & a_{12} & \cdots & a_{1n} \\ a_{21} & a_{22} & \cdots & a_{2n} \\ \vdots & \vdots & \vdots & \vdots \\ a_{n1} & a_{n2} & \cdots & a_{nn} \end{bmatrix} \tag{3.18}$$

We claim that $\mathbf{A}$ is the $n \times n$ identity matrix.

We have from Theorem 1 that $\mathbf{v}_{\mathbb{D}} = \mathbf{P}_{\mathbb{C}}^{\mathbb{D}} \mathbf{P}_{\mathbb{D}}^{\mathbb{C}}$. Substituting into the right side of this equation the expression for $\mathbf{v}_{\mathbb{C}}$ given by (3.17), we obtain

$$\mathbf{v}_{\mathbb{D}} = \left(\mathbf{P}_{\mathbb{C}}^{\mathbb{D}} \mathbf{P}_{\mathbb{D}}^{\mathbb{C}}\right) \mathbf{v}_{\mathbb{D}} = \mathbf{A}\mathbf{v}_{\mathbb{D}} \tag{3.19}$$

Equation (3.18) is valid for any n-tuple representation with respect to the $\mathbb{D}$ basis. For the special case, $\mathbf{v}_{\mathbb{D}} = [1\ 0\ 0 \ldots 0]^{\mathrm{T}}$, equation (3.19) reduces to

$$\begin{bmatrix} 1 \\ 0 \\ 0 \\ 0 \\ \vdots \\ 0 \end{bmatrix} = \begin{bmatrix} a_{11} & a_{12} & \cdots & a_{1n} \\ a_{21} & a_{22} & \cdots & a_{2n} \\ a_{31} & a_{32} & \cdots & a_{3n} \\ \vdots & \vdots & \ddots & \vdots \\ a_{n1} & a_{n2} & \cdots & a_{nn} \end{bmatrix} \begin{bmatrix} 1 \\ 0 \\ 0 \\ 0 \\ \vdots \\ 0 \end{bmatrix}$$

or

$$\begin{bmatrix} 1 \\ 0 \\ 0 \\ \vdots \\ 0 \end{bmatrix} = \begin{bmatrix} a_{11} \\ a_{21} \\ a_{31} \\ \vdots \\ a_{n1} \end{bmatrix}$$

which defines the first column of the product matrix in (3.18). For the special case, $\mathbf{v}_{\mathbb{D}} = [0\ 1\ 0 \ldots 0]^{\mathrm{T}}$, equation (3.19) reduces to

$$\begin{bmatrix} 0 \\ 1 \\ 0 \\ \vdots \\ 0 \end{bmatrix} = \begin{bmatrix} a_{11} & a_{12} & \cdots & a_{1n} \\ a_{21} & a_{22} & \cdots & a_{2n} \\ a_{31} & a_{32} & \cdots & a_{3n} \\ \vdots & \vdots & \ddots & \vdots \\ a_{n1} & a_{n2} & \cdots & a_{nn} \end{bmatrix} \begin{bmatrix} 0 \\ 1 \\ 0 \\ \vdots \\ 0 \end{bmatrix}$$

or

$$\begin{bmatrix} 0 \\ 1 \\ 0 \\ \vdots \\ 0 \end{bmatrix} = \begin{bmatrix} a_{12} \\ a_{22} \\ a_{32} \\ \vdots \\ a_{n2} \end{bmatrix}$$

which defines the second column of $\mathbf{A}$. Successively, substituting for $\mathbf{v}_{\mathbb{D}}$ the various vectors in the standard basis, we find that

$$\mathbf{P}_{\mathbb{C}}^{\mathbb{D}} \mathbf{P}_{\mathbb{D}}^{\mathbb{C}} = 1$$

from which we conclude that $\mathbf{P}_{\mathbb{C}}^{\mathbb{D}}$ and $\mathbf{P}_{\mathbb{D}}^{\mathbb{C}}$ are inverses of one another.

Example 3 Find transition matrices between the two bases $\mathbb{G} = \{t + 1, t - 1\}$ and $\mathbb{H} = \{2t + 1, 3t + 1\}$ for $\mathbb{P}^1$ and verify the results for the coordinate representations of the polynomial $3t+5$ with respect to each basis.

Solution: Setting $\mathbf{v} = 3t + 5$, we may express $\mathbf{v}$ as a linear combination of vectors in either basis. We have

$$3t + 5 = [4](t + 1) + [-1](t - 1)$$

and

$$3t + 5 = [12](2t + 1) + [-7](3t + 1)$$

so the coordinate representations of $\mathbf{v}$ with respect to these bases are

$$\mathbf{v}_{\mathbb{G}} = \begin{bmatrix} 4 \\ -1 \end{bmatrix}_{\mathbb{G}} \quad \text{and} \quad \mathbf{v}_{\mathbb{H}} = \begin{bmatrix} 12 \\ -7 \end{bmatrix}_{\mathbb{H}}$$

Now writing each vector in the $\mathbb{H}$ basis as a linear combination of the vectors in the $\mathbb{G}$ basis, we obtain

$$2t + 1 = [1.5](t+1) + [0.5](t-1) \leftrightarrow \begin{bmatrix} 1.5 \\ 0.5 \end{bmatrix}_{\mathbb{G}}$$

and

$$3t + 1 = [2](t+1) + [1](t-1) \leftrightarrow \begin{bmatrix} 2 \\ 1 \end{bmatrix}_{\mathbb{G}}$$

Consequently, the transition matrix from the $\mathbb{H}$ basis to the $\mathbb{G}$ basis is

$$\mathbf{P}_{\mathbb{H}}^{\mathbb{G}} = \begin{bmatrix} 1.5 & 2 \\ 0.5 & 1 \end{bmatrix}$$

while the transition matrix from the $\mathbb{G}$ basis to the $\mathbb{H}$ basis is

$$\mathbf{P}_{\mathbb{H}}^{\mathbb{G}} = \left(\mathbf{P}_{\mathbb{H}}^{\mathbb{G}}\right)^{-1} = \begin{bmatrix} 2 & -4 \\ -1 & 3 \end{bmatrix}$$

Then

$$\mathbf{P}_{\mathbb{H}}^{\mathbb{G}}\mathbf{v}_{\mathbb{H}} = \begin{bmatrix} 1.5 & 2 \\ 0.5 & 1 \end{bmatrix}\begin{bmatrix} 12 \\ -7 \end{bmatrix} = \begin{bmatrix} 4 \\ -1 \end{bmatrix} = \mathbf{v}_{\mathbb{G}}$$

and

$$\mathbf{P}_{\mathbb{G}}^{\mathbb{H}}\mathbf{v}_{\mathbb{G}} = \begin{bmatrix} 2 & -4 \\ -1 & 3 \end{bmatrix}\begin{bmatrix} 4 \\ -1 \end{bmatrix} = \begin{bmatrix} 12 \\ 7 \end{bmatrix} = \mathbf{v}_{\mathbb{H}}$$

If we graph the standard basis in $\mathbb{R}^2$ in the x-y plane, we have the directed line segments $\mathbf{e}_1$ and $\mathbf{e}_2$ shown in Figure 3.13. Another basis for $\mathbb{R}^2$ is obtained

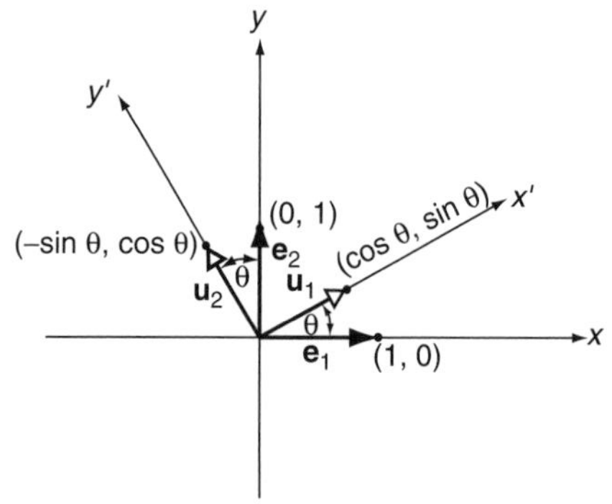

FIGURE 3.13

by *rotating* these two vectors counterclockwise about the origin by an angle θ, resulting in the directed line segments $\mathbf{u}_1$ and $\mathbf{u}_2$ graphed in Figure 3.13. The magnitudes of all four directed line segments are one. It then follows from elementary trigonometry that the arrowhead for $\mathbf{u}_1$ falls on the point ($\cos\theta$, $\sin\theta$) while that for $\mathbf{u}_2$ falls on the point ($-\sin\theta$, $\cos\theta$). Setting $\mathbb{S} = \{\mathbf{e}_1, \mathbf{e}_2\}$ and $\mathbb{R} = \{\mathbf{u}_1, \mathbf{e}_2\}$, we have

$$\begin{bmatrix} \cos\theta \\ \sin\theta \end{bmatrix} = \cos\theta \begin{bmatrix} 1 \\ 0 \end{bmatrix} + \sin\theta \begin{bmatrix} 0 \\ 1 \end{bmatrix} \leftrightarrow \begin{bmatrix} \cos\theta \\ \sin\theta \end{bmatrix}_{\mathbb{S}}$$

and

$$\begin{bmatrix} -\sin\theta \\ \cos\theta \end{bmatrix} = -\sin\theta \begin{bmatrix} 1 \\ 0 \end{bmatrix} + \cos\theta \begin{bmatrix} 0 \\ 1 \end{bmatrix} \leftrightarrow \begin{bmatrix} -\sin\theta \\ \cos\theta \end{bmatrix}_{\mathbb{S}}$$

The transition matrix from the $\mathbb{R}$ basis to the $\mathbb{S}$ basis is

$$\mathbf{P}_{\mathbb{R}}^{\mathbb{S}} = \begin{bmatrix} \cos\theta & -\sin\theta \\ \sin\theta & \cos\theta \end{bmatrix}$$

Hence, the transition matrix from the $\mathbb{S}$ basis to the $\mathbb{R}$ basis is

$$\mathbf{P}_{\mathbb{S}}^{\mathbb{R}} = \left(\mathbf{P}_{\mathbb{R}}^{\mathbb{S}}\right)^{-1} = \begin{bmatrix} \cos\theta & \sin\theta \\ -\sin\theta & \cos\theta \end{bmatrix}$$

Consequently, if

$$\mathbf{v}_{\mathbb{S}} \begin{bmatrix} x \\ y \end{bmatrix}_{\mathbb{S}} \quad \text{and} \quad \mathbf{v}_{\mathbb{R}} \begin{bmatrix} x' \\ y' \end{bmatrix}_{\mathbb{R}}$$

denote, respectively, the coordinate representation of the vector $\mathbf{v}$ with respect to the standard basis $\mathbb{S}$ and the coordinate representation of $\mathbf{v}$ with respect to the $\mathbb{R}$ basis, then

$$\begin{bmatrix} x' \\ y' \end{bmatrix} = \mathbf{v}_{\mathbb{R}} = \mathbf{P}_{\mathbb{S}}^{\mathbb{R}} \mathbf{v}_{\mathbb{S}} = \begin{bmatrix} \cos\theta & \sin\theta \\ -\sin\theta & \cos\theta \end{bmatrix} \begin{bmatrix} x \\ y \end{bmatrix} = \begin{bmatrix} x\cos\theta + y\sin\theta \\ -x\sin\theta + y\cos\theta \end{bmatrix}$$

Equating components, we have the well-known transformations for a rotation of the coordinate axis in the x-y plane by an angle of θ in the counterclockwise direction:

$$\begin{aligned} x' &= x\cos\theta + y\sin\theta \\ y' &= -x\sin\theta + y\cos\theta \end{aligned}$$

We showed in Section 3.3 that a linear transformation from one finite-dimensional vector space to another can be represented by a matrix. Such a matrix, however, is basis dependent; as the basis for either the domain or range is changed, the matrix changes accordingly.

In general, a linear transformation has many matrix representations, a different matrix for each pair of bases in the domain and range.

Example 4 Find matrix representations for the linear transformation $T: \mathbb{R}^2 \to \mathbb{R}^2$ defined by

$$T\begin{bmatrix} a \\ b \end{bmatrix} = \begin{bmatrix} 11a + 3b \\ -5a - 5b \end{bmatrix}$$

(a) with respect to the standard basis $\mathbb{C} = \{[1\ 0]^T, [0\ 1]^T\}$, (b) with respect to the basis $\mathbb{D} = \left\{[1\ 1]^T, [1\ -1]^T\right\}$, and (c) with respect to the basis $\mathbb{E} = \left\{[3\ -1]^T, [1\ -5]^T\right\}$.

Solution: (a) Using the standard basis, we have

$$T\begin{bmatrix} 1 \\ 0 \end{bmatrix} = \begin{bmatrix} 11 \\ -5 \end{bmatrix} = 11\begin{bmatrix} 1 \\ 0 \end{bmatrix} - 5\begin{bmatrix} 0 \\ 1 \end{bmatrix} \leftrightarrow \begin{bmatrix} 11 \\ -5 \end{bmatrix}_{\mathbb{C}}$$

$$T\begin{bmatrix} 0 \\ 1 \end{bmatrix} = \begin{bmatrix} 3 \\ -5 \end{bmatrix} = 3\begin{bmatrix} 1 \\ 0 \end{bmatrix} - 5\begin{bmatrix} 0 \\ 1 \end{bmatrix} \leftrightarrow \begin{bmatrix} 3 \\ -5 \end{bmatrix}_{\mathbb{C}}$$

and

$$T \leftrightarrow \begin{bmatrix} 11 & 3 \\ -5 & -5 \end{bmatrix}_{\mathbb{C}}^{\mathbb{C}} = \mathbf{A}_{\mathbb{C}}^{\mathbb{C}}$$

(b) Using the $\mathbb{B}$ basis, we have

$$T\begin{bmatrix} 1 \\ 1 \end{bmatrix} = \begin{bmatrix} 14 \\ -10 \end{bmatrix} = 2\begin{bmatrix} 1 \\ 1 \end{bmatrix} - 12\begin{bmatrix} 1 \\ -1 \end{bmatrix} \leftrightarrow \begin{bmatrix} 2 \\ 12 \end{bmatrix}_{\mathbb{D}}$$

$$T\begin{bmatrix} 1 \\ -1 \end{bmatrix} = \begin{bmatrix} 8 \\ 0 \end{bmatrix} = 4\begin{bmatrix} 1 \\ 1 \end{bmatrix} + 4\begin{bmatrix} 1 \\ -1 \end{bmatrix} \leftrightarrow \begin{bmatrix} 4 \\ 4 \end{bmatrix}_{\mathbb{D}}$$

and

$$T \leftrightarrow \begin{bmatrix} 2 & 4 \\ 12 & 4 \end{bmatrix}_{\mathbb{D}}^{\mathbb{D}} = \mathbf{A}_{\mathbb{D}}^{\mathbb{D}}$$

(c) Using the $\mathbb{E}$ basis, we obtain

$$T\begin{bmatrix} 3 \\ -1 \end{bmatrix} = \begin{bmatrix} 30 \\ -10 \end{bmatrix} = 10\begin{bmatrix} 3 \\ -1 \end{bmatrix} + 0\begin{bmatrix} 1 \\ -5 \end{bmatrix} \leftrightarrow \begin{bmatrix} 10 \\ 0 \end{bmatrix}_{\mathbb{E}}$$

$$T\begin{bmatrix} 1 \\ -5 \end{bmatrix} = \begin{bmatrix} -4 \\ 20 \end{bmatrix} = 0\begin{bmatrix} 3 \\ -1 \end{bmatrix} - 4\begin{bmatrix} 1 \\ -5 \end{bmatrix} \leftrightarrow \begin{bmatrix} 0 \\ -4 \end{bmatrix}_{\mathbb{E}}$$

and

$$T \leftrightarrow \begin{bmatrix} 10 & 0 \\ 0 & -4 \end{bmatrix}_{\mathbb{E}}^{\mathbb{E}} = \mathbf{A}_{\mathbb{E}}^{\mathbb{E}}$$

It is natural to ask whether *different* matrices representing the *same* linear transformation are related. We limit ourselves to linear transformations from a vector space into *itself*, that is, linear transformations of the form $T: \mathbb{V} \to \mathbb{V}$, because these are the transformations that will interest us the most. When the domain and range are identical, both have the same dimension, and any matrix representation of T must be square. The more general case of transformations that map from one vector space $\mathbb{V}$ into a *different* vector space $\mathbb{W}$ is addressed in Problem 40.

Let $T: \mathbb{V} \to \mathbb{V}$ be a linear transformation on an n-dimensional vector space $\mathbb{V}$ with $\mathbf{w} = T(\mathbf{v})$. If $\mathbb{C}$ is a basis for a vector space $\mathbb{V}$, then the n-tuple representation for $\mathbf{w}$ with respect to $\mathbb{C}$, denoted by $\mathbf{w}_{\mathbb{C}}$, can be obtained indirectly (see Section 3.3), by first determining the n-tuple representation for $\mathbf{v}$ with respect to $\mathbb{C}$, denoted by $\mathbf{v}_{\mathbb{C}}$, then determining the matrix representation for T with respect to the $\mathbb{C}$ basis, denoted by $\mathbf{A}_{\mathbb{C}}^{\mathbb{C}}$, and finally calculating the product $\mathbf{A}_{\mathbb{C}}^{\mathbb{C}}\mathbf{v}_{\mathbb{C}}$. That is,

$$\mathbf{w}_{\mathbb{C}} = \mathbf{A}_{\mathbb{C}}^{\mathbb{C}}\mathbf{v}_{\mathbb{C}} \tag{3.20}$$

If we use a different basis, denoted by $\mathbb{D}$, then we also have

$$\mathbf{w}_{\mathbb{D}} = \mathbf{A}_{\mathbb{D}}^{\mathbb{D}}\mathbf{v}_{\mathbb{D}} \tag{3.21}$$

Since $\mathbf{v}_{\mathbb{C}}$ and $\mathbf{v}_{\mathbb{D}}$ are n-tuple representations for the same vector $\mathbf{v}$, but with respect to different bases, it follows from Theorem 1 that there exists a transition matrix $\mathbf{P}_{\mathbb{C}}^{\mathbb{D}}$ for which

$$\mathbf{v}_{\mathbb{D}} = \mathbf{P}_{\mathbb{C}}^{\mathbb{D}}\mathbf{v}_{\mathbb{C}} \tag{3.22}$$

Because (3.22) is true for any vector in $\mathbb{V}$, it is also true for $\mathbf{w}$, hence

$$\mathbf{w}_{\mathbb{D}} = \mathbf{P}_{\mathbb{C}}^{\mathbb{D}}\mathbf{w}_{\mathbb{C}} \tag{3.23}$$

Now, (3.21) and (3.22) imply that

$$\mathbf{w}_{\mathbb{D}} = \mathbf{A}_{\mathbb{D}}^{\mathbb{D}}\mathbf{v}_{\mathbb{D}} = \mathbf{A}_{\mathbb{D}}^{\mathbb{D}}\mathbf{P}_{\mathbb{C}}^{\mathbb{D}}\mathbf{v}_{\mathbb{C}} \tag{3.24}$$

while (3.23) and (3.20) imply that

$$\mathbf{w}_{\mathbb{D}} = \mathbf{P}_{\mathbb{C}}^{\mathbb{D}}\mathbf{v}_{\mathbb{C}} = \mathbf{P}_{\mathbb{C}}^{\mathbb{D}}\mathbf{P}_{\mathbb{C}}^{\mathbb{C}}\mathbf{v}_{\mathbb{C}} \tag{3.25}$$

It follows from (3.24) and (3.25) that

$$\mathbf{P}_{\mathbb{C}}^{\mathbb{D}}\mathbf{A}_{\mathbb{C}}^{\mathbb{C}}\mathbf{v}_{\mathbb{C}} = \mathbf{A}_{\mathbb{D}}^{\mathbb{D}}\mathbf{P}_{\mathbb{C}}^{\mathbb{D}}\mathbf{v}_{\mathbb{C}}$$

This equality is valid for all n-tuples $\mathbf{v}_{\mathbb{C}}$ with respect to the $\mathbb{C}$ basis. If we successively take $\mathbf{v}_{\mathbb{C}}$ to be the vector having 1 as its first component with all other components equal to zero, then the vector having 1 in its second component with all other components equal to zero, and so on through the entire standard basis, we conclude that

$$\mathbf{P}_{\mathbb{C}}^{\mathbb{D}}\mathbf{A}_{\mathbb{C}}^{\mathbb{C}} = \mathbf{A}_{\mathbb{D}}^{\mathbb{D}}\mathbf{P}_{\mathbb{C}}^{\mathbb{D}}$$

We know from Theorem 2 that the transition matrix is invertible, so we may rewrite this last equation as

$$\mathbf{A}_{\mathbb{C}}^{\mathbb{C}} = \left(\mathbf{P}_{\mathbb{C}}^{\mathbb{D}}\right)^{-1}\mathbf{A}_{\mathbb{D}}^{\mathbb{D}}\mathbf{P}_{\mathbb{C}}^{\mathbb{D}} \tag{3.26}$$

Conversely, the same reasoning shows that if (3.26) is valid, then $\mathbf{A}_{\mathbb{C}}^{\mathbb{C}}$ and $\mathbf{A}_{\mathbb{D}}^{\mathbb{D}}$ are matrix representations for the same linear transformations with respect to the $\mathbb{C}$ basis and $\mathbb{D}$ basis, respectively, where these two bases are related by the transition matrix $\mathbf{P}_{\mathbb{C}}^{\mathbb{D}}$. If we simplify our notation by omitting the subscripts and superscripts and using different letters to distinguish different matrices, we have proven:

▶THEOREM 3

*Two $n \times n$ matrices **A** and **B** represent the same linear transformation if and only if there exists an invertible matrix **P** such that*

$$\mathbf{A} = \mathbf{P}^{-1}\mathbf{B}\mathbf{P} \tag{3.27}$$ ◀

Although equation (3.27) is notationally simpler, equation (3.26) is more revealing because it explicitly exhibits the dependencies on the different bases.

Example 5 Verify equation (3.26) for the matrix representations obtained in parts (a) and (b) of Example 4.

Solution: From Example 4,

$$\mathbf{A}_{\mathbb{S}}^{\mathbb{S}} = \begin{bmatrix} 11 & 3 \\ -5 & -5 \end{bmatrix}, \mathbf{A}_{\mathbb{B}}^{\mathbb{B}} = \begin{bmatrix} 2 & 4 \\ 12 & 4 \end{bmatrix}$$

and from Example 1,

$$\mathbf{P}_{\mathbb{S}}^{\mathbb{B}} = \begin{bmatrix} \frac{1}{2} & \frac{1}{2} \\ \frac{1}{2} & -\frac{1}{2} \end{bmatrix}$$

Therefore,

$$\left(\mathbf{P}_{\mathbb{S}}^{\mathbb{B}}\right)^{-1}\mathbf{A}_{\mathbb{B}}^{\mathbb{B}}\mathbf{P}_{\mathbb{S}}^{\mathbb{B}} = \begin{bmatrix} 1 & 1 \\ 1 & -1 \end{bmatrix}\begin{bmatrix} 2 & 4 \\ 12 & 4 \end{bmatrix}\begin{bmatrix} \frac{1}{2} & \frac{1}{2} \\ \frac{1}{2} & -\frac{1}{2} \end{bmatrix}$$
$$= \begin{bmatrix} 11 & 3 \\ -5 & -5 \end{bmatrix} = \mathbf{A}_{\mathbb{S}}^{\mathbb{S}}$$

Example 6 Verify equation (3.26) for the matrix representations obtained in parts (a) and (c) of Example 4.

Solution: Here the bases are $\mathbb{S} = \left\{[1\ 0]^{\mathrm{T}}, [0\ 1]^{\mathrm{T}}\right\}$ and $\mathbb{E} = \left\{[3 - 1]^{\mathrm{T}}, [1 - 5]^{\mathrm{T}}\right\}$, so equation (3.26) takes the notational form

$$\mathbf{A}_{\mathbb{S}}^{\mathbb{S}} = \left(\mathbf{P}_{\mathbb{S}}^{\mathbb{E}}\right)^{-1}\mathbf{A}_{\mathbb{E}}^{\mathbb{E}}\mathbf{P}_{\mathbb{S}}^{\mathbb{E}}$$

From Example 4,

$$\mathbf{A}_{\mathbb{C}}^{\mathbb{C}} = \begin{bmatrix} 11 & 3 \\ -5 & -5 \end{bmatrix}, \quad \text{and} \quad \mathbf{A}_{\mathbb{E}}^{\mathbb{E}} = \begin{bmatrix} 10 & 0 \\ 0 & -4 \end{bmatrix}$$

Writing each vector in the $\mathbb{S}$ basis as a linear combination of vectors in the $\mathbb{E}$ basis, we find that

$$\begin{bmatrix} 1 \\ 0 \end{bmatrix} = \frac{5}{14}\begin{bmatrix} 3 \\ -1 \end{bmatrix} - \frac{1}{14}\begin{bmatrix} 1 \\ -5 \end{bmatrix} \leftrightarrow \begin{bmatrix} 5/14 \\ -1/14 \end{bmatrix}_{\mathbb{E}}$$

$$\begin{bmatrix} 0 \\ 1 \end{bmatrix} = \frac{1}{14}\begin{bmatrix} 3 \\ -1 \end{bmatrix} - \frac{3}{14}\begin{bmatrix} 1 \\ -5 \end{bmatrix} \leftrightarrow \begin{bmatrix} 1/14 \\ -3/14 \end{bmatrix}_{\mathbb{E}}$$

whereupon

$$\mathbf{P}_{\mathbb{S}}^{\mathbb{E}} = \begin{bmatrix} 5/14 & 1/14 \\ -1/14 & -3/14 \end{bmatrix}$$

Matrices **A** and **B** are similar if they represent the same linear transformation, in which case there exists a transition matrix **P** such that $\mathbf{A} = \mathbf{P}^{-1}\mathbf{B}\mathbf{P}$.

Therefore,

$$\begin{aligned}\left(\mathbf{P}_{\mathbb{S}}^{\mathbb{E}}\right)^{-1}\mathbf{A}_{\mathbb{E}}^{\mathbb{E}}\mathbf{P}_{\mathbb{S}}^{\mathbb{E}} &= \begin{bmatrix} 3 & 1 \\ -1 & -5 \end{bmatrix}\begin{bmatrix} 10 & 0 \\ 0 & -4 \end{bmatrix}\begin{bmatrix} 5/14 & 1/14 \\ -1/14 & -3/14 \end{bmatrix} \\ &= \begin{bmatrix} 11 & 3 \\ -5 & -5 \end{bmatrix} = \mathbf{A}_{\mathbb{S}}^{\mathbb{S}}\end{aligned}$$

We say that two matrices are *similar* if they represent the same linear transformation. It follows from equation (3.27) that similar matrices satisfy the matrix equation

$$\mathbf{A} = \mathbf{P}^{-1}\mathbf{B}\mathbf{P} \qquad (3.27 \text{ repeated})$$

If we premultiply equation (3.27) by **P**, it follows that **A** is similar to **B** if and only if there exists a nonsingular matrix **P** such that

$$\mathbf{PA} = \mathbf{BP} \qquad (3.28)$$

Of all the similar matrices that can represent a particular linear transformation, some will be simpler in structure than others and one may be the simplest of all. In Example 4, we identified three different matrix representations for the same linear transformation. We now know all three of these matrices are similar. One, in particular, is a diagonal matrix, which is in many respects the simplest possible structure for a matrix. Could we have known this in advance? Could we have known in advance what basis would result in the simplest matrix representation? The answer is yes in both cases, and we will spend much of Chapters 4 and 6 developing methods for producing the appropriate bases and their related matrices.

Problems 3.4

In Problems 1 through 13, find the transition matrix from the first listed basis to the second.

(1) $\mathbb{B} = \left\{[1\ 0]^T, [1\ 1]^T\right\}, \mathbb{C} = \left\{[0\ 1]^T, [1\ 1]^T\right\}.$

(2) $\mathbb{B} = \left\{[1\ 0]^T, [1\ 1]^T\right\}, \mathbb{D} = \left\{[1\ 1]^T, [1\ 2]^T\right\}.$

(3) $\mathbb{C} = \left\{[0\ 1]^T, [1\ 1]^T\right\}, \mathbb{D} = \left\{[1\ 1]^T, [0\ 2]^T\right\}.$

(4) Same as Problem 3 but with $\mathbb{D}$ listed first.

(5) $\mathbb{E} = \left\{[1\ 2]^T, [1\ 3]^T\right\}, \mathbb{F} = \left\{[-1\ 1]^T, [0\ 1]^T\right\}.$

(6) Same as Problem 5 but with $\mathbb{F}$ listed first.

(7) $\mathbb{G} = \left\{[10\ 20]^T, [10 - 20^T\}, \mathbb{F} = \{-1\ 1^T], [0\ 1]^T\right\}.$

(8) $\mathbb{S} = \left\{[1\ 0\ 0]^T, [0\ 1\ 0]^T, [0\ 0\ 1]^T\right\}, \mathbb{T} = \left\{[1\ 1\ 0]^T, [0\ 1\ 1]^T, [1\ 0\ 1]^T\right\}.$

(9) $\mathbb{S} = \left\{[1\ 0\ 0]^T, [0\ 1\ 0]^T, [0\ 0\ 1]^T\right\}, \mathbb{U} = \left\{[1\ 0\ 0]^T, [1\ 1\ 0]^T, [1\ 1\ 1]^T\right\}.$

(10) Same as Problem 9 but with $\mathbb{U}$ listed first.

(11) $\mathbb{U} = \left\{[1\ 0\ 0]^T, [1\ 1\ 0]^T, [1\ 1\ 1]^T\right\}, \mathbb{T} = \left\{[1\ 1\ 0]^T, [0\ 1\ 1]^T, [1\ 0\ 1]^T\right\}.$

(12) $\mathbb{V} = \left\{[1\ 1\ 0]^T, [0\ 1\ 1]^T, [1\ 3\ 1]^T\right\}, \mathbb{T} = \left\{[1\ 1\ 0]^T, [0\ 1\ 1]^T, [1\ 0\ 1]^T\right\}.$

(13) $\mathbb{V} = \left\{[1\ 1\ 0]^T, [0\ 1\ 1]^T, [1\ 3\ 1]^T\right\}, \mathbb{U} = \left\{[1\ 0\ 1]^T,, [1\ 1\ 0]^T, [1\ 1\ 1]^T\right\}.$

In Problems 14 through 25, a linear transformation is defined and two bases are specified. Find (a) the matrix representation for $T\colon \mathbb{V} \to \mathbb{V}$ with respect to the first listed bases, (b) the matrix representation for the linear transformation with respect to the second listed basis, and (c) verify equation (3.26) using the results of parts (a) and (b) with a suitable transition matrix.

(14) $T\begin{bmatrix} a \\ b \end{bmatrix} = \begin{bmatrix} 2a+b \\ a-3b \end{bmatrix}$; $\mathbb{B}$ and $\mathbb{C}$ as given in Problem 1.

(15) $T\begin{bmatrix} a \\ b \end{bmatrix} = \begin{bmatrix} 2a+b \\ a-3b \end{bmatrix}$; $\mathbb{E}$ and $\mathbb{F}$ as given in Problem 5.

(16) $T\begin{bmatrix} a \\ b \end{bmatrix} = \begin{bmatrix} 8a-3b \\ 6a-b \end{bmatrix}$; $\mathbb{B}$ and $\mathbb{D}$ as given in Problem 2.

(17) $T\begin{bmatrix} a \\ b \end{bmatrix} = \begin{bmatrix} 2a \\ 3a - b \end{bmatrix}$; $\mathbb{B}$ and $\mathbb{C}$ as given in Problem 1.

(18) $T\begin{bmatrix} a \\ b \end{bmatrix} = \begin{bmatrix} 11a - 4b \\ 24a - 9b \end{bmatrix}$; $\mathbb{E}$ and $\mathbb{F}$ as given in Problem 5.

(19) $T\begin{bmatrix} a \\ b \end{bmatrix} = \begin{bmatrix} 11a - 4b \\ 24a - 9b \end{bmatrix}$; $\mathbb{B}$ and $\mathbb{D}$ as given in Problem 2.

(20) $T\begin{bmatrix} a \\ b \end{bmatrix} = \begin{bmatrix} a \\ b \end{bmatrix}$; $\mathbb{E}$ and $\mathbb{F}$ as given in Problem 5.

(21) $T\begin{bmatrix} a \\ b \end{bmatrix} = \begin{bmatrix} 0 \\ 0 \end{bmatrix}$; $\mathbb{C}$ and $\mathbb{D}$ as given in Problem 3.

(22) $T\begin{bmatrix} a \\ b \\ c \end{bmatrix} = \begin{bmatrix} 3a - b + c \\ 2a - 2c \\ 3a - 3b + c \end{bmatrix}$; $\mathbb{S}$ and $\mathbb{T}$ as given in Problem 8.

(23) $T\begin{bmatrix} a \\ b \\ c \end{bmatrix} = \begin{bmatrix} 3a - b + c \\ 2a - 2c \\ 3a - 3b + c \end{bmatrix}$; $\mathbb{S}$ and $\mathbb{U}$ as given in Problem 9.

(24) $T\begin{bmatrix} a \\ b \\ c \end{bmatrix} = \begin{bmatrix} a - b \\ 2b \\ a - 3c \end{bmatrix}$; $\mathbb{S}$ and $\mathbb{T}$ as given in Problem 8.

(25) $T\begin{bmatrix} a \\ b \\ c \end{bmatrix} = \begin{bmatrix} a \\ 2b \\ -3c \end{bmatrix}$; $\mathbb{S}$ and $\mathbb{U}$ as given in Problem 9.

(26) Show directly that $\mathbf{A} = \begin{bmatrix} 2 & 0 \\ 0 & 2 \end{bmatrix}$ and $\mathbf{B} = \begin{bmatrix} 2 & 1 \\ 0 & 2 \end{bmatrix}$ are not similar.

Hint: Set $\mathbf{P} = \begin{bmatrix} a & b \\ c & d \end{bmatrix}$ and show that no elements of this matrix exist that make equation (3.27) valid.

(27) Show directly that there does exist an invertible matrix **P** this satisfies equation (3.27) for $\mathbf{A} = \begin{bmatrix} 4 & 3 \\ -2 & -1 \end{bmatrix}$ and $\mathbf{B} = \begin{bmatrix} 5 & -4 \\ 3 & -2 \end{bmatrix}$.

(28) Prove that if **A** is similar to **B** then **B** is similar to **A**.

(29) Prove that if **A** is similar to **B** and **B** is similar to **C**, then **A** is similar to **C**.

(30) Prove that if **A** is similar to **B**, then $\mathbf{A}^2$ is similar to $\mathbf{B}^2$.

(31) Prove that if **A** is similar to **B**, then $\mathbf{A}^3$ is similar to $\mathbf{B}^3$.

(32) Prove that if **A** is similar to **B**, then $\mathbf{A}^{\mathrm{T}}$ is similar to $\mathbf{B}^{\mathrm{T}}$.

(33) Prove that every square matrix is similar to itself.

(34) Prove that if $\mathbf{A}$ is similar to $\mathbf{B}$, then $k\mathbf{A}$ is similar to $k\mathbf{B}$ for any constant k.

(35) Prove that if $\mathbf{A}$ is similar to $\mathbf{B}$ and if $\mathbf{A}$ is invertible, then $\mathbf{B}$ is also invertible and $\mathbf{A}^{-1}$ is similar to $\mathbf{B}^{-1}$.

(36) Show that there are many $\mathbf{P}$ matrices that make equation (3.26) valid for the two matrix representations obtained in Problem 20.

(37) Show that there are many $\mathbf{P}$ matrices that make equation (3.26) valid for the two matrix representations obtained in Problem 21.

(38) Let $\mathbb{C}=\{\mathbf{v}_1, \mathbf{v}_2, \ldots, \mathbf{v}_n\}$ and let $\mathbb{D}=\{\mathbf{v}_2, \mathbf{v}_3, \ldots, \mathbf{v}_n, \mathbf{v}_1\}$ be a re-ordering of the $\mathbb{C}$ basis by listing $\mathbf{v}_1$ last instead of first. Find the transition matrix from the $\mathbb{C}$ basis to the $\mathbb{D}$ basis.

(39) Let $\mathbb{S}$ be the standard basis for $\mathbb{R}^n$ written as column vectors. Show that if $\mathbb{B}=\{\mathbf{v}_1, \mathbf{v}_2, \ldots, \mathbf{v}_n\}$ is any other basis of column vectors for $\mathbb{R}^n$, then the columns of the transition matrix from $\mathbb{B}$ to $\mathbb{S}$ are the vectors in $\mathbb{B}$.

(40) Let $\mathbb{C}$ and $\mathbb{E}$ be two bases for a vector space $\mathbb{V}$, $\mathbb{D}$, and $\mathbb{F}$ be two bases for a vector space $\mathbb{W}$, and $\mathbf{T}\colon \mathbb{V} \to \mathbb{W}$ be a linear transformation. Verify the following:

(i) For any vector $\mathbf{v}$ in $\mathbb{V}$ there exists a transition matrix $\mathbf{P}$ such that $\mathbf{v}_{\mathbb{C}} = \mathbf{P}_{\mathbb{E}}^{\mathbb{C}}\mathbf{v}_{\mathbb{E}}$.

(ii) For any vector $\mathbf{w}$ in $\mathbb{W}$ there exists a transition matrix $\mathbf{Q}$ such that $\mathbf{w}_{\mathbb{D}} = \mathbf{Q}_{\mathbb{F}}^{\mathbb{D}}\mathbf{v}_{\mathbb{F}}$.

(iii) If $\mathbf{A}$ is a matrix representation of $\mathbf{T}$ with respect to the $\mathbb{C}$ and $\mathbb{D}$ bases, then $\mathbf{w}_{\mathbb{D}} = \mathbf{A}_{\mathbb{C}}^{\mathbb{D}}\mathbf{v}_{\mathbb{C}}$.

(iv) If $\mathbf{A}$ is a matrix representation of $\mathbf{T}$ with respect to the $\mathbb{E}$ and $\mathbb{F}$ bases, then $\mathbf{w}_{\mathbb{F}} = \mathbf{A}_{\mathbb{E}}^{\mathbb{F}}\mathbf{v}_{\mathbb{E}}$.

(v) $\mathbf{w}_{\mathbb{D}} = \mathbf{A}_{\mathbb{C}}^{\mathbb{D}}\mathbf{P}_{\mathbb{E}}^{\mathbb{C}}\mathbf{v}_{\mathbb{E}}$.

(vi) $\mathbf{w}_{\mathbb{D}} = \mathbf{Q}_{\mathbb{F}}^{\mathbb{D}}\mathbf{A}_{\mathbb{E}}^{\mathbb{F}}\mathbf{v}_{\mathbb{E}}$.

(vii) $\mathbf{A}_{\mathbb{C}}^{\mathbb{D}}\mathbf{P}_{\mathbb{E}}^{\mathbb{C}} = \mathbf{Q}_{\mathbb{F}}^{\mathbb{D}}\mathbf{A}_{\mathbb{E}}^{\mathbb{F}}$.

(viii) $\mathbf{A}_{\mathbb{E}}^{\mathbb{F}} = \left(\mathbf{Q}_{\mathbb{F}}^{\mathbb{D}}\right)^{-1}\mathbf{A}_{\mathbb{C}}^{\mathbb{D}}\mathbf{P}_{\mathbb{E}}^{\mathbb{C}}$.

3.5 PROPERTIES OF LINEAR TRANSFORMATIONS

Because a linear transformation from one finite-dimensional vector space to another can be represented by a matrix, we can use our understanding of matrices to gain a broader understanding of linear transformations. Alternatively, because matrices are linear transformations, we can transport properties of linear transformations to properties of matrices. Sometimes it will be easier to discover properties dealing with matrices, because the structure of a matrix is so concrete. Other times, it will be easier to work directly with linear transformations in the abstract, because their structures are so simple. In either case, knowledge about one, either linear transformations or matrices, provides an understanding about the other.

> **▶THEOREM 1**
>
> *If* $\mathbf{T}$: $\mathbb{V} \to \mathbb{W}$ is a linear transformation, then $\mathbf{T(0)}=\mathbf{0}$.◀

Proof: We have from Theorem 1 of Section 2.1 that $0\mathbf{0}=\mathbf{0}$. In addition, $T(\mathbf{0})$ is a vector in $\mathbb{W}$, so $0T(\mathbf{0})=\mathbf{0}$. Combining these results with the properties of linear transformations, we conclude that

$$T(0) = T(00) = 0T(0) = 0$$

Note how simple Theorem 1 was to prove using the properties of vector spaces and linear transformations. To understand Theorem 1 in the context of matrices, we first note that regardless of the basis $\mathbb{B} = \{\mathbf{u}_1, \mathbf{u}_2, \ldots, \mathbf{u}_p\}$ selected for a vector space, the zero vector has the form

$$(0) = 0\mathbf{u}_1 + 0\mathbf{u}_2 + \cdots + 0\mathbf{u}_p$$

The zero vector is unique (Theorem 4 of Section 2.1) and can be written only one way as a linear combination of basis vectors (Theorem 5 of Section 2.4), hence the coordinate representation of the zero vector is a zero column matrix. Thus, in terms of matrices, Theorem 1 simply states that the product of a matrix with a zero column matrix is again a zero column matrix. Theorem 1 is obvious in the context of matrices, but only after we set it up. In contrast, the theorem was not so obvious in the context of linear transformations, but much simpler to prove. In a nutshell, that is the advantage (and disadvantage) of each approach.

Theorem 1 states that a linear transformation always maps the zero vector in the domain into the zero vector in $\mathbb{W}$. This may not, however, be the only vector mapped into the zero vector; there may be many more. The projection L: $\mathbb{R}^2 \to \mathbb{R}^2$ defined in Example 7 of Section 3.2 as

$$L[a\ b] = [a\ 0]$$

generates the mappings $L[0\ 1]=[0\ 0]=\mathbf{0}$, $L[0\ 2]=[0\ 0]=\mathbf{0}$, and, in general, $L[0\ k]=0$ for any real number k. This projection maps infinitely many different vectors in the domain into the zero vector. In contrast, the identity mapping $I(\mathbf{v})=\mathbf{v}$ maps only the zero vector into the zero vector. We define the *kernel (or null space)* of a linear transformation T:$\mathbb{V} \to \mathbb{W}$, denoted by *ker*$(T)$, as the set of all vectors $\mathbf{v} \in \mathbb{V}$ that are mapped by T into the zero vector in $\mathbb{W}$; that is, all $\mathbf{v}$ for which $T(\mathbf{v})=\mathbf{0}$. It follows from Theorem 1 that *ker*(T) always contains the zero vector from the domain, so the kernel is never an empty set. We can say even more.

The kernel of a linear transformation $\boldsymbol{T}$ is the set of all vectors $\mathbf{v}$ in the domain for which $\boldsymbol{T}(\mathbf{v})=\mathbf{0}$.

▸THEOREM 2

The kernel of a linear transformation is a subspace of the domain.◂

Proof: Let **u** and **v** be any two vectors in the kernel of a linear transformation T, where $T(\mathbf{u})=\mathbf{0}$ and $T(\mathbf{v})=\mathbf{0}$. Then for any two scalars α and β, it follows from the properties of a linear transformation that

$$T(\alpha\mathbf{u}+\beta\mathbf{v}) = \alpha T(\mathbf{u}) + \beta T(\mathbf{v}) = \alpha\mathbf{0} + \beta\mathbf{0} = \mathbf{0}+\mathbf{0} = \mathbf{0}$$

Thus, $\alpha\mathbf{u}+\beta\mathbf{v}$ is also in the kernel and the kernel is a subspace.

The set of vectors that satisfy the homogeneous matrix equation $\mathbf{Ax}=\mathbf{0}$ is a subspace called the *kernel* of **A**.

In terms of a specific matrix **A**, the kernel is the set of column vectors **x** that satisfy the matrix equation $\mathbf{Ax}=\mathbf{0}$. That is, *ker*(**A**) is the set of all solutions to the system of homogeneous equations $\mathbf{Ax}=\mathbf{0}$. Theorem 2 implies that this set is a subspace.

Example 1 Determine the kernel of the matrix $\mathbf{A}=\begin{bmatrix}1 & 1 & 5\\ 2 & -1 & 1\end{bmatrix}$.

Solution: The kernel of **A** is the set of all three-dimensional column matrices $\mathbf{x}=[x\ y\ z]^{\mathrm{T}}$ that satisfy the matrix equation

$$\begin{bmatrix}1 & 1 & 5\\ 2 & -1 & 1\end{bmatrix}\begin{bmatrix}x\\ y\\ z\end{bmatrix}=\begin{bmatrix}0\\ 0\\ 0\end{bmatrix}$$

or, equivalently, the system of linear equations

$$\begin{aligned} x+y+5z&=0\\ 2x-y+z&=0\end{aligned}$$

The solution to this system is found by Gaussian elimination to be $x=-2z$, $y=-3z$, with z arbitrary. Thus, $\mathbf{x}\in ker(\mathbf{A})$ if and only if

$$\mathbf{x}=\begin{bmatrix}x\\ y\\ z\end{bmatrix}3=z\begin{bmatrix}-2\\ -3\\ 1\end{bmatrix}$$

where z is an arbitrary real number. The kernel of **A** is a one-dimensional subspace of the domain $\mathbb{R}^3$; a basis for *ker*(**K**) consists of the single vector $[-2\ -3\ 1]^{\mathrm{T}}$.

The image of a linear transformation T is the set of all vectors **w** in the range for which there is a vector **v** in the domain satisfying $T(\mathbf{v})=\mathbf{w}$.

The image of a transformation $T\colon \mathbb{V}\rightarrow\mathbb{W}$ is the set of vectors in $\mathbb{W}$ that are matched with at least one vector in $\mathbb{V}$; that is, **w** is in the image of T if and only if there exists at least one vector **v** in the domain for which $T(\mathbf{v})=\mathbf{w}$. We shall denote the image of T by $Im(T)$. If T is linear, it follows from Theorem 1 that $Im(T)$ always contains the zero vector in $\mathbb{W}$, because the zero vector in **V** is mapped into the zero vector in $\mathbb{W}$. We can say even more.

> **▶THEOREM 3**
>
> *The image of a linear transformation* $\mathbf{T}: \mathbb{V} \to \mathbb{W}$ *is a sub space of* $\mathbb{W}$.◀

Proof: Let $\mathbf{w}_1$ and $\mathbf{w}_2$ be any two vectors in the image of a linear transformation T. Then there must exist vectors $\mathbf{v}_1$ and $\mathbf{v}_2$ in the domain having the property that $T(\mathbf{v}_1)=\mathbf{w}_1$ and $T(\mathbf{v}_2)=\mathbf{w}_2$. For any two scalars α and β, it follows from the properties of a linear transformation that

$$(\alpha\mathbf{w}_1 + \beta\mathbf{w}_2) = \alpha T(\mathbf{v}_1) + \beta T(\mathbf{v}_2) = T(\alpha\mathbf{v}_1 + \alpha\mathbf{v}_1)$$

Because $\mathbb{V}$ is a vector space, $\alpha\mathbf{v}_1 + \beta\mathbf{v}_2$ is in the domain, and because this linear combination maps into $\alpha\mathbf{w}_1 + \beta\mathbf{w}_2$, it follows that $\alpha\mathbf{w}_1 + \beta\mathbf{w}_2$ is in the image of T. Consequently, $Im(T)$ is a subspace.

In terms of a specific matrix $\mathbf{A}$, the image is the set of column matrices $\mathbf{y}$ that satisfy the matrix equation $\mathbf{Ax}=\mathbf{y}$. That is, $Im(\mathbf{A})$ is the set of products $\mathbf{Ax}$ for any vector $\mathbf{x}$ in the domain. Theorem 3 implies that this set is a subspace. Denote the columns of $\mathbf{A}$ by $\mathbf{A}_1, \mathbf{A}_2, \ldots, \mathbf{A}_n$, respectively, and a column matrix $\mathbf{x}$ as $\mathbf{x}= [x_1\ x_2 \ldots x_n]^{\mathrm{T}}$. Then

$$\mathbf{Ax} = x_1\mathbf{A}_1 + x_2\mathbf{A}_2 + \cdots + x_n\mathbf{A}_n$$

That is, the image of $\mathbf{A}$ is the span of the columns of $\mathbf{A}$, which is the column space of $\mathbf{A}$.

The image of a matrix is its column space.

Example 2 Determine the image of the matrix $\mathbf{A} = \begin{bmatrix} 1 & 1 & 5 \\ 2 & -1 & 1 \end{bmatrix}$.

Solution: The column space of $\mathbf{A}$ is identical to the row space of $\mathbf{A}^{\mathrm{T}}$. Using elementary row operations to transform $\mathbf{A}^{\mathrm{T}}$ to row-reduced form, we obtain

$$\begin{bmatrix} 1 & 2 \\ 0 & 1 \\ 0 & 0 \end{bmatrix}$$

This matrix has two nonzero rows; hence, its rank is 2. Thus the rank of $\mathbf{A}^{\mathrm{T}}$, as well as the rank of $\mathbf{A}$, is 2. $\mathbf{A}$ is a 2×3 matrix mapping $\mathbb{R}^3$ into $\mathbb{R}^2$. The range $\mathbb{R}^2$ has dimension 2, and since the image also has dimension 2, the image must be the entire range. Thus, $Im(\mathbf{A}) = \mathbb{R}^2$.

Example 3 Identify the kernel and the image of the linear transformation $T: \mathbb{P}^2 \to \mathbb{M}_{2\times 2}$ defined by

$$T\left(at^2 + bt + c\right)\begin{bmatrix} a & 2b \\ 0 & a \end{bmatrix}$$

for all real numbers a, b, and c.

Solution: This transformation maps polynomials in t of degree 2 or less into 2×2 matrices. In particular,

$$T(3t^2 + 4t + 5)\begin{bmatrix} 3 & 8 \\ 0 & 3 \end{bmatrix}$$

and

$$T(-t^2 + 5t + 2) = \mathrm{T}(-t^2 + 5t - 8) = \begin{bmatrix} -1 & 10 \\ 0 & -1 \end{bmatrix}$$

A polynomial in the domain is mapped into the zero matrix if and only if $a = b = 0$, so the kernel is the set of all polynomials of the form $0t^2 + 0t + c$; that is, the subspace of all zero-degree polynomials. A basis for *ker*(*T*) is $\{1\}$. Thus, the kernel is a one-dimensional subspace of $\mathbb{P}^2$.

$\mathbb{M}_{2\times 2}$ is a four-dimensional vector space. The image of *T* is the subspace containing all matrices of the form

$$\begin{bmatrix} a & 2b \\ 0 & a \end{bmatrix} = a\begin{bmatrix} 1 & 0 \\ 0 & 1 \end{bmatrix} + b\begin{bmatrix} 0 & 2 \\ 0 & 0 \end{bmatrix}$$

which is spanned by the two matrices

$$\begin{bmatrix} 1 & 0 \\ 0 & 1 \end{bmatrix} \quad \text{and} \quad \begin{bmatrix} 0 & 2 \\ 0 & 0 \end{bmatrix}$$

It is a simple matter to prove that these two matrices are linearly independent, so they form a basis for the image of *T*. Thus, *Im*(*T*) is a two-dimensional subspace of $\mathbb{M}_{2\times 2}$.

It is important to recognize that the kernel and image of a linear transformation $T\colon \mathbb{V} \to \mathbb{W}$ are conceptually different subspaces: the kernel is a subspace of the domain $\mathbb{V}$ while the image is a subspace of the range $\mathbb{W}$. Figure 3.14 is a schematic rendition of these concepts. The vector space $\mathbb{V}$ is depicted by the palette on the left, the vector space $\mathbb{W}$ by the palette on the right, and because these vector spaces can be different, the palettes are drawn differently. Each point in the interior of a palette denotes a vector in its respective vector space.

Needless to say, both palettes are just symbolic representations of vector spaces and not true geometrical renditions of either the domain or range.

The palettes in Figure 3.14 are partitioned into two sections, one shaded and one not. The shaded portion of the left palette represents *ker*(*T*), and, as such, every point in it must be mapped into the zero vector in $\mathbb{W}$. This is shown symbolically by the vector $\mathbf{v}_1$. Vectors in the unshaded portion of the left palette, illustrated by

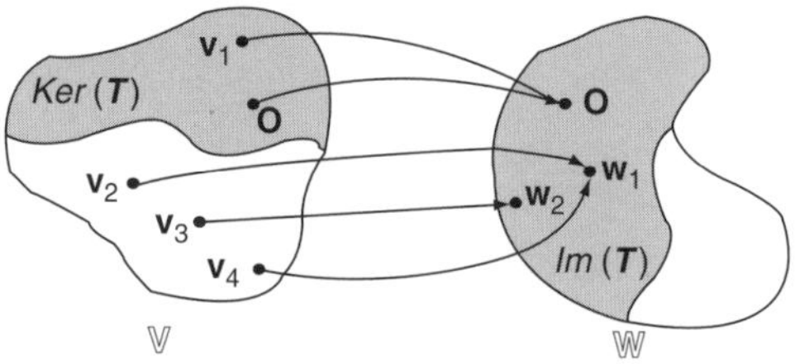

FIGURE 3.14

the vectors $\mathbf{v}_2$, $\mathbf{v}_3$, and $\mathbf{v}_4$, are mapped into other vectors in $\mathbb{W}$. The zero vector in $\mathbb{V}$ is mapped into the zero vector in $\mathbb{W}$ as a consequence of Theorem 1.

The shaded portion of the right palette represents the image of T. Any vector $\mathbf{w}$ in this region has associated with it a vector $\mathbf{v}$ in the left palette for which $\mathbf{w}=T(\mathbf{v})$. The unshaded portion of the right palette is not in the image of T and vectors in it are not matched with any vectors in domain represented by the left palette.

Even though the kernel and image of a linear transformation are conceptually different, their bases are related.

▶THEOREM 4

Let $\mathbf{T}$ be a linear transformation from an n-dimensional vector space $\mathbb{V}$ into $\mathbb{W}$ and let $\{\mathbf{v}_1, \mathbf{v}_2, \ldots, \mathbf{v}_k\}$ be a basis for the kernel of $\mathbf{T}$. If this basis is extended to a basis $\{\mathbf{v}_1, \mathbf{v}_2, \ldots, \mathbf{v}_k, \mathbf{v}_{k+1}, \ldots, \mathbf{v}_n\}$ for $\mathbb{V}$, then $\{\mathbf{T}(\mathbf{v}_{k+1}), \mathbf{T}(\mathbf{v}_{k+2}), \ldots, \mathbf{T}(\mathbf{v}_n)\}$ is a basis for the image of $\mathbf{T}$. ◀

Proof: We must show that $\{T(\mathbf{v}_{k+1}), T(\mathbf{v}_{k+2}), \ldots, T(\mathbf{v}_n)\}$ is a linearly independent set that spans the image of T. To prove linear independence, we form the equation

$$c_{k+1}T(\mathbf{v}_{k+1}) + c_{k+2}T(\mathbf{v}_{k+2}) + \cdots + c_nT(\mathbf{v}_n) = 0 \tag{3.29}$$

and show that the only solution to this equation is $c_{k+1}=c_{k+2}= = \cdots = c_n = 0$. Because T is linear, equation (3.29) can be rewritten as

$$T(c_{k+1}\mathbf{v}_{k+1} + c_{k+2}\mathbf{v}_{k+2} + \cdots + c_n\mathbf{v}_n) = 0$$

which implies that the sum $c_{k+1}\mathbf{v}_{k+1}+c_{k+2}\mathbf{v}_{k+2}+\cdots c_n\mathbf{v}_n$ in a vector in the kernel of T. Every vector in the kernel can be expressed as a unique linear combination of its basis vectors (Theorem 5 of Section 2.5), so there must exist a unique set of scalars $c_1, c_2, \ldots, c_k$ such that

$$c_{k+1}\mathbf{v}_{k+1} + c_{k+2}\mathbf{v}_{k+2} + \cdots + c_n\mathbf{v}_n = c_1\mathbf{v}_1 + c_2\mathbf{v}_2 + \cdots + c_k\mathbf{v}_k$$

which can be rewritten as

$$-c_1\mathbf{v}_1 - c_2\mathbf{v}_2 - \cdots - c_k\mathbf{v}_k + c_{k+1}\mathbf{v}_{k+1} + c_{k+2}\mathbf{v}_{k+2} + \cdots + c_n\mathbf{v}_n = 0 \tag{3.30}$$

But $\{\mathbf{v}_1, \mathbf{v}_2, \ldots, \mathbf{v}_n\}$ is basis for $\mathbb{V}$; consequently, it is linearly independent and the only solution to equation (3.30) is $-c_1=-c_2=\cdots=-c_k=c_{k+1}=c_{k+2}=\cdots=c_n=0$. Thus, $c_{k+1}=c_{k+2}=\cdots=c_n=0$ is the only solution to equation (3.29), and $\{T(\mathbf{v}_{k+1}), T(\mathbf{v}_{k+2}), \ldots, T(\mathbf{v}_n)\}$ is linearly independent.

It remains to show that $\{T(\mathbf{v}_{k+1}), T(\mathbf{v}_{k+2}), \ldots, T(\mathbf{v}_n)\}$ spans the image of T. Let $\mathbf{w}$ denote an arbitrary vector in the image. Then there must be at least one vector $\mathbf{v}$ in the domain having the property that $T(\mathbf{v})=\mathbf{w}$. Writing $\mathbf{v}$ as a linear combination of basis vectors, we have $\mathbf{v}=d_1\mathbf{v}_1+d_2\mathbf{v}_2+\cdots+d_k\mathbf{v}_k+d_{k+1}\mathbf{v}_{k+1}+d_{k+2}\mathbf{v}_{k+2}+\cdots+d_n\mathbf{v}_n$ for a unique set of scalars $d_1, d_2, \ldots, d_n$. Then

$$\begin{aligned}\mathbf{w} = T(\mathbf{v}) &= T(d_1\mathbf{v}_1 + d_2\mathbf{v}_2 + \cdots + d_k\mathbf{v}_k + d_{k+1}\mathbf{v}_{k+1} + d_{k+2}\mathbf{v}_{k+2} + \cdots + d_n\mathbf{v}_n) \\ &= d_1T(\mathbf{v}_1) + d_2T(\mathbf{v}_2) + \cdots + d_kT(\mathbf{v}_k) + d_{k+1}T(\mathbf{v}_{k+1}) + d_{k+2}T(\mathbf{v}_{k+2}) \\ &\quad + \cdots + d_nT(\mathbf{v}_n) \\ &= d_1 0 + d_2 0 + \cdots + d_k 0 + d_{k+1}T(\mathbf{v}_{k+1}) + d_{k+2}T(\mathbf{v}_{k+2}) + \cdots + d_nT(\mathbf{v}_n) \\ &= d_{k+1}T(\mathbf{v}_{k+1}) + d_{k+2}T(\mathbf{v}_{k+2}) + \cdots + d_nT(\mathbf{v}_n)\end{aligned}$$

because $\mathbf{v}_1, \mathbf{v}_2, \ldots, \mathbf{v}_k$ are (basis) vectors in the kernel of T and all vectors in $ker(T)$ map into the zero vector. We conclude that every vector $\mathbf{w}$ in the image of T can be written as a linear combination of $\{T(\mathbf{v}_{k+1}), T(\mathbf{v}_{k+2}), \ldots, T(\mathbf{v}_n)\}$, so this set spans the image.

We have shown that $\{T(\mathbf{v}_{k+1}), T(\mathbf{v}_{k+2}), \ldots, T(\mathbf{v}_n)\}$ is a linearly independent set that spans the image of T; hence, it is a basis for that image.

Example 4 Apply Theorem 4 to the linear transformation given in Example 3.

Solution: A basis for the kernel was found to be the set $\{1\}$ while a basis for the domain is $\{1, t, t^2\}$. Theorem 4 states that

$$T(t^2) = T(1t^2 + 0t + 0) = \begin{bmatrix} 1 & 0 \\ 0 & 1 \end{bmatrix}$$

and

$$T(t) = T(0t^2 + 1t + 0) = \begin{bmatrix} 0 & 2 \\ 0 & 0 \end{bmatrix}$$

form a basis for the image of T, which is precisely the same result obtained in Example 3.

Example 5 Apply Theorem 4 to the linear transformation $T: \mathbb{R}^4 \to \mathbb{R}^3$ defined by

$$T\begin{bmatrix} a \\ b \\ c \\ d \end{bmatrix} = \begin{bmatrix} a + b \\ b + c + d \\ a - c - d \end{bmatrix}$$

Solution: A vector in $\mathbb{R}^4$ is in the kernel of T if and only if its components a, b, c, and d satisfy the system of equations

$$\begin{aligned} a + b &= 0 \\ b + c + d &= 0 \\ a - c - d &= 0 \end{aligned}$$

Using Gaussian elimination on this system, we obtain as its solution $a = c + d$, $b = -c - d$ with c and d arbitrary, which takes the vector form

$$\begin{bmatrix} a \\ b \\ c \\ d \end{bmatrix} = \begin{bmatrix} c + b \\ -c - d \\ c \\ d \end{bmatrix} = c\begin{bmatrix} 1 \\ -1 \\ 1 \\ 0 \end{bmatrix} + d\begin{bmatrix} 1 \\ -1 \\ 0 \\ 1 \end{bmatrix}$$

Every vector of this form is in the kernel of *T*. It is clear that the two vectors on the right side of this last equation span the kernel of *T*. It is also easy to show that these two vectors are linearly independent, so they form a basis for *ker*(*T*).

This basis for *ker*(*T*) can be extended to the set

$$\left\{\begin{bmatrix}1\\-1\\1\\0\end{bmatrix}, \begin{bmatrix}1\\-1\\0\\1\end{bmatrix}, \begin{bmatrix}1\\0\\0\\0\end{bmatrix}, \begin{bmatrix}0\\1\\0\\0\end{bmatrix}\right\}$$

which forms a basis for $\mathbb{R}^4$. It now follows that

$$T\begin{bmatrix}1\\0\\0\\0\end{bmatrix} = \begin{bmatrix}1\\0\\1\end{bmatrix} \text{ and } T\begin{bmatrix}0\\1\\0\\0\end{bmatrix} = \begin{bmatrix}1\\1\\0\end{bmatrix}$$

form a basis for the image of *T*.

The nullity and rank of a linear transformation are, respectively, the dimensions of its kernel and image.

Because the kernel and image of a linear transformation $T: \mathbb{V} \to \mathbb{W}$ are subspaces, each has a dimension. The dimension of the kernel is its *nullity*, denoted by $\nu(T)$; the dimension of the image is its *rank*, denoted by $r(T)$. Assume that $dim(\mathbb{V})\text{I} = n$. It follows from Theorem 4 that if there are k vectors in the basis $\{\mathbf{v}_1, \mathbf{v}_2, \ldots, \mathbf{v}_k\}$ for the kernel of *T*, so that $\nu(T) = k$, then a basis for the image of *T* given by $\{T(\mathbf{v}_{k+1}), T(\mathbf{v}_{k+2}), \ldots, T(\mathbf{v}_n)\}$ contains $n-k$ vectors and $r(T) = n-k$. Together, $r(T) + \nu(T) = (n-k) + k = n$, the dimension of $\mathbb{V}$.

The proof of Theorem 4 assumes that $1 \le k < n$. If $k = 0$, then *ker*(*T*) contains just the zero vector, which has dimension 0. In this case, we let $\{\mathbf{v}_1, \mathbf{v}_1, \ldots, \mathbf{v}_n\}$ be any basis for $\mathbb{V}$, and with minor modifications the proof of Theorem 4 can be adapted to show that $\{T(\mathbf{v}_1), T(\mathbf{v}_2), \ldots, T(\mathbf{v}_n)\}$ is a basis for the image of *T*. Once again, $r(T) + \nu(T) = n + 0 = n$. Finally, if $\nu(T) = n$, then *ker*(*T*) must be all of the domain, all vectors in $\mathbb{V}$ map into $\mathbf{0}$, the image of *T* is just the zero vector, $r(T) = 0$, and $r(T) + \nu(T) = 0 + n = n$. We have, therefore, proven one of the more fundamental results of linear algebra.

►COROLLARY 1

*For any linear transformation **T** from an n-dimensional vector space $\mathbb{V}$ to $\mathbb{W}$, the rank of **T** plus the nullity of **T** equals n, the dimension of the domain. That is,*

$$r(\mathbf{T}) + \nu(\mathbf{T}) = n. ◄$$

The startling aspect of Corollary 1 is that the dimension of $\mathbb{W}$ is of no consequence. Although the image of *T* is a subspace of $\mathbb{W}$, its dimension when summed with the dimension of the null space of T is the dimension of the domain.

Example 6 Verify Corollary 1 for the linear transformation $T: \mathbb{P}^2 \to \mathbb{M}_{2\times 2}$ defined by

$$T(at^2 + bt + c) = \begin{bmatrix} a & 2b \\ 0 & a \end{bmatrix}$$

for all real numbers a, b, and c.

Solution: The domain $\mathbb{P}^2$ has dimension 3. We showed in Example 3 that a basis for the kernel contains a single vector and a basis for the image of T contains two elements. Thus, $r(T)=2$, $v(T)=1$, and $r(T)+v(T)=2+1=3$, the dimension of the domain.

Example 7 Verify Corollary 1 for the linear transformation $T: \mathbb{R}^4 \to \mathbb{R}^3$ defined by

$$T\begin{bmatrix} a \\ b \\ c \\ d \end{bmatrix} = \begin{bmatrix} a+b \\ b+c+d \\ a-c-d \end{bmatrix}$$

Solution: The domain $\mathbb{R}^4$ has dimension four. We showed in Example 5 that bases for both the kernel and the image contain two vectors, so $r(T)+v(T)=2+2=4$, the dimension of the domain.

If we restrict our attention to an $n \times p$ matrix $\mathbf{A}$, then the kernel of $\mathbf{A}$ is the subspace of all solutions to the homogeneous system of equation $\mathbf{Ax}=\mathbf{0}$ and the dimension of this subspace is $v(\mathbf{A})$, the nullity of $\mathbf{A}$. The image of $\mathbf{A}$ is the column space of $\mathbf{A}$ and its dimension is the column rank of $\mathbf{A}$, which is the rank of the matrix. Thus, Corollary 1 is simply an alternate formulation of Theorem 3 of Section 2.6.

A linear transformation is one-to-one if it maps different vectors in the domain into different vectors in the range.

A linear transformation $T: \mathbb{V} \to \mathbb{W}$ is *one-to-one* if the equality $T(\mathbf{u})=T(\mathbf{v})$ implies $\mathbf{u}=\mathbf{v}$. A one-to-one linear transformation maps different vectors in $\mathbb{V}$ into different vectors in $\mathbb{W}$, as illustrated in Figure 3.15a. If two different vectors $\mathbf{u}$ and $\mathbf{v}$ in $\mathbb{V}$ map into the same vector in $\mathbb{W}$, as illustrated in Figure 3.15b, then $T(\mathbf{u})=T(\mathbf{v})$ with $\mathbf{u}\neq\mathbf{v}$, and the transformation is not one-to-one.

Example 8 Determine whether the linear transformation $T: \mathbb{P}^2 \to \mathbb{M}_{2\times 2}$ defined by

$$T(at^2 + bt + c) = \begin{bmatrix} a & 2b \\ 0 & a \end{bmatrix}$$

is one-to-one.

Solution: Here

$$T(-t^2 + 5t + 2) = T(-t^2 + 5t - 8) = \begin{bmatrix} -1 & 10 \\ 0 & -1 \end{bmatrix}$$

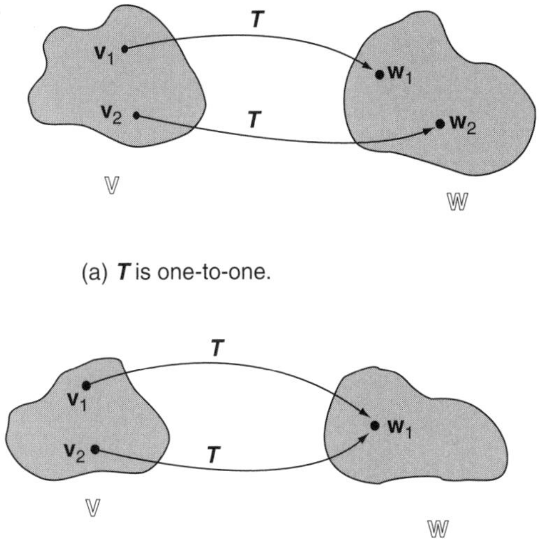

(a) ***T*** is one-to-one.

(b) ***T*** is not one-to-one.

FIGURE 3.15

Setting $\mathbf{u} = -t^2 + 5t + 2$ and $\mathbf{v} = -t^2 + 5t - 8$, we have $T(\mathbf{u}) = T(\mathbf{v})$ with $\mathbf{u} \neq \mathbf{v}$, hence T is *not* one-to-one.

Example 9 Determine whether the linear transformation $T : \mathbb{R}^2 \to \mathbb{R}^3$ defined by

$$T\begin{bmatrix} a \\ b \end{bmatrix} = \begin{bmatrix} a+b \\ a-b \\ 2a+3b \end{bmatrix}$$

is one-to-one.

Solution: Setting $\mathbf{u} = [a\,b]^{\mathrm{T}}$, $\mathbf{v} = [c\,d]^{\mathrm{T}}$, and $T(\mathbf{u}) = T(\mathbf{v})$, we obtain the vector equation

$$\begin{bmatrix} a+b \\ a-b \\ 2a+3b \end{bmatrix} = \begin{bmatrix} c+d \\ c-d \\ 2c+3d \end{bmatrix}$$

which is equivalent to the system of equations

$$\begin{aligned} a+b &= c+d \\ a-b &= c-d \\ 2a+3b &= 2c+3d \end{aligned}$$

Solving this system by Gaussian elimination for the variables a and b, thinking of c and d as fixed constants, we generate the single solution $a = c$ and $b = d$. Therefore, the equality $T(\mathbf{u}) = T(\mathbf{v})$ implies that $\mathbf{u} = \mathbf{v}$, and T is one-to-one.

Often, the easiest way to show whether a linear transformation is one-to-one is to use the following:

▶THEOREM 5

*A linear transformation $T : \mathbb{V} \to \mathbb{W}$ is one-to-one if and only if the kernel of **T** contains just the zero vector, i.e., $\nu(T) = 0$.*◀

Proof: Assume that T is one-to-one. If $\mathbf{v} \in ker(T)$, then $T(\mathbf{v}) = \mathbf{0}$. We know from Theorem 1 that $T(\mathbf{0}) = \mathbf{0}$. Consequently, $T(\mathbf{v}) = T(\mathbf{0})$, which implies that $\mathbf{v} = \mathbf{0}$, because T is one-to-one. Thus, if $\mathbf{v} \in ker(T)$, then $\mathbf{v} = \mathbf{0}$, from which we conclude that the kernel of T contains just the zero vector.

Conversely, assume that the kernel of T contains just the zero vector. If $\mathbf{u}$ and $\mathbf{v}$ are vectors in the domain for which $T(\mathbf{u}) = T(\mathbf{v})$, then $T(\mathbf{u}) - T(\mathbf{v}) = \mathbf{0}$ and $T(\mathbf{u} - \mathbf{v}) = \mathbf{0}$, which implies that the vector $\mathbf{u} - \mathbf{v}$ is in the kernel of T. Since this kernel contains only the zero vector, it follows that $\mathbf{u} - \mathbf{v} = \mathbf{0}$ and $\mathbf{u} = \mathbf{v}$. Thus, the equality $T(\mathbf{u}) = T(\mathbf{v})$ implies $\mathbf{u} = \mathbf{v}$, from which we conclude that T is one-to-one.

Example 10 Determine whether the linear transformation $T : \mathbb{R}^4 \to \mathbb{R}^3$ defined by

$$T\begin{bmatrix} a \\ b \\ c \\ d \end{bmatrix} = \begin{bmatrix} a + b \\ b + c + d \\ a - c - d \end{bmatrix}$$

is one-to-one.

Solution: We showed in Example 5 that a basis for the kernel of T contained two vectors. Thus, $\nu(T) = 2 \neq 0$, and the transformation is not one-to-one.

A linear transformation is onto if its image is its range.

A linear transformation $T : \mathbb{V} \to \mathbb{W}$ is *onto* if the image of T is all of $\mathbb{W}$; that is, if the image equals the range. The dimension of the image of T is the rank of T. Thus, T is onto if and only if the rank of T equals the dimension of $\mathbb{W}$. This provides a straightforward algorithm for testing whether a linear transformation is onto.

Example 11 Determine whether the linear transformation $T : \mathbb{P}^2 \to \mathbb{M}_{2\times 2}$ defined by

$$T\left(at^2 + bt + c\right) = \begin{bmatrix} a & 2b \\ 0 & a \end{bmatrix}$$

is onto.

Solution: We showed in Example 3 that a basis for the kernel of the transformation is the set $\{1\}$, hence $\nu(T) = 1$. The dimension of the domain $\mathbb{P}^2$ is 3, so it follows from Corollary 1 that $r(T) + 1 = 3$ and $r(T) = 2$. Here $\mathbb{W} = \mathbb{M}_{2\times 2}$ has dimension 4. Since $r(T) = 2 \neq 4 = dim(\mathbb{W})$, the transformation is *not* onto.

Example 12 Determine whether the linear transformation $T: \mathbb{M}_{2\times 2} \to \mathbb{R}^3$ defined by

$$\mathbf{T}\begin{bmatrix} a & b \\ c & d \end{bmatrix} = \begin{bmatrix} a+b \\ b+c \\ c+d \end{bmatrix}$$

is onto.

Solution: A matrix in $\mathbb{M}_{2\times 2}$ is in the kernel of T if and only if its components a, b, c, and d satisfy the system of equations

$$\begin{aligned} a+b&=0 \\ b+c&=0 \\ c+d&=0 \end{aligned}$$

The solution to this system is found immediately by back substitution to be $a=-d$, $b=d$, $c=-d$, with d arbitrary. Thus, a matrix in $ker(T)$ must have the form

$$\begin{bmatrix} -d & d \\ -d & d \end{bmatrix} = d\begin{bmatrix} -1 & 1 \\ -1 & 1 \end{bmatrix}$$

which implies that the kernel of T is spanned by the matrix

$$\begin{bmatrix} -1 & 1 \\ -1 & 1 \end{bmatrix}$$

This matrix is nonzero. It follows from Theorem 2 of Section 2.4 that, by itself, this matrix is a linearly independent set. Consequently, this matrix forms a basis for $ker(T)$, and $\nu(T)=1$. The dimension of the domain $\mathbb{V} = \mathbb{M}_{2\times 2}$ is 4, so it follows from Corollary 1 that $r(T)+1=4$ and $r(T)=3$. The dimension of the range $\mathbb{R}^3$ is also 3, hence the transformation is onto.

Alternatively, we may show that the matrix representation of T with respect to the standard bases in both $\mathbb{M}_{2\times 2}$ and $\mathbb{R}^3$ is

$$\mathbf{A} = \begin{bmatrix} 1 & 1 & 0 & 0 \\ 0 & 1 & 1 & 0 \\ 0 & 0 & 1 & 1 \end{bmatrix}$$

A is in row-reduced form and has rank 3. Therefore, $r(T) = r(\mathbf{A}) = 3 = dim(\mathbb{R}^3)$, and we once again conclude that the transformation is onto.

In general, the attributes of one-to-one and onto are quite distinct. A linear transformation can be one-to-one and onto, or one-to-one and not onto, onto but not one-to-one, or neither one-to-one nor onto. All four possibilities exist. There is one situation, however, when one-to-one implies onto and vice versa.

▶THEOREM 6

*Let a linear transformation **T**:$\mathbb{V} \to \mathbb{W}$ have the property that the dimension of $\mathbb{V}$ equals the dimension of $\mathbb{W}$. Then **T** is one-to-one if and only if **T** is onto.*◀

Proof: T is one-to-one if and only if (from Theorem 5) $v(T)=0$, which is true if and only if (Corollary 1) $r(T) = dim(\mathbb{V})$. But $dim(\mathbb{V}) = dim(\mathbb{W})$; hence, T is one-to-one if and only if $r(T) = dim(\mathbb{W})$, which is valid if and only if T is onto.

Problems 3.5

(1) Define $T : \mathbb{R}^3 \to \mathbb{R}^2$ by $\mathbf{T}[a\ b\ c] = [a+b\ c]$. Determine whether any of the following vectors are in the kernel of **T**.

(a) $[1 -1\ \ 3]$, (b) $[1 -1\ \ 0]$,
(c) $[2 -2\ \ 0]$, (d) $[1\,25\,1\ \ 0]$.

(2) Define $S : \mathbb{R}^3 \to \mathbb{R}^2$ by $\mathbf{S}[a\ b\ c] = [a - c\ c - b]$. Determine whether any of the following vectors are in the kernel of **S**.

(a) $[1\ -1\ 1]$, (b) $[1\ 1\ 1]$,
(c) $[-2\ -2\ -2]$, (d) $[1\ 1\ 0]$.

(3) Define $L : \mathbb{R}^3 \to \mathbb{R}^2$, $\mathbf{L}[a\ b\ c] = [a + 2b - 3c\ 0]$. Determine whether any of the following vectors are in the kernel of **L**.

(a) $[1\ 1\ 1]$, (b) $[5\ -1\ 1]$,
(c) $[-1\ 2\ 1]$, (d) $[-1\ 5\ 3]$.

(4) Define $P : \mathbb{M}_{2\times 2} \to \mathbb{M}_{2\times 2}, P\begin{bmatrix} a & b \\ c & d \end{bmatrix} = \begin{bmatrix} a+b & 0 \\ 0 & c-d \end{bmatrix}$. Determine whether any of the following matrices are in the kernel of **P**.

(a) $\begin{bmatrix} 1 & 1 \\ 1 & 1 \end{bmatrix}$, (b) $\begin{bmatrix} 1 & -1 \\ 1 & 1 \end{bmatrix}$, (c) $\begin{bmatrix} 1 & 1 \\ 1 & -1 \end{bmatrix}$, (d) $\begin{bmatrix} 1 & -1 \\ -1 & -1 \end{bmatrix}$.

(5) Define $T : \mathbb{P}^2 \to \mathbb{P}^2$ by $\mathbf{T}(a_2t^2 + a_1t + a_0) = (a_2 - a_1)t^2 + (a_1 - a_0)t$. Determine whether any of the following vectors are in the kernel of T.

(a) $2t^2 - 3t + 4$, (b) $t^2 + t$, (c) $3t + 3$, (d) $-t^2 - t - 1$.

(6) Determine whether any of the following vectors are in the image of the linear transformation defined in Problem 1. For each one that is, produce an element in the domain that maps into it.

(a) $[1\ 1]$, (b) $[1\ -1]$, (c) $[2\ 0]$, (d) $[1\ 2]$.

(7) Determine whether any of the following vectors are in the image of the linear transformation defined in Problem 3. For each one that is, produce an element in the domain that maps into it.

(a) [1 1], (b) [1 0], (c) [2 0], (d) [1 2].

(8) Determine whether any of the following matrices are in the image of the linear transformation defined in Problem 4. For each one that is, produce an element in the domain that maps into it.

(a) $\begin{bmatrix} 1 & 1 \\ 1 & 1 \end{bmatrix}$, (b) $\begin{bmatrix} 1 & 0 \\ 0 & 0 \end{bmatrix}$, (c) $\begin{bmatrix} 0 & 1 \\ 0 & 0 \end{bmatrix}$, (d) $\begin{bmatrix} 3 & 0 \\ 0 & -5 \end{bmatrix}$.

(9) Redo Problem 8 for $\mathbf{P}\colon \mathbb{M}_{2\times 2} \to \mathbb{M}_{2\times 2}$, by $\mathbf{P}\begin{bmatrix} a & b \\ c & d \end{bmatrix} = \begin{bmatrix} c & a \\ d & b \end{bmatrix}$.

(10) Determine whether any of the following vectors are in the image of the linear transformation defined in Problem 5. For each one that is, produce an element in the domain that maps into it.

(a) $2t^2-3t+4$, (b) t^2+2t, (c) $3t$, (d) $2t-1$.

In Problems 11 through 30, find the nullity and rank of the given linear transformations, and determine which are one-to-one and which are onto.

(11) $T\colon \mathbb{R}^2 \to \mathbb{R}^2, T[a\ b] = [2a 3b]$.

(12) $T\colon \mathbb{R}^2 \to \mathbb{R}^2, T[a\ b] = [aa + \mathrm{b}]$.

(13) $T\colon \mathbb{R}^2 \to \mathbb{R}^2, T[a\ b] = [a0]$.

(14) $S\colon \mathbb{R}^3 \to \mathbb{R}^2, S[a\ b\ c] = [a + b\ c]$.

(15) $S\colon \mathbb{R}^3 \to \mathbb{R}^2, S[a\ b\ c] = [a - c\ c - b]$.

(16) $S\colon \mathbb{R}^3 \to \mathbb{R}^2, S[a\ b\ c] = [a + 2b - 3c0]$.

(17) $S\colon \mathbb{R}^2 \to \mathbb{R}^3, S[a\ b] = [a + b2a + b\ a]$.

(18) $S\colon \mathbb{R}^2 \to \mathbb{R}^3, S[a\ b] = [a0b]$.

(19) $N\colon \mathbb{R}^2 \to \mathbb{R}^3, N[a\ b] = [a + b2a + b\ b]$.

(20) $N\colon \mathbb{R}^2 \to \mathbb{R}^3, N[a\ b] = [0\ 0\ 2a - 5b]$.

(21) $T\colon \mathbb{R}^2 \to \mathbb{R}^3, T[a\ b] = [a \quad -a \quad -8a]$.

(22) $T\colon \mathbb{R}^3 \to \mathbb{R}^1, T[a\ b\ c] = a - c$.

(23) $L\colon \mathbb{R}^3 \to \mathbb{R}^1, L[a\ b\ c] = 0$.

(24) $P\colon \mathbb{M}_{2\times 2} \to \mathbb{M}_{2\times 2}, \mathbf{P}\begin{bmatrix} a & b \\ c & d \end{bmatrix} = \begin{bmatrix} c & a \\ d & b \end{bmatrix}$.

(25) $P\colon \mathbb{M}_{2\times 2} \to \mathbb{M}_{2\times 2}$, $P\begin{bmatrix} a & b \\ c & d \end{bmatrix} = \begin{bmatrix} a+b & 0 \\ 0 & c-d \end{bmatrix}$.

(26) $T\colon \mathbb{M}_{2\times 2} \to \mathbb{M}_{2\times 2}$, $T\begin{bmatrix} a & b \\ c & d \end{bmatrix} = \begin{bmatrix} 2d & 0 \\ 0 & 0 \end{bmatrix}$.

(27) $R\colon \mathbb{M}_{2\times 2} \to \mathbb{R}^1$, $R\begin{bmatrix} a & b \\ c & d \end{bmatrix} = b + 2c - 3d$.

(28) $L\colon \mathbb{P}^2 \to \mathbb{P}^2$, $L(a_2t^2 + a_1t + a_0) = a_0t$.

(29) $T\colon \mathbb{P}^2 \to \mathbb{P}^2$, $T(a_2t^2 + a_1t + a_0) = (a_2 - a_1)t^2 + (a_1 - a_0)t$.

(30) $S\colon \mathbb{P}^2 \to \mathbb{P}^2$, $S(a_2t^2 + a_1t + a_0) = 0$.

(31) Determine whether any of the following vectors are in the image of $\mathbf{A} = \begin{bmatrix} 1 & 3 \\ 0 & 0 \end{bmatrix}$.

(a) $\begin{bmatrix} 2 \\ 6 \end{bmatrix}$, (b) $\begin{bmatrix} 2 \\ 0 \end{bmatrix}$, (c) $\begin{bmatrix} 0 \\ 2 \end{bmatrix}$, (d) $\begin{bmatrix} 0 \\ 0 \end{bmatrix}$.

(32) Redo the previous problem for the matrix $\mathbf{A} = \begin{bmatrix} 1 & 3 \\ 0 & 0 \end{bmatrix}$.

(33) Determine whether any of the following vectors are in the image of $\mathbf{A} = \begin{bmatrix} 1 & 0 \\ 1 & 1 \\ 1 & 1 \end{bmatrix}$.

(a) $\begin{bmatrix} 1 \\ 0 \\ 1 \end{bmatrix}$, (b) $\begin{bmatrix} 2 \\ 0 \\ 0 \end{bmatrix}$, (c) $\begin{bmatrix} 4 \\ 3 \\ 3 \end{bmatrix}$, (d) $\begin{bmatrix} 4 \\ 4 \\ 3 \end{bmatrix}$.

In Problems 34 through 42, find a basis for the kernels and a basis for the image of the given matrices.

(34) $\mathbf{A} = \begin{bmatrix} 1 & 2 \\ 2 & 4 \end{bmatrix}$.

(35) $\mathbf{B} = \begin{bmatrix} 1 & 2 \\ 2 & 5 \end{bmatrix}$.

(36) $\mathbf{C} = \begin{bmatrix} 1 & -1 & 0 \\ -1 & 1 & 0 \end{bmatrix}$.

(37) $\mathbf{D} = \begin{bmatrix} 1 & 0 & 2 \\ 3 & 0 & 4 \end{bmatrix}$.

(38) $\mathbf{E} = \begin{bmatrix} 1 & 0 & 1 \\ 2 & 1 & 3 \\ 3 & 1 & 4 \end{bmatrix}$.

(39) $\mathbf{F} = \begin{bmatrix} 1 & 1 & 1 \\ 1 & 1 & 1 \\ 1 & 1 & 1 \end{bmatrix}$.

(40) $\mathbf{G} = \begin{bmatrix} 1 & 1 & 0 \\ 1 & 0 & 1 \\ 0 & 1 & 1 \end{bmatrix}$.

(41) $\mathbf{H} = \begin{bmatrix} 1 \\ 2 \\ 3 \end{bmatrix}$.

(42) $K = [1\ 1\ 2\ 2]$.

(43) What can be said about the ranks of similar matrices?

(44) Prove that if a linear transformation $T: \mathbb{V} \to \mathbb{W}$ is onto, then the dimension of $\mathbb{W}$ cannot be greater than the dimension of $\mathbb{V}$.

(45) Use the results of the previous problem to show directly that the transformation defined in Example 3 is not onto.

(46) Use the results of Problem 44 to show directly that the transformation defined in Example 9 is not onto.

(47) Prove that if $\{\mathbf{w}_1, \mathbf{w}_2, \ldots, \mathbf{w}_k\}$ are linearly independent vectors in the image of a linear transformation $L: \mathbb{V} \to \mathbb{W}$, and if $\mathbf{w}_1 = T(\mathbf{v}_i)$ $(i = 1, 2, \ldots, k)$, then $\{\mathbf{v}_1, \mathbf{v}_2, \ldots, \mathbf{v}_k\}$ is also linearly independent.

(48) Prove that a linear transformation $T: \mathbb{V} \to \mathbb{W}$ cannot be one-to-one if the dimension of $\mathbb{W}$ is less than the dimension of $\mathbb{V}$.

(49) Use the result of the previous problem to show directly that the transformation defined in Example 5 cannot be one-to-one.

(50) Use the result of Problem 48 to show directly that the transformation defined in Example 12 cannot be one-to-one.

(51) Let $\{\mathbf{v}_1, \mathbf{v}_2, \ldots, \mathbf{v}_p\}$ be a spanning set for $\mathbb{V}$ and let $T: \mathbb{V} \to \mathbb{W}$ be a linear transformation. Prove that $\{T(\mathbf{v}_1), T(\mathbf{v}_2), \ldots, T(\mathbf{v}_p)\}$ is a spanning set for the image of T.

(52) Prove that a linear transformation $T: \mathbb{V} \to \mathbb{W}$ is one-to-one if and only if the image of every linearly independent set of vectors in $\mathbb{V}$ is a linearly independent set of vectors in $\mathbb{W}$.

(53) Let $T: \mathbb{V} \to \mathbb{W}$ be a linear transformation having the property that the dimension of $\mathbb{V}$ is the same as the dimension of $\mathbb{W}$. Prove that T is one-to-one if the image of any basis of $\mathbb{V}$ is a basis for $\mathbb{W}$.

(54) Prove that a matrix representation of a linear transformation $T: \mathbb{V} \to \mathbb{V}$ has an inverse if and only if T is one-to-one.

(55) Prove that a matrix representation of a linear transformation $T: \mathbb{V} \to \mathbb{V}$ has an inverse if and only if T is onto.

CHAPTER 3 REVIEW

Important Terms

coordinate representation
dilation
domain
function
image
projection onto the x-axis
projection onto the y-axis
range
rank
reflection across the x-axis

kernel
linear transformation
nullity
null space
one-to-one
onto
reflection across the y-axis
rotations in the x-y plane
similar matrices
transformation
transition matrix
zero transformation

Important Concepts

Section 3.1

- A function is a rule of correspondence between two sets, a domain and range, that assigns to each element in the domain exactly one element (but not necessarily a different one) in the range.

Section 3.2

- A transformation T is a rule of correspondence between two vector spaces, a domain $\mathbb{V}$ and a range $\mathbb{W}$, that assigns to each element in $\mathbb{V}$ exactly one element (but not necessarily a different one) in $\mathbb{W}$.
- A transformation is linear if it preserves linear combinations.
- Every matrix defines a linear transformation.

Section 3.3

- A linear transformation is described completely by its actions on a basis for the domain.
- Every linear transformation from one finite-dimensional vector space to another can be represented by a matrix that is basis dependent.

Section 3.4

- In general, a vector has many coordinate representations, a different one for each basis.
- The transition matrix from $\mathbb{C}$ to $\mathbb{D}$, where both $\mathbb{C}$ and $\mathbb{D}$ are bases for the same finite-dimensional vector space, is invertible and its inverse is the transition matrix from $\mathbb{D}$ to $\mathbb{C}$.
- If $\mathbf{v}_{\mathbb{C}}$ and $\mathbf{v}_{\mathbb{D}}$ are the coordinate representations of the same vector with respect to the bases $\mathbb{C}$ and $\mathbb{D}$, respectively, then $\mathbf{v}_{\mathbb{D}} = \mathbf{P}\mathbf{v}_{\mathbb{C}}$ where $\mathbf{P}$ is the transition matrix from $\mathbb{C}$ to $\mathbb{D}$.
- In general, a linear transformation may be represented by many matrices, a different one for each basis.
- Two square matrices $\mathbf{A}$ and $\mathbf{B}$ represent the same linear transformation if and only if there exists a transition matrix $\mathbf{P}$ such that $\mathbf{A} = \mathbf{P}^{-1}\mathbf{B}\mathbf{P}$.

Section 3.5

- A linear transformation always maps the zero vector in the domain to the zero vector in the range.
- The kernel of a linear transformation is a nonempty subspace of the domain; the image of a linear transformation is a nonempty subspace of the range.
- The kernel of the linear transformation defined by a matrix **A** is the set of all solutions to the system of homogeneous equations $\mathbf{Ax}=\mathbf{0}$; the image of the linear transformation is the column space of **A**.
- If $\{\mathbf{v}_1, \mathbf{v}_2, \ldots, \mathbf{v}_k\}$ is a basis for the kernel of a linear transformation T and if this basis is extended to a basis $\{\mathbf{v}_1, \mathbf{v}_2, \ldots, \mathbf{v}_k, \mathbf{v}_{k+1}, \ldots, \mathbf{v}_n\}$ for the domain, then $\{T(\mathbf{v}_{k+1}), T(\mathbf{v}_{k+2}), \ldots, T(\mathbf{v}_n)\}$ is a basis for the image of T.
- The rank plus the nullity of a linear transformation from one finite-dimensional vector space to another equals the dimension of the domain.
- A linear transformation is one-to-one if and only if its kernel contains just the zero vector.
- A linear transformation is onto if and only if its rank equals the dimension of the range.
- A linear transformation $T : \mathbb{V} \rightarrow \mathbb{W}$, having the property that $dim(\mathbb{V}) = dim(\mathbb{W})$, is one-to-one if and only if the transformation is onto.

CHAPTER 4

Eigenvalues, Eigenvectors, and Differential Equations

Chapter Outline

4.1 EIGENVECTORS AND EIGENVALUES

Many of the uses and applications of linear algebra are especially evident by considering *diagonal matrices*. In addition to the fact that they are easy to multiply, a number of other properties readily emerge: their determinants (see Appendix A) are trivial to compute, we can quickly determine whether such matrices have inverses and, when they do, their inverses are easy to obtain. Thus, diagonal matrices are simple matrix representations for linear transformations from a finite-dimensional vector space $\mathbb{V}$ to itself (see Section 3.4). Unfortunately, not all linear transformations from $\mathbb{V}$ to $\mathbb{V}$ can be represented by diagonal matrices. In this section and Section 4.3, we determine which linear transformations have diagonal matrix representations and which bases generate those representations.

To gain insight into the conditions needed to produce a diagonal matrix representation, we consider a linear transformation $T\colon \mathbb{R}^3 \to \mathbb{R}^3$ having the diagonal matrix representation

$$\mathbf{D} = \begin{bmatrix} \lambda_1 & 0 & 0 \\ 0 & \lambda_2 & 0 \\ 0 & 0 & \lambda_3 \end{bmatrix}$$

with respect to the basis $\mathbb{B} = \{\mathbf{x}_1, \mathbf{x}_2, \mathbf{x}_3\}$. The first column of $\mathbf{D}$ is the coordinate representation of $T(\mathbf{x}_1)$ with respect to $\mathbb{B}$, the second column of $\mathbf{D}$ is the coordinate representation of $T(\mathbf{x}_2)$ with respect to $\mathbb{B}$, and the third column of **D** is the coordinate representation of $T(\mathbf{x}_3)$ with respect to $\mathbb{B}$. That is,

$$T(\mathbf{x}_1) = \lambda_1\mathbf{x}_1 + 0\mathbf{x}_2 + 0\mathbf{x}_3 = \lambda_1\mathbf{x}_1$$

$$T(\mathbf{x}_2) = 0\mathbf{x}_1 + \lambda_2\mathbf{x}_2 + 0\mathbf{x}_3 = \lambda_2\mathbf{x}_2$$

$$T(\mathbf{x}_3) = 0\mathbf{x}_1 + 0\mathbf{x}_2 + \lambda_3\mathbf{x}_3 = \lambda_3\mathbf{x}_3$$

Mapping the basis vectors $\mathbf{x}_1$, $\mathbf{x}_2$, or $\mathbf{x}_3$ from the domain of T to the range of T is equivalent to simply multiplying each vector by the scalar λ_1, λ_2, or λ_3, respectively.

We say that a nonzero vector **x** is an *eigenvector* of a linear transformation T if there exists a scalar λ such that

$$T(\mathbf{x}) = \lambda\mathbf{x} \tag{4.1}$$

In terms of a matrix representation **A** for T, we define a nonzero vector **x** to be an *eigenvector* of **A** if there exists a nonzero scalar λ such that

$$\mathbf{A}\mathbf{x} = \lambda\mathbf{x} \tag{4.2}$$

A nonzero vector **x** is an eigenvector of a square matrix **A** if there exists a scalar λ, called an eigenvalue, such that $\mathbf{A}\mathbf{x} = \lambda\mathbf{x}$.

The scalar λ in Equation (4.1) is an *eigenvalue* of the linear transformation T; the scalar λ in Equation (4.2) is an *eigenvalue* of the matrix **A**. Note that an eigenvector must be nonzero; eigenvalues, however, may be zero.

Eigenvalues and eigenvectors have an interesting geometric interpretation in $\mathbb{R}^2$ or $\mathbb{R}^3$ when the eigenvalues are real. As described in Section 2.1, multiplying a vector in either vector space by a real number λ results in an elongation of the vector by a factor of $|\lambda|$ when $|\lambda| > 1$, or a contraction of the vector by a factor of $|\lambda|$ when $|\lambda| < 1$, followed by no rotation when λ is positive, or a rotation of 180° when λ is negative. These four possibilities are illustrated in Figure 4.1 for the vector **u** in $\mathbb{R}^2$ with $\lambda = 1/2$ and $\lambda = -1/2$, and for the vector **v** in $\mathbb{R}^2$ with $\lambda = 3$ and $\lambda = -2$. Thus, an eigenvector **x** of a linear transformation T in $\mathbb{R}^2$ or $\mathbb{R}^3$ is always mapped into a vector $T(\mathbf{x})$ that is parallel to **x**.

Not every linear transformation has real eigenvalues. Under the rotation transformation R described in Example 7 of Section 3.2, each vector is rotated around the origin by an angle θ in the counterclockwise direction (see Figure 4.2). As long as θ is not an integral multiple of 180°, *no* nonzero vector is mapped into another vector parallel to itself.

Example 1 The vector $\mathbf{x} = \begin{bmatrix} -1 \\ 1 \end{bmatrix}$ is an eigenvector of $\mathbf{A} = \begin{bmatrix} 1 & 2 \\ 4 & 3 \end{bmatrix}$ because

$$\mathbf{A}\mathbf{x} = \begin{bmatrix} 1 & 2 \\ 4 & 3 \end{bmatrix}\begin{bmatrix} -1 \\ 1 \end{bmatrix} = \begin{bmatrix} -1 \\ 1 \end{bmatrix} = (-1)\begin{bmatrix} -1 \\ 1 \end{bmatrix} = (-1)\mathbf{x}$$

The corresponding eigenvalue is $\lambda = -1$.

Example 2 The vector $\mathbf{x} = \begin{bmatrix} 4 \\ 1 \\ -2 \end{bmatrix}$ is an eigenvector of $\mathbf{A} = \begin{bmatrix} 1 & 2 & 3 \\ 2 & 4 & 6 \\ 3 & 6 & 9 \end{bmatrix}$ because

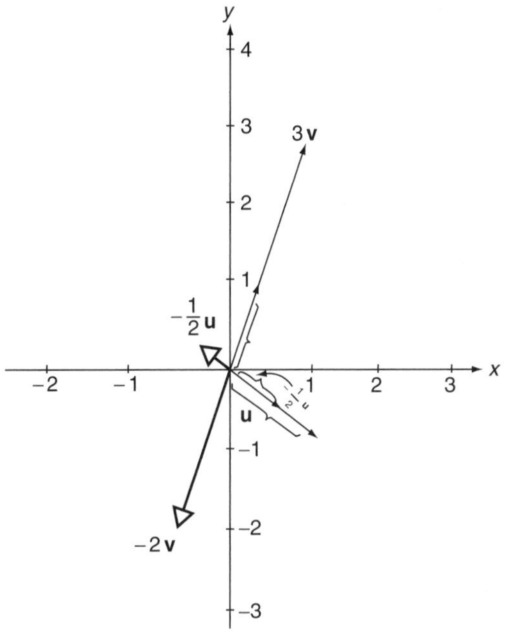

FIGURE 4.1

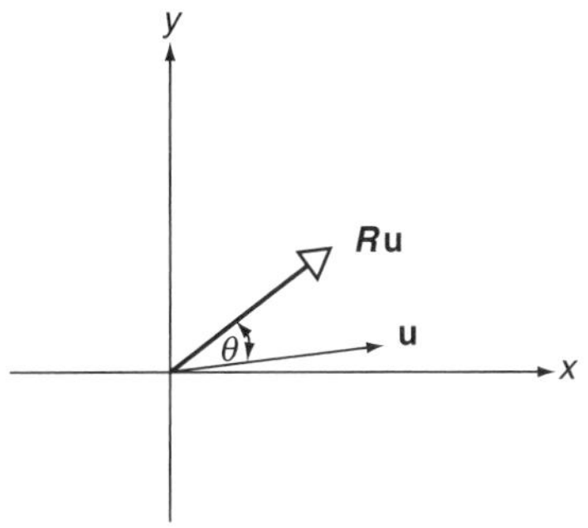

FIGURE 4.2

$$\mathbf{Ax} = \begin{bmatrix} 1 & 2 & 3 \\ 2 & 4 & 6 \\ 3 & 6 & 9 \end{bmatrix} \begin{bmatrix} 4 \\ 1 \\ -2 \end{bmatrix} = \begin{bmatrix} 0 \\ 0 \\ 0 \end{bmatrix} = 0 \begin{bmatrix} 4 \\ 1 \\ -2 \end{bmatrix} = 0\mathbf{x}$$

The corresponding eigenvalue is $\lambda = 0$.

Eigenvectors and eigenvalues come in pairs. If $\mathbf{x}$ is an eigenvector of a matrix $\mathbf{A}$, then there must exist an eigenvalue λ such that $\mathbf{Ax} = \lambda\mathbf{x}$, which is equivalent to the equation $\mathbf{Ax} - \lambda\mathbf{x} = \mathbf{0}$ or

$$(\mathbf{A} - \lambda\mathbf{I})\mathbf{x} = 0 \tag{4.3}$$

Note that we cannot write Equation (4.3) as $(\mathbf{A}-\lambda)\mathbf{x}=\mathbf{0}$ because subtraction between a scalar λ and a matrix $\mathbf{A}$ is undefined. In contrast, $\mathbf{A}-\lambda\mathbf{I}$ is the difference between two matrices, which is defined when $\mathbf{A}$ and $\mathbf{I}$ have the same order.

Equation (4.3) is a linear homogeneous equation for the vector $\mathbf{x}$. If $(\mathbf{A}-\lambda\mathbf{I})^{-1}$ exists, we can solve Equation (4.3) for $\mathbf{x}$, obtaining $\mathbf{x}=(\mathbf{A}-\lambda\mathbf{I})^{-1}\mathbf{0}=\mathbf{0}$, which violates the condition that an eigenvector be nonzero. It follows that $\mathbf{x}$ is an eigenvector for $\mathbf{A}$ corresponding to the eigenvalue λ if and only if $(\mathbf{A}-\lambda\mathbf{I})$ does *not* have an inverse. Alternatively, because a square matrix has an inverse if and only if its determinant is nonzero, we may conclude that $\mathbf{x}$ is an eigenvector for $\mathbf{A}$ corresponding to the eigenvalue λ if and only if

To find eigenvalues and eigenvectors for a matrix **A**, first solve the characteristic equation, Equation (4.4), for the eigenvalues and then for each eigenvalue solve Equation (4.3) for the corresponding eigenvectors.

$$\det(\mathbf{A}-\lambda\mathbf{I})=0 \tag{4.4}$$

Equation (4.4) is the *characteristic equation of* $\mathbf{A}$. If $\mathbf{A}$ has order $n\times n$, then det $(\mathbf{A}-\lambda\mathbf{I})$ is an nth degree polynomial in λ and the characteristic equation of $\mathbf{A}$ has exactly n roots, which are the eigenvalues of $\mathbf{A}$. Once an eigenvalue is located, corresponding eigenvectors are obtained by solving Equation (4.3).

Example 3 Find the eigenvalues and eigenvectors of $\mathbf{A}=\begin{bmatrix}1 & 2\\ 4 & 3\end{bmatrix}$.

Solution:

$$\mathbf{A}-\lambda\mathbf{I}=\begin{bmatrix}1 & 2\\ 4 & 3\end{bmatrix}-\lambda\begin{bmatrix}1 & 0\\ 0 & 1\end{bmatrix}=\begin{bmatrix}1-\lambda & 2\\ 4 & 3-\lambda\end{bmatrix}$$

with det $(\mathbf{A}-\lambda\mathbf{I})=(1-\lambda)(3-\lambda)-8=\lambda^2-4\lambda-5$. The characteristic equation of $\mathbf{A}$ is $\lambda^2-4\lambda-5=0$, having as its roots $\lambda=-1$ and $\lambda=5$. These two roots are the eigenvalues of $\mathbf{A}$.

Eigenvectors of $\mathbf{A}$ have the form $\mathbf{x}=[x\ y]^{\mathrm{T}}$. With $\lambda=-1$, Equation (4.3) becomes

$$(\mathbf{A}-\lambda\mathbf{I})\mathbf{x}=\left\{\begin{bmatrix}1 & 2\\ 4 & 3\end{bmatrix}-(-1)\begin{bmatrix}1 & 0\\ 0 & 1\end{bmatrix}\right\}\begin{bmatrix}x\\ y\end{bmatrix}=\begin{bmatrix}0\\ 0\end{bmatrix}$$

or

$$\begin{bmatrix}2 & 2\\ 4 & 4\end{bmatrix}\begin{bmatrix}x\\ y\end{bmatrix}=\begin{bmatrix}0\\ 0\end{bmatrix}$$

The solution to this homogeneous matrix equation is $x=-y$, with y arbitrary. The eigenvectors corresponding to $\lambda=-1$ are

$$\mathbf{x}=\begin{bmatrix}x\\ y\end{bmatrix}=\begin{bmatrix}-y\\ y\end{bmatrix}=y\begin{bmatrix}-1\\ 1\end{bmatrix}$$

for any nonzero scalar y. We restrict y to be nonzero to insure that the eigenvectors are nonzero.

With $\lambda=5$, Equation (4.3) becomes

$$(\mathbf{A}-\lambda\mathbf{I})\mathbf{x}=\left\{\begin{bmatrix}1 & 2\\ 4 & 3\end{bmatrix}-5\begin{bmatrix}1 & 0\\ 0 & 1\end{bmatrix}\right\}\begin{bmatrix}x\\ y\end{bmatrix}=\begin{bmatrix}0\\ 0\end{bmatrix}$$

or

$$\begin{bmatrix}-4 & 2\\ 4 & -2\end{bmatrix}\begin{bmatrix}x\\ y\end{bmatrix}=\begin{bmatrix}0\\ 0\end{bmatrix}$$

The solution to this homogeneous matrix equation is $x=y/2$, with y arbitrary. The eigenvectors corresponding to $\lambda=5$ are

$$\mathbf{x}=\begin{bmatrix}x\\ y\end{bmatrix}=\begin{bmatrix}y/2\\ y\end{bmatrix}=\frac{y}{2}\begin{bmatrix}1\\ 2\end{bmatrix}$$

for any nonzero scalar y.

Example 4 Find the eigenvalues and eigenvectors of $\mathbf{A}=\begin{bmatrix}2 & -1 & 0\\ 3 & -2 & 0\\ 0 & 0 & 1\end{bmatrix}$.

Solution:

$$\mathbf{A}-\lambda\mathbf{I}=\begin{bmatrix}2 & -1 & 0\\ 3 & -2 & 0\\ 0 & 0 & 1\end{bmatrix}-\lambda\begin{bmatrix}1 & 0 & 0\\ 0 & 1 & 0\\ 0 & 0 & 1\end{bmatrix}=\begin{bmatrix}2-\lambda & -1 & 0\\ 3 & -2-\lambda & 0\\ 0 & 0 & 1-\lambda\end{bmatrix}$$

Using expansion by cofactors with the last row, we find that

$$\det(\mathbf{A}-\lambda\mathbf{I})=(1-\lambda)[(2-\lambda)(-2-\lambda)+3]=(1-\lambda)\left(\lambda^2-1\right)$$

The characteristic equation of $\mathbf{A}$ is $(1-\lambda)(\lambda^2-1)=0$; hence, the eigenvalues of $\mathbf{A}$ are $\lambda_1=\lambda_2=1$ and $\lambda_3=-1$.

Eigenvectors of $\mathbf{A}$ have the form $\mathbf{x}=[x\ y\ z]^{\mathrm{T}}$. With $\lambda=1$, Equation (4.3) becomes

$$(\mathbf{A}-\lambda\mathbf{I})\mathbf{x}=\left\{\begin{bmatrix}2 & -1 & 0\\ 3 & -2 & 0\\ 0 & 0 & 1\end{bmatrix}-(1)\begin{bmatrix}1 & 0 & 0\\ 0 & 1 & 0\\ 0 & 0 & 1\end{bmatrix}\right\}\begin{bmatrix}x\\ y\\ z\end{bmatrix}=\begin{bmatrix}0\\ 0\\ 0\end{bmatrix}$$

or

$$\begin{bmatrix}1 & -1 & 0\\ 3 & -3 & 0\\ 0 & 0 & 0\end{bmatrix}\begin{bmatrix}x\\ y\\ z\end{bmatrix}=\begin{bmatrix}0\\ 0\\ 0\end{bmatrix}$$

The solution to this homogeneous matrix equation is $x=y$, with both y and z arbitrary. The eigenvectors corresponding to $\lambda=1$ are

$$\mathbf{x}=\begin{bmatrix}x\\y\\z\end{bmatrix}=\begin{bmatrix}y\\y\\z\end{bmatrix}=y\begin{bmatrix}1\\1\\0\end{bmatrix}+z\begin{bmatrix}0\\0\\1\end{bmatrix}$$

for y and z arbitrary, but not both zero to insure that the eigenvectors are nonzero.

With $\lambda=-1$, Equation (4.3) becomes

$$(\mathbf{A}-\lambda\mathbf{I})\mathbf{x}=\left\{\begin{bmatrix}2&-1&0\\3&-2&0\\0&0&1\end{bmatrix}-(-1)\begin{bmatrix}1&0&0\\0&1&0\\0&0&1\end{bmatrix}\right\}\begin{bmatrix}x\\y\\z\end{bmatrix}=\begin{bmatrix}0\\0\\0\end{bmatrix}$$

or

$$\begin{bmatrix}3&-1&0\\3&-1&0\\0&0&2\end{bmatrix}\begin{bmatrix}x\\y\\z\end{bmatrix}=\begin{bmatrix}0\\0\\0\end{bmatrix}$$

The solution to this homogeneous matrix equation is $x=y/3$ and $z=0$, with y arbitrary. The eigenvectors corresponding to $\lambda=-1$ are

$$\mathbf{x}=\begin{bmatrix}x\\y\\z\end{bmatrix}=\begin{bmatrix}y/3\\y\\0\end{bmatrix}=\frac{y}{3}\begin{bmatrix}1\\3\\0\end{bmatrix}$$

for any nonzero scalar y.

The roots of a characteristic equation can be repeated. If $\lambda_1=\lambda_2=\lambda_3=\cdots\lambda_k$, the eigenvalue is said to be of *multiplicity k*. Thus, in Example 4, $\lambda=1$ is an eigenvalue of multiplicity 2 while $\lambda=-1$ is an eigenvalue of multiplicity 1.

Locating eigenvalues is a matrix-based process. To find the eigenvalues of a more general linear transformation, we could identify a matrix representation for the linear transformation and then find the eigenvalues of that matrix. Because a linear transformation has many matrix representations, in general a different one for each basis, this approach would be useless if different matrix representations of the same linear transformation yielded different eigenvalues. Fortunately, this cannot happen. We know from Theorem 3 of Section 3.4 that two different matrix representations of the same linear transformation are similar. To this we now add:

►THEOREM 1

Similar matrices have the same characteristic equation (and, therefore, the same eigenvalues).◄

Proof: Let **A** and **B** be similar matrices. Then there must exist a nonsingular matrix **P** such that $\mathbf{A}=\mathbf{P}^{-1}\,\mathbf{BP}$. Since

$$\lambda\mathbf{I} = \lambda\mathbf{P}^{-1}\mathbf{P} = \mathbf{P}^{-1}\lambda\mathbf{P} = \mathbf{P}^{-1}\lambda\mathbf{IP}$$

it follows that

$$\begin{aligned} |\mathbf{A}-\lambda\mathbf{I}| &= |\mathbf{P}^{-1}\mathbf{BP}-\mathbf{P}^{-1}\lambda\mathbf{IP}| = |\mathbf{P}^{-1}(\mathbf{B}-\lambda\mathbf{I})\mathbf{P}| \\ &= |\mathbf{P}^{-1}||\mathbf{B}-\lambda\mathbf{I}||\mathbf{P}| && \text{Theorem 1 of Appendix A} \\ &= \frac{1}{|P|}|B-\lambda I||P| && \text{Theorem 8 of Appendix A} \\ &= |\mathbf{B}-\lambda\mathbf{I}| \end{aligned}$$

Thus the characteristic equation of **A**, namely $|\mathbf{A}-\lambda\mathbf{I}|=0$, is identical to the characteristic of **B**, namely $|\mathbf{B}-\lambda\mathbf{I}|=0$.

It follows from Theorem 1 that if two matrices do *not* have the same characteristic equations then the matrices *cannot* be similar. It is important to note, however, that Theorem 1 makes *no* conclusions about matrices with the same characteristic equation. Such matrices may or may not be similar.

If two matrices do not have the same characteristic equations, then they are not similar.

Example 5 Determine whether $\mathbf{A} = \begin{bmatrix} 1 & 2 \\ 4 & 3 \end{bmatrix}$ is similar to $\mathbf{B} = \begin{bmatrix} 1 & 2 \\ 4 & 3 \end{bmatrix}$.

Solution: The characteristic equation of **A** is $\lambda^2-4\lambda-5=0$ while that of **B** is $\lambda^2-3\lambda-10=0$. Because these equations are *not* identical, **A** *cannot* be similar to **B**.

The eigenvectors **x** corresponding to the eigenvalue λ of a matrix **A** are all nonzero solutions of the matrix Equation $(\mathbf{A}-\lambda\mathbf{I})\mathbf{x}=\mathbf{0}$. This matrix equation defines the kernel of $(\mathbf{A}-\lambda\mathbf{I})$, a vector space known as the *eigenspace* of **A** for the eigenvalue λ. The nonzero vectors of an eigenspace are the eigenvectors. Because basis vectors must be nonzero, the eigenvectors corresponding to a particular eigenvalue are described most simply by just listing a basis for the corresponding eigenspace.

Example 6 Find bases for the eigenspaces of $\mathbf{A} = \begin{bmatrix} 2 & -1 & 0 \\ 3 & -2 & 0 \\ 0 & 0 & 1 \end{bmatrix}$.

Solution: We have from Example 4 that the eigenvalues of **A** are 1 and -1. Vectors in the kernel of $\mathbf{A}-(1)\mathbf{I}$ have the form

$$\mathbf{x} = y\begin{bmatrix} 1 \\ 1 \\ 0 \end{bmatrix} + z\begin{bmatrix} 0 \\ 0 \\ 1 \end{bmatrix}$$

An eigenspace of **A** for the eigenvalue λ is the kernel of $\mathbf{A}-\lambda\mathbf{I}$. Nonzero vectors of this vector space are eigenvectors of **A**.

with y and z arbitrary, but not both zero. Clearly $[1 \quad 1 \quad 0]^{\mathrm{T}}$ and $[0 \quad 0 \quad 1]^{\mathrm{T}}$ span the eigenspace of **A** for $\lambda = 1$, and because these two vectors are linearly independent they form a basis for that eigenspace.

Vectors in the kernel of $\mathbf{A}-(-1)\mathbf{I}$ have the form

$$\mathbf{x} = \frac{y}{3}\begin{bmatrix} 1 \\ 3 \\ 0 \end{bmatrix}$$

Because every vector in the eigenspace of **A** for $\lambda = -1$ is a scalar multiple of $[1 \quad 3 \quad 0]^{\mathrm{T}}$, this vector serves as a basis for that eigenspace.

If $\mathbf{A}_{\mathbb{C}}^{\mathbb{C}}$ is a matrix representation of a linear transformation with respect to a basis **C** and if $\mathbf{A}_{\mathbb{D}}^{\mathbb{D}}$ is a matrix representation of the same linear transformation but with respect to a basis $\mathbb{D}$, then it follows from Equation (3.26) of Section 3.4 that

$$\mathbf{A}_{\mathbb{C}}^{\mathbb{C}} = \left(\mathbf{P}_{\mathbb{C}}^{\mathbb{D}}\right)^{-1} \mathbf{A}_{\mathbb{D}}^{\mathbb{D}} \mathbf{P}_{\mathbb{C}}^{\mathbb{D}}$$

where $\mathbf{P}_{\mathbb{C}}^{\mathbb{D}}$ denotes a transition matrix from $\mathbb{C}$ to $\mathbb{D}$. Let λ be an eigenvalue of $\mathbf{A}_{\mathbb{C}}^{\mathbb{C}}$ with a corresponding eigenvalue **x**. Then

$$\mathbf{A}_{\mathbb{C}}^{\mathbb{C}}\mathbf{x} = \lambda\mathbf{x}$$

$$\left(\mathbf{P}_{\mathbb{C}}^{\mathbb{D}}\right)^{-1} \mathbf{A}_{\mathbb{D}}^{\mathbb{D}} \mathbf{P}_{\mathbb{C}}^{\mathbb{D}}\mathbf{x} = \lambda\mathbf{x}$$

and

$$\mathbf{A}_{\mathbb{D}}^{\mathbb{D}} \mathrm{P}_{\mathbb{C}}^{\mathbb{D}}\mathbf{x} = \mathrm{P}_{\mathbb{C}}^{\mathbb{D}}(\lambda\mathbf{x}) = \lambda\mathrm{P}_{\mathbb{C}}^{\mathbb{D}}\mathbf{x}$$

If we set

$$\mathbf{y} = \mathbf{P}_{\mathbb{C}}^{\mathbb{D}}\mathbf{x} \tag{4.5}$$

we have

$$\mathbf{A}_{\mathbb{D}}^{\mathbb{D}}\mathbf{y} = \lambda\mathbf{y}$$

which implies that **y** is an eigenvector of $\mathbf{A}_{\mathbb{D}}^{\mathbb{D}}$. But it follows from Theorem 1 of Section 3.4 that **y** is the same vector as **x**, just expressed in a different basis. Thus, once we identify an eigenvector for a matrix representation of a linear transformation T, that eigenvector is a coordinate representation for an eigenvector of T, in the same basis used to create the matrix.

To find the eigenvalues and eigenvectors for a linear transformation $T\colon \mathbb{V} \to \mathbb{V}$, find the eigenvalues and eigenvectors of any matrix representation for T.

We now have a procedure for finding the eigenvalues and eigenvectors of a linear transformation T from one finite-dimensional vector space to itself. We first identify a matrix representation **A** for T and then determine the eigenvalues and eigenvectors of **A**. Any matrix representation will do, although a standard basis is used normally when one is available. The eigenvalues of **A** are the eigenvalues T (see Theorem 1). The eigenvectors of **A** are coordinate representations for the eigenvectors of T, with respect to the basis used to generate $\mathbf{A}$.

Example 7 Determine the eigenvalues and a basis for the eigenspaces of $T: \mathbb{P}^1 \to \mathbb{P}^1$ defined by

$$T(at+b) = (a+2b)t + (4a+3b)$$

Solution: A standard basis for $\mathbb{P}^1$ is $\mathbb{B} = \{t, 1\}$. With respect to this basis

$$T(t) = t + 4 = (1)t + 4(1) \leftrightarrow \begin{bmatrix} 1 \\ 4 \end{bmatrix}$$

$$T(1) = 2t + 3 = (2)t + 3(1) \leftrightarrow \begin{bmatrix} 2 \\ 3 \end{bmatrix}$$

so the matrix representation of T with respect to $\mathbb{B}$ is

$$\mathbf{A} = \begin{bmatrix} 1 & 2 \\ 4 & 3 \end{bmatrix}$$

We have from Example 3 that the eigenvalues of this matrix are -1 and 5, which are also the eigenvalues of T. The eigenvectors of $\mathbf{A}$ are, respectively,

$$y\begin{bmatrix} -1 \\ 1 \end{bmatrix} \text{ and } \frac{y}{2}\begin{bmatrix} 1 \\ 2 \end{bmatrix}$$

with y arbitrary but nonzero.

The eigenspace of $\mathbf{A}$ for $\lambda = -1$ is spanned by $[-1 \quad 1]^{\mathrm{T}}$, hence this vector serves as a basis for that eigenspace. Similarly, the eigenspace of $\mathbf{A}$ for $\lambda = 5$ is spanned by $[1 \quad 2]^{\mathrm{T}}$, so this vector serves as a basis for that eigenspace. These 2-tuples are coordinate representations for

$$\begin{bmatrix} -1 \\ 1 \end{bmatrix} \leftrightarrow (-1)t + (1)1 = -t + 1$$

and

$$\begin{bmatrix} 1 \\ 2 \end{bmatrix} \leftrightarrow (1)t + (2)1 = t + 2$$

Therefore, the polynomial $-t+1$ is a basis for the eigenspace of T for the eigenvalue -1 while the polynomial $t+2$ is a basis for the eigenspace of T for the eigenvalue 5. As a check, we note that

$$T(-t+1) = t - 1 = -1(-t+1)$$
$$T(t+2) = 5t + 10 = 5(t+2)$$

The characteristic equation of a real matrix may have complex roots, and these roots are *not* eigenvalues for linear transformations on real-valued vector spaces.

If a matrix is real, then eigenvectors corresponding to complex eigenvalues have complex components and such vectors are *not* elements of real vector space. Thus, there are *no* vectors in a real-valued vector space that satisfy $\mathbf{Ax} = \lambda\mathbf{x}$ when λ is complex.

Example 8 Determine the eigenvalues of $T: \mathbb{R}^3 \to \mathbb{R}^3$ defined by

$$T\begin{bmatrix} a \\ b \\ c \end{bmatrix} = \begin{bmatrix} 2a \\ 2b + 5c \\ -b - 2c \end{bmatrix}$$

Solution: Using the standard basis for $\mathbb{R}^3$, we have

$$T\begin{bmatrix} 1 \\ 0 \\ 0 \end{bmatrix} = \begin{bmatrix} 2 \\ 0 \\ 0 \end{bmatrix} = 2\begin{bmatrix} 1 \\ 0 \\ 0 \end{bmatrix} + 0\begin{bmatrix} 0 \\ 1 \\ 0 \end{bmatrix} + 0\begin{bmatrix} 0 \\ 0 \\ 1 \end{bmatrix} \leftrightarrow \begin{bmatrix} 2 \\ 0 \\ 0 \end{bmatrix}$$

$$T\begin{bmatrix} 0 \\ 1 \\ 0 \end{bmatrix} = \begin{bmatrix} 0 \\ 2 \\ -1 \end{bmatrix} = 0\begin{bmatrix} 1 \\ 0 \\ 0 \end{bmatrix} + 2\begin{bmatrix} 0 \\ 1 \\ 0 \end{bmatrix} + (-1)\begin{bmatrix} 0 \\ 0 \\ 1 \end{bmatrix} \leftrightarrow \begin{bmatrix} 0 \\ 2 \\ -1 \end{bmatrix}$$

$$T\begin{bmatrix} 0 \\ 0 \\ 1 \end{bmatrix} = \begin{bmatrix} 0 \\ 5 \\ -2 \end{bmatrix} = 0\begin{bmatrix} 1 \\ 0 \\ 0 \end{bmatrix} + 5\begin{bmatrix} 0 \\ 1 \\ 0 \end{bmatrix} + (-2)\begin{bmatrix} 0 \\ 0 \\ 1 \end{bmatrix} \leftrightarrow \begin{bmatrix} 0 \\ 5 \\ -2 \end{bmatrix}$$

where (as always when using this basis) the coordinate representation for any vector in $\mathbb{R}^3$ is the vector itself. The matrix representation for T with respect to the standard basis is

$$\mathbf{A} = \begin{bmatrix} 2 & 0 & 0 \\ 0 & 2 & 5 \\ 0 & -1 & -2 \end{bmatrix}$$

Here

$$\mathbf{A} - \lambda\mathbf{I} = \begin{bmatrix} 2-\lambda & 0 & 0 \\ 0 & 2-\lambda & 5 \\ 0 & -1 & -2-\lambda \end{bmatrix}$$

Using expansion by cofactors with the first row, we find that

$$\det(\mathbf{A} - \lambda\mathbf{I}) = (2-\lambda)[(2-\lambda)(-2-\lambda) + 5] = (2-\lambda)(\lambda^2 + 1)$$

The characteristic equation of **A** is $(2-\lambda)(\lambda^2+1) = 0$ with roots $\lambda_1 = 2$, $\lambda_2 = i$, and $\lambda_3 = -i$. The only real root is 2, which is the only eigenvalue for the given linear transformation.

Once an eigenvalue of a matrix is known, it is straightforward to identify the corresponding eigenspace. Unfortunately, determining the eigenvalues of a matrix, especially a square matrix with more than 10 rows, is difficult. Even some square matrices with just a few rows, such as

$$\mathbf{A} = \begin{bmatrix} 10 & 7 & 8 & 7 \\ 7 & 5 & 6 & 5 \\ 8 & 6 & 10 & 9 \\ 7 & 5 & 9 & 10 \end{bmatrix}$$

can be problematic. In most applications, numerical techniques (see Sections 4.4, 5.4, and Appendix D) are used to approximate the eigenvalues.

Problems 4.1

(1) Determine by direct multiplication which of the following vectors are eigenvectors for $\mathbf{A} = \begin{bmatrix} 1 & 2 \\ -4 & 7 \end{bmatrix}$.

(a) $\begin{bmatrix} 1 \\ 1 \end{bmatrix}$, (b) $\begin{bmatrix} 1 \\ -1 \end{bmatrix}$, (c) $\begin{bmatrix} 2 \\ 1 \end{bmatrix}$,

(d) $\begin{bmatrix} 1 \\ 2 \end{bmatrix}$, (e) $\begin{bmatrix} 2 \\ 2 \end{bmatrix}$, (f) $\begin{bmatrix} 0 \\ 0 \end{bmatrix}$,

(g) $\begin{bmatrix} -4 \\ -4 \end{bmatrix}$, (h) $\begin{bmatrix} 4 \\ -4 \end{bmatrix}$, (i) $\begin{bmatrix} 2 \\ 4 \end{bmatrix}$.

(2) What are the eigenvalues that correspond to the eigenvectors found in Problem 1?

(3) Determine by direct multiplication which of the following vectors are eigenvectors for $\mathbf{A} = \begin{bmatrix} 2 & 0 & -1 \\ 1 & 2 & 1 \\ -1 & 0 & 2 \end{bmatrix}$.

(a) $\begin{bmatrix} 1 \\ 0 \\ 0 \end{bmatrix}$, (b) $\begin{bmatrix} 0 \\ 1 \\ 0 \end{bmatrix}$, (c) $\begin{bmatrix} 1 \\ -2 \\ 1 \end{bmatrix}$,

(d) $\begin{bmatrix} -3 \\ 6 \\ -3 \end{bmatrix}$, (e) $\begin{bmatrix} -1 \\ 0 \\ 1 \end{bmatrix}$, (f) $\begin{bmatrix} 1 \\ 0 \\ 1 \end{bmatrix}$,

(g) $\begin{bmatrix} 2 \\ 0 \\ -2 \end{bmatrix}$, (h) $\begin{bmatrix} 1 \\ 1 \\ 1 \end{bmatrix}$, (i) $\begin{bmatrix} 0 \\ 0 \\ 0 \end{bmatrix}$.

(4) What are the eigenvalues that correspond to the eigenvectors found in Problem 3?

(5) Determine by direct evaluation which of the following matrices are eigenvectors for the linear transformation $T\colon \mathbb{M}_{2\times 2} \to \mathbb{M}_{2\times 2}$ defined by $\mathbf{T}\begin{bmatrix} a & b \\ c & d \end{bmatrix} = \begin{bmatrix} a+3b & a-b \\ c+2d & 4c+3d \end{bmatrix}$.

(a) $\begin{bmatrix} 1 & -1 \\ 0 & 0 \end{bmatrix}$, (b) $\begin{bmatrix} 0 & 0 \\ 1 & -1 \end{bmatrix}$, (c) $\begin{bmatrix} 1 & 0 \\ 0 & -1 \end{bmatrix}$,

(d) $\begin{bmatrix} 3 & 1 \\ 0 & 0 \end{bmatrix}$, (e) $\begin{bmatrix} 0 & 0 \\ 0 & 0 \end{bmatrix}$, (f) $\begin{bmatrix} 1 & 1 \\ 0 & 0 \end{bmatrix}$.

(6) What are the eigenvalues that correspond to the eigenvectors found in Problem 5?

(7) Determine by direct evaluation which of the following polynomials are eigenvectors for the linear transformation $T: \mathbb{P}^1 \rightarrow \mathbb{P}^1$ defined by $T(at+b) = (3a+5b)t - (2a+4b)$.

(a) $t-1$, (b) t^2+1, (c) $5t-5$,

(d) $5t-2$, (e) $5t$, (f) $-10t+2$.

(8) What are the eigenvalues that correspond to the eigenvectors found in Problem 7?

In Problems 9 through 32, find the eigenvalues and a basis for the eigenspace associated with each eigenvalue for the given matrices.

(9) $\begin{bmatrix} 1 & 2 \\ -1 & 4 \end{bmatrix}$. **(10)** $\begin{bmatrix} 2 & 1 \\ 2 & 3 \end{bmatrix}$. **(11)** $\begin{bmatrix} 2 & 3 \\ 4 & 6 \end{bmatrix}$.

(12) $\begin{bmatrix} 3 & 6 \\ 9 & 6 \end{bmatrix}$. **(13)** $\begin{bmatrix} 1 & 2 \\ 4 & -1 \end{bmatrix}$. **(14)** $\begin{bmatrix} 2 & 5 \\ -1 & -2 \end{bmatrix}$.

(15) $\begin{bmatrix} 3 & 1 \\ 0 & 3 \end{bmatrix}$. **(16)** $\begin{bmatrix} 3 & 0 \\ 0 & 3 \end{bmatrix}$. **(17)** $\begin{bmatrix} 0 & t \\ 2t & -t \end{bmatrix}$.

(18) $\begin{bmatrix} 4\theta & 2\theta \\ -\theta & \theta \end{bmatrix}$. **(19)** $\begin{bmatrix} 1 & 0 & 3 \\ 1 & 2 & 1 \\ 3 & 0 & 1 \end{bmatrix}$. **(20)** $\begin{bmatrix} 2 & 0 & -1 \\ 2 & 2 & 2 \\ -1 & 0 & 2 \end{bmatrix}$.

(21) $\begin{bmatrix} 3 & 0 & -1 \\ 2 & 3 & 2 \\ -1 & 0 & 3 \end{bmatrix}$. **(22)** $\begin{bmatrix} 2 & 1 & 1 \\ 0 & 1 & 0 \\ 1 & 1 & 2 \end{bmatrix}$. **(23)** $\begin{bmatrix} 2 & 1 & 1 \\ 0 & 1 & 0 \\ 1 & 2 & 2 \end{bmatrix}$.

(24) $\begin{bmatrix} 1 & 2 & 3 \\ 2 & 4 & 6 \\ 3 & 6 & 9 \end{bmatrix}$. **(25)** $\begin{bmatrix} 0 & 1 & 0 \\ 0 & 0 & 1 \\ 27 & -27 & 9 \end{bmatrix}$. **(26)** $\begin{bmatrix} 4 & 2 & 1 \\ 2 & 7 & 2 \\ 1 & 2 & 4 \end{bmatrix}$.

(27) $\begin{bmatrix} 5 & -7 & 7 \\ 4 & -3 & 4 \\ 4 & -1 & 2 \end{bmatrix}$. **(28)** $\begin{bmatrix} 3 & 1 & -1 \\ 1 & 3 & -1 \\ -1 & -1 & 5 \end{bmatrix}$. **(29)** $\begin{bmatrix} 0 & 1 & 0 & 0 \\ 0 & 0 & 1 & 0 \\ 0 & 0 & 0 & 1 \\ -1 & 4 & -6 & 4 \end{bmatrix}$.

(30) $\begin{bmatrix} 1 & 0 & 0 & 0 \\ 0 & 0 & 1 & 0 \\ 0 & 0 & 0 & 1 \\ 0 & 1 & -3 & 3 \end{bmatrix}$. **(31)** $\begin{bmatrix} 1 & 0 & 0 & 0 \\ 1 & 2 & 1 & 1 \\ 1 & 1 & 2 & 1 \\ 1 & 1 & 1 & 2 \end{bmatrix}$. **(32)** $\begin{bmatrix} 3 & 1 & 1 & 2 \\ 0 & 3 & 1 & 1 \\ 0 & 0 & 2 & 0 \\ 0 & 0 & 0 & 2 \end{bmatrix}$.

In Problems 33 through 37, find a basis of unit eigenvectors for the eigenspaces associated with each eigenvalue of the following matrices.

(33) The matrix in Problem 9.

(34) The matrix in Problem 10.

(35) The matrix in Problem 11.

(36) The matrix in Problem 19.

(37) The matrix in Problem 20.

In Problems 38 through 53, find the eigenvalues and a basis for the eigenspace associated with each eigenvalue for the given linear transformations.

(38) $T\colon \mathbb{P}^1 \to \mathbb{P}^1$ such that $T(at+b)=(3a+5b)t+(5a-3b)$.

(39) $T\colon \mathbb{P}^1 \to \mathbb{P}^1$ such that $T(at+b)=(3a+5b)t-(2a+4b)$.

(40) $T\colon \mathbb{P}^2 \to \mathbb{P}^2$ such that $T(at^2+bt+c)=(2a-c)t^2+(2a+b-2c)t+(-a+2c)$.

(41) $T\colon \mathbb{R}^2 \to \mathbb{R}^2$ such that $T\begin{bmatrix} a \\ b \end{bmatrix} = \begin{bmatrix} 2a-b \\ a+4b \end{bmatrix}$.

(42) $T\colon \mathbb{R}^2 \to \mathbb{R}^2$ such that $T\begin{bmatrix} a \\ b \end{bmatrix} = \begin{bmatrix} 4a+10b \\ 9a-5b \end{bmatrix}$.

(43) $T\colon \mathbb{R}^3 \to \mathbb{R}^3$ such that $T\begin{bmatrix} a \\ b \\ c \end{bmatrix} = \begin{bmatrix} a+b-c \\ 0 \\ a+2b+3c \end{bmatrix}$.

(44) $T\colon \mathbb{R}^3 \to \mathbb{R}^3$ such that $T\begin{bmatrix} a \\ b \\ c \end{bmatrix} = \begin{bmatrix} 3a-b+c \\ -a+3b-c \\ a-b+3c \end{bmatrix}$.

(45) $T\colon \mathbb{V} \to \mathbb{V}$, where $\mathbb{V}$ is the set of all 2×2 real upper triangular matrices, such that

$$T\begin{bmatrix} a & b \\ 0 & c \end{bmatrix} = \begin{bmatrix} b & c \\ 0 & a-3b+3c \end{bmatrix}.$$

(46) $T\colon \mathbb{P}^1 \to \mathbb{P}^1$ such that $T=d/dt$; that is, $T(at+b) = \dfrac{d}{dt}(at+b) = a$.

(47) $T\colon \mathbb{P}^2 \to \mathbb{P}^2$ such that $T=d/dt$; that is, $T(at^2+bt+c) = \dfrac{d}{dt}(at^2+bt+c)$ $= 2at+b$.

(48) $T\colon \mathbb{P}^2 \to \mathbb{P}^2$ such that $T=d^2/dt^2$; that is, $T(at^2+bt+c) = \dfrac{d^2}{dt^2}(at^2+bt+c)$ $= 2a$.

(49) $T\colon \mathbb{V} \to \mathbb{V}$ such that $T=d/dt$ and $\mathbb{V} = span\{e^{3t}, e^{-3t}\}$.

(50) $T\colon \mathbb{V} \to \mathbb{V}$ such that $T=d^2/dt^2$ and $\mathbb{V} = span\{e^{3t}, e^{-3t}\}$.

(51) T: $\mathbb{V} \to \mathbb{V}$ such that $T = d/dt$ and $\mathbb{V} = span\{\sin t, \cos t\}$.

(52) T: $\mathbb{V} \to \mathbb{V}$ such that $T = d^2/dt^2$ and $\mathbb{V} = span\{\sin t, \cos t\}$.

(53) T: $\mathbb{V} \to \mathbb{V}$ such that $T = d^2/dt^2$ and $\mathbb{V} = span\{\sin 2t, \cos 2t\}$.

(54) Consider the matrix

$$\mathbf{C} = \begin{bmatrix} 0 & 1 & 0 & \cdots & 0 \\ 0 & 0 & 1 & \cdots & 0 \\ \vdots & \vdots & \vdots & \ddots & \vdots \\ 0 & 0 & 0 & \vdots & 1 \\ -a_0 & -a_1 & -a_2 & \cdots & -a_{n-1} \end{bmatrix}.$$

Use mathematical induction to prove that

$$\det(\mathbf{C} - \lambda\mathbf{I}) = (-1)^n\left(\lambda^n + a_{n-1}\lambda^{n-1} + \cdots + a_2\lambda^2 + a_1\lambda + a_0\right).$$

Deduce that the characteristic equation for this matrix is

$$\lambda^n + a_{n-1}\lambda^{n-1} + \cdots + a_2\lambda^2 + a_1\lambda + a_0 = 0.$$

The matrix **C** is called the *companion matrix* for this characteristic equation.

4.2 PROPERTIES OF EIGENVALUES AND EIGENVECTORS

The eigenvalues of a linear transformation T from a finite-dimensional vector space to itself are identical to the eigenvalues of any matrix representation for T. Consequently, we discover information about one by studying the other.

The kernel of $\mathbf{A} - \lambda\mathbf{I}$ is a vector space for any square matrix **A**, and all nonzero vectors of this kernel are eigenvectors of **A**. A vector space is closed under scalar multiplication, so $k\mathbf{x}$ is an eigenvector of **A** for any nonzero scalar k whenever **x** is an eigenvector. Thus, in general, a matrix has a finite number of eigenvalues but infinitely many eigenvectors. A vector space is also closed under vector addition, so if **x** and **y** are two eigenvectors corresponding to the *same* eigenvalue λ, then so too is $\mathbf{x} + \mathbf{y}$, providing this sum is not the zero vector.

The *trace* of a square matrix **A**, designated by $tr(\mathbf{A})$, is the sum of the elements on the main diagonal of **A**. In particular, the trace of

$$\mathbf{A} = \begin{bmatrix} -1 & 2 & 0 \\ -3 & 6 & 8 \\ 5 & 4 & -2 \end{bmatrix}$$

is $tr(\mathbf{A}) = -1 + 6 + (-2) = 3$.

▶THEOREM 1

The sum of the eigenvalues of a matrix equals the trace of the matrix.◀

We leave the proof of Theorem 1 as an exercise (see Problem 21). This result provides a useful check on the accuracy of computed eigenvalues. If the sum of the computed eigenvalues of a matrix do *not* equal the trace of the matrix, there is an error! Beware, however, that Theorem 1 only provides a necessary condition on eigenvalues, not a sufficient condition. That is, no conclusions can be drawn from Theorem 1 if the sum of a set of eigenvalues equals the trace. Eigenvalues of a matrix can be computed incorrectly and still have their sum equal the trace of the matrix.

Example 1 Determine whether $\lambda_1 = 12$ and $\lambda_2 = -4$ are eigenvalues for

$$\mathbf{A} = \begin{bmatrix} 11 & 3 \\ -5 & -5 \end{bmatrix}$$

Solution: Here $tr(\mathbf{A}) = 11 + (-5) = 6 \neq 8 = \lambda_1 + \lambda_2$, so these numbers are *not* the eigenvalues of **A**. The eigenvalues for this matrix are 10 and −4, and their sum is the trace of **A**.

The determinant of an upper (or lower) triangular matrix is the product of elements on the main diagonal, so it follows immediately that

►THEOREM 2

The eigenvalues of an upper or lower triangular matrix are the elements on the main diagonal.◄

Example 2 The matrix $\begin{bmatrix} 1 & 0 & 0 \\ 2 & 1 & 0 \\ 3 & 4 & -1 \end{bmatrix}$ is lower triangular, so its eigenvalues are $\lambda_1 = \lambda_2 = 1$ and $\lambda_3 = -1$.

Once the eigenvalues of a matrix are known, one can determine immediately whether the matrix is singular.

►THEOREM 3

A matrix is singular if and only if it has a zero eigenvalue.◄

Proof: A matrix **A** has a zero eigenvalue if and only if $\det(\mathbf{A} - 0\mathbf{I}) = 0$, or (since $0\mathbf{I} = \mathbf{0}$) if and only if $\det(\mathbf{A}) = 0$, which is true (see Theorem 11 of Appendix A) if and only if **A** is singular.

A nonsingular matrix and its inverse have reciprocal eigenvalues and identical eigenvectors.

▶THEOREM 4

If $\mathbf{x}$ *is an eigenvector of an invertible matrix* $\mathbf{A}$ *corresponding to the eigenvalue* λ*, then* $\mathbf{x}$ *is also an eigenvector of* $\mathbf{A}^{-1}$ *corresponding to the eigenvalue* $1/\lambda$.◀

Proof: Since $\mathbf{A}$ is invertible, Theorem 3 implies that $\lambda \neq 0$; hence $1/\lambda$ exists. We have that $\mathbf{Ax} = \lambda\mathbf{x}$. Premultiplying both sides of this equation by $\mathbf{A}^{-1}$, we obtain

$$\mathbf{x} = \lambda\mathbf{A}^{-1}\mathbf{x} \quad \text{or} \quad \mathbf{A}^{-1}\mathbf{x} = (1/\lambda)\mathbf{x}$$

Thus, $\mathbf{x}$ is an eigenvector of $\mathbf{A}^{-1}$ with corresponding eigenvalue $1/\lambda$.

We may combine Theorem 3 with Theorem 10 of Appendix A and Theorems 5 and 6 of Section 2.6 to obtain the following result.

▶THEOREM 5

The following statements are equivalent for an $n \times n$ *matrix* $\mathbf{A}$:

(i) $\mathbf{A}$ *has an inverse.*
(ii) $\mathbf{A}$ *has rank* n.
(iii) $\mathbf{A}$ *can be transformed by elementary row operations to an upper triangular matrix with only unity elements on the main diagonal.*
(iv) $\mathbf{A}$ *has a nonzero determinant.*
(v) *Every eigenvalue of* $\mathbf{A}$ *is nonzero.*◀

Multiplying the equation $\mathbf{Ax} = \lambda\mathbf{x}$ by a scalar k, we obtain $(k\mathbf{A})\mathbf{x} = (k\lambda)\mathbf{x}$. Thus we have proven Theorem 6.

▶THEOREM 6

If $\mathbf{x}$ *is an eigenvector of* $\mathbf{A}$ *corresponding to the eigenvalue* λ*, then* $k\lambda$ *and* $\mathbf{x}$ *are a corresponding pair of eigenvalues and eigenvectors of* $k\mathbf{A}$*, for any nonzero scalar* k.◀

Theorem 1 provides a relationship between the sum of the eigenvalues of a matrix and its trace. There is also a relationship between the product of those eigenvalues and the determinant of the matrix. The proof of the next theorem is left as an exercise (see Problem 22).

▶THEOREM 7

The product of all the eigenvalues of a matrix (counting multiplicity) equals the determinant of the matrix.◀

Example 3 The eigenvalues of $A = \begin{bmatrix} 11 & 3 \\ -5 & -5 \end{bmatrix}$ are $\lambda_1 = 10$ and $\lambda_2 = -4$. Here $\det(A) = -55 + 15 = -40 = \lambda_1\lambda_2$.

▶THEOREM 8

If $\mathbf{x}$ is an eigenvector of $\mathbf{A}$ corresponding to the eigenvalue λ, then λ^n and $\mathbf{x}$ are a corresponding pair of eigenvalues and eigenvectors of $\mathbf{A}^n$, for any positive integer n.◀

Proof: We are given that $\mathbf{Ax} = \lambda\mathbf{x}$ and we need to show that

$$\mathbf{A}^n\mathbf{x} = \lambda^n\mathbf{x} \tag{4.6}$$

We prove this last equality by mathematical induction on the power n. Equation (4.6) is true for $n = 1$ as a consequence of the hypothesis of the theorem. Now assume that the proposition is true for $n = k - 1$. Then

$$\mathbf{A}^{k-1}\mathbf{x} = \lambda^{k-1}\mathbf{x}$$

Premultiplying this equation by A, we have

$$\mathbf{A}(\mathbf{A}^{k-1}\mathbf{x}) = \mathbf{A}\left(\lambda^{k-1}\mathbf{x}\right)$$

or

$$\mathbf{A}^k\mathbf{x} = \lambda^{k-1}(\mathbf{Ax})$$

It now follows from the hypothesis of the theorem that

$$\mathbf{A}^k\mathbf{x} = \lambda^{k-1}(\lambda\mathbf{x})$$

or

$$\mathbf{A}^k\mathbf{x} = \lambda^k\mathbf{x}$$

which implies that the proposition is true for $n = k$. Thus, Theorem 8 is proved by mathematical induction.

The proofs of the next two results are left as exercises for the reader (see Problems 16 and 17).

▶THEOREM 9

If $\mathbf{x}$ is an eigenvector of $\mathbf{A}$ corresponding to the eigenvalue λ, then for any scalar c, $\lambda - c$ and $\mathbf{x}$ are a corresponding pair of eigenvalues and eigenvectors of $\mathbf{A} - c\mathbf{I}$.◀

▶THEOREM 10

If λ is an eigenvalue of $\mathbf{A}$, then λ also an eigenvalue of $\mathbf{A}^{\mathrm{T}}$.◀

Problems 4.2

(1) One eigenvalue of the matrix $\mathbf{A} = \begin{bmatrix} 8 & 2 \\ 3 & 3 \end{bmatrix}$ is known to be 2.

Determine the second eigenvalue by inspection.

(2) One eigenvalue of the matrix $\mathbf{A} = \begin{bmatrix} 8 & 3 \\ 3 & 2 \end{bmatrix}$ is known to be 0.7574 rounded to four decimal places. Determine the second eigenvalue by inspection.

(3) Two eigenvalues of a 3×3 matrix are known to be 5 and 8. What can be said about the third eigenvalue if the trace of the matrix is -4?

(4) Redo Problem 3 if -4 is the determinant of the matrix instead of its trace.

(5) The determinant of a 4×4 matrix is 144 and two of its eigenvalues are known to be -3 and 2. What can be said about the remaining eigenvalues?

(6) A 2×2 matrix $\mathbf{A}$ is known to have the eigenvalues -3 and 4. What are the eigenvalues of

(a) $2\mathbf{A}$, (b) $5\mathbf{A}$, (c) $\mathbf{A} - 3\mathbf{I}$, (d) $\mathbf{A} + 4\mathbf{I}$.

(7) A 3×3 matrix $\mathbf{A}$ is known to have the eigenvalues -2, 2, and 4. What are the eigenvalues of

(a) $\mathbf{A}^2$, (b) $\mathbf{A}^3$, (c) $-3\mathbf{A}$, (d) $\mathbf{A} + 3\mathbf{I}$.

(8) A 2×2 matrix $\mathbf{A}$ is known to have the eigenvalues -1 and 1. Find a matrix in terms of $\mathbf{A}$ that has for its eigenvalues

(a) -2 and 2, (b) -5 and 5, (c) 1 and 1, (d) 2 and 4.

(9) A 3×3 matrix $\mathbf{A}$ is known to have the eigenvalues 2, 3, and 4. Find a matrix in terms of $\mathbf{A}$ that has for its eigenvalues

(a) 4, 6, and 8, (b) 4, 9, and 16, (c) 8, 27, and 64, (d) 0, 1, and 2.

(10) Verify Theorems 1 and 7 for $\mathbf{A} = \begin{bmatrix} 8 & 3 \\ 3 & 2 \end{bmatrix}$.

(11) Verify Theorems 1 and 7 for $\mathbf{A} = \begin{bmatrix} 1 & 3 & 6 \\ -1 & 2 & -1 \\ 2 & 1 & 7 \end{bmatrix}$.

(12) What are the eigenvalues of $\mathbf{A}^{-1}$ for the matrices defined in Problems 10 and 11?

(13) Show by example that, in general, an eigenvalue of $\mathbf{A} + \mathbf{B}$ is *not* the sum of an eigenvalue of $\mathbf{A}$ with an eigenvalue of $\mathbf{B}$.

(14) Show by example that, in general, an eigenvalue of $\mathbf{AB}$ is *not* the product of an eigenvalue of $\mathbf{A}$ with an eigenvalue of $\mathbf{B}$.

(15) Show by example that an eigenvector of $\mathbf{A}$ need *not* be an eigenvector of $\mathbf{A}^{\mathrm{T}}$.

(16) Prove Theorem 9.

(17) Prove Theorem 10.

(18) The determinant of $\mathbf{A} - \lambda\mathbf{I}$ is known as the *characteristic polynomial* of **A**. For an $n \times n$ matrix **A** it has the form

$$\det(\mathbf{A} - \lambda\mathbf{I}) = (-1)^n\left(\lambda^n + a_{n-1}\lambda^{n-1} + a_{n-2}\lambda^{n-2} + \cdots + a_2\lambda^2 + a_1\lambda + a_0\right),$$

where $a_{n-1}, a_{n-2}, \ldots, a_2, a_1,$ and a_0 are constants that depend on the elements of **A**. Show that $(-1)^n a_0 = \det(\mathbf{A})$.

(19) (Problem 18 continued.) Convince yourself by considering arbitrary 2×2, 3×3, and 4×4 matrices that $(-1)a_{n-1} = tr(\mathbf{A})$.

(20) Consider an $n \times n$ matrix **A** with eigenvalues $\lambda_1, \lambda_2, \ldots, \lambda_n$, where some or all of the eigenvalues may be equal. Each eigenvalue $\lambda_i (i = 1, 2, \ldots, n)$ is a root of the characteristic polynomial; hence $(\lambda - \lambda_i)$ must be a factor of that polynomial. Deduce that $\det(\mathbf{A} - \lambda\mathbf{I}) = (-1)^n(\lambda - \lambda_1)(\lambda - \lambda_2)\ldots(\lambda - \lambda_n)$.

(21) Use the results of Problems 19 and 20 to prove Theorem 1.

(22) Use the results of Problems 18 and 20 to prove Theorem 7.

(23) The Cayley-Hamilton theorem states that every square matrix **A** satisfies its own characteristic equation. That is, if the characteristic equation of **A** is

$$\lambda^n + a_{n-1}\lambda^{n-1} + a_{n-2}\lambda^{n-2} + \cdots + a_2\lambda^2 + a_1\lambda + a_0 = 0,$$

then

$$A^n + a_{n-1}A^{n-1} + a_{n-2}A^{n-2} + \cdots + a_2A^2 + a_1A + a_0I = 0.$$

Verify the Cayley-Hamilton theorem for

(a) $\begin{bmatrix} 1 & 2 \\ 3 & 4 \end{bmatrix}$, (b) $\begin{bmatrix} 1 & 2 \\ 2 & 4 \end{bmatrix}$, (c) $\begin{bmatrix} 2 & 0 & 1 \\ 4 & 0 & 2 \\ 0 & 0 & -1 \end{bmatrix}$,

(d) $\begin{bmatrix} 1 & -1 & 2 \\ 0 & 3 & 2 \\ 2 & 1 & 2 \end{bmatrix}$, (e) $\begin{bmatrix} 1 & 0 & 0 & 0 \\ 0 & -1 & 0 & 0 \\ 0 & 0 & -1 & 0 \\ 0 & 0 & 0 & 1 \end{bmatrix}$.

(24) Let the characteristic equation of a square matrix **A** be as given in Problem 23. Use the results of Problem 18 to prove that **A** is invertible if and only if $a_0 \neq 0$.

(25) Let the characteristic equation of a square matrix **A** be given as in Problem 23. Use the Cayley-Hamilton theorem to show that

$$A^{-1} = \frac{-1}{a_0}\left(A^{n-1} + a_{n-1}A^{n-2} + \cdots + a_2A + a_1I\right)$$

when $a_0 \neq 0$.

(26) Use the result of Problem 25 to find the inverses, when they exist, for the matrices defined in Problem 23.

4.3 DIAGONALIZATION OF MATRICES

We are ready to answer the question that motivated this chapter: Which linear transformations can be represented by diagonal matrices and what bases generate such representations? Recall that different matrices represent the same linear transformation if and only if those matrices are similar (Theorem 3 of Section 3.4). Therefore, a linear transformation has a diagonal matrix representation if and only if any matrix representation of the transformation is similar to a diagonal matrix.

To establish whether a linear transformation T has a diagonal matrix representation, we first create one matrix representation for the transformation and then determine whether that matrix is similar to a diagonal matrix. If it is, we say the matrix is *diagonalizable*, in which case T has a diagonal matrix representation.

If a matrix **A** is similar to a diagonal matrix **D**, then the form of **D** is determined. Both **A** and **D** have identical eigenvalues, and the eigenvalues of a diagonal matrix (which is both upper and lower triangular) are the elements on its main diagonal. Consequently, the main diagonal of **D** must be the eigenvalues of **A**. If, for example,

A matrix is diagonalizable if it is similar to a diagonal matrix.

$$\mathbf{A} = \begin{bmatrix} 1 & 2 \\ 4 & 3 \end{bmatrix}$$

with eigenvalues -1 and 5, is diagonalizable, then **A** must be similar to either

$$\begin{bmatrix} -1 & 0 \\ 0 & 5 \end{bmatrix} \quad \text{or} \quad \begin{bmatrix} 5 & 0 \\ 0 & -1 \end{bmatrix}$$

Now let **A** be an $n \times n$ matrix with n *linearly independent eigenvectors* $\mathbf{x}_1, \mathbf{x}_2, \ldots, \mathbf{x}_n$ corresponding to the eigenvalues $\lambda_1, \lambda_2, \ldots, \lambda_n$, respectively. Therefore,

$$\mathbf{A}\mathbf{x}_j = \lambda_j \mathbf{x}_j \tag{4.7}$$

for $j = 1, 2, \ldots, n$. There are no restrictions on the multiplicity of the eigenvalues, so some or all of them may be equal. Set

$$\mathbf{M} = \begin{bmatrix} \mathbf{x}_1 & \mathbf{x}_2 & \ldots & \mathbf{x}_n \end{bmatrix} \text{ and}$$

$$\mathbf{D} = \begin{bmatrix} \lambda_1 & 0 & \ldots & 0 \\ 0 & \lambda_2 & \ldots & 0 \\ \vdots & \vdots & \ddots & \vdots \\ 0 & 0 & \ldots & \lambda_n \end{bmatrix}$$

Here M is called a *modal matrix* for A and D a *spectral matrix* for A. Now

$$\begin{aligned} \mathbf{AM} &= \mathbf{A}[\mathbf{x}_1 \quad \mathbf{x}_2 \quad \dots \quad \mathbf{x}_n] \\ &= [\mathbf{Ax}_1 \quad \mathbf{Ax}_2 \quad \dots \quad \mathbf{Ax}_n] \\ &= [\lambda_1\mathbf{x}_1 \quad \lambda_2\mathbf{x}_2 \quad \dots \quad \lambda_n\mathbf{x}_n] \\ &= [\mathbf{x}_1 \quad \mathbf{x}_2 \quad \dots \quad \mathbf{x}_n]\mathbf{D} \\ &= \mathbf{MD} \end{aligned} \tag{4.8}$$

Because the columns of M are linearly independent, the column rank of M is n, the rank of M is n, and $\mathbf{M}^{-1}$ exists. Premultiplying Equation (4.8) by $\mathbf{M}^{-1}$, we obtain

$$\mathbf{D} = \mathbf{M}^{-1}\mathbf{AM} \tag{4.9}$$

Postmultiplying Equation (4.8) by $\mathbf{M}^{-1}$, we have

$$\mathbf{A} = \mathbf{MDM}^{-1} \tag{4.10}$$

Thus, A is similar to D. We can retrace our steps and show that if Equation (4.10) is satisfied, then M must be an invertible matrix having as its columns a set of eigenvectors of A. We have proven the following result.

►THEOREM 1

An $n \times n$ matrix is diagonalizable if and only if the matrix possesses n linearly independent eigenvectors.◄

Example 1 Determine whether $\mathbf{A} = \begin{bmatrix} 1 & 2 \\ 4 & 3 \end{bmatrix}$ is diagonalizable.

Solution: Using the results of Example 3 of Section 4.1, we have $\lambda_1 = -1$ and $\lambda_2 = 5$ as the eigenvalues of A with corresponding eigenspaces spanned by the vectors

$$\mathbf{x}_1 = \begin{bmatrix} -1 \\ 1 \end{bmatrix} \quad \text{and} \quad \mathbf{x}_2 = \begin{bmatrix} 1 \\ 2 \end{bmatrix}$$

respectively. These two vectors are linearly independent, so A is diagonalizable. We can choose either

$$\mathbf{M} = \begin{bmatrix} -1 & 1 \\ 1 & 2 \end{bmatrix} \quad \text{or} \quad \mathbf{M} = \begin{bmatrix} 1 & -1 \\ 2 & 1 \end{bmatrix}$$

Making the first choice, we find

$$\mathbf{D} = \mathbf{M}^{-1}\mathbf{AM} = \frac{1}{3}\begin{bmatrix} -2 & 1 \\ 1 & 1 \end{bmatrix}\begin{bmatrix} 1 & 2 \\ 4 & 3 \end{bmatrix}\begin{bmatrix} -1 & 1 \\ 1 & 2 \end{bmatrix} = \begin{bmatrix} -1 & 0 \\ 0 & 5 \end{bmatrix}$$

Making the second choice, we find

$$\mathbf{D} = \mathbf{M}^{-1}\mathbf{AM} = \frac{1}{3}\begin{bmatrix} 1 & 1 \\ -2 & 1 \end{bmatrix}\begin{bmatrix} 1 & 2 \\ 4 & 3 \end{bmatrix}\begin{bmatrix} 1 & -1 \\ 2 & 1 \end{bmatrix} = \begin{bmatrix} 5 & 0 \\ 0 & -1 \end{bmatrix}$$

In general, neither the modal matrix **M** nor the spectral matrix **D** is unique. However, once **M** is selected, then **D** is fully determined. The element of **D** located in the jth row and jth column must be the eigenvalue corresponding to the eigenvector in the jth column of **M**. In particular,

$$\mathbf{M} = [\mathbf{x}_2 \quad \mathbf{x}_1 \quad \mathbf{x}_3 \quad \ldots \quad \mathbf{x}_n]$$

is matched with

$$\mathbf{D} = \begin{bmatrix} \lambda_2 & 0 & 0 & \ldots & 0 \\ 0 & \lambda_1 & 0 & \ldots & 0 \\ 0 & 0 & \lambda_3 & \ldots & 0 \\ \vdots & \vdots & \vdots & \ddots & \vdots \\ 0 & 0 & 0 & \ldots & \lambda_n \end{bmatrix}$$

while

$$\mathbf{M} = [\mathbf{x}_n \quad \mathbf{x}_{n-1} \quad \ldots \quad \mathbf{x}_1]$$

is matched with

$$\mathbf{D} = \begin{bmatrix} \lambda_n & 0 & \ldots & 0 \\ 0 & \lambda_{n-1} & \ldots & 0 \\ \vdots & \vdots & \ddots & \vdots \\ 0 & 0 & \ldots & \lambda_1 \end{bmatrix}$$

Example 2 Determine whether $\mathbf{A} = \begin{bmatrix} 2 & -1 & 0 \\ 3 & -2 & 0 \\ 0 & 0 & 1 \end{bmatrix}$ is diagonalizable.

Solution: Using the results of Example 6 of Section 4.1, we have

$$\mathbf{x}_1 = \begin{bmatrix} 1 \\ 1 \\ 0 \end{bmatrix} \quad \text{and} \quad \mathbf{x}_2 = \begin{bmatrix} 0 \\ 0 \\ 1 \end{bmatrix}$$

as a basis for the eigenspace corresponding to eigenvalue $\lambda = 1$ of multiplicity 2 and

$$\mathbf{x}_3 = \begin{bmatrix} 1 \\ 3 \\ 0 \end{bmatrix}$$

as a basis corresponding to eigenvalue $\lambda = -1$ of multiplicity 1. These three vectors are linearly independent, so **A** is diagonalizable. If we choose

$$\mathbf{M} = \begin{bmatrix} 1 & 0 & 1 \\ 1 & 0 & 3 \\ 0 & 1 & 0 \end{bmatrix}, \quad \text{then} \quad \mathbf{M}^{-1}\mathbf{A}\mathbf{M} = \begin{bmatrix} 1 & 0 & 0 \\ 0 & 1 & 0 \\ 0 & 0 & -1 \end{bmatrix}$$

The process of determining whether a given set of eigenvectors is linearly independent is simplified by the following two results.

▶THEOREM 2

Eigenvectors of a matrix corresponding to distinct eigenvalues are linearly independent. ◀

Proof: Let $\lambda_1, \lambda_2, \ldots, \lambda_k$ denote the *distinct* eigenvalues of an $n \times n$ matrix $\mathbf{A}$ with corresponding eigenvectors $\mathbf{x}_1, \mathbf{x}_2, \ldots, \mathbf{x}_k$. If all the eigenvalues have multiplicity 1, then $k=n$, otherwise $k<n$. We use mathematical induction to prove that $\{\mathbf{x}_1, \mathbf{x}_2, \ldots, \mathbf{x}_k\}$ is a linearly independent set.

For $k=1$, the set $\{\mathbf{x}_1\}$ is linearly independent because the eigenvector $\mathbf{x}_1$ cannot be 0. We now assume that the set $\{\mathbf{x}_1, \mathbf{x}_2, \ldots, \mathbf{x}_{k-1}\}$ is linearly independent and use this to show that the set $\{\mathbf{x}_1, \mathbf{x}_2, \ldots, \mathbf{x}_{k-1}, \mathbf{x}_k\}$ is linearly independent. This is equivalent to showing that the only solution to the vector equation

$$c_1\mathbf{x}_1 + c_2\mathbf{x}_2 + \cdots + c_{k-1}\mathbf{x}_{k-1} + c_k\mathbf{x}_k = 0 \tag{4.11}$$

is $c_1=c_2=\cdots=c_{k-1}=c_k=0$.

Multiplying Equation (4.11) on the left by $\mathbf{A}$ and using the fact that $\mathbf{A}\mathbf{x}_j=\lambda_j\mathbf{x}_j$ for $j=1,2, \ldots, k$, we obtain

$$c_1\lambda_1\mathbf{x}_1 + c_2\lambda_2\mathbf{x}_2 + \cdots + c_{k-1}\lambda_{k-1}\mathbf{x}_{k-1} + c_k\lambda_k\mathbf{x}_k = 0 \tag{4.12}$$

Multiplying Equation (4.11) by λ_k, we obtain

$$c_1\lambda_k\mathbf{x}_1 + c_2\lambda_k\mathbf{x}_2 + \cdots + c_{k-1}\lambda_k\mathbf{x}_{k-1} + c_k\lambda_k\mathbf{x}_k = 0 \tag{4.13}$$

Subtracting Equation (4.13) from (4.12), we have

$$c_1(\lambda_1 - \lambda_k)\mathbf{x}_1 + c_2(\lambda_2 - \lambda_k)\mathbf{x}_2 + \cdots + c_{k-1}(\lambda_{k-1} - \lambda_k)\mathbf{x}_{k-1} = 0$$

But the vectors $\{\mathbf{x}_1, \mathbf{x}_2, \ldots, \mathbf{x}_{k-1}\}$ are linearly independent by the induction hypothesis, hence the coefficients in the last equation must all be 0; that is,

$$c_1(\lambda_1 - \lambda_k) = c_2(\lambda_2 - \lambda_k) = \cdots = c_{k-1}(\lambda_{k-1} - \lambda_k) = 0$$

from which we imply that $c_1=c_2=\cdots=c_{k-1}=0$, because the eigenvalues are distinct. Equation (4.11) reduces to $c_k\mathbf{x}_k=\mathbf{0}$ and because $\mathbf{x}_k$ is an eigenvector, and therefore nonzero, we also conclude that $c_k=0$, and the proof is complete.

It follows from Theorems 1 and 2 that any $n \times n$ real matrix having n distinct real roots of its characteristic equation, that is a matrix having n eigenvalues all of multiplicity 1, must be diagonalizable (see, in particular, Example 1).

Example 3 Determine whether $\mathbf{A} = \begin{bmatrix} 2 & 0 & 0 \\ -3 & 3 & 0 \\ 2 & -1 & 4 \end{bmatrix}$ is diagonalizable.

Solution: The matrix is lower triangular so its eigenvalues are the elements on the main diagonal, namely 2, 3, and 4. Every eigenvalue has multiplicity 1, hence **A** is diagonalizable.

▶THEOREM 3

If λ is an eigenvalue of multiplicity k of an $n \times n$ matrix $\mathbf{A}$, then the number of linearly independent eigenvectors of $\mathbf{A}$ associated with λ is $n - r(\mathbf{A} - \lambda\mathbf{I})$, where r denotes rank.◀

Proof: The eigenvectors of **A** corresponding to the eigenvalue λ are all nonzero solutions of the vector Equation $(\mathbf{A} - \lambda\mathbf{I})\mathbf{x} = \mathbf{0}$. This homogeneous system is consistent, so by Theorem 3 of Section 2.6 the solutions will be in terms of $n - r(\mathbf{A} - \lambda\mathbf{I})$ arbitrary unknowns. Since these unknowns can be picked independently of each other, they generate $n - r(\mathbf{A} - \lambda\mathbf{I})$ linearly independent eigenvectors.

In Example 2, **A** is a 3×3 matrix ($n = 3$) and $\lambda = 1$ is an eigenvalue of multiplicity 2. In this case,

$$\mathbf{A} - (1)\mathbf{I} = \mathbf{A} - \mathbf{I} = \begin{bmatrix} 1 & -1 & 0 \\ 3 & -3 & 0 \\ 0 & 0 & 0 \end{bmatrix}$$

can be transformed into row-reduced form (by adding to the second row -3 times the first row)

$$\begin{bmatrix} 1 & -1 & 0 \\ 0 & 0 & 0 \\ 0 & 0 & 0 \end{bmatrix}$$

having rank 1. Thus, $n - r(\mathbf{A} - \mathbf{I}) = 3 - 1 = 2$ and **A** has two linearly independent eigenvectors associated with $\lambda = 1$. Two such vectors are exhibited in Example 2.

Example 4 Determine whether $\mathbf{A} = \begin{bmatrix} 2 & 1 \\ 0 & 2 \end{bmatrix}$ is diagonalizable.

Solution: The matrix is upper triangular so its eigenvalues are the elements on the main diagonal, namely, 2 and 2. Thus, **A** is 2×2 matrix with one eigenvalue of multiplicity 2. Here

$$\mathbf{A} - 2\mathbf{I} = \begin{bmatrix} 0 & 1 \\ 0 & 0 \end{bmatrix}$$

has a rank of 1. Thus, $n - r(\mathbf{A} - 2\mathbf{I}) = 2 - 1 = 1$ and **A** has only *one* linearly independent eigenvector associated with its eigenvalues, not two as needed. Matrix **A** is *not* diagonalizable.

We saw in the beginning of Section 4.1 that if a linear transformation $\mathbf{T}: \mathbb{V} \to \mathbb{V}$ is represented by a diagonal matrix, then the basis that generates such a representation is a basis of eigenvectors. To this we now add that a linear transformation $T: \mathbb{V} \to \mathbb{V}$, where $\mathbb{V}$ is n-dimensional, can be represented by a diagonal matrix if and only if T possesses n-linearly independent eigenvectors. When such a set exists, it is a basis for $\mathbb{V}$.

If $\mathbb{V}$ is an n-dimensional vector space, then a linear transformation $T: \mathbb{V} \to \mathbb{V}$ may be represented by a diagonal matrix if and only if T possesses a basis of eigenvectors.

Example 5 Determine whether the linear transformation $T: \mathbb{P}^1 \to \mathbb{P}^1$ defined by

$$T(at + b) = (a + 2b)t + (4a + 3b)$$

can be represented by a diagonal matrix.

Solution: A standard basis for $\mathbb{P}^1$ is $\mathbb{B} = \{t, 1\}$, and we showed in Example 7 of Section 4.1 that a matrix representation for T with respect to this basis is

$$\mathbf{A} = \begin{bmatrix} 1 & 2 \\ 4 & 3 \end{bmatrix}$$

It now follows from Example 1 that this matrix is diagonalizable; hence T can be represented by a diagonal matrix $\mathbf{D}$, in fact, either of the two diagonal matrices produced in Example 1.

Furthermore, we have from Example 7 of Section 4.1 that $-t+1$ is an eigenvector of T corresponding to $\lambda_1 = -1$ while $5t+10$ is an eigenvector corresponding $\lambda_2 = 5$. Since both polynomials correspond to distinct eigenvalues, the vectors are linearly independent and, therefore, constitute a basis. Setting $\mathbb{C} = \{-t + 1, 5t + 10\}$, we have the matrix representation of T with respect to $\mathbb{C}$ as

$$\mathbf{A}_{\mathbb{C}}^{\mathbb{C}} = \mathbf{D} = \begin{bmatrix} -1 & 0 \\ 0 & 5 \end{bmatrix}$$

Example 6 Let $\mathbb{U}$ be the set of all 2×2 real upper triangular matrices. Determine whether the linear transformation $T: \mathbb{U} \to \mathbb{U}$ defined by

$$T\begin{bmatrix} a & b \\ 0 & c \end{bmatrix} = \begin{bmatrix} 3a + 2b + c & 2b \\ 0 & a + 2b + 3c \end{bmatrix}$$

can be represented by a diagonal matrix and, if so, produce a basis that generates such a representation.

Solution: $\mathbb{U}$ is closed under addition and scalar multiplication, so it is a sub-space of $\mathbb{M}_{2\times2}$. A simple basis for $\mathbb{U}$ is given by

$$\mathbb{B} = \left\{ \begin{bmatrix} 1 & 0 \\ 0 & 0 \end{bmatrix}, \begin{bmatrix} 0 & 1 \\ 0 & 0 \end{bmatrix}, \begin{bmatrix} 0 & 0 \\ 0 & 1 \end{bmatrix} \right\}$$

With respect to these basis vectors,

$$T\begin{bmatrix} 1 & 0 \\ 0 & 0 \end{bmatrix} = \begin{bmatrix} 3 & 0 \\ 0 & 1 \end{bmatrix} = 3\begin{bmatrix} 1 & 0 \\ 0 & 0 \end{bmatrix} + 0\begin{bmatrix} 0 & 1 \\ 0 & 0 \end{bmatrix} + 1\begin{bmatrix} 0 & 0 \\ 0 & 1 \end{bmatrix} \leftrightarrow \begin{bmatrix} 3 \\ 0 \\ 1 \end{bmatrix}$$

$$T\begin{bmatrix} 0 & 1 \\ 0 & 0 \end{bmatrix} = \begin{bmatrix} 2 & 2 \\ 0 & 2 \end{bmatrix} = 2\begin{bmatrix} 1 & 0 \\ 0 & 0 \end{bmatrix} + 2\begin{bmatrix} 0 & 1 \\ 0 & 0 \end{bmatrix} + 2\begin{bmatrix} 0 & 0 \\ 0 & 1 \end{bmatrix} \leftrightarrow \begin{bmatrix} 2 \\ 2 \\ 2 \end{bmatrix}$$

$$T\begin{bmatrix} 0 & 0 \\ 0 & 1 \end{bmatrix} = \begin{bmatrix} 1 & 0 \\ 0 & 3 \end{bmatrix} = 1\begin{bmatrix} 1 & 0 \\ 0 & 0 \end{bmatrix} + 0\begin{bmatrix} 0 & 1 \\ 0 & 0 \end{bmatrix} + 3\begin{bmatrix} 0 & 0 \\ 0 & 1 \end{bmatrix} \leftrightarrow \begin{bmatrix} 1 \\ 0 \\ 3 \end{bmatrix}$$

and a matrix representation for T is

$$\mathbf{A} = \begin{bmatrix} 3 & 2 & 1 \\ 0 & 2 & 0 \\ 1 & 2 & 3 \end{bmatrix}$$

The eigenvalues of this matrix are 2, 2, and 4. Even though the eigenvalues are not all distinct, the matrix still has three linearly independent eigenvectors, namely,

$$\mathbf{x}_1 = \begin{bmatrix} -2 \\ 1 \\ 0 \end{bmatrix}, \quad \mathbf{x}_2 = \begin{bmatrix} -1 \\ 0 \\ 1 \end{bmatrix}, \quad \text{and} \quad \mathbf{x}_3 = \begin{bmatrix} 1 \\ 0 \\ 1 \end{bmatrix}$$

Thus, $\mathbf{A}$ is diagonalizable and, therefore, T has a diagonal matrix representation. Setting

$$\mathbf{M} = \begin{bmatrix} -2 & -1 & 1 \\ 1 & 0 & 0 \\ 0 & 1 & 1 \end{bmatrix}, \quad \text{we have} \quad \mathbf{D} = \mathbf{M}^{-1}\mathbf{A}\mathbf{M} = \begin{bmatrix} 2 & 0 & 0 \\ 0 & 2 & 0 \\ 0 & 0 & 4 \end{bmatrix}$$

which is one diagonal representation for T.

The vectors $\mathbf{x}_1$, $\mathbf{x}_2$, and $\mathbf{x}_3$ are coordinate representations with respect to the $\mathbb{B}$ basis for

$$\begin{bmatrix} -2 \\ 1 \\ 0 \end{bmatrix} \leftrightarrow (-2)\begin{bmatrix} 1 & 0 \\ 0 & 0 \end{bmatrix} + 1\begin{bmatrix} 0 & 1 \\ 0 & 0 \end{bmatrix} + 0\begin{bmatrix} 0 & 0 \\ 0 & 1 \end{bmatrix} = \begin{bmatrix} -2 & 1 \\ 0 & 0 \end{bmatrix}$$

$$\begin{bmatrix} -1 \\ 0 \\ 1 \end{bmatrix} \leftrightarrow (-1)\begin{bmatrix} 1 & 0 \\ 0 & 0 \end{bmatrix} + 0\begin{bmatrix} 0 & 1 \\ 0 & 0 \end{bmatrix} + 1\begin{bmatrix} 0 & 0 \\ 0 & 1 \end{bmatrix} = \begin{bmatrix} -1 & 0 \\ 0 & 1 \end{bmatrix}$$

$$\begin{bmatrix} 1 \\ 0 \\ 1 \end{bmatrix} \leftrightarrow 1\begin{bmatrix} 1 & 0 \\ 0 & 0 \end{bmatrix} + 0\begin{bmatrix} 0 & 1 \\ 0 & 0 \end{bmatrix} + 1\begin{bmatrix} 0 & 0 \\ 0 & 1 \end{bmatrix} = \begin{bmatrix} 1 & 0 \\ 0 & 1 \end{bmatrix}$$

The set

$$\mathbb{C} = \left\{ \begin{bmatrix} -2 & 1 \\ 0 & 0 \end{bmatrix}, \begin{bmatrix} -1 & 0 \\ 0 & 1 \end{bmatrix}, \begin{bmatrix} 1 & 0 \\ 0 & 1 \end{bmatrix} \right\}$$

is a basis of eigenvectors of T for the vector space $\mathbb{U}$. A matrix representation of T with respect to the $\mathbb{C}$ basis is the diagonal matrix $\mathbf{D}$.

Problems 4.3

In Problems 1 through 11, determine whether the matrices are diagonalizable. If they are, identify a modal matrix M and calculate $M^{-1}AM$.

(1) $A = \begin{bmatrix} 2 & -3 \\ 1 & -2 \end{bmatrix}$.

(2) $A = \begin{bmatrix} 4 & 3 \\ 3 & -4 \end{bmatrix}$.

(3) $A = \begin{bmatrix} 3 & 1 \\ -1 & 5 \end{bmatrix}$.

(4) $A = \begin{bmatrix} 1 & 1 & 1 \\ 0 & 1 & 0 \\ 0 & 0 & 1 \end{bmatrix}$.

(5) $A = \begin{bmatrix} 1 & 0 & 0 \\ 2 & -3 & 3 \\ 1 & 2 & 2 \end{bmatrix}$.

(6) $A = \begin{bmatrix} 5 & 1 & 2 \\ 0 & 3 & 0 \\ 2 & 1 & 5 \end{bmatrix}$.

(7) $A = \begin{bmatrix} 1 & 2 & 3 \\ 2 & 4 & 6 \\ 3 & 6 & 9 \end{bmatrix}$.

(8) $A = \begin{bmatrix} 3 & -1 & 1 \\ -1 & 3 & -1 \\ 1 & -1 & 3 \end{bmatrix}$.

(9) $A = \begin{bmatrix} 7 & 3 & 3 \\ 0 & 1 & 0 \\ -3 & -3 & 1 \end{bmatrix}$.

(10) $A = \begin{bmatrix} 3 & 1 & 0 \\ 0 & 3 & 1 \\ 0 & 0 & 3 \end{bmatrix}$.

(11) $A = \begin{bmatrix} 3 & 0 & 0 \\ 0 & 3 & 1 \\ 0 & 0 & 3 \end{bmatrix}$.

In Problems 12 through 21, determine whether the linear transformations can be represented by diagonal matrices and, if so, produce bases that will generate such representations.

(12) $T: \mathbb{P}^1 \to \mathbb{P}^1$ defined by $T(at+b) = (2a-3b)t + (a-2b)$.

(13) $T: \mathbb{P}^1 \to \mathbb{P}^1$ defined by $T(at+b) = (4a+3b)t + (3a-4b)$.

(14) $T: \mathbb{P}^2 \to \mathbb{P}^2$ defined by $T(at^2+bt+c) = at^2 + (2a-3b+3c)t + (a+2b+2c)$.

(15) $T: \mathbb{P}^2 \to \mathbb{P}^2$ defined by $T(at^2+bt+c) = (5a+b+2c)t^2 + 3bt + (2a+b+5c)$.

(16) $T: \mathbb{P}^2 \to \mathbb{P}^2$ defined by $T(at^2+bt+c) = (3a+b)t^2 + (3b+c)t + 3c$.

(17) $T: \mathbb{U} \to \mathbb{U}$ where $\mathbb{U}$ is the set of all 2×2 real upper triangular matrices and

$$T\begin{bmatrix} a & b \\ 0 & c \end{bmatrix} = \begin{bmatrix} a+2b+3c & 2a+4b+6c \\ 0 & 3a+6b+9c \end{bmatrix}.$$

(18) $T: \mathbb{U} \to \mathbb{U}$ where $\mathbb{U}$ is the set of all 2×2 real upper triangular matrices and

$$T\begin{bmatrix} a & b \\ 0 & c \end{bmatrix} = \begin{bmatrix} 7a+3b+3c & b \\ 0 & -3a-3b+c \end{bmatrix}.$$

(19) T: $\mathbb{W} \to \mathbb{W}$ where $\mathbb{W}$ is the set of all 2×2 real lower triangular matrices and

$$T\begin{bmatrix} a & 0 \\ b & c \end{bmatrix} = \begin{bmatrix} 3a - b + c & 0 \\ -a + 3b - c & a - b + 3c \end{bmatrix}.$$

(20) T: $\mathbb{R}^3 \to \mathbb{R}^3$ defined by $T\begin{bmatrix} a \\ b \\ c \end{bmatrix} = \begin{bmatrix} c \\ a \\ b \end{bmatrix}$.

(21) T: $\mathbb{R}^3 \to \mathbb{R}^3$ defined by $T\begin{bmatrix} a \\ b \\ c \end{bmatrix} = \begin{bmatrix} 3a + b \\ 3b + c \\ c \end{bmatrix}$.

4.4 THE EXPONENTIAL MATRIX

In this section and the next section (Section 4.5), we will use eigenvalues and eigenvectors extensively and conclude our chapter with sections dealing with differential equations.

One of the most important functions in the calculus is the exponential function e^x. It should not be surprising, therefore, to find that the "exponentials of matrices" are equally useful and important.

To develop this idea, we extend the idea of Maclaurin series to include matrices. As we further our discussion, we will make reference to the Jordan canonical form (see Appendix A).

We recall that this function can be written as a Maclaurin series:

$$e^x = \sum_{k=0}^{\infty} \frac{x^k}{k!} = 1 + x + \frac{x^2}{2!} + \frac{x^3}{2!} + \cdots \tag{4.14}$$

The exponential of a square matrix **A** is defined by the infinite series

$$e^{\mathrm{A}} = \sum_{k=0}^{\infty} \frac{\mathrm{A}^k}{k!} = \mathrm{I} + \frac{\mathrm{A}}{1!} + \frac{\mathrm{A}^2}{2!} + \frac{\mathrm{A}^3}{3!} + \cdots$$

Then we can use this expansion to define the exponential of a square matrix **A** as

$$e^{\mathbf{A}} = \sum_{k=0}^{\infty} \frac{\mathbf{A}^k}{k!} = \mathbf{I} + \frac{\mathbf{A}}{1!} + \frac{\mathbf{A}^2}{2!} + \frac{\mathbf{A}^3}{3!} + \cdots \tag{4.15}$$

Equation (4.14) converges for all values of the variable x; analogously, it can-be shown that Equation (4.15) converges for all square matrices **A**, although the actual proof is well beyond the scope of this book. Using Equation (4.14), we can easily sum the right side of Equation (4.15) for any diagonal matrix.

Example 1 For $\mathbf{A} = \begin{bmatrix} 2 & 0 \\ 0 & -0.3 \end{bmatrix}$, we have

$$e^{\mathbf{A}} = \begin{bmatrix} 1 & 0 \\ 0 & 1 \end{bmatrix} + \frac{1}{1!}\begin{bmatrix} 2 & 0 \\ 0 & -0.3 \end{bmatrix} + \frac{1}{2!}\begin{bmatrix} 2 & 0 \\ 0 & -0.3 \end{bmatrix}^2 + \frac{1}{3!}\begin{bmatrix} 2 & 0 \\ 0 & -0.3 \end{bmatrix}^3 + \cdots$$

$$= \begin{bmatrix} 1 & 0 \\ 0 & 1 \end{bmatrix} + \begin{bmatrix} 2/1! & 0 \\ 0 & (-0.3)/1! \end{bmatrix} + \begin{bmatrix} (2)^2/2! & 0 \\ 0 & (-0.3)^2/2! \end{bmatrix}$$

$$+ \begin{bmatrix} (2)^3/3! & 0 \\ 0 & (-0.3)^3/3! \end{bmatrix} + \ldots$$

$$= \begin{bmatrix} \sum_{k=0}^{\infty} \frac{2^k}{k!} & 0 \\ 0 & \sum_{k=0}^{\infty} \frac{(-0.3)^k}{k!} \end{bmatrix} = \begin{bmatrix} e^2 & 0 \\ 0 & e^{-0.3} \end{bmatrix}$$

In general, if **D** is the diagonal matrix

$$\mathbf{D} = \begin{bmatrix} \lambda_1 & 0 & \ldots & 0 \\ 0 & \lambda_2 & \ldots & 0 \\ \vdots & \vdots & \ddots & \vdots \\ 0 & 0 & \ldots & \lambda_n \end{bmatrix}$$

To calculate the exponential of a diagonal matrix, replace each diagonal element by the exponential of that diagonal element.

Then

$$e^{\mathbf{D}} = \begin{bmatrix} e^{\lambda_1} & 0 & \ldots & 0 \\ 0 & e^{\lambda_2} & \ldots & 0 \\ \vdots & \vdots & \ddots & \vdots \\ 0 & 0 & \ldots & e^{\lambda_n} \end{bmatrix} \tag{4.16}$$

Example 2 Find $e^{\mathbf{D}}$ for $\mathbf{D} = \begin{bmatrix} 1 & 0 & 0 \\ 0 & 2 & 0 \\ 0 & 0 & 2 \end{bmatrix}$.

Solution:

$$e^{\mathbf{D}} = \begin{bmatrix} e^1 & 0 & 0 \\ 0 & e^2 & 0 \\ 0 & 0 & e^2 \end{bmatrix}$$

If a square matrix **A** is not diagonal, but diagonalizable, then we know from our work in Section 4.3 that there exists a modal matrix **M** such that

$$\mathbf{A} = \mathbf{MDM}^{-1} \tag{4.17}$$

where **D** is a diagonal matrix. It follows that

$$\begin{aligned}\mathbf{A}^2 &= \mathbf{AA} = \left(\mathbf{MDM}^{-1}\right)\left(\mathbf{MDM}^{-1}\right) = \mathbf{MD}\left(\mathbf{M}^{-1}\mathbf{M}\right)\mathbf{DM}^{-1}\\ &= \mathbf{MD(I)DM}^{-1} = \mathbf{MD}^2\mathbf{M}^{-1}\end{aligned}$$

$$\begin{aligned}\mathbf{A}^3 &= \mathbf{A}^2\mathbf{A} = \left(\mathbf{MD}^2\mathbf{M}^{-1}\right)\left(\mathbf{MDM}^{-1}\right) = \mathbf{MD}^2\left(\mathbf{M}^{-1}\mathbf{M}\right)\mathbf{DM}^{-1} = \mathbf{MD}^2\mathbf{(I)DM}^{-1}\\ &= \mathbf{MD}^3\mathbf{M}^{-1}\end{aligned}$$

and, in general,

$$\mathbf{A}^n = \mathbf{MD}^n\mathbf{M}^{-1} \tag{4.18}$$

for any positive integer n. Consequently,

$$e^{\mathbf{A}} = \sum_{k=0}^{\infty}\frac{\mathbf{A}^k}{k!} = \sum_{k=0}^{\infty}\frac{\mathbf{MD}^k\mathbf{M}^{-1}}{k!} = \mathbf{M}\left(\sum_{k=0}^{\infty}\frac{\mathbf{D}^k}{k!}\right)\mathbf{M}^{-1} = \mathbf{M}e^{\mathbf{D}}\mathbf{M}^{-1} \tag{4.19}$$

Example 3 Find $e^{\mathbf{A}}$ for $\mathbf{A} = \begin{bmatrix}1 & 2\\4 & 3\end{bmatrix}$.

Solution: The eigenvalues of **A** are -1 and 5 with corresponding eigenvectors $\begin{bmatrix}1\\-1\end{bmatrix}$ and $\begin{bmatrix}1\\2\end{bmatrix}$. Here,

$$\mathbf{M} = \begin{bmatrix}1 & 1\\-1 & 2\end{bmatrix}, \mathbf{M}^{-1} = \begin{bmatrix}2/3 & -1/3\\1/3 & 1/3\end{bmatrix}, \quad \text{and} \quad \mathbf{D} = \begin{bmatrix}-1 & 0\\0 & 5\end{bmatrix}.$$

It follows first from Equation (4.19) and then from (4.16) that

$$\begin{aligned}e^{\mathbf{A}} = \mathbf{M}e^{\mathbf{D}}\mathbf{M}^{-1} &= \begin{bmatrix}1 & 1\\-1 & 2\end{bmatrix}\begin{bmatrix}e^{-1} & 0\\0 & e^5\end{bmatrix}\begin{bmatrix}2/3 & -1/3\\1/3 & 1/3\end{bmatrix}\\ &= \frac{1}{3}\begin{bmatrix}2e^{-1}+e^5 & -e^{-1}+e^5\\-2e^{-1}+2e^5 & e^{-1}+2e^5\end{bmatrix}\end{aligned}$$

If **A** is similar to a matrix **J** in Jordan canonical form, so that $\mathbf{A} = \mathbf{MJM}^{-1}$ for a generalized modal matrix **M**, then $e^{\mathbf{A}} = \mathbf{M}e^{\mathbf{J}}\mathbf{M}^{-1}$ (see Appendix A).

Even if a matrix **A** is not diagonalizable, it is still similar to a matrix **J** in Jordan canonical form (see Appendix B). That is, there exists a generalized modal matrix **M** such that

$$\mathbf{A} = \mathbf{MJM}^{-1} \tag{4.20}$$

Repeating the derivation of (4.18) and (4.19), with **J** replacing **D**, we obtain

$$e^{\mathbf{A}} = \mathbf{M}e^{\mathbf{J}}\mathbf{M}^{-1} \tag{4.21}$$

Thus, once we know how to calculate $e^{\mathbf{J}}$ for a matrix **J** in Jordan canonical form, we can use Equation (4.21) to find $e^{\mathbf{A}}$ for any square matrix **A**.

A matrix **J** in Jordan canonical form has the block diagonal pattern

$$\mathbf{J} = \begin{bmatrix} J_1 & 0 & \dots & 0 \\ 0 & J_2 & \dots & 0 \\ \vdots & \vdots & \ddots & \vdots \\ 0 & 0 & \dots & J_r \end{bmatrix} \tag{4.22}$$

with each $\mathbf{J}_i(i=1, 2, \dots, r)$ being a Jordan block of the form

$$\mathbf{J}_i = \begin{bmatrix} \lambda_i & 1 & 0 & \dots & 0 & 0 \\ 0 & \lambda_i & 1 & \dots & 0 & 0 \\ \vdots & \vdots & \vdots & \ddots & \ddots & \ddots \\ 0 & 0 & 0 & \dots & \lambda_i & 1 \\ 0 & 0 & 0 & \dots & 0 & \lambda_i \end{bmatrix} \tag{4.23}$$

Powers of a matrix in Jordan canonical form are relatively easy to calculate.

$$\mathbf{J}^2 = \mathbf{J}\mathbf{J} = \begin{bmatrix} J_1 & 0 & \cdots & 0 \\ 0 & J_2 & \cdots & 0 \\ \vdots & \vdots & \ddots & \vdots \\ 0 & 0 & \cdots & J_r \end{bmatrix} \begin{bmatrix} J_1 & 0 & \cdots & 0 \\ 0 & J_2 & \cdots & 0 \\ \vdots & \vdots & \ddots & \vdots \\ 0 & 0 & \cdots & J_r \end{bmatrix} = \begin{bmatrix} J_1^2 & 0 & \cdots & 0 \\ 0 & J_2^2 & \cdots & 0 \\ \vdots & \vdots & \ddots & \vdots \\ 0 & 0 & \cdots & J_r^2 \end{bmatrix}$$

$$\mathbf{J}^3 = \mathbf{J}\mathbf{J}^2 = \begin{bmatrix} J_1 & 0 & \dots & 0 \\ 0 & J_2 & \dots & 0 \\ \vdots & \vdots & \ddots & \vdots \\ 0 & 0 & \dots & J_r \end{bmatrix} \begin{bmatrix} J_1^2 & 0 & \dots & 0 \\ 0 & J_2^2 & \dots & 0 \\ \vdots & \vdots & \ddots & \vdots \\ 0 & 0 & \dots & J_r^2 \end{bmatrix} = \begin{bmatrix} J_1^3 & 0 & \dots & 0 \\ 0 & J_2^3 & \dots & 0 \\ \vdots & \vdots & \ddots & \vdots \\ 0 & 0 & \dots & J_r^3 \end{bmatrix}$$

and, in general,

$$\mathbf{J}^k = \begin{bmatrix} J_1^k & 0 & \dots & 0 \\ 0 & J_2^k & \dots & 0 \\ \vdots & \vdots & \ddots & \vdots \\ 0 & 0 & \dots & J_r^k \end{bmatrix}$$

for any positive integer value of k. Consequently,

$$e^{\mathbf{J}} = \sum_{k=0}^{\infty} \frac{J^k}{k!} = \sum_{k=0}^{\infty} \frac{1}{k!} \begin{bmatrix} J_1^k & 0 & \dots & 0 \\ 0 & J_2^k & \dots & 0 \\ \vdots & \vdots & \ddots & \vdots \\ 0 & 0 & \dots & J_r^k \end{bmatrix} = \begin{bmatrix} \sum_{k=0}^{\infty} \frac{J_1^k}{k!} & 0 & \dots & 0 \\ 0 & \sum_{k=0}^{\infty} \frac{J_2^k}{k!} & \dots & 0 \\ \vdots & \vdots & \ddots & \vdots \\ 0 & 0 & \dots & \sum_{k=0}^{\infty} \frac{J_r^k}{k!} \end{bmatrix} \tag{4.24}$$

$$= \begin{bmatrix} e^{J_1} & 0 & \dots & 0 \\ 0 & e^{J_2} & \dots & 0 \\ \vdots & \vdots & \ddots & \vdots \\ 0 & 0 & \dots & e^{J_r} \end{bmatrix}$$

Thus, once we know how to calculate the exponential of a Jordan block, we can use Equation (4.24) to find $e^{\mathbf{J}}$ for a matrix $\mathbf{J}$ in Jordan canonical form and then Equation (4.21) to obtain $e^{\mathbf{A}}$ for a square matrix $\mathbf{A}$.

A 1×1 Jordan block has the form $[\lambda]$ for some scalar λ. Such a matrix is a diagonal matrix, indeed all 1×1 matrices are, by default, diagonal matrices, and it follows directly from Equation (4.16) that $e^{[\lambda]} = [e^{\lambda}]$. All other Jordan blocks have superdiagonal elements, which are all ones. For $p \times p$ Jordan block in the form of Equation (4.23), we can show by direct calculations that each successive power has one additional diagonal of nonzero entries, until all elements above the main diagonal become nonzero. On each diagonal, the entries are identical. If we designate the nth power of a Jordan block as the matrix $[a_{ij}^n]$, then the entries can be expressed compactly in terms of derivatives as

$$a_{i,\, i+j}^n = \begin{cases} \frac{1}{j!} \frac{d^j}{d\lambda_i^j} (\lambda_i^n) & \text{for } j = 0, 1, \dots, n \\ 0 & \text{otherwise} \end{cases}$$

The exponential of a matrix in Jordan canonical form (Equation 4.22) has block diagonal form (Equation 4.24), with the exponential of each Jordan block given by Equation (4.25).

Equation (4.15) then reduces to

$$e^{J_i} = e^{\lambda_i} \begin{bmatrix} 1 & \frac{1}{1!} & \frac{1}{2!} & \frac{1}{3!} & \dots & \frac{1}{(p-1)!} \\ 0 & 1 & \frac{1}{1!} & \frac{1}{2!} & \dots & \frac{1}{(p-2)!} \\ 0 & 0 & 1 & \frac{1}{1!} & \dots & \frac{1}{(p-3)!} \\ \dots & \dots & \dots & \dots & \dots & \dots \\ 0 & 0 & 0 & 0 & \dots & 1 \end{bmatrix} \tag{4.25}$$

Example 4 Find $e^{\mathbf{J}}$ for $\mathbf{J} = \begin{bmatrix} 2t & 1 & 0 \\ 0 & 2t & 1 \\ 0 & 0 & 2t \end{bmatrix}$.

Solution: **J** is a single Jordan block with diagonal elements $\lambda_i = 2t$. For this matrix, Equation (4.25) becomes

$$e^{\mathbf{J}} = e^{2t} \begin{bmatrix} 1 & 1 & 1/2 \\ 0 & 1 & 1 \\ 0 & 0 & 1 \end{bmatrix}$$

Example 5 Find $e^{\mathbf{J}}$ for $\mathbf{J} = \begin{bmatrix} 2 & 0 & 0 & 0 & 0 & 0 \\ 0 & 3 & 0 & 0 & 0 & 0 \\ 0 & 0 & 1 & 1 & 0 & 0 \\ 0 & 0 & 0 & 1 & 1 & 0 \\ 0 & 0 & 0 & 0 & 1 & 1 \\ 0 & 0 & 0 & 0 & 0 & 1 \end{bmatrix}$

Solution: **J** is in the Jordan canonical form

$$\mathbf{J} = \begin{bmatrix} \mathbf{J}_1 & 0 & 0 \\ 0 & \mathbf{J}_2 & 0 \\ 0 & 0 & \mathbf{J}_3 \end{bmatrix}$$

with $\mathbf{J}_1 = [2]$ and $\mathbf{J}_2 = [3]$ both of order 1×1, and

$$\mathbf{J}_3 = \begin{bmatrix} 1 & 1 & 0 & 0 \\ 0 & 1 & 1 & 0 \\ 0 & 0 & 1 & 1 \\ 0 & 0 & 0 & 1 \end{bmatrix}$$

Here,

$$e^{J_1} = \left[e^2\right], \quad e^{J_2} = \left[e^3\right], \quad \text{and}$$

$$e^{J_3} = e^1 \begin{bmatrix} 1 & 1 & 1/2 & 1/6 \\ 0 & 1 & 1 & 1/2 \\ 0 & 0 & 1 & 1 \\ 0 & 0 & 0 & 1 \end{bmatrix} = \begin{bmatrix} e & e & e/2 & e/6 \\ 0 & e & e & e/2 \\ 0 & 0 & e & e \\ 0 & 0 & 0 & e \end{bmatrix}$$

Then,

$$e^{\mathbf{J}} = \begin{bmatrix} e^2 & 0 & 0 & 0 & 0 & 0 \\ 0 & e^3 & 0 & 0 & 0 & 0 \\ 0 & 0 & e & e & e/2 & e/6 \\ 0 & 0 & 0 & e & e & e/2 \\ 0 & 0 & 0 & 0 & e & e \\ 0 & 0 & 0 & 0 & 0 & e \end{bmatrix}$$

Example 6 Find $e^{\mathbf{A}}$ for $\mathbf{A} = \begin{bmatrix} 0 & 4 & 2 \\ -3 & 8 & 3 \\ 4 & -8 & -2 \end{bmatrix}$.

Solution: A canonical basis for this matrix has one chain of length 2: $\mathbf{x}_2 = [0 \;\; 0 \;\; 1]^T$ and $\mathbf{x}_1 = [2 \;\; 3 \;\; -4]^T$, and one chain of length 1: $\mathbf{y}_1 = [2 \;\; 1 \;\; 0]^T$, each corresponding to the eigenvalue 2. Setting

$$\mathbf{M} = \begin{bmatrix} 2 & 2 & 0 \\ 1 & 3 & 0 \\ 0 & -4 & 1 \end{bmatrix} \quad \text{and} \quad \mathbf{J} = \begin{bmatrix} 2 & 0 & 0 \\ 0 & 2 & 1 \\ 0 & 0 & 2 \end{bmatrix}$$

we have $\mathbf{A} = \mathbf{MJM}^{-1}$. Here $\mathbf{J}$ contains two Jordan blocks, the 1×1 matrix $\mathbf{J}_1 = [2]$ and the 2×2 matrix $\mathbf{J}_2 = \begin{bmatrix} 2 & 1 \\ 0 & 2 \end{bmatrix}$. We have,

$$e^{J_1} = [e^2], e^{J_2} = e^2 \begin{bmatrix} 1 & 1 \\ 0 & 1 \end{bmatrix} = \begin{bmatrix} e^2 & e^2 \\ 0 & e^2 \end{bmatrix}$$

$$e^{\mathbf{J}} = 65 = \begin{bmatrix} e^2 & 0 & 0 \\ 0 & e^2 & e^2 \\ 0 & 0 & e^2 \end{bmatrix}$$

$$e^{\mathbf{A}} = \mathbf{M}e^{\mathbf{J}}\mathbf{M}^{-1} = \begin{bmatrix} 2 & 2 & 0 \\ 1 & 3 & 0 \\ 0 & -4 & 1 \end{bmatrix} \begin{bmatrix} e^2 & 0 & 0 \\ 0 & e^2 & e^2 \\ 0 & 0 & e^2 \end{bmatrix} \begin{bmatrix} 3/4 & -1/2 & 0 \\ -1/4 & 1/2 & 0 \\ -1 & 2 & 1 \end{bmatrix}$$

$$= e^2 \begin{bmatrix} -1 & 4 & 2 \\ -3 & 7 & 3 \\ 4 & -8 & -3 \end{bmatrix}$$

Two important properties of the exponential of a matrix are given in the next theorems.

> **▸THEOREM 1**
>
> $e^{\mathbf{0}} = \mathbf{I}$, *where* $\mathbf{0}$ *is the* $n \times n$ *zero matrix and* $\mathbf{I}$ *is the* $n \times n$ *identity matrix.*◂

Proof: In general,

$$e^{\mathbf{A}} = \sum_{k=0}^{\infty} \frac{A^k}{k!} = I + \sum_{k=1}^{\infty} \frac{A^k}{k!} \tag{4.26}$$

With $\mathbf{A} = \mathbf{0}$, we have

$$e^0 = I + \sum_{k=1}^{\infty} \frac{0^k}{k!} = \mathbf{I}$$

> **▸THEOREM 2**
>
> $(e^{\mathbf{A}})^{-1} = e^{-\mathbf{A}}$.◂

Proof:

$$e^{\mathbf{A}}e^{-\mathbf{A}} = \left[\sum_{k=0}^{\infty}\frac{A^k}{k!}\right]\left[\sum_{k=0}^{\infty}\frac{(-A)^k}{k!}\right]$$

$$= \left[I + A + \frac{A^2}{2!} + \frac{A^3}{3!} + \cdots\right]\left[I + A + \frac{A^2}{2!} + \frac{A^3}{3!} + \cdots\right]$$

$$= \mathbf{II} + \mathbf{A}[1-1] + \mathbf{A}^2\left[\frac{1}{2!} - 1 + \frac{1}{2!}\right] + \mathbf{A}^3\left[-\frac{1}{3!} + \frac{1}{2!} - \frac{1}{2!} + \frac{1}{3!}\right] + \cdots$$

$$= \mathbf{I}$$

Thus, $e^{-\mathbf{A}}$ is the inverse of $e^{\mathbf{A}}$.

To calculate $e^{\mathbf{A}t}$, where **A** is a square constant matrix and *t* is a variable, set $\mathbf{B} = \mathbf{A}t$ and calculate $e^{\mathbf{B}}$.

We conclude from Theorem 2 that $e^{\mathbf{A}}$ is *always invertible* even when **A** is not. To calculate $e^{-\mathbf{A}}$ directly, set $\mathbf{B} = -\mathbf{A}$, and then determine $e^{\mathbf{B}}$.

A particularly useful matrix function for solving differential equations is $e^{\mathbf{A}t}$, where **A** is a square constant matrix (that is, all of its elements are constants) and *t* is a variable, usually denoting time. This function may be obtained directly by setting $\mathbf{B} = \mathbf{A}t$ and then calculating $e^{\mathbf{B}}$.

Example 7 Find $e^{\mathbf{A}t}$ for $\mathbf{A} = \begin{bmatrix} 3 & 0 & 4 \\ 1 & 2 & 1 \\ -1 & 0 & -2 \end{bmatrix}$.

Solution: Set $\mathbf{B} = \mathbf{A}t = \begin{bmatrix} 3t & 0 & 4t \\ t & 2t & t \\ -t & 0 & -2t \end{bmatrix}$.

A canonical basis for **B** contains one chain of length 1, corresponding to the eigenvalue $-t$ of multiplicity 1, and one chain of length 2, corresponding to the eigenvalue $2t$ of multiplicity 2. A generalized modal matrix for **B** is

$$\mathbf{M} = \begin{bmatrix} 1 & 0 & 4 \\ 0 & 3t & 0 \\ -1 & 0 & -1 \end{bmatrix}$$

Then,

$$\mathbf{J} = \mathbf{M}^{-1}\mathbf{B}\mathbf{M} = \begin{bmatrix} -t & 0 & 0 \\ 0 & 2t & 1 \\ 0 & 0 & 2t \end{bmatrix}, \; e^{\mathbf{J}} = \begin{bmatrix} e^{-t} & 0 & 0 \\ 0 & e^{2t} & e^{2t} \\ 0 & 0 & e^{2t} \end{bmatrix}$$

$$e^{\mathbf{A}t} = e^{\mathbf{B}} = \mathbf{M}e^{\mathbf{J}}\mathbf{M}^{-1} = \begin{bmatrix} 1 & 0 & 4 \\ 0 & 3t & 0 \\ -1 & 0 & -1 \end{bmatrix}\begin{bmatrix} e^{-t} & 0 & 0 \\ 0 & e^{2t} & e^{2t} \\ 0 & 0 & e^{2t} \end{bmatrix}\begin{bmatrix} -1/3 & 0 & -4/3 \\ 0 & 1/(3t) & 0 \\ 1/3 & 0 & 1/3 \end{bmatrix}$$

$$= \frac{1}{3}\begin{bmatrix} -e^{-t} + 4e^{2t} & 0 & -4e^{-t} + 4e^{2t} \\ 3te^{2t} & 3e^{2t} & 3te^{2t} \\ e^{-t} - e^{2t} & 0 & 4e^{-t} - e^{2t} \end{bmatrix}$$

Observe that this derivation may not be valid for $t=0$ because $\mathbf{M}^{-1}$ is undefined there. Considering the case $t=0$ separately, we find that $e^{\mathbf{A}0}=e^{0}=\mathbf{I}$. Our answer also reduces to the identity matrix at $t=0$, so our answer is correct for all t.

The roots of the characteristic equation of $\mathbf{B}=\mathbf{A}t$ may be complex. As noted in Section 4.1, such a root is *not* an eigenvalue when the underlying vector space is $\mathbb{R}^n$, because there is no corresponding eigenvector with real-valued components. Complex roots of a characteristic equation *are* eigenvalues when the underlying vector space is the set of all n-tuples with complex-valued components. When calculating matrix exponentials, it is convenient to take the underlying vector space to be complex-valued n-tuples and to accept each root of a characteristic equation as an eigenvalue. Consequently, a generalized modal matrix $\mathbf{M}$ may contain complex-valued elements.

If $\mathbf{A}$ is a real matrix and t a real-valued variable, then $\mathbf{B}t$ is real-valued. Because all integral powers of matrices with real elements must also be real, it follows from Equation (4.26) that $e^{\mathbf{B}}$ must be real. Thus, even if $\mathbf{J}$ and $\mathbf{M}$ have complex-valued elements, the product $e^{\mathbf{B}}=\mathbf{M}e^{\mathbf{J}}\mathbf{M}^{-1}$ must be real. Complex roots of the characteristic equation of a real matrix must appear in conjugate pairs, which often can be combined into real-valued quantities by using Euler's relations:

$$\cos\theta = \frac{e^{i\theta}+e^{-i\theta}}{2} \quad \text{and} \quad \sin\theta = \frac{e^{i\theta}-e^{-i\theta}}{2i}$$

Example 8 Find $e^{\mathbf{A}t}$ for $\mathbf{A} = \begin{bmatrix} 0 & 1 \\ -1 & 0 \end{bmatrix}$.

Solution: Set $\mathbf{B} = \mathbf{A}t = \begin{bmatrix} 0 & t \\ -t & 0 \end{bmatrix}$.

The eigenvalues of $\mathbf{B}$ are $\lambda_1=it$ and $\lambda_2=-it$, with corresponding eigenvectors $[1 \;\; i]^{\mathrm{T}}$ and $[1 \;\; -i]^{\mathrm{T}}$, respectively. Thus,

$$\mathbf{M} = \begin{bmatrix} 1 & 1 \\ i & -i \end{bmatrix}, \quad \mathbf{J} = \begin{bmatrix} it & 0 \\ 0 & -it \end{bmatrix}$$

and

$$e^{\mathbf{A}t} = e^{\mathbf{B}} = \begin{bmatrix} 1 & 1 \\ i & -i \end{bmatrix} \begin{bmatrix} e^{it} & 0 \\ 0 & e^{-it} \end{bmatrix} \begin{bmatrix} 1/2 & -i/2 \\ 1/2 & i/2 \end{bmatrix}$$

$$= \begin{bmatrix} \dfrac{e^{it}+e^{-it}}{2} & \dfrac{e^{it}-e^{-it}}{2i} \\[2ex] -\dfrac{e^{it}-e^{-it}}{2i} & \dfrac{e^{it}+e^{-it}}{2} \end{bmatrix} = \begin{bmatrix} \cos t & \sin t \\ -\sin t & \cos t \end{bmatrix}$$

If the eigenvalues of $\mathbf{B}=\mathbf{A}t$ are not pure imaginary but rather complex numbers of the form $\beta+i\theta$ and $\beta-i\theta$, then the algebraic operations needed to simplify $e^{\mathbf{B}}$

are more tedious. Euler's relations remain applicable, but as part of the following identities:

$$\frac{e^{\beta+i\theta}+e^{\beta-i\theta}}{2}=\frac{e^{\beta}e^{i\theta}+e^{\beta}e^{-i\theta}}{2}=\frac{e^{\beta}\left(e^{i\theta}+e^{-i\theta}\right)}{2}=e^{\beta}\cos\theta$$

and

$$\frac{e^{\beta+i\theta}-e^{\beta-i\theta}}{2i}=\frac{e^{\beta}e^{i\theta}-e^{\beta}e^{-i\theta}}{2i}=\frac{e^{\beta}\left(e^{i\theta}-e^{-i\theta}\right)}{2i}=e^{\beta}\sin\theta$$

The exponential of a matrix is useful in matrix calculus for the same reason the exponential function is so valuable in the calculus: the derivative of $e^{\mathbf{A}t}$ is closely related to the function itself. The *derivative of a matrix* is obtained by differentiating each element in the matrix. Thus, a matrix $\mathbf{C}=[c_{ij}]$ has a derivative if and only if each element c_{ij} has a derivative, in which case, we write

The derivative of a matrix is obtained by differentiating each element in the matrix.

$$\dot{\mathbf{C}}(t)=\frac{d\mathbf{C}(t)}{dt}=\left[\frac{dc_{ij}(t)}{dt}\right] \tag{427}$$

Example 9 If $\mathbf{C}(t)=\begin{bmatrix} t^2 & \sin t \\ \ln t & e^{t^2}\end{bmatrix}$, then

$$\dot{\mathbf{C}}(t)=\frac{d\mathbf{C}(t)}{dt}=\begin{bmatrix}\dfrac{d(t^2)}{dt} & \dfrac{d(\sin t)}{dt} \\ \dfrac{d(\ln t)}{dt} & \dfrac{d\left(e^{t^2}\right)}{dt}\end{bmatrix}=\begin{bmatrix} 2t & \cos t \\ 1/t & 2te^{t^2}\end{bmatrix}$$

▶THEOREM 3

*If **A** is a constant matrix, then* $\frac{de^{\mathbf{A}t}}{dt}=\mathbf{A}e^{\mathbf{A}t}=e^{\mathbf{A}t}\mathbf{A}$. ◀

Proof:

$$\frac{de^{\mathbf{A}t}}{dt}=\frac{d}{dt}\left(\sum_{k=0}^{\infty}\frac{(\mathbf{A}t)^k}{k!}\right)=\frac{d}{dt}\left(\sum_{k=0}^{\infty}\frac{\mathbf{A}^k t^k}{k!}\right)=\sum_{k=0}^{\infty}\frac{d}{dt}\left(\frac{\mathbf{A}^k t^k}{k!}\right)=\sum_{k=0}^{\infty}\frac{k\mathbf{A}^k t^{k-1}}{k!}$$

$$=0+\sum_{k=1}^{\infty}\frac{\mathbf{A}\mathbf{A}^{k-1}t^{k-1}}{(k-1)!}=\mathbf{A}\left(\sum_{k=1}^{\infty}\frac{\mathbf{A}^{k-1}t^{k-1}}{(k-1)!}\right)$$

$$=\mathbf{A}\left(\sum_{j=0}^{\infty}\frac{\mathbf{A}^j t^j}{j!}\right)=\mathbf{A}\left(\sum_{j=0}^{\infty}\frac{(\mathbf{A}t)^j}{j!}\right)=\mathbf{A}e^{\mathbf{A}t}$$

If we factor **A** on the right, instead of the left, we obtain the other identity.

By replacing **A** with −**A** in Theorem 3, we obtain:

> ▶**COROLLARY 1**
>
> If **A** is a constant matrix, then $\dfrac{de^{-\mathbf{A}t}}{dt} = -\mathbf{A}e^{\mathbf{A}t} = -e^{\mathbf{A}t}\mathbf{A}$. ◀

Problems 4.4

In Problems 1 through 29, find the exponential of each matrix.

(1) $\begin{bmatrix} -1 & 0 \\ 0 & 4 \end{bmatrix}$. **(2)** $\begin{bmatrix} 2 & 0 \\ 0 & 3 \end{bmatrix}$. **(3)** $\begin{bmatrix} -7 & 0 \\ 0 & -7 \end{bmatrix}$.

(4) $\begin{bmatrix} 0 & 0 \\ 0 & 0 \end{bmatrix}$. **(5)** $\begin{bmatrix} -7 & 1 \\ 0 & -7 \end{bmatrix}$. **(6)** $\begin{bmatrix} 2 & 1 \\ 0 & 2 \end{bmatrix}$.

(7) $\begin{bmatrix} 3 & 1 \\ 0 & 3 \end{bmatrix}$. **(8)** $\begin{bmatrix} 0 & 1 \\ 0 & 0 \end{bmatrix}$. **(9)** $\begin{bmatrix} 2 & 0 & 0 \\ 0 & 3 & 0 \\ 0 & 0 & 4 \end{bmatrix}$.

(10) $\begin{bmatrix} 1 & 0 & 0 \\ 0 & -5 & 0 \\ 0 & 0 & -1 \end{bmatrix}$. **(11)** $\begin{bmatrix} 2 & 0 & 0 \\ 0 & 2 & 0 \\ 0 & 0 & 2 \end{bmatrix}$. **(12)** $\begin{bmatrix} 2 & 1 & 0 \\ 0 & 2 & 1 \\ 0 & 0 & 2 \end{bmatrix}$.

(13) $\begin{bmatrix} -1 & 1 & 0 \\ 0 & -1 & 1 \\ 0 & 0 & -1 \end{bmatrix}$. **(14)** $\begin{bmatrix} 0 & 1 & 0 \\ 0 & 0 & 1 \\ 0 & 0 & 0 \end{bmatrix}$. **(15)** $\begin{bmatrix} -1 & 0 & 0 \\ 0 & -1 & 1 \\ 0 & 0 & -1 \end{bmatrix}$.

(16) $\begin{bmatrix} 2 & 0 & 0 \\ 0 & 2 & 1 \\ 0 & 0 & 2 \end{bmatrix}$. **(17)** $\begin{bmatrix} 1 & 0 & 0 & 0 \\ 0 & 5 & 0 & 0 \\ 0 & 0 & -5 & 0 \\ 0 & 0 & 0 & 3 \end{bmatrix}$. **(18)** $\begin{bmatrix} -5 & 0 & 0 & 0 \\ 0 & -5 & 0 & 0 \\ 0 & 0 & -5 & 0 \\ 0 & 0 & 0 & -5 \end{bmatrix}$.

(19) $\begin{bmatrix} -5 & 0 & 0 & 0 \\ 0 & -5 & 0 & 0 \\ 0 & 0 & -5 & 1 \\ 0 & 0 & 0 & -5 \end{bmatrix}$. **(20)** $\begin{bmatrix} -5 & 0 & 0 & 0 \\ 0 & -5 & 1 & 0 \\ 0 & 0 & -5 & 1 \\ 0 & 0 & 0 & -5 \end{bmatrix}$.

(21) $\begin{bmatrix} -5 & 1 & 0 & 0 \\ 0 & -5 & 0 & 0 \\ 0 & 0 & -5 & 1 \\ 0 & 0 & 0 & -5 \end{bmatrix}$. **(22)** $\begin{bmatrix} -5 & 1 & 0 & 0 \\ 0 & -5 & 1 & 0 \\ 0 & 0 & -5 & 1 \\ 0 & 0 & 0 & -5 \end{bmatrix}$.

(23) $\begin{bmatrix} 2 & 0 & 1 \\ 0 & 2 & 0 \\ 0 & 0 & 2 \end{bmatrix}$. **(24)** $\begin{bmatrix} 1 & 3 \\ 4 & 2 \end{bmatrix}$. **(25)** $\begin{bmatrix} 4 & -1 \\ 1 & 2 \end{bmatrix}$.

(26) $\begin{bmatrix} 1 & 1 & 2 \\ -1 & 3 & 4 \\ 0 & 0 & 2 \end{bmatrix}$. **(27)** $\begin{bmatrix} \pi & \pi/3 & -\pi \\ 0 & \pi & \pi/2 \\ 0 & 0 & \pi \end{bmatrix}$. **(28)** $\begin{bmatrix} 2 & 1 & 0 \\ 0 & 2 & 2 \\ 0 & 0 & 2 \end{bmatrix}$.

(29) $\begin{bmatrix} 1 & 0 & 0 \\ 2 & 3 & -1 \\ 1 & 1 & 1 \end{bmatrix}$.

(30) Verify Theorem 2 for $\mathbf{A} = \begin{bmatrix} 1 & 3 \\ 0 & 1 \end{bmatrix}$.

(31) Verify Theorem 2 for $\mathbf{A} = \begin{bmatrix} 0 & 1 \\ -64 & 0 \end{bmatrix}$.

(32) Verify Theorem 2 for $\mathbf{A} = \begin{bmatrix} 0 & 1 & 0 \\ 0 & 0 & 1 \\ 0 & 0 & 0 \end{bmatrix}$. What is the inverse of $\mathbf{A}$?

(33) Find $e^{\mathbf{A}}e^{\mathbf{B}}$, $e^{\mathbf{B}}e^{\mathbf{A}}$, and $e^{\mathbf{A}+\mathbf{B}}$ when

$$\mathbf{A} = \begin{bmatrix} 1 & 1 \\ 0 & 0 \end{bmatrix} \quad \text{and} \quad \mathbf{B} = \begin{bmatrix} 0 & 1 \\ 0 & 1 \end{bmatrix},$$

and show that $e^{\mathbf{A}+\mathbf{B}} \neq e^{\mathbf{A}}e^{\mathbf{B}} \neq e^{\mathbf{B}}e^{\mathbf{A}}$.

(34) Find two matrices $\mathbf{A}$ and $\mathbf{B}$ such that $e^{\mathbf{A}}e^{\mathbf{B}} = e^{\mathbf{A}+\mathbf{B}}$.

(35) Using Equation (4.15) directly, prove that $e^{\mathbf{A}}e^{\mathbf{B}} = e^{\mathbf{A}+\mathbf{B}}$ when $\mathbf{A}$ and $\mathbf{B}$ commute.

In Problems 36 through 54, find $e^{\mathbf{A}t}$ for the given matrix $\mathbf{A}$.

(36) $\begin{bmatrix} 4 & 4 \\ 3 & 5 \end{bmatrix}$.

(37) $\begin{bmatrix} 2 & 1 \\ -1 & -2 \end{bmatrix}$.

(38) $\begin{bmatrix} 4 & 1 \\ -1 & 2 \end{bmatrix}$.

(39) $\begin{bmatrix} 0 & 1 \\ -14 & 9 \end{bmatrix}$.

(40) $\begin{bmatrix} -3 & 2 \\ 2 & -6 \end{bmatrix}$.

(41) $\begin{bmatrix} -10 & 6 \\ 6 & -10 \end{bmatrix}$.

(42) $\begin{bmatrix} 2 & 1 \\ 0 & 2 \end{bmatrix}$.

(43) $\begin{bmatrix} 0 & 1 & 0 \\ 0 & 0 & 1 \\ 0 & 0 & 0 \end{bmatrix}$.

(44) $\begin{bmatrix} 1 & 0 & 0 \\ 4 & 1 & 2 \\ -1 & 4 & -1 \end{bmatrix}$.

(45) $\begin{bmatrix} -1 & 1 & 0 \\ 0 & -1 & 1 \\ 0 & 0 & -1 \end{bmatrix}$.

(46) $\begin{bmatrix} 4 & 1 & 0 \\ 0 & 4 & 0 \\ 0 & 0 & 4 \end{bmatrix}$.

(47) $\begin{bmatrix} 2 & 1 & 0 \\ 0 & 2 & 0 \\ 0 & 0 & -1 \end{bmatrix}$.

(48) $\begin{bmatrix} 2 & 3 & 0 \\ -1 & -2 & 0 \\ 1 & 1 & 1 \end{bmatrix}$.

(49) $\begin{bmatrix} 3 & 1 & 0 \\ -1 & 1 & 0 \\ 1 & 2 & 2 \end{bmatrix}$.

(50) $\begin{bmatrix} 5 & -2 & 2 \\ 2 & 0 & 1 \\ -7 & 5 & -2 \end{bmatrix}$.

(51) $\begin{bmatrix} 0 & 1 \\ -64 & 0 \end{bmatrix}$.

(52) $\begin{bmatrix} 2 & 5 \\ -1 & -2 \end{bmatrix}$.

(53) $\begin{bmatrix} 0 & 1 \\ -25 & -8 \end{bmatrix}$.

(54) $\begin{bmatrix} 3 & 1 \\ -2 & 5 \end{bmatrix}$.

(55) Verify Theorem 3 for the matrix $\mathbf{A}$ given in Example 7.

(56) Verify Theorem 3 for the matrix $\mathbf{A}$ given in Example 8.

(57) Using the formula

$$\frac{d[A(t)B(t)]}{dt} = \left(\frac{dA(t)}{dt}\right)B(t) + A(t)\left(\frac{dB(t)}{dt}\right),$$

derive a formula for differentiating $\mathbf{A}^2(t)$. Use this formula to find $d\mathbf{A}^2(t)/dt$ when

$$\mathbf{A}(t) = \begin{bmatrix} t & 2t^2 \\ 4t^3 & e^t \end{bmatrix},$$

and show that $d\mathbf{A}^2(t)/dt \neq 2\mathbf{A}(t)d\mathbf{A}(t)/dt$. Therefore, the power rule of differentiation does *not* hold for matrices unless a matrix commutes with its derivative.

4.5 POWER METHODS

The analytic methods described in Section 4.1 are impractical for calculating the eigenvalues and eigenvectors of matrices of large order. Determining the characteristic equations for such matrices involves enormous effort, and finding its roots algebraically is usually impossible. Instead, iterative methods that lend themselves to computer implementation are used. Ideally, each iteration yields a new approximation, which converges to an eigenvalue and the corresponding eigenvector.

The dominant eigenvalue of a matrix is the one having the largest absolute value.

The *dominant* eigenvalue of a matrix is the eigenvalue with the largest absolute value. Thus, if the eigenvalues of a matrix are 2, 5, and −13, then −13 is the dominant eigenvalue because it is the largest in absolute value. The *power method* is an algorithm for locating the dominant eigenvalue and a corresponding eigenvector for a matrix of real numbers when the following two conditions exist:

Condition 1. The dominant eigenvalue of a matrix is real (not complex) and is strictly greater in absolute value than all other eigenvalues.

Condition 2. If the matrix has order $n \times n$, then it possesses n linearly independent eigenvectors.

Denote the eigenvalues of a given square matrix $\mathbf{A}$ satisfying Conditions 1 and 2 by $\lambda_1, \lambda_2, \ldots, \lambda_n$, and a set of corresponding eigenvectors by $\mathbf{v}_1, \mathbf{v}_2, \ldots, \mathbf{v}_n$, respectively. Assume the indexing is such that

$$|\lambda_1| > |\lambda_2| \geq |\lambda_3| \geq \cdots \geq |\lambda_n|$$

Any vector $\mathbf{x}_0$ can be expressed as a linear combination of the eigenvectors of $\mathbf{A}$, so we may write

$$\mathbf{x}_0 = c_1\mathbf{v}_1 + c_2\mathbf{v}_2 + \cdots + c_n\mathbf{v}_n$$

Multiplying this equation by $\mathbf{A}^k$, for some large, positive integer k, we get

$$\begin{aligned} \mathbf{A}^k\mathbf{x}_0 &= \mathbf{A}^k(c_1\mathbf{v}_1 + c_2\mathbf{v}_2 + \cdots + c_n\mathbf{v}_n) \\ &= c_1\mathbf{A}^k\mathbf{v}_1 + c_2\mathbf{A}^k\mathbf{v}_2 + \cdots + c_n\mathbf{A}^k\mathbf{v}_n \end{aligned}$$

It follows from Theorem 8 of Section 4.2 that

$$\begin{aligned}\mathbf{A}^k\mathbf{x}_0 &= c_1\lambda_1^k\mathbf{v}_1 + c_1\lambda_2^k\mathbf{v}_2 + \cdots + c_n\lambda_n^k\mathbf{v}_n \\ &= \lambda_1^k\left[c_1\mathbf{v}_1 + c_2\left(\frac{\lambda_2}{\lambda_1}\right)^k\mathbf{v}_2 + \cdots + c_n\left(\frac{\lambda_n}{\lambda_1}\right)^k\mathbf{v}_n\right] \\ &\approx \lambda_1^k c_1\mathbf{v}_1 \quad \text{for large } k\end{aligned}$$

This last pseudo-equality follows from noting that each quotient of eigenvalues is less than unity in absolute value, as a result of indexing the first eigenvalue as the dominant one, and therefore tends to 0 as that quotient is raised to successively higher powers.

Thus, $\mathbf{A}^k\mathbf{x}_0$ approaches a scalar multiple of $\mathbf{v}_1$. But any nonzero scalar multiple of an eigenvector is itself an eigenvector, so $\mathbf{A}^k\mathbf{x}_0$ approaches a scalar multiple of $\mathbf{v}_1$, which is itself an eigenvector of **A** corresponding to the dominant eigenvalue, providing c_1 is not 0. The scalar c_1 will be 0 only if $\mathbf{x}_0$ is a linear combination of $\{\mathbf{v}_2, \mathbf{v}_3, \ldots, \mathbf{v}_n\}$.

The power method begins with an initial vector $\mathbf{x}_0$, usually the vector having all ones for its components, and then iteratively calculates the vectors

$$\begin{aligned}\mathbf{x}_1 &= \mathbf{A}\mathbf{x}_0 \\ \mathbf{x}_2 &= \mathbf{A}\mathbf{x}_1 = \mathbf{A}^2\mathbf{x}_0 \\ \mathbf{x}_3 &= \mathbf{A}\mathbf{x}_2 = \mathbf{A}^3\mathbf{x}_0 \\ &\vdots \\ \mathbf{x}_k &= \mathbf{A}\mathbf{x}_{k-1} = A^k\mathbf{x}_0\end{aligned}$$

As k gets larger, $\mathbf{x}_k$ approaches an eigenvector of **A** corresponding to its dominant eigenvalue.

THE POWER METHOD

Step 1. Begin with an initial guess $\mathbf{x}_0$ for an eigenvector of a matrix **A**, having the property that the largest component of $\mathbf{x}_0$ in absolute value is one. Set a counter k equal to 1.

Step 2. Calculate $\mathbf{x}_k = \mathbf{A}\mathbf{x}_{k-1}$.

Step 3. Set λ equal to the largest component of $\mathbf{x}_k$ in absolute value and use λ as an estimate for the dominant eigenvalue.

Step 4. Rescale $\mathbf{x}_k$ by dividing each of its components by λ. Relabel the resulting vector as $\mathbf{x}_k$.

Step 5. If λ is an adequate estimate for the dominant eigenvalue, with $\mathbf{x}_k$ as a corresponding eigenvector, stop; otherwise increment k by one and return to Step 2.

We can even determine the dominant eigenvalue. If k is large enough so the $\mathbf{x}_k$ is a good approximation to the eigenvector to within acceptable roundoff error, then it follows that $\mathbf{A}\mathbf{x}_k = \lambda_1 \mathbf{x}_k$. If $\mathbf{x}_k$ is scaled so that its largest component in absolute value is 1, then the component of $\mathbf{x}_{k+1} = \mathbf{A}\mathbf{x}_k = \lambda_1 \mathbf{x}_k$ that has the largest absolute value must be λ_1. We can now formalize the power method.

Example 1 Find the dominant eigenvalue and a corresponding eigenvector for

$$\mathbf{A} = \begin{bmatrix} 1 & 2 \\ 4 & 3 \end{bmatrix}$$

Solution: We initialize $\mathbf{x}_0 = [1 \quad 1]^{\mathrm{T}}$. Then, for the first iteration,

$$\mathbf{x}_1 = \mathbf{A}\mathbf{x}_0 = \begin{bmatrix} 1 & 2 \\ 4 & 3 \end{bmatrix} \begin{bmatrix} 1 \\ 1 \end{bmatrix} = \begin{bmatrix} 3 \\ 7 \end{bmatrix}$$

$$\lambda \approx 7$$

$$\mathbf{x}_1 \leftarrow \frac{1}{7}[3 \quad 7]^{\mathrm{T}} = [0.428571 \quad 1]^{\mathrm{T}}$$

For the second iteration,

$$\mathbf{x}_2 = \mathbf{A}\mathbf{x}_1 = \begin{bmatrix} 1 & 2 \\ 4 & 3 \end{bmatrix} \begin{bmatrix} 0.428571 \\ 1 \end{bmatrix} = \begin{bmatrix} 2.428571 \\ 4.714286 \end{bmatrix}$$

$$\lambda \approx 4.714286$$

$$\mathbf{x}_2 \leftarrow \frac{1}{4.714286}[2.428571 \quad 4.714286]^{\mathrm{T}} = [0.515152 \quad 1]^{\mathrm{T}}$$

For the third iteration,

$$\mathbf{x}_3 = \mathbf{A}\mathbf{x}_2 = \begin{bmatrix} 1 & 2 \\ 4 & 3 \end{bmatrix} \begin{bmatrix} 0.515152 \\ 1 \end{bmatrix} = \begin{bmatrix} 2.515152 \\ 5.060606 \end{bmatrix}$$

$$\lambda \approx 5.060606$$

$$\mathbf{x}_3 \leftarrow \frac{1}{5.060606}[2.515152 \quad 5.060606]^{\mathrm{T}} = [0.497006 \quad 1]^{\mathrm{T}}$$

For the fourth iteration,

$$\mathbf{x}_4 = \mathbf{A}\mathbf{x}_3 = \begin{bmatrix} 1 & 2 \\ 4 & 3 \end{bmatrix} \begin{bmatrix} 0.497006 \\ 1 \end{bmatrix} = \begin{bmatrix} 2.497006 \\ 4.988024 \end{bmatrix}$$

$$\lambda \approx 4.988024$$

$$\mathbf{x}_4 \leftarrow \frac{1}{4.988024}[2.497006 \quad 4.988024]^{\mathrm{T}} = [0.500600 \quad 1]^{\mathrm{T}}$$

The method is converging to the eigenvalue 5 and its corresponding eigenvector $[0.5 \quad 1]^{\mathrm{T}}$.

Example 2 Find the dominant eigenvalue and a corresponding eigenvector for

$$\mathbf{A} = \begin{bmatrix} 0 & 1 & 0 \\ 0 & 0 & 1 \\ 18 & -1 & -7 \end{bmatrix}$$

Solution: We initialize $\mathbf{x}_0 = [1 \quad 1 \quad 1]^T$. Then, for the first iteration,

$$\mathbf{x}_1 = \mathbf{A}\mathbf{x}_0 = \begin{bmatrix} 0 & 1 & 0 \\ 0 & 0 & 1 \\ 18 & -1 & -7 \end{bmatrix} \begin{bmatrix} 1 \\ 1 \\ 1 \end{bmatrix} = \begin{bmatrix} 1 \\ 1 \\ 10 \end{bmatrix}$$

For the second iteration,

$$\mathbf{x}_2 = \mathbf{A}\mathbf{x}_1 = \begin{bmatrix} 0 & 1 & 0 \\ 0 & 0 & 1 \\ 18 & -1 & -7 \end{bmatrix} \begin{bmatrix} 0.1 \\ 0.1 \\ 1 \end{bmatrix} = \begin{bmatrix} 0.1 \\ 1 \\ -5.3 \end{bmatrix}$$

$$\lambda \approx -5.3$$

$$\mathbf{x}_2 \leftarrow \frac{1}{-5.3}[0.1 \quad 1 \quad -5.3]^T = [-0.018868 \quad -0.188679 \quad 1]^T$$

For the third iteration,

$$\mathbf{x}_3 = \mathbf{A}\mathbf{x}_2 = \begin{bmatrix} 0 & 1 & 0 \\ 0 & 0 & 1 \\ 18 & -1 & -7 \end{bmatrix} \begin{bmatrix} -0.018868 \\ -0.188679 \\ 1 \end{bmatrix} = \begin{bmatrix} -0.188679 \\ 1 \\ -7.150943 \end{bmatrix}$$

$$\lambda \approx -7.150943$$

$$\mathbf{x}_3 \leftarrow \frac{1}{-7.150943}[-0.188679 \quad 1 \quad -7.150943]^T$$

$$= [0.026385 \quad -0.139842 \quad 1]^T$$

Continuing in this manner, we generate Table 4.1, where all entries are rounded to four decimal places. The algorithm is converging through six decimal places to the eigenvalue -6.405125 and its corresponding eigenvector

$$[0.024375 \quad -0.156125 \quad 1]^T$$

Although effective when it converges, the power method has deficiencies. It does not converge to the dominant eigenvalue when that eigenvalue is complex, and it may not converge when there is more than one equally dominant eigenvalue (see Problem 12). Furthermore, the method, in general, cannot be used to locate all the eigenvalues.

Table 4.1

Iteration	Eigenvector Components			Eigenvalue
0	1.0000	1.0000	1.0000	
1	0.1000	0.1000	1.0000	10.0000
2	−0.0189	−0.1887	1.0000	−5.3000
3	0.0264	−0.1398	1.0000	−7.1509
4	0.0219	−0.1566	1.0000	−6.3852
5	0.0243	−0.1551	1.0000	−6.4492
6	0.0242	−0.1561	1.0000	−6.4078
7	0.0244	−0.1560	1.0000	−6.4084
8	0.0244	−0.1561	1.0000	−6.4056

The inverse power method is the power method applied to the inverse of a matrix **A**; in general, the inverse power method converges to the smallest eigenvalue of **A** in absolute value.

A more powerful numerical method is the *inverse power method*, which is the power method applied to the inverse of a matrix. This, of course, adds another assumption: The inverse must exist, or equivalently, the matrix must not have any zero eigenvalues. Since a nonsingular matrix and its inverse share identical eigenvectors and reciprocal eigenvalues (see Theorem 4 of Section 4.4), once we know the eigenvalues and eigenvectors of the inverse of a matrix, we have the analogous information about the matrix itself.

The power method applied to the inverse of a matrix **A** will generally converge to the dominant eigenvalue of $\mathbf{A}^{-1}$. Its reciprocal will be the eigenvalue of **A** having the smallest absolute value. The advantages of the inverse power method are that it converges more rapidly than the power method, and it often can be used to find all real eigenvalues of **A**; a disadvantage is that it deals with $\mathbf{A}^{-1}$, which is laborious to calculate for matrices of large order. Such a calculation, however, can be avoided using **LU** decomposition.

The power method generates the sequence of vectors

$$\mathbf{x}_k = \mathbf{A}\mathbf{x}_{k-1}$$

The inverse power method will generate the sequence

$$\mathbf{x}_k = \mathbf{A}^{-1}\mathbf{x}_{k-1}$$

which may be written as

$$\mathbf{A}\mathbf{x}_k = \mathbf{x}_{k-1}$$

We solve for the unknown vector $\mathbf{x}_k$ using **LU** decomposition (see Section 1.7).

Example 3 Use the inverse power method to find an eigenvalue for

$$\mathbf{A} = \begin{bmatrix} 2 & 1 \\ 2 & 3 \end{bmatrix}$$

Solution: We initialize $\mathbf{x}_0 = [1 \quad 1]^T$. The **LU** decomposition for **A** has $\mathbf{A} = \mathbf{LU}$ with

$$\mathbf{L} = \begin{bmatrix} 1 & 0 \\ 1 & 1 \end{bmatrix} \quad \text{and} \quad \mathbf{U} = \begin{bmatrix} 2 & 1 \\ 0 & 2 \end{bmatrix}$$

For the first iteration, we solve the system $\mathbf{LUx}_1 = \mathbf{x}_0$ by first solving the system $\mathbf{Ly} = \mathbf{x}_0$ for $\mathbf{y}$, and then solving the system $\mathbf{Ux}_1 = \mathbf{y}$ for $\mathbf{x}_1$. Set $\mathbf{y} = [y_1 \quad y_2]^T$ and $\mathbf{x}_1 = [a \quad b]^T$. The first system is

$$\begin{aligned} y_1 + 0y_2 &= 1 \\ y_1 + y_2 &= 1 \end{aligned}$$

which has as its solution $y_1 = 1$ and $y_2 = 0$. The system $\mathbf{Ux}_1 = \mathbf{y}$ becomes

$$\begin{aligned} 2a + b &= 1 \\ 2b &= 0 \end{aligned}$$

which admits the solution $a = 0.5$ and $b = 0$. Thus,

$$\begin{aligned} &\mathbf{x}_1 = \mathbf{A}^{-1}\mathbf{x}_0 = [0.5 \quad 0]^T \\ &\lambda \approx 0.5 \quad \text{(an approximation to an eigenvalue for } \mathbf{A}^{-1}) \\ &\mathbf{x}_1 \leftarrow \frac{1}{0.5}[0.5 \quad 0]^T = [1 \quad 0]^T \end{aligned}$$

For the second iteration, we solve the system $\mathbf{LUx}_2 = \mathbf{x}_1$ by first solving the system $\mathbf{Ly} = \mathbf{x}_1$ for $\mathbf{y}$, and then solving the system $\mathbf{Ux}_2 = \mathbf{y}$ for $\mathbf{x}_2$. Set $\mathbf{y} = [y_1 \quad y_2]^T$ and $\mathbf{x}_2 = [a \; b]^T$. The first system is

$$\begin{aligned} y_1 + 0y_2 &= 1 \\ y_1 + y_2 &= 0 \end{aligned}$$

which has as its solution $y_1 = 1$ and $y_2 = -1$. The system $\mathbf{Ux}_2 = \mathbf{y}$ becomes

$$\begin{aligned} 2a + b &= 1 \\ 2b &= -1 \end{aligned}$$

which admits the solution $a = 0.75$ and $b = -0.5$. Thus,

$$\begin{aligned} &\mathbf{x}_2 = \mathbf{A}^{-1}\mathbf{x}_1 = [0.75 \quad -0.5]^T \\ &\lambda \approx 0.75 \\ &\mathbf{x}_2 \leftarrow \frac{1}{0.75}[0.75 \quad -0.5]^T = [1 \quad -0.666667]^T \end{aligned}$$

For the third iteration, we first solve $\mathbf{Ly} = \mathbf{x}_2$ to obtain $\mathbf{y} = [1 \quad -1.666667]^T$, and then $\mathbf{Ux}_3 = \mathbf{y}$ to obtain $\mathbf{x}_3 = [0.916667 \quad -0.833333]^T$ Then,

$$\begin{aligned} &\lambda \approx 0.916667 \\ &\mathbf{x}_3 \leftarrow \frac{1}{0.916667}[0.916667 \quad -0.833333]^T = [1 \quad -0.909091]^T \end{aligned}$$

Continuing, we converge to the eigenvalue 1 for $\mathbf{A}^{-1}$ and its reciprocal $1/1 = 1$ for $\mathbf{A}$. The vector approximations are converging to $[1 \quad -1]^T$, which is an eigenvector for both $\mathbf{A}^{-1}$ and $\mathbf{A}$.

Example 4 Use the inverse power method to find an eigenvalue for

$$\mathbf{A} = \begin{bmatrix} 7 & 2 & 0 \\ 2 & 1 & 6 \\ 0 & 6 & 7 \end{bmatrix}$$

Solution: We initialize $\mathbf{x}_0 = [1 \quad 1 \quad 1]^T$. The **LU** decomposition for **A** has $\mathbf{A} = \mathbf{LU}$ with

$$\mathbf{L} = \begin{bmatrix} 1 & 0 & 0 \\ 0.285714 & 1 & 0 \\ 0 & 14 & 1 \end{bmatrix} \quad \text{and} \quad \mathbf{U} = \begin{bmatrix} 7 & 2 & 0 \\ 0 & 0.428571 & 6 \\ 0 & 0 & -77 \end{bmatrix}$$

For the first iteration, set $\mathbf{y} = [y_1 \quad y_2 \quad y_3]^T$ and $\mathbf{x}_1 = [a \quad b \quad c]^T$. The first system is

$$\begin{aligned} y_1 + 0y_2 + 0y_3 &= 1 \\ 0.285714y_1 + y_2 + 0y_3 &= 1 \\ 0y_1 + 14y_2 + y_3 &= 1 \end{aligned}$$

which has as its solution $y_1 = 1$, $y_2 = 0.714286$, and $y_3 = -9$. The system $\mathbf{Ux}_1 = \mathbf{y}$ becomes

$$\begin{aligned} 7a + 2b &= 1 \\ 0.428571b + 6c &= 0.714286 \\ -77c &= -9 \end{aligned}$$

which admits the solution $a = 0.134199$, $b = 0.030303$, and $c = 0.116883$. Thus,

$$\begin{aligned} \mathbf{x}_1 &= \mathbf{A}^{-1}\mathbf{x}_0 = [0.134199 \quad 0.030303 \quad 0.116833]^T \\ \lambda &\approx 0.134199 \quad (\text{an approximation to an eigenvalue for} \mathbf{A}^{-1}) \\ \mathbf{x}_1 &\leftarrow \frac{1}{0.134199}[0.134199 \quad 0.030303 \quad 0.116833]^T \\ &= [1 \quad 0.225806 \quad 0.870968]^T \end{aligned}$$

For the second iteration, solving the system $\mathbf{Ly} = \mathbf{x}_1$ for **y**, we obtain

$$\mathbf{y} = [1 \quad -0.059908 \quad 1.709677]^T$$

Then, solving the system $\mathbf{Ux}_2 = \mathbf{y}$ for $\mathbf{x}_2$, we get

$$\mathbf{x}_2 = [0.093981 \quad 0.171065 \quad -0.022204]^T$$

Therefore,

$$\lambda \approx 0.171065$$

$$\mathbf{x}_2 \leftarrow \frac{1}{0.171065}[0.093981 \quad 0.171065 \quad -0.022204]^{\mathrm{T}}$$

$$= [0.549388 \quad 1 \quad -0.129796]^{\mathrm{T}}$$

For the third iteration, solving the system $\mathbf{Ly}=\mathbf{x}_2$ for $\mathbf{y}$, we obtain

$$\mathbf{y} = [0.549388 \quad 0.843032 \quad -11.932245]^{\mathrm{T}}$$

Then, solving the system $\mathbf{Ux}_3=\mathbf{y}$ for $\mathbf{x}_3$, we get

$$\mathbf{x}_3 = [0.136319 \quad -0.202424 \quad 0.154964]^{\mathrm{T}}$$

Therefore,

$$\lambda \approx -0.202424$$

$$\mathbf{x}_3 \leftarrow \frac{1}{-0.202424}[0.136319 \quad -0.202424 \quad 0.154964]^{\mathrm{T}}$$

$$= [-0.673434 \quad 1 \quad -0.765542]^{\mathrm{T}}$$

Continuing in this manner, we generate Table 4.2, where all entries are rounded to four decimal places. The algorithm is converging to the eigenvalue $-1/3$ for $\mathbf{A}^{-1}$ and its reciprocal -3 for $\mathbf{A}$. The vector approximations are converging to $[-0.2 \quad 1 \quad -0.6]^{\mathrm{T}}$, which is an eigenvector for both $\mathbf{A}^{-1}$ and $\mathbf{A}$.

Table 4.2

Iteration	Eigenvector Components			Eigenvalue
0	1.0000	1.0000	1.0000	
1	1.0000	0.2258	0.8710	0.1342
2	0.5494	1.0000	−0.1298	0.1711
3	−0.6734	1.0000	−0.7655	−0.2024
4	−0.0404	1.0000	−0.5782	−0.3921
5	−0.2677	1.0000	−0.5988	−0.3197
6	−0.1723	1.0000	−0.6035	−0.3372
7	−0.2116	1.0000	−0.5977	−0.3323
8	−0.1951	1.0000	−0.6012	−0.3336
9	−0.2021	1.0000	−0.5994	−0.3333
10	−0.1991	1.0000	−0.6003	−0.3334
11	−0.2004	1.0000	−0.5999	−0.3333
12	−0.1998	1.0000	−0.6001	−0.3333

We can use Theorem 9 of Section 4.2 in conjunction with the inverse power method to develop a procedure for finding all eigenvalues and a set of corresponding eigenvectors for a matrix, providing that the eigenvalues are real and distinct, and estimates of their locations are known. The algorithm is known as the *shifted inverse power method.*

If c is an estimate for an eigenvalue of **A**, then $\mathbf{A}-c\mathbf{I}$ will have an eigenvalue near 0 and its reciprocal will be the dominant eigenvalue of $(\mathbf{A}-c\mathbf{I})^{-1}$. We use the inverse power method with an **LU** decomposition of $\mathbf{A}-c\mathbf{I}$ to calculate the dominant eigenvalue λ and its corresponding eigenvector **x** for $(\mathbf{A}-c\mathbf{I})^{-1}$. Then $1/\lambda$ and **x** are an eigenvalue and eigenvector pair for $\mathbf{A}-c\mathbf{I}$ while $c+(1/\lambda)$ and **x** are an eigenvalue and eigenvector pair for **A**.

THE SHIFTED INVERSE POWER METHOD

Step 1. Begin with an initial guess $\mathbf{x}_0$ for an eigenvector of a matrix **A**, having the property that the largest component of $\mathbf{x}_0$ in absolute value is one. Set a counter k equal to 1 and choose a value for the constant c (preferably an estimate for an eigenvalue if such an estimate is available).

Step 2. Calculate $\mathbf{x}_k=(\mathbf{A}-c\mathbf{I})\mathbf{x}_{k-1}$.

Step 3. Set λ equal to the largest component of $\mathbf{x}_k$ in absolute value.

Step 4. Rescale $\mathbf{x}_k$ by dividing each of its components by λ. Relabel the resulting vector as $\mathbf{x}_k$.

Step 5. If $c+(1/\lambda)$ is an adequate estimate for an eigenvalue of **A**, with $\mathbf{x}_k$ as a corresponding eigenvector, stop; otherwise increment k by one and return to Step 2.

Example 5 Find a second eigenvalue for the matrix given in Example 4.

Solution: Since we do not have an estimate for any of the eigenvalues, we arbitrarily choose $c=15$. Then

$$\mathbf{A}-c\mathbf{I}=\begin{bmatrix}-8 & 2 & 0\\ 2 & -14 & 6\\ 0 & 6 & -8\end{bmatrix}$$

which has an **LU** decomposition with

$$\mathbf{L}=\begin{bmatrix}1 & 0 & 0\\ -0.25 & 1 & 0\\ 0 & -0.444444 & 1\end{bmatrix} \quad \text{and} \quad \mathbf{U}=\begin{bmatrix}-8 & 2 & 0\\ 0 & -13.5 & 6\\ 0 & 0 & -5.333333\end{bmatrix}$$

Applying the inverse power method to $\mathbf{A}-15\mathbf{I}$, we generate Table 4.3, which is converging to $\lambda=-0.25$ and $\mathbf{x}=\begin{bmatrix}\frac{1}{3} & \frac{2}{3} & 1\end{bmatrix}^{\mathrm{T}}$. The corresponding eigenvalue of **A** is $(1/-0.25)+15=11$, with the same eigenvector.

Table 4.3

Iteration	Eigenvector Components			Eigenvalue
0	1.0000	1.0000	1.0000	
1	0.6190	0.7619	1.0000	−0.2917
2	0.4687	0.7018	1.0000	−0.2639
3	0.3995	0.6816	1.0000	−0.2557
4	0.3661	0.6736	1.0000	−0.2526
5	0.3496	0.6700	1.0000	−0.2513
6	0.3415	0.6683	1.0000	−0.2506
7	0.3374	0.6675	1.0000	−0.2503
8	0.3354	0.6671	1.0000	−0.2502
9	0.3343	0.6669	1.0000	−0.2501
10	0.3338	0.6668	1.0000	−0.2500
11	0.3336	0.6667	1.0000	−0.2500

Using the results of Examples 4 and 5, we have two eigenvalues, $\lambda_1 = -3$ and $\lambda_2 = 11$, of the 3×3 matrix defined in Example 4. Since the trace of a matrix equals the sum of the eigenvalues (Theorem 1 of Section 4.2), we know $7 + 1 + 7 = -3 + 11 + \lambda_3$, so the last eigenvalue is $\lambda_3 = 7$.

Problems 4.5

In Problems 1 through 10, use the power method to locate the dominant eigenvalue and a corresponding eigenvector for the given matrices. Stop after five iterations.

(1) $\begin{bmatrix} 2 & 1 \\ 2 & 3 \end{bmatrix}$. **(2)** $\begin{bmatrix} 2 & 3 \\ 4 & 6 \end{bmatrix}$. **(3)** $\begin{bmatrix} 3 & 6 \\ 9 & 6 \end{bmatrix}$.

(4) $\begin{bmatrix} 0 & 1 \\ -4 & 6 \end{bmatrix}$. **(5)** $\begin{bmatrix} 8 & 2 \\ 3 & 3 \end{bmatrix}$. **(6)** $\begin{bmatrix} 8 & 3 \\ 3 & 2 \end{bmatrix}$.

(7) $\begin{bmatrix} 3 & 0 & 0 \\ 2 & 6 & 4 \\ 2 & 3 & 5 \end{bmatrix}$. **(8)** $\begin{bmatrix} 7 & 2 & 0 \\ 2 & 1 & 6 \\ 0 & 6 & 7 \end{bmatrix}$. **(9)** $\begin{bmatrix} 3 & 2 & 3 \\ 2 & 6 & 6 \\ 3 & 6 & 11 \end{bmatrix}$.

(10) $\begin{bmatrix} 2 & -17 & 7 \\ -17 & -4 & 1 \\ 7 & 1 & -14 \end{bmatrix}$.

(11) Use the power method on

$$\mathbf{A} = \begin{bmatrix} 2 & 0 & -1 \\ 2 & 2 & 2 \\ -1 & 0 & 2 \end{bmatrix}$$

and explain why it does not converge to the dominant eigenvalue $\lambda = 3$.

(12) Use the power method on

$$A = \begin{bmatrix} 3 & 5 \\ 5 & -3 \end{bmatrix}$$

and explain why it does not converge.

(13) Shifting can also be used with the power method to locate the next most dominant eigenvalue, if it is real and distinct, once the dominant eigenvalue has been determined. Construct $\mathbf{A} - \lambda\mathbf{I}$, where λ is the dominant eigenvalue of $\mathbf{A}$, and apply the power method to the shifted matrix. If the algorithm converges to μ, and $\mathbf{x}$, then $\mu + \lambda$ is an eigenvalue of $\mathbf{A}$ with the corresponding eigenvector $\mathbf{x}$. Apply this shifted power method algorithm to the matrix in Problem 1. Use the result of Problem 1 to determine the appropriate shift.

(14) Use the shifted power method as described in Problem 13 on the matrix in Problem 9. Use the results of Problem 9 to determine the appropriate shift.

(15) Use the inverse power method on the matrix defined in Example 1. Stop after five iterations.

(16) Use the inverse power method on the matrix defined in Problem 3. Take $\mathbf{x}_0 = [1 \quad -0.5]^T$ and stop after five iterations.

(17) Use the inverse power method on the matrix defined in Problem 5. Stop after five iterations.

(18) Use the inverse power method on the matrix defined in Problem 6. Stop after five iterations.

(19) Use the inverse power method on the matrix defined in Problem 9. Stop after five iterations.

(20) Use the inverse power method on the matrix defined in Problem 10. Stop after five iterations.

(21) Use the inverse power method on the matrix defined in Problem 11. Stop after five iterations.

(22) Use the inverse power method on the matrix defined in Problem 4. Explain the difficulty and suggest a way to avoid it.

(23) Use the inverse power method on the matrix defined in Problem 2. Explain the difficulty and suggest a way to avoid it.

(24) Can the power method converge to a dominant eigenvalue if that eigenvalue is not distinct?

(25) Apply the shifted inverse power method to the matrix defined in Problem 9, with a shift constant of 10.

(26) Apply the shifted inverse power method to the matrix defined in Problem 10, with a shift constant of −25.

CHAPTER 4 REVIEW

Important Terms

characteristic equation
determinant
derivative of a matrix
diagonalizable matrix
dominant eigenvalue
$e^{\mathbf{A}t}$
eigenspace
eigenvalue
eigenvector
Euler's relations
exponential of a matrix
inverse power method
modal matrix
model
power method
shifted inverse power method
spectral matrix
trace

Important Concepts

Section 4.1

- A nonzero vector **x** is an eigenvector of a square matrix **A** if there exists a scalar λ, called an *eigenvalue*, such that $\mathbf{Ax} = \lambda\mathbf{x}$.
- Similar matrices have the same characteristic equation (and, therefore, the same eigenvalues).
- Nonzero vectors in the eigenspace of the matrix **A** for the eigenvalue λ are eigenvectors of **A**.
- Eigenvalues and eigenvectors for a linear transformation $T: \mathbb{V} \rightarrow \mathbb{V}$ are determined by locating the eigenvalues and eigenvectors of any matrix representation for T; the eigenvectors of the matrix are coordinate representations of the eigenvector of T.

Section 4.2

- Any nonzero scalar multiple of an eigenvector is again an eigenvector; the nonzero sum of two eigenvectors corresponding to the same eigenvalue is again an eigenvector
- The sum of the eigenvalues of a matrix equals the trace of the matrix.
- The eigenvalues of an upper (lower) triangular matrix are the elements on the main diagonal of the matrix.
- The product of all the eigenvalues of a matrix (counting multiplicity) equals the determinant of the matrix.
- A matrix is singular if and only if it has a zero eigenvalue.
- If **x** is an eigenvector of **A** corresponding to the eigenvalue λ, then
  - for any nonzero scalar k, $k\lambda$ and **x** are a corresponding pair of eigenvalues and eigenvectors of $k\mathbf{A}$,
  - λ^n and **x** are a corresponding pair of eigenvalues and eigenvectors of $\mathbf{A}^n$, for any positive integer n,
  - for any scalar c, $\lambda - c$ and **x** are a corresponding pair of eigenvalues and eigenvectors of $\mathbf{A} - c\mathbf{I}$,
  - $1/\lambda$ and **x** are a corresponding pair of eigenvalues and eigenvectors of $\mathbf{A}^{-1}$, providing the inverse exists,
  - λ is an eigenvalue of $\mathbf{A}^{\mathrm{T}}$.

Section 4.3

- An $n \times n$ matrix is diagonalizable if and only if it has n linearly independent eigenvectors.
- Eigenvectors of a matrix corresponding to distinct eigenvalues are linearly independent.
- If λ is an eigenvalue of multiplicity k of an $n \times n$ matrix $\mathbf{A}$, then the number of linearly independent eigenvectors of $\mathbf{A}$ associated with λ is $n - r(\mathbf{A} - \lambda\mathbf{I})$, where r denotes rank.
- If $\mathbb{V}$ is an n-dimensional vector space, then a linear transformation $T: \mathbb{V} \to \mathbb{V}$ may be represented by a diagonal matrix if and only if T possesses a basis of eigenvectors.

Section 4.4

- To calculate the exponential of a diagonal matrix, replace each diagonal element by the exponential of that diagonal element.
- If $\mathbf{A}$ is similar to a matrix $\mathbf{J}$ in Jordan canonical form, so that $\mathbf{A} = \mathbf{MJM}^{-1}$ for a generalized modal matrix $\mathbf{M}$, then $e^{\mathbf{A}} = \mathbf{M}e^{\mathbf{J}}\mathbf{M}^{-1}$.
- $e^{\mathbf{0}} = \mathbf{I}$, where $\mathbf{0}$ is the $n \times n$ zero matrix and $\mathbf{I}$ is the $n \times n$ identity matrix.

Section 4.5

- The power method is a numerical method for estimating the dominant eigenvalue and a corresponding eigenvector for a matrix.
- The inverse power method is the power method applied to the inverse of a matrix $\mathbf{A}$. In general, the inverse power method converges to the smallest eigenvalue in absolute value of $\mathbf{A}$.

CHAPTER 5

Applications of Eigenvalues

Chapter Outline

5.1 DIFFERENTIAL EQUATIONS

A *differential equation* is an equation involving an unknown function and one or more of its derivatives. For the next few sections of this chapter, we will take advantage of some of the concepts we have thus far developed and apply them to solving differential equations.

5.2 DIFFERENTIAL EQUATIONS IN FUNDAMENTAL FORM

An important application of Jordan canonical forms (see Appendix A), in general, and the exponential of a matrix, in particular, occurs in the solution of differential equations with constant coefficients. A working knowledge of the integral calculus and a familiarity with differential equations is required to understand the scope of this application. In this section, we show how to transform many systems of differential equations into a matrix differential equation. In the next section, we show how to solve such systems using the exponential of matrix.

A differential equation in the unknown functions $x_1(t)$, $x_2(t)$, ..., $x_n(t)$ is an equation that involves these functions and one or more of their derivatives. We shall be interested in systems of first-order differential equations of the form

$$
\begin{aligned}
\frac{dx_1(t)}{dt} &= a_{11}x_1(t) + a_{12}x_2(t) + \cdots + a_{1n}x_n(t) + f_1(t) \\
\frac{dx_2(t)}{dt} &= a_{21}x_1(t) + a_{22}x_2(t) + \cdots + a_{2n}x_n(t) + f_2(t) \\
&\vdots \\
\frac{dx_n(t)}{dt} &= a_{11}x_1(t) + a_{12}x_2(t) + \cdots + a_{1n}x_n(t) + f_1(t)
\end{aligned}
\tag{5.1}
$$

Here, a_{ij} $(i, j = 1,2, \ldots, n)$ is restricted to be a constant and $f_i(t)$ is presumed to be a known function of the variable t. If we define,

$$
\mathbf{x}(t) = \begin{bmatrix} x_1(t) \\ x_2(t) \\ \vdots \\ x_n(t) \end{bmatrix}, \quad \mathbf{A} = \begin{bmatrix} a_{11} & a_{12} & \cdots & a_{1n} \\ a_{21} & a_{22} & \cdots & a_{2n} \\ \vdots & \vdots & \ddots & \vdots \\ a_{n1} & a_{n2} & \cdots & a_{nn} \end{bmatrix}, \quad \text{and} \quad \mathbf{f}(t) = \begin{bmatrix} f_1(t) \\ f_2(t) \\ \vdots \\ f_n(t) \end{bmatrix} \tag{5.2}
$$

then Equation (5.1) is equivalent to the single matrix equation

$$
\frac{dx(t)}{dt} = \mathbf{A}\mathbf{x}(t) + \mathbf{f}(t) \tag{5.3}
$$

Example 1 The system of equations

$$
\begin{aligned}
\frac{dx(t)}{dt} &= 2x(t) + 3y(t) + 4z(t) + (t^2 - 1) \\
\frac{dy(t)}{dt} &= 5y(t) + 6z(t) + e^t \\
\frac{dz(t)}{dt} &= 7x(t) - 8y(t) - 9z(t)
\end{aligned}
$$

is equivalent to the matrix equation

$$
\begin{bmatrix} dx(t)/dt \\ dy(t)/dt \\ dz(t)/dt \end{bmatrix} = \begin{bmatrix} 2 & 3 & 4 \\ 0 & 5 & 6 \\ 7 & -8 & -9 \end{bmatrix} \begin{bmatrix} x(t) \\ y(t) \\ z(t) \end{bmatrix} + \begin{bmatrix} t^2 - 1 \\ e^t \\ 0 \end{bmatrix}
$$

This matrix equation is in form (4.30) with

$$
\mathbf{x}(t) = \begin{bmatrix} x(t) \\ y(t) \\ z(t) \end{bmatrix}, \quad \mathbf{A} = \begin{bmatrix} 2 & 3 & 4 \\ 0 & 5 & 6 \\ 7 & -8 & -9 \end{bmatrix}, \quad \text{and} \quad \mathbf{f}(t) = \begin{bmatrix} t^2 - 1 \\ e^t \\ 0 \end{bmatrix}
$$

In this example, $x_1(t) = x(t)$, $x_2(t) = y(t)$, and $x_3(t) = z(t)$.

We solve Equation (5.3) in the interval $a \le t \le b$ by identifying a column matrix $\mathbf{x}(t)$ that when substituted into Equation (5.3) makes the equation true for all values of t in the given interval. Often, however, we need to solve more than just a set of differential equations. Often, we seek functions $x_1(t), x_2(t), \ldots, x_n(t)$ that satisfy all the differential equations in Equation (5.1) or, equivalently, Equation (5.3) and also a set of *initial conditions* of the form

$$x_1(t_0) = c_1, x_2(t_0) = c_2, \ldots, x_n(t_0) = c_0 \tag{5.4}$$

where $c_1, c_2, \ldots, c_n$ are all constants, and t_0 is a specific value of the variable t inside the interval of interest. Upon defining

$$\mathbf{c} = \begin{bmatrix} c_1 \\ c_2 \\ \vdots \\ c_n \end{bmatrix}$$

it follows that

$$\mathbf{x}(t_0) = \begin{bmatrix} x_1(t_0) \\ x_2(t_0) \\ \vdots \\ x_n(t_0) \end{bmatrix} = \begin{bmatrix} c_1 \\ c_2 \\ \vdots \\ c_n \end{bmatrix} = \mathbf{c}$$

Thus, initial conditions (Equation 4.31) have the matrix form

$$\mathbf{x}(t_0) = \mathbf{c} \tag{5.5}$$

We say that a system of differential equations is in *fundamental form* if it is given by the matrix equations

$$\frac{dx(t)}{dt} = \mathbf{A}\mathbf{x}(t) + \mathbf{f}(t)$$
$$\mathbf{x}(t_0) = \mathbf{c} \tag{5.6}$$

A system of differential equations is in fundamental form if it is given by the matrix equations

$$\frac{dx(t)}{dt} = \mathbf{A}\mathbf{x}(t) + \mathbf{f}(t)$$

$$\mathbf{x}(t_0) = \mathbf{c}.$$

Example 2 The system of equations

$$\frac{dr(t)}{dt} = 2r(t) - 3s(t)$$

$$\frac{ds(t)}{dt} = 4r(t) + 5s(t)$$

$$r(\pi) = 10, s(\pi) = -20$$

is equivalent to the matrix equations

$$\begin{bmatrix} dr(t)/dt \\ ds(t)/dt \end{bmatrix} = \begin{bmatrix} 2 & -3 \\ 4 & 5 \end{bmatrix} \begin{bmatrix} r(t) \\ s(t) \end{bmatrix} + \begin{bmatrix} 0 \\ 0 \end{bmatrix}$$

$$\begin{bmatrix} r(\pi) \\ s(\pi) \end{bmatrix} = \begin{bmatrix} 10 \\ -20 \end{bmatrix}$$

This set of equations is in fundamental form (4.33) with

$$\mathbf{x}(t) = \begin{bmatrix} r(t) \\ s(t) \end{bmatrix}, \mathbf{A} = \begin{bmatrix} 2 & -3 \\ 4 & 5 \end{bmatrix}, \mathbf{f}(t) = \begin{bmatrix} 0 \\ 0 \end{bmatrix}, \text{ and } \mathbf{c} = \begin{bmatrix} 10 \\ -20 \end{bmatrix}$$

In this example, $x_1(t) = r(t)$ and $x_2(t) = s(t)$.

A system of differential equations in fundamental form is homogeneous when $\mathbf{f}(t) = \mathbf{0}$.

A system of differential equations in fundamental form is *homogeneous* when $\mathbf{f}(t) = \mathbf{0}$ and *nonhomogeneous* when $\mathbf{f}(t) \neq \mathbf{0}$ (i.e., when at least one element of $\mathbf{f}(t)$ is not zero). The system in Example 2 is homogeneous; the system in Example 1 is nonhomogeneous.

Generally, systems of differential equations do not appear in fundamental form. However, many such systems can be transformed into fundamental form by appropriate reduction techniques. One such group are initial-value problems of the form

$$\begin{aligned} &a_n \frac{d^n x(t)}{dt^n} + a_{n-1} \frac{d^{n-1} x(t)}{dt^{n-1}} + \ldots + a_1 \frac{dx(t)}{dt} + a_0 x(t) = f(t) \\ &x(t_0) = c_1, \ \frac{dx(t_0)}{dt} = c_2 \ldots, \ \frac{d^{n-1} x(t_0)}{dt^{n-1}} = c_{n-1} \end{aligned} \tag{5.7}$$

This is a system containing a single nth-order, linear differential equation with constant coefficients along with $n-1$ initial conditions at t_0. The coefficients $a_0, a_1, \ldots, a_n$ are restricted to be constants and the function $f(t)$ is presumed to be known and continuous on some interval centered around t_0.

A method of reduction for transforming system (5.5) into fundamental form is given by the following six steps.

Step 1. Solve system (5.5) for the nth derivative of $x(t)$.

$$\frac{d^n x(t)}{dt^n} = -\left(\frac{a_{n-1}}{a_n}\right) \frac{d^{n-1} x(t)}{dt^{n-1}} - \cdots - \left(\frac{a_1}{a_n}\right) \frac{dx(t)}{dt} - \left(\frac{a_0}{a_n}\right) x(t) + \frac{f(t)}{a_n}$$

Step 2. Define n new variables (the same number as the order of the differential equations) $x_1(t), x_2(t), \ldots, x_n(t)$ by the equations

$$x_1 = x(t), \quad x_2 = \frac{dx}{dt}, x_3 = \frac{d^2 x}{dt^2}, \ldots, x_{n-1} = \frac{d^{n-2} x}{dt^{n-2}}, \quad x_n = \frac{d^{n-1} x}{dt^{n-1}} \tag{5.8}$$

Here, we simplified $x_j(t)$ $(j = 1, 2, \ldots, n)$ to x_j. By differentiating the last equation in system (5.8), we obtain

$$\frac{dx_n}{dt} = \frac{d^n x}{dt^n} \tag{5.9}$$

Step 3. Substitute Equations (5.8) and (5.9) into the equation obtained in Step 1, thereby obtaining an equation for dx_n/dt in terms of the new variables. The result is

$$\frac{dx_n}{dt} = -\left(\frac{a_{n-1}}{a_n}\right)x_n - \ldots - \left(\frac{a_1}{a_n}\right)x_2 - \left(\frac{a_0}{a_n}\right)x_1 + \frac{f(t)}{a_n} \tag{5.10}$$

Step 4. Using Equations (5.8) and (5.10), construct a system of n first-order differential equations for $x_1, x_2, \ldots, x_n$. The system is

$$\frac{dx_1}{dt} = x_2$$

$$\frac{dx_2}{dt} = x_3$$

$$\vdots$$

$$\frac{dx_{n-1}}{dt} = x_n$$

$$\frac{dx_n}{dt} = -\left(\frac{a_0}{a_n}\right)x_1 - \left(\frac{a_1}{a_n}\right)x_2 - \ldots - \left(\frac{a_{n-1}}{a_n}\right)x_n + \frac{f(t)}{a_n} \tag{5.11}$$

In this last equation, the order of the terms in Equation (5.10) was rearranged so that x_1 appears before x_2, which appears before x_3 and so on. This was done to simplify the next step.

Step 5. Write system (5.11) as a single matrix differential equation. Define

$$\mathbf{x}(t) = \begin{bmatrix} x_1 \\ x_2 \\ \vdots \\ x_{n-1} \\ x_n \end{bmatrix}, \mathbf{f}(t) = \begin{bmatrix} 0 \\ 0 \\ \vdots \\ 0 \\ f(t)/a_n \end{bmatrix}$$

$$\mathbf{A} = \begin{bmatrix} 0 & 1 & 0 & 0 & \ldots & 0 \\ 0 & 0 & 1 & 0 & \ldots & 0 \\ 0 & 0 & 0 & 1 & \ldots & 0 \\ \vdots & \vdots & \vdots & \vdots & & \vdots \\ 0 & 0 & 0 & 0 & \ldots & 1 \\ -\dfrac{a_0}{a_n} & -\dfrac{a_1}{a_n} & -\dfrac{a_2}{a_n} & -\dfrac{a_3}{a_n} & \ldots & -\dfrac{a_{n-1}}{a_n} \end{bmatrix}$$

Then Equation (5.11) is equivalent to the matrix equation $\dfrac{dx(t)}{dt} = \mathbf{A}\mathbf{x}(t) + \mathbf{f}(t)$.

Step 6. Write the initial conditions as a matrix equation. Define $\mathbf{c} = [c_1\ c_2 \ldots c_n]^{\mathrm{T}}$. Then,

$$\mathbf{x}(t_0) = \begin{bmatrix} x_1(t_0) \\ x_2(t_0) \\ \vdots \\ x_n(t_0) \end{bmatrix} = \begin{bmatrix} x(t_0) \\ dx(t_0)/dt \\ \vdots \\ d^{n-1}x(t_0)/dt^{n-1} \end{bmatrix} = \begin{bmatrix} c_1 \\ c_2 \\ \vdots \\ c_n \end{bmatrix} = \mathbf{c}$$

The results of Steps 5 and 6 are a matrix system in fundamental form.

Example 3 Write the initial-value problem

$$\frac{d^2x(t)}{dt^2} + x(t) = 2; \quad x(\pi) = 0, \ \frac{dx(\pi)}{dt} = -1$$

in fundamental form.

Solution: The differential equation may be rewritten as

$$\frac{d^2x(t)}{dt} = -x(t) + 2$$

This is a second-order differential equation, so we define two new variables $x_1 = x(t)$ and $x_2 = \frac{dx}{dt}$. Thus, $\frac{dx_2}{dt} = \frac{d^2x}{dt^2}$ and the original differential equation becomes $\frac{dx_2}{dt} = -x_1 + 2$. A first-order system for the new variables is

$$\frac{dx_1}{dt} = x_2 = 0x_1 + 1x_2$$

$$\frac{dx_2}{dt} = -x_1 + 2 = -1x_1 + 0x_2 + 2$$

Define $\mathbf{x}(t) = \begin{bmatrix} x_1 \\ x_2 \end{bmatrix}$, $\mathbf{A} = \begin{bmatrix} 0 & 1 \\ -1 & 0 \end{bmatrix}$, $\mathbf{f}(t) = \begin{bmatrix} 0 \\ 2 \end{bmatrix}$, and $\mathbf{c} = \begin{bmatrix} 0 \\ -1 \end{bmatrix}$. Then, the initial-value problem is equivalent to the fundamental form

$$\frac{d\mathbf{x}(t)}{dt} = \mathbf{A}\mathbf{x}(t) + \mathbf{f}(t); \quad \mathbf{x}(\pi) = \mathbf{c}$$

Example 4 Write the initial-value problem

$$2\frac{d^4x}{dt^4} - 4\frac{d^3x}{dt^3} + 16\frac{d^2x}{dt^2} - \frac{dx}{dt} + 2x = \sin t$$

$$x(0) = 1, \quad \frac{dx(0)}{dt} = 2, \quad \frac{d^2x(0)}{dt^2} = -1, \quad \frac{d^3x(0)}{dt^3} = 0$$

in fundamental form.

Solution: The differential equation may be rewritten as

$$\frac{d^4x}{dt^4} = 2\frac{d^3x}{dt^3} - 8\frac{d^2x}{dt^2} + \frac{1}{2}\frac{dx}{dt} - x + \frac{1}{2}\sin t$$

This is a fourth-order differential equation, so we define four new variables

$$x_1 = x(t), \quad x_2 = \frac{dx}{dt}, \quad x_3 = \frac{d^2x}{dt^2}, \quad \text{and} \quad x_4 = \frac{d^3x}{dt^3}$$

Thus, $\frac{dx_4}{dt} = \frac{d^4x}{dt^4}$ and the original differential equation becomes

$$\frac{dx_4}{dt} = 2x_4 - 8x_3 + \frac{1}{2}x_2 - x_1 + \frac{1}{2}\sin t$$

A first-order system for the new variables is

$$\frac{dx_1}{dt} = x_2$$

$$\frac{dx_2}{dt} = x_3$$

$$\frac{dx_3}{dt} = x_4$$

$$\frac{dx_4}{dt} = -x_1 + \frac{1}{2}x_2 - 8x_3 + 2x_4 + \frac{1}{2}\sin t.$$

Define $\mathbf{x}(t) = \begin{bmatrix} x_1 \\ x_2 \\ x_3 \\ x_4 \end{bmatrix}$, $\mathbf{A} = \begin{bmatrix} 0 & 1 & 0 & 0 \\ 0 & 0 & 1 & 0 \\ 0 & 0 & 0 & 1 \\ -1 & \frac{1}{2} & -8 & 2 \end{bmatrix}$, $\mathbf{f}(t) = \begin{bmatrix} 0 \\ 0 \\ 0 \\ \frac{1}{2}\sin t \end{bmatrix}$, and $\mathbf{c} = \begin{bmatrix} 1 \\ 2 \\ -1 \\ 0 \end{bmatrix}$.

Then, the initial-value problem is equivalent to the fundamental form

$$\frac{dx(t)}{dt} = \mathbf{A}\mathbf{x}(t) + \mathbf{f}(t); \ \mathbf{x}(0) = \mathbf{c}$$

Problems 5.2

Put the following initial-value problems into fundamental form

(1)
$$\frac{dx(t)}{dt} = 2x(t) + 3y(t)$$
$$\frac{dy(t)}{dt} = 4x(t) + 5y(t)$$
$$x(0) = 6, \quad y(0) = 7$$

(2)
$$\frac{dy(t)}{dt} = 3y(t) + 2z(t)$$
$$\frac{dz(t)}{dt} = 4y(t) + z(t)$$
$$y(0) = 1, \quad z(0) = 1$$

(3)
$$\frac{dx(t)}{dt} = -3x(t) + 3y(t) + 1$$
$$\frac{dy(t)}{dt} = 4x(t) - 4y(t) - 1$$
$$x(0) = 0, \quad y(0) = 0$$

(4)
$$\frac{dx(t)}{dt} = 3x(t) + t$$
$$\frac{dy(t)}{dt} = 2x(t) + t + 1$$
$$x(0) = 1, \quad y(0) - 1$$

(5)
$$\frac{dx(t)}{dt} = 3x(t) + 7y(t) + 2$$
$$\frac{dy(t)}{dt} = x(t) + y(t) + 2t$$
$$x(1) = 2, \quad y(1) = -3$$

(6)
$$\frac{du(t)}{dt} = u(t) + v(t) + w(t)$$
$$\frac{dv(t)}{dt} = u(t) - 3v(t) + w(t)$$
$$\frac{dw(t)}{dt} = v(t) + w(t)$$
$$u(4) = 0, v(4) = 1, w(4) = -1$$

(7)
$$\frac{dx(t)}{dt} = 6y(t) + z(t)$$
$$\frac{dy(t)}{dt} = x(t) - 3z(t)$$
$$\frac{dz(t)}{dt} = -2y(t)$$
$$x(0) = 10, y(0) = 10, z(0) = 20$$

(8)
$$\frac{dr(t)}{dt} = r(t) - 3s(t) - u(t) + \sin t$$
$$\frac{ds(t)}{dt} = r(t) - s(t) + t^2 + 1$$
$$\frac{dt(t)}{dt} = 2r(t) + s(t) - u(t) + \cos t$$
$$r(1) = 4, s(1) = -2, u(1) = 5$$

(9) $\frac{d^2x(t)}{dt^2} - 2\frac{dx(t)}{dt} - 3x(t) = 0$

$x(0) = 4, \frac{dx(0)}{dt} = 5$

(10) $\frac{d^2x(t)}{dt^2} + \frac{dx(t)}{dt} - x(t) = 0$

$x(1) = 2, \frac{dx(1)}{dt} = 0$

(11) $\frac{d^2x(t)}{dt^2} - x(t) = t^2$

$x(0) = -3, \frac{dx(0)}{dt} = 0$

(12) $\frac{d^2x(t)}{dt^2} - 2\frac{dx(t)}{dt} - 3x(t) = 2$

$x(0) = 0, \frac{dx(0)}{dt} = 0$

(13) $\frac{d^2x(t)}{dt^2} - 3\frac{dx(t)}{dt} + 2x(t) = e^{-t}$

$x(1) = 2, \frac{dx(1)}{dt} = 2$

(14) $\frac{d^3x(t)}{dt^3} + \frac{d^2(t)}{dt^2} - x(t) = 0$

$x(-1) = 2, \frac{dx(-1)}{dt} = 1, \frac{d^2x(t)}{dt^2} = -205$

(15) $\frac{d^4x}{dt^4} + \frac{d^2x}{dt^2} = 1 + \frac{dx}{dt}$

$x(0) = 1, \frac{dx(0)}{dt} = 2, \frac{d^2x(0)}{dt^2} = \pi, \frac{d^3x(0)}{dt^3} = e^3$

(16) $\frac{d^6x}{dt^6} + 4\frac{d^4x}{dt^4} = t^2 - t$

$x(\pi) = 2, \frac{dx(\pi)}{dt} = 1, \frac{d^2x(\pi)}{dt^2} = 0, \frac{d^3x(\pi)}{dt^3} = 2$

$\frac{d^4x(\pi)}{dt^4} = 1, \frac{d^5x(\pi)}{dt^5} = 0$

5.3 SOLVING DIFFERENTIAL EQUATIONS IN FUNDAMENTAL FORM

We demonstrated in Section 5.2 how various systems of differential equations could be transformed into the fundamental matrix form

$$\begin{aligned} \frac{d\mathbf{x}(t)}{dt} &= \mathbf{A}\mathbf{x}(t) + \mathbf{f}(t) \\ \mathbf{x}(t_0) &= \mathbf{c} \end{aligned} \tag{5.12}$$

The matrix **A** is assumed to be a matrix of constants, as is the column matrix **c**. In contrast, the column matrix $\mathbf{f}(t)$ may contain known functions of the variable t. Such differential equations can be solved in terms of $e^{\mathbf{A}t}$.

The matrix differential equation in Equation (5.12) can be rewritten as

$$\frac{d\mathbf{x}(t)}{dt} - \mathbf{A}\mathbf{x}(t) = \mathbf{f}(t)$$

If we premultiply each side of this equation by $e^{-\mathbf{A}t}$, we obtain

$$e^{-\mathbf{A}t}\left[\frac{d\mathbf{x}(t)}{dt} - \mathbf{A}\mathbf{x}(t)\right] = e^{-\mathbf{A}t}\mathbf{f}(t)$$

which may be rewritten as (see Corollary 1 of Section 4.4)

$$\frac{d}{dt}\left[e^{-\mathbf{A}t}\mathbf{x}(t)\right] = e^{-\mathbf{A}t}\mathbf{f}(t)$$

Integrating this last equation between the limits of t_0 and t, we have

$$\int_{t_0}^{t} \frac{d}{dt}\left[e^{-\mathbf{A}t}\mathbf{x}(t)\right]dt = \int_{t_0}^{t} e^{-\mathbf{A}t}\mathbf{f}(t)dt$$

or

$$e^{-\mathbf{A}t}\mathbf{x}(t)\Big|_{t_0}^{t} = \int_{t_0}^{t} e^{-\mathbf{A}s}\mathbf{f}(s)ds \tag{5.13}$$

Note that we have replaced the dummy variable t by the dummy variable s in the right-side of Equation (5.13), which has *no* effect on the definite integral (see Problem 1). Evaluating the left side of Equation (5.13), we obtain

$$e^{-\mathbf{A}t}\mathbf{x}(t) - e^{\mathbf{A}t_0}\mathbf{x}(t_0) = \int_{t_0}^{t} e^{-\mathbf{A}s}\mathbf{f}(s)ds$$

or

$$e^{-\mathbf{A}t}\mathbf{x}(t) = e^{\mathbf{A}t_0}\mathbf{c} + \int_{t_0}^{t} e^{-\mathbf{A}s}\mathbf{f}(s)ds \tag{5.14}$$

where we substituted for $\mathbf{x}(t_0)$ the initial condition $\mathbf{x}(t_0) = \mathbf{c}$. We solve explicitly for $\mathbf{x}(t)$ by premultiplying both sides of Equation (5.14) by $(e^{-\mathbf{A}t})^{-1}$, whence

$$\mathbf{x}(t) = \left(e^{-\mathbf{A}t}\right)^{-1}e^{\mathbf{A}t_0}\mathbf{c} + \left(e^{-\mathbf{A}t}\right)^{-1}\int_{t_0}^{t} e^{-\mathbf{A}s}\mathbf{f}(s)ds \tag{5.15}$$

But $(e^{-\mathbf{A}t})^{-1} = e^{\mathbf{A}t}$ (see Theorem 2 of Section 4.4). Also, $\mathbf{A}t$ commutes with $\mathbf{A}t_0$, so $e^{\mathbf{A}t}e^{\mathbf{A}t_0} = e^{\mathbf{A}(t-t_0)}$ (see Problem 36 of Section 4.4). Equation (5.15) may be simplified to

$$x(t) = e^{-\mathbf{A}(t-t_0)}c + e^{\mathbf{A}t}\int_{t_0}^{t} e^{-\mathbf{A}s}f(s)ds \tag{5.16}$$

and we have proven

▶THEOREM 1

The solution to the system $\dfrac{dx(t)}{dt} = \mathbf{Ax}(t) + \mathbf{f}(t); \mathbf{x}(t_0) = \mathbf{c}$ *in fundamental form is* $x(t) = e^{\mathbf{A}(t-t_0)}c + e^{\mathbf{A}t}\int_{t_0}^{t} e^{-\mathbf{A}s}f(s)ds$. ◀

A simple technique for calculating the matrices $e^{\mathbf{A}(t-t_0)}$ and $e^{-\mathbf{A}s}$ is to first find $e^{\mathbf{A}t}$ and then replace the variable t wherever it appears by the quantities $(t-t_0)$ and $(-s)$, respectively.

Example 1 $e^{\mathbf{A}t} = \begin{bmatrix} e^{-t} & te^{-t} \\ 0 & e^{-t} \end{bmatrix}$ for $\mathbf{A} = \begin{bmatrix} -1 & 1 \\ 0 & -1 \end{bmatrix}$. Consequently, $e^{\mathbf{A}(t-t_0)} = \begin{bmatrix} e^{-(t-t_0)} & (t-t_0)e^{-(t-t_0)} \\ 0 & e^{-(t-t_0)} \end{bmatrix}$ and $e^{-\mathbf{A}s} = \begin{bmatrix} e^{s} & -se^{s} \\ 0 & e^{s} \end{bmatrix}$.

Note that when t is replaced by $(t-t_0)$ in e^{-t}, the result is $e^{-(t-t_0)} = e^{-t+t_0}$ and *note* e^{-t-t_0}. That is, we replace the *quantity* t by the *quantity* $(t-t_0)$; we do not simply add $-t_0$ to the variable t wherever t appeared.

Example 2 Use matrix methods to solve

$$\frac{du(t)}{dt} = u(t) + 2v(t) + 1$$

$$\frac{dv(t)}{dt} = 4u(t) + 3v(t) - 1$$

$$u(0) = 1, v(0) = 2$$

Solution: This system can be transformed into fundamental form if we define

$$\mathbf{x}(t) = \begin{bmatrix} u(t) \\ v(t) \end{bmatrix}, \quad \mathbf{A} = \begin{bmatrix} 1 & 2 \\ 4 & 3 \end{bmatrix}, \quad \mathbf{f}(t) = \begin{bmatrix} 1 \\ -1 \end{bmatrix}, \quad \text{and} \quad \mathbf{c} = \begin{bmatrix} 1 \\ 2 \end{bmatrix}$$

and take $t_0 = 0$. For this $\mathbf{A}$, we calculate

$$e^{\mathbf{A}t} = \frac{1}{6}\begin{bmatrix} 2e^{5t} + 4e^{-t} & 2e^{5t} - 2e^{-t} \\ 4e^{5t} - 4e^{-t} & 4e^{5t} + 2e^{-t} \end{bmatrix}$$

Hence,

$$e^{-\mathbf{A}s} = \frac{1}{6}\begin{bmatrix} 2e^{-5s}+4e^{s} & 2e^{-5s}-2e^{s} \\ 4e^{-5s}-4e^{s} & 4e^{-5s}+2e^{s} \end{bmatrix}$$

and

$$e^{\mathbf{A}(t-t_0)} = e^{\mathbf{A}t}$$

since $t_0=0$. Thus,

$$e^{\mathbf{A}(t-t_0)}\mathbf{c} = \frac{1}{6}\begin{bmatrix} 2e^{5t}+4e^{-t} & 2e^{5t}-2e^{-t} \\ 4e^{5t}-4e^{-t} & 4e^{5t}+2e^{-t} \end{bmatrix}\begin{bmatrix} 1 \\ 2 \end{bmatrix}$$

$$= \frac{1}{6}\begin{bmatrix} 1[2e^{5t}+4e^{-t}]+2[2e^{5t}-2e^{-t}] \\ 1[4e^{5t}-4e^{-1}]+2[4e^{5t}+2e^{-t}] \end{bmatrix}$$

$$= \begin{bmatrix} e^{5t} \\ 2e^{5t} \end{bmatrix}. \tag{5.17}$$

$$e^{-\mathbf{A}s}\mathbf{f}(s) = \frac{1}{6}\begin{bmatrix} 2e^{-5s}+4e^{s} & 2e^{-5s}-2e^{s} \\ 4e^{-5s}-4e^{s} & 4e^{-5s}+2e^{s} \end{bmatrix}\begin{bmatrix} 1 \\ -1 \end{bmatrix}$$

$$= \frac{1}{6}\begin{bmatrix} 1[2e^{-5s}+4e^{s}]-1[2e^{-5s}-2e^{s}] \\ 1[4e^{-5t}-4e^{s}]-1[4e^{-5s}+2e^{s}] \end{bmatrix} = \begin{bmatrix} e^{s} \\ -e^{s} \end{bmatrix}.$$

Hence,

$$\int_{t_0}^{t} e^{-\mathbf{A}s}\mathbf{f}(s)ds = \begin{bmatrix} \int_0^t e^{s}ds \\ \int_0^t -e^{s}ds \end{bmatrix} = \begin{bmatrix} e^{s}|_0^t \\ -e^{s}|_0^t \end{bmatrix} = \begin{bmatrix} e^{t}-1 \\ -e^{t}+1 \end{bmatrix}$$

$$e^{\mathbf{A}t}\int_{t_0}^{t} e^{-\mathbf{A}s}\mathbf{f}(s)ds = \frac{1}{6}\begin{bmatrix} 2e^{5t}+4e^{-t} & 2e^{5t}-2e^{-t} \\ 4e^{5t}-4e^{-t} & 4e^{5t}+2e^{-t} \end{bmatrix}\begin{bmatrix} (e^{t}-1) \\ (1-e^{t}) \end{bmatrix} \tag{5.18}$$

$$= \frac{1}{6}\begin{bmatrix} [2e^{5t}+4e^{-t}][e^{t}-1]+[2e^{5t}-2e^{-t}][1-e^{t}] \\ [4e^{5t}-4e^{-t}][e^{t}-1]+[4e^{5t}+2e^{-t}][1-e^{t}] \end{bmatrix}$$

$$= \begin{bmatrix} (1-e^{-t}) \\ (-1+e^{-t}) \end{bmatrix}$$

Substituting Equations (5.17) and (5.18) into Equation (5.16), we have

$$\begin{bmatrix} u(t) \\ v(t) \end{bmatrix} = \mathbf{x}(t) = \begin{bmatrix} e^{5t} \\ 2e^{5t} \end{bmatrix} + \begin{bmatrix} 1-e^{-t} \\ -1+e^{-t} \end{bmatrix} = \begin{bmatrix} e^{5t}+1-e^{-t} \\ 2e^{5t}-1+e^{-t} \end{bmatrix}$$

or

$$u(t) = e^{5t} - e^{-t} + 1$$
$$v(t) = 2e^{5t} + e^{-t} - 1$$

Example 3 Use matrix methods to solve

$$\frac{d^2y}{dt^2} - 3\frac{dy}{dt} + 2y = e^{-3t}$$

$$y(1) = 1, \quad \frac{dy(1)}{dt} = 0$$

Solution: This system can be transformed into fundamental form if we define

$$\mathbf{x}(t) = \begin{bmatrix} x_1(t) \\ x_2(t) \end{bmatrix}, \quad \mathbf{A} = \begin{bmatrix} 0 & 1 \\ -2 & 3 \end{bmatrix}, \quad \mathbf{f}(t) = \begin{bmatrix} 0 \\ e^{-3t} \end{bmatrix}, \text{ and } \mathbf{c} = \begin{bmatrix} 1 \\ 0 \end{bmatrix}$$

and take $t_0 = 0$. For this **A**, we calculate

$$e^{\mathbf{A}t} = \begin{bmatrix} -e^{2t}+2e^{t} & e^{2t}-e^{t} \\ -2e^{2t}+2e^{t} & 2e^{2t}-e^{t} \end{bmatrix}$$

Thus,

$$\begin{aligned} e^{\mathbf{A}(t-t_0)}\mathbf{c} &= \begin{bmatrix} -e^{2(t-1)}+2e^{(t-1)} & e^{2(t-1)}-e^{(t-1)} \\ -2e^{2(t-1)}+2e^{(t-1)} & 2e^{2(t-1)}-e^{(t-1)} \end{bmatrix} \begin{bmatrix} 1 \\ 0 \end{bmatrix} \\ &= \begin{bmatrix} -e^{2(t-1)}+2e^{(t-1)} \\ -2e^{2(t-1)}+2e^{(t-1)} \end{bmatrix} \end{aligned} \tag{5.19}$$

Now

$$\mathbf{f}(t) = \begin{bmatrix} 0 \\ e^{-3t} \end{bmatrix}, \quad \mathbf{f}(s) = \begin{bmatrix} 0 \\ e^{-3s} \end{bmatrix}$$

$$\begin{aligned} e^{-\mathbf{A}s}\mathbf{f}(s) &= \begin{bmatrix} -e^{-2s}+2e^{-s} & e^{-2s}-e^{-s} \\ -2e^{-2s}+2e^{-s} & 2e^{-2s}-e^{-s} \end{bmatrix} \begin{bmatrix} 0 \\ e^{-3s} \end{bmatrix} \\ &= \begin{bmatrix} e^{-5s}-e^{-4s} \\ 2e^{-5s}-e^{-4s} \end{bmatrix} \end{aligned}$$

Hence,

$$\int_{t_0}^{t} e^{-\mathbf{A}s}\mathbf{f}(s)ds = \begin{bmatrix} \int_1^t (e^{-5s} - e^{-4s})ds \\ \int_1^1 (2e^{-5s} - e^{-4s})ds \end{bmatrix}$$

$$= \begin{bmatrix} \left(-\frac{1}{5}\right)e^{-5t} + \left(\frac{1}{4}\right)e^{-4t} + \left(\frac{1}{5}\right)e^{-5} - \left(\frac{1}{4}\right)e^{-4} \\ \left(-\frac{2}{5}\right)e^{-5t} + \left(\frac{1}{4}\right)e^{-4t} + \left(\frac{2}{5}\right)e^{-5} - \left(\frac{1}{4}\right)e^{-4} \end{bmatrix}$$

$$e^{\mathbf{A}t}\int_{t_0}^{t} e^{-\mathbf{A}s}\mathbf{f}(s)ds = \begin{bmatrix} (-e^{2t} + 2e^{t}) & (e^{2t} - e^{t}) \\ (-2e^{2t} + 2e^{t}) & (2e^{2t} - e^{t}) \end{bmatrix}$$

$$\times \begin{bmatrix} \left(-\frac{1}{5}e^{-5t} + \frac{1}{4}e^{-4t} + \frac{1}{5}e^{-5} - \frac{1}{4}e^{-4}\right) \\ \left(-\frac{2}{5}e^{-5t} + \frac{1}{4}e^{-4t} + \frac{2}{5}e^{-5} - \frac{1}{4}e^{-4}\right) \end{bmatrix}$$

$$= \begin{bmatrix} \frac{1}{20}e^{-3t} + \frac{1}{5}e^{(2t-5)} - \frac{1}{4}e^{t-4} \\ -\frac{3}{20}e^{-3t} + \frac{2}{5}e^{(2t-5)} - \frac{1}{4}e^{t-4} \end{bmatrix} \tag{5.20}$$

Substituting Equations (5.19) and (5.20) into Equation (5.16), we have that

$$\mathbf{x}(t) = \begin{bmatrix} x_1(t) \\ x_2(t) \end{bmatrix} = \begin{bmatrix} -e^{2(t-1)} + 2e^{t-1} \\ -2e^{2(t-1)} + 2e^{t-1} \end{bmatrix} + \begin{bmatrix} \frac{1}{20}e^{-3t} + \frac{1}{3}e^{(2t-5)} - \frac{1}{4}e^{t-4} \\ \frac{1}{20}e^{-3t} + \frac{2}{5}e^{(2t-5)} - \frac{1}{4}e^{t-4} \end{bmatrix}$$

$$= \begin{bmatrix} -e^{2(t-1)} + 2e^{t-1} + \frac{1}{20}e^{-3t} + \frac{1}{5}e^{(2t-5)} - \frac{1}{4}e^{t-4} \\ -2e^{2(t-1)} + 2e^{t-1} + \frac{3}{20}e^{-3t} + \frac{2}{5}e^{(2t-5)} - \frac{1}{4}e^{t-4} \end{bmatrix}$$

It follows that the solution to the original initial-value problem is

$$y(t) = x_1(t) = -e^{2(t-1)} + 2e^{t-1}\left(\frac{1}{20}\right)e^{(2t-5)} - \frac{1}{4}e^{t-4}$$

The most tedious step in Example 3 was multiplying the matrix $e^{\mathbf{A}t}$ by the column matrix $\int_{t_0}^{t} e^{-\mathbf{A}s}f(s)ds$. This step can be eliminated if we are willing to tolerate a slightly more complicated integral. The integration in Equation (5.16) is with respect to the dummy variable s. If we bring the matrix $e^{\mathbf{A}t}$, appearing in front of the integral, inside the integral, we may rewrite Equation (5.16) as

An alternate form of the solution to a matrix differential equation in fundamental form is $x(t) = e^{\mathbf{A}(t-t_0)}c + \int_{t_0}^{t} e^{\mathbf{A}(t-s)}f(s)ds.$

$$x(t) = e^{\mathbf{A}(t-t_0)}c + \int_{t_0}^{t} e^{-\mathbf{A}t}e^{-\mathbf{A}s}f(s)ds \tag{5.21}$$

But $\mathbf{A}t$ and $-\mathbf{A}s$ commute, so $e^{\mathbf{A}t}e^{-\mathbf{A}s} = e^{\mathbf{A}(t-s)}$ and Equation (5.21) becomes

$$x(t) = e^{\mathbf{A}(t-t_0)}c + \int_{t_0}^{t} e^{\mathbf{A}(t-s)}f(s)ds \tag{5.22}$$

The matrix $e^{\mathbf{A}(t-s)}$ is obtained by replacing the variable t in $e^{\mathbf{A}t}$ by the quantity $(t-s)$.

Example 4 Use matrix methods to solve

$$\frac{d^2x}{dt^2} + x = 2$$

$$x(\pi) = 0, \quad \frac{dx(\pi)}{dt} = -1$$

Solution: This system can be transformed into fundamental form if we define

$$\mathbf{x}(t) = \begin{bmatrix} x_1(t) \\ x_2(t) \end{bmatrix}, \quad \mathbf{A} = \begin{bmatrix} 0 & 1 \\ -1 & 0 \end{bmatrix}, \quad \mathbf{f}(t) = \begin{bmatrix} 0 \\ 2 \end{bmatrix}, \quad \text{and} \quad \mathbf{c} = \begin{bmatrix} 0 \\ -1 \end{bmatrix}$$

and take $t_0 = \pi$. The solution to this initial-value problem is given by either Equation (5.16) or (5.22). In this example, we shall evaluate Equation (5.22), thereby saving one matrix multiplication. For this $\mathbf{A}$, $e^{\mathbf{A}t}$ was determined in Example 8 of Section 4.4 to be

$$e^{\mathbf{A}t} = \begin{bmatrix} \cos t & \sin t \\ -\sin t & \cos t \end{bmatrix}$$

Thus,

$$
\begin{aligned}
e^{\mathbf{A}(t-t_0)_c} &= \begin{bmatrix} \cos(t-\pi) & \sin(t-\pi) \\ -\sin(t-\pi) & \cos(t-\pi) \end{bmatrix} \begin{bmatrix} 0 \\ -1 \end{bmatrix} \\
&= \begin{bmatrix} -\sin(t-\pi) \\ -\cos(t-\pi) \end{bmatrix} \\
e^{\mathbf{A}(t-s)}\mathbf{f}(s) &= \begin{bmatrix} \cos(t-s) & \sin(t-s) \\ -\sin(t-s) & \cos(t-s) \end{bmatrix} \begin{bmatrix} 0 \\ 2 \end{bmatrix} \\
&= \begin{bmatrix} 2\sin(t-s) \\ 2\cos(t-s) \end{bmatrix}
\end{aligned}
\tag{5.23}
$$

Hence,

$$
\begin{aligned}
\int_{t_0}^{t} e^{\mathbf{A}(t-s)}\mathbf{f}(s)ds &= \begin{bmatrix} \int_{\pi}^{t} 2\sin(t-s)ds \\ \int_{\pi}^{t} 2\cos t(t-s)ds \end{bmatrix} \\
&= \begin{bmatrix} 2-2\cos(t-\pi) \\ 2\sin(t-\pi) \end{bmatrix}
\end{aligned}
\tag{5.24}
$$

Substituting Equations (5.23) and (5.24) into Equation (5.22) and using the trigonometric identities $\sin(t-\pi)=-\sin\ t$ and $\cos(t-\pi)=-\cos\ t$, we have

$$
\begin{aligned}
\begin{bmatrix} x_1(t) \\ x_2(t) \end{bmatrix} = \mathbf{x}(t) &= \begin{bmatrix} -\sin(t-\pi) \\ -\cos(t-\pi) \end{bmatrix} + \begin{bmatrix} 2-2\cos(t-\pi) \\ 2\sin(t-\pi) \end{bmatrix} \\
&= \begin{bmatrix} \sin t + 2\cos t + 2 \\ \cos t - 2\sin t \end{bmatrix}
\end{aligned}
$$

Thus, since $x(t)=x_1(t)$, it follows that the solution to the initial-value problem is given by

$$x(t) = \sin t + 2\cos t + 2$$

A great simplification to both Equation (5.16) and Equation (5.22) is effected when the differential equation is homogeneous, that is, when $\mathbf{f}(t)=\mathbf{0}$. In both formulas, the integral becomes a zero-column matrix, and the solution reduces to

$$x(t) = e^{\mathbf{A}(t-t_0)}c \tag{5.25}$$

Occasionally, one needs to solve a differential equation by itself, and not an entire initial-value problem. In such cases, the general solution is (see Problem 2)

$$x(t) = e^{\mathbf{A}t}\mathbf{k} + e^{\mathbf{A}t}\int e^{-\mathbf{A}t}f(t)dt \tag{5.26}$$

where $\mathbf{k}$ is an arbitrary column matrix of suitable dimension. The general solution to a homogeneous differential equation by itself is

$$x(t) = e^{\mathbf{A}t}\mathbf{k} \tag{5.27}$$

The solution to the homogeneous system

$$\frac{dx(t)}{dt} = \mathbf{A}\mathbf{x}(t);$$

$\mathbf{x}(t_0) = \mathbf{c}$ is $x(t) = e^{\mathbf{A}(t-t_0)}c$.

Example 5 Use matrix methods to solve

$$\frac{du(t)}{dt} = u(t) + 2v(t)$$

$$\frac{dv(t)}{dt} = 4u(t) + 3v(t)$$

Solution: This system can be transformed into fundamental form if we define

$$\mathrm{x}(t) = \begin{bmatrix} u(t) \\ v(t) \end{bmatrix}, \quad \mathbf{A} = \begin{bmatrix} 1 & 2 \\ 4 & 3 \end{bmatrix}, \quad \text{and} \quad \mathbf{f}(t) = \begin{bmatrix} 0 \\ 0 \end{bmatrix}$$

This is a homogeneous system with no initial conditions specified; the general solution is given in Equation (5.27). For this $\mathbf{A}$, we have

$$e^{\mathbf{A}t} = \frac{1}{6}\begin{bmatrix} 2e^{5t} + 4e^{-t} & 2e^{5t} - 2e^{-t} \\ 4e^{5t} - 4e^{-t} & 4e^{5t} + 2e^{-t} \end{bmatrix}$$

Thus,

$$\begin{aligned} e^{\mathbf{A}t}\mathbf{k} &= \frac{1}{6}\begin{bmatrix} 2e^{5t} + 4e^{-t} & 2e^{5t} - 2e^{-t} \\ 4e^{5t} - 4e^{-t} & 4e^{5t} + 2e^{-t} \end{bmatrix}\begin{bmatrix} k_1 \\ k_2 \end{bmatrix} \\ &= \frac{1}{6}\begin{bmatrix} k_1[2e^{5t} + 4e^{-t}] + k_2[2e^{5t} - 2e^{-t}] \\ k_1[4e^{5t} - 4e^{-t}] + k_2[4e^{5t} + 2e^{-t}] \end{bmatrix} \\ &= \frac{1}{6}\begin{bmatrix} e^{5t}(2k_1 + 2k_2) + e^{-t}(4k_1 - 2k_2) \\ e^{5t}(4k_1 + 4k_2) + e^{-t}(-4k_1 + 2k_2) \end{bmatrix} \end{aligned} \tag{5.28}$$

Substituting Equation (5.28) into Equation (5.27), we have that

$$\begin{bmatrix} u(t) \\ v(t) \end{bmatrix} = \mathbf{x}(t) = \frac{1}{6}\begin{bmatrix} e^{5t}(2k_1 + 2k_2) + e^{-t}(4k_1 - 2k_2) \\ e^{5t}(4k_1 + 4k_2) + e^{-t}(-4k_1 + 2k_2) \end{bmatrix}$$

or

$$
\begin{aligned}
u(t) &= \left(\frac{2k_1 + 2k_2}{6}\right)e^{5t} + \left(\frac{4k_1 - 2k_2}{6}\right)e^{-t} \\
v(t) &= 2\left(\frac{2k_1 + 2k_2}{6}\right)e^{5t} + \left(\frac{-4k_1 + 2k_2}{6}\right)e^{-t}
\end{aligned} \tag{5.29}
$$

We can simplify the expressions for $u(t)$ and $v(t)$ if we introduce two new arbitrary constants k_3, and k_4 defined by

$$k_3 = \frac{2k_1 + 2k_2}{6}, \quad k_4 = \frac{4k_1 - 2k_2}{6}$$

Substituting these values into Equation (5.29), we obtain

$$
\begin{aligned}
u(t) &= k_3 e^{5t} + k_4 e^{-t} \\
v(t) &= 2k_3 e^{5t} - k_4 e^{-t}
\end{aligned}
$$

Problems 5.3

(1) Show by direct integration that

$$\int_{t_0}^{t} t^2 dt = \int_{t_0}^{t} s^2 ds = \int_{t_0}^{t} p^2 dp$$

In general, show that if $f(t)$ is integrable on the interval $[a, b]$, then

$$\int_a^b f(t)dt = \int_a^b f(s)ds$$

Hint: Assume $\int f(t)dt = F(t) + c$. Hence, $\int f(s)ds = F(s) + c$. Then use the fundamental theorem of integral calculus.

(2) Derive Equation (5.26). *Hint:* Follow the derivation of Equation (5.16) using indefinite integration, rather than definite integration, and note that

$$\int \frac{d}{dt}\left[e^{-\mathbf{A}t}\mathbf{x}(t)\right]dt = e^{-\mathbf{A}t}\mathbf{x}(t) + \mathbf{k}$$

where $\mathbf{k}$ is an arbitrary column matrix of integration.

(3) Find (a) e^{-At}, (b) $e^{A(t-2)}$, (c) $e^{A(t-s)}$, (d) $e^{-A(t-2)}$, if

$$e^{At} = e^{3t}\begin{bmatrix} 1 & t & t^2/2 \\ 0 & 1 & t \\ 0 & 0 & 1 \end{bmatrix}$$

(4) Find (a) e^{-At}, (b) e^{-As}, (c) $e^{A(t-3)}$, if

$$e^{At} = \frac{1}{6}\begin{bmatrix} 2e^{5t}+4e^{-t} & 2e^{5t}-2e^{-t} \\ 4e^{5t}-4e^{-t} & 4e^{5t}+2e^{-t} \end{bmatrix}$$

(5) Find (a) e^{-At}, (b) e^{-As}, (c) $e^{-A(t-s)}$, if

$$e^{At} = \frac{1}{3}\begin{bmatrix} -\sin 3t + 3\cos t & 5\sin 3t \\ -2\sin 3t & \sin 3t + 3\cos 3t \end{bmatrix}$$

(6) Determine which of the following column vectors **x** are solutions to the system

$$\frac{d}{dt}\begin{bmatrix} x_1(t) \\ x_2(t) \end{bmatrix} = \begin{bmatrix} 0 & 1 \\ -1 & 0 \end{bmatrix}\begin{bmatrix} x_1(t) \\ x_2(t) \end{bmatrix}; \quad \begin{bmatrix} x_1(0) \\ x_2(0) \end{bmatrix} = \begin{bmatrix} 1 \\ 0 \end{bmatrix}$$

$$\text{(a)}\begin{bmatrix} \sin t \\ \cos t \end{bmatrix}, \quad \text{(b)}\begin{bmatrix} e^t \\ 0 \end{bmatrix}, \quad \text{(c)}\begin{bmatrix} \cos t \\ -\sin t \end{bmatrix}.$$

(7) Determine which of the following column vectors **x** are solutions to the system

$$\frac{d}{dt}\begin{bmatrix} x_1(t) \\ x_2(t) \end{bmatrix} = \begin{bmatrix} 1 & 2 \\ 4 & 3 \end{bmatrix}\begin{bmatrix} x_1(t) \\ x_2(t) \end{bmatrix}; \quad \begin{bmatrix} x_1(0) \\ x_2(0) \end{bmatrix} = \begin{bmatrix} 1 \\ 2 \end{bmatrix}$$

$$\text{(a)}\begin{bmatrix} e^{-t} \\ -e^{-t} \end{bmatrix}, \quad \text{(b)}\begin{bmatrix} e^{-t} \\ 2e^{-t} \end{bmatrix}, \quad \text{(c)}\begin{bmatrix} e^{5t} \\ 2e^{5t} \end{bmatrix}.$$

(8) Determine which of the following column vectors **x** are solutions to the system

$$\frac{d}{dt}\begin{bmatrix} x_1(t) \\ x_2(t) \end{bmatrix} = \begin{bmatrix} 0 & 1 \\ -2 & 3 \end{bmatrix}\begin{bmatrix} x_1(t) \\ x_2(t) \end{bmatrix}; \quad \begin{bmatrix} x_1(1) \\ x_2(1) \end{bmatrix} = \begin{bmatrix} 1 \\ 0 \end{bmatrix}$$

$$\text{(a)}\begin{bmatrix} -e^{2t}+2e^{t} \\ -2e^{2t}+2e^{t} \end{bmatrix}, \quad \text{(b)}\begin{bmatrix} -e^{2(t-1)}+2e^{(t-1)} \\ -2e^{2(t-1)}+2e^{(t-1)} \end{bmatrix}, \quad \text{(c)}\begin{bmatrix} e^{2(t-1)} \\ 0 \end{bmatrix}.$$

Solve the systems described in Problems 9 through 16 by matrix methods. Note that Problems 9 through 12 have the same coefficient matrix.

(9) $$\frac{dx(t)}{dx} = -2x(t) + 3y(t)$$
$$\frac{dy(t)}{dt} = -x(t) + 2y(t)$$
$$x(2) = 2, \quad y(2) = 4$$

(10) $$\frac{dx(t)}{dt} = -2x(t) + 3y(t) + 1$$
$$\frac{dy(t)}{dt} = -x(t) + 2y(t) + 1$$
$$x(1) = 1, \quad y(1) = 1$$

(11) $$\frac{dx(t)}{dt} = -2x(t) + 3y(t)$$
$$\frac{dy(t)}{dt} = -x(t) + 2y(t)$$

(12) $$\frac{dx(t)}{dt} = -2x(t) + 3y(t) + 1$$
$$\frac{dy(t)}{dt} = -x(t) + 2y(t) + 1$$

(13) $$\frac{d^2x}{dt^2} + 4x = \sin t; \quad x(0) = 1, \frac{dx(0)}{dt} = 0$$

(14) $$\frac{d^3x}{dt^3} = t; \quad x(1) = 1, \frac{dx(1)}{dt} = 2, \frac{d^2x(1)}{dt^2} = 3$$

(15) $$\frac{d^2x}{dt^2} - \frac{dx}{dt} - 2x = e^{-e}; \quad x(0) = 1, \frac{dx(0)}{dt} = 0$$

(16) $$\frac{d^2x}{dt^2} = 2\frac{dx}{dt} + 5y + 3$$
$$\frac{dy}{dt} = -\frac{dx}{dt} - 2y$$
$$x(0) = 0, \frac{dx(0)}{dt} = 0, y(0) = 1.$$

5.4 MODELING AND DIFFERENTIAL EQUATIONS

Mathematical models are used in virtually all branches of science, technology, and engineering. Many models are presented in terms of differential equations. There is a delicate balance between making sure a model is "reflective enough" to govern or mirror a situation, and—at the same time—"easy enough" to solve the associated equations.

In this section, we consider a mixing problem which will be modeled by a system of differential equations. In our discussion, we will make various assumptions and then "tweak" the model by changing various parameters.

Consider Figure 5.1. A saline solution, of concentration 2 pounds of salt/gal, is introduced into Tank 1 at a rate of 5 gal/min. As we can see from the diagram, the tanks are connected by a system of pipes.

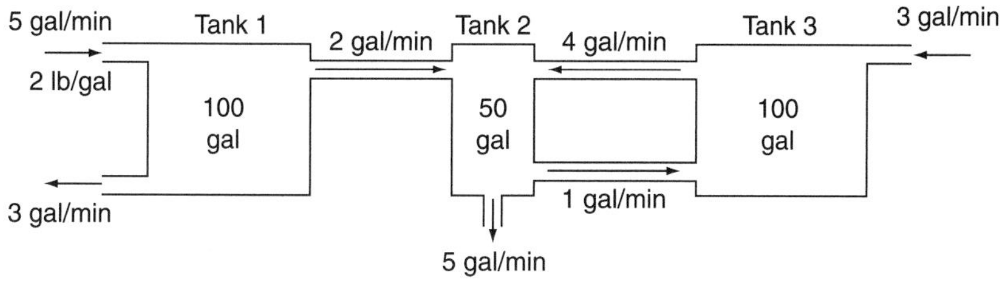

FIGURE 5.1

Assuming that the salt is distributed uniformly in the solution, we will model the problem with the following variables:

$$t = \text{time (min)}$$
$$S_1(t) = \text{amount of salt in Tank 1 at time } t \text{ (pounds)}$$
$$S_2(t) = \text{amount of salt in Tank 2 at time } t \text{ (pounds)}$$
$$S_3(t) = \text{amount of salt in Tank 3 at time } t \text{ (pounds)}$$
$$\frac{dS_k}{dt} = \text{rate of change of salt in Tank } k \text{ (pounds/min)}, \ k = 1,2,3$$

Let us now consider Tank 1. Because there are three pipes connected to the tank, the rate of change of the salt in this tank will have three terms:

$$\frac{dS_1}{dt} = \frac{5\,\text{gal}}{\text{min}} \times \frac{2\,\text{lbs}}{\text{gal}} - \frac{S_1\,\text{lb}}{100\,\text{gal}} \times \frac{2\,\text{gal}}{\text{min}} - \frac{S_1\,\text{lb}}{100\,\text{gal}} \times \frac{3\,\text{gal}}{\text{min}} \tag{5.30}$$

We note in this equation the consistency of units (lbs/min) and the division by the capacity of Tank 1 (100 gal).

The two other tanks are modeled as follows:

$$\frac{dS_2}{dt} = \frac{S_1\,\text{lb}}{100\,\text{gal}} \times \frac{2\,\text{gal}}{\text{min}} + \frac{S_3\,\text{lb}}{100\,\text{gal}} \times \frac{4\,\text{gal}}{\text{min}} - \frac{S_2\,\text{lb}}{50\,\text{gal}} \times \frac{1\,\text{gal}}{\text{min}} - \frac{S_2\,\text{lb}}{50\,\text{gal}} \times \frac{3\,\text{gal}}{\text{min}} \tag{5.31}$$

$$\frac{dS_3}{dt} = \frac{S_2\,\text{lb}}{50\,\text{gal}} \times \frac{1\,\text{gal}}{\text{min}} - \frac{S_3\,\text{lb}}{100\,\text{gal}} \times \frac{4\,\text{gal}}{\text{min}} - \frac{0\,\text{lb}}{\text{gal}} \times \frac{3\,\text{gal}}{\text{min}} \tag{5.32}$$

We note here that the last term of Equation (5.32) is 0, because there is no salt in the incoming solution from the right.

Finally, let us assume that initially there is no salt in any tank. That is,

$$S_1(0) = S_2(0) = S_3(0) = 0$$

We now will rewrite our problem in matrix notation

$$\frac{d}{dt}\begin{bmatrix} S_1 \\ S_2 \\ S_3 \end{bmatrix} = \begin{bmatrix} \frac{-5}{100} & 0 & 0 \\ \frac{2}{100} & \frac{-6}{50} & \frac{4}{100} \\ 0 & \frac{1}{50} & \frac{-4}{100} \end{bmatrix} \begin{bmatrix} S_1 \\ S_2 \\ S_3 \end{bmatrix} + \begin{bmatrix} 10 \\ 0 \\ 0 \end{bmatrix} \tag{5.33}$$

We can now expand on the techniques discussed in Sections 5.2 and 5.3 to solve this problem. However, in this case, the use of technological methods is preferred (see Appendix D). This is primarily due to the fact that we have a 3-by-3 coefficient matrix instead of a 2-by-2 matrix.

We end our discussion with the following observations and ask the following questions:

We note that the system was "closed"; that is, the amount of solution coming in (8 gal) is equal to the amount going out (8 gal). What if this was not the case?

We assumed no salt was initially present. What if this was not the case?

If the salt in the solution was not uniformly distributed, the modeling of our problem becomes much more difficult. The same is true if the solution is not introduced continuously. In these cases, our approach must be radically altered and a numerical approach might be more useful.

Problems 5.4

(1) Assume vat V_1 is placed above vat V_2 and that both vats have a capacity of 100 l. If 7 l of a sucrose solution (5 kg sugar/l) is poured into V_1 every minute, how much sugar is in each vat at time t, if V_1 drains into V_2 at the rate of 7 l/min, while V_2 drains off at the same rate and there is no sugar in either vat initially?

(2) Consider the previous problem. If vat V_2 drains off at a rate of 8 l/min, how much sugar will it contain in the long run, realizing that it will eventually be empty?

(3) Consider the previous problem. If vat V_2 drains off at a rate of 6 l/min, how much sugar will it contain in the long run, realizing that it will eventually overflow?

(4) Solve problem 1 if $V_1(0)=5$ and $V_2(0)=12$.

(5) Suppose two lakes (x and y) are connected by a series of canals in such a way that the rate of change of the pollution in each lake can be modeled by the following matrix equation:

$$\frac{d}{dt}\begin{bmatrix} x \\ y \end{bmatrix} = \begin{bmatrix} -2 & 3 \\ 4 & -3 \end{bmatrix} \begin{bmatrix} x \\ y \end{bmatrix} + \begin{bmatrix} 1 \\ 0 \end{bmatrix}$$

where $x(t)$ and $y(t)$ represent the amount of pollution (in tons) at time t (months). If both lakes are initially clean, find the amount of pollution at time t, along with the long-range pollution in each lake.

(6) Do the previous problem if the model is given by

$$\frac{d}{dt}\begin{bmatrix} x \\ y \end{bmatrix} = \begin{bmatrix} -2 & -3 \\ -4 & -3 \end{bmatrix}\begin{bmatrix} x \\ y \end{bmatrix} + \begin{bmatrix} 1 \\ 0 \end{bmatrix}$$

(7) Suppose Problem 5 is modeled by

$$\frac{d}{dt}\begin{bmatrix} x \\ y \end{bmatrix} = \begin{bmatrix} -2 & -3 \\ -4 & -3 \end{bmatrix}\begin{bmatrix} x \\ y \end{bmatrix}$$

with $x(0) = 100$, and $y(0) = 300$. Find the long-range pollution of each lake.

5.5 A BRIEF INTRODUCTION TO GRAPHS AND NETWORKS

One area of Mathematics that has a definite starting point is the field of Graph Theory, which can trace its origin to Leonhard Euler's 1736 solution to the Königsberg Bridge Problem (for more information, see Hopkins, Brian, and Robin J. Wilson, (2004). "The bridges of Königsberg." *The College Mathematics Journal* 35.3: 198-207). A *graph* can be thought of as a picture that shows a set of points, some of which are related. The relationship is indicated by the placement of lines between the points. In graphs, the lines have no direction, so traversal between the points can occur in either direction along the line. If the order mattered, they would be called *directed graphs*, but they are not under consideration here.

The points are called *vertices*, and lines are called *edges*. If a pair of vertices is joined by more than one edge, the edge is called a *multiple edge*, and the graph is called a *multigraph*. Graphs without multiple edges are called *simple graphs*. When a graph on n vertices has an edge between every pair of vertices, the graph is called a *complete graph on n vertices*, denoted K_n. The number of edges incident on a vertex is the *degree* of the vertex, and if all the vertices have equal degree r, the graph is *regular of degree r*.

If in a graph, one can begin at a particular vertex, traverse through several other vertices via incident edges, never repeated a vertex or edge, and return to the starting vertex, then the part of the graph just described is called a *cycle*. Figure 5.6 of Section 5.6 depicts the graph C_4, a cycle on four vertices. Not every graph contains a cycle, or cycles. In some graphs, it might not be possible to find a sequence of vertices and edges between every pair of vertices. If there is a pair of vertices for which such a sequence does not exist, the graph is said to be *disconnected*. Otherwise, the graph is *connected*, and there is a sequence of vertices and edges between every pair of vertices. If such a sequence exists and is unique for every pair of vertices, then the graph is a *tree*. Trees also are acyclic, that is, they have no cycles. A tree which includes every vertex of a graph is a *spanning tree*.

In Figure 5.2 of Section 5.6, the vertex sequence (and incident edges) formed by 1-2-3 is a tree, while 1-2-3-1 is a cycle (as is 1-2-3-4-1), and 1-2-3-4 is a spanning tree.

Graphs can be used to model different types of networks, such as transportation networks, communications networks, or computer networks. The actual behavior of such a network can be modeled more completely by including some assumptions about the vertices and edges, and matrices play a critical role in this analysis. One such model assumes that the vertices (i.e., the landmasses, telephones, or computers) are always "operational", while the edges (i.e., the bridges, telephone lines, or computer cables) fail with some numerical probability. This is an example of a problem in *network reliability*: can any pair of vertices communicate with each other via a path through the surviving links? Our way of framing the question is: does failure of certain links still lead to a surviving graph that has a spanning tree? If so, how many spanning trees does it have?

5.6 THE ADJACENCY MATRIX

Figure 5.2 shows a graph; its vertices are labeled 1, 2, 3, and 4 arbitrarily, and its edges are labeled using the end vertices in numerical order, although such an order does not matter for undirected graphs. We define Adjacency Matrix of a simple graph $\mathbf{A}(G)$ to be an $n \times n$ matrix with entry a_{ij} denoting the number of edges from v_i to v_j. For simple graphs these entries are always either 0 or 1. The adjacency matrix for the graph in Figure 5.2 would, therefore, have first row [0,1,1,0] because there are no edges labeled "(1,1) " or "(1,4)", but one edge each has label "(1,2)" and "(1,3)." Note that, if edge "(1,1)" were to exist, it would be a so-called self-loop from vertex 1 to itself. The existence of at least one self-loop means that the graph would be termed a *pseudograph*. The full adjacency matrix for the graph in Figure 5.2 would therefore be $\mathbf{A} = \begin{bmatrix} 0 & 1 & 1 & 0 \\ 1 & 0 & 1 & 1 \\ 1 & 1 & 0 & 1 \\ 0 & 1 & 1 & 0 \end{bmatrix}$.

Note that all zeroes down the main diagonal indicate that the graph has no

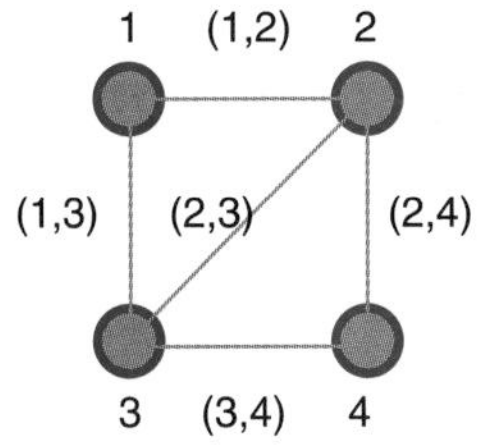

FIGURE 5.2

self-loops. We note that **A** is symmetric about the main diagonal, that is, the (x,y) entry is equal to the (y,x) entry, where x and y represent a row or column number.

We will define the left hand side of the characteristic equation, $\det(\mathbf{A}-\lambda\mathbf{I})=0$ (4.4), as the *characteristic polynomial* for the matrix in question. The characteristic polynomial for the adjacency matrix of a graph contains some very important information about the graph, as we see in our next theorem.

▶THEOREM 2

Let G be a graph having adjacency matrix **A** *and characteristic polynomial* $\det(\mathbf{A}-\lambda\mathbf{I})=a_0+a_1\lambda+a_2\lambda^2+\ldots+a_{n-2}\lambda^{n-2}+a_{n-1}\lambda^{n-1}+\lambda^n$. *Then the coefficients of the characteristic polynomial give the following information about the graph:*

(i) $-a_{n-2}$ *is the number of edges in G;*
(ii) $-a_{n-3}$ *is twice the number of triangles in G;*
(iii) *the number of edge sequences of length k joining the vertices v_i and v_j of a graph G is equal to the ij-th entry of the matrix* $\mathbf{A}(G)^k$. ◀

We will explore the proof of part (iii) of Theorem 1 in the exercises, and observe that part (iii) is demonstrated when the exponent $k=1$, that is, any edge is itself an edge sequence of length one between its end vertices. We further observe that the numbering convention we select is not absolute, that is, we would obtain the same characteristic polynomial and information if the nodes were numbered differently, and the matrix entries altered accordingly. It is beyond the scope of this text to prove this notion.

Example 1 For the graph in Figure 5.2,

$$\det(\mathbf{A}-\lambda\mathbf{I})=\det\left(\begin{bmatrix}-\lambda & 1 & 1 & 0\\ 1 & -\lambda & 1 & 1\\ 1 & 1 & -\lambda & 1\\ 0 & 1 & 1 & -\lambda\end{bmatrix}\right)=-4\lambda-5\lambda^2+\lambda^4. \quad \text{Here,} \quad n=4,$$

indicating that there are $-a_{n-2}=-a_{n-2}=-a_2=-(-5)=5$ edges in the graph, and $-a_{n-3}=-a_{4-3}=-a_1=-(-4)=4=2(2)$, so G has two triangles (namely, formed by vertices 1-2-3 and their incident edges, and vertices 2-3-4 and their incident edges). If we raise **A** to the second power, we will find out the number of paths of length two between each pair of vertices, including those that begin and end at the same vertex, so $\mathbf{A}^2=\mathbf{A}\times\mathbf{A}=\begin{bmatrix}0&1&1&0\\1&0&1&1\\1&1&0&1\\0&1&1&0\end{bmatrix}\times$

$$\begin{bmatrix}0&1&1&0\\1&0&1&1\\1&1&0&1\\0&1&1&0\end{bmatrix}=\begin{bmatrix}2&1&1&2\\1&3&2&1\\1&2&3&1\\2&1&1&2\end{bmatrix}$$ Thus, for example, the (1,4) (and, of course,

(4,1)) entry of $\mathbf{A}^2$ is 2, so there are two paths of length 2 between that pair of vertices in G, that is, 1-2-4 and 1-3-4, and no others. The (2,2) entry is 3, indicating 3-4-3, 3-2-3, and 3-1-3 are the only paths of length 2 from vertex 2 to itself.

As was demonstrated in Chapter 4, the zeros of the characteristic polynomial are the eigenvalues for the matrix, and the eigenvalues of adjacency matrices contain more information about the graph's structure. The list of a matrix's eigenvalues is its *spectrum*. Theorem 3 gives some details about the spectra of adjacency matrices.

▶THEOREM 3

Let G be a (nonpseudo) graph having adjacency matrix **A**, *whose characteristic polynomial* $\det(\mathbf{A}-\lambda\mathbf{I}) = a_0 + a_1\lambda + a_2\lambda^2 + \ldots + a_{n-2}\lambda^{n-2} + a_{n-1}\lambda^{n-1} + \lambda^n$ *has factorization* $(\lambda-\alpha_1)(\lambda-\alpha_2)\cdots(\lambda-\alpha_{n-1})(\lambda-\alpha_n)$, *where* $\alpha_1 \le \alpha_2 \le \cdots \le \alpha_n$. *Then*

(i) $\sqrt{\dfrac{2(-a_{n-2})(n-1)}{n}} \ge \alpha_n$;

(ii) $\sum_{i=1}^{n} \alpha_i = 0$;

(iii) $\sum_{i=1}^{n} (\alpha_i)^2 = 2(-a_{n-2})$. ◀

Example 2 For the graph given in Figure 5.2, its characteristic polynomial $\lambda^4 - 5\lambda^2 - 4\lambda$ has roots $\frac{1}{2} - \frac{\sqrt{17}}{2}, -1, 0, \frac{1}{2} + \frac{\sqrt{17}}{2}$. Clearly, the upper bound on the largest eigenvalue for this graph is $\frac{\sqrt{2(-a_{n-2})(n-1)}}{n} = \frac{\sqrt{2(s)(4-1)}}{4} = \sqrt{\frac{30}{4}} = \sqrt{\frac{15}{2}} \ge \alpha_4 = \frac{1}{2} + \frac{\sqrt{17}}{2}$, satisfying (i), while $\frac{1}{2} - \frac{\sqrt{17}}{2} + (-1) + 0 + \frac{1}{2} + \frac{\sqrt{17}}{2} = 0$, demonstrating (ii), and $\left(\frac{1}{2} - \frac{\sqrt{17}}{2}\right)^2, +(-1)^2 + 0^2 + \left(\frac{1}{2} + \frac{\sqrt{17}}{2}\right)^2 = 10 = 2(5)$, or twice the number of edges of, as per (iii).

As stated earlier, graphs that have vertices of all the same degree, say, r, are called regular graphs. We present the following theorem regarding the eigenvalues of regular graphs.

▶THEOREM 4

Let G be a graph that is regular of degree r having adjacency matrix **A**. *Then*

(i) *r is an eigenvalue of* **A**;

(ii) *r is the largest magnitude of an eigenvalue for* **A**, *that is,* $|\alpha_i| \le r$ *for all i.* ◀

Problems 5.6

We recommend the use of computer software to assist in the computation of characteristic polynomials and eigenvalues in the next two sections.

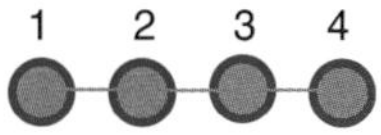

FIGURE 5.3

Problems 1-6 refer to the graph in Figure 5.3. The graph is a path on four vertices and will be referred to as P_4.

(1) Find the adjacency matrix for the graph P_4.

(2) Find the characteristic polynomial for the adjacency matrix for P_4.

(3) Verify Theorem 2, parts (i) and (ii) for P_4.

(4) (a) Find A^3 for P_4.
(b) How many paths of length 3 are there between vertex "1" and each of the other vertices in P_4?

(5) Find the eigenvalues for the characteristic polynomial found in problem 2.

(6) Verify Theorem 3, parts (i)-(iii), for the characteristic polynomial found in problem 2.

(7) Prove Theorem 2, part (iii).

Problems 8-13 refer to the graph in Figure 5.4. The graph is on four vertices and will be referred to as G_1.

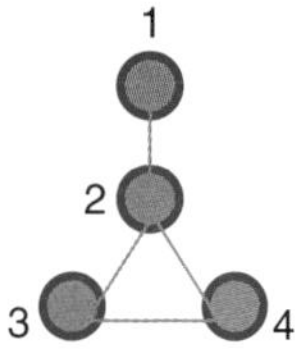

FIGURE 5.4

(8) Find the adjacency matrix for the graph G_1.

(9) Find the characteristic polynomial for the adjacency matrix for G_1.

(10) Verify Theorem 2, parts (i) and (ii) for G_1.

(11) (a) Find A^3 for G_1.
(b) How many paths of length 3 are there between vertex "2" and each of the other vertices in G_1?

(12) Find the eigenvalues for the characteristic polynomial found in problem 9.

(13) Verify Theorem 3, parts (i)-(iii), for the characteristic polynomial found in problem 9.

Problems 14-17 refer to the graph in Figure 5.5. The graph is on four vertices and will be referred to as G_2.

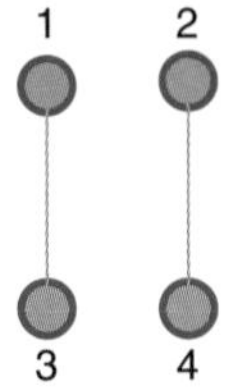

FIGURE 5.5

(14) Find the adjacency matrix for the graph G_2.

(15) Find the characteristic polynomial for the adjacency matrix for G_2.

(16) Find the eigenvalues for the characteristic polynomial found in problem 15.

(17) Verify Theorem 3, parts (i)-(iii), for the characteristic polynomial found in problem 15.

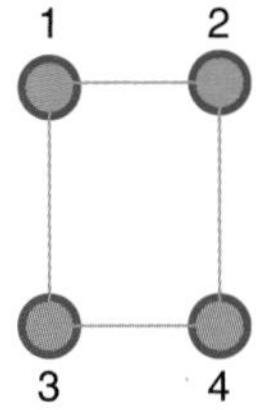

FIGURE 5.6

Problems 18-22 refer to the graph in Figure 5.6. The graph is on four vertices and will be referred to as C_4.

(18) Find the adjacency matrix for the graph C_4.

(19) Find the characteristic polynomial for the adjacency matrix for C_4.

(20) Find the eigenvalues for the characteristic polynomial found in problem 19.

(21) Verify Theorem 3, parts (i)-(iii), for the characteristic polynomial found in problem 19.

(22) Verify Theorem 4, part (i), for the characteristic polynomial found in problem 19.

(23) Prove Theorem 4, part (i).

5.7 THE LAPLACIAN MATRIX

For example 1, we consider the graph in Figure 5.2.

Example 1 Since the degree of vertices 1, 2, 3, and 4 are, respectively, 2, 3, 3, and 2, we can form a diagonal matrix using the degrees of the corresponding vertices as the diagonal entries. For the graph in Figure 5.2, the matrix is $\mathbf{D} = \begin{bmatrix} 2 & 0 & 0 & 0 \\ 0 & 3 & 0 & 0 \\ 0 & 0 & 3 & 0 \\ 0 & 0 & 0 & 2 \end{bmatrix}$.

Recall that the graph had adjacency matrix $\mathbf{A} = \begin{bmatrix} 0 & 1 & 1 & 0 \\ 1 & 0 & 1 & 1 \\ 1 & 1 & 0 & 1 \\ 0 & 1 & 1 & 0 \end{bmatrix}$. We can form the Laplacian matrix for the graph, denoted $\mathbf{L}$, where $\mathbf{L}=\mathbf{D}-\mathbf{A}$, by

$$\mathbf{L} = \begin{bmatrix} 2 & 0 & 0 & 0 \\ 0 & 3 & 0 & 0 \\ 0 & 0 & 3 & 0 \\ 0 & 0 & 0 & 2 \end{bmatrix} - \begin{bmatrix} 0 & 1 & 1 & 0 \\ 1 & 0 & 1 & 1 \\ 1 & 1 & 0 & 1 \\ 0 & 1 & 1 & 0 \end{bmatrix} = \begin{bmatrix} 2 & -1 & -1 & 0 \\ -1 & 3 & -1 & -1 \\ -1 & -1 & 3 & -1 \\ 0 & -1 & -1 & 2 \end{bmatrix}.$$

We remark that the Laplacian matrix is sometimes referred to as the nodal admittance matrix in Electrical Engineering applications. Further, this matrix's main diagonal is comprised of the degrees of its corresponding vertices, it has a "-1" wherever there is an edge between the two associated vertices, and each row and column sum up to "0."

In 1847, Gustav Robert Kirchhoff published a paper, the title of which translates to "On the solution of the equations obtained from the investigation of the linear distribution of galvanic currents," *Annalen der Physik und Chemie*, in which his work led to the study of spanning trees of connected graphs.

►THEOREM 5. KIRCHHOFF'S MATRIX-TREE THEOREM

All cofactors of **L** *are equal and their common value is the number of spanning trees in the associated graph.* ◄

The proof of this well-known theorem involves some advanced matrix theory that is beyond the scope of this text.

Theorem 5 can be employed to prove a more useful way to determine the number of spanning trees of a graph. It appeared in the paper by A. K. Kelmans and V. M. Chelnokov, (1974). "A certain polynomial of graph and graphs with an extremal number of trees," *Journal of Combinatorial Theory B* 16: 197-214 and we state it here.

▶THEOREM 6

The number of spanning trees t(G) of the Laplacian matrix of a graph is related to the eigenvalues as follows: $t(G) = \frac{1}{n}\prod_{i=2}^{n} \lambda_i(G)$, $0 = \lambda_1 \leq \lambda_2 \leq \ldots \leq \lambda_n$ ◀

Example 2 For the graph in Figure 5.2, we have $L = \begin{bmatrix} 2 & -1 & -1 & 0 \\ -1 & 3 & -1 & -1 \\ -1 & -1 & 3 & -1 \\ 0 & -1 & -1 & 2 \end{bmatrix}$ and

$$\det(\mathbf{L} - \lambda\mathbf{I}) = \begin{vmatrix} 2-\lambda & -1 & -1 & 0 \\ -1 & 3-\lambda & -1 & -1 \\ -1 & -1 & 3-\lambda & -1 \\ 0 & -1 & -1 & 2-\lambda \end{vmatrix} = \lambda^4 - 10\lambda^3 + 32\lambda^2 - 32\lambda$$

$$= \lambda(\lambda-2)(\lambda-4)^2.$$

This translates to $\frac{1}{4}(2)(4)(4) = 8$ spanning trees for the graph in Figure 5.2. One such spanning tree is 1-2-3-4; yet another is 1-2-4-3. In the exercises, you will be asked to determine the remaining six spanning trees for the graph.

We note that the eigenvalue product to determine the number of spanning trees of a graph eliminates the first (smallest) eigenvalue, which is always zero. The number of Laplacian eigenvalues that equal zero indicates the number of connected pieces (called components) that constitute the graph. Therefore, a connected graph will have a single eigenvalue equal to zero, and a graph that has three distinct pieces, between which there are not any links, will have three eigenvalues equaling zero. Since the formula only removes one such eigenvalue, there will be zeros in the spanning tree product, which indicates zero spanning trees for a disconnected graph (which, of course, is the case).

We illustrate this fact with an example.

Example 3 For the graph in Figure 5.5, we have

$$\mathbf{L} = \mathbf{D} - \mathbf{A} = \begin{bmatrix} 1 & 0 & 0 & 0 \\ 0 & 1 & 0 & 0 \\ 0 & 0 & 1 & 0 \\ 0 & 0 & 0 & 1 \end{bmatrix} - \begin{bmatrix} 0 & 0 & 1 & 0 \\ 0 & 0 & 0 & 1 \\ 1 & 0 & 0 & 0 \\ 0 & 1 & 0 & 0 \end{bmatrix} = \begin{bmatrix} 1 & 0 & -1 & 0 \\ 0 & 1 & 0 & -1 \\ -1 & 0 & 1 & 0 \\ 0 & -1 & 0 & 1 \end{bmatrix}$$

and

$$\det(\mathbf{L} - \lambda\mathbf{I}) = \begin{vmatrix} 1-\lambda & 0 & -1 & 0 \\ 0 & 1-\lambda & 0 & -1 \\ -1 & 0 & 1-\lambda & 0 \\ 0 & -1 & 0 & 1-\lambda \end{vmatrix} = \lambda^4 - 4\lambda^3 + 4\lambda^2 = \lambda^2(\lambda-2)^2.$$

This translates to $\frac{1}{4}(0)(2)(2) = 0$ spanning trees.

We conclude this section with some useful results about a graph's Laplacian eigenvalues.

▶THEOREM 7

*Let G be a graph on n vertices and e edges, and **L** its associated Laplacian matrix. Then,*

(i) *for all i,* $\lambda_i \geq 0$;
(ii) $\lambda_n \leq n$;
(iii) *trace*(**L**) $= 2e = \sum_{i=1}^{n} \lambda_t$ ◀

Problems 5.7

We recommend the use of computer software to assist in the computation of characteristic polynomials and eigenvalues in this section and the previous section.

(1) List the spanning trees for the graph in Figure 5.2 of the previous section.

Problems 2-6 refer to the graph P_4 in Figure 5.3 of the previous section.

(2) Find the Laplacian matrix for the graph P_4.

(3) Find the characteristic polynomial for the Laplacian matrix for P_4.

(4) Find the eigenvalues for the characteristic polynomial found in problem 3.

(5) Apply Theorem 6 to determine the number of spanning trees for P_4.

(6) Verify Theorem 7 for the eigenvalues of the Laplacian matrix associated with P_4.

Problems 7-12 refer to the graph G_1 depicted in Figure 5.4 of the previous section.

(7) List the spanning trees for the graph in Figure 5.4.

(8) Find the Laplacian matrix for the graph G_1.

(9) Find the characteristic polynomial for the Laplacian matrix for G_1.

(10) Find the eigenvalues for the characteristic polynomial found in problem 9.

(11) Apply Theorem 6 to determine the number of spanning trees for G_1.

(12) Verify Theorem 7 for the eigenvalues for the Laplacian matrix associated with G_1.

Recall: When a graph on n vertices has an edge between every pair of vertices, the graph is a *complete graph on n vertices*, denoted K_n. Exercises 13-17 ask you to work with complete graphs and conjecture a general formula for their number of spanning trees.

(13) (i) Draw K_3.
(ii) Find the Laplacian matrix for K_3.
(iii) Find the eigenvalues for K_3.

(14) (i) Draw K_4.
(ii) Find the Laplacian matrix for K_4.
(iii) Find the eigenvalues for K_4.

(15) (i) Draw K_5.
(ii) Find the Laplacian matrix for K_5.
(iii) Find the eigenvalues for K_5.

(16) What pattern do you see in your responses to questions 13c, 14c, and 15c? Can you formulate a conjecture about the Laplacian eigenvalues for any complete graph K_n?

(17) Using your response to question 16, and using Theorem 6, can you generalize a formula for the number of spanning trees for any graph K_n? This result is known as Cayley's Theorem, after Arthur Cayley.

CHAPTER 5 REVIEW

Important Terms

fundamental form of differential equations
homogeneous differential equation
initial conditions
model
nonhomogeneous differential equation
regular graph
complete graph
adjacency matrix
Laplacian matrix
spanning trees

Important Concepts

Section 5.1

- A differential equation in the unknown functions $x_1(t), x_2(t), \ldots, x_n(t)$ is an equation that involves these functions and one or more of their derivatives.

Section 5.2

- The solution to the system $\dfrac{dx(t)}{dt} = \mathbf{A}\mathbf{x}(t) + \mathbf{f}(t); \quad \mathbf{x}(t_0) = \mathbf{c}$ is

$$\mathbf{x}(t) = e^{\mathbf{A}(t-t_0)}\mathbf{c} + e^{\mathbf{A}t}\int_{t_0}^{t} e^{-\mathbf{A}s}\mathbf{f}(s)ds$$

$$= e^{\mathbf{A}(t-t_0)}\mathbf{c} + \int_{t_0}^{t} e^{-\mathbf{A}(t-s)}\mathbf{f}(s)ds$$

- The solution to the homogenous equation

$$\frac{dx(t)}{dt} = \mathbf{Ax}(t); \quad \mathbf{x}(t_0) = \mathbf{c} \text{ is}$$

$$\mathbf{x}(t) = e^{\mathbf{A}(t-t_0)}\mathbf{c}$$

Section 5.3

- Models are useful in everyday life.

Section 5.4

- Graphs can be used to model different types of networks, such as transportation networks, communications networks, or computer networks.
- Matrices can play a critical role in analysis of networks represented by graphs.

Section 5.5

- The coefficients in the characteristic polynomial and the eigenvalues of the adjacency matrix give information about the corresponding graph.

Section 5.6

- The Laplacian matrix of a graph can be formed from the adjacency matrix.
- The eigenvalues of the Laplacian matrix can be used to determine the number of spanning trees in the corresponding network represented by the graph.

CHAPTER 6

Euclidean Inner Product

Chapter Outline

6.1 ORTHOGONALITY

Perpendicularity is such a useful concept in Euclidean geometry that we want to extend the notion to all finite dimensional vector spaces. This is relatively easy for vector spaces of two or three dimensions, because such vectors have graphical representations. Each vector in a two-dimensional vector space can be written as a 2-tuple and graphed as a directed line segment (arrow) in the plane. Similarly, each vector in a three-dimensional vector space can be written as a 3-tuple and graphed as a directed line segment in space. Using geometrical principles on such graphs, we can determine whether directed line segments from the same vector space meet at right angles. However, to extend the concept of perpendicularity to $\mathbb{R}^n$, $n>3$, we need a different approach.

The *Euclidean inner product* of two column matrices $\mathbf{x}=[x_1\ x_2\ x_3 \cdots x_n]^{\mathrm{T}}$ and $\mathbf{y}=[y_1\ y_2\ y_3 \cdots y_n]^{\mathrm{T}}$ in $\mathbb{R}^n$, denoted by $\langle \mathbf{x}, \mathbf{y} \rangle$ is

$$\langle \mathbf{x}, \mathbf{y} \rangle = x_1y_1 + x_2y_2 + x_3y_3 + \cdots + x_ny_n \tag{6.1}$$

To calculate the Euclidean inner product, we multiply corresponding components of two column matrices in $\mathbb{R}^n$ and sum the resulting products. Although we will work exclusively in this chapter with n-tuples written as column matrices, the Euclidean inner product is equally applicable to row matrices. Either way, the

Euclidean inner product of two vectors in $\mathbb{R}^n$ is a real number and *not* another vector in $\mathbb{R}^n$. In terms of column matrices,

The inner product of two vectors **x** and **y** in $\mathbb{R}^n$ is a real number determined by multiplying corresponding components of **x** and **y** and then summing the resulting products.

$$\langle \mathbf{x}, \mathbf{y} \rangle = \mathbf{x}^{\mathrm{T}}\mathbf{y} \tag{6.2}$$

Example 1 The Euclidean inner product of $\mathbf{x} = \begin{bmatrix} 1 \\ 2 \\ 3 \end{bmatrix}$ and $\mathbf{y} = \begin{bmatrix} 4 \\ -5 \\ 6 \end{bmatrix}$ in $\mathbb{R}^3$ is

$$\langle \mathbf{x}, \mathbf{y} \rangle = 1(4) + 2(-5) + 3(6) = 12$$

while the Euclidean inner product of $\mathbf{u} = \begin{bmatrix} 20 \\ -4 \\ 30 \\ 10 \end{bmatrix}$ and $\mathbf{v} = \begin{bmatrix} 10 \\ -5 \\ -8 \\ -6 \end{bmatrix}$ in $\mathbb{R}^4$ is

$$\langle \mathbf{u}, \mathbf{v} \rangle = 20(10) + (-4)(-5) + 30(-8) + 10(-6) = -80$$

▶THEOREM 1

If **x**, **y**, *and* **z** *are vectors in* $\mathbb{R}^n$, *then*

(a) $\langle \mathbf{x}, \mathbf{x} \rangle$ *is positive if* $\mathbf{x} \neq 0$; $\langle \mathbf{x}, \mathbf{x} \rangle = \mathbf{0}$ *if and only if* $\mathbf{x} = \mathbf{0}$.
(b) $\langle \mathbf{x}, \mathbf{y} \rangle = \langle \mathbf{y}, \mathbf{x} \rangle$.
(c) $\langle \lambda\mathbf{x}, \mathbf{y} \rangle = \lambda\langle \mathbf{x}, \mathbf{y} \rangle$, *for any real number* λ.
(d) $\langle \mathbf{x}+\mathbf{z}, \mathbf{y} \rangle = \langle \mathbf{x}, \mathbf{y} \rangle + \langle \mathbf{z}, \mathbf{y} \rangle$.
(e) $\langle 0, y \rangle = 0$.◀

Proof: We prove parts (a) and (b) here and leave the proofs of the other parts as exercises (see Problems 28 through 30). With $\mathbf{x} = [x_1\ x_2\ x_3 \ldots x_n]^{\mathrm{T}}$, we have

$$\langle \mathbf{x}, \mathbf{x} \rangle = (x_1)^2 + (x_2)^2 + (x_3)^3 + \ldots + (x_n)^2$$

This sum of squares is zero if and only if $x_1 = x_2 = x_3 = \ldots = x_n = 0$, which in turn implies that $\mathbf{x} = \mathbf{0}$. If any component is not zero, that is, if **x** is not the zero vector in $\mathbb{R}^n$, then the sum of the squares must be positive.

For part (b), we set $\mathbf{y} = [y_1\ y_2\ y_3 \ldots y_n]^{\mathrm{T}}$. Then

$$\begin{aligned} \langle \mathbf{x}, \mathbf{y} \rangle &= x_1y_1 + x_2y_2 + x_3y_3 + \ldots + x_ny_n \\ &= y_1x_1 + y_2x_2 + y_3x_3 + \ldots + y_nx_n \\ &= \langle \mathbf{y}, \mathbf{x} \rangle \end{aligned}$$

The magnitude of an *n*-tuple **x** (see Section 2.1) is related to the Euclidean inner product by the formula

$$\|\mathbf{x}\| = \sqrt{\langle \mathbf{x}, \mathbf{x} \rangle} = \sqrt{\mathrm{x}_1^2 + \mathrm{x}_2^2 + \mathrm{x}_3^2 + \ldots + \mathrm{x}_n^2}. \tag{6.3}$$

Example 2 The magnitude of $\mathbf{x}=[2 \quad -3 \quad -4]^{\mathrm{T}}$ in $\mathbb{R}^3$ is

$$\|\mathbf{x}\| = \sqrt{\langle \mathbf{x}, \mathbf{x}\rangle} = \sqrt{(2)^2 + (-3)^2 + (-4)^2} = \sqrt{29}$$

while the magnitude of $\mathbf{y}=[1 \quad -1 \quad 1 \quad -1]^{\mathrm{T}}$ in $\mathbb{R}^4$ is

$$\|\mathbf{y}\| = \sqrt{\langle \mathbf{y}, \mathbf{y}\rangle} = \sqrt{(1)^2 + (-1)^2 + (1)^2 + (-1)^2} = 2$$

The magnitude of a vector $\mathbf{x}$ in $\mathbb{R}^n$ is the square root of the inner product of $\mathbf{x}$ with itself.

A *unit vector* is a vector having a magnitude of 1. A nonzero vector $\mathbf{x}$ is *normalized* if it is divided by its magnitude. It follows that

$$\begin{aligned}
\left\langle \frac{1}{\|\mathbf{x}\|}\mathbf{x}, \frac{1}{\|\mathbf{x}\|}\mathbf{x} \right\rangle &= \frac{1}{\|\mathbf{x}\|}\left\langle \mathbf{x}, \frac{1}{\|\mathbf{x}\|}\mathbf{x} \right\rangle \quad \text{Part (c) of Theorem 1} \\
&= \frac{1}{\|\mathbf{x}\|}\left\langle \frac{1}{\|\mathbf{x}\|}\mathbf{x}, \mathbf{x} \right\rangle \quad \text{Part (b) of Theorem 1} \\
&= \left(\frac{1}{\|\mathbf{x}\|}\right)^2 \langle \mathbf{x}, \mathbf{x}\rangle \quad \text{Part (c) of Theorem 1} \\
&= \left(\frac{1}{\|\mathbf{x}\|}\right)^2 \|\mathbf{x}\|^2 \\
&= 1
\end{aligned}$$

Thus, a normalized vector is always a unit vector.

As with other vector operations, the Euclidean inner product has a geometrical interpretation in two or three dimensions. For simplicity, we consider two-dimensional vectors here; the extension to three dimensions is straightforward.

Let $\mathbf{u}$ and $\mathbf{v}$ be two nonzero vectors in $\mathbb{R}^2$ represented by directed line segments in the plane, each emanating from the origin. The *angle between* $\mathbf{u}$ *and* $\mathbf{v}$ is the angle θ between the two line segments, with $0^\circ \leq \theta \leq 180^\circ$ as illustrated in Figure 6.1. The

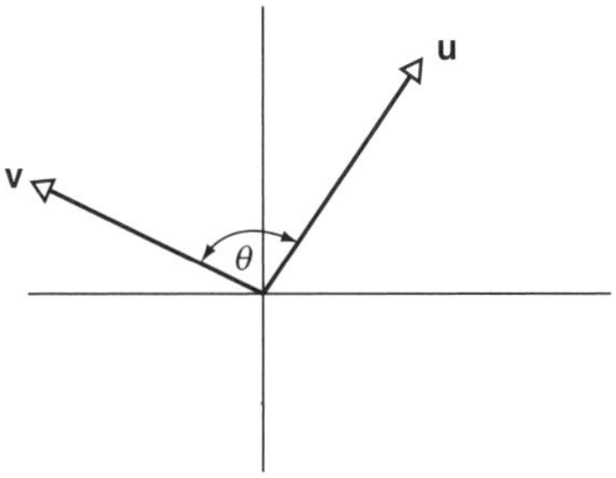

FIGURE 6.1

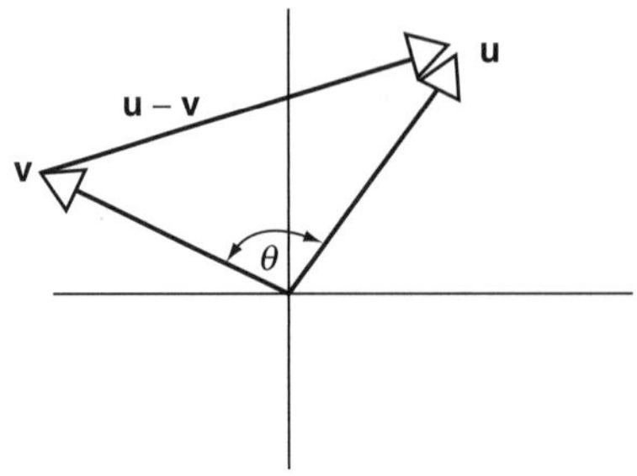

FIGURE 6.2

vectors **u** and **v**, along with their difference **u** − **v**, form a triangle (see Figure 6.2) having sides $\|\mathbf{u}\|$, $\|\mathbf{v}\|$, and $\|\mathbf{u}-\mathbf{v}\|$. It follows from the law of cosines that

$$\|\mathbf{u}-\mathbf{v}\|^2 = \|\mathbf{u}\|^2 + \|\mathbf{v}\|^2 - 2\|\mathbf{u}\|\|\mathbf{v}\|\cos\theta$$

where upon

$$\begin{aligned}\|\mathbf{u}\|\|\mathbf{v}\|\cos\theta &= \frac{1}{2}\left(\|\mathbf{u}\|^2 + \|\mathbf{v}\|^2 - \|\mathbf{u}-\mathbf{v}\|^2\right)\\ &= \frac{1}{2}(\langle\mathbf{u},\mathbf{u}\rangle + \langle\mathbf{v},\mathbf{v}\rangle - \langle\mathbf{u}-\mathbf{v},\mathbf{u}-\mathbf{v}\rangle)\\ &= \frac{1}{2}(\langle\mathbf{u},\mathbf{u}\rangle + \langle\mathbf{v},\mathbf{v}\rangle - [\langle\mathbf{u},\mathbf{u}\rangle - 2\langle\mathbf{u},\mathbf{v}\rangle + \langle\mathbf{v},\mathbf{v}\rangle])\\ &= \langle\mathbf{u},\mathbf{v}\rangle\end{aligned} \tag{6.4}$$

and

$$\cos\theta = \frac{\langle\mathbf{u},\mathbf{v}\rangle}{\|\mathbf{u}\|\|\mathbf{v}\|} \tag{6.5}$$

We use Equation (6.5) to calculate the angle between two directed line segments in $\mathbb{R}^2$.

Example 3 Find the angle between the vectors $\mathbf{u} = \begin{bmatrix}2\\5\end{bmatrix}$ and $\mathbf{v} = \begin{bmatrix}-3\\4\end{bmatrix}$.

Solution:

$$\langle\mathbf{u},\mathbf{v}\rangle = 2(-3)(-3) + 5(4) = 14,\ \|\mathbf{u}\| = \sqrt{4+25} = \sqrt{29},\ \|\mathbf{v}\| = \sqrt{9+16} = 5,$$

$$\textit{so, } \cos\theta = \frac{14}{5\sqrt{29}} \approx 0.5199, \text{ and } \theta \approx 58.7°.$$

If **u** is a nonzero vector in $\mathbb{R}^2$, we have from Theorem 1 that **(u, u)** is positive and then, from Equation (6.3), that $\|\mathbf{u}\| > 0$. Similarly, if **v** is a nonzero vector in $\mathbb{R}^2$, then $\|\mathbf{v}\| > 0$. Because

$$\langle \mathbf{u}, \mathbf{v} \rangle = \|\mathbf{u}\| \|\mathbf{v}\| \cos \theta \qquad (6.4 \text{ repeated})$$

We see that the inner product of two nonzero vectors in $\mathbb{R}^2$ is 0 if and only if $\cos \theta = 0$. The angle θ is the angle between the two directed line segments representing **u** and **v** (see Figure 6.1) with $0° \leq \theta < 180°$. Thus, $\cos \theta = 0$ if and only if $\theta = 90°$, from which we conclude that the inner product of two nonzero vectors in $\mathbb{R}^2$ is 0 if and only if their directed line segments form a right angle. Here now is a characteristic of perpendicularity we can extend to n-tuples of all dimensions! We use the word orthogonal instead of perpendicularity for generalizations to higher dimensions, and say that two vectors in the same vector space are *orthogonal* if their inner product is 0.

Two vectors in the same vector space are orthogonal if their Euclidean inner product is zero.

Example 4 For the vectors $\mathbf{x} = \begin{bmatrix} 1 \\ 2 \\ 3 \end{bmatrix}$, $\mathbf{y} = \begin{bmatrix} -3 \\ -6 \\ 5 \end{bmatrix}$, and $\mathbf{z} = \begin{bmatrix} 0 \\ 5 \\ 6 \end{bmatrix}$ in $\mathbb{R}^3$ we have that **x** is orthogonal to **y** and **y** is orthogonal to **z**, because

$$\langle \mathbf{x}, \mathbf{y} \rangle = 1(-3) + 2(-6) + 3(5) = 0$$

and

$$\langle \mathbf{y}, \mathbf{z} \rangle = (-3)(0) + (-6)(5) + 5(6) = 0$$

but **x** is *not* orthogonal to **z**, because

$$\langle \mathbf{x}, \mathbf{z} \rangle = 1(0) + 2(5) + 3(6) = 28 \neq 0$$

As a direct consequence of Theorem 1, part (e), we have that the zero vector in $\mathbb{R}^n$ is orthogonal to every vector in $\mathbb{R}^n$.

▸THEOREM 2. (GENERALIZED THEOREM OF PYTHAGORAS)

*If **u** and **v** are orthogonal vectors in $\mathbb{R}^n$, then $\|\mathbf{u} - \mathbf{v}\|^2 = \|\mathbf{u}\|^2 + \|\mathbf{v}\|^2$.* ◂

Proof: In the special case of $\mathbb{R}^2$, this result reduces directly to Pythagoras's theorem when we consider the right triangle bounded by the directed line segments representing **u**, **v** and $\mathbf{u} - \mathbf{v}$ (see Figure 6.3). More generally, if **u** and **v** are orthogonal, then $\langle \mathbf{u}, \mathbf{v} \rangle = 0$ and

$$\begin{aligned} \|\mathbf{u} - \mathbf{v}\|^2 &= \langle \mathbf{u} - \mathbf{v}, \mathbf{u} - \mathbf{v} \rangle \\ &= \langle \mathbf{u}, \mathbf{u} \rangle - 2\langle \mathbf{u}, \mathbf{v} \rangle + \langle \mathbf{v}, \mathbf{v} \rangle \\ &= \langle \mathbf{u}, \mathbf{u} \rangle - 2(0) + \langle \mathbf{v}, \mathbf{v} \rangle \\ &= \|\mathbf{u}\|^2 + \|\mathbf{v}\|^2 \end{aligned}$$

▸THEOREM 3. (CAUCHY-SCHWARZ INEQUALITY)

*If **u** and **v** are vectors in $\mathbb{R}^n$, then $|\langle \mathbf{u}, \mathbf{v} \rangle| \leq \|\mathbf{u}\| \|\mathbf{v}\|$.* ◂

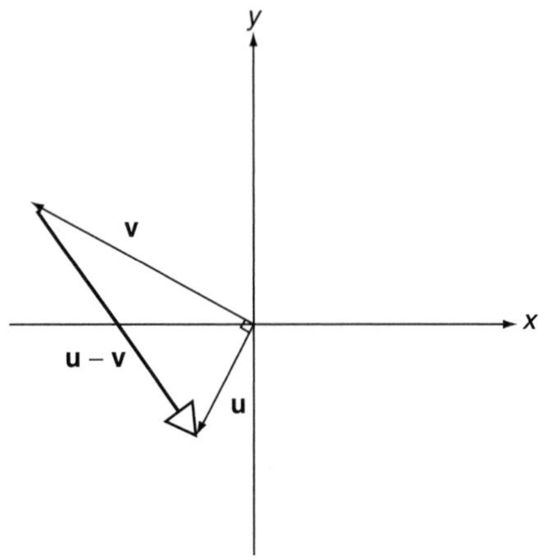

FIGURE 6.3

Proof: In the special case of $\mathbb{R}^2$, we have

$$\langle \mathbf{u}, \mathbf{v} \rangle = \|\mathbf{u}\| \|\mathbf{v}\| \cos\theta, \qquad \text{(6.4 repeated)}$$

hence

$$\begin{aligned} |\langle \mathbf{u}, \mathbf{v} \rangle| &= |\|\mathbf{u}\| \|\mathbf{v}\| \cos\theta| \\ &= |\|\mathbf{u}\| \|\mathbf{v}\| \cos\theta| \\ &\leq \|\mathbf{u}\| \|\mathbf{v}\| \end{aligned}$$

because $|\cos\theta| \leq 1$ for any angle θ. The proof for more general vector spaces is left as an exercise (see Problems 35 and 36).

Matrices can form a Euclidean inner product, but not every combination of matrices produces a Euclidean inner product.

Example 5 Show that $\langle \mathbf{A}, \mathbf{B} \rangle = \det(\mathbf{AB})$ does not represent a Euclidean inner product in the vector space $\mathbf{M}_{2\times2}$.

Solution: Let $\mathbf{A} = \begin{bmatrix} x & x \\ x & x \end{bmatrix}$ and $\mathbf{B} = \begin{bmatrix} x & x \\ x & x \end{bmatrix}$. Then $\det(\mathbf{AB}) = \det\begin{bmatrix} 2x^2 & 2x^2 \\ 2x^2 & 2x^2 \end{bmatrix} = 0$, but $\mathbf{A} \neq 0$, so part (a) of Theorem 1 is violated.

The Euclidean inner product in $\mathbb{R}^n$ induces an inner product on pairs of vectors in other n-dimensional vector spaces. A vector in an n-dimensional vector space $\mathbb{V}$ has a coordinate representation with respect to an underlying basis (see Section 2.5). We define an inner product on two vectors $\mathbf{x}$ and $\mathbf{y}$ in $\mathbb{V}$ by forming the Euclidean inner product on the coordinate representations of both vectors with respect to the *same* underlying basis.

Example 6 Calculate $\langle \mathbf{A}, \mathbf{B} \rangle$ for $\mathbf{A} = \begin{bmatrix} 4 & 3 \\ 6 & 2 \end{bmatrix}$ and $\mathbb{B} = \begin{bmatrix} 1 & 2 \\ 1 & 2 \end{bmatrix}$ in the vector space $\mathbb{M}_{2\times 2}$ with respect to the standard basis

$$\mathbb{S} = \left\{ \begin{bmatrix} 1 & 0 \\ 0 & 0 \end{bmatrix}, \begin{bmatrix} 0 & 1 \\ 0 & 0 \end{bmatrix}, \begin{bmatrix} 0 & 0 \\ 1 & 0 \end{bmatrix}, \begin{bmatrix} 0 & 0 \\ 0 & 1 \end{bmatrix} \right\}$$

Solution: The coordinate representations with respect to this basis are

An inner product is basis dependent. Two vectors can be orthogonal with respect to one basis and not orthogonal with respect to another basis.

$$\begin{bmatrix} 4 & 3 \\ 6 & 2 \end{bmatrix} \leftrightarrow \begin{bmatrix} 4 \\ 3 \\ 6 \\ 2 \end{bmatrix} \quad \text{and} \quad \begin{bmatrix} 1 & 2 \\ 1 & 2 \end{bmatrix} \leftrightarrow \begin{bmatrix} 1 \\ 2 \\ 1 \\ 2 \end{bmatrix}$$

The induced inner product is

$$\langle \mathbf{A}, \mathbf{B} \rangle = 4(1) + 3(2) + 6(1) + 2(2) = 20$$

With respect to the standard basis, the induced inner product of two matrices *of the same order* is obtained by multiplying corresponding elements of both matrices and summing the results.

Example 7 Redo Example 6 with respect to the basis

$$\mathbb{B} = \left\{ \begin{bmatrix} 0 & 1 \\ 1 & 1 \end{bmatrix}, \begin{bmatrix} 1 & 0 \\ 1 & 1 \end{bmatrix}, \begin{bmatrix} 1 & 1 \\ 0 & 1 \end{bmatrix}, \begin{bmatrix} 1 & 1 \\ 1 & 0 \end{bmatrix} \right\}$$

Solution: The coordinate representations with respect to this basis is (see Example 13 of Section 2.5)

$$\mathbf{x} = \begin{bmatrix} 4 & 3 \\ 6 & 2 \end{bmatrix} \leftrightarrow \begin{bmatrix} 1 \\ 2 \\ -1 \\ 3 \end{bmatrix}_{\mathbb{B}} \quad \text{and} \quad \mathbf{y} = \begin{bmatrix} 1 & 2 \\ 1 & 2 \end{bmatrix} \leftrightarrow \begin{bmatrix} 1 \\ 0 \\ 1 \\ 0 \end{bmatrix}_{\mathbb{B}}$$

The induced inner product is now

$$\langle \mathbf{A}, \mathbf{B} \rangle = 1(1) + 2(0) + (-1)(1) + 3(0) = 0$$

which is different from the inner product calculated in Example 6.

An inner product is basis dependent. Two vectors can be orthogonal with respect to one basis and not orthogonal with respect to another basis.

It follows from the previous two examples that an inner product depends on the underlying basis; different bases can induce different inner products. Consequently, two vectors can be orthogonal with respect to one basis, as in Example 6, and notorthogonal with respect to another basis, as in Example 5. We can see this distinction graphically, by considering the vectors

$$\mathbf{x} = \begin{bmatrix} 1 \\ 1 \end{bmatrix} \text{ and } \mathbf{y} = \begin{bmatrix} 1 \\ -1 \end{bmatrix}$$

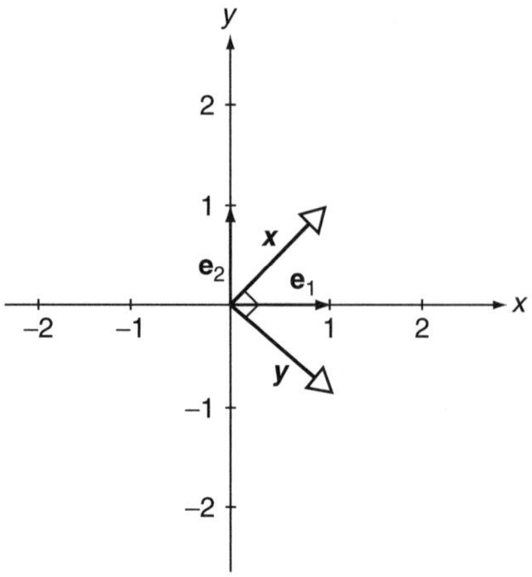

FIGURE 6.4

With respect to the standard basis

$$\mathbb{S} = \left\{ \mathbf{e}_1 = \begin{bmatrix} 1 \\ 0 \end{bmatrix}, \mathbf{e}_2 = \begin{bmatrix} 0 \\ 1 \end{bmatrix} \right\}$$

$\langle \mathbf{x}, \mathbf{y} \rangle = 0$, and $\mathbf{x}$ is perpendicular to $\mathbf{y}$, as illustrated in Figure 6.4. If, instead, we take as the basis

$$\mathbb{D} = \left\{ \mathbf{d}_1 = \begin{bmatrix} 2 \\ 1 \end{bmatrix}, \mathbf{d}_2 = \begin{bmatrix} 5 \\ 2 \end{bmatrix} \right\}$$

then we have as coordinate representations in the $\mathbb{D}$ basis,

$$\mathbf{x} = \begin{bmatrix} 1 \\ 1 \end{bmatrix} = (3)\begin{bmatrix} 2 \\ 1 \end{bmatrix} + (-1)\begin{bmatrix} 5 \\ 2 \end{bmatrix} \leftrightarrow \begin{bmatrix} 3 \\ -1 \end{bmatrix}_{\mathbb{D}}$$

$$\mathbf{y} = \begin{bmatrix} 1 \\ -1 \end{bmatrix} = (-7)\begin{bmatrix} 2 \\ 1 \end{bmatrix} + (3)\begin{bmatrix} 5 \\ 2 \end{bmatrix} \leftrightarrow \begin{bmatrix} -7 \\ 3 \end{bmatrix}_{\mathbb{D}}$$

$$\mathbf{d}_1 = \begin{bmatrix} 2 \\ 1 \end{bmatrix} = (1)\begin{bmatrix} 2 \\ 1 \end{bmatrix} + (0)\begin{bmatrix} 5 \\ 2 \end{bmatrix} \leftrightarrow \begin{bmatrix} 1 \\ 0 \end{bmatrix}_{\mathbb{D}}$$

$$\mathbf{d}_2 = \begin{bmatrix} 5 \\ 2 \end{bmatrix} = (0)\begin{bmatrix} 2 \\ 1 \end{bmatrix} + (1)\begin{bmatrix} 5 \\ 2 \end{bmatrix} \leftrightarrow \begin{bmatrix} 0 \\ 1 \end{bmatrix}_{\mathbb{D}}$$

Graphing the coordinate representations in the $\mathbb{D}$ basis, we generate Figure 6.5. Note that $\mathbf{x}$ and $\mathbf{y}$ are no longer perpendicular. Indeed, $\langle \mathbf{x}, \mathbf{y} \rangle = 3(-7) + (-1)(3) = 24 \neq 0$. Furthermore, $\langle \mathbf{x}, \mathbf{x} \rangle = (3)^2 + (-1)^2 = 10$, $\langle \mathbf{y}, \mathbf{y} \rangle = (-7)^2 + (3)^2 = 58$, and it follows from Equation (6.5) that the angle between $\mathbf{x}$ and $\mathbf{y}$ is

$$\theta = \arccos \frac{-24}{\sqrt{10}\sqrt{58}} \approx 175°$$

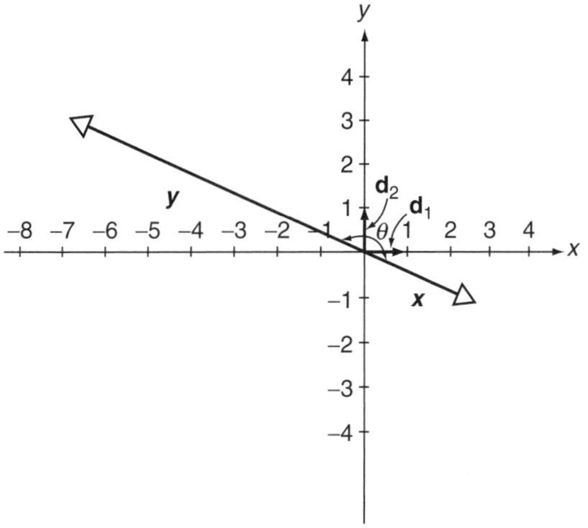

FIGURE 6.5

Example 8 Calculate $\langle p(t), q(t)\rangle$ with respect to the standard basis in $\mathbb{P}^2$ for

$$p(t) = 3t^2 - t + 5 \text{ and } q(t) = -2t^2 + 4t + 2$$

Solution: Using the standard basis $\mathbb{S} = \{t^2, t, 1\}$, we have the coordinate representations

$$3t^2 - t + 5 \leftrightarrow \begin{bmatrix} 3 \\ -1 \\ 5 \end{bmatrix} \text{ and } -2t^2 + 4t + 2 \leftrightarrow \begin{bmatrix} -2 \\ 4 \\ 2 \end{bmatrix}$$

An induced inner product of two polynomials is obtained by multiplying the coefficients of like powers of the variable and summing the results.

The induced inner product is

$$\langle p(t), q(t)\rangle = 3(-2) + (-1)(4) + 5(2) = 0$$

and the polynomials are orthogonal. With respect to the standard basis, the induced inner product of two polynomials is obtained by multiplying the coefficients of like powers of the variable and summing the results.

Problems 6.1

In Problems 1 through 17, (a) find $\langle \mathbf{x}, \mathbf{y}\rangle$, (b) find $|\mathbf{x}|$, and (c) determine whether $\mathbf{x}$ and $\mathbf{y}$ are orthogonal.

(1) $\mathbf{x} = [1 \quad 2]^T$, $\mathbf{y} = [3 \quad 4]^T$.

(2) $\mathbf{x} = [1 \quad 1]^T$, $\mathbf{y} = [-4 \quad 4]^T$.

(3) $\mathbf{x} = [-5 \quad 7]^T$, $\mathbf{y} = [3 \quad -5]^T$.

(4) $\mathbf{x} = [-2 \quad -8]^T$, $\mathbf{y} = [20 \quad -5]^T$.

(5) $\mathbf{x} = [-3 \quad 4]^T$, $\mathbf{y} = [0 \quad 0]^T$.

(6) $\mathbf{x} = [2 \quad 0 \quad 1]^T$, $\mathbf{y} = [1 \quad 2 \quad 4]^T$.

(7) $\mathbf{x} = [-2 \quad 2 \quad -4]^T$, $\mathbf{y} = [-4 \quad 3 \quad -3]^T$.

(8) $\mathbf{x} = [-3 \quad -2 \quad 5]^T$, $\mathbf{y} = [6 \quad -4 \quad -4]^T$.

(9) $\mathbf{x} = [10 \quad 20 \quad 30]^T$, $\mathbf{y} = [5 \quad -7 \quad 3]^T$.

(10) $\mathbf{x} = \left[\frac{1}{4} \quad \frac{1}{2} \quad \frac{1}{8}\right]^T$, $\mathbf{y} = \left[\frac{1}{3} \quad \frac{1}{3} \quad \frac{1}{3}\right]^T$.

(11) $\mathbf{x} = [1 \quad 0 \quad 1 \quad 1]^T$, $\mathbf{y} = [1 \quad 1 \quad 0 \quad 1]^T$.

(12) $\mathbf{x} = [1 \quad 0 \quad 1 \quad -1]^T$, $\mathbf{y} = [1 \quad 1 \quad 0 \quad 1]^T$.

(13) $\mathbf{x} = [1 \quad 0 \quad 1 \quad 0]^T$, $\mathbf{y} = [0 \quad 1 \quad 0 \quad 1]^T$.

(14) $\mathbf{x} = \left[\frac{1}{2} \quad \frac{1}{2} \quad \frac{1}{2} \quad \frac{1}{2}\right]^T$, $\mathbf{y} = [1 \; 2 \; 3 \; -4]^T$.

(15) $\mathbf{x} = \left[\frac{1}{2} \quad \frac{1}{2} \quad 0 \quad \frac{1}{2}\right]^T$, $\mathbf{y} = \left[\frac{1}{3} \quad \frac{1}{2} \quad 1 \quad \frac{-2}{3}\right]^T$.

(16) $\mathbf{x} = [1 \quad 2 \quad 3 \quad 4 \quad 5]^T$, $\mathbf{y} = [1 \quad 2 \quad -3 \quad 4 \quad -5]^T$.

(17) $\mathbf{x} = [1 \quad 2 \quad 3 \quad 4 \quad 5 \quad 6]^T$, $\mathbf{y} = [1 \quad 2 \quad 3 \quad 4 \quad 5]^T$.

(18) Normalize the following vectors:

(a) $\mathbf{y}$ as defined in Problem 1.

(b) $\mathbf{y}$ as defined in Problem 4.

(c) $\mathbf{y}$ as defined in Problem 6.

(d) $\mathbf{y}$ as defined in Problem 7.

(e) $\mathbf{y}$ as defined in Problem 10.

(f) $\mathbf{y}$ as defined in Problem 17.

In Problems 19 through 26, find the angle between the given vectors.

(19) $\mathbf{x} = [1 \quad 2]^T$, $\mathbf{y} = [2 \quad 1]^T$.

(20) $\mathbf{x} = [1 \quad 1]^T$, $\mathbf{y} = [3 \quad 5]^T$.

(21) $\mathbf{x} = [3 \quad -2]^T$, $\mathbf{y} = [3 \quad 3]^T$.

(22) $\mathbf{x} = [4 \quad -1]^T$, $\mathbf{y} = [2 \quad 8]^T$.

(23) $\mathbf{x} = [-7 \quad -2]^T$, $\mathbf{y} = [2 \quad 9]^T$.

(24) $\mathbf{x} = [2 \quad 1 \quad 0]^T$, $\mathbf{y} = [2 \quad 0 \quad 2]^T$.

(25) $\mathbf{x} = [1 \quad 1 \quad 0]^T$, $\mathbf{y} = [2 \quad 2 \quad 1]^T$.

(26) $\mathbf{x} = [0 \quad 3 \quad 4]^T$, $\mathbf{y} = [2 \quad 5 \quad 5]^T$.

(27) Show that, for real numbers a,b, $\langle a,b\rangle = |a+b|$ does not form a Euclidean inner product over the set of real numbers.

(28) Prove that if $\mathbf{x}$ and $\mathbf{y}$ are vectors in $\mathbb{R}^n$, then $\langle \lambda\mathbf{x}, \mathbf{y}\rangle = \lambda\langle \mathbf{x}, \mathbf{y}\rangle$ for any real number λ.

(29) Prove that if $\mathbf{x}$, $\mathbf{y}$ and $\mathbf{z}$ are vectors in $\mathbb{R}^n$, then $\langle \mathbf{x}+\mathbf{z}, \mathbf{y}\rangle = \langle \mathbf{x}, \mathbf{y}\rangle + \langle \mathbf{z}, \mathbf{y}\rangle$.

(30) Prove for any vector $\mathbf{y}$ in $\mathbb{R}^n$ that $\langle \mathbf{0}, \mathbf{y}\rangle = 0$.

(31) Prove that if $\mathbf{x}$ and $\mathbf{y}$ are orthogonal vectors in $\mathbb{R}^n$, then $\|\mathbf{x}+\mathbf{y}\|^2 = \|\mathbf{x}\|^2 + \|\mathbf{y}\|^2$.

(32) Prove the following: $\|\mathbf{x}+\mathbf{y}\| = \|\mathbf{x}-\mathbf{y}\|$ if and only if $\mathbf{x}$ and $\mathbf{y}$ are orthogonal.

(33) Prove the *parallelogram law* for any two vectors $\mathbf{x}$ and $\mathbf{y}$ in $\mathbb{R}^n$:

$$\|\mathbf{x} + \mathbf{y}\|^2 + \|\mathbf{x} - \mathbf{y}\|^2 = 2\|\mathbf{x}\|^2 + 2\|\mathbf{y}\|^2.$$

(34) Prove that for any two vectors $\mathbf{x}$ and $\mathbf{y}$ in $\mathbb{R}^n$:

$$\|\mathbf{x} + \mathbf{y}\|^2 - \|\mathbf{x} - \mathbf{y}\|^2 = 4\langle \mathbf{x}, \mathbf{y}\rangle.$$

(35) Let $\mathbf{x}$, $\mathbf{y}$ and $\mathbf{z}$ be vectors in $\mathbb{R}^n$. Show that if $\mathbf{x}$ is orthogonal to $\mathbf{y}$ and if $\mathbf{x}$ is orthogonal to $\mathbf{z}$ then $\mathbf{x}$ is also orthogonal to all linear combinations of the vectors $\mathbf{y}$ and $\mathbf{z}$.

(36) (a) Prove that, for any scalar $\boldsymbol{\lambda}$,

$$0 \leq \|\boldsymbol{\lambda}\mathbf{x} - \mathbf{y}\|^2 = \boldsymbol{\lambda}^2\|\mathbf{x}\|^2 - 2\boldsymbol{\lambda}\langle \mathbf{x}, \mathbf{y}\rangle + \|\mathbf{y}\|^2.$$

(36) (b) Take $\boldsymbol{\lambda} = \langle \mathbf{x}, \mathbf{y}\rangle / \|\mathbf{x}\|^2$ and show that

$$0 \leq \frac{-\langle \mathbf{x}, \mathbf{y}\rangle^2}{\|\mathbf{x}\|^2} + \|\mathbf{y}\|^2$$

From this deduce that

$$\langle \mathbf{x}, \mathbf{y}\rangle^2 \leq \|\mathbf{x}\|^2\|\mathbf{y}\|^2$$

and then the Cauchy-Schwarz inequality.

(37) Prove that the Cauchy-Schwarz inequality is an equality in $\mathbb{R}^2$ if and only if one vector is a scalar multiple of the other.

(38) Use the Cauchy-Schwarz inequality to show that

$$-1 \leq \frac{\langle \mathbf{u}, \mathbf{v} \rangle}{\|\mathbf{u}\| \|\mathbf{v}\|} \leq 1.$$

Thus, Equation (6.5) can be used to define the cosine of the angle between any two vectors in $\mathbb{R}^n$. Use Equation (6.5) to find the cosine of the angle between the following **x** and **y** vectors

(a). $\mathbf{x} = [0 \quad 1 \quad 1 \quad 1]^{\mathrm{T}}$, $\mathbf{y} = [1 \quad 1 \quad 1 \quad 0]^{\mathrm{T}}$,

(b) $\mathbf{x} = [1 \quad 2 \quad 3 \quad 4]^{\mathrm{T}}$, $\mathbf{y} = [1 \quad -2 \quad 0 \quad -1]^{\mathrm{T}}$,

(c) $\mathbf{x} = [\frac{1}{2} \ \frac{1}{2} \ \frac{1}{2} \ \frac{1}{2}]^{\mathrm{T}}$, $\mathbf{y} = [-1 - 1 - 1 - 1]^{\mathrm{T}}$,

(d) $\mathbf{x} = [1 \quad 1 \quad 2 \quad 2 \quad 3]^{\mathrm{T}}$, $\mathbf{y} = [1 \quad 2 \quad 3 \quad 2 \quad 1]^{\mathrm{T}}$,

(e) $\mathbf{x} = [1 \quad 2 \quad 3 \quad 4 \quad 5 \quad 6]^{\mathrm{T}}$, $\mathbf{y} = [1 \quad 1 \quad 1 \quad 1 \quad 1 \quad 1]^{\mathrm{T}}$.

(39) Verify the following relationships:

$$\begin{aligned} \|\mathbf{x} + \mathbf{y}\|^2 &= \|\mathbf{x}\|^2 + 2\langle \mathbf{x}, \mathbf{y} \rangle + \|\mathbf{y}\|^2 \\ &\leq \|\mathbf{x}\|^2 + 2\|\mathbf{x}\|\|\mathbf{y}\| + \|\mathbf{y}\|^2 \\ &= (\|\mathbf{x}\| + \|\mathbf{y}\|)^2 \end{aligned}$$

and then, using the Cauchy-Schwarz inequality, deduce the triangle inequality

$$\|\mathbf{x} + \mathbf{y}\| \leq \|\mathbf{x}\| + \|\mathbf{y}\|$$

(40) Calculate induced inner products for the following pairs of matrices with respect to standard bases:

(a) $\mathbf{A} = \begin{bmatrix} 1 & 5 \\ 6 & 2 \end{bmatrix}$ and $\mathbf{B} = \begin{bmatrix} 5 & 5 \\ 1 & 4 \end{bmatrix}$ in $\mathbb{M}_{2\times 2}$,

(b) $\mathbf{A} = \begin{bmatrix} 1 & -2 \\ 0 & 4 \end{bmatrix}$ and $\mathbf{B} = \begin{bmatrix} 3 & -3 \\ 2 & -8 \end{bmatrix}$ in $\mathbb{M}_{2\times 2}$,

(c) $\mathbf{A} = \begin{bmatrix} -2 & 7 \\ 1 & 1 \end{bmatrix}$ and $\mathbf{B} = \begin{bmatrix} 2 & -3 \\ 2 & 6 \end{bmatrix}$ in $\mathbb{M}_{2\times 2}$,

(d) $\mathbf{A} = \begin{bmatrix} 4 & 2 \\ 1 & -3 \\ 3 & -5 \end{bmatrix}$ and $\mathbf{B} = \begin{bmatrix} 1 & 2 \\ 3 & 4 \\ 5 & 6 \end{bmatrix}$ in $\mathbb{M}_{3\times 2}$,

(e) $\mathbf{A} = \begin{bmatrix} 1 & 2 & 3 \\ 4 & 5 & 6 \end{bmatrix}$ and $\mathbf{B} = \begin{bmatrix} 1 & 1 & 2 \\ -3 & 2 & -3 \end{bmatrix}$ in $\mathbb{M}_{2\times 3}$,

(f) $\mathbf{A} = \begin{bmatrix} 1 & 2 & 3 \\ 4 & 5 & 6 \\ 7 & 8 & 9 \end{bmatrix}$ and $\mathbf{B} = \begin{bmatrix} -3 & 4 & 1 \\ 2 & 0 & -4 \\ 5 & 1 & 2 \end{bmatrix}$ in $\mathbb{M}_{3\times 3}$.

(41) Redo parts (a), (b), and (c) of Problem 40 with respect to the basis

$$\mathbb{C} = \left\{ \begin{bmatrix} 1 & 1 \\ 0 & 0 \end{bmatrix}, \begin{bmatrix} 1 & -1 \\ 0 & 0 \end{bmatrix}, \begin{bmatrix} 0 & 0 \\ 1 & 1 \end{bmatrix}, \begin{bmatrix} 0 & 0 \\ 1 & -1 \end{bmatrix} \right\}.$$

(42) A generalization of the inner product for n-dimensional column matrices with real components is

$$\langle \mathbf{x}, \mathbf{y} \rangle_{\mathbf{A}} = \langle \mathbf{Ax}, \mathbf{Ay} \rangle$$

where the inner product on the right is the Euclidean inner product between $\mathbf{Ax}$ and $\mathbf{Ay}$ for a given $n \times n$ real, nonsingular matrix $\mathbf{A}$. Show that $\langle \mathbf{x}, \mathbf{y} \rangle_{\mathbf{A}}$ satisfies all the properties of Theorem 1.

(43) Calculate $\langle \mathbf{x}, \mathbf{y} \rangle_{\mathbf{A}}$ for the vectors in Problem 1 when $\mathbf{A} = \begin{bmatrix} 2 & 3 \\ 1 & -1 \end{bmatrix}$.

(44) Calculate $\langle \mathbf{x}, \mathbf{y} \rangle_{\mathbf{A}}$ for the vectors in Problem 7 when $\mathbf{A} = \begin{bmatrix} 1 & 1 & 0 \\ 1 & 0 & 1 \\ 0 & 1 & 1 \end{bmatrix}$.

(45) Redo Problem 44 with $\mathbf{A} = \begin{bmatrix} 1 & -1 & 1 \\ 0 & 1 & -1 \\ 1 & 1 & 1 \end{bmatrix}$.

(46) Show that $\langle \mathbf{x}, \mathbf{y} \rangle_{\mathbf{A}}$ is the Euclidean inner product when $\mathbf{x}$ and $\mathbf{y}$ are coordinate representations with respect to a basis $\mathbb{B}$ made up of the columns of $\mathbf{A}$ and $\mathbf{A}$ is the transition matrix from the $\mathbb{B}$ basis to the standard basis.

(47) Calculate induced inner products for the following pairs of polynomials with respect to standard bases:

(a) $p(t) = t^2 + 2t + 3$ and $q(t) = t^2 + 3t - 5$ in $\mathbb{P}^2$,

(b) $p(t) = 10t^2 - 5t + 1$ and $q(t) = 2t^2 - t - 30$ in $\mathbb{P}^2$,

(c) $p(t) = t^2 + 5$ and $q(t) = 2t^2 - 2t + 1$ in $\mathbb{P}^2$,

(d) $p(t) = 2t^2 + 3t$ and $q(t) = t + 8$ in $\mathbb{P}^2$,

(e) $p(t) = 3t^3 + 2t^2 - t + 4$ and $q(t) = t^3 + t$ in $\mathbb{P}^3$,

(f) $p(t) = t^3 - t^2 + 2t$ and $q(t) = t^2 + t + 1$ in $\mathbb{P}^3$.

(48) Redo parts (a) through (d) of Problem 47 with respect to the basis

$$\mathbb{B} = \{t^2, \ t + 1, \ t\}.$$

(49) A different inner product on $\mathbb{P}^n$ is defined by

$$\langle p(t), q(t)\rangle = \int_a^b p(t)q(t)dt$$

for polynomials $p(t)$ and $q(t)$ and real numbers a and b with $b>a$. Show that this inner product satisfies all the properties of Theorem 1.

(50) Redo Problem 47 with the inner product defined in Problem 48, taking $a=0$ and $b=1$.

6.2 PROJECTIONS AND GRAM-SCHMIDT ORTHONORMALIZATION

An important problem in the applied sciences is to write a given nonzero vector **x** in $\mathbb{R}^2$ or $\mathbb{R}^3$ as the sum of two vectors $\mathbf{u}+\mathbf{v}$ where **u** is parallel to a known reference vector **a** and **v** is perpendicular to **a** (see Figure 6.6). In physics, **u** is called the *parallel component* of **x** and **v** is called the *perpendicular component* of **x**, where parallel and perpendicular are relative to the reference vector **a**.

If **u** is to be parallel to **a**, it must be a scalar multiple of **a**; that is, $\mathbf{u}=\lambda\mathbf{a}$ for some value of the scalar λ. If $\mathbf{x}=\mathbf{u}+\mathbf{v}$, then necessarily $\mathbf{v}=\mathbf{x}-\mathbf{u}=\mathbf{x}-\lambda\mathbf{a}$. If **u** and **v** are to be perpendicular, then

$$\begin{aligned} 0 = \langle \mathbf{u}, \mathbf{v}\rangle &= \langle \lambda\mathbf{a}, \mathbf{x} - \lambda\mathbf{a}\rangle \\ &= \lambda\langle \mathbf{a}, \mathbf{x}\rangle - \lambda^2\langle \mathbf{a}, \mathbf{a}\rangle \\ &= \lambda[\langle \mathbf{a}, \mathbf{x}\rangle - \lambda\langle \mathbf{a}, \mathbf{a}\rangle] \end{aligned}$$

Either $\lambda=0$ or $\lambda=\langle \mathbf{a}, \mathbf{x}\rangle/\langle \mathbf{a}, \mathbf{a}\rangle$. If $\lambda=0$, then $\mathbf{u}=\lambda\mathbf{a}=0\mathbf{a}=\mathbf{0}$, and $\mathbf{x}=\mathbf{u}+\mathbf{v}=\mathbf{v}$, from which we conclude that **x** and **a**, the given vector and the reference vector, are perpendicular and $\langle \mathbf{a}, \mathbf{x}\rangle=0$. Thus, $\lambda=\langle \mathbf{a}, \mathbf{x}\rangle/\langle \mathbf{a}, \mathbf{a}\rangle$ is always true and

$$\mathbf{u} = \frac{\langle \mathbf{a}, \mathbf{x}\rangle}{\langle \mathbf{a}, \mathbf{a}\rangle}\mathbf{a} \quad \text{and} \quad \mathbf{v} = \mathbf{x} - \frac{\langle \mathbf{a}, \mathbf{x}\rangle}{\langle \mathbf{a}, \mathbf{a}\rangle}\mathbf{a}$$

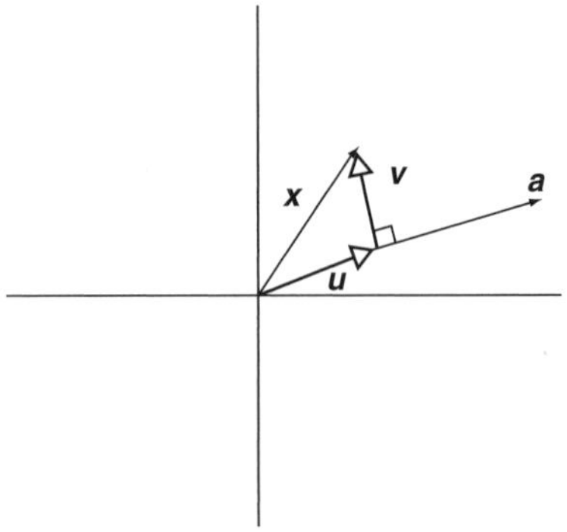

FIGURE 6.6

In this context, **u** is the *projection of* **x** *onto* **a** and **v** is the *orthogonal complement.*

Example 1 Write the vector $\mathbf{x} = \begin{bmatrix} 2 \\ 7 \end{bmatrix}$ as the sum of two vectors, one parallel to $\mathbf{a} = \begin{bmatrix} -3 \\ 4 \end{bmatrix}$ and one perpendicular to **a**.

Solution:

$$\langle \mathbf{a}, \mathbf{x} \rangle = -3(2) + 4(7) = 22$$

$$\langle \mathbf{a}, \mathbf{a} \rangle = (-3)^2 + (4)^2 = 25,$$

$$= \frac{\langle \mathbf{a}, \mathbf{x} \rangle}{\langle \mathbf{a}, \mathbf{a} \rangle} \mathbf{a} = \frac{22}{25} \begin{bmatrix} -3 \\ 4 \end{bmatrix} = \begin{bmatrix} -2.64 \\ 3.52 \end{bmatrix}$$

$$\mathbf{v} = \mathbf{x} - \mathbf{u} = \begin{bmatrix} 2 \\ 7 \end{bmatrix} - \begin{bmatrix} -2.64 \\ 3.52 \end{bmatrix} = \begin{bmatrix} 4.64 \\ 3.48 \end{bmatrix}$$

Then, $\mathbf{x} = \mathbf{u} + \mathbf{v}$, with **u** parallel to **a** and **v** perpendicular to **a**.

Example 2 Find the point on the line $x + 4y = 0$ closest to $(-3, -1)$.

Solution: One point on the line is $(4, -1)$, so $\mathbf{a} = [4 - 1]^{\mathrm{T}}$ is a reference vector in the plane parallel to the line. The given point $(-3, -1)$ is associated with the vector $\mathbf{x} = [-3 \ \ -1]^{\mathrm{T}}$, and we seek the coordinates of the point P (see Figure 6.7) on the line $x + 4y = 0$. The vector **u** that begins at the origin and terminates at P is the projection of **x** onto **a**. Therefore,

$$\langle \mathbf{a}, \mathbf{x} \rangle = 4(-3) + (-1)(-1) = -11$$

$$\langle \mathbf{a}, \mathbf{a} \rangle = (4)^2 + (-1)^2 = 17$$

$$\mathbf{u} = \frac{\langle \mathbf{a}, \mathbf{x} \rangle}{\langle \mathbf{a}, \mathbf{a} \rangle} \mathbf{a} = \frac{-11}{17} \begin{bmatrix} 4 \\ -1 \end{bmatrix} = \begin{bmatrix} -44/17 \\ 11/17 \end{bmatrix}$$

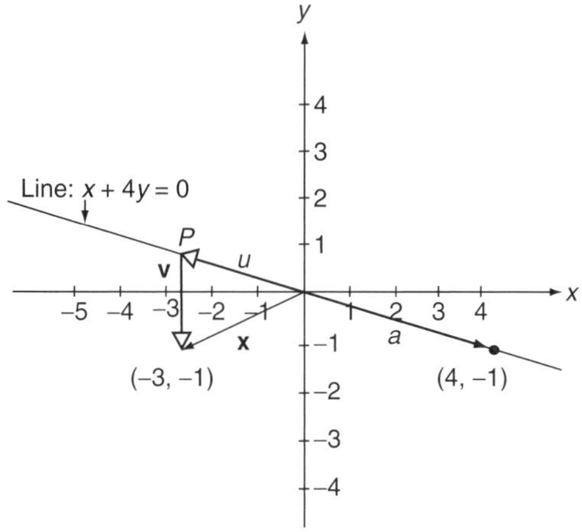

FIGURE 6.7

$$P = (-44/17, 11/17)$$

The concepts of projections and orthogonal complements in $\mathbb{R}^2$ can be extended to any finite dimensional vector space $\mathbb{V}$ with an inner product. Given a nonzero vector $\mathbf{x}$ and a reference vector $\mathbf{a}$, both in $\mathbb{V}$, we define the *projections of* $\mathbf{x}$ *onto* $\mathbf{a}$ as

$$\text{proj}_{\mathbf{a}}\mathbf{x} = \frac{\langle \mathbf{a}, \mathbf{x} \rangle}{\langle \mathbf{a}, \mathbf{a} \rangle}\mathbf{a} \tag{6.6}$$

It then follows (see Problem 34) that

$$\mathbf{x} - \frac{\langle \mathbf{a}, \mathbf{x} \rangle}{\langle \mathbf{a}, \mathbf{a} \rangle}\mathbf{a} \text{ is orthogonal to } \mathbf{a} \tag{6.7}$$

Subtracting from a nonzero vector $\mathbf{x}$ the projection $\mathbf{x}$ onto another nonzero vector $\mathbf{a}$ leaves a vector that is orthogonal to both $\mathbf{a}$ and the projection of $\mathbf{x}$ onto $\mathbf{a}$.

If $\mathbf{x}$ is a nonzero vector, then $\mathbf{x}$ minus its projection onto another nonzero vector $\mathbf{a}$ yields a vector that is orthogonal to both $\mathbf{a}$ and the projection of $\mathbf{x}$ onto $\mathbf{a}$.

Example 3 Write the polynomial $x(t) = 2t^2 + 3t + 4$ in $\mathbb{P}^2$ as the sum of two polynomials, one that is the projection of $x(t)$ onto $a(t) = 5t^2 + 6$ and one that is orthogonal to $a(t)$ under the inner product induced by the Euclidean inner product in $\mathbb{R}^3$.

Solution: The induced inner product between two polynomials is obtained by multiplying the coefficients of like powers of t and summing the resulting products (see Example 7 of Section 6.1). Thus,

$$\langle a(t), x(t) \rangle = 5(2) + 0(3) + 6(4) = 34$$

$$\langle a(t), a(t) \rangle = (5)^2 + (0)^2 + (6)^2 = 61$$

$$u(t) = \frac{\langle a(t), x(t) \rangle}{\langle a(t), a(t) \rangle} a(t) = \frac{34}{61}\left(5t^2 + 6\right) = \frac{170}{61}t^2 + \frac{204}{61}$$

is the projection of $x(t)$ onto $a(t)$.

$$v(t) = x(t) - u(t) = -\frac{48}{61}t^2 + 3t + \frac{40}{61}$$

is orthogonal to $a(t)$, and $x(t) = u(t) + v(t)$.

A set of vectors is called an *orthogonal set* if each vector in the set is orthogonal to every other vector in the set.

Example 4 The vectors $\{\mathbf{x}, \mathbf{y}, \mathbf{z}\}$ in $\mathbb{R}^3$ defined by

$$\mathbf{x} = \begin{bmatrix} 1 \\ 1 \\ 1 \end{bmatrix}, \mathbf{y} = \begin{bmatrix} 1 \\ 1 \\ -2 \end{bmatrix}, \mathbf{z} = \begin{bmatrix} 1 \\ -1 \\ 0 \end{bmatrix}$$

are an orthogonal set of vectors because $\langle \mathbf{x}, \mathbf{y} \rangle = \langle \mathbf{x}, \mathbf{z} \rangle = \langle \mathbf{y}, \mathbf{z} \rangle = 0$. In contrast, the set of vectors $\{\mathbf{a}, \mathbf{b}, \mathbf{c}\}$ in $\mathbb{R}^4$ defined by

$$\mathbf{a} = [1 \quad 1 \quad 0 \quad 1]^{\mathrm{T}}, \mathbf{b} = [-1 \quad 1 \quad 2 \quad 0]^{\mathrm{T}}, \mathbf{c} = [1 \quad 1 \quad 0 \quad 2]^{\mathrm{T}}$$

is *not* an orthogonal set because $\langle \mathbf{a}, \mathbf{c} \rangle \neq 0$. If $\mathbf{c}$ is redefined as

$$\mathbf{c} = [1 \quad 1 \quad 0 \quad -2]^{\mathrm{T}}$$

then $\{\mathbf{a}, \mathbf{b}, \mathbf{c}\}$ is orthogonal, because now $\langle \mathbf{a}, \mathbf{b}\rangle = \langle \mathbf{a}, \mathbf{c}\rangle = \langle \mathbf{b}, \mathbf{c}\rangle = 0$.

An orthogonal set of unit vectors (vectors all having magnitude 1) is called an *orthonormal* set. Using the *Kronecker delta* notation,

An orthonormal set of vectors is an orthogonal set of unit vectors.

$$\delta_{ij} = \begin{cases} 1 & \text{if } i = j \\ 0 & \text{if } i \neq j \end{cases} \tag{6.8}$$

We say that a set of vectors $\{\mathbf{x}_1, \mathbf{x}_2, \ldots, \mathbf{x}_n\}$ is orthonormal if and only if

$$\langle \mathbf{x}_i, \mathbf{x}_j \rangle = \delta_{ij} \; (i, j = 1, 2, \ldots, m) \tag{6.9}$$

Example 5 The set of vectors $\{\mathbf{u}, \mathbf{v}, \mathbf{w}\}$ in $\mathbb{R}^3$ defined by

$$\mathbf{u} = \begin{bmatrix} 1/\sqrt{2} \\ 1/\sqrt{2} \\ 0 \end{bmatrix}, \mathbf{v} = \begin{bmatrix} 1/\sqrt{2} \\ -1/\sqrt{2} \\ 0 \end{bmatrix}, \mathbf{w} = \begin{bmatrix} 0 \\ 0 \\ 1 \end{bmatrix}$$

is an orthonormal set of vectors because each vector is orthogonal to the other two *and* each vector is a unit vector.

Any orthogonal set of *nonzero* vectors can be transformed into an orthonormal set by dividing each vector by its magnitude. It follows from Example 4 that the vectors

$$\mathbf{x} = \begin{bmatrix} 1 \\ 1 \\ 1 \end{bmatrix}, \mathbf{y} = \begin{bmatrix} 1 \\ 1 \\ -2 \end{bmatrix}, \mathbf{z} = \begin{bmatrix} 1 \\ -1 \\ 0 \end{bmatrix}$$

form an orthogonal set. Dividing each vector by its magnitude, we generate

$$\left\{ \frac{\mathbf{x}}{\|\mathbf{x}\|}, \frac{\mathbf{y}}{\|\mathbf{y}\|}, \frac{\mathbf{z}}{\|\mathbf{z}\|} \right\} = \left\{ \begin{bmatrix} 1/\sqrt{3} \\ 1/\sqrt{3} \\ 1/\sqrt{3} \end{bmatrix}, \begin{bmatrix} 1/\sqrt{6} \\ 1/\sqrt{6} \\ -2/\sqrt{6} \end{bmatrix}, \begin{bmatrix} 1/\sqrt{2} \\ -1/\sqrt{2} \\ 0 \end{bmatrix} \right\}$$

as an orthonormal set.

> **▶THEOREM 1**
>
> *An orthonormal set of a finite number of vectors is linearly independent.*◀

Proof: Let $\{\mathbf{x}_1, \mathbf{x}_2, \ldots, \mathbf{x}_n\}$ be an orthonormal set and consider the vector equation

$$c_1\mathbf{x}_1 + c_2\mathbf{x}_2 + \ldots + c_n\mathbf{x}_n = \mathbf{0} \tag{6.10}$$

where $c_j (j = 1, 2, \ldots, n)$ is a scalar. This set of vectors is linearly independent if and only if the only solution to Equation (6.10) is $c_1 = c_2 = \ldots = c_n = 0$. Taking the inner product of both sides of Equation (6.10) with $\mathbf{x}_j$, we have

$$\langle c_1\mathbf{x}_1 + c_2\mathbf{x}_2 + \ldots + c_j\mathbf{x}_j + \ldots + c_n\mathbf{x}_n, \mathbf{x}_j\rangle = \langle \mathbf{0}, \mathbf{x}_j\rangle$$

Using parts (c), (d), and (e) of Theorem 1 of Section 6.1, we rewrite this last equation as

$$c_1\langle\mathbf{x}_1, \mathbf{x}_j\rangle + c_2\langle\mathbf{x}_2, \mathbf{x}_j\rangle + \ldots + c_j\langle\mathbf{x}_j, \mathbf{x}_j\rangle + \ldots + c_n\langle\mathbf{x}_n, \mathbf{x}_j\rangle = \mathbf{0}$$

or

$$\sum_{i=1}^{n} c_i\langle\mathbf{x}_i, \mathbf{x}_j\rangle = 0$$

As a consequence of Equation (6.9),

$$\sum_{i=1}^{n} c_i\delta_{ij} = 0$$

or $c_j = 0$ $(j = 1, 2, \ldots, n)$.

If $\mathbb{B} = \{\mathbf{x}_1, \mathbf{x}_2, \ldots, \mathbf{x}_n\}$ is a basis for $\mathbb{V}$, then any vector $\mathbf{x}$ in $\mathbb{V}$ can be written as a linear combination of the basis vectors in one and only one way (see Theorem 5 of Section 2.5). That is,

$$\mathbf{x} = c_1\mathbf{x}_1 + c_2\mathbf{x}_2 + \ldots + c_n\mathbf{x}_n = \sum_{i=1}^{n} c_i x_i$$

with each $c_j(i = 1, 2, \ldots, n)$ uniquely determined by the choice of the basis. If the basis is orthonormal, we can use the additional structure of an inner product to say more. In particular,

$$\begin{aligned}\langle\mathbf{x}, \mathbf{x}_j\rangle &= \left\langle \sum_{i=1}^{n} c_i\mathbf{x}_i, \mathbf{x}_j \right\rangle \\ &= \sum_{i=1}^{n} \langle c_i\mathbf{x}_i, \mathbf{x}_j\rangle \\ &= \sum_{i=1}^{n} c_i\langle\mathbf{x}_i, \mathbf{x}_j\rangle \\ &= \sum_{i=1}^{n} c_i\delta_{ij} = c_j.\end{aligned}$$

We have proven Theorem 2.

▶THEOREM 2

If $\{\mathbf{x}_1, \mathbf{x}_2, \ldots, \mathbf{x}_n\}$ is orthonormal basis for a vector space $\mathbb{V}$, then for any vector $\mathbf{x}$ in $\mathbb{V}$,
$\mathbf{x} = \langle\mathbf{x},\mathbf{x}_1\rangle\mathbf{x}_1 + \langle\mathbf{x},\mathbf{x}_2\rangle\mathbf{x}_2 + \ldots + \langle\mathbf{x},\mathbf{x}_n\rangle\mathbf{x}_n$ ◀

Theorem 2 is one of those wonderful results that saves time and effort. In general, to write a vector in an n-dimensional vector space in terms of a given basis, we must solve a set n simultaneous linear equations (see Example 11, Section 2.5). If, however, the basis is orthonormal, the work is reduced to taking n-inner products and solving *no* simultaneous equations.

Example 6 Write $\mathbf{x} = [1\ 2\ 3]^T$ as a linear combination of the vectors

$$\mathbf{q}_1 = \begin{bmatrix} 1/\sqrt{3} \\ 1/\sqrt{3} \\ 1/\sqrt{3} \end{bmatrix}, \mathbf{q}_2 = \begin{bmatrix} 1/\sqrt{6} \\ 1/\sqrt{6} \\ -2/\sqrt{6} \end{bmatrix}, \mathbf{q}_3 = \begin{bmatrix} 1/\sqrt{2} \\ -1/\sqrt{2} \\ 0 \end{bmatrix}$$

Proof: The set $\{\mathbf{q}_1, \mathbf{q}_2, \mathbf{q}_3\}$ is an orthonormal basis for $\mathbb{R}^3$. Consequently,

$$\langle \mathbf{x}, \mathbf{q}_1 \rangle = 1\left(\frac{1}{\sqrt{3}}\right) + 2\left(\frac{1}{\sqrt{3}}\right) + 3\left(\frac{1}{\sqrt{3}}\right) = \frac{6}{\sqrt{3}}$$

$$\langle \mathbf{x}, \mathbf{q}_2 \rangle = 1\left(\frac{1}{\sqrt{6}}\right) + 2\left(\frac{1}{\sqrt{6}}\right) + 3\left(-\frac{2}{\sqrt{6}}\right) = \frac{-3}{\sqrt{6}}$$

$$\langle \mathbf{x}, \mathbf{q}_3 \rangle = 1\left(\frac{1}{\sqrt{2}}\right) + 2\left(\frac{-1}{\sqrt{2}}\right) + 3(0) = \frac{-1}{\sqrt{2}}$$

$$\begin{bmatrix} 1 \\ 2 \\ 3 \end{bmatrix} = \frac{6}{\sqrt{3}} \begin{bmatrix} 1/\sqrt{3} \\ 1/\sqrt{3} \\ 1/\sqrt{3} \end{bmatrix} + \left(\frac{-3}{\sqrt{6}}\right) \begin{bmatrix} 1/\sqrt{6} \\ 1/\sqrt{6} \\ -2/\sqrt{6} \end{bmatrix} + \left(\frac{-1}{\sqrt{2}}\right) \begin{bmatrix} 1/\sqrt{2} \\ -1/\sqrt{2} \\ 0 \end{bmatrix}$$

Example 7 Write $\mathbf{A} = \begin{bmatrix} 1 & 2 \\ 3 & 4 \end{bmatrix}$ as a linear combination of the four matrices

$$\mathbf{Q}_1 = \begin{bmatrix} 1/\sqrt{3} & 1/\sqrt{3} \\ -1/\sqrt{3} & 0 \end{bmatrix},$$

$$\mathbf{Q}_2 = \begin{bmatrix} 0 & -1/\sqrt{3} \\ -1/\sqrt{3} & 1/\sqrt{3} \end{bmatrix},$$

$$\mathbf{Q}_3 = \begin{bmatrix} 1/\sqrt{3} & 0 \\ 1/\sqrt{3} & 1/\sqrt{3} \end{bmatrix},$$

$$\mathbf{Q}_4 = \begin{bmatrix} -1/\sqrt{3} & 1/\sqrt{3} \\ 0 & 1/\sqrt{3} \end{bmatrix},$$

Solution: The set $\{\mathbf{Q}_1, \mathbf{Q}_2, \mathbf{Q}_3, \mathbf{Q}_4\}$ is an orthonormal basis for $\mathbb{M}_{2\times 2}$ under the induced inner product (see Example 5 of Section 6.1) defined by multiplying corresponding elements and summing the resulting products. Consequently,

$$\langle \mathbf{A}, \mathbf{Q}_1\rangle = \left\langle \begin{bmatrix} 1 & 2 \\ 3 & 4 \end{bmatrix}, \begin{bmatrix} 1/\sqrt{3} & 1/\sqrt{3} \\ -1/\sqrt{3} & 0 \end{bmatrix} \right\rangle$$

$$= 1\left(\frac{1}{\sqrt{3}}\right) + 2\left(\frac{1}{\sqrt{3}}\right) + 3\left(\frac{-1}{\sqrt{3}}\right) + 4(0) = 0$$

$$\langle \mathbf{A}, \mathbf{Q}_2\rangle = \left\langle \begin{bmatrix} 1 & 2 \\ 3 & 4 \end{bmatrix}, \begin{bmatrix} 0 & -1/\sqrt{3} \\ -1/\sqrt{3} & 1/\sqrt{3} \end{bmatrix} \right\rangle$$

$$= 1(0) + 2\left(\frac{-1}{\sqrt{3}}\right) + 3\left(\frac{-1}{\sqrt{3}}\right) + 4\left(\frac{1}{\sqrt{3}}\right) = \frac{-1}{\sqrt{3}}$$

$$\langle \mathbf{A}, \mathbf{Q}_3\rangle = \left\langle \begin{bmatrix} 1 & 2 \\ 3 & 4 \end{bmatrix}, \begin{bmatrix} 1/\sqrt{3} & 0 \\ 1/\sqrt{3} & 1/\sqrt{3} \end{bmatrix} \right\rangle$$

$$= 1\left(\frac{1}{\sqrt{3}}\right) + 2(0) + 3\left(\frac{1}{\sqrt{3}}\right) + 4\left(\frac{1}{\sqrt{3}}\right) = \frac{8}{\sqrt{3}}$$

$$\langle \mathbf{A}, \mathbf{Q}_4\rangle = \left\langle \begin{bmatrix} 1 & 2 \\ 3 & 4 \end{bmatrix}, \begin{bmatrix} -1/\sqrt{3} & 1/\sqrt{3} \\ 0 & 1/\sqrt{3} \end{bmatrix} \right\rangle$$

$$= 1\left(\frac{-1}{\sqrt{3}}\right) + 2\left(\frac{1}{\sqrt{3}}\right) + 3(0) + 4\left(\frac{1}{\sqrt{3}}\right) = \frac{5}{\sqrt{3}}$$

and

$$\begin{bmatrix} 1 & 2 \\ 3 & 4 \end{bmatrix} = (0)\begin{bmatrix} 1/\sqrt{3} & 1/\sqrt{3} \\ -1/\sqrt{3} & 0 \end{bmatrix} + \left(\frac{-1}{\sqrt{3}}\right)\begin{bmatrix} 0 & -1/\sqrt{3} \\ -1/\sqrt{3} & 1/\sqrt{3} \end{bmatrix}$$
$$+ \left(\frac{8}{\sqrt{3}}\right)\begin{bmatrix} 1/\sqrt{3} & 0 \\ 1/\sqrt{3} & 1/\sqrt{3} \end{bmatrix} + \left(\frac{5}{\sqrt{3}}\right)\begin{bmatrix} -1/\sqrt{3} & 1/\sqrt{3} \\ 0 & 1/\sqrt{3} \end{bmatrix}.$$

An inner product space is a vector space with an inner product defined between pairs of vectors.

An *inner product space* is a vector space with an inner product defined between pairs of vectors. Using projections, we can transform any basis for a finite dimensional inner product space $\mathbb{V}$ into an orthonormal basis for $\mathbb{V}$. To see how, let $\{\mathbf{x}_1, \mathbf{x}_2, \mathbf{x}_3\}$ be a basis for $\mathbb{R}^3$. Taking $\mathbf{x}_1$ as our reference vector, it follows from Equation (6.7), with $\mathbf{x}_2$ replacing $\mathbf{x}$, that

$$\mathbf{x}_4 = \mathbf{x}_2 - \frac{\langle \mathbf{x}_1, \mathbf{x}_2\rangle}{\langle \mathbf{x}_1, \mathbf{x}_1\rangle}\mathbf{x}_1 \quad \text{is orthogonal to } \mathbf{x}_1$$

Similarly, it follows from Equation (6.7), with $\mathbf{x}_3$ replacing $\mathbf{x}$, that

$$\mathbf{x}_5 = \mathbf{x}_3 - \frac{\langle \mathbf{x}_1, \mathbf{x}_3 \rangle}{\langle \mathbf{x}_1, \mathbf{x}_1 \rangle}\mathbf{x}_1 \quad \text{is orthogonal to } \mathbf{x}_1$$

These formulas may be simplified when $\mathbf{x}_1$ is a unit vector, because $\langle \mathbf{x}_1, \mathbf{x}_1 \rangle = \|\mathbf{x}_1\|^2 = 1$. We can guarantee that the first vector in any basis be a unit vector by dividing that vector by its magnitude. Assuming this has been done and noting that $\langle \mathbf{x}_1, \mathbf{x}_2 \rangle = \langle \mathbf{x}_2, \mathbf{x}_1 \rangle$ and $\langle \mathbf{x}_1, \mathbf{x}_3 \rangle = \langle \mathbf{x}_3, \mathbf{x}_1 \rangle$, we have that

$$\mathbf{x}_4 = \mathbf{x}_2 - \langle \mathbf{x}_2, \mathbf{x}_1 \rangle \mathbf{x}_1 \quad \text{is orthogonal to } \mathbf{x}_1$$

and

$$\mathbf{x}_5 = \mathbf{x}_3 - \langle \mathbf{x}_3, \mathbf{x}_1 \rangle \mathbf{x}_1 \quad \text{is orthogonal to } \mathbf{x}_1$$

Furthermore, $\mathbf{x}_4 \neq \mathbf{0}$ because it is a linear combination of $\mathbf{x}_1$ and $\mathbf{x}_2$, which are linearly independent, with the coefficient of $\mathbf{x}_2$ equal to 1. The only way for a linear combination of linearly independent vectors to be 0 is for all the coefficients of the vectors to be 0. Similarly $\mathbf{x}_5 \neq \mathbf{0}$ because it is a linear combination of $\mathbf{x}_1$ and $\mathbf{x}_3$ with the coefficient of $\mathbf{x}_3$ set to 1. Thus, the set $\{\mathbf{x}_1, \mathbf{x}_4, \mathbf{x}_5\}$ has the property that $\mathbf{x}_1$ is a unit vector orthogonal to both nonzero vectors $\mathbf{x}_4$ and $\mathbf{x}_5$. The vectors $\mathbf{x}_4$ and $\mathbf{x}_5$ are not necessarily unit vectors and may not be orthogonal, but we have made progress in our attempt to create an orthonormal set. Now, taking $\mathbf{x}_4$ as our reference vector, it follows from Equation (6.7), with $\mathbf{x}_5$ replacing $\mathbf{x}$, that

$$\mathbf{x}_6 = \mathbf{x}_5 - \frac{\langle \mathbf{x}_4, \mathbf{x}_5 \rangle}{\langle \mathbf{x}_4, \mathbf{x}_4 \rangle}\mathbf{x}_4 \quad \text{is orthogonal to } \mathbf{x}_4$$

This formula may be simplified if $\mathbf{x}_4$ is a unit vector, a condition we can force by dividing $\mathbf{x}_4$ by its magnitude. Assuming this has been done and noting that $\langle \mathbf{x}_4, \mathbf{x}_5 \rangle = \langle \mathbf{x}_5, \mathbf{x}_4 \rangle$, we have that

$$\mathbf{x}_6 = \mathbf{x}_5 - \langle \mathbf{x}_5, \mathbf{x}_4 \rangle \mathbf{x}_4 \quad \text{is orthogonal to } \mathbf{x}_4$$

Also,

$$\begin{aligned}
\langle \mathbf{x}_6, \mathbf{x}_1 \rangle &= \langle \mathbf{x}_5 - \langle \mathbf{x}_5, \mathbf{x}_4 \rangle \mathbf{x}_4, \mathbf{x}_1 \rangle \\
&= \langle \mathbf{x}_5, \mathbf{x}_1 \rangle - \langle \langle \mathbf{x}_5, \mathbf{x}_4 \rangle \mathbf{x}_4, \mathbf{x}_1 \rangle \\
&= \langle \mathbf{x}_5, \mathbf{x}_1 \rangle - \langle \mathbf{x}_5, \mathbf{x}_4 \rangle \langle \mathbf{x}_4, \mathbf{x}_1 \rangle \\
&= 0
\end{aligned}$$

because $\mathbf{x}_1$ is orthogonal to both $\mathbf{x}_4$ and $\mathbf{x}_5$. Thus, $\mathbf{x}_1$ is orthogonal to both $\mathbf{x}_4$ and $\mathbf{x}_6$ and these last two vectors are themselves orthogonal. Furthermore, $\mathbf{x}_6 \neq \mathbf{0}$, because it can be written as a linear combination of the linearly independent vectors $\mathbf{x}_1$, $\mathbf{x}_2$, and $\mathbf{x}_3$ with the coefficient of $\mathbf{x}_3$ set to one. If $\mathbf{x}_6$ is not a unit vector, we may force it to become a unit vector by dividing $\mathbf{x}_6$ by its magnitude. Assuming this is done, we have that $\{\mathbf{x}_1, \mathbf{x}_4, \mathbf{x}_6\}$ is an orthonormal set.

If we apply this construction to arbitrary n-dimensional inner product spaces, and use $\mathbf{q}_i$ to denote the ith vector in an orthonormal set, we have Theorem 3.

▶THEOREM 3. (THE GRAM-SCHMIDT ORTHONORMALIZATION PROCESS)

Let $\{\mathbf{x}_1, \mathbf{x}_2, \ldots, \mathbf{x}_n\}$ be a basis for an inner product space $\mathbb{V}$. For $k=1, 2, \ldots, n$, do iteratively:

Step 1. Calculate $r_{kk}=|\mathbf{x}_k|$.

Step 2. Set $\mathbf{q}_k=(1/r_{kk})\mathbf{x}_k$.

Step 3. For $j=k+1, k+2, \ldots, n$, calculate $r_{kj}=\langle \mathbf{x}_j, \mathbf{q}_k \rangle$.

Step 4. For $j=k+1, k+2, \ldots, n$, replace $\mathbf{x}_j$ with $\mathbf{y}_j=\mathbf{x}_j-r_{kj}\mathbf{q}_k$; that is, $\mathbf{x}_j \leftarrow \mathbf{x}_j-r_{kj}\mathbf{q}_k$.

After the kth iteration ($k=1, 2, \ldots, n$), $\{\mathbf{q}_1, \mathbf{q}_2, \ldots, \mathbf{q}_k\}$ is an ortho-normal set, the span of $\{\mathbf{q}_1, \mathbf{q}_2, \ldots, \mathbf{q}_k\}$ equals the span of $\{\mathbf{x}_1, \mathbf{x}_2, \ldots, \mathbf{x}_R\}$, and each new $\mathbf{x}_j$ ($j=k+1, k+2, \ldots, n$) is a nonzero vector orthogonal to each $\mathbf{q}_i$ ($i=1, 2, \ldots, k$)◀

Proof: (by mathematical induction on the iterations). Setting $\mathbf{q}_1=\mathbf{x}_1/\|\mathbf{x}_1\|$, we have $span\{\mathbf{q}_1\}=span\{\mathbf{x}_1\}$ and $\|\mathbf{q}_1\|=1$. Furthermore, it follows from Equation (6.7) that $\mathbf{x}_j-r_{1j}\mathbf{q}_1$, ($j=2, 3, \ldots, n$) is orthogonal to $\mathbf{q}_1$. Thus, the proposition is true for $n=1$.

Assume that the proposition is true for $n=k$. Then $\mathbf{x}_{k+1}$ is nonzero and orthogonal to $\mathbf{q}_1, \mathbf{q}_2, \ldots, \mathbf{q}_k$, hence $\mathbf{q}_{k+1}=\mathbf{x}_{k+1}/\|\mathbf{x}_{k+1}\|$ is a unit vector and $\{\mathbf{q}_1, \mathbf{q}_2, \ldots, \mathbf{q}_k, \mathbf{q}_{k+1}\}$ is an orthonormal set. From the induction hypothesis,

$$span\{\mathbf{q}_1, \mathbf{q}_2, \ldots, \mathbf{q}_k\} = span\{\mathbf{x}_1, \mathbf{x}_2, \ldots, \mathbf{x}_k\}, \text{ so}$$

$$\begin{aligned} span\{\mathbf{q}_1, \mathbf{q}_2, \ldots, \mathbf{q}_k, \mathbf{q}_{k+1}\} &= span\{\mathbf{x}_1, \mathbf{x}_2, \ldots, \mathbf{x}_k, \mathbf{q}_{k+1}\} \\ &= span\{\mathbf{x}_1, \mathbf{x}_2, \ldots, \mathbf{x}_k, \mathbf{x}_{k+1}/\|\mathbf{x}_{k+1}\|\} \\ &= span\{\mathbf{x}_1, \mathbf{x}_2, \ldots, \mathbf{x}_k, \mathbf{x}_{k+1}\}. \end{aligned}$$

For $j=k+2, k+3, \ldots, n$, we construct $\mathbf{y}_j=\mathbf{x}_j-r_{k+1,\,j}\mathbf{q}_{k+1}$. It follows from Equation (6.7) that each $\mathbf{y}_j$ vector is orthogonal to $\mathbf{q}_{k+1}$. In addition, for $i=1, 2, \ldots, k$,

$$\begin{aligned} \left\langle \mathbf{y}_j, \mathbf{q}_i \right\rangle &= \langle \mathbf{x}_j - r_{k+1,j}\mathbf{q}_{k+1}, \mathbf{q}_i \rangle \\ &= \langle \mathbf{x}_j, \mathbf{q}_i \rangle - r_{k+1,j}\langle \mathbf{q}_{k+1}, \mathbf{q}_i \rangle \\ &= 0 \end{aligned}$$

Here $\langle \mathbf{x}_j, \mathbf{q}_i \rangle=0$ as a result of the induction hypothesis and $\langle \mathbf{q}_{k+1}, \mathbf{q}_i \rangle=0$ because $\{\mathbf{q}_1, \mathbf{q}_2, \ldots, \mathbf{q}_k, \mathbf{q}_{k+1}\}$ is an orthonormal set. Letting $\mathbf{x}_j \leftarrow \mathbf{y}_j$, $j=k+2, k+3, \ldots, n$, we have that each new $\mathbf{x}_j$ is orthogonal to each $\mathbf{q}_i$, $i=1, 2, \ldots, k+1$. Thus, Theorem 3 is proved by mathematical induction (see Appendix A).

The first two steps in the orthonormalization process create unit vectors; the third and fourth steps subtract projections from vectors, thereby generating orthogonality. These four steps are also known as the *revised* (*or modified*) *Gram-Schmidt algorithm*.

Example 8 Use the Gram-Schmidt orthonormalization process to construct an orthonormal set of vectors from the linearly independent set $\{\mathbf{x}_1, \mathbf{x}_2, \mathbf{x}_3\}$, where

$$\mathbf{x}_1 = \begin{bmatrix} 1 \\ 1 \\ 0 \end{bmatrix}, \mathbf{x}_2 = \begin{bmatrix} 0 \\ 1 \\ 1 \end{bmatrix}, \mathbf{x}_3 = \begin{bmatrix} 1 \\ 0 \\ 1 \end{bmatrix}$$

Solution: For the first iteration ($k=1$),

$$r_{11} = \sqrt{\langle \mathbf{x}_1, \mathbf{x}_1 \rangle} = \sqrt{2}$$

$$\mathbf{q}_1 = \frac{1}{r_{11}}\mathbf{x}_1 = \frac{1}{\sqrt{2}}\begin{bmatrix} 1 \\ 1 \\ 0 \end{bmatrix} = \begin{bmatrix} 1/\sqrt{2} \\ 1/\sqrt{2} \\ 0 \end{bmatrix}$$

$$r_{12} = \langle \mathbf{x}_2, \mathbf{q}_1 \rangle = \frac{1}{\sqrt{2}}$$

$$r_{13} = \langle \mathbf{x}_3, \mathbf{q}_1 \rangle = \frac{1}{\sqrt{2}}$$

$$\mathbf{x}_2 \leftarrow \mathbf{x}_2 - r_{12}\mathbf{q}_1 = \begin{bmatrix} 0 \\ 1 \\ 1 \end{bmatrix} - \frac{1}{\sqrt{2}}\begin{bmatrix} 1/\sqrt{2} \\ 1/\sqrt{2} \\ 0 \end{bmatrix} = \begin{bmatrix} -1/2 \\ 1/2 \\ 1 \end{bmatrix}$$

$$\mathbf{x}_3 \leftarrow \mathbf{x}_3 - r_{13}\mathbf{q}_1 = \begin{bmatrix} 1 \\ 0 \\ 1 \end{bmatrix} - \frac{1}{\sqrt{2}} - \frac{1}{\sqrt{2}}\begin{bmatrix} 1/\sqrt{2} \\ 1/\sqrt{2} \\ 0 \end{bmatrix} = \begin{bmatrix} 1/2 \\ -1/2 \\ 1 \end{bmatrix}$$

Note that both $\mathbf{x}_2$ and $\mathbf{x}_3$ are now orthogonal to $\mathbf{q}_1$.

For the second iteration ($k=2$), using vectors from the first iteration, we compute

$$r_{22} = \sqrt{\langle \mathbf{x}_2, \mathbf{x}_2 \rangle} = \sqrt{3/2}$$

$$\mathbf{q}_2 = \frac{1}{r_{22}}\mathbf{x}_2 = \frac{1}{\sqrt{3/2}}\begin{bmatrix} -1/2 \\ 1/2 \\ 1 \end{bmatrix} = \begin{bmatrix} -1/\sqrt{6} \\ 1/\sqrt{6} \\ 2/\sqrt{6} \end{bmatrix}$$

$$r_{23} = \langle \mathbf{x}_3, \mathbf{q}_2 \rangle = \frac{1}{\sqrt{6}}$$

$$\mathbf{x}_3 \leftarrow \mathbf{x}_3 - r_{23}\mathbf{q}_2 = \begin{bmatrix} 1/2 \\ -1/2 \\ 1 \end{bmatrix} - \frac{1}{\sqrt{6}}\begin{bmatrix} -1/\sqrt{6} \\ 1/\sqrt{6} \\ 2/\sqrt{6} \end{bmatrix} = \begin{bmatrix} 2/3 \\ -2/3 \\ 2/3 \end{bmatrix}$$

For the third iteration ($k = 3$), using vectors from the second iteration, we compute

$$r_{33} = \sqrt{\langle \mathbf{x}_3, \mathbf{x}_3 \rangle} = \frac{2}{\sqrt{3}}$$

$$\mathbf{q}_3 = \frac{1}{r_{33}}\mathbf{x}_3 = \frac{1}{2/\sqrt{3}}\begin{bmatrix} 2/3 \\ -2/3 \\ 2/3 \end{bmatrix} = \begin{bmatrix} 1/\sqrt{3} \\ -1/\sqrt{3} \\ 2/\sqrt{3} \end{bmatrix}$$

The orthonormal set is $\{\mathbf{q}_1, \mathbf{q}_2, \mathbf{q}_3\}$.

Example 9 Use the Gram-Schmidt orthonormalization process to construct an orthonormal set of vectors from the linearly independent set $\{\mathbf{x}_1, \mathbf{x}_2, \mathbf{x}_3, \mathbf{x}_4\}$, where

$$\mathbf{x}_1 = \begin{bmatrix} 1 \\ 1 \\ 0 \\ 1 \end{bmatrix}, \mathbf{x}_2 = \begin{bmatrix} 1 \\ 2 \\ 1 \\ 0 \end{bmatrix}, \mathbf{x}_3 = \begin{bmatrix} 0 \\ 1 \\ 2 \\ 1 \end{bmatrix}, \mathbf{x}_4 = \begin{bmatrix} 1 \\ 0 \\ 1 \\ 1 \end{bmatrix}$$

Solution: Carrying eight significant figures through all computations but rounding to four decimals for presentation purposes, we get

For the first iteration ($k = 1$)

$$r_{11} = \sqrt{\langle \mathbf{x}_1, \mathbf{x}_1 \rangle} = \sqrt{3} = 1.7321,$$

$$\mathbf{q}_1 = \frac{1}{r_{11}}\mathbf{x}_1 = \frac{1}{\sqrt{3}}\begin{bmatrix} 1 \\ 1 \\ 0 \\ 1 \end{bmatrix} = \begin{bmatrix} 0.5774 \\ 0.5774 \\ 0.0000 \\ 0.5774 \end{bmatrix},$$

$$r_{12} = \langle \mathbf{x}_2, \mathbf{q}_1 \rangle = 1.7321,$$

$$r_{13} = \langle \mathbf{x}_3, \mathbf{q}_1 \rangle = 1.1547,$$

$$r_{14} = \langle \mathbf{x}_4, \mathbf{q}_1 \rangle = 1.1547,$$

$$\mathbf{x}_2 \leftarrow \mathbf{x}_2 - r_{12}\mathbf{q}_1 = \begin{bmatrix} 1 \\ 2 \\ 1 \\ 0 \end{bmatrix} - 1.7321\begin{bmatrix} 0.5774 \\ 0.5774 \\ 0.0000 \\ 0.5774 \end{bmatrix} = \begin{bmatrix} 0.0000 \\ 1.0000 \\ 1.0000 \\ -1.0000 \end{bmatrix},$$

$$\mathbf{x}_3 \leftarrow \mathbf{x}_3 - r_{13}\mathbf{q}_1 = \begin{bmatrix} 0 \\ 1 \\ 2 \\ 1 \end{bmatrix} - 1.1547\begin{bmatrix} 0.5774 \\ 0.5774 \\ 0.0000 \\ 0.5774 \end{bmatrix} = \begin{bmatrix} -0.6667 \\ 0.3333 \\ 2.0000 \\ 0.3333 \end{bmatrix},$$

$$\mathbf{x}_4 \leftarrow \mathbf{x}_4 - r_{14}\mathbf{q}_1 = \begin{bmatrix} 1 \\ 0 \\ 1 \\ 1 \end{bmatrix} - 1.1547 \begin{bmatrix} 0.5774 \\ 0.5774 \\ 0.0000 \\ 0.5774 \end{bmatrix} = \begin{bmatrix} 0.3333 \\ -0.6667 \\ 1.0000 \\ 0.3333 \end{bmatrix}.$$

For the second iteration ($k=2$), using vectors from the first iteration, we compute

$$\left(\begin{aligned} r_{22} &= \sqrt{\langle \mathbf{x}_2, \mathbf{x}_2 \rangle} = 1.7321, \\ \mathbf{q}_2 &= \frac{1}{r_{22}}\mathbf{x}_2 = \frac{1}{1.7321}\begin{bmatrix} 0.0000 \\ 1.0000 \\ 1.0000 \\ -1.0000 \end{bmatrix} = \begin{bmatrix} 0.0000 \\ 0.5774 \\ 0.5774 \\ -0.5774 \end{bmatrix}, \\ r_{23} &= \langle \mathbf{x}_3, \mathbf{q}_2 \rangle = 1.1547, \\ r_{24} &= \langle \mathbf{x}_4, \mathbf{q}_2 \rangle = 0.0000, \end{aligned}\right)$$

$$\mathbf{x}_3 \leftarrow \mathbf{x}_3 - r_{23}\mathbf{q}_2 = \begin{bmatrix} -0.6667 \\ 0.3333 \\ 2.0000 \\ 0.3333 \end{bmatrix} - 1.1547 \begin{bmatrix} 0.0000 \\ 0.5774 \\ 0.5774 \\ -0.5774 \end{bmatrix} = \begin{bmatrix} -0.6667 \\ -0.3333 \\ 1.3333 \\ 1.0000 \end{bmatrix},$$

$$\mathbf{x}_4 \leftarrow \mathbf{x}_4 - r_{24}\mathbf{q}_2 = \begin{bmatrix} 0.3333 \\ -0.6667 \\ 1.0000 \\ 0.3333 \end{bmatrix} - 0.0000 \begin{bmatrix} 0.0000 \\ 0.5774 \\ 0.5774 \\ -0.5774 \end{bmatrix} = \begin{bmatrix} 0.3333 \\ -0.6667 \\ 1.0000 \\ 0.3333 \end{bmatrix}.$$

For the third iteration ($k=3$), using vectors from the second iteration, we compute

$$\begin{aligned} r_{33} &= \sqrt{\langle \mathbf{x}_3, \mathbf{x}_3 \rangle} = 1.8257, \\ \mathbf{q}_3 &= \frac{1}{r_{33}}\mathbf{x}_3 = \frac{1}{1.8257}\begin{bmatrix} -0.6667 \\ -0.3333 \\ 1.3333 \\ 1.0000 \end{bmatrix} = \begin{bmatrix} -0.3651 \\ -0.1826 \\ 0.7303 \\ 0.5477 \end{bmatrix}, \\ r_{34} &= \langle \mathbf{x}_4, \mathbf{q}_3 \rangle = 0.9129, \end{aligned}$$

$$\mathbf{x}_4 \leftarrow \mathbf{x}_4 - r_{34}\mathbf{q}_3 = \begin{bmatrix} 0.3333 \\ -0.6667 \\ 1.0000 \\ 0.3333 \end{bmatrix} - 0.9129 \begin{bmatrix} -0.3651 \\ -0.1826 \\ 0.7303 \\ 0.5477 \end{bmatrix} = \begin{bmatrix} 0.6667 \\ -0.5000 \\ 0.3333 \\ -0.1667 \end{bmatrix}.$$

For the fourth iteration ($k=4$), using vectors from the third iteration, we compute

$$r_{44} = \sqrt{\langle \mathbf{x}_4, \mathbf{x}_4 \rangle} = 0.9129,$$

$$\mathbf{q}_4 = \frac{1}{r_{44}}\mathbf{x}_4 = \frac{1}{0.9129} \begin{bmatrix} 0.6667 \\ -0.5000 \\ 0.3333 \\ -0.1667 \end{bmatrix} = \begin{bmatrix} 0.7303 \\ -0.5477 \\ 0.3651 \\ -0.1826 \end{bmatrix}.$$

The orthonormal set is $\{\mathbf{q}_1, \mathbf{q}_2, \mathbf{q}_3, \mathbf{q}_4\}$.

If $\mathbb{B} = \{\mathbf{x}_1, \mathbf{x}_2, \ldots, \mathbf{x}_p\}$ is a linearly independent set of vectors in an inner product space $\mathbb{U}$, and not necessarily a basis, then the Gram-Schmidt orthonormalization process can be applied directly on $\mathbb{B}$ to transform it into an orthonormal set of vectors with the same span as $\mathbb{B}$. This follows immediately from Theorem 3 because $\mathbb{B}$ is a basis for the subspace $\mathbb{V} = span\{\mathbf{x}_1, \mathbf{x}_2, \ldots, \mathbf{x}_p\}$.

Problems 6.2

In Problems 1 through 10, determine (a) the projection of $\mathbf{x}_1$ onto $\mathbf{x}_2$, and (b) the orthogonal complement.

(1) $\mathbf{x}_1 = \begin{bmatrix} 1 \\ 2 \end{bmatrix}, \mathbf{x}_2 = \begin{bmatrix} 2 \\ 1 \end{bmatrix}.$

(2) $\mathbf{x}_1 = \begin{bmatrix} 1 \\ 1 \end{bmatrix}, \mathbf{x}_2 = \begin{bmatrix} 3 \\ 5 \end{bmatrix}.$

(3) $\mathbf{x}_1 = \begin{bmatrix} 3 \\ -2 \end{bmatrix}, \mathbf{x}_2 = \begin{bmatrix} 3 \\ 3 \end{bmatrix}.$

(4) $\mathbf{x}_1 = \begin{bmatrix} 4 \\ -1 \end{bmatrix}, \mathbf{x}_2 = \begin{bmatrix} 2 \\ 8 \end{bmatrix}.$

(5) $\mathbf{x}_1 = \begin{bmatrix} -7 \\ -2 \end{bmatrix}, \mathbf{x}_2 = \begin{bmatrix} 2 \\ 9 \end{bmatrix}.$

(6) $\mathbf{x}_1 = \begin{bmatrix} 2 \\ 1 \\ 0 \end{bmatrix}, \mathbf{x}_2 = \begin{bmatrix} 2 \\ 0 \\ 2 \end{bmatrix}.$

(7) $\mathbf{x}_1 = \begin{bmatrix} 1 \\ 1 \\ 0 \end{bmatrix}, \mathbf{x}_2 = \begin{bmatrix} 2 \\ 2 \\ 1 \end{bmatrix}.$

(8) $\mathbf{x}_1 = \begin{bmatrix} 0 \\ 3 \\ 4 \end{bmatrix}, \mathbf{x}_2 = \begin{bmatrix} 2 \\ 5 \\ 5 \end{bmatrix}.$

(9) $\mathbf{x}_1 = \begin{bmatrix} 0 \\ 1 \\ 1 \\ 1 \end{bmatrix}, \mathbf{x}_2 = \begin{bmatrix} 1 \\ 1 \\ 1 \\ 0 \end{bmatrix}.$

(10) $\mathbf{x}_1 = \begin{bmatrix} 1 \\ 2 \\ 3 \\ 4 \end{bmatrix}, \mathbf{x}_2 = \begin{bmatrix} 1 \\ -2 \\ 0 \\ -1 \end{bmatrix}.$

In Problems 11 through 23, show that the set $\mathbb{B}$ is an orthonormal basis (under the Euclidean inner product or the inner product induced by the Euclidean inner product) for the given vector space and then write $\mathbf{x}$ as a linear combination of those basis vectors.

(11) $\mathbb{B} = \left\{ \begin{bmatrix} 3/5 \\ 4/5 \end{bmatrix}, \begin{bmatrix} 4/5 \\ -3/5 \end{bmatrix} \right\}$ in $\mathbb{R}^2$; $\mathbf{x} = \begin{bmatrix} 3 \\ 5 \end{bmatrix}$.

(12) $\mathbb{B} = \left\{ \begin{bmatrix} 1/\sqrt{2} \\ 1/\sqrt{2} \end{bmatrix}, \begin{bmatrix} 1/\sqrt{2} \\ -1/\sqrt{2} \end{bmatrix} \right\}$ in $\mathbb{R}^2$; $\mathbf{x} = \begin{bmatrix} 3 \\ 5 \end{bmatrix}$.

(13) $\mathbb{B} = \left\{ \begin{bmatrix} 1/\sqrt{5} \\ 2/\sqrt{5} \end{bmatrix}, \begin{bmatrix} -2/\sqrt{5} \\ 1/\sqrt{5} \end{bmatrix} \right\}$ in $\mathbb{R}^2$; $\mathbf{x} = \begin{bmatrix} 2 \\ -3 \end{bmatrix}$.

(14) $\mathbb{B} = \left\{ \begin{bmatrix} 3/5 \\ 4/5 \\ 0 \end{bmatrix}, \begin{bmatrix} 4/5 \\ -3/5 \\ 0 \end{bmatrix}, \begin{bmatrix} 0 \\ 0 \\ 1 \end{bmatrix} \right\}$ in $\mathbb{R}^3$; $\mathbf{x} = \begin{bmatrix} 1 \\ 2 \\ 3 \end{bmatrix}$.

(15) $\mathbb{B} = \left\{ \begin{bmatrix} 3/5 \\ 4/5 \\ 0 \end{bmatrix}, \begin{bmatrix} 4/5 \\ -3/5 \\ 0 \end{bmatrix}, \begin{bmatrix} 0 \\ 0 \\ 1 \end{bmatrix} \right\}$ in $\mathbb{R}^3$; $\mathbf{x} = \begin{bmatrix} 10 \\ 0 \\ -20 \end{bmatrix}$.

(16) $\mathbb{B} = \left\{ \begin{bmatrix} 1/\sqrt{2} \\ 1/\sqrt{2} \\ 0 \end{bmatrix}, \begin{bmatrix} -1/\sqrt{6} \\ 1/\sqrt{6} \\ 2/\sqrt{6} \end{bmatrix}, \begin{bmatrix} 1/\sqrt{3} \\ -1/\sqrt{3} \\ 1/\sqrt{3} \end{bmatrix} \right\}$ in $\mathbb{R}^3$; $\mathbf{x} = \begin{bmatrix} 10 \\ 0 \\ -20 \end{bmatrix}$.

(17) $\mathbb{B} = \left\{ \begin{bmatrix} -1/\sqrt{2} \\ 1/\sqrt{2} \\ 0 \end{bmatrix}, \begin{bmatrix} 1/\sqrt{6} \\ 1/\sqrt{6} \\ -2/\sqrt{6} \end{bmatrix}, \begin{bmatrix} -1/\sqrt{3} \\ -1/\sqrt{3} \\ -1/\sqrt{3} \end{bmatrix} \right\}$ in $\mathbb{R}^3$; $\mathbf{x} = \begin{bmatrix} 10 \\ 0 \\ -20 \end{bmatrix}$.

(18) $\mathbb{B} = \{0.6t - 0.8,\ 0.8t + 0.6\}$ in $\mathbb{P}^1$; $\mathbf{x} = 2t + 1$.

(19) $\mathbb{B} = \{0.6t^2 - 0.8,\ 0.8t^2 + 0.6,\ t\}$ in $\mathbb{P}^2$; $\mathbf{x} = t^2 + 2t + 3$.

(20) $\mathbb{B} = \{0.6t^2 - 0.8,\ 0.8t^2 + 0.6,\ t\}$ in $\mathbb{P}^2$; $\mathbf{x} = t^2 - 1$.

(21) $\mathbb{B} = \left\{ \begin{bmatrix} 1/\sqrt{3} & 1/\sqrt{3} \\ -1/\sqrt{3} & 0 \end{bmatrix}, \begin{bmatrix} 0 & -1/\sqrt{3} \\ -1/\sqrt{3} & 1/\sqrt{3} \end{bmatrix}, \begin{bmatrix} 1/\sqrt{3} & 0 \\ 1/\sqrt{3} & 1/\sqrt{3} \end{bmatrix}, \begin{bmatrix} -1/\sqrt{3} & 1/\sqrt{3} \\ 0 & 1/\sqrt{3} \end{bmatrix} \right\}$

in $\mathbb{M}_{2\times 2}$; $\mathbf{x} = \begin{bmatrix} 1 & 1 \\ -1 & 2 \end{bmatrix}$.

(22) $\mathbb{B} = \left\{ \begin{bmatrix} 3/5 & 4/5 \\ 0 & 0 \end{bmatrix}, \begin{bmatrix} 4/5 & -3/5 \\ 0 & 0 \end{bmatrix}, \begin{bmatrix} 0 & 0 \\ 3/5 & -4/5 \end{bmatrix}, \begin{bmatrix} 0 & 0 \\ -4/5 & -3/5 \end{bmatrix} \right\}$;

in $\mathbb{M}_{2\times 2}$; $\mathbf{x} = \begin{bmatrix} 1 & 2 \\ 3 & 4 \end{bmatrix}$.

(23) $\mathbb{B} = \left\{ \begin{bmatrix} 1/2 & 1/2 \\ 1/\sqrt{2} & 0 \end{bmatrix}, \begin{bmatrix} -1/2 & -1/2 \\ 1/\sqrt{2} & 0 \end{bmatrix}, \begin{bmatrix} -1/2 & 1/2 \\ 0 & 1/\sqrt{2} \end{bmatrix}, \begin{bmatrix} 1/2 & -1/2 \\ 0 & 1/\sqrt{2} \end{bmatrix} \right\}$

in $\mathbb{M}_{2\times 2}$; $\mathbf{x} = \begin{bmatrix} 4 & 5 \\ -6 & 7 \end{bmatrix}$

In Problems 24 through 32, use the Gram-Schmidt orthonormalization process to construct an orthonormal set from the given set of linearly independent vectors.

(24) The vectors in Problem 1.

(25) The vectors in Problem 2.

(26) The vectors in Problem 3.

(27) $\mathbf{x}_1 = \begin{bmatrix} 1 \\ 2 \\ 1 \end{bmatrix}, \mathbf{x}_2 = \begin{bmatrix} 1 \\ 0 \\ 1 \end{bmatrix}, \mathbf{x}_3 = \begin{bmatrix} 1 \\ 0 \\ 2 \end{bmatrix}.$

(28) $\mathbf{x}_1 = \begin{bmatrix} 2 \\ 1 \\ 0 \end{bmatrix}, \mathbf{x}_2 = \begin{bmatrix} 0 \\ 1 \\ 1 \end{bmatrix}, \mathbf{x}_3 = \begin{bmatrix} 2 \\ 0 \\ 2 \end{bmatrix}.$

(29) $\mathbf{x}_1 = \begin{bmatrix} 1 \\ 1 \\ 0 \end{bmatrix}, \mathbf{x}_2 = \begin{bmatrix} 2 \\ 0 \\ 1 \end{bmatrix}, \mathbf{x}_3 = \begin{bmatrix} 2 \\ 2 \\ 1 \end{bmatrix}.$

(30) $\mathbf{x}_1 = \begin{bmatrix} 0 \\ 3 \\ 4 \end{bmatrix}, \mathbf{x}_2 = \begin{bmatrix} 3 \\ 5 \\ 0 \end{bmatrix}, \mathbf{x}_3 = \begin{bmatrix} 2 \\ 5 \\ 5 \end{bmatrix}.$

(31) $\mathbf{x}_1 = \begin{bmatrix} 0 \\ 1 \\ 1 \\ 1 \end{bmatrix}, \mathbf{x}_2 = \begin{bmatrix} 1 \\ 0 \\ 1 \\ 1 \end{bmatrix}, \mathbf{x}_3 = \begin{bmatrix} 1 \\ 1 \\ 0 \\ 1 \end{bmatrix}, \mathbf{x}_4 = \begin{bmatrix} 1 \\ 1 \\ 1 \\ 0 \end{bmatrix}.$

(32) $\mathbf{x}_1 = \begin{bmatrix} 1 \\ 1 \\ 0 \\ 0 \end{bmatrix}, \mathbf{x}_2 = \begin{bmatrix} 0 \\ 1 \\ -1 \\ 0 \end{bmatrix}, \mathbf{x}_3 = \begin{bmatrix} 1 \\ 0 \\ -1 \\ 0 \end{bmatrix}, \mathbf{x}_4 = \begin{bmatrix} 1 \\ 0 \\ 0 \\ -1 \end{bmatrix}.$

(33) The vectors $\mathbf{x}_1 = \begin{bmatrix} 1 \\ 1 \\ 0 \end{bmatrix}, \mathbf{x}_2 = \begin{bmatrix} 0 \\ 1 \\ 1 \end{bmatrix}, \mathbf{x}_3 = \begin{bmatrix} 1 \\ 0 \\ -1 \end{bmatrix}.$

are linearly dependent. Apply the Gram-Schmidt orthonormalization process to them and use the results to deduce what occurs when the process is applied to a linearly dependent set of vectors.

(34) Prove directly that $\mathbf{x} - \dfrac{\langle \mathbf{a}, \mathbf{x} \rangle}{\langle \mathbf{a}, \mathbf{a} \rangle}\mathbf{a}$ is orthogonal to $\mathbf{a}$.

(35) Prove that if $\mathbf{x}$ and $\mathbf{y}$ are orthonormal, then $||s\mathbf{x} + t\mathbf{y}||^2 = s^2 + t^2$ for any two scalars s and t.

(36) Let $\mathbf{Q}$ be any $n \times n$ real matrix having columns that, when considered as n-dimensional vectors, form an orthonormal set. What can one say about the product $\mathbf{Q}^{\mathrm{T}}\mathbf{Q}$?

(37) Prove that if $\langle \mathbf{y}, \mathbf{x} \rangle = 0$ for every n-dimensional vector $\mathbf{y}$, then $\mathbf{x} = \mathbf{0}$.

(38) Let $\mathbf{A}$ be an $n \times n$ real matrix and $\mathbf{p}$ be a real n-dimensional column matrix. Show that if $\mathbf{p}$ is orthogonal to the columns of $\mathbf{A}$, then $\langle \mathbf{A}y, \mathbf{p} \rangle = 0$ for every n-dimensional real column matrix $\mathbf{y}$.

(39) Prove that if $\mathbb{B}$ is an orthonormal set of vectors that span a vector space $\mathbb{U}$, then $\mathbb{B}$ is a basis for $\mathbb{U}$.

6.3 THE QR ALGORITHM

To express a matrix in a more convenient form as a product of two other matrices is called a *factorization*. One of the more useful of these is the **QR** factorization, which is based on Gram-Schmidt orthonormalization. The **QR** algorithm is a robust numerical method for computing eigenvalues of real matrices. In contrast to the power methods described in Section 4.5, which converge to a single dominant real eigenvalue, the **QR** algorithm generally locates *all* eigenvalues of a matrix, both real and complex, regardless of multiplicity.

To use the algorithm, we must factor a given matrix **A** into the matrix product

$$\mathbf{A} = \mathbf{QR} \tag{6.11}$$

where **R** is an upper (or right) triangular matrix and the columns of **Q**, considered as individual column matrices, form an orthonormal set. Equation (6.11) is a **QR** decomposition of **A**. Such a decomposition is always possible when the columns of **A** are linearly independent.

Example 1 Two **QR** decompositions are

$$\begin{bmatrix} 1 & 3 \\ 1 & 5 \end{bmatrix} = \begin{bmatrix} 1/\sqrt{2} & -1/\sqrt{2} \\ 1/\sqrt{2} & 1/\sqrt{2} \end{bmatrix} \begin{bmatrix} \sqrt{2} & 4/\sqrt{2} \\ 0 & \sqrt{2} \end{bmatrix}$$

and

$$\begin{bmatrix} 1 & 2 \\ 2 & 2 \\ 2 & 1 \end{bmatrix} = \begin{bmatrix} 1/3 & 10/\sqrt{153} \\ 2/3 & 2/\sqrt{153} \\ 2/3 & -7/\sqrt{153} \end{bmatrix} \begin{bmatrix} 3 & 8/3 \\ 0 & \sqrt{153}/9 \end{bmatrix}$$

It is apparent from Example 1 that **QR** decompositions exist for square and rectangular matrices. The orders of **A** and **Q** are the same and **R** is a square matrix having the same number of columns as **A**. For the remainder of this section, we restrict **A** to be square because we are interested in locating eigenvalues, and eigenvalues are defined only for square matrices. Then both **Q** and **R** are square and have the same order as **A**.

A **QR** decomposition of a matrix **A** comes directly from the Gram-Schmidt orthonormalization process (see Theorem 3 of Section 6.2) applied to the linearly independent columns of **A**. The elements of $\mathbf{R} = [r_{ij}]$ are the scalars from Steps 1 and 3 of the orthonormalization process, and the columns of **Q** are the orthonormal column matrices constructed in Step 2 of that process. To see why, we let

In a **QR** decomposition of a matrix **A**, the elements of $\mathbf{R} = [r_{ij}]$ are the scalars from Steps 1 and 3 of the Gram-Schmidt orthonormalization process applied to the linearly independent columns of **A**, while the columns of **Q** are the orthonormal column matrices constructed in Step 2 of the Gram-Schmidt process.

$x_j^{(i)}$ denote $\mathbf{x}_j$ after the *ith* iteration of the Gram-Schmidt process ($j>i$). Thus, $x_j^{(1)}$ is the new value of $\mathbf{x}_j$ after the first iteration of the orthonormalization process, $x_j^{(2)}$ is value of $\mathbf{x}_j$ after the second iteration, and so on. In this context, $x_j^{(0)}$ is the initial value of $\mathbf{x}_j$.

▸THEOREM 1

After the ith iteration of the Gram-Schmidt orthonormalization process, $x_j^{(i)}=x_j^{(0)}-r_{1,j}\,\mathbf{q}_1-r_{2,j}\,\mathbf{q}_2-\ldots-r_{i,j}\,\mathbf{q}_i$ ◂

Proof: (by mathematical induction on the iterations): After the first iteration, we have from Step 4 of the process that $x_j^{(1)}=x_j^{(0)}-r_{1,j}\,\mathbf{q}_1$ for $j=2, 3, \ldots, n$, so the proposition is true for $n=1$.

Assume the proposition is true for $n=i$. Then after the $i+1$ iteration, it follows from Step 4 that for $j=i+2, i+3, \ldots, n$.

$$x_j^{(i+1)}=x_j^{(i)}-r_{i+1,j}\;\mathbf{q}_{i+1}$$

and then from the induction hypothesis that

$$x_j^{(i+1)}=\left[x_j^{(0)}-r_{1,j}\mathbf{q}_1-r_{2,j}\mathbf{q}_2-\ldots-r_{i,j}\mathbf{q}_i\right]-r_{i+1,j}\mathbf{q}_{i+1}$$

which is of the required form. Therefore, Theorem 1 is proved by mathematical induction.

Designate the columns of an $n\times n$ matrix $\mathbf{A}$ as $\mathbf{x}_1, \mathbf{x}_2, \ldots, \mathbf{x}_n$, respectively, so that $\mathbf{A}=[\mathbf{x}_1 \quad \mathbf{x}_2 \quad \ldots \quad \mathbf{x}_n]$. Set

$$\mathbf{Q}=[q_1\ q_2\cdots q_n]. \tag{6.12}$$

and

$$\mathbf{R}=\begin{bmatrix} r_{11} & r_{12} & r_{13} & \cdots & r_{1n} \\ 0 & r_{22} & r_{23} & \cdots & r_{2n} \\ 0 & 0 & r_{33} & \cdots & r_{3n} \\ \vdots & \vdots & \vdots & \vdots & \vdots \\ 0 & 0 & 0 & \cdots & r_{nn} \end{bmatrix} \tag{6.13}$$

Then it follows from Theorem 1 that $\mathbf{A}=\mathbf{QR}$.

Example 2 Construct a **QR** decomposition for $\mathbf{A}=\begin{bmatrix} 1 & 0 & 1 \\ 1 & 1 & 0 \\ 0 & 1 & 1 \end{bmatrix}$.

Solution: The columns of **A** are

$$\mathbf{x}_1 = \begin{bmatrix} 1 \\ 1 \\ 0 \end{bmatrix}, \mathbf{x}_2 = \begin{bmatrix} 0 \\ 1 \\ 1 \end{bmatrix}, \mathbf{x}_3 = \begin{bmatrix} 1 \\ 0 \\ 1 \end{bmatrix}$$

Using the results of Example 8 of Section 6.2, we have immediately that

$$\mathbf{Q} = \begin{bmatrix} 1/\sqrt{2} & -1/\sqrt{6} & 1/\sqrt{3} \\ 1/\sqrt{2} & 1/\sqrt{6} & -1/\sqrt{3} \\ 0 & 2/\sqrt{6} & 1/\sqrt{3} \end{bmatrix}, \mathbf{R} = \begin{bmatrix} \sqrt{2} & 1/\sqrt{2} & 1/\sqrt{2} \\ 0 & \sqrt{3/2} & 1/\sqrt{6} \\ 0 & 0 & 2/\sqrt{3} \end{bmatrix}$$

from which **A**=**QR**.

Example 3 Construct a **QR** decomposition for $\mathbf{A} = \begin{bmatrix} 1 & 1 & 0 & 1 \\ 1 & 2 & 1 & 0 \\ 0 & 1 & 2 & 1 \\ 1 & 0 & 1 & 1 \end{bmatrix}$.

Solution: The columns of **A** are

$$\mathbf{x}_1 = \begin{bmatrix} 1 \\ 1 \\ 0 \\ 1 \end{bmatrix}, \mathbf{x}_2 = \begin{bmatrix} 1 \\ 2 \\ 1 \\ 0 \end{bmatrix}, \mathbf{x}_3 = \begin{bmatrix} 0 \\ 1 \\ 2 \\ 1 \end{bmatrix}, \mathbf{x}_4 = \begin{bmatrix} 1 \\ 0 \\ 1 \\ 1 \end{bmatrix}$$

Using the results of Example 9 of Section 6.2, we have immediately that

$$\mathbf{Q} = \begin{bmatrix} 0.5774 & 0.0000 & -0.3651 & 0.7303 \\ 0.5774 & 0.5774 & -0.1826 & -0.5477 \\ 0.0000 & 0.5774 & 0.7303 & 0.3651 \\ 0.5774 & -0.5774 & 0.5477 & -0.1826 \end{bmatrix},$$

$$\mathbf{R} = \begin{bmatrix} 1.7321 & 1.7321 & 1.1547 & 1.1547 \\ 0 & 1.7321 & 1.1547 & 0.0000 \\ 0 & 0 & 1.8257 & 0.9129 \\ 0 & 0 & 0 & 0.9129 \end{bmatrix}$$

from which **A**=**QR** to within round-off error.

The **QR** algorithm uses **QR** decompositions to identify the eigenvalues of a square matrix. The algorithm involves many arithmetic calculations, making it unattractive for hand computations but ideal for implementation on a computer. Although a proof of the **QR** algorithm is beyond the scope of this book, the algorithm itself is deceptively simple.

We begin with a square real matrix $\mathbf{A}_0$ having linearly independent columns. To determine its eigenvalues, we create a sequence of new matrices $\mathbf{A}_1, \mathbf{A}_2, \ldots, \mathbf{A}_{k-1}, \mathbf{A}_k, \ldots$, having the property that each new matrix has the same eigenvalues as $\mathbf{A}_0$, and that these eigenvalues become increasingly obvious as the sequence progresses. To calculate $\mathbf{A}_k$ ($k=1, 2, \ldots$) once $\mathbf{A}_{k-1}$ is known, we construct a **QR** decomposition of $\mathbf{A}_{k-1}$:

$$\mathbf{A}_{k-1} = \mathbf{Q}_{k-1}\mathbf{R}_{k-1}$$

and then reverse the order of the product to define

$$\mathbf{A}_k = \mathbf{R}_{k-1}\mathbf{Q}_{k-1} \tag{6.14}$$

Each matrix in the sequence $\{\mathbf{A}_k\}$ has identical eigenvalues (see Problem 29), and the sequence generally converges to one of the following two partitioned forms:

$$\left[\begin{array}{ccccc|c} & & \mathbf{S} & & & \mathbf{T} \\ \hline 0 & 0 & 0 & \cdots & 0 & a \end{array}\right] \tag{6.15}$$

or

$$\left[\begin{array}{ccccc|cc} & & \mathbf{U} & & & & \mathbf{V} \\ \hline 0 & 0 & 0 & \cdots & 0 & b & c \\ 0 & 0 & 0 & \cdots & 0 & d & e \end{array}\right] \tag{6.16}$$

If matrix (6.15) occurs, then the element a is an eigenvalue, and the remaining eigenvalues are found by applying the **QR** algorithm anew to the submatrix **S**. If, on the other hand, matrix (6.16) occurs, then two eigenvalues are determined by solving for the roots of the characteristic equation of the 2×2 matrix in the lower right partition, namely,

$$\lambda^2 - (b+e)\lambda + (be - cd) = 0$$

The remaining eigenvalues are found by applying the **QR** algorithm anew to the submatrix **U**.

Convergence of the algorithm is accelerated by performing a shift at each iteration. If the orders of all matrices are $n \times n$, we denote the element in the (n, n) position of the matrix $\mathbf{A}_{k-1}$ as w_{k-1}, and construct a **QR** decomposition for the shifted matrix $\mathbf{A}_{k-1} - w_{k-1}\mathbf{I}$. That is,

$$\mathbf{A}_{k-1} - w_{k-1}\mathbf{I} = \mathbf{Q}_{k-1}\mathbf{R}_{k-1} \tag{6.17}$$

We define

$$\mathbf{A}_k = \mathbf{R}_{k-1}\mathbf{Q}_{k-1} + w_{k-1}\mathbf{I} \tag{6.18}$$

Example 4 Find the eigenvalues of

$$\mathbf{A}_0 = \begin{bmatrix} 0 & 1 & 0 \\ 0 & 0 & 1 \\ 18 & -1 & -7 \end{bmatrix}$$

Solution: Using the **QR** algorithm with shifting, carrying all calculations to eight significant figures but rounding to four decimals for presentation, we compute

$$\mathbf{A}_0-(-7)\mathbf{I} = \begin{bmatrix} 7 & 1 & 0 \\ 0 & 7 & 1 \\ 18 & -1 & 0 \end{bmatrix}$$

$$= \begin{bmatrix} 0.3624 & 0.1695 & -0.9165 \\ 0.0000 & 0.9833 & 0.1818 \\ 0.9320 & -0.0659 & 0.3564 \end{bmatrix} \begin{bmatrix} 19.3132 & -0.5696 & 0.0000 \\ 0.0000 & 7.1187 & 0.9833 \\ 0.0000 & 0.0000 & 0.1818 \end{bmatrix}$$

$$= \mathbf{Q}_0\mathbf{R}_0$$

$$\mathbf{A}_1 = \mathbf{R}_0\mathbf{Q}_0 + (-7)\mathbf{I}$$

$$= \begin{bmatrix} 19.3132 & -0.5696 & 0.0000 \\ 0.0000 & 7.1187 & 0.9833 \\ 0.0000 & 0.0000 & 0.1818 \end{bmatrix} \begin{bmatrix} 0.3624 & 0.1695 & -0.9165 \\ 0.0000 & 0.9833 & 0.1818 \\ 0.9320 & -0.0659 & 0.3564 \end{bmatrix} + \begin{bmatrix} -7 & 0 & 0 \\ 0 & -7 & 0 \\ 0 & 0 & -7 \end{bmatrix}$$

$$= \begin{bmatrix} 0.0000 & 2.7130 & -17.8035 \\ 0.9165 & 6.8704 & 1.6449 \\ 0.1695 & -0.0120 & -6.9352 \end{bmatrix}$$

$$\mathbf{A}_1 - (-6.9352)\mathbf{I}$$

$$= \begin{bmatrix} 6.9352 & 2.7130 & -178035 \\ 0.9165 & 6.8704 & 1.6449 \\ 0.1695 & -0.0120 & 0.0000 \end{bmatrix}$$

$$= \begin{bmatrix} 0.9911 & -0.1306 & -0.0260 \\ 0.1310 & 0.9913 & 0.0120 \\ 0.0242 & -0.0153 & 0.9996 \end{bmatrix} \begin{bmatrix} 6.9975 & 3.5884 & -17.4294 \\ 0.0000 & 6.4565 & 3.9562 \\ 0.0000 & 0.0000 & 0.4829 \end{bmatrix}$$

$$= \mathbf{Q}_1\mathbf{R}_1$$

$$\mathbf{A}_2 = \mathbf{R}_1\mathbf{Q}_1 + (-6.9352)\mathbf{I} = \begin{bmatrix} 0.0478 & 2.9101 & -17.5612 \\ 0.9414 & -0.5954 & 4.0322 \\ 0.0117 & -0.0074 & -6.4525 \end{bmatrix}$$

Continuing in this manner, we generate sequentially

$$\mathbf{A}_3 = \begin{bmatrix} 0.5511 & 2.7835 & -16.8072 \\ 0.7826 & -1.1455 & 6.5200 \\ 0.0001 & -0.0001 & -6.4056 \end{bmatrix}$$

$$\mathbf{A}_4 = \begin{bmatrix} 0.9259 & 2.5510 & -15.9729 \\ 0.5497 & -1.5207 & 8.3583 \\ 0.0000 & -0.0000 & -6.4051 \end{bmatrix}$$

$\mathbf{A}_4$ has form (6.15) with

$$\mathbf{S} = \begin{bmatrix} 0.9259 & 2.5510 \\ 0.5497 & -1.5207 \end{bmatrix} \text{ and } a = -6.4051$$

One eigenvalue is -6.4051, which is identical to the value obtained in Example 2 of Section 4.6. In addition, the characteristic equation of $\mathbf{R}$ is $\lambda^2 + 0.5948\lambda - 2.8103 = 0$, which admits both -2 and 1.4052 as roots. These are the other two eigenvalues of $\mathbf{A}_0$.

Example 5 Find the eigenvalues of

$$\mathbf{A}_0 = \begin{bmatrix} 0 & 0 & 0 & -25 \\ 1 & 0 & 0 & 30 \\ 0 & 0 & 1 & -18 \\ 0 & 0 & 1 & 6 \end{bmatrix}$$

Solution: Using the **QR** algorithm with shifting, carrying all calculations to eight significant figures but rounding to four decimals for presentation, we compute

$$\mathbf{A}_0 - (6)\mathbf{I} = \begin{bmatrix} -6 & 0 & 0 & -25 \\ 1 & -6 & 0 & 30 \\ 0 & 1 & -6 & -18 \\ 0 & 0 & 1 & 0 \end{bmatrix}$$

$$= \begin{bmatrix} -0.9864 & -0.1621 & -0.0270 & -0.0046 \\ 0.1644 & -0.9726 & -0.1620 & -0.0274 \\ 0.0000 & 0.1666 & -0.9722 & -0.1643 \\ 0.0000 & 0.0000 & 0.1667 & -0.9860 \end{bmatrix}$$

$$\times \begin{bmatrix} 6.0828 & -0.9864 & 0.0000 & 29.5918 \\ 0.0000 & 6.0023 & -0.9996 & -28.1246 \\ 0.0000 & 0.0000 & 6.0001 & 13.3142 \\ 0.0000 & 0.0000 & 0.0000 & 2.2505 \end{bmatrix}$$

$$= \mathbf{Q}_0\mathbf{R}_0$$

$$\mathbf{A}_1 = \mathbf{R}_0\mathbf{Q}_0 + (6)\mathbf{I} = \begin{bmatrix} -0.1622 & -0.0266 & 4.9275 & -29.1787 \\ 0.9868 & -0.0044 & -4.6881 & 27.7311 \\ 0.0000 & 0.9996 & 2.3858 & -14.1140 \\ 0.0000 & 0.0000 & 0.3751 & 3.7810 \end{bmatrix}$$

$$\mathbf{A}_1 - (3.7810)\mathbf{I} = \begin{bmatrix} -3.9432 & -0.0266 & 4.9275 & -29.1787 \\ 0.9868 & -3.7854 & -4.6881 & 27.7311 \\ 0.0000 & 0.9996 & -1.3954 & -14.1140 \\ 0.0000 & 0.0000 & 0.3751 & 0.0000 \end{bmatrix}$$

$$= \begin{bmatrix} -0.9701 & -0.2343 & -0.0628 & -0.0106 \\ 0.2428 & -0.9361 & -0.2509 & -0.0423 \\ 0.0000 & 0.2622 & -0.9516 & -0.1604 \\ 0.0000 & 0.0000 & 0.1662 & -0.9861 \end{bmatrix}$$

$$\times \begin{bmatrix} 4.0647 & -0.8931 & -5.9182 & 35.0379 \\ 0.0000 & 3.8120 & 2.8684 & -22.8257 \\ 0.0000 & 0.0000 & 2.2569 & 8.3060 \\ 0.0000 & 0.0000 & 0.0000 & 1.3998 \end{bmatrix}$$

$$= \mathbf{Q}_1\mathbf{R}_1$$

$$\mathbf{A}_2 = \mathbf{R}_1\mathbf{Q}_1 + (3.7810)\mathbf{I}$$

$$= \begin{bmatrix} -0.3790 & -1.6681 & 11.4235 & -33.6068 \\ 0.9254 & 0.9646 & -7.4792 & 21.8871 \\ 0.0000 & 0.5918 & 3.0137 & -8.5524 \\ 0.0000 & 0.0000 & 0.2326 & 2.4006 \end{bmatrix}$$

Continuing in this manner, we generate, after 25 iterations,

$$\mathbf{A}_{25} = \begin{bmatrix} 4.8641 & -4.4404 & 18.1956 & -28.7675 \\ 4.2635 & -2.8641 & 13.3357 & -21.3371 \\ 0.0000 & 0.0000 & 2.7641 & -4.1438 \\ 0.0000 & 0.0000 & 0.3822 & 1.2359 \end{bmatrix}$$

which has form (6.16) with

$$\mathbf{U} = \begin{bmatrix} 4.8641 & -4.4404 \\ 4.2635 & -2.8641 \end{bmatrix} \text{ and } \begin{bmatrix} b & c \\ d & e \end{bmatrix} = \begin{bmatrix} 2.7641 & -4.1438 \\ 0.3822 & 1.2359 \end{bmatrix}$$

The characteristic equation of $\mathbf{U}$ is $\lambda^2-2\lambda+5=0$, which has as its roots $1\pm 2i$; the characteristic equation of the other 2×2 matrix is $\lambda^2-4\lambda+4.9999=0$, which has as its roots $2\pm i$. These roots are the four eigenvalues of $\mathbf{A}_0$.

Problems 6.3

(1) Given the matrix $\mathbf{A}=\begin{bmatrix}1 & 2\\ 2 & 0\\ 0 & 2\end{bmatrix}$ and matrix $\mathbf{Q}=\begin{bmatrix}\frac{\sqrt{5}}{5} & \frac{4\sqrt{5}}{15} & -\frac{2}{3}\\ \frac{2\sqrt{5}}{5} & -\frac{2\sqrt{5}}{15} & \frac{1}{3}\\ 0 & \frac{\sqrt{5}}{3} & \frac{2}{3}\end{bmatrix}$,

find an upper triangular matrix $\mathbf{R}$ such that $\mathbf{A}=\mathbf{QR}$.

In Problems 2 through 12, construct $\mathbf{QR}$ decompositions for the given matrices.

(2) $\begin{bmatrix}1 & 2\\ 2 & 1\end{bmatrix}$.

(3) $\begin{bmatrix}1 & 3\\ 1 & 5\end{bmatrix}$.

(4) $\begin{bmatrix}3 & 3\\ -2 & 3\end{bmatrix}$.

(5) $\begin{bmatrix}1 & 2\\ 2 & 2\\ 2 & 1\end{bmatrix}$.

(6) $\begin{bmatrix}1 & 1\\ 1 & 0\\ 3 & 5\end{bmatrix}$.

(7) $\begin{bmatrix}3 & 1\\ -2 & 1\\ 1 & 1\\ -1 & 1\end{bmatrix}$.

(8) $\begin{bmatrix}2 & 0 & 2\\ 1 & 1 & 0\\ 0 & 1 & 2\end{bmatrix}$.

(9) $\begin{bmatrix}1 & 2 & 2\\ 1 & 0 & 2\\ 0 & 1 & 1\end{bmatrix}$.

(10) $\begin{bmatrix}0 & 3 & 2\\ 3 & 5 & 5\\ 4 & 0 & 5\end{bmatrix}$.

(11) $\begin{bmatrix}0 & 1 & 1\\ 1 & 0 & 1\\ 1 & 1 & 0\\ 1 & 1 & 1\end{bmatrix}$.

(12) $\begin{bmatrix}1 & 0 & 1\\ 1 & 1 & 0\\ 0 & -1 & -1\\ 0 & 0 & 0\end{bmatrix}$.

(13) Use one iteration of the **QR** algorithm to calculate $\mathbf{A}_1$ for

$$\mathbf{A}_0 = \begin{bmatrix} 0 & 1 & 0 \\ 0 & 0 & 1 \\ 18 & -1 & 7 \end{bmatrix}.$$

Note that this matrix differs from the one in Example 4 by a single sign.

(14) Use one iteration of the **QR** algorithm to calculate $\mathbf{A}_1$ for

$$\mathbf{A}_0 = \begin{bmatrix} 2 & -17 & 7 \\ -17 & -4 & 1 \\ 7 & 1 & -14 \end{bmatrix}.$$

(15) Use one iteration of the **QR** algorithm to calculate $\mathbf{A}_1$ for

$$\mathbf{A}_0 = \begin{bmatrix} 0 & 0 & 0 & -13 \\ 1 & 0 & 0 & 4 \\ 0 & 1 & 0 & -14 \\ 0 & 0 & 1 & 4 \end{bmatrix}.$$

In Problems 16 through 24, use the **QR** algorithm to calculate the eigenvalues of the given matrices.

(16) The matrix in Problem 13.

(17) The matrix in Problem 14.

(18) $\begin{bmatrix} 3 & 0 & 0 \\ 2 & 6 & 4 \\ 2 & 3 & 5 \end{bmatrix}.$

(19) $\begin{bmatrix} 7 & 2 & 0 \\ 2 & 1 & 6 \\ 0 & 6 & 7 \end{bmatrix}.$

(20) $\begin{bmatrix} 3 & 2 & 3 \\ 2 & 6 & 6 \\ 3 & 6 & 11 \end{bmatrix}.$

(21) $\begin{bmatrix} 1 & 1 & 0 \\ 0 & 1 & 1 \\ 5 & -9 & 6 \end{bmatrix}.$

(22) The matrix in Problem 15.

(23) $\begin{bmatrix} 0 & 3 & 2 & -1 \\ 1 & 0 & 2 & -3 \\ 3 & 1 & 0 & -1 \\ 2 & -2 & 1 & 1 \end{bmatrix}.$

(24) $\begin{bmatrix} 10 & 7 & 8 & 7 \\ 7 & 5 & 6 & 5 \\ 8 & 6 & 10 & 9 \\ 7 & 5 & 9 & 10 \end{bmatrix}.$

(25) Prove that **R** is nonsingular in a **QR** decomposition.

(26) Evaluate $\mathbf{Q}^T\mathbf{Q}$ for any *square* matrix **Q** in a **QR** decomposition, and then prove that **Q** is nonsingular.

(27) Using Problem 25, show that $\mathbf{A}_k$ is similar to $\mathbf{A}_{k-1}$ in the **QR** algorithm and deduce that both matrices have the same eigenvalues.

6.4 LEAST SQUARES

Analyzing data to interpret and predict events is common to business, engineering, and the physical and social sciences. If such data are plotted, as in Figure 6.8, they constitute a *scatter diagram*, which may provide insight into the underlying relationship between system variables. Figure 6.8 could represent a relationship between advertising expenditures and sales in a business environment, or between time and velocity in physics, or between formal control and deterrence in sociology.

Small random variations from expected patterns are called noise.

The data in Figure 6.8 appears to follow a straight line relationship, but with minor random distortions. Such distortions, called *noise*, are expected when data are obtained experimentally. To understand why, assume you are asked to ride a bicycle on a painted line down the middle of a straight path. A paint pot with a mechanism that releases a drop of paint intermittently is attached to the bicycle to check your accuracy. If you ride flawlessly, the paint spots will all fall on the line you are to follow. A perfect ride, however, is not likely. Wind, road imperfections, fatigue, and other random events will move the bicycle slightly away from its intended path. Repeat this experiment three times, and the paint spots from all three rides would look like the data points in Figure 6.8.

Generally, we have a set of data points obtained experimentally from a process of interest, such as those in Figure 6.8, and we want the equation of the underlying theoretical relationship. For example, we have the spots left by a bicycle, and we want the equation of the path the rider followed. In this section, we limit ourselves to relationships that appear linear.

A straight line in the variables x and y satisfying the equation

$$y = mx + c \tag{6.19}$$

where m and c are constants, will have one y value on the line for each value of x. This y value may not agree with the data at each value of x where data exists (see Figure 6.9). The difference between the y value of the data point at x and the y value defined by Equation (6.19) for this same value of x is known as the *residual at x*, which we denote $e(x)$.

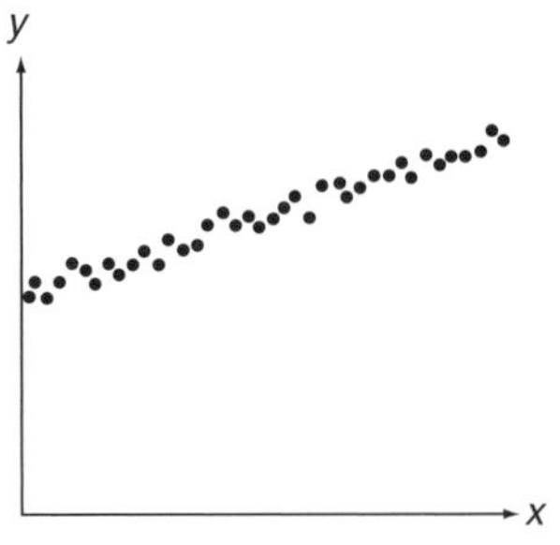

FIGURE 6.8

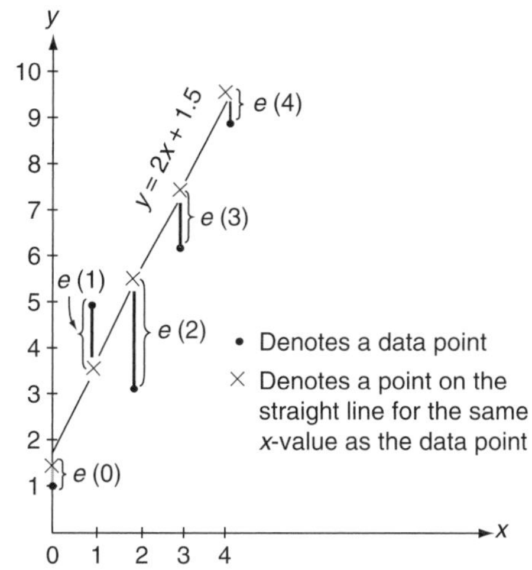

FIGURE 6.9

Example 1 Calculate the residuals between the five data points in Figure 6.9 and their corresponding points on the line defined by $y = 2x + 1.5$.

Solution: Data points are provided at $x = 0$, $x = 1$, $x = 2$, $x = 3$, and $x = 4$. Evaluating the equation $y = 2x + 1.5$ at these values of x, we generate Table 6.1. The residuals are

$$e(0) = 1 - 1.5 = -0.5$$
$$e(1) = 5 - 3.5 = 1.5$$
$$e(2) = 3 - 5.5 = -2.5$$
$$e(3) = 6 - 7.5 = -1.5$$
$$e(4) = 9 - 9.5 = -0.5$$

Note that these residuals can be read directly from Figure 6.9.

In general, we have N data points at (x_1, y_1), (x_2, y_2), (x_3, y_3), ... (x_N, y_N) with residuals $e(x_1)$, $e(x_2)$, $e(x_3)$, ... , $e(x_N)$ between the data points and a straight line

Table 6.1

Given Data		Evaluated from $y = 2x + 1.5$
x	y	y
0	1	1.5
1	5	3.5
2	3	5.5
3	6	7.5
4	9	9.5

approximation to the data. Residuals may be positive, negative, or 0, with a zero residual occurring only when a data point is on the straight line approximation. The *least-squares error* E is the sum of the squares of the individual residuals. That is,

$$E = [e(x_1)]^2 + [e(x_2)]^2 + [e(x_3)]^2 + \ldots + [e(x_N)]^2$$

The least-squares error is 0 if and only if all the residuals are 0.

Example 2 Calculate the least-squares error made in approximating the data in Figure 6.9 by the straight line defined by $y = 2x + 1.5$.

Solution: Using the residuals determined in Example 1, we have

$$\begin{aligned} E &= [e(0)]^2 + [e(1)]^2 + [e(2)]^2 + [e(3)]^2 + [e(4)]^2 \\ &= (-0.5)^2 + (1.5)^2 + (-2.5)^2 + (-1.5)^2 + (-0.5)^2 \\ &= 0.25 + 2.25 + 6.25 + 2.25 + 0.25 \\ &= 11.25 \end{aligned}$$

The least-squares error is the sum of the squares of the individual residuals, and the least-squares straight line is the line that minimizes the least-squares error.

Corresponding to every straight line approximation to a given set of data is a set of residuals and a least-squares error. Different straight lines can produce different least-squares errors, and we define the *least-squares straight line* to be the line that minimizes the least-squares error. A nonvertical straight line satisfies the equation

$$y = mx + c \qquad (6.19 \text{ repeated})$$

and has residuals

$$e(x_i) = y_i - (mx_i + c)$$

at $x_i (i = 1, 2, \ldots, N)$. We seek values of m and c that minimize

$$E = \sum_{i=1}^{N} (y_i - mx_i - c)^2$$

This occurs when

$$\frac{\partial E}{\partial m} = \sum_{i=1}^{N} 2(y_i - mx_i - c)(-x_i) = 0$$

$$\frac{\partial E}{\partial c} = \sum_{i=1}^{N} 2(y_i - mx_i - c)(-1) = 0$$

or, upon simplifying, when

$$\begin{aligned} \left(\sum_{i=1}^{N} x_i^2\right) m + \left(\sum_{i=1}^{N} x_i\right) c &= \sum_{i=1}^{N} x_i y_i \\ \left(\sum_{i=1}^{N} x_i\right) m + Nc &= \sum_{i=1}^{N} y_i \end{aligned} \qquad (6.20)$$

Table 6.2

x_i	y_i	$(x_i)^2$	x_iy_i
0	1	0	0
1	5	1	5
2	3	4	6
3	6	9	18
4	9	16	36
$\sum_{i=1}^{5} x_i = 10$	$\sum_{i=1}^{5} y_i = 24$	$\sum_{i=1}^{5} (x_i)^2 = 30$	$\sum_{i=1}^{5} x_iy_i = 65$

System (6.20) makes up the *normal equations* for a least-squares fit in two variables.

Example 3 Find the least-squares straight line for the following xy data:

x	0	1	2	3	4
y	1	5	3	6	9

Solution: Table 6.2 contains the required summations. For this data, the normal equations become

$$30m + 10c = 65$$
$$10m + 5c = 24$$

which has as its solution $m = 1.7$ and $c = 1.4$. The least-squares straight line is $y = 1.7x + 1.4$.

The normal equations have a simple matrix representation. Ideally, we would like to choose m and c for Equation (6.19) so that,

$$y_i = mx_i + c$$

for all data pairs (x_i, y_i), $i = 1, 2, \ldots, N$. That is, we want the constants m and c to solve the system

$$\begin{aligned} mx_1 + c &= y_1 \\ mx_2 + c &= y_2 \\ mx_3 + c &= y_3 \\ &\vdots \\ mx_N + c &= y_N \end{aligned}$$

or, equivalently, the matrix equation

$$\begin{bmatrix} x_1 & 1 \\ x_2 & 1 \\ x_3 & 1 \\ \vdots & \vdots \\ x_N & 1 \end{bmatrix} \begin{bmatrix} m \\ c \end{bmatrix} = \begin{bmatrix} y_1 \\ y_2 \\ y_3 \\ \vdots \\ y_N \end{bmatrix}$$

This system has the standard form $\mathbf{Ax}=\mathbf{b}$, where $\mathbf{A}$ is defined as a matrix having two columns, the first being the data vector $[x_1 \quad x_2 \quad x_3 \quad \ldots \quad x_N]^{\mathrm{T}}$, and the second containing all ones, $x=[m \quad c]^{\mathrm{T}}$, and $\mathbf{b}$ is the data vector $[y_1 \quad y_2 \quad y_3 \quad \ldots \quad y_N]^{\mathrm{T}}$. In this context, $\mathbf{Ax}=\mathbf{b}$ has a solution for $\mathbf{x}$ if and only if the data falls on a straight line. If not, then the matrix system is inconsistent, and we seek the least-squares solution. That is, we seek the vector $\mathbf{x}$ that minimizes the least-squares error having the matrix form

$$E = \|\mathbf{Ax} - \mathbf{b}\|^2 \tag{6.21}$$

The solution is the vector $\mathbf{x}$ satisfying the normal equations, which take the matrix form

$$\mathbf{A}^{\mathrm{T}}\mathbf{Ax} = \mathbf{A}^{\mathrm{T}}\mathbf{b} \tag{6.22}$$

System (6.22) is identical to system (6.20) when $\mathbf{A}$ and $\mathbf{b}$ are as just defined.

We now generalize to all linear systems of the form $\mathbf{Ax}=\mathbf{b}$. We are primarily interested in cases where the system is inconsistent (rendering the methods developed in Chapter 1 useless), and this generally occurs when $\mathbf{A}$ has more rows than columns. We place no restrictions on the number of columns in $\mathbf{A}$, but we will assume that *the columns are linearly independent*. We seek the vector $\mathbf{x}$ that minimizes the least-squares error defined by Equation (6.21).

▶THEOREM 1

If $\mathbf{x}$ has the property that $\mathbf{Ax}-\mathbf{b}$ is orthogonal to the columns of $\mathbf{A}$, then $\mathbf{x}$ minimizes $\|\mathbf{Ax}-\mathbf{b}\|^2$.◀

Proof: For any vector $\mathbf{x}_0$ of appropriate dimension,

$$\begin{aligned}
\|\mathbf{Ax}_0 - \mathbf{b}\|^2 &= \|(\mathbf{Ax}_0 - \mathbf{Ax}) + (\mathbf{Ax} - \mathbf{b})\|^2 \\
&= \langle(\mathbf{Ax}_0 - \mathbf{Ax}) + (\mathbf{Ax} - \mathbf{b}), (\mathbf{Ax}_0 - \mathbf{Ax}) + (\mathbf{Ax} - \mathbf{b})\rangle \\
&= \langle(\mathbf{Ax}_0 - \mathbf{Ax}), (\mathbf{Ax}_0 - \mathbf{Ax})\rangle + \langle(\mathbf{Ax} - \mathbf{b}), (\mathbf{Ax} - \mathbf{b})\rangle \\
&\quad + 2\langle(\mathbf{Ax}_0 - \mathbf{Ax}), (\mathbf{Ax} - \mathbf{b})\rangle \\
&= \|(\mathbf{Ax}_0 - \mathbf{Ax})\|^2 + \|(\mathbf{Ax} - \mathbf{b})\|^2 + 2\langle\mathbf{Ax}_0, (\mathbf{Ax} - \mathbf{b})\rangle - 2\langle\mathbf{Ax}, (\mathbf{Ax} - \mathbf{b})\rangle
\end{aligned}$$

It follows directly from Problem 38 of Section 6.2 that the last two inner products are both 0 (take $\mathbf{p}=\mathbf{Ax}-\mathbf{b}$). Therefore,

$$\begin{aligned}
\|\mathbf{Ax}_0 - \mathbf{b}\|^2 &= \|(\mathbf{Ax}_0 - \mathbf{Ax})\|^2 + \|(\mathbf{Ax} - \mathbf{b})\|^2 \\
&\geq \|(\mathbf{Ax} - \mathbf{b})\|^2
\end{aligned}$$

and $\mathbf{x}$ minimizes Equation (6.21).

As a consequence of Theorem 1, we seek a vector $\mathbf{x}$ having the property that $\mathbf{Ax}-\mathbf{b}$ is orthogonal to the columns of A. Denoting the columns of A as $\mathbf{A}_1, \mathbf{A}_2, \ldots, \mathbf{A}_n$, respectively, we require

$$\langle \mathbf{A}_i, \mathbf{Ax}-\mathbf{b}\rangle = 0 \quad (i = 1, 2, \ldots, n)$$

If $\mathbf{y}=[y_1 \quad y_2 \quad \ldots \quad y_n]^T$ denotes an arbitrary vector of appropriate dimension, then

$$\mathbf{Ay} = \mathbf{A}_1 y_1 + \mathbf{A}_2 y_2 + \cdots + \mathbf{A}_n y_n = \sum_{i=1}^{n} \mathbf{A}_i y_i$$

$$\begin{aligned} \langle \mathbf{Ay}, (\mathbf{Ax}-\mathbf{b})\rangle &= \left\langle \sum_{i=1}^{n} \mathbf{A}_i y_i (\mathbf{Ax}-\mathbf{b}) \right\rangle \\ &= \sum_{i=1}^{n} \langle \mathbf{A}_i y_i, (\mathbf{Ax}-\mathbf{b})\rangle \\ &= \sum_{i=1}^{n} y_i \langle \mathbf{A}_i, (\mathbf{Ax}-\mathbf{b})\rangle \\ &= 0 \end{aligned} \tag{6.23}$$

It follows from

$$\langle \mathbf{x}, \mathbf{y}\rangle = \mathbf{x}^T\mathbf{y} \qquad \text{(6.2 repeated)}$$

that

$$\begin{aligned} \langle \mathbf{Ay}, (\mathbf{Ax}-\mathbf{b})\rangle &= (\mathbf{Ay})^T(\mathbf{Ax}-\mathbf{b}) \\ &= (\mathbf{y}^T\mathbf{A}^T)(\mathbf{Ax}-\mathbf{b}) \\ &= \mathbf{y}^T(\mathbf{A}^T\mathbf{Ax}-\mathbf{A}^T\mathbf{b}) \\ &= \langle \mathbf{y}, (\mathbf{A}^T\mathbf{Ax}-\mathbf{A}^T\mathbf{b})\rangle \end{aligned} \tag{6.24}$$

Equations (6.23) and (6.24) imply that $\langle \mathbf{y}, (\mathbf{A}^T\mathbf{Ax}-\mathbf{A}^T\mathbf{b})\rangle = 0$ for any $\mathbf{y}$. Using the results of Problem 37 of Section 6.2, we conclude that $(\mathbf{A}^T\mathbf{Ax}-\mathbf{A}^T\mathbf{b})=\mathbf{0}$ or that $\mathbf{A}^T\mathbf{Ax}=\mathbf{A}^T\mathbf{b}$, *which has the same form as* Equation (6.22)*!* thus, we have Theorem 2.

▶THEOREM 2

A vector $\mathbf{x}$ *is the least-squares solution to* $\mathbf{Ax}=\mathbf{b}$ *if and only if* $\mathbf{x}$ *is a solution to the normal equations* $\mathbf{A}^T\mathbf{Ax}=\mathbf{A}^T\mathbf{b}$.◀

The set of normal equations has a unique solution whenever the columns of **A** are linearly independent, and these normal equations may be solved using any of the methods presented in the previous chapters for solving systems of simultaneous linear equations.

Example 4 Find the least-squares solution to

$$\begin{aligned} x + 2y + z &= 1 \\ 3x - y &= 2 \\ 2x + y - z &= 2 \\ x + 2y + 2z &= 1 \end{aligned}$$

Solution: This system takes the matrix form $\mathbf{Ax}=\mathbf{b}$, with

$$\mathbf{A} = \begin{bmatrix} 1 & 2 & 1 \\ 3 & -1 & 0 \\ 2 & 1 & -1 \\ 1 & 2 & 2 \end{bmatrix}, \mathbf{x} = \begin{bmatrix} x \\ y \\ z \end{bmatrix}, \text{ and } \mathbf{b} = \begin{bmatrix} 1 \\ 2 \\ 2 \\ 1 \end{bmatrix}$$

Then,

$$\mathbf{A}^{\mathrm{T}}\mathbf{A} = \begin{bmatrix} 15 & 3 & 1 \\ 3 & 10 & 5 \\ 1 & 5 & 6 \end{bmatrix} \text{ and } \mathbf{A}^{\mathrm{T}}\mathbf{b} = \begin{bmatrix} 12 \\ 4 \\ 1 \end{bmatrix}$$

and the normal equations become

$$\begin{bmatrix} 15 & 3 & 1 \\ 3 & 10 & 5 \\ 1 & 5 & 6 \end{bmatrix} \begin{bmatrix} x \\ y \\ z \end{bmatrix} = \begin{bmatrix} 12 \\ 4 \\ 1 \end{bmatrix}$$

Using Gaussian elimination, we obtain as the unique solution to this set of equations $x=0.7597$, $y=0.2607$, and $\mathbf{z}=-0.1772$, rounded to four decimals, which is also the least-squares solution to the original system.

Example 5 Find the least-squares solution to

$$\begin{aligned} 0x + 3y &= 180 \\ 2x + 5y &= 100 \\ 5x - 2y &= 60 \\ -x + 8y &= 130 \\ 10x - y &= 150 \end{aligned}$$

Solution: This system takes the matrix form $\mathbf{Ax}=\mathbf{b}$, with

$$\mathbf{A}=\begin{bmatrix} 1 & 3 \\ 2 & 5 \\ 5 & -2 \\ -1 & 8 \\ 10 & -1 \end{bmatrix}, \mathbf{x}=\begin{bmatrix} x \\ y \end{bmatrix}, \quad \text{and} \quad \mathbf{b}=\begin{bmatrix} 80 \\ 100 \\ 60 \\ 130 \\ 150 \end{bmatrix}$$

Then,

$$\mathbf{A}^{\mathrm{T}}\mathbf{A}=\begin{bmatrix} 131 & -15 \\ -15 & 103 \end{bmatrix} \quad \text{and} \quad \mathbf{A}^{\mathrm{T}}\mathbf{b}=\begin{bmatrix} 1950 \\ 1510 \end{bmatrix}$$

and the normal equations become

$$\begin{bmatrix} 131 & -15 \\ -15 & 103 \end{bmatrix}\begin{bmatrix} x \\ y \end{bmatrix}=\begin{bmatrix} 1950 \\ 1510 \end{bmatrix}$$

The unique solution to this set of equations is $x=16.8450$ and $y=17.1134$, rounded to four decimals, which is also the least-squares solution to the original system.

Problems 6.4

In Problems 1 through 8, find the least-squares solution to the given systems of equations.

(1) $2x+3y=8,$
$3x-y=5,$
$x+y=6.$

(2) $2x+y=8,$
$y=4,$
$-x+y=0,$
$3x+y=13.$

(3) $x+3y=65,$
$2x-y=0,$
$3x+y=50,$
$2x+2y=55.$

(4) $2x+y=6,$
$x+y=8,$
$-2x+y=11,$
$-x+y=8,$
$3x+y=4.$

(5) $2x+3y-4z=1,$
$x-2y+3z=3,$
$x+4y+2z=6,$
$2x+y-3z=1.$

(6) $2x+3y+2z=25,$
$2x-y+3z=30,$
$3x+4y-2z=20,$
$3x+5y+4z=55.$

(7) $x+y-z=90,$
$2x+y+z=200,$
$x+2y+2z=320,$
$3x-2y-4z=10,$
$3x+2y-3z=220.$

(8) $x+2y+2z=1$
$2x+3y+2z=2,$
$2x+4y+4z=-2,$
$3x+5y+4z=1,$
$x+3y+2z=-1.$

(9) The monthly sales figures (in thousands of dollars) for a newly opened shoe store are

Month	1	2	3	4	5
Sales	9	16	14	15	21

(a) Plot a scatter diagram for this data.

(b) Find the least-squares straight line that best fits this data.

(c) Use this line to predict sales revenue for month 6.

(10) Major League Baseball attendance for every ten years since 1960 is

Year	Attendance (in millions)
1960	19.9
1970	28.7
1980	43.0
1990	54.8
2000	72.7
2010	73.1

Source: www.ballparksofbaseball.com

(a) Find the least-squares straight line that best fits this data.

(b) Use this line to predict total major league baseball attendance in 2020.

(11) Annual rainfall data (in inches) for a given town over the last seven years are

Year	1	2	3	4	5	6	7
Rainfall	10.5	10.8	10.9	11.7	11.4	11.8	12.2

(a) Find the least-squares straight line that best fits this data.

(b) Use this line to predict next year's rainfall.

(12) Solve system (6.20) algebraically and explain why the solution would be susceptible to round-off error.

(13) (Coding) To minimize the round-off error associated with solving the normal equations for a least-squares straight line fit, the (x_i, y_i) data are coded before using them in calculations. Each x_i value is replaced by the difference between x_i and the average of all x_i data. That is, if

$$X = \frac{1}{N}\sum_{i=1}^{N} x_i$$

then set $x_i' = x_i - \overline{X}$ and fit a straight line to the (x_i', y_i) data instead.

Explain why this coding scheme avoids the round-off errors associated with un-coded data.

(14) (a) Code the data given in Problem 9 using the procedure described in Problem 13.

(b) Find the least-squares straight line fit for this coded data.

(15) Census figures for the population (in millions of people) for a particular region of the country are as follows:

Year	1950	1960	1970	1980	1990
Population	25.3	23.5	20.6	18.7	17.8

(a) Code this data using the procedure described in Problem 13, and then find the least-squares straight line that best fits it.

(b) Use this line to predict the population in 2000.

(16) Show that if $\mathbf{A}=\mathbf{QR}$ is a $\mathbf{QR}$ decomposition of $\mathbf{A}$, then the normal equations given by Equation (6.22) can be written as $\mathbf{R}^{\mathrm{T}}\mathbf{R}\mathbf{x}=\mathbf{R}^{\mathrm{T}}\mathbf{Q}^{\mathrm{T}}\mathbf{b}$, which reduces to $\mathbf{R}\mathbf{x}=\mathbf{Q}^{\mathrm{T}}\mathbf{b}$. This is a numerically stable set of equations to solve, not subject to the same round-off errors associated with solving the normal equations directly.

(17) Use the procedure described in Problem 16 to solve Problem 1.

(18) Use the procedure described in Problem 16 to solve Problem 2.

(19) Use the procedure described in Problem 16 to solve Problem 5.

(20) Use the procedure described in Problem 16 to solve Problem 6.

(21) Determine the column matrix of residuals associated with the least-squares solution of Problem 1, and then calculate the inner product of this vector with each of the columns of the coefficient matrix associated with the given set of equations.

(22) Determine the column matrix of residuals associated with the least-squares solution of Problem 5, and then calculate the inner product of this vector with each of the columns of the coefficient matrix associated with the given set of equations.

6.5 ORTHOGONAL COMPLEMENTS

Two vectors in the same inner product space are orthogonal if their inner product is zero. More generally, we say that the two subspaces $\mathbb{U}$ and $\mathbb{W}$ of an inner product space $\mathbb{V}$ are *orthogonal*, written $\mathbb{U}\perp\mathbb{W}$, if $\langle \mathbf{u},\mathbf{w}\rangle=0$ for every $\mathbf{u}\in\mathbb{U}$ and every $\mathbf{w}\in\mathbb{W}$.

Two subspaces $\mathbb{U}$ and $\mathbb{W}$ of the inner product space $\mathbb{V}$ are orthogonal if $\langle u, w\rangle=0$ for every $\mathbf{u}\in\mathbb{U}$ and every $\mathbf{w}\in\mathbb{W}$.

Example 1 The subspaces

$$\mathbb{U}=\left\{at^2+bt+c\in\mathbb{P}^2 \middle| b=0\right\} \text{ and } \mathbb{W}=\left\{at^2+bt+c\in\mathbb{P}^2 \middle| a=c=0\right\}$$

are orthogonal with respect to the induced Euclidean inner product. If $p(t)\in\mathbb{U}$, then $p(t)=at^2+c$, for some choice of the real numbers a and c. If $q(t)\in\mathbb{W}$, then $q(t)=bt$ for some choice of the real number b. Then

$$\langle p(t), q(t)\rangle = \langle at^2 + c, bt\rangle = \langle at^2 + 0t + c, 0t^2 + bt + 0\rangle$$
$$= a(0) + 0(b) + c(0) = 0$$

Example 2 The subspaces $\mathbb{U} = span\{[1 \quad 1 \quad 1]^T, [1 \quad -1 \quad 0]^T\}$ and $\mathbb{W} = span\{[1 \quad 1 \quad -2]^T\}$ in $\mathbb{R}^3$ are orthogonal with respect to the Euclidean inner product. Every vector in $\mathbf{u} \in \mathbb{U}$ must have the form

$$\mathbf{u} = a\begin{bmatrix} 1 \\ 1 \\ 1 \end{bmatrix} + b\begin{bmatrix} 1 \\ -1 \\ 0 \end{bmatrix} = \begin{bmatrix} a+b \\ a-b \\ a \end{bmatrix}$$

for some choice of scalars a and b, while every vector in $\mathbf{w} \in \mathbb{W}$ must have the form

$$\mathbf{w} = c\begin{bmatrix} 1 \\ 1 \\ -2 \end{bmatrix} = \begin{bmatrix} c \\ c \\ -2c \end{bmatrix}$$

for some choice of scalar c. Here,

$$\langle \mathbf{u}, \mathbf{w}\rangle = (a+b)(c) + (a-b)(c) + a(-2c) = 0$$

Orthogonal subspaces in $\mathbb{R}^3$ do not always agree with our understanding of perpendicularity. The *xy*-plane is perpendicular to the *yz*-plane, as illustrated in Figure 6.10, but the two planes are not orthogonal. The *xy*-plane is the subspace defined by

$$\mathbb{U} = \left\{[x \quad y \quad z]^T \in \mathbb{R}^3 | z = 0\right\}$$

Therefore, $\mathbf{u} = [1 \quad 1 \quad 0]^T$ is a vector in $\mathbb{U}$. The *yz*-plane is the subspace defined by

$$\mathbb{W} = \left\{[x \quad y \quad z]^T \in \mathbb{R}^3 | x = 0\right\}$$

and $\mathbf{w} = [0 \quad 1 \quad 1]^T$ is in $\mathbb{W}$. Here,

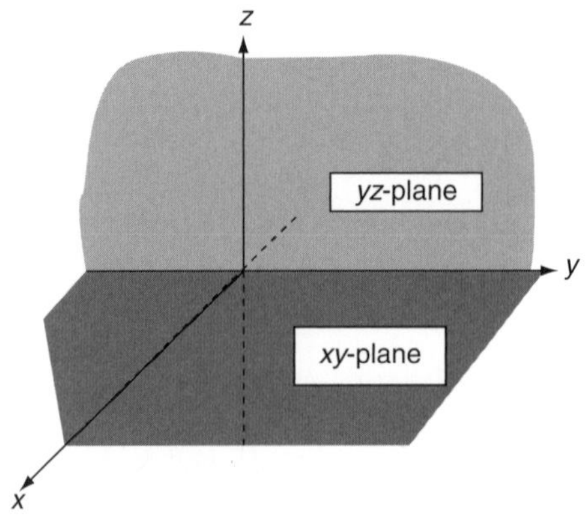

FIGURE 6.10

$$\langle \mathbf{u}, \mathbf{v} \rangle = 1(0) + 1(1) + 0(1) = 1 \neq 0$$

If $\mathbb{U}$ is a subspace of an inner product space $\mathbb{V}$, we define the *orthogonal complement of* $\mathbb{U}$, denoted as $\mathbb{U}^{\perp}$, as the set of all vectors in $\mathbb{V}$ that are orthogonal to every vector in $\mathbb{U}$, that is,

$$\mathbb{U}^{\perp} = \{\mathbf{v} \in \mathbb{V} | \langle \mathbf{u}, \mathbf{v} \rangle = 0 \text{ for every } \mathbf{u} \in \mathbb{U}\} \tag{6.25}$$

If $\mathbb{U}$ is a subspace of an inner product space $\mathbb{V}$, then $\mathbb{U}^{\perp}$, the orthogonal complement of $\mathbb{U}$, is the set of all vectors in $\mathbb{V}$ that are orthogonal to every vector in $\mathbb{U}$.

Example 3 In $\mathbb{R}^3$, the orthogonal complement of the *z*-axis is the *xy*-plane. The z-axis is the subspace

$$\mathbb{Y} = \left\{ [x\, y\, z]^{\mathrm{T}} \in \mathbb{R}^3 | x = y = 0 \right\}$$

so any vector in this subspace has the form $[0 \;\; 0 \;\; a]^{\mathrm{T}}$ for some choice of the scalar a. A general vector in $\mathbb{R}^3$ has the form $[x \;\; y \;\; z]^{\mathrm{T}}$ for any choice of the scalars x, y, and z. If

$$\left\langle \begin{bmatrix} x \\ y \\ z \end{bmatrix}, \begin{bmatrix} 0 \\ 0 \\ 0 \end{bmatrix} \right\rangle = za$$

is to be zero for every choice of the scalar a, then $z = 0$. Thus, the orthogonal complement of the *z*-axis is the set

$$\left\{ [x \quad y \quad z]^{\mathrm{T}} \in \mathbb{R}^3 | z = 0 \right\}$$

which defines the *xy*-plane.

▶THEOREM 1

If $\mathbb{U}$ is a subspace of an inner product space $\mathbb{V}$, then so too is the orthogonal complement of $\mathbb{U}$.◀

Proof: Let $\mathbf{x}$ and $\mathbf{y}$ be elements of $\mathbb{U}^{\perp}$, and let $\mathbf{u} \in \mathbb{U}$. Then $(\mathbf{x}, \mathbf{u}) = 0$, $(\mathbf{y}, \mathbf{u}) = 0$, and for any two scalars α and β

$$\langle \alpha\mathbf{x} + \beta\mathbf{y}, \mathbf{u} \rangle = \langle \alpha\mathbf{x}, \mathbf{u} \rangle + \langle \beta\mathbf{y}, \mathbf{u} \rangle = \alpha\langle \mathbf{x}, \mathbf{u} \rangle + \beta\langle \mathbf{y}, \mathbf{u} \rangle = \alpha(0) + \beta(0) = 0.$$

Thus, $\alpha\mathbf{x} + \beta\mathbf{y} \in \mathbb{U}^{\perp}$ and $\mathbb{U}^{\perp}$ is a subspace of $\mathbb{V}$.

▶THEOREM 2

If $\mathbb{U}$ is a subspace of an inner product space $\mathbb{V}$, then the only vector common to both $\mathbb{U}$ and $\mathbb{U}^{\perp}$ is the zero vector.◀

Proof: Let $\mathbf{x}$ be a vector in both $\mathbb{U}$ and $\mathbb{U}^{\perp}$. Since $\mathbf{x} \in \mathbb{U}^{\perp}$, it must be orthogonal to every vector in $\mathbb{U}$, hence $\mathbf{x}$ must be orthogonal to itself, because $\mathbf{x} \in \mathbb{U}$. Thus, $(\mathbf{x}, \mathbf{x}) = 0$, and it follows immediately from Theorem 1 of Section 6.1 that $\mathbf{x} = 0$.

Identifying the orthogonal complement of subspaces $\mathbb{U}$ of $\mathbb{R}^n$ is straightforward when we know a spanning set $\mathbb{S} = \{\mathbf{u}_1, \mathbf{u}_2, \ldots, \mathbf{u}_k\}$ for $\mathbb{U}$. We define a matrix **A** to be

$$\mathbf{A} = \begin{bmatrix} \boldsymbol{u}_1^{\mathrm{T}} \\ \boldsymbol{u}_2^{\mathrm{T}} \\ \vdots \\ \boldsymbol{u}_k^{\mathrm{T}} \end{bmatrix} \tag{6.26}$$

where the column matrices in $\mathbb{S}$ become the *rows* of **A**. We then transform **A** to row-reduced form using elementary row operations, obtaining

$$\mathbf{A} \rightarrow \begin{bmatrix} \boldsymbol{v}_1^{\mathrm{T}} \\ \boldsymbol{v}_2^{\mathrm{T}} \\ \vdots \\ \boldsymbol{v}_k^{\mathrm{T}} \end{bmatrix}$$

The nonzero rows of this row-reduced matrix are a basis for $\mathbb{U}$. Any vector $\mathbf{x} \in \mathbb{U}^{\perp}$ must be orthogonal to each basis vector in $\mathbb{U}$, so

$$\langle \mathbf{v}_j, \mathbf{x} \rangle = 0 \quad (j = 1, 2, \ldots, k). \tag{6.27}$$

Equation (6.27) yields a set of k; equations (some of which will be $0 = 0$ when the rank of **A** is less than k) for the components of **x**. These equations define all vectors in the orthogonal complement of $\mathbb{U}$. But Equation (6.27) also defines the kernel of the matrix **A** in Equation (6.26), so we have proven Theorem 3.

▶THEOREM 3

If $\mathbb{S}$ is a spanning set for a subspace $\mathbb{U}$ of $\mathbb{R}^n$ (considered as column matrices) and if a matrix $\mathbb{A}$ is created so that each row of $\mathbb{A}$ is the transpose of the vectors in $\mathbb{S}$, then $\mathbb{U}^{\perp} = ker(\mathbf{A})$. ◀

Example 4 Find the orthogonal complement of the sub space in $\mathbb{R}^4$ spanned by

$$\mathbb{S} = \left\{ \begin{bmatrix} 1 \\ 3 \\ 1 \\ -1 \end{bmatrix}, \begin{bmatrix} 2 \\ 7 \\ 2 \\ 1 \end{bmatrix}, \begin{bmatrix} 1 \\ 4 \\ 1 \\ 2 \end{bmatrix} \right\}$$

Solution: For these vectors, matrix (6.26) becomes

$$\mathbf{A} = \begin{bmatrix} 1 & 3 & 1 & -1 \\ 2 & 7 & 2 & 1 \\ 1 & 4 & 1 & 2 \end{bmatrix}$$

which is transformed by elementary row operation to the row-reduced form

$$\rightarrow \begin{bmatrix} 1 & 3 & 1 & -1 \\ 0 & 1 & 0 & 3 \\ 0 & 1 & 0 & 3 \end{bmatrix}$$
$$\rightarrow \begin{bmatrix} 1 & 3 & 1 & -1 \\ 0 & 1 & 0 & 3 \\ 0 & 0 & 0 & 0 \end{bmatrix}$$

A basis for $\mathbb{U}$ is $\{[1 \quad 3 \quad 1 \quad -1]^{\mathrm{T}}, [0 \quad 1 \quad 0 \quad 3]^{\mathrm{T}}\}$, hence $\mathbb{U}$ is a two-dimensional subspace of $\mathbb{R}^4$. If we let $\mathbf{x} = [x_1, x_2, x_3, x_1]^{\mathrm{T}}$, denote an arbitrary element in the kernel of **A**, then

$$\begin{aligned} x_1 + 3x_2 + x_3 - x_4 &= 0 \\ x_2 + 3x_4 &= 0 \\ 0 &= 0 \end{aligned}$$

whence, $x_1 = -x_3 + 10x_4$, $x_2 = -3x_4$ with x_3 and x_4 arbitrary. Thus the kernel of **A** is

$$\left\{ \begin{bmatrix} -x_3 + 10x_4 \\ -3x_4 \\ x_3 \\ x_4 \end{bmatrix} = x_3 \begin{bmatrix} -1 \\ 0 \\ 1 \\ 0 \end{bmatrix} + x_4 \begin{bmatrix} 10 \\ -3 \\ 0 \\ 1 \end{bmatrix} \; x_3 \text{ and } x_4 \text{ are arbitrary} \right\}$$

A basis for and $\mathbb{U}^{\perp}$ is $\{[-1 \quad 0 \quad 1 \quad 0]^{\mathrm{T}}, [10 \quad -3 \quad 0 \quad 1]^{\mathrm{T}}\}$, and $\mathbb{U}^{\mathrm{T}}$ also a two-dimensional subspace of $\mathbb{R}^4$.

►THEOREM 4

*If **U** is a subspace of $\mathbb{R}^n$, then dim(**U**) + dim($\mathbf{U}^{\perp}$) = n.*◄

Proof: The proposition is true when $\mathbb{U} = \{0\}$, because $\langle 0, \mathbf{y} \rangle = 0$ for every $\mathbf{y} \in \mathbb{R}^n$ and $(\mathbb{U}^{\perp} = \mathbb{R}^n$. In all other cases, let $\mathbb{S} = \{\mathbf{u}_1, \mathbf{u}_2, \ldots, \mathbf{u}_k\}$ be a basis for $\mathbb{U}$, and construct **A** as in Equation (6.26). Then **A** is a linear transformation from $\mathbb{R}^n$ to $\mathbb{R}^k$. Because $\mathbb{S}$ is a basis, $r(\mathbf{A}) = dim(\mathbb{U}) = k$, where $r(\mathbf{A})$ denotes the rank of **A.** The nullity of **A**, $v(\mathbf{A})$, is the dimension of the kernel of **A**, hence $v(\mathbf{A})$ is the dimension of $\mathbb{U}^{\perp}$. But $r(\mathbf{A}) + v(\mathbf{A}) = n$ (Corollary 1 of Section 3.5), so Theorem 4 is immediate.

►THEOREM 5

If $\mathbb{U}$ is a subspace of $\mathbb{R}^n$, then $(\mathbb{U}^{\perp})^{\perp} = \mathbf{U}$.◄

Proof: If $\mathbf{u} \in \mathbb{U}$, then **u** is orthogonal to every vector in $\mathbb{U}^{\perp}$, so $\mathbf{u} \in (\mathbb{U}^{\perp})^{\perp}$ and $\mathbb{U}$ is a subset of $(\mathbb{U}^{\perp})^{\perp}$. Denote the dimension of $\mathbb{U}$ as k. It follows from Theorem 4 that

$dim(\mathbb{U}^{\perp}) = n - k$. But it also follows from Theorem 4 that $dim(\mathbb{U}^{\perp}) + dim((\mathbb{U}^{\perp})^{\perp}) = n$, whereupon $dim((\mathbb{U}^{\perp})^{\perp}) = n - (n-k) = k = dim(\mathbb{U})$. Thus, $\mathbb{U} \subseteq (\mathbb{U}^{\perp})^{\perp}$ with each subspace having the same dimension, hence $\mathbb{U} = (\mathbb{U}^{\perp})^{\perp}$.

We began Section 6.2 by writing a vector $\mathbf{x} \in \mathbb{R}^2$ as the sum of two vectors $\mathbf{u} + \mathbf{v}$, which were orthogonal to one another. We now do even more.

▶THEOREM 6

If $\mathbb{U}$ is a subspace of $\mathbb{R}^n$, then each vector $\mathbf{x} \in \mathbb{R}^n$ can be written uniquely as $\mathbf{x} = \mathbf{u} + \mathbf{u}^{\perp}$, where $\mathbf{u} \in \mathbb{U}$ and $\mathbf{u}^{\perp} \in \mathbb{U}^{\perp}$. ◀

Proof: If $\mathbb{U} = \mathbb{R}^n$, then $\mathbb{U}^{\perp} = \{0\}$, and, conversely, if $\mathbb{U} = \{0\}$, then $\mathbb{U}^{\perp} = \mathbb{R}^n$, because $\mathbf{x} = \mathbf{x} + \mathbf{0} = \mathbf{0} + \mathbf{x}$. In all other cases, let $\{\mathbf{u}_1, \mathbf{u}_2, \ldots, \mathbf{u}_k\}$ be a basis for $\mathbb{U}(k < n)$, and let $\{\mathbf{u}_{k+1}, \mathbf{u}_{k+2}, \ldots, \mathbf{u}_n\}$ be a basis for $\mathbb{U}^{\perp}$. We first claim that the set $\mathbf{S} = \{\mathbf{u}_1, \mathbf{u}_2, \ldots, \mathbf{u}_k, \mathbf{u}_{k+1}, \mathbf{u}_{k+2}, \ldots, \mathbf{u}_n\}$ is linearly independent, which is equivalent to showing that the only solution to

$$c_1\mathbf{u}_1 + c_2\mathbf{u}_2 + \cdots + c_k\mathbf{u}_k + c_{k+1}\mathbf{u}_{k+1} + c_{k+2}\mathbf{u}_{k+2} + \cdots + c_n\mathbf{u}_n = \mathbf{0} \qquad (6.28)$$

is $c_1 = c_2 = \ldots = c_n = 0$. If we rewrite Equation (6.28) as

$$c_1\mathbf{u}_1 + c_2\mathbf{u}_2 + \cdots + c_k\mathbf{u}_k = -c_{k+1}\mathbf{u}_{k+1} - c_{k+2}\mathbf{u}_{k+2} - \cdots - c_n\mathbf{u}_n$$

we see that the left side of this equation is a vector in $\mathbb{U}$ while the right side is a vector in $\mathbb{U}^{\perp}$. Since the vectors are equal, they represent a vector in both $\mathbb{U}$ and $\mathbb{U}^{\perp}$ that, from Theorem 2, must be the zero vector. Thus,

$$c_1\mathbf{u}_1 + c_2\mathbf{u}_2 + \cdots + c_k\mathbf{u}_k = 0$$

and since $\{\mathbf{u}_1, \mathbf{u}_2, \ldots, \mathbf{u}_k\}$ is a linearly independent set (it is a basis for $\mathbb{U}$), we conclude that $c_1 = c_2 = \ldots = c_k = 0$. Similarly

$$c_{k+1}\mathbf{u}_{k+1} + c_{k+2}\mathbf{u}_{k+2} + \cdots + c_n\mathbf{u}_n = 0$$

and since $\{\mathbf{u}_k +_1, \mathbf{u}_k +_2, \ldots, \mathbf{u}_n\}$ is a linearly independent set (it is as basis for $\mathbb{U}^{\perp}$), we conclude that $c_k +_1 = c_k +_2 = \ldots = c_n = 0$. Thus, **S** is linearly independent as claimed.

Since the dimension of $\mathbb{R}^n$ is n and $\mathbb{S}$ is a linearly independent set of n-vectors in $\mathbb{R}^n$, it follows that $\mathbb{S}$ is a basis for $\mathbb{R}^n$. We now have (see Theorem 5 of Section 2.5) that each vector in $\mathbb{R}^n$ can be written *uniquely* as a linear combination of the vectors in $\mathbb{S}$. That is, if $\mathbf{x} \in \mathbb{R}^n$, then there exists a unique set of scalars $d_1 = d_2 = \ldots = d_n$ such that

$$\mathbf{x} = d_1\mathbf{u}_1 + d_2\mathbf{u}_2 + \cdots + d_k\mathbf{u}_k + d_{k+1}\mathbf{u}_{k+1} + d_{k+2}\mathbf{u}_{k+2} + \cdots + d_n\mathbf{u}_n$$

Setting $\mathbf{u} = d_1\mathbf{u}_1 + d_2\mathbf{u}_2 + \ldots + d_k\mathbf{u}_k$ and $\mathbf{u}^{\perp} = d_{k+1}\mathbf{u}_{k+1} + d_{k+2}\mathbf{u}_{k+2} + \ldots + d_n\mathbf{u}_n$, we have $\mathbf{u} \in \mathbb{U}$, $\mathbf{u}^{\perp} \in \mathbb{U}^{\perp}$, and $\mathbf{x} = \mathbf{u} + \mathbf{u}^{\perp}$.

Example 5 Decompose $\mathbf{x}=[-14 \quad -10 \quad 12]^{\mathrm{T}}$ into the sum of two vectors, one in the subspace $\mathbb{U}$ spanned by $\{[1 \quad 1 \quad 5]^{\mathrm{T}}, [2 \quad -1 \quad 1]^{\mathrm{T}}\}$ and the other in $\mathbb{U}^{\perp}$.

Solution: The vectors $\mathbf{u}_1=[1 \quad 1 \quad 5]^{\mathrm{T}}$ and $u_2=[2 \quad -1 \quad 1]^{\mathrm{T}}$ are linearly independent, so they form a basis for $\mathbb{U}$. We set

$$\mathbf{A}=\begin{bmatrix} 1 & 1 & 5 \\ 2 & -1 & 1 \end{bmatrix}$$

and then determine that $\mathbf{u}_3=[-2 \quad -2 \quad 1]^{\mathrm{T}}$ is a basis for $ker(\mathbf{A})$ (see Example 1 of Section 3.5) and, therefore, a basis for $\mathbb{U}^{\perp}$. Thus, $\mathbb{B}=\{\mathbf{u}_1, \mathbf{u}_2, \mathbf{u}_3\}$ is a basis for $\mathbb{R}^3$.

We want the coordinates of the given vector $\mathbf{x}=[-14 \quad -10 \quad 12]^{\mathrm{T}}$ with respect to the $\mathbb{B}$ basis; that is, we want the values of the scalars d_1, d_2, and d_3 so that

$$d_1\begin{bmatrix} 1 \\ 1 \\ 5 \end{bmatrix} + d_2\begin{bmatrix} 2 \\ -1 \\ 5 \end{bmatrix} + d_3\begin{bmatrix} -2 \\ -3 \\ 1 \end{bmatrix} = \begin{bmatrix} -14 \\ -10 \\ 12 \end{bmatrix}.$$

Solving the associated system of simultaneous linear equations by Gaussian elimination, we find $d_1=2$, $d_2=-3$, and $d_3=5$. Finally setting

$$\mathbf{u} = 2\mathbf{u}_1 + (-3)\mathbf{u}_2 = 2\begin{bmatrix} 1 \\ 1 \\ 5 \end{bmatrix} + (-3)\begin{bmatrix} 2 \\ -1 \\ 5 \end{bmatrix} = \begin{bmatrix} -4 \\ 5 \\ 7 \end{bmatrix}$$

$$\mathbf{u}^{\perp} = 5\mathbf{u}_3 = \begin{bmatrix} -10 \\ -15 \\ 5 \end{bmatrix}$$

we have $\mathbf{u}\in\mathbb{U}$, $\mathbf{u}^{\perp} \perp \mathbb{U}^{\perp}$, and $\mathbf{x}=\mathbf{u}+\mathbf{u}^{\perp}$.

Whenever we have a decomposition of a given vector $\mathbf{x}$ into the sum of two vectors as described in Theorem 6, $\mathbf{x}=\mathbf{u}+\mathbf{u}^{\perp}$, then $\mathbf{u}$ is called the *projection of* $\mathbf{x}$ *on* $\mathbb{U}$. In the special case where $\mathbb{U}$ is a one-dimensional subspace spanned by a single vector $\mathbf{a}$, the projection of $\mathbf{x}$ on $\mathbb{U}$ is obtained most easily by Equation (6.6) in Section 6.2.

Example 6 Using the results of Example 5, we have that $\mathbf{u}=[-4 \quad 5 \quad 7]^{\mathrm{T}}$ is the projection of the vector $\mathbf{x}=[-14 \quad -10 \quad 12]^{\mathrm{T}}$ on the subspace $\mathbb{U}$ spanned by $\{[1 \quad 1 \quad 5]^{\mathrm{T}}, [2 \quad -1 \quad 1]^{\mathrm{T}}\}$.

A vector space $\mathbb{V}$ is the *direct sum* of two subspaces $\mathbb{U}$ and $\mathbb{W}$, written $\mathbb{V}=\mathbb{U}\oplus\mathbb{W}$, if each vector in $\mathbb{V}$ can be written *uniquely* as the sum $\mathbf{u}+\mathbf{v}$, where $\mathbf{u}\in\mathbb{U}$ and $\mathbf{v}\in\mathbb{V}$. It follows from Theorem 6 that $\mathbb{R}^n=\mathbb{U}\oplus\mathbb{U}^{\perp}$ for each subspace $\mathbb{U}$ of $\mathbb{R}^n$.

A vector space $\mathbb{V}$ is the direct sum of two subspaces $\mathbb{U}$ and $\mathbb{W}$ if each vector in $\mathbb{V}$ can be written uniquely as the sum of a vector in $\mathbb{U}$ and a vector in $\mathbb{V}$.

Problems 6.5

In Problems 1 through 10, (a) find the orthogonal complement for the subspace $\mathbb{U}$ of $\mathbb{R}^3$ spanned by the given set of vectors, and then (b) find the projection of $\mathbf{x}=[1 \quad 1 \quad 0]^T$ on $\mathbb{U}$.

(1) $\{[0 \quad 1 \quad 1]^T\}$.

(2) $\{[1 \quad 1 \quad 1]^T\}$.

(3) $\{[2 \quad 1 \quad 1]^T\}$.

(4) $\{[1 \quad 1 \quad 1]^T, [0 \quad 1 \quad 2]^T\}$.

(5) $\{[2 \quad 1 \quad 1]^T, [0 \quad 1 \quad 2]^T\}$.

(6) $\{[0 \quad 1 \quad 1]^T, [0 \quad 1 \quad 2]^T\}$.

(7) $\{[1 \quad 1 \quad 1]^T, [2 \quad 2 \quad 0]^T\}$.

(8) $\{[1 \quad 1 \quad 1]^T, [2 \quad 2 \quad 2]^T\}$.

(9) $\{[1 \quad 1 \quad 1]^T, [0 \quad 1 \quad 1]^T, [3 \quad 2 \quad 2]^T\}$.

(10) $\{[1 \quad 1 \quad 1]^T, [1 \quad 0 \quad 1]^T, [1 \quad 1 \quad 0]^T\}$.

In Problems 11 through 20, (a) find the orthogonal complement for the subspace $\mathbb{U}$ of $\mathbb{R}^4$ spanned by the given set of vectors, and then (b) find the projection of $\mathbf{x}=[1 \quad 0 \quad 1 \quad 0]^T$ on $\mathbb{U}$.

(11) $\{[0 \quad 0 \quad 1 \quad 1]^T\}$.

(12) $\{[0 \quad 1 \quad 1 \quad 1]^T\}$.

(13) $\{[0 \quad 0 \quad 1 \quad 1]^T, [0 \quad 1 \quad 1 \quad 1]^T\}$.

(14) $\{[0 \quad 1 \quad 0 \quad 1]^T, [0 \quad 1 \quad 0 \quad 2]^T\}$.

(15) $\{[1 \quad 1 \quad 1 \quad 0]^T, [1 \quad 1 \quad 0 \quad 1]^T\}$.

(16) $\{[1 \quad 1 \quad 1 \quad 0]^T, [1 \quad 1 \quad 0 \quad 1]^T, [1 \quad 0 \quad 1 \quad 1]^T\}$.

(17) $\{[1 \quad 1 \quad 1 \quad 0]^T, [1 \quad 1 \quad 1 \quad 1]^T, [1 \quad 1 \quad 1 \quad 2]^T\}$.

(18) $\{[1 \quad 1 \quad 0 \quad 0]^T, [0 \quad 1 \quad 0 \quad 1]^T, [1 \quad 0 \quad 1 \quad 0]^T\}$.

(19) $\{[1 \quad 1 \quad 1 \quad 0]^T, [1 \quad 1 \quad 0 \quad 1]^T, [1 \quad 0 \quad 1 \quad 1]^T, [0 \quad 1 \quad 1 \quad 1]^T\}$.

(20) $\{[1 \quad 1 \quad 1 \quad 0]^T, [1 \quad 1 \quad 0 \quad 1]^T, [1 \quad 2 \quad 1 \quad 1]^T, [3 \quad 4 \quad 2 \quad 2]^T\}$.

(21) Is it possible for $\mathbf{x}=[1 \quad 1 \quad 0]^T$ to be in the kernel of a 3×3 matrix $\mathbf{A}$ and also for $\mathbf{y}=[1 \quad 0 \quad 1]^T$ to be in the row space of $\mathbf{A}$?

(22) Show that if $\mathbf{x}=\mathbf{u}+\mathbf{u}^{\perp}$ as in Theorem 6, then $\|x\| = \sqrt{\|u^2\| + \|u^{\perp}\|^2}$.

(23) Let $\mathbb{U}$ be a subspace of a finite-dimensional vector space $\mathbb{V}$ with a basis $\mathbb{B}$, and let $\mathbb{W}$ be subspace of $\mathbb{V}$ with basis $\mathbb{C}$. Show that if $\mathbb{V}=\mathbb{U}\oplus\mathbb{W}$, then $\mathbb{B}\cup\mathbb{C}$ is a basis for $\mathbb{V}$.

(24) Prove that if $\mathbb{U}$ and $\mathbb{W}$ are subspaces of a finite-dimensional vector space $\mathbb{V}$ with $\mathbb{V}=\mathbb{U}\oplus\mathbb{W}$, then the only vector common to both $\mathbb{U}$ and $\mathbb{W}$ is 0.

(25) Prove that if $\mathbb{U}$ and $\mathbb{W}$ are subspaces of a finite-dimensional vector space $\mathbb{V}$ with $\mathbb{V}=\mathbb{U}\oplus\mathbb{W}$, then $dim(\mathbb{U})+dim(\mathbb{W})=dim(\mathbb{V})$.

CHAPTER 6 REVIEW

Important Terms

angle between n-tuples
Cauchy-Schwarz Inequality
direct sum
Euclidean inner product
Gram-Schmidt orthonormalization process
induced inner product
inner product space
Kronecker delta
least-squares error
least-squares straight line
magnitude of an n-tuple
noise
normal equations
normalized vector
orthogonal complement
orthogonal vectors
orthonormal set of vectors
orthogonal subspaces
projection
QR algorithm
QR decomposition
residual
scatter diagram
unit vector

Important Concepts

Section 6.1

- The Euclidean inner product of two vectors **x** and **y** in $\mathbb{R}^n$ is a real number obtained by multiplying corresponding components of **x** and **y** and then summing the resulting products.
- The inner product of a vector with itself is positive, unless the vector is the zero vector, in which case the inner product is zero.
- The inner product of a vector with the zero vector yields the zero scalar.
- $\langle \mathbf{x}, \mathbf{y}\rangle = \langle \mathbf{y}, \mathbf{x}\rangle = \langle \mathbf{y}, \mathbf{x}\rangle$ for vectors **x** and **y** in R^n.
- $\langle \lambda\mathbf{x}, \mathbf{y}\rangle = \lambda\langle \mathbf{x}, \mathbf{y}\rangle$, for any real number λ.
- $\langle \mathbf{x}+\mathbf{z}, \mathbf{y}\rangle = \langle \mathbf{x}, \mathbf{y}\rangle + \langle \mathbf{z}, \mathbf{y}\rangle$.
- The magnitude of a vector $\mathbf{x}\in\mathbb{R}^n$ is the square root of the inner product of **x** with itself.
- If **u** and **v** are vectors in $\mathbb{R}^n$, then $|(\text{u,v})| \le \|\mathbf{u}\|\ \|\mathbf{v}\|$.
- An induced inner product on two matrices of the same order is obtained by multiplying corresponding elements of both matrices and summing the results.

- An induced inner product of two polynomials is obtained by multiplying the coefficients of like powers of the variable and summing the results.
- Two vectors can be orthogonal with respect to one basis and not orthogonal with respect to another basis.

Section 6.2

- Subtracting from a nonzero vector $\mathbf{x}$ its projection onto another nonzero vector a yields a vector that is orthogonal to both a and the projection of $\mathbf{x}$ onto a.
- An orthonormal set of vectors is an orthogonal set of unit vectors.
- An orthonormal set of a finite number of vectors is linearly independent.
- If $\{\mathbf{x}_1, \mathbf{x}_2, \ldots, \mathbf{x}_n\}$ is orthonormal basis for a vector space $\mathbb{V}$, then for any vector $\mathbf{x} \in \mathbb{V}$, $\mathbf{x} = \langle \mathbf{x}, \mathbf{x}_1 \rangle \mathbf{x}_1 + \langle \mathbf{x}, \mathbf{x}_2 \rangle \mathbf{x}_2 + \cdots + \langle \mathbf{x}, \mathbf{x}_n \rangle \mathbf{x}_n$.
- Every set of linearly independent vectors in an inner product space can be transformed into an orthonormal set of vectors that spans the same subspace.

Section 6.3

- If the columns of a matrix $\mathbf{A}$ are linearly independent, then $\mathbf{A}$ can be factored into the product of a matrix $\mathbf{Q}$, having columns that form an orthonormal set, and another matrix $\mathbf{R}$, that is upper triangular.
- The **QR** algorithm is a numerical method of locating all eigenvalues of a real matrix.

Section 6.4

- The least-squares straight line is the line that minimizes the least-squares error for a given set of data.
- A vector $\mathbf{x}$ is the least-squares solution to $\mathbf{Ax} = \mathbf{b}$ if and only if $\mathbf{x}$ is a solution to the normal equation $\mathbf{A}^T\mathbf{Ax} = \mathbf{A}^T\mathbf{b}$.

Section 6.5

- If $\mathbb{U}$ is a subspace of an inner product space $\mathbb{V}$, then so too is the orthogonal complement of $\mathbb{U}$.
- If $\mathbb{U}$ is a subspace of an inner product space $\mathbb{V}$, then the only vector common to both $\mathbb{U}$ and $\mathbb{U}^\perp$ is the zero vector.
- If $\mathbb{S}$ is a spanning set for a subspace $\mathbb{U}$ of $\mathbb{R}^n$ (considered as column matrices) and if a matrix $\mathbb{A}$ is created so that each row of $\mathbf{A}$ is the transpose of the vectors in $\mathbb{S}$, then $\mathbb{U}^\perp = ker(\mathbf{A})$.
- If $\mathbb{U}$ is a subspace of $\mathbb{R}^n$, then $dim(\mathbb{U}) + dim(\mathbb{U}^\perp) = n$.
- If $\mathbb{U}$ is a subspace of $\mathbb{R}^n$, then each vector $\mathbf{x} \in \mathbb{R}^n$ can be written uniquely as $\mathbf{x} = \mathbf{u} + \mathbf{u}^\perp$, where $\mathbf{u} \in \mathbb{U}$ and $\mathbf{u}^\perp \in \mathbb{U}^\perp$.

APPENDIX A

Jordan Canonical Forms

In Chapter 4, we began identifying bases that generate simple matrix representations for linear transformations of the form $T: \mathbb{V} \to \mathbb{V}$, when $\mathbb{V}$ is an n-dimensional vector space. Every basis for $\mathbb{V}$ contains n-vectors, and every matrix representation of T has order $n \times n$. We concluded (see Section 4.3) that T may be represented by a *diagonal* matrix if and only if T possesses n linearly independent eigenvectors.

Eigenvectors for a linear transformation T are found by first producing a matrix representation for T, generally the matrix with respect to a standard basis, and then calculating the eigenvectors of that matrix. Let **A** denotes a matrix representation of T. Eigenvectors of **A** are coordinate representations for the eigenvectors of T. If **A** has n linearly independent eigenvectors, then so does T, and T can be represented by a diagonal matrix that is similar to **A**. If **A** does not have n linearly independent eigenvectors, then neither does T, and T does *not* have a diagonal matrix representation.

In this appendix, we focus on identifying simple matrix representations for all linear transformations from a finite-dimensional vector space back to itself. We classify a matrix representation as simple if it contains many zeros. The more zeros, the simpler the matrix. By this criterion, the simplest matrix is the zero matrix. The zero matrix represents the zero transformation $\mathit{0}$, having the property $\mathit{0}(\mathbf{v}) = \mathbf{0}$ for every vector $\mathbf{v} \in \mathbf{V}$. The next simplest class of matrices is diagonal matrices, because they have zeros for all elements not on the main diagonal. These matrices represent linear transformations having sufficiently many linearly independent eigenvectors. The elements on the main diagonal are the eigenvalues.

The more zeros a matrix has, the simpler it is as a matrix representation for a linear transformation.

Another simple class of matrices are block diagonal matrices having the partitioned form

$$\mathbf{A} = \begin{bmatrix} \mathbf{A}_1 & & & 0 \\ & \mathbf{A}_2 & & \\ & & \ddots & \\ 0 & & & \mathbf{A}_k \end{bmatrix} \tag{A.1}$$

We will show that every linear transformation from a finite-dimensional vector space $\mathbb{V}$ back to itself can be represented by a matrix in block diagonal form. To do so, we must develop the concepts of direct sums and invariant subspaces.

Direct sums were introduced in Section 6.5. A vector space $\mathbb{V}$ is the direct sum of two subspaces $\mathbb{U}$ and $\mathbb{W}$, written $\mathbb{V} = \mathbb{U} \oplus \mathbb{W}$, if each vector in $\mathbb{V}$ can be written *uniquely* as the sum of a vector in $\mathbb{U}$ and a vector in $\mathbb{W}$. We know from our work in the last chapter that if $\mathbb{V}$ is an inner product space and if $\mathbb{U}$ is any subspace of $\mathbb{V}$, then $\mathbb{V} = \mathbb{U} \oplus \mathbb{U}^{\perp}$. However, there are many other direct sums available to us.

▶THEOREM 1

Let $\mathbb{M}$ and $\mathbb{N}$ be subspaces of a finite dimensional vector space $\mathbb{V}$, with $\mathbb{B}$ being a basis for $\mathbb{M}$ and $\mathbb{C}$ being a basis for $\mathbb{N}$. $\mathbb{V} = \mathbb{M} \oplus \mathbb{N}$ if and only if $\mathbb{B} \cup \mathbb{C}$ is a basis for $\mathbb{V}$. ◀

Proof: Assume that $\mathbb{V} = \mathbb{M} \oplus \mathbb{N}$. If $\mathbf{x} \in \mathbb{V}$, then $\mathbf{x}$ can be written uniquely as the sum $\mathbf{y}+\mathbf{z}$ with $\mathbf{y} \in \mathbb{M}$ and $\mathbf{z} \in \mathbb{N}$. Let $\mathbb{B} = \{\mathbf{m}_1, \mathbf{m}_2, \ldots, \mathbf{m}_r\}$. Since $\mathbb{B}$ is a basis for $\mathbb{M}$, there exist scalars $c_1, c_2, \ldots, c_r$ such that

$$\mathbf{y} = c_1\mathbf{m}_1 + c_2\mathbf{m}_2 + \cdots + c_r\mathbf{m}_r \tag{A.2}$$

Let $\mathbb{C} = \{\mathbf{n}_1, \mathbf{n}_2, \ldots, \mathbf{n}_s\}$. Since $\mathbb{C}$ is a basis for $\mathbb{N}$, there exist scalars and $d_1, d_2, \ldots, d_s$ such that

$$\mathbf{z} = d_1\mathbf{n}_1 + d_2\mathbf{n}_2 + \cdots + d_s\mathbf{n}_s \tag{A.3}$$

Therefore,

$$\mathbf{x} = \mathbf{y} + \mathbf{z} = c_1\mathbf{m}_1 + c_2\mathbf{m}_2 + \cdots + c_r\mathbf{m}_r + d_1\mathbf{n}_1 + d_2\mathbf{n}_2 + \cdots + d_s\mathbf{n}_s \tag{A.4}$$

and $\mathbb{B} \cup \mathbb{C}$ is a spanning set for $\mathbb{V}$.

To show that $\mathbb{B} \cup \mathbb{C}$ is a linearly independent set of vectors, we consider the vector equation

$$0 = (c_1\mathbf{m}_1 + c_2\mathbf{m}_2 + \cdots + c_r\mathbf{m}_r) + (d_1\mathbf{n}_1 + d_2\mathbf{n}_2 + \cdots + d_s\mathbf{n}_s)$$

Clearly,

$$0 = (0\mathbf{m}_1 + 0\mathbf{m}_2 + \cdots + 0\mathbf{m}_r) + (0\mathbf{n}_1 + 0\mathbf{n}_2 + \cdots + 0\mathbf{n}_s)$$

The last two equations are two representations of the vector $\mathbf{0}$ as the sum of a vector in $\mathbb{M}$ (the terms in the first set of parentheses of each equation) and a vector in $\mathbb{N}$ (the terms in the second set of parentheses of each equation). Since $\mathbb{V} = \mathbb{M} \oplus \mathbb{N}$, the zero vector can only be represented *one* way as a vector in $\mathbb{M}$ with a vector in $\mathbb{N}$, so it must be the case that $c_j = 0$ for $j = 1, 2, \ldots, r$ and $d_k = 0$ for $k = 1, 2, \ldots, s$. Thus, $\mathbb{B} \cup \mathbb{C}$ is a linearly independent set of vectors. A linearly independent spanning set of vectors is a basis, hence $\mathbb{B} \cup \mathbb{C}$ is a basis for $\mathbb{V}$. Conversely, assume that $\mathbb{B} \cup \mathbb{C}$ is a basis for $\mathbb{V}$. If $\mathbf{x} \in \mathbb{V}$, then there exists a *unique* set of scalars $c_1, c_2, \ldots, c_r$ and $d_1, d_2, \ldots, d_s$ such that Eq. (A.4) is satisfied. If we now use Eqs. (A.2) and (A.3) to define $\mathbf{y}$ and $\mathbf{z}$, we have $\mathbf{x}$ written uniquely as the sum of a vector $\mathbf{y} \in \mathbb{M}$ and a vector $\mathbf{z} \in \mathbb{N}$. Therefore, $\mathbb{V} = \mathbb{M} \oplus \mathbb{N}$.

Example 1 $\mathbb{D} = \left\{ \mathbf{x}_1 = \begin{bmatrix} 1 \\ 1 \\ 0 \\ 0 \end{bmatrix}, \mathbf{x}_2 = \begin{bmatrix} 5 \\ 2 \\ 0 \\ 0 \end{bmatrix}, \mathbf{x}_3 = \begin{bmatrix} 1 \\ 0 \\ -1 \\ 2 \end{bmatrix}, \mathbf{x}_4 = \begin{bmatrix} -1 \\ 0 \\ 1 \\ 3 \end{bmatrix} \right\}$ is a basis for $\mathbb{R}^4$. If we set $\mathbb{B} = \{\mathbf{x}_1, \mathbf{x}_2\}$, $\mathbb{C} = \{\mathbf{x}_3, \mathbf{x}_4\}$, $\mathbb{M} = span\{\mathbf{B}\}$, and $\mathbb{N} = span\{\mathbb{C}\}$, then it follows from Theorem 1 that $\mathbb{R}^4 = \mathbb{M} \oplus \mathbb{N}$. Alternatively, if we set $\mathbb{Q} = span\{\mathbf{x}_2, \mathbf{x}_3\}$ and $\mathbb{S} = span\{\mathbf{x}_1, \mathbf{x}_4\}$, then $\mathbb{R}^4 = \mathbb{Q} \oplus \mathbb{S}$. Still a third possibility is to set $\mathbb{U} = span\{\mathbf{x}_1, \mathbf{x}_2, \mathbf{x}_3\}$ and $\mathbb{W} = span\{\mathbf{x}_4\}$, in which case $\mathbb{R}^4 = \mathbb{U} \oplus \mathbb{W}$.

A subspace $\mathbb{U}$ of an n-dimensional vector space $\mathbb{V}$ is invariant under a linear transformation $T\colon \mathbb{V} \to \mathbb{V}$ if $T(\mathbf{u}) \in \mathbb{U}$ whenever $\mathbf{u} \in \mathbb{U}$. That is, T maps vectors in $\mathbb{U}$ back into vectors in $\mathbb{U}$.

Example 2 The subspace $ker\{T\}$ is invariant under T because T maps every vector in the kernel into the zero vector, which is itself in the kernel. The subspace $Im(T)$ is invariant under T because $T(\mathbf{u}) \in Im(T)$ for every vector in $\mathbb{V}$, including those in $Im(T)$. If $\mathbf{x}$ is an eigenvector of T corresponding to the eigenvalue λ, then $span\{\mathbf{x}\}$ is invariant under T if $\mathbf{u} \in span\{\mathbf{x}\}$, then $\mathbf{u} = \alpha\mathbf{x}$, for some choice of the scalar α, and

$$T(\mathbf{u}) = T(\alpha\mathbf{x}) = \alpha T(\mathbf{x}) = \alpha(\lambda\mathbf{x}) = (\alpha\lambda)\mathbf{x} \in span\{\mathbf{x}\}.$$

▸THEOREM 2

Let $\mathbb{B} = \{\mathbf{u}_1, \mathbf{u}_2, \ldots, \mathbf{u}_m\}$ be basis for a subspace $\mathbb{U}$ of an n-dimensional vector space $\mathbb{V}$. $\mathbb{U}$ is an invariant subspace under the linear transformation $\mathbf{T}\colon \mathbb{V} \to \mathbb{V}$ if and only if $\mathbf{T}(\mathbf{u}_j) \in \mathbb{U}$ for $j = 1, 2, \ldots, m$. ◂

Proof: If $\mathbb{U}$ is an invariant subspace under T, then $T(\mathbf{u}) \in \mathbb{U}$ for every vector $\mathbf{u} \in \mathbb{U}$. Since the basis vectors $\mathbf{u}_j (j = 1, 2, \ldots, m)$ are vectors in $\mathbb{U}$, it follows that $T(\mathbf{u}_j) \in \mathbf{U}$. Conversely, if $\mathbf{u} \in \mathbb{U}$, then there exist scalars $c_1, c_2, \ldots, c_m$ such that

$$\mathbf{u} = c_1\mathbf{u}_1 + c_2\mathbf{u}_2 + \cdots + c_m\mathbf{u}_m$$

Now

$$\begin{aligned} T(\mathbf{u}) &= T(c_1\mathbf{u}_1 + c_2\mathbf{u}_2 + \cdots + c_m\mathbf{u}_m) \\ &= c_1 T(\mathbf{u}_1) + c_2 T(\mathbf{u}_2) + \cdots + c_m T(\mathbf{u}_m) \end{aligned}$$

Thus, $T(\mathbf{u})$ is a linear combination of the vectors $T(\mathbf{u}_j)$ for $j = 1, 2, \ldots, m$. Since each vector $T(\mathbf{u}_j) \in \mathbb{U}$, it follows that $T(\mathbf{u}) \in \mathbb{U}$ and that $\mathbb{U}$ in invariant under T.

Example 3 Determine whether the subspace

$$\mathbb{M} = span\left\{ \begin{bmatrix} 1 \\ 1 \\ 0 \\ 0 \end{bmatrix}, \begin{bmatrix} 5 \\ 2 \\ 0 \\ 0 \end{bmatrix} \right\} \text{ is invariant under } T \begin{bmatrix} a \\ b \\ c \\ d \end{bmatrix} = \begin{bmatrix} a + b - d \\ b \\ c + d \\ d \end{bmatrix}$$

Solution: The two vectors that span $\mathbb{M}$ are linearly independent and, therefore, are a basis for $\mathbb{M}$. Here,

$$T\begin{bmatrix}1\\1\\0\\0\end{bmatrix}=\begin{bmatrix}2\\1\\0\\0\end{bmatrix}=\left(\frac{1}{3}\right)\begin{bmatrix}1\\1\\0\\0\end{bmatrix}+\left(\frac{1}{3}\right)\begin{bmatrix}5\\2\\0\\0\end{bmatrix}\in\mathbb{M}$$

$$T\begin{bmatrix}5\\2\\0\\0\end{bmatrix}=\begin{bmatrix}7\\2\\0\\0\end{bmatrix}=\left(-\frac{4}{3}\right)\begin{bmatrix}1\\1\\0\\0\end{bmatrix}+\left(\frac{5}{3}\right)\begin{bmatrix}5\\2\\0\\0\end{bmatrix}\in\mathbb{M}$$

It follows from Theorem 2 that $\mathbb{M}$ is an invariant subspace of $\mathbb{R}^4$ under T.

Example 4 Determine whether the subspace

$$\mathbf{N}=span\left\{\begin{bmatrix}1\\0\\-1\\2\end{bmatrix},\begin{bmatrix}-1\\0\\1\\3\end{bmatrix}\right\}$$

is invariant under the linear transformation defined in Example 3.

Solution: The two vectors that span $\mathbb{N}$ are linearly independent and, therefore, are a basis for $\mathbb{N}$. Here,

$$T\begin{bmatrix}1\\0\\-1\\2\end{bmatrix}=\begin{bmatrix}-1\\0\\1\\2\end{bmatrix}=\left(-\frac{1}{5}\right)\begin{bmatrix}1\\0\\-1\\2\end{bmatrix}+\left(\frac{4}{5}\right)\begin{bmatrix}-1\\0\\1\\3\end{bmatrix}\in\mathbb{N}$$

$$T\begin{bmatrix}-1\\0\\1\\3\end{bmatrix}=\begin{bmatrix}-4\\0\\4\\3\end{bmatrix}=\left(-\frac{9}{5}\right)\begin{bmatrix}1\\0\\-1\\2\end{bmatrix}+\left(\frac{11}{5}\right)\begin{bmatrix}-1\\0\\1\\3\end{bmatrix}\in\mathbb{N}$$

It follows from Theorem 2 that $\mathbb{N}$ is an invariant subspace of $\mathbb{R}^4$ under T.

The next result establishes a link between direct sums of invariant subspaces and matrix representations in block diagonal form.

►THEOREM 3

If $\mathbb{M}$ and $\mathbb{N}$ are invariant subspaces of a finite-dimensional vector space $\mathbb{V}$ with $\mathbb{V}=\mathbb{M}\oplus\mathbb{N}$, and if $\mathbf{T}\colon\mathbb{V}\to\mathbb{V}$ is linear, then $\mathbf{T}$ has a matrix representation of the form

$$\mathbf{A}=\begin{bmatrix}\mathbf{B}&0\\0&\mathbf{C}\end{bmatrix}$$

*where **B** and **C** are square matrices having as many rows (and columns) as the dimensions of $\mathbb{M}$ and $\mathbb{N}$, respectively.*◄

Proof: Let $\mathbb{B} = \{\mathbf{m}_1, \mathbf{m}_2, \ldots, \mathbf{m}_r\}$ be a basis for $\mathbb{M}$ and let $\mathbb{C} = \{\mathbf{n}_1, \mathbf{n}_2, \ldots, \mathbf{n}_s\}$ be a basis for $\mathbb{N}$. Then, because $\mathbb{V}$ is the direct sum of $\mathbb{M}$ and $\mathbb{N}$, it follows that

$$\mathbb{D} = \mathbb{B} \cup \mathbb{C} = \{\mathbf{m}_1, \mathbf{m}_2, \ldots, \mathbf{m}_r, \mathbf{n}_1, \mathbf{n}_2, \ldots, \mathbf{n}_s\}$$

is a basis for $\mathbb{V}$ (see Theorem 1). $\mathbb{M}$ is given to be an invariant subspace of T, so all vectors in $\mathbb{M}$, in particular the basis vectors themselves, map into vectors in $\mathbb{M}$. Every vector in $\mathbb{M}$ can be written uniquely as a linear combination of the basis vectors for $\mathbb{M}$. Thus, for jth basis vector in $\mathbb{M}$ $(j = 1, 2, \ldots, r)$, we have

$$\begin{aligned} T(\mathbf{m}_j) &= b_{1j}\mathbf{m}_1 + b_{2j}\mathbf{m}_2 + \cdots + b_{rj}\mathbf{m}_r \\ &= b_{1j}\mathbf{m}_1 + b_{2j}\mathbf{m}_2 + \cdots + b_{rj}\mathbf{m}_r + 0\mathbf{n}_1 + 0\mathbf{n}_2 + \cdots + 0\mathbf{n}_s \end{aligned}$$

for some choice of the scalars $b_{1j}, b_{2j}, \ldots, b_{rj}$. $T(\mathbf{m}_j)$ has the coordinate representation

$$T(\mathbf{m}_j) \leftrightarrow \left[b_{1j}b_{2j}\ldots b_{rj}00\ldots0\right]^{\mathrm{T}}.$$

Similarly, $\mathbb{N}$ is an invariant subspace of T, so all vectors in $\mathbb{N}$, in particular the basis vectors themselves, map into vectors in $\mathbb{N}$. Every vector in $\mathbb{N}$ can be written uniquely as a linear combination of the basis vectors for $\mathbb{N}$. Thus, for kth basis vector in $\mathbb{N}$ $(k = 1, 2, \ldots, s)$, we have

$$\begin{aligned} T(\mathbf{n}_k) &= c_{1k}\mathbf{n}_1 + c_{2k}\mathbf{n}_2 + \cdots + c_{sk}\mathbf{n}_s \\ &= 0\mathbf{m}_1 + 0\mathbf{m}_2 = \cdots + 0\mathbf{m}_r + c_{1k}\mathbf{n}_1 + c_{2k}\mathbf{n}_2 + \cdots + c_{sk}\mathbf{n}_s \end{aligned}$$

for some choice of the scalars $c_{1k}, c_{2k}, \ldots, c_{sk}$. $T(\mathbf{n}_k)$ has the coordinate representation

$$T(\mathbf{n}_k) \leftrightarrow [0\,0\ldots0c_{1k}\,c_{2k}\ldots c_{sk}]^{\mathrm{T}}.$$

These coordinate representations for $T(\mathbf{m}_j)$ $(j = 1, 2, \ldots, r)$ and $T(\mathbf{n}_k)$ $(k = 1, 2, \ldots, s)$ become columns of the matrix representation for T with respect to the $\mathbb{D}$ basis. That is,

$$T \leftrightarrow \mathbf{A}_{\mathbb{D}}^{\mathbb{D}} = \begin{bmatrix} b_{11} & b_{12} & b_{1r} & 0 & 0 & 0 \\ b_{21} & b_{22} & b_{2r} & 0 & 0 & 0 \\ b_{r1} & b_{r2} & b_{rr} & 0 & 0 & 0 \\ 0 & 0 & 0 & c_{11} & c_{12} & c_{13} \\ 0 & 0 & 0 & c_{21} & c_{22} & c_{23} \\ 0 & 0 & 0 & c_{s1} & c_{s2} & c_{ss} \end{bmatrix}$$

which is the form claimed in Theorem 3.

Example 5 We showed in Example 1 that $\mathbb{R}^4 = \mathbb{M} \oplus \mathbb{N}$ when

$$\mathbb{M} = span\left\{\begin{bmatrix}1\\1\\0\\0\end{bmatrix}, \begin{bmatrix}5\\2\\0\\0\end{bmatrix}\right\} \text{ and } \mathbb{N} = span\left\{\begin{bmatrix}1\\0\\-1\\2\end{bmatrix}, \begin{bmatrix}-1\\0\\1\\3\end{bmatrix}\right\}$$

We established in Examples 3 and 4 that both $\mathbb{M}$ and $\mathbb{N}$ are invariant subspaces under

$$T\begin{bmatrix}a\\b\\c\\d\end{bmatrix} = \begin{bmatrix}a+b-d\\b\\c+d\\d\end{bmatrix}$$

It now follows from Theorem 3 and its proof that T has a matrix representation in block diagonal form with respect to the basis

$$\mathbb{D} = \left\{\begin{bmatrix}1\\1\\0\\0\end{bmatrix}, \begin{bmatrix}5\\2\\0\\0\end{bmatrix}, \begin{bmatrix}1\\0\\-1\\2\end{bmatrix}, \begin{bmatrix}-1\\0\\1\\3\end{bmatrix}\right\}$$

for $\mathbb{R}^4$. Here,

$$T\begin{bmatrix}1\\1\\0\\0\end{bmatrix} = \begin{bmatrix}2\\1\\0\\0\end{bmatrix} = \left(\frac{1}{3}\right)\begin{bmatrix}1\\1\\0\\0\end{bmatrix} + \left(\frac{1}{3}\right)\begin{bmatrix}5\\2\\0\\0\end{bmatrix} + (0)\begin{bmatrix}1\\0\\-1\\2\end{bmatrix} + (0)\begin{bmatrix}-1\\0\\1\\3\end{bmatrix} \leftrightarrow \begin{bmatrix}1/3\\1/3\\0\\0\end{bmatrix}_{\mathbb{D}}$$

$$T\begin{bmatrix}5\\2\\0\\0\end{bmatrix} = \begin{bmatrix}7\\2\\0\\0\end{bmatrix} = \left(-\frac{4}{3}\right)\begin{bmatrix}1\\1\\0\\0\end{bmatrix} + \left(\frac{5}{3}\right)\begin{bmatrix}5\\2\\0\\0\end{bmatrix} + (0)\begin{bmatrix}1\\0\\-1\\2\end{bmatrix} + (0)\begin{bmatrix}-1\\0\\1\\3\end{bmatrix} \leftrightarrow \begin{bmatrix}-4/3\\5/3\\0\\0\end{bmatrix}_{\mathbb{D}}$$

$$T\begin{bmatrix}1\\0\\-1\\2\end{bmatrix} = \begin{bmatrix}-1\\0\\1\\2\end{bmatrix} = (0)\begin{bmatrix}1\\1\\0\\0\end{bmatrix} + (0)\begin{bmatrix}5\\2\\0\\0\end{bmatrix} + \left(-\frac{1}{5}\right)\begin{bmatrix}1\\0\\-1\\2\end{bmatrix} + \left(\frac{4}{5}\right)\begin{bmatrix}-1\\0\\1\\3\end{bmatrix} \leftrightarrow \begin{bmatrix}0\\0\\-1/5\\4/5\end{bmatrix}_{\mathbb{D}}$$

$$T\begin{bmatrix}-1\\0\\1\\3\end{bmatrix} = \begin{bmatrix}-4\\0\\4\\3\end{bmatrix} = (0)\begin{bmatrix}1\\1\\0\\0\end{bmatrix} + (0)\begin{bmatrix}5\\2\\0\\0\end{bmatrix} + \left(-\frac{9}{5}\right)\begin{bmatrix}1\\0\\-1\\2\end{bmatrix} + \left(\frac{11}{5}\right)\begin{bmatrix}-1\\0\\1\\3\end{bmatrix} \leftrightarrow \begin{bmatrix}0\\0\\-9/5\\11/5\end{bmatrix}_{\mathbb{D}}$$

The matrix representation of T with respect to the $\mathbb{D}$ basis is

$$\mathbf{A}_{\mathbb{D}}^{\mathbb{D}} = \begin{bmatrix} 1/3 & -4/3 & 0 & 0 \\ 1/3 & 5/3 & 0 & 0 \\ 0 & 0 & -1/5 & -9/5 \\ 0 & 0 & 4/5 & 11/5 \end{bmatrix}$$

Theorem 3 deals with two invariant subspaces, but that result is easily generalized to any finite number of subspaces. If $\mathbb{M}_1, \mathbb{M}_2, \ldots, \mathbb{M}_k$ are invariant subspaces of a linear transformation $T\colon \mathbb{V} \to \mathbb{V}$ with $\mathbb{V} = \mathbb{M}_1 \oplus \mathbb{M}_2 \oplus \cdots \oplus \mathbb{M}_k$, then the union of bases for each subspace is a basis for $\mathbb{V}$. A matrix representation of T with respect to this basis for $\mathbb{V}$ has the block diagonal form displayed in Eq. (A.1). Thus, the key to developing block diagonal matrix representations for linear transformations is to identify invariant subspaces.

The span of any set of eigenvectors of a linear transformation generates an invariant subspace for that transformation (see Problem 35), but there may not be enough linearly independent eigenvectors to form a basis for the entire vector space. A vector $\mathbf{x}_m$ is a *generalized eigenvector of type m* for the linear transformation T corresponding to the eigenvalue λ if

$$(T - \lambda\mathbf{I})^m(\mathbf{x}_m) = 0 \quad \text{and} \quad (T - \lambda\mathbf{I})^{m-1}(\mathbf{x}_m) \neq 0 \tag{A.5}$$

As was the case with eigenvectors, it is often easier to find generalized eigenvectors for a matrix representation for a linear transformation than for the linear transformations, *per se*. A vector $\mathbf{x}_m$ is a generalized eigenvector of type m corresponding to the eigenvalue λ for the matrix $\mathbf{A}$ if

A vector $\mathbf{x}_m$ is a generalized eigenvector of type m corresponding to the eigenvalue λ for the matrix $\mathbf{A}$ if $(\mathbf{A} - \lambda\mathbf{I})^m(\mathbf{x}_m) = \mathbf{0}$ and $(\mathbf{A} - \lambda\mathbf{I})^{m-1}(\mathbf{x}_m) \neq \mathbf{0}$.

$$(\mathbf{A} - \lambda\mathbf{I})^m\mathbf{x}_m = 0 \quad \text{and} \quad (\mathbf{A} - \lambda\mathbf{I})^{m-1}\mathbf{x}_m \neq 0 \tag{A.6}$$

Example 6 $\mathbf{x}_3 = [0 \quad 0 \quad 1]^{\mathrm{T}}$ is a generalized eigenvector of type 3 corresponding to $\lambda = 2$ for

$$\mathbf{A} = \begin{bmatrix} 2 & 1 & -1 \\ 0 & 2 & 1 \\ 0 & 0 & 2 \end{bmatrix}$$

because

$$(\mathbf{A} - 2\mathbf{I})^3\mathbf{x}_3 = \begin{bmatrix} 0 & 0 & 0 \\ 0 & 0 & 0 \\ 0 & 0 & 0 \end{bmatrix}\begin{bmatrix} 0 \\ 0 \\ 1 \end{bmatrix} = \begin{bmatrix} 0 \\ 0 \\ 0 \end{bmatrix}$$

while

$$(\mathbf{A} - 2\mathbf{I})^2\mathbf{x}_3 = \begin{bmatrix} 0 & 0 & 0 \\ 0 & 0 & 0 \\ 0 & 0 & 0 \end{bmatrix}\begin{bmatrix} 0 \\ 0 \\ 1 \end{bmatrix} = \begin{bmatrix} 1 \\ 0 \\ 0 \end{bmatrix} \neq \mathbf{0}$$

Also, $\mathbf{x}_2 = [-1 \quad 1 \quad 0]^{\mathrm{T}}$ is a generalized eigenvector of type 2 corresponding to $\lambda = 2$ for this same matrix because

$$(\mathbf{A} - 2\mathbf{I})^3 \mathbf{x}_3 = \begin{bmatrix} 0 & 0 & 1 \\ 0 & 0 & 0 \\ 0 & 0 & 0 \end{bmatrix} \begin{bmatrix} -1 \\ 1 \\ 0 \end{bmatrix} = \begin{bmatrix} 0 \\ 0 \\ 0 \end{bmatrix}$$

while

$$(\mathbf{A} - 2\mathbf{I})^1 \mathbf{x}_2 = \begin{bmatrix} 0 & 0 & -1 \\ 0 & 0 & 1 \\ 0 & 0 & 0 \end{bmatrix} \begin{bmatrix} -1 \\ 1 \\ 0 \end{bmatrix} = \begin{bmatrix} 1 \\ 0 \\ 0 \end{bmatrix} \neq \mathbf{0}$$

Furthermore, $\mathbf{x}_1 = [1 \quad 0 \quad 0]^{\mathrm{T}}$ is a generalized eigenvector of type 1 corresponding to $\lambda = 2$ for $\mathbf{A}$ because $(\mathbf{A} - 2\mathbf{I})^1 \mathbf{x}_1 = \mathbf{0}$ but $(\mathbf{A} - 2\mathbf{I})^0 \mathbf{x}_1 = \mathbf{I}\mathbf{x}_1 = \mathbf{x}_1 \neq \mathbf{0}$.

Example 7 It is known, and we shall see why later, that the matrix

$$\mathbf{A} = \begin{bmatrix} 5 & 1 & -2 & 4 \\ 0 & 5 & 2 & 2 \\ 0 & 0 & 5 & 3 \\ 0 & 0 & 0 & 4 \end{bmatrix}$$

has a generalized eigenvector of type 3 corresponding to $\lambda = 5$. Find it.

Solution: We seek a vector $\mathbf{x}_3$ such that

$$(\mathbf{A} - 5\mathbf{I})^3 \mathbf{x}_3 = \mathbf{0} \quad \text{and} \quad (\mathbf{A} - 5\mathbf{I})^2 \mathbf{x} \neq \mathbf{0}$$

Set $\mathbf{x}_3 = [w \quad x \quad y \quad z]^{\mathrm{T}}$. Then

$$(\mathbf{A} - 5\mathbf{I})^3 \mathbf{x}_3 = \begin{bmatrix} 0 & 0 & 0 & 14 \\ 0 & 0 & 0 & -4 \\ 0 & 0 & 0 & 3 \\ 0 & 0 & 0 & -1 \end{bmatrix} \begin{bmatrix} w \\ x \\ y \\ z \end{bmatrix} = \begin{bmatrix} 14z \\ -4z \\ 3z \\ z \end{bmatrix}$$

$$(\mathbf{A} - 5\mathbf{I})^2 \mathbf{x}_3 = \begin{bmatrix} 0 & 0 & 2 & -8 \\ 0 & 0 & 0 & 4 \\ 0 & 0 & 0 & -3 \\ 0 & 0 & 0 & 1 \end{bmatrix} \begin{bmatrix} w \\ x \\ y \\ z \end{bmatrix} = \begin{bmatrix} 2y - 8z \\ 4z \\ -3z \\ z \end{bmatrix}$$

To satisfy the condition $(\mathbf{A} - 5\mathbf{I})^3 \mathbf{x}_3 = \mathbf{0}$, we must have $z = 0$. To satisfy the condition $(\mathbf{A} - 5\mathbf{I})^2 \mathbf{x}_3 \neq \mathbf{0}$, with $z = 0$, we must have $y \neq 0$. No restrictions are placed on w and x. By choosing $w = x = z = 0$, $y = 1$, we obtain $\mathbf{x}_3 = [0 \quad 0 \quad 1 \quad 0]^{\mathrm{T}}$ as a generalized eigenvector of type 3 corresponding to $\lambda = 5$. There are infinitely many other generalized eigenvector of type 3, each obtained by selecting other values for w, x, and y $(y \neq 0)$ with $z = 0$. In particular, the values w $= -1$, $x = 2$, $y = 15$, $z = 0$ lead to $\mathbf{x}_3 = [- \quad 1 \quad 2 \quad 15 \quad 0]^{\mathrm{T}}$. Our first choice, however, is the simplest.

Generalized eigenvectors are the building blocks for invariant subspaces. Each generalized eigenvector propagates a *chain* of vectors that serves as a basis for

The *chain* propagated by $\mathbf{x}_m$, a generalized eigenvector of type m corresponding to the eigenvalue λ for a matrix $\mathbf{A}$, is the set of vectors $\{\mathbf{x}_m, \mathbf{x}_{m-1}, \ldots, \mathbf{x}_1\}$ defined sequentially by $\mathbf{x}_j = (\mathbf{A} - \lambda\mathbf{I})\mathbf{x}_{j+1}$ for $j = 1, 2, \ldots, m-1$.

an invariant subspace. The chain propagated by $\mathbf{x}_m$, a generalized eigenvector of type m corresponding to the eigenvalue λ for $\mathbf{A}$, is the set of vectors $\{\mathbf{x}_m, \mathbf{x}_{m-1}, \ldots, \mathbf{x}_1\}$ given by

$$\begin{aligned}
\mathbf{x}_{m-1} &= (\mathbf{A} - \lambda\mathbf{I})\mathbf{x}_m \\
\mathbf{x}_{m-2} &= (\mathbf{A} - \lambda\mathbf{I})^2\mathbf{x}_m = (\mathbf{A} - \lambda\mathbf{I})\mathbf{x}_{m-1} \\
\mathbf{x}_{m-3} &= (\mathbf{A} - \lambda\mathbf{I})^3\mathbf{x}_m = (\mathbf{A} - \lambda\mathbf{I})\mathbf{x}_{m-2} \\
&\vdots \\
\mathbf{x}_1 &= (\mathbf{A} - \lambda\mathbf{I})^{m-1}\mathbf{x}_m = (\mathbf{A} - \lambda\mathbf{I})\mathbf{x}_2
\end{aligned} \tag{A.7}$$

In general, for $j = 1, 2, \ldots, m-1$,

$$\mathbf{x}_j = (\mathbf{A} - \lambda\mathbf{I})^{m-j}\mathbf{x}_m = (\mathbf{A} - \lambda\mathbf{I})\mathbf{x}_{j+1} \tag{A.8}$$

▶THEOREM 4

The jth vector in a chain, $\mathbf{x}_j$, as defined by equation (B.8), is a generalized eigenvector of type j corresponding to the same matrix and eigenvalue associated with the generalized eigenvector of type m that propagated the chain. ◀

Proof: Let $\mathbf{x}_m$ be a generalized eigenvector of type m for a matrix $\mathbf{A}$ with eigenvalue λ. Then, $(\mathbf{A}-\lambda\mathbf{I})^m\mathbf{x}_m = \mathbf{0}$ and $(\mathbf{A}-\lambda\mathbf{I})^{m-1}\mathbf{x}_m \neq \mathbf{0}$. Using equation (A.8), we conclude that

$$(\mathbf{A} - \lambda\mathbf{I})^j\mathbf{x}_j = (\mathbf{A} - \lambda\mathbf{I})^j\left[(\mathbf{A} - \lambda\mathbf{I})^{m-j}\mathbf{x}_m\right] = (\mathbf{A} - \lambda\mathbf{I})^m\mathbf{x}_m = \mathbf{0}$$

and

$$(\mathbf{A} - \lambda\mathbf{I})^{j-1}\mathbf{x}_j = (\mathbf{A} - \lambda\mathbf{I})^{j-1}\left[(\mathbf{A} - \lambda\mathbf{I})^{m-j}\mathbf{x}_m\right] = (\mathbf{A} - \lambda\mathbf{I})^{m-1}\mathbf{x}_m \neq \mathbf{0}$$

Thus, $\mathbf{x}_j$ is a generalized eigenvector of type j corresponding to the eigenvalue λ for $\mathbf{A}$.

It follows from Theorem 4 that once we have a generalized eigenvector of type m, for any positive integer m, we can use Eq. (A.8) to produce other generalized eigenvectors of type less than m.

Example 8 In Example 7, we showed that $\mathbf{x}_3 = [0 \quad 0 \quad 1 \quad 0]^T$ is a generalized eigenvector of type 3 for

$$\mathrm{A} = \begin{bmatrix} 5 & 1 & -2 & 4 \\ 0 & 5 & 2 & 2 \\ 0 & 0 & 5 & 3 \\ 0 & 0 & 0 & 4 \end{bmatrix}$$

corresponding to $\lambda=5$. Using Theorem 4, we now can state that

$$(\mathbf{A}-5\mathbf{I})\mathbf{x}_3 = \begin{bmatrix} 0 & 1 & -2 & 4 \\ 0 & 0 & 2 & 2 \\ 0 & 0 & 0 & 3 \\ 0 & 0 & 0 & -1 \end{bmatrix} \begin{bmatrix} 0 \\ 0 \\ 1 \\ 0 \end{bmatrix} = \begin{bmatrix} -2 \\ 2 \\ 0 \\ 0 \end{bmatrix}$$

is a generalized eigenvector of type 2 for **A** corresponding to $\lambda=5$, while

$$(\mathbf{A}-5\mathbf{I})\mathbf{x}_2 = \begin{bmatrix} 0 & 1 & -2 & 4 \\ 0 & 0 & 2 & 2 \\ 0 & 0 & 0 & 3 \\ 0 & 0 & 0 & -1 \end{bmatrix} \begin{bmatrix} -2 \\ 2 \\ 0 \\ 0 \end{bmatrix} = \begin{bmatrix} -2 \\ 2 \\ 0 \\ 0 \end{bmatrix}$$

is a generalized eigenvector of type 1, and, therefore, an eigenvector of **A** corresponding to $\lambda=5$. The set

$$\{\mathbf{x}_3, \mathbf{x}_2, \mathbf{x}_1\} = \left\{ \begin{bmatrix} 0 \\ 0 \\ 1 \\ 0 \end{bmatrix}, \begin{bmatrix} -2 \\ 2 \\ 0 \\ 0 \end{bmatrix}, \begin{bmatrix} 2 \\ 0 \\ 0 \\ 0 \end{bmatrix} \right\}$$

is the chain propagated by the $\mathbf{x}_3$.

The relationship between chains of generalized eigenvectors and invariant subspaces is established by the next two theorems.

▶THEOREM 5

A chain is a linearly independent set of vectors. ◀

Proof: Let $\{\mathbf{x}_m, \mathbf{x}_{m-1}, \ldots, \mathbf{x}_1\}$ be a chain propagated from $\mathbf{x}_m$, a generalized eigenvector of type m corresponding to the eigenvalue λ for **A**. We consider the vector equation

$$c_m\mathbf{x}_m + c_{m-1}\mathbf{x}_{m-1} + \cdots + c_1\mathbf{x}_1 = \mathbf{0} \tag{A.9}$$

To prove that this chain is linearly independent, we must show that the only solution to Eq. (A.9) is the trivial solution $c_m=c_{m-1}=\ldots=c_1=0$. We shall do this iteratively. First, we multiply both sides of Eq. (A.9) by $(\mathbf{A}-\lambda\mathbf{I})^{m-1}$. Note that for $j=1, 2, \ldots, m-1$,

$$\begin{aligned}(\mathbf{A}-\lambda\mathbf{I})^{m-1}c_j\mathbf{x}_j &= c_j(\mathbf{A}-\lambda\mathbf{I})^{m-j-1}\left[(\mathbf{A}-\lambda\mathbf{I})^j\mathbf{x}_m\right] \\ &= c_j(\mathbf{A}-\lambda\mathbf{I})^{m-j-1}[\mathbf{0}] \quad \text{because } x_j \text{ is a generalized eigenvector of type } j \\ &= \mathbf{0}\end{aligned}$$

Thus, Eq. (A.9) becomes $c_m(\mathbf{A}-\lambda\mathbf{I})^{m-1}\,\mathbf{x}_m=\mathbf{0}$. But $\mathbf{x}_m$ is a generalized eigenvector of type m, so the vector $(\mathbf{A}-\lambda\mathbf{I})^{m-1}\,\mathbf{x}_m \neq 0$. It then follows (see Section 2.2) that

$c_m = 0$. Substituting $c_m = 0$ into Eq. (A.9) and then multiplying the resulting equation by $(\mathbf{A} - \lambda\mathbf{I})^{m-2}$, we find, by similar reasoning, that $c_{m-1} = 0$. Continuing this process, we find iteratively that $c_m = c_{m-1} = \cdots = c_1 = 0$, which implies that the chain is linearly independent.

▶THEOREM 6

The span of a set of vectors that forms a chain of generalized eigenvectors for a matrix **A** *corresponding to an eigenvalue* **A** *is an invariant subspace for* **A**. ◀

Proof: The span of any set of vectors in a vector space is a subspace, so it only remains to show that the subspace is invariant under **A**. Let $\{\mathbf{x}_m, \mathbf{x}_{m-1}, \ldots, \mathbf{x}_1\}$ be a chain propagated from $\mathbf{x}_m$, a generalized eigenvector of type m for **A** corresponding to the eigenvalue λ. It follows that

$$\mathbf{x}_j = (\mathbf{A} - \lambda\mathbf{I})\mathbf{x}_{j+1} (j = 1, 2, \ldots, m-1) \qquad \text{(A.8 repeated)}$$

This equation may be rewritten as

$$\mathbf{A}\mathbf{x}_{j+1} = \lambda\mathbf{x}_{j+1} (j = 1, 2, \ldots, m-1) \qquad \text{(A.10)}$$

A generalized eigenvector of type 1 is an eigenvector, so we also have

$$\mathbf{A}\mathbf{x}_1 = \lambda\mathbf{x}_1 \qquad \text{(A.11)}$$

If $\mathbf{v} \in span\{\mathbf{x}_m, \mathbf{x}_{m-1}, \ldots, \mathbf{x}_2, \mathbf{x}_1\}$, then there exists a set of scalars $d_m, d_{m-1}, \ldots, d_2, d_1$ such that

$$\mathbf{v} = d_m\mathbf{x}_m + d_{m-1}\mathbf{x}_{m-1} + \cdots + d_2\mathbf{x}_2 + d_1\mathbf{x}_1$$

Multiplying this equation by **A** and then using (A.10) and (A.11), we have

$$\begin{aligned}
\mathbf{A}\mathbf{v} &= d_m\mathbf{A}\mathbf{x}_m + d_{m-1}\mathbf{A}\mathbf{x}_{m-1} + \cdots + d_2\mathbf{A}\mathbf{x}_2 + d_1\mathbf{A}\mathbf{x}_1 \\
&= d_m(\lambda\mathbf{x}_m + \mathbf{x}_{m-1}) + d_{m-1}(\lambda\mathbf{x}_{m-1} + \mathbf{x}_{m-2}) + \cdots d_2(\lambda\mathbf{x}_2 + \mathbf{x}_1) + d_1(\lambda\mathbf{x}_1) \\
&= (\lambda d_m)\mathbf{x}_m + (d_m + \lambda d_{m-1})\mathbf{x}_{m-1} + (d_{m-1} + \lambda d_{m-2})\mathbf{x}_{m-2} + \cdots \\
&\quad + (d_3 + \lambda d_2)\mathbf{x}_2 + (d_2 + \lambda d_1)\mathbf{x}_1
\end{aligned}$$

which is also a linear combination of the vectors in the chain and, therefore, in the subspace spanned by the vectors in the chain. Thus, if $\mathbf{v} \in span\{\mathbf{x}_m, \mathbf{x}_{m-1}, \ldots, \mathbf{x}_2, \mathbf{x}_1\}$, then $\mathbf{A}\mathbf{v} \in span\{\mathbf{x}_m, \mathbf{x}_{m-1}, \ldots, \mathbf{x}_2, \mathbf{x}_1\}$ and $span\{\mathbf{x}_m, \mathbf{x}_{m-1}, \ldots, \mathbf{x}_2, \mathbf{x}_1\}$ is an invariant subspace of **A**.

It follows from Theorems 5 and 6 that a chain of generalized eigenvectors is a basis for the invariant subspace spanned by that chain.

We now have the mathematical tools to produce a simple matrix representation for a linear transformation $T : \mathbb{V} \to \mathbb{V}$ on a finite-dimensional vector space $\mathbb{V}$. A linear transformation T may not have enough linearly independent eigenvectors to serve as a basis for $\mathbb{V}$ and, therefore, as a basis for a diagonal matrix representation of T. We shall see shortly that a linear transformation always has enough generalized eigenvectors to form a basis for $\mathbb{V}$, and the matrix representation of T with respect to such a basis is indeed simple.

A generalized eigenvector $\mathbf{x}_j$ of type j in the chain propagated by $\mathbf{x}_m$ is related to its immediate ancestor, the generalized eigenvector $\mathbf{x}_{j+1}$ of type $j+1$, by the formula

$$\mathbf{x}_j = [T - \lambda\mathbf{I}](\mathbf{x}_{j+1}) = T(\mathbf{x}_{j+1}) - \lambda\mathbf{x}_{j+1}$$

which may be rewritten as

$$T(\mathbf{x}_{j+1}) = \lambda\mathbf{x}_{j+1} + \mathbf{x}_j \quad (j = 1, 2, \ldots, m-1) \tag{A.12}$$

Since a generalized eigenvector of type 1 is an eigenvector, we also have

$$T(\mathbf{x}_1) = \lambda\mathbf{x}_1 \tag{A.13}$$

Now let $\mathbb{U}$ be the invariant subspace of $\mathbb{V}$ spanned by the chain propagated by $\mathbf{x}_m$. This chain forms a basis for $\mathbb{U}$. If we extend this chain into a basis for $\mathbb{V}$, say

$$\mathbb{B} = \{\mathbf{x}_1, \mathbf{x}_2, \ldots, \mathbf{x}_{m-1}, \mathbf{x}_m, \mathbf{v}_1, \mathbf{v}_2, \ldots, \mathbf{v}_{n-m}\}$$

and define $\mathbb{W} = span\{\mathbf{v}_1, \mathbf{v}_2, \ldots, \mathbf{v}_{n-m}\}$, then it follows from Theorem 1 that $\mathbb{V} = \mathbb{U} \oplus \mathbb{W}$. If $\mathbb{W}$ is also an invariant subspace of T, then we have from Theorem 3 that a matrix representation of T with respect to the $\mathbb{B}$ basis has the block diagonal form

$$\mathbf{A} = \begin{bmatrix} \mathbf{B} & 0 \\ 0 & \mathbf{C} \end{bmatrix} \tag{A.14}$$

But now we can say even more.

Using (A.12) and (A.13), we have

$$T(\mathbf{x}_1) = \lambda\mathbf{x}_1 = \lambda\mathbf{x}_1 + 0\mathbf{x}_2 + 0\mathbf{x}_3 + \cdots + 0\mathbf{x}_{m-1} + 0\mathbf{x}_m + 0\mathbf{v}_1 + 0\mathbf{v}_2 + \cdots + 0\mathbf{v}_{n-m}$$

with a coordinate representation with respect to the $\mathbb{B}$ basis of

$$T(\mathbf{x}_1) \leftrightarrow [\lambda\ 0\ 0 \ldots 0]^{\mathrm{T}}$$

$$T(\mathbf{x}_2) = \lambda\mathbf{x}_2 + \mathbf{x}_1 = \lambda\mathbf{x}_1 + \lambda\mathbf{x}_2 + 0\mathbf{x}_3 + \cdots + 0\mathbf{x}_{m-1} + 0\mathbf{x}_m + 0\mathbf{v}_1 + 0\mathbf{v}_2 + \cdots + 0\mathbf{v}_{n-m}$$

with a coordinate representation of

$$T(\mathbf{x}_2) \leftrightarrow [1\ \lambda\ 0 \ldots 0]^{\mathrm{T}}$$

$$T(\mathbf{x}_3) = \lambda\mathbf{x}_3 + \mathbf{x}_2 = 0\mathbf{x}_1 + 1\mathbf{x}_2 + \lambda\mathbf{x}_3 + \cdots + 0\mathbf{x}_{m-1} + 0\mathbf{x}_m + 0\mathbf{v}_1 + 0\mathbf{v}_2 + \cdots + 0\mathbf{v}_{n-m}$$

with a coordinate representation of

$$T(\mathbf{x}_3) \leftrightarrow [0\ 1\ \lambda\ 0 \ldots 0]^{\mathrm{T}}$$

This pattern continues through $T(\mathbf{x}_m)$. In particular,

$$T(\mathbf{x}_4) \leftrightarrow [0\ 0\ 1\ \lambda\ 0 \ldots 0]^{\mathrm{T}}$$

$$T(\mathbf{x}_2) \leftrightarrow [0\ 0\ 0\ 1\ \lambda\ 0 \ldots 0]^{\mathrm{T}}$$

and so on. The resulting coordinate representations become the first m columns **A** as given by (A.14). Because the basis for **U** is a chain, the submatrix **B** in (A.13) has the upper triangular form

$$\mathbf{B} = \begin{bmatrix} \lambda & 1 & 0 & \cdots & 0 & 0 \\ 0 & \lambda & 1 & \cdots & 0 & 0 \\ \vdots & \vdots & \vdots & \ddots & \ddots & \vdots \\ 0 & 0 & 0 & \cdots & \lambda & 1 \\ 0 & 0 & 0 & \cdots & 0 & \lambda \end{bmatrix} \tag{A.15}$$

with all of its diagonal elements equal to λ, all elements on its *superdiagonal* (i.e., all elements directly above the diagonal elements) equal to 1, and all of its other elements equal to 0.

To emphasize, a matrix is in Jordan canonical form if it is a block diagonal matrix in which every diagonal block is a Jordan block.

We call matrices having form (A.15) *Jordan blocks*. Jordan blocks contain many zeros and are simple building blocks for matrix representations of linear transformations. A matrix representation is in *Jordan canonical form* if it is a block diagonal matrix in which every diagonal block is a Jordan block.

Example 9 The linear transformation $T: \mathbb{R}^4 \to \mathbb{R}^4$ defined by

$$T\begin{bmatrix} a \\ b \\ c \\ d \end{bmatrix} = \begin{bmatrix} 4a - c - d \\ -4a + 2b + 2c + 2d \\ 2a + b + 2c \\ 2a - b - 2c \end{bmatrix}$$

has a matrix representation with respect to the standard basis of

$$\mathbf{G} = \begin{bmatrix} 4 & 0 & -1 & -1 \\ -4 & 2 & 2 & 2 \\ 2 & 1 & 2 & 0 \\ 2 & -1 & -2 & 0 \end{bmatrix}$$

which is not simple. We will show in Example 11 that **G** has two linearly independent generalized eigenvectors of type 2 corresponding to the eigenvalue 2. Using the techniques previously discussed, we find that two such vectors are

$$\mathbf{x}_2 = \begin{bmatrix} 1 \\ 0 \\ 0 \\ 0 \end{bmatrix} \quad \text{and} \quad \mathbf{v}_2 = \begin{bmatrix} 1 \\ 0 \\ 0 \\ 0 \end{bmatrix}$$

Creating chains from each of these two vectors, we obtain

$$\mathbf{x}_1 = (\mathbf{G} - 2\mathbf{I})\mathbf{x}_2 = \begin{bmatrix} 2 & 0 & -1 & -1 \\ -4 & 0 & 2 & 2 \\ 2 & 1 & 0 & 0 \\ 2 & -1 & -2 & -2 \end{bmatrix}\begin{bmatrix} 1 \\ 0 \\ 0 \\ 0 \end{bmatrix} = \begin{bmatrix} 2 \\ -4 \\ 2 \\ 2 \end{bmatrix}$$

$$\mathbf{v}_1 = (\mathbf{G} - 2\mathbf{I})\mathbf{v}_2 = \begin{bmatrix} 2 & 0 & -1 & -1 \\ -4 & 0 & 2 & 2 \\ 2 & 1 & 0 & 0 \\ 2 & -1 & -2 & -2 \end{bmatrix}\begin{bmatrix} 0 \\ 1 \\ 0 \\ 0 \end{bmatrix} = \begin{bmatrix} 0 \\ 0 \\ 1 \\ -1 \end{bmatrix}$$

Setting $\mathbb{U} = span\{\mathbf{x}_1, \mathbf{x}_2\}$ and $\mathbb{W} = span\{\mathbf{v}_1, \mathbf{v}_2\}$, we have two invariant subspaces of $\mathbb{R}^4$, each having as a basis a single chain. Thus, we expect the matrix representation of T with respect to the basis $\mathbb{B} = \{\mathbf{x}_1, \mathbf{x}_2, \mathbf{v}_1, \mathbf{v}_2\}$ to contain two Jordan blocks. Using this basis, we have

$$T\begin{bmatrix}2\\-4\\2\\2\end{bmatrix} = \begin{bmatrix}4\\-8\\4\\4\end{bmatrix} = (2)\begin{bmatrix}2\\-4\\2\\1\end{bmatrix} + (0)\begin{bmatrix}1\\0\\0\\0\end{bmatrix} + (0)\begin{bmatrix}0\\0\\1\\-1\end{bmatrix} + (0)\begin{bmatrix}0\\1\\0\\0\end{bmatrix} \leftrightarrow \begin{bmatrix}2\\0\\0\\0\end{bmatrix}_{\mathbb{B}}$$

$$T\begin{bmatrix}1\\0\\0\\0\end{bmatrix} = \begin{bmatrix}4\\-4\\2\\2\end{bmatrix} = (1)\begin{bmatrix}2\\-4\\2\\1\end{bmatrix} + (2)\begin{bmatrix}1\\0\\0\\0\end{bmatrix} + (0)\begin{bmatrix}0\\0\\1\\-1\end{bmatrix} + (0)\begin{bmatrix}0\\1\\0\\0\end{bmatrix} \leftrightarrow \begin{bmatrix}1\\2\\0\\0\end{bmatrix}_{\mathbb{B}}$$

$$T\begin{bmatrix}1\\0\\0\\0\end{bmatrix} = \begin{bmatrix}0\\0\\2\\-2\end{bmatrix} = (0)\begin{bmatrix}2\\-4\\2\\2\end{bmatrix} + (0)\begin{bmatrix}1\\0\\0\\0\end{bmatrix} + (2)\begin{bmatrix}0\\0\\1\\-1\end{bmatrix} + (0)\begin{bmatrix}0\\1\\0\\0\end{bmatrix} \leftrightarrow \begin{bmatrix}0\\0\\2\\0\end{bmatrix}_{\mathbb{B}}$$

$$T\begin{bmatrix}0\\1\\0\\0\end{bmatrix} = \begin{bmatrix}0\\2\\1\\-1\end{bmatrix} = (0)\begin{bmatrix}2\\-4\\2\\2\end{bmatrix} + (0)\begin{bmatrix}1\\0\\0\\0\end{bmatrix} + (1)\begin{bmatrix}0\\0\\1\\-1\end{bmatrix} + (2)\begin{bmatrix}0\\1\\0\\0\end{bmatrix} \leftrightarrow \begin{bmatrix}0\\0\\1\\2\end{bmatrix}_{\mathbb{B}}$$

The matrix representation of T with respect to the $\mathbb{B}$ basis is

$$\mathbf{A} = \begin{bmatrix}2 & 1 & 0 & 0\\0 & 2 & 0 & 0\\0 & 0 & 2 & 1\\0 & 0 & 0 & 2\end{bmatrix}$$

A 1×1 Jordan block has only a single diagonal element. Therefore, a diagonal matrix is a matrix in Jordan canonical form in which every diagonal block is a 1×1 Jordan block.

In Example 9, we wrote the domain $\mathbb{R}^4$ of a linear transformation as the direct sum of two invariant subspaces, with each subspace having a single chain as a basis. Perhaps it is possible to always write the domain of a linear transformation $T\colon \mathbb{V} \to \mathbb{V}$ as the direct sum of a finite number of subspaces, say $\mathbb{V} = \mathbb{U}_1 \oplus \mathbb{U}_2 \oplus \cdots \oplus \mathbb{U}_p$, where each subspace is invariant under T, and each subspace has as a basis a single chain of generalized eigenvectors for T. If so, we could produce a matrix representation of T that is in Jordan canonical form.

When finding eigenvalues and eigenvectors, we generally work with matrix representations of linear transformations rather than with the linear transformations *per se* because it is easier to do so. Either we begin with a matrix or we construct a matrix representation for a given linear transformation, generally a matrix with respect to a standard basis as we did with the matrix $\mathbf{G}$ in Example 9.

A generalized eigenvector $\mathbf{x}_m$ of rank m corresponding to an eigenvalue λ of an $n \times n$ matrix $\mathbf{A}$ has the property that

$$(\mathbf{A} - \lambda\mathbf{I})^m \mathbf{x}_m = 0 \quad \text{and} \quad (\mathbf{A} - \lambda\mathbf{I})^{m-1} \mathbf{x}_m \neq \mathbf{0} \qquad \text{(A.6 repeated)}$$

Thus, $\mathbf{x}_m$ is in the kernel of $(\mathbf{A}-\lambda\mathbf{I})^m$ but not in the kernel of $(\mathbf{A}-\lambda\mathbf{I})^{m-1}$. Clearly, if $\mathbf{x} \in ker[(\mathbf{A}-\lambda\mathbf{I})^{m-1}]$, then $\mathbf{x} \in ker[(\mathbf{A}-\lambda\mathbf{I})^m]$. Consequently, the dimension of $ker[(\mathbf{A}-\lambda\mathbf{I})^{m-1}] < ker[(\mathbf{A}-\lambda\mathbf{I})^m]$ or, in terms of rank (see Corollary 1 of Section 3.5),

$$r\left[(\mathbf{A} - \lambda\mathbf{I})^{m-1}\right] > r[(\mathbf{A} - \lambda\mathbf{I})^m] \qquad \text{(A.16)}$$

The converse is also true. If (A.16) is valid, then there must exist a vector $\mathbf{x}_m$ that satisfies (A.6), in which case $\mathbf{x}_m$ is a generalized eigenvector of type m corresponding to $\mathbf{A}$ and λ. The difference

$$\rho_m = r\left[(\mathbf{A} - \lambda\mathbf{I})^{m-1}\right] - r[(\mathbf{A} - \lambda\mathbf{I})^m] \qquad \text{(A.17)}$$

is the number of linearly independent generalized eigenvectors of type m corresponding to $\mathbf{A}$ and its eigenvalue λ. The differences p_m, $m=1, 2, \ldots$ are called *index numbers*.

Example 10 The matrix

$$\mathbf{A} = \begin{bmatrix} 2 & 1 & -1 & 0 & 0 & 0 \\ 0 & 2 & 1 & 0 & 0 & 0 \\ 0 & 0 & 2 & 0 & 0 & 0 \\ 0 & 0 & 0 & 2 & 1 & 0 \\ 0 & 0 & 0 & 0 & 2 & 1 \\ 0 & 0 & 0 & 0 & 0 & 4 \end{bmatrix}$$

has an eigenvalue 4 of multiplicity 1 and an eigenvalue 2 of multiplicity 5. Here,

$$\mathbf{A} - 2\mathbf{I} = \begin{bmatrix} 0 & 1 & -1 & 0 & 0 & 0 \\ 0 & 0 & 1 & 0 & 0 & 0 \\ 0 & 0 & 0 & 0 & 0 & 0 \\ 0 & 0 & 0 & 0 & 1 & 0 \\ 0 & 0 & 0 & 0 & 0 & 1 \\ 0 & 0 & 0 & 0 & 0 & 2 \end{bmatrix}$$

has rank 4.

$$(\mathbf{A} - 2\mathbf{I})^2 = \begin{bmatrix} 0 & 0 & 1 & 0 & 0 & 0 \\ 0 & 0 & 0 & 0 & 0 & 0 \\ 0 & 0 & 0 & 0 & 0 & 0 \\ 0 & 0 & 0 & 0 & 0 & 1 \\ 0 & 0 & 0 & 0 & 0 & 2 \\ 0 & 0 & 0 & 0 & 0 & 4 \end{bmatrix}$$

has rank 2.

$$(\mathbf{A}-2\mathbf{I})^3 = \begin{bmatrix} 0 & 0 & 0 & 0 & 0 & 0 \\ 0 & 0 & 0 & 0 & 0 & 0 \\ 0 & 0 & 0 & 0 & 0 & 0 \\ 0 & 0 & 0 & 0 & 0 & 2 \\ 0 & 0 & 0 & 0 & 0 & 4 \\ 0 & 0 & 0 & 0 & 0 & 8 \end{bmatrix}$$

has rank 1.

$$(\mathbf{A}-2\mathbf{I})^4 = \begin{bmatrix} 0 & 0 & 0 & 0 & 0 & 0 \\ 0 & 0 & 0 & 0 & 0 & 0 \\ 0 & 0 & 0 & 0 & 0 & 0 \\ 0 & 0 & 0 & 0 & 0 & 4 \\ 0 & 0 & 0 & 0 & 0 & 8 \\ 0 & 0 & 0 & 0 & 0 & 16 \end{bmatrix}$$

also has rank 1. Therefore, we have the index numbers

$$\rho_1 = r\left[(\mathbf{A}-2\mathbf{I})^0\right] - r\left[(\mathbf{A}-2\mathbf{I})^1\right] = r(\mathbf{I}) - 4 = 6 - 4 = 2$$

$$\rho_2 = r\left[(\mathbf{A}-2\mathbf{I})^1\right] - r\left[(\mathbf{A}-2\mathbf{I})^2\right] = 4 - 2 = 2$$

$$\rho_3 = r\left[(\mathbf{A}-2\mathbf{I})^2\right] - r\left[(\mathbf{A}-2\mathbf{I})^3\right] = 2 - 1 = 1$$

$$\rho_4 = r\left[(\mathbf{A}-2\mathbf{I})^3\right] - r\left[(\mathbf{A}-2\mathbf{I})^4\right] = 1 - 1 = 0$$

Corresponding to $\lambda = 2$, **A** has two linearly independent generalized eigenvectors of type 1 (which are eigenvectors), two linearly independent generalized eigenvectors of type 2, one linearly independent generalized eigenvector of type 3, and no generalized eigenvectors of type 4. There are also no generalized eigenvectors of type greater than 4 because if one existed we could create a chain from it and produce a generalized eigenvector of type 4. The eigenvalue 4 has multiplicity 1 and only one linearly independent eigenvector associated with it.

Example 11 The matrix

$$\mathbf{G} = \begin{bmatrix} 4 & 0 & -1 & -1 \\ -4 & 2 & 2 & 2 \\ 2 & 1 & 2 & 0 \\ 2 & -1 & -2 & 0 \end{bmatrix}$$

has an eigenvalue 2 of multiplicity 4. Here,

$$\mathbf{G}-2\mathbf{I} = \begin{bmatrix} 2 & 0 & -1 & -1 \\ -4 & 0 & 2 & 2 \\ 2 & 1 & 0 & 0 \\ 2 & -1 & -2 & -2 \end{bmatrix}$$

has rank 2.

$$(\mathbf{G}-2\mathbf{I})^2 = \begin{bmatrix} 0 & 0 & 0 & 0 \\ 0 & 0 & 0 & 0 \\ 0 & 0 & 0 & 0 \\ 0 & 0 & 0 & 0 \end{bmatrix}$$

has rank 0, as will every power of $\mathbf{G}-2\mathbf{I}$ greater than 2. The associated index numbers are

$$\rho_1 = r\left[(\mathbf{G}-2\mathbf{I})^0\right] - r\left[(\mathbf{G}-2\mathbf{I})^1\right] = r(\mathbf{I}) - 2 = 4 - 2 = 2$$

$$\rho_2 = r\left[(\mathbf{G}-2\mathbf{I})^1\right] - r\left[(\mathbf{G}-2\mathbf{I})^2\right] = 2 - 0 = 2$$

$$\rho_3 = r\left[(\mathbf{G}-2\mathbf{I})^2\right] - r\left[(\mathbf{G}-2\mathbf{I})^3\right] = 0 - 0 = 0$$

Corresponding to $\lambda = 2$, $\mathbf{G}$ has two linearly independent generalized eigenvectors of type 1 (eigenvectors) and two linearly independent generalized eigenvectors of type 2.

Once we have a generalized eigenvector $\mathbf{x}_m$ of type m, we can identify a sequence of generalized eigenvectors of decreasing types by constructing the chain propagated by $\mathbf{x}_m$. An $n \times n$ matrix $\mathbf{A}$ may not have enough linearly independent eigenvectors to constitute a basis for $\mathbb{R}^n$, but $\mathbf{A}$ will always have n linearly independent generalized eigenvectors that can serve as a basis. If these generalized eigenvectors are chains, then they form invariant subspaces.

We define a *canonical basis* for an $n \times n$ matrix to be a set of n linearly independent generalized eigenvectors composed entirely of chains. Therefore, once we have determined that a generalized eigenvector $\mathbf{x}_m$ of type m is part of a canonical basis, then so too are the vectors $\mathbf{x}_{m-1}, \mathbf{x}_{m-2}, \ldots, \mathbf{x}_1$ that are in the chain propagated by $\mathbf{x}_m$. The following result, the proof of which is beyond the scope of this book, summarizes the relevant theory.

A canonical basis for an $n \times n$ matrix is a set of n linearly independent generalized eigenvectors composed entirely of chains.

▶THEOREM 7

Every $n \times n$ matrix possesses a canonical basis in $\mathbb{R}^n$. ◀

In terms of a linear transformation $T: \mathbb{V} \to \mathbb{V}$, where $\mathbb{V}$ is an n-dimensional vector space, Theorem 1 states that $\mathbb{V}$ has a basis consisting entirely of chains of generalized eigenvectors of T. With respect to such a basis, a matrix representation of T will be in Jordan canonical form. This is as simple a matrix representation as we can get for any linear transformation. The trick is to identify a canonical basis. It is one thing to know such a basis exists, and it is another matter entirely to find it.

If $\mathbf{x}_m$ is a generalized eigenvector of type m corresponding to the eigenvalue λ for the matrix $\mathbf{A}$, then

$$(\mathbf{A}-\lambda\mathbf{I})^m \mathbf{x}_m = \mathbf{0} \quad \text{and} \quad (\mathbf{A}-\lambda\mathbf{I})^{m-1} \mathbf{x}_m \neq \mathbf{0} \qquad \text{(A.6 repeated)}$$

This means that $\mathbf{x}_m$ is in the kernel of $(\mathbf{A}-\lambda\mathbf{I})^m$ *and* in the range of $(\mathbf{A}-\lambda\mathbf{I})^{m-1}$. If we find a basis for the range of $(\mathbf{A}-\lambda\mathbf{I})^{m-1}$ composed only of vectors that are also in the kernel of $(\mathbf{A}-\lambda\mathbf{I})^m$, we will then have a maximal set of linearly independent generalized eigenvectors of type m. This number will equal the index number p_m. Let us momentarily assume that $p_m=r$, and let us designate these generalized eigenvectors of type m as $\mathbf{v}_1, \mathbf{v}_2, \ldots, \mathbf{v}_r$. These r vectors are linearly independent vectors in the range of $(\mathbf{A}-\lambda\mathbf{I})^{m-1}$, so the only constants that satisfy the equation

$$c_1(\mathbf{A}-\lambda\mathbf{I})^{m-1}\mathbf{v}_1 + c_2(\mathbf{A}-\lambda\mathbf{I})^{m-1}\mathbf{v}_2 + \cdots + c_r(\mathbf{A}-\lambda\mathbf{I})^{m-1}\mathbf{v}_r = 0 \qquad \text{(A.18)}$$

are $c_1=c_2=\cdots=c_r=0$. It follows that $\{\mathbf{v}_1, \mathbf{v}_2, \ldots, \mathbf{v}_r\}$ is a linearly independent set, because if we multiply the equation

$$c_1\mathbf{v}_1 + c_2\mathbf{v}_2 + \ldots + c_n\mathbf{v}_r = \mathbf{0}$$

by $(\mathbf{A}-\lambda\mathbf{I})^{m-1}$, we obtain (A.18) and conclude that $c_1=c_2=\ldots=c_r=0$. It also follows that the set $\{(\mathbf{A}-\lambda\mathbf{I})\mathbf{v}_1, (\mathbf{A}-\lambda\mathbf{I})\mathbf{v}_2, \ldots, (\mathbf{A}-\lambda\mathbf{I})\mathbf{v}_r\}$ of generalized eigenvectors of type $m-1$ is also linearly independent, because if we multiply the equation

$$c_1(\mathbf{A}-\lambda\mathbf{I})\mathbf{v}_1 + c_2(\mathbf{A}-\lambda\mathbf{I})\mathbf{v}_2 + \cdots + c_n(\mathbf{A}-\lambda\mathbf{I})\mathbf{v}_r = \mathbf{0}$$

by $(\mathbf{A}-\lambda\mathbf{I})^{m-2}$, we again obtain (A.18) and conclude that $c_1=c_2=\ldots=c_r=0$. Thus, we have proved Theorem 8.

►THEOREM 8

If $\mathbb{S}=\{\mathbf{v}_1, \mathbf{v}_2, \ldots, \mathbf{v}_r\}$ is a set of generalized eigenvectors of type m such that $\{(\mathbf{A}-\lambda\mathbf{I})^{m-1}\mathbf{v}_1, (\mathbf{A}-\lambda\mathbf{I})^{m-1}\mathbf{v}_2, \ldots, (\mathbf{A}-\lambda\mathbf{I})^{m-1}\mathbf{v}_r\}$ is a linearly independent set, then $\mathbb{S}$ itself is a linearly independent set as is the set $\{(\mathbf{A}-\lambda\mathbf{I})\mathbf{v}_1, (\mathbf{A}-\lambda\mathbf{I})\mathbf{v}_2, \ldots, (\mathbf{A}-\lambda\mathbf{I})\mathbf{v}_r\}$ of generalized eigenvectors of type $m-1$. ◄

Example 12 The linear transformation T: $\mathbb{R}^6 \to \mathbb{R}^6$ defined by

$$T = \begin{bmatrix} a \\ b \\ c \\ d \\ e \\ f \end{bmatrix} = \begin{bmatrix} 5a+b+c \\ 5b+c \\ 5c \\ 5d+e-f \\ 5e+f \\ 5f \end{bmatrix}$$

has as its matrix representation with respect to the standard basis

$$\mathbf{A} = \begin{bmatrix} 5 & 1 & 1 & 0 & 0 & 0 \\ 0 & 5 & 1 & 0 & 0 & 0 \\ 0 & 0 & 5 & 0 & 0 & 0 \\ 0 & 0 & 0 & 5 & 1 & -1 \\ 0 & 0 & 0 & 0 & 5 & 1 \\ 0 & 0 & 0 & 0 & 0 & 5 \end{bmatrix}$$

This matrix (as well as T) has one eigenvalue 5 of multiplicity 6. Here,

$$\mathbf{A}-5\mathbf{I}=\begin{bmatrix}0&1&1&0&0&0\\0&0&1&0&0&0\\0&0&0&0&0&0\\0&0&0&0&1&-1\\0&0&0&0&0&1\\0&0&0&0&0&0\end{bmatrix}$$

has rank 4,

$$(\mathbf{A}-5\mathbf{I})^2=\begin{bmatrix}0&0&1&0&0&0\\0&0&0&0&0&0\\0&0&0&0&0&0\\0&0&0&0&0&1\\0&0&0&0&0&0\\0&0&0&0&0&0\end{bmatrix}$$

has rank 2, and all higher powers equal the zero matrix with rank 0. The index numbers are

$$\rho_1=r\left[(\mathbf{A}-5\mathbf{I})^0\right]-r\left[(\mathbf{A}-5\mathbf{I})^1\right]=r(\mathbf{I})-4=6-4=2$$

$$\rho_2=r\left[(\mathbf{A}-5\mathbf{I})^1\right]-r\left[(\mathbf{A}-5\mathbf{I})^2\right]=4-2=2$$

$$\rho_3=r\left[(\mathbf{A}-5\mathbf{I})^2\right]-r\left[(\mathbf{A}-5\mathbf{I})^3\right]=2-0=2$$

$$\rho_4=r\left[(\mathbf{A}-5\mathbf{I})^3\right]-r\left[(\mathbf{A}-5\mathbf{I})^4\right]=0-0=0$$

A has two generalized eigenvectors of type 3, two generalized eigenvectors of type 2, and two generalized eigenvectors of type 1. Generalized eigenvectors of type 3 must satisfy the two conditions $(\mathbf{A}-5\mathbf{I})^3\mathbf{x}=\mathbf{0}$ and $(\mathbf{A}-5\mathbf{I})^2\mathbf{x}\neq\mathbf{0}$. Here, $(\mathbf{A}-5\mathbf{I})^3=\mathbf{0}$, so the first condition places no restrictions on **x**. If we let $\mathbf{x}=[a\quad b\quad c\quad d\quad e\quad f]^{\mathrm{T}}$, then

$$(\mathbf{A}-5\mathbf{I})^2\mathbf{x}=\begin{bmatrix}c\\0\\0\\f\\0\\0\end{bmatrix}$$

and, this will be 0 if either c or f is nonzero. If we first take $c=1$ with $a=b=d=e=f=0$ and then take $f=1$ with $a=b=c=d=e=0$, we generate

$$\mathbf{x}_3=\begin{bmatrix}a\\b\\c\\d\\e\\f\end{bmatrix}=\begin{bmatrix}0\\0\\1\\0\\0\\0\end{bmatrix}\quad\text{and}\quad\mathbf{y}_3=\begin{bmatrix}a\\b\\c\\d\\e\\f\end{bmatrix}=\begin{bmatrix}0\\0\\0\\0\\0\\1\end{bmatrix}$$

as two generalized eigenvectors of type 3. It is important to note that $\mathbf{x}_3$ and $\mathbf{y}_3$ were *not* chosen to be linearly independent; they were chosen so that

$$\mathbf{x}_1 = (\mathbf{A} - 5\mathbf{I})^2\mathbf{x}_3 = \begin{bmatrix} 1 \\ 0 \\ 0 \\ 0 \\ 0 \\ 0 \end{bmatrix} \quad \text{and} \quad \mathbf{y}_1 = (\mathbf{A} - 5\mathbf{I})^2\mathbf{y}_3 = \begin{bmatrix} 0 \\ 0 \\ 0 \\ 1 \\ 0 \\ 0 \end{bmatrix}$$

are linearly independent. It follows from Theorem 2 that $\mathbf{x}_3$ and $\mathbf{y}_3$ are linearly independent, as are

$$\mathbf{x}_2 = (\mathbf{A} - 5\mathbf{I})^2\mathbf{x}_3 = \begin{bmatrix} 1 \\ 1 \\ 0 \\ 0 \\ 0 \\ 0 \end{bmatrix} \quad \text{and} \quad \mathbf{y}_2 = (\mathbf{A} - 5\mathbf{I})\mathbf{y}_3 = \begin{bmatrix} 0 \\ 0 \\ 0 \\ -1 \\ 1 \\ 0 \end{bmatrix}$$

The vectors $\mathbf{x}_1, \mathbf{x}_2, \mathbf{x}_3$ form a chain as do the vectors $\mathbf{y}_1, \mathbf{y}_2, \mathbf{y}_3$. A canonical basis is $\{\mathbf{x}_1, \mathbf{x}_2, \mathbf{x}_3, \mathbf{y}_1, \mathbf{y}_2, \mathbf{y}_3\}$, and with respect to this basis a matrix representation of T is in the Jordan canonical form

$$\mathbf{J} = \begin{bmatrix} 5 & 1 & 0 & 0 & 0 & 0 \\ 0 & 5 & 1 & 0 & 0 & 0 \\ 0 & 0 & 5 & 0 & 0 & 0 \\ 0 & 0 & 0 & 5 & 1 & 0 \\ 0 & 0 & 0 & 0 & 5 & 1 \\ 0 & 0 & 0 & 0 & 0 & 5 \end{bmatrix}$$

Theorem 8 provides the foundation for obtaining canonical bases. We begin with a set of index numbers for an eigenvalue. Let m denotes the highest type of generalized eigenvector. We first find a set of generalized eigenvectors of type m, $\{\mathbf{v}_1, \mathbf{v}_2, \ldots, \mathbf{v}_r\}$, such that $\{(\mathbf{A}-\lambda\mathbf{I})^{m-1}\mathbf{v}_1, (\mathbf{A}-\lambda\mathbf{I})^{m-1}\mathbf{v}_2, \ldots, (\mathbf{A}-\lambda\mathbf{I})^{m}\mathbf{v}_r\}$ is a basis for the range of $(\mathbf{A}-\lambda\mathbf{I})^{m-1}$. The vectors $\{\mathbf{w}_1 = (\mathbf{A}-\lambda\mathbf{I})\mathbf{v}_1, \mathbf{w}_2 = (\mathbf{A}-\lambda\mathbf{I})\mathbf{v}_2, \ldots, \mathbf{w}_r = (\mathbf{A}-\lambda\mathbf{I})\mathbf{v}_r\}$ are a linearly independent set of generalized eigenvectors of type $m-1$. If more generalized eigenvectors of type $m-1$ are needed, we find them. That is, if $p_{m-1} = s > r$, then we find $s-r$ additional generalized eigenvectors, $\mathbf{w}_{r+1}, \mathbf{w}_{r+2}, \ldots, \mathbf{w}_s$, such that

$$\{(\mathbf{A} - \lambda\mathbf{I})^{m-2}\mathbf{w}_1, (\mathbf{A} - \lambda\mathbf{I})^{m-2}, \mathbf{w}_2, \ldots (\mathbf{A} - \lambda\mathbf{I})^{m-2}\mathbf{w}_r,$$
$$(\mathbf{A} - \lambda\mathbf{I})^{m-2}\mathbf{w}_{r+1}, \ldots, (\mathbf{A} - \lambda\mathbf{I})^{m-2}\mathbf{w}_s\}$$

is a basis for the range of $(\mathbf{A}-\lambda\mathbf{I})^{m-2}$. It follows from Theorem 8 that

$$\{(\mathbf{A} - \lambda\mathbf{I})\mathbf{w}_1, (\mathbf{A} - \lambda\mathbf{I})\mathbf{w}_2, \ldots, (\mathbf{A} - \lambda\mathbf{I})\mathbf{w}_r, (\mathbf{A} - \lambda\mathbf{I})\mathbf{w}_{r+1}, \ldots, (\mathbf{A} - \lambda\mathbf{I})\mathbf{w}_s\}$$

is a linearly independent set of generalized eigenvectors of type $m-2$. Now the process is repeated sequentially, in decreasing order, through all types of generalized eigenvectors.

TO CREATE A CANONICAL BASIS

For each distinct eigenvalue of a matrix **A**, do the following:

Step 1. Using the index numbers, determine the number of linearly independent generalized eigenvectors of highest type, say type m, corresponding to λ. Determine one such set, $\{\mathbf{v}_1, \mathbf{v}_2, \ldots, \mathbf{v}_r\}$, so that the product of each of these vectors with $(\mathbf{A}-\lambda\mathbf{I})^{m-1}$ forms a basis for the range of $(\mathbf{A}-\lambda\mathbf{I})^{m-1}$. Call the set of **v** vectors the *current set*.

Step 2. If $m=1$, stop; otherwise continue.

Step 3. For *each* vector **v** in the current set of vectors, calculate $(\mathbf{A}-\lambda\mathbf{I})\mathbf{v}$, the next vector in its chain.

Step 4. Using the index numbers, determine the number of linearly independent generalized eigenvectors of the type $m-1$. If this number coincides with the number of vectors obtained in Step 3, call this new set of vectors the *current set* and go to Step 6; otherwise continue.

Step 5. Find additional generalized eigenvectors of type $m-1$ so that when these new vectors are adjoined to the current set, the product of each vector in the newly expanded set with $(\mathbf{A}-\lambda\mathbf{I})^{m-2}$ forms a basis for the range of $(\mathbf{A}-\lambda\mathbf{I})^{m-2}$. Call this newly expanded set the *current set* of vectors.

Step 6. Decrement m by 1 and return to Step 2.

Example 13 Find a matrix representation in Jordan canonical form for the linear transformation $T: \mathbb{R}^6 \to \mathbb{R}^6$ defined by

$$T = \begin{bmatrix} a \\ b \\ c \\ d \\ e \\ f \end{bmatrix} = \begin{bmatrix} 2a+b-c \\ 2b+c \\ 2c \\ 2d+e \\ 2e+f \\ 4f \end{bmatrix}$$

Solution: The matrix representation of T with respect to the standard basis is the matrix **A** exhibited in Example 10. It follows from Example 10 that **A** has one eigenvalue 2 of multiplicity 5 and one eigenvalue 4 of multiplicity 1. Associated with the eigenvalue 2 are one generalized eigenvector of type 3, two generalized eigenvectors of type 2, and two generalized eigenvectors of type 1. A generalized eigenvector of type 3 is

$$\mathbf{x}_3 = \begin{bmatrix} 0 \\ 0 \\ 1 \\ 0 \\ 0 \\ 0 \end{bmatrix}$$

Then,

$$\mathbf{x}_2 = (\mathbf{A} - 2\mathbf{I})\mathbf{x}_3 \begin{bmatrix} -1 \\ 1 \\ 0 \\ 0 \\ 0 \\ 0 \end{bmatrix}$$

is a generalized eigenvector of type 2. We still need another generalized eigenvector of type 2, so we set $\mathbf{y}_2 = [a \quad b \quad c \quad d \quad e \quad f]^{\mathrm{T}}$, and choose the components so that $\mathbf{y}_2$ is in the kernel of $(\mathbf{A} - 2\mathbf{I})^2$ and also so that $(\mathbf{A} - 2\mathbf{I})\mathbf{y}_2$ and $(\mathbf{A} - 2\mathbf{I})\mathbf{x}_2$ constitute a basis for the range of $(\mathbf{A} - 2\mathbf{I})$. If $\mathbf{y}_2$ is to be in the kernel of $(\mathbf{A} - 2\mathbf{I})^2$, then $c = f = 0$. Furthermore,

$$(\mathbf{A} - 2\mathbf{I})\mathbf{y}_2 (\mathbf{A} - 2\mathbf{I}) \begin{bmatrix} a \\ b \\ 0 \\ d \\ e \\ 0 \end{bmatrix} = \begin{bmatrix} b \\ 0 \\ 0 \\ e \\ 0 \\ 0 \end{bmatrix}, (\mathbf{A} - 2\mathbf{I})\mathbf{x}_2 = \begin{bmatrix} 1 \\ 0 \\ 0 \\ 0 \\ 0 \\ 0 \end{bmatrix}$$

and $\mathbf{y}_2$ must be chosen so that these two vectors are linearly independent. A simple choice is $b = 0$ and $e = 1$. There are many choices for $\mathbf{y}_2 = [a \quad 0 \quad 0 \quad d \quad 1 \quad 0]^{\mathrm{T}}$, depending how a and d are selected. The simplest is to take $a = d = 0$, whereupon

$$\mathbf{y}_2 = \begin{bmatrix} 0 \\ 0 \\ 0 \\ 0 \\ 1 \\ 0 \end{bmatrix}$$

Next,

$$\mathbf{y}_1 = (\mathbf{A} - 2\mathbf{I})\mathbf{y}_2 = \begin{bmatrix} 0 \\ 0 \\ 0 \\ 1 \\ 0 \\ 0 \end{bmatrix} \quad \text{and} \quad \mathbf{x}_1 = (\mathbf{A} - 2\mathbf{I})\mathbf{x}_2 = \begin{bmatrix} 1 \\ 0 \\ 0 \\ 0 \\ 0 \\ 0 \end{bmatrix}$$

are the required generalized eigenvectors of type 1. There is only one linearly independent generalized eigenvector associated with the eigenvalue 4. A suitable candidate is

$$\mathbf{z}_1 = \begin{bmatrix} 0 \\ 0 \\ 0 \\ 1 \\ 2 \\ 4 \end{bmatrix}$$

We take our canonical basis to be $\{\mathbf{z}_1, \mathbf{y}_1, \mathbf{y}_2, \mathbf{x}_1, \mathbf{x}_2, \mathbf{x}_3\}$. With respect to this basis, T is represented by the matrix in Jordan canonical form

$$\mathbf{J} = \begin{bmatrix} 4 & 0 & 0 & 0 & 0 & 0 \\ 0 & 2 & 1 & 0 & 0 & 0 \\ 0 & 0 & 2 & 0 & 0 & 0 \\ 0 & 0 & 0 & 2 & 1 & 0 \\ 0 & 0 & 0 & 0 & 2 & 1 \\ 0 & 0 & 0 & 0 & 0 & 2 \end{bmatrix}$$

The Jordan canonical form found in Example 13 contained a 1×1 Jordan block with the eigenvalue 4 on the main diagonal, a 2×2 Jordan block with the eigenvalue 2 on the main diagonal, and a 3×3 Jordan block again with the eigenvalue 2 on the main diagonal. The 1×1 Jordan block corresponds to the single element chain $\mathbf{z}_1$ in the canonical basis, the 2×2 Jordan block corresponds to the two element chain $\mathbf{y}_1$, $\mathbf{y}_2$ in the canonical basis, while the 3×3 Jordan block corresponds to the three element chain in the canonical basis. If we rearrange the ordering of the chains in the canonical basis, then the Jordan blocks in the Jordan canonical form will be rearranged in a corresponding manner. In particular, if we take the canonical basis to be $\{\mathbf{x}_1, \mathbf{x}_2, \mathbf{x}_3, \mathbf{y}_1, \mathbf{y}_2, \mathbf{z}_1\}$, then the corresponding Jordan canonical form becomes

$$\mathbf{J} = \begin{bmatrix} 2 & 1 & 0 & 0 & 0 & 0 \\ 0 & 2 & 1 & 0 & 0 & 0 \\ 0 & 0 & 2 & 0 & 0 & 0 \\ 0 & 0 & 0 & 2 & 1 & 0 \\ 0 & 0 & 0 & 0 & 2 & 0 \\ 0 & 0 & 0 & 0 & 0 & 4 \end{bmatrix}$$

If, instead, we take the ordering of the canonical basis to be $\{\mathbf{x}_1, \mathbf{x}_2, \mathbf{x}_3, \mathbf{z}_1, \mathbf{y}_1, \mathbf{y}_2\}$, then the corresponding Jordan canonical form becomes

$$\mathbf{J} = \begin{bmatrix} 2 & 1 & 1 & 0 & 0 & 0 \\ 0 & 2 & 1 & 0 & 0 & 0 \\ 0 & 0 & 2 & 0 & 0 & 0 \\ 0 & 0 & 0 & 4 & 0 & 0 \\ 0 & 0 & 0 & 0 & 0 & 1 \\ 0 & 0 & 0 & 0 & 0 & 2 \end{bmatrix}$$

In a canonical basis, all vectors from the same chain are grouped together, and generalized eigenvectors in each chain are ordered by increasing type.

Two criteria must be observed if a canonical basis is to generate a matrix in Jordan canonical form. First, all vectors in the same chain must be grouped together (not separated by vectors from other chains), and second, each chain must be ordered by increasing type (so that the generalized eigenvector of type 1 appears before the generalized eigenvector of type 2 of the same chain, which appears before the generalized eigenvector of type 3 of the same chain, and so on). If either criterion is violated, then the ones will not appear, in general, on the superdiagonal. In particular, if vectors are ordered by decreasing type, then all the ones appear on the *subdiagonal*, the diagonal just below the main diagonal.

Let **A** denote a matrix representation of a linear transformation $T: \mathbb{V} \to \mathbb{V}$ with respect to a basis $\mathbb{B}$ (perhaps the standard basis), and let **J** be a matrix representation in Jordan canonical form for T. **J** is the matrix representation with respect to a canonical basis $\mathbb{C}$. Since **J** and **A** are two matrix representations of the same linear transformation, with respect to different basis, they must be similar. Using the notation developed in Section 3.4 we may write

$$\mathbf{J}_{\mathbb{C}}^{\mathbb{C}} = \left(\mathbf{P}_{\mathbb{C}}^{\mathbb{B}}\right)^{-1}\mathbf{A}_{\mathbb{B}}^{\mathbb{B}}\mathbf{P}_{\mathbb{C}}^{\mathbb{B}} \tag{A.19}$$

where $\mathbf{P}_{\mathbb{C}}^{\mathbb{B}}$ is the transition matrix from the $\mathbb{B}$ basis to the $\mathbb{C}$ basis.

Let $\{\mathbf{x}_1, \mathbf{x}_2, \ldots, \mathbf{x}_n\}$ be a canonical basis of generalized eigenvectors for **A**. A *generalized modal matrix* is a matrix **M** whose columns are the vectors in the canonical basis, that is,

$$\mathbf{M} = [\mathbf{x}_1\ \mathbf{x}_2 \cdots \mathbf{x}_n] \tag{A.20}$$

If $\mathbf{x}_{j+1}$ is a direct ancestor of $\mathbf{x}_y$, in the same chain corresponding to the eigenvalue λ, then

$$\mathbf{A}\mathbf{x}_{j+1} = \lambda\mathbf{x}_{j+1} + \mathbf{x}_j \qquad \text{(A.10 repeated)}$$

If $\mathbf{x}_1$ is an eigenvector corresponding to λ, then

$$\mathbf{A}\mathbf{x}_1 = \lambda\mathbf{x}_1 \qquad \text{(A.11 repeated)}$$

Using these relationships, it is a simple matter to show that $\mathbf{AM} = \mathbf{MJ}$. Since the columns of **M** are linearly independent, **M** has an inverse. Therefore,

$$\mathbf{J} = \mathbf{M}^{-1}\mathbf{A}\mathbf{M} \tag{A.21}$$

$$\mathbf{A} = \mathbf{M}\mathbf{J}\mathbf{M}^{-1} \tag{A.22}$$

Comparing (A.21) with (A.22), we see that the generalized modal matrix is just the transition matrix from the canonical basis $\mathbb{C}$ to the **B** basis. It then follows that $\mathbf{M}^{-1}$ is the transition matrix from the **B** basis to the $\mathbb{C}$ basis.

PROBLEMS APPENDIX A

(1) Let $L: \mathbb{R}^2 \to \mathbb{R}^2$ be defined by $T\begin{bmatrix} a \\ b \end{bmatrix} = \begin{bmatrix} a+2b \\ 4a+3b \end{bmatrix}$. Determine whether the subspaces spanned by the following sets of vectors are invariant subspaces of L.

(a) $\mathbb{A} = \left\{ \begin{bmatrix} 1 \\ 1 \end{bmatrix}, \begin{bmatrix} 1 \\ -1 \end{bmatrix} \right\}$,

(b) $\mathbb{B} = \left\{ \begin{bmatrix} 1 \\ 1 \end{bmatrix}, \begin{bmatrix} -1 \\ -1 \end{bmatrix} \right\}$,

(c) $\mathbb{C} = \left\{ \begin{bmatrix} 2 \\ 1 \end{bmatrix} \right\}$,

(d) $\mathbb{D} = \left\{ \begin{bmatrix} 1 \\ 2 \end{bmatrix} \right\}$,

(e) $\mathbb{E} = \left\{ \begin{bmatrix} 0 \\ 0 \end{bmatrix} \right\}$,

(f) $\mathbb{F} = \left\{ \begin{bmatrix} 0 \\ 0 \end{bmatrix}, \begin{bmatrix} 1 \\ -1 \end{bmatrix} \right\}$,

(2) Let T: $\mathbb{R}^3 \to \mathbb{R}^3$ be defined by $T\begin{bmatrix} a \\ b \\ c \end{bmatrix} = \begin{bmatrix} 4b + 2c \\ -3a + 8b + 3c \\ 4a - 8b - 2c \end{bmatrix}$. Determine whether the subspaces spanned by the following sets of vectors are invariant subspaces of T.

(a) $\mathbb{A} = \left\{ \begin{bmatrix} 2 \\ 1 \\ 0 \end{bmatrix}, \begin{bmatrix} 2 \\ 3 \\ -4 \end{bmatrix} \right\}$,

(b) $\mathbb{B} = \left\{ \begin{bmatrix} 0 \\ -2 \\ 4 \end{bmatrix}, \begin{bmatrix} 4 \\ 4 \\ -4 \end{bmatrix} \right\}$,

(c) $\mathbb{C} = \left\{ \begin{bmatrix} 2 \\ 1 \\ 0 \end{bmatrix}, \begin{bmatrix} 0 \\ 0 \\ 1 \end{bmatrix} \right\}$,

(d) $\mathbb{D} = \left\{ \begin{bmatrix} 0 \\ 0 \\ 1 \end{bmatrix}, \begin{bmatrix} 2 \\ 3 \\ -4 \end{bmatrix} \right\}$,

(e) $\mathbb{E} = \left\{ \begin{bmatrix} 0 \\ 0 \\ 1 \end{bmatrix} \right\}$,

(f) $\mathbb{F} = \left\{ \begin{bmatrix} 2 \\ 3 \\ -4 \end{bmatrix} \right\}$,

(3) Let $\mathbf{R}$: $\mathbb{R}^4 \to \mathbb{R}^4$ be defined by $T\begin{bmatrix} a \\ b \\ c \\ d \end{bmatrix} = \begin{bmatrix} 2a + b - d \\ 2b + c + d \\ 2c \\ 2d \end{bmatrix}$. Determine whether the subspaces spanned by the following sets of vectors are invariant subspaces of $\mathbf{R}$.

(a) $\mathbf{A} = \left\{ \begin{bmatrix} 1 \\ 0 \\ 0 \\ 0 \end{bmatrix}, \begin{bmatrix} -1 \\ 1 \\ 0 \\ 0 \end{bmatrix} \right\}$,

(b) $\mathbf{B} = \left\{ \begin{bmatrix} 1 \\ 0 \\ 0 \\ 0 \end{bmatrix}, \begin{bmatrix} 0 \\ 0 \\ 0 \\ 1 \end{bmatrix} \right\}$,

(c) $\mathbf{C} = \left\{ \begin{bmatrix} 1 \\ 0 \\ 0 \\ 0 \end{bmatrix}, \begin{bmatrix} 0 \\ -1 \\ 1 \\ -1 \end{bmatrix} \right\}$,

(d) $\mathbf{D} = \left\{ \begin{bmatrix} 0 \\ 0 \\ 0 \\ 1 \end{bmatrix}, \begin{bmatrix} 0 \\ -1 \\ 1 \\ -1 \end{bmatrix} \right\}$,

(e) $\mathbf{E} = \left\{ \begin{bmatrix} 1 \\ 0 \\ 0 \\ 0 \end{bmatrix}, \begin{bmatrix} -1 \\ 1 \\ 0 \\ 0 \end{bmatrix}, \begin{bmatrix} 0 \\ 0 \\ 0 \\ 1 \end{bmatrix} \right\}$,

(f) $\mathbf{F} = \left\{ \begin{bmatrix} 1 \\ 0 \\ 0 \\ 0 \end{bmatrix}, \begin{bmatrix} -1 \\ 1 \\ 0 \\ 0 \end{bmatrix}, \begin{bmatrix} 0 \\ -1 \\ 1 \\ -1 \end{bmatrix} \right\}$,

(4) Determine whether the subspaces spanned by the following sets of vectors are invariant subspaces of $\mathbf{A} = \begin{bmatrix} 3 & 1 \\ -1 & 5 \end{bmatrix}$.

(a) $\mathbb{A} = \left\{ \begin{bmatrix} 0 \\ 1 \end{bmatrix} \right\}$,

(b) $\mathbb{B} = \left\{ \begin{bmatrix} 1 \\ 1 \end{bmatrix} \right\}$,

(c) $\mathbb{C} = \left\{ \begin{bmatrix} 1 \\ 2 \end{bmatrix} \right\}$,

(d) $\mathbb{D} = \left\{ \begin{bmatrix} 1 \\ 1 \end{bmatrix}, \begin{bmatrix} 0 \\ 1 \end{bmatrix} \right\}$,

(e) $\mathbb{E} = \left\{ \begin{bmatrix} 1 \\ 1 \end{bmatrix}, \begin{bmatrix} 2 \\ 2 \end{bmatrix} \right\}$,

(f) $\mathbb{F} = \left\{ \begin{bmatrix} 1 \\ 1 \end{bmatrix}, \begin{bmatrix} 1 \\ 2 \end{bmatrix} \right\}$,

(5) Determine whether the subspaces spanned by the following sets of vectors are invariant subspaces of $\mathbf{A} = \begin{bmatrix} 5 & 1 & -1 \\ 0 & 5 & 2 \\ 0 & 0 & 5 \end{bmatrix}$.

(a) $\mathbb{A} = \left\{ \begin{bmatrix} 0 \\ 0 \\ 1 \end{bmatrix}, \begin{bmatrix} 2 \\ 3 \\ -4 \end{bmatrix} \right\}$,

(b) $\mathbb{B} = \left\{ \begin{bmatrix} 2 \\ 0 \\ 0 \end{bmatrix}, \begin{bmatrix} -1 \\ 2 \\ 0 \end{bmatrix} \right\}$,

(c) $\mathbb{C} = \left\{ \begin{bmatrix} 2 \\ 0 \\ 0 \end{bmatrix}, \begin{bmatrix} 0 \\ 0 \\ 1 \end{bmatrix} \right\}$,

(d) $\mathbb{D} = \left\{ \begin{bmatrix} 2 \\ 0 \\ 0 \end{bmatrix} \right\}$,

(e) $\mathbb{E} = \left\{ \begin{bmatrix} 0 \\ 0 \\ 1 \end{bmatrix} \right\}$,

(f) $\mathbb{F} = \left\{ \begin{bmatrix} -1 \\ 2 \\ 0 \end{bmatrix} \right\}$,

(6) Determine whether the subspaces spanned by the following sets of vectors are invariant subspaces of $\mathbf{A} = \begin{bmatrix} 3 & 1 & 0 & -1 \\ 0 & 3 & 1 & 0 \\ 0 & 0 & 4 & 1 \\ 0 & 0 & 0 & 4 \end{bmatrix}$.

(a) $\mathbb{A} = \left\{ \begin{bmatrix} 1 \\ 0 \\ 0 \\ 0 \end{bmatrix}, \begin{bmatrix} 0 \\ 1 \\ 0 \\ 0 \end{bmatrix} \right\}$,

(b) $\mathbb{B} = \left\{ \begin{bmatrix} 1 \\ 0 \\ 0 \\ 0 \end{bmatrix}, \begin{bmatrix} 1 \\ 1 \\ 1 \\ 0 \end{bmatrix} \right\}$,

(c) $\mathbb{C} = \left\{ \begin{bmatrix} 1 \\ 0 \\ 0 \\ 0 \end{bmatrix}, \begin{bmatrix} 0 \\ 0 \\ 1 \\ 1 \end{bmatrix} \right\}$,

(d) $\mathbb{D} = \left\{ \begin{bmatrix} 0 \\ 1 \\ 0 \\ 0 \end{bmatrix}, \begin{bmatrix} 1 \\ 1 \\ 1 \\ 0 \end{bmatrix} \right\}$,

(e) $\mathbb{E} = \left\{ \begin{bmatrix} 1 \\ 0 \\ 0 \\ 0 \end{bmatrix}, \begin{bmatrix} 1 \\ 1 \\ 1 \\ 0 \end{bmatrix}, \begin{bmatrix} 3 \\ 1 \\ 0 \\ -1 \end{bmatrix} \right\}$,

(f) $\mathbb{F} = \left\{ \begin{bmatrix} 1 \\ 0 \\ 0 \\ 0 \end{bmatrix}, \begin{bmatrix} 0 \\ 1 \\ 0 \\ 0 \end{bmatrix}, \begin{bmatrix} 1 \\ 1 \\ 1 \\ 0 \end{bmatrix} \right\}$,

(7) Using the information provided in Problem 1, determine which of the following statements are true:

(a) $\mathbb{R}^2 = span\{\mathbb{A}\} \oplus span\{\mathbb{B}\}$ (c) $\mathbb{R}^2 = span\{\mathbb{C}\} \oplus span\{\mathbb{D}\}$
(b) $\mathbb{R}^2 = span\{\mathbb{A}\} \oplus span\{\mathbb{C}\}$ (d) $\mathbb{R}^2 = span\{\mathbb{D}\} \oplus span\{\mathbb{E}\}$

(8) Using the information provided in Problem 2, determine which of the following statements are true:

(a) $\mathbb{R}^3 = span\{\mathbb{A}\} \oplus span\{\mathbb{B}\}$ (c) $\mathbb{R}^3 = span\{\mathbb{B}\} \oplus span\{\mathbb{F}\}$
(b) $\mathbb{R}^3 = span\{\mathbb{A}\} \oplus span\{\mathbb{E}\}$ (d) $\mathbb{R}^3 = span\{\mathbb{C}\} \oplus span\{\mathbb{F}\}$

(9) Using the information provided in Problem 4, determine which of the following statements are true:

(a) $\mathbb{R}^2 = span\{\mathbb{A}\} \oplus span\{\mathbb{B}\}$ (c) $\mathbb{R}^2 = span\{\mathbb{B}\} \oplus span\{\mathbb{C}\}$
(b) $\mathbb{R}^2 = span\{\mathbb{A}\} \oplus span\{\mathbb{C}\}$ (d) $\mathbb{R}^2 = span\{\mathbb{E}\} \oplus span\{\mathbb{F}\}$

(10) Using the information provided in Problem 5, determine which of the following statements are true:

(a) $\mathbb{R}^3 = span\{\mathbb{A}\} \oplus span\{\mathbb{D}\}$ (c) $\mathbb{R}^3 = span\{\mathbb{B}\} \oplus span\{\mathbb{E}\}$
(b) $\mathbb{R}^3 = span\{\mathbb{B}\} \oplus span\{\mathbb{D}\}$ (d) $\mathbb{R}^3 = span\{\mathbb{D}\} \oplus span\{\mathbb{E}\}$

(11) Characterize the subspace $\mathbb{U} = span\{\mathbb{D})\} \oplus span\{\mathbb{E}\}$ for the sets $\mathbb{D}$ and $\mathbb{E}$ described in Problem 5.

(12) Let $T : \mathbb{R}^4 \to \mathbb{R}^4$ be defined by $T\begin{bmatrix} a \\ b \\ c \\ d \end{bmatrix} = \begin{bmatrix} 3a+b-d \\ 3b+c \\ 4c+d \\ 4d \end{bmatrix}$. Set $\mathbb{B} = \left\{ \begin{bmatrix} 1 \\ 1 \\ 0 \\ 0 \end{bmatrix}, \begin{bmatrix} 1 \\ -1 \\ 0 \\ 0 \end{bmatrix} \right\}$,

$\mathbb{C} = \left\{ \begin{bmatrix} -1 \\ -1 \\ -1 \\ 0 \end{bmatrix}, \begin{bmatrix} 3 \\ 1 \\ 0 \\ -1 \end{bmatrix} \right\}$, $\mathbb{M} = span(\mathbb{B})$, and $\mathbb{N} = span(\mathbb{C})$.

(a) Show that $\mathbb{M}$ and $\mathbb{N}$ are both invariant subspaces of T with $\mathbb{R}^4 = \mathbb{M} \oplus \mathbb{N}$.
(b) Show that T has a matrix representation in the block diagonal form with respect to the basis $\mathbb{B} \cup \mathbb{C}$.

(13) Let $T : \mathbb{R}^4 \to \mathbb{R}^4$ be defined by $T\begin{bmatrix} a \\ b \\ c \\ d \end{bmatrix} = \begin{bmatrix} 2a+b-d \\ 2b+c+d \\ 2c \\ 2d \end{bmatrix}$. Set $\mathbb{B} = \left\{ \begin{bmatrix} 0 \\ 1 \\ -1 \\ 1 \end{bmatrix} \right\}$,

$\mathbb{C} = \left\{ \begin{bmatrix} 1 \\ 1 \\ 0 \\ 0 \end{bmatrix}, \begin{bmatrix} 1 \\ -1 \\ 0 \\ 0 \end{bmatrix}, \begin{bmatrix} 0 \\ 0 \\ 0 \\ 1 \end{bmatrix} \right\}$, $\mathbb{M} = span(\mathbb{B})$, and $\mathbb{N} = span(\mathbb{C})$.

(a) Show that $\mathbb{M}$ and $\mathbb{N}$ are both invariant subspaces of T with $\mathbb{R}^4 = \mathbb{M} \oplus \mathbb{N}$.
(b) Show that T has a matrix representation in the block diagonal form with respect to the basis $\mathbb{B} \cup \mathbb{C}$.

(14) Let $T: \mathbb{R}^4 \to \mathbb{R}^4$ be defined by $T\begin{bmatrix} a \\ b \\ c \\ d \end{bmatrix} = \begin{bmatrix} 4a+c \\ 2a+2b+3c \\ -a+2c \\ 4a+c+2d \end{bmatrix}$. Set $\mathbb{B} = \left\{ \begin{bmatrix} 0 \\ 1 \\ 0 \\ 1 \end{bmatrix}, \begin{bmatrix} 0 \\ 0 \\ 0 \\ 1 \end{bmatrix} \right\}$,

$\mathbb{C} = \left\{ \begin{bmatrix} 1 \\ -1 \\ -1 \\ 3 \end{bmatrix}, \begin{bmatrix} 1 \\ 3 \\ 0 \\ 1 \end{bmatrix} \right\}$, $\mathbb{M} = span(\mathbb{B})$, and $\mathbb{N} = span(\mathbb{C})$.

(a) Show that $\mathbb{M}$ and $\mathbb{N}$ are both invariant subspaces of T with $\mathbb{R}^4 = \mathbb{M} \oplus \mathbb{N}$.
(b) Show that T has a matrix representation in the block diagonal form with respect to the basis $\mathbb{B} \cup \mathbb{C}$.

(15) Determine whether the following vectors are generalized eigenvectors of type 3 corresponding to the eigenvalue $\lambda = 2$ for the matrix

$$\mathbf{A} = \begin{bmatrix} 2 & 2 & 1 & 1 \\ 0 & 2 & -1 & 0 \\ 0 & 0 & 2 & 0 \\ 0 & 0 & 0 & 1 \end{bmatrix}$$

(a) $\begin{bmatrix} 1 \\ 1 \\ 1 \\ 0 \end{bmatrix}$, (b) $\begin{bmatrix} 0 \\ 1 \\ 0 \\ 0 \end{bmatrix}$, (c) $\begin{bmatrix} 0 \\ 0 \\ 1 \\ 0 \end{bmatrix}$, (d) $\begin{bmatrix} 2 \\ 0 \\ 3 \\ 0 \end{bmatrix}$, (e) $\begin{bmatrix} 0 \\ 0 \\ 0 \\ 1 \end{bmatrix}$, (f) $\begin{bmatrix} 0 \\ 0 \\ 0 \\ 0 \end{bmatrix}$.

For the matrices in Problems 16 through 20, find a generalized eigenvector of type 2 corresponding the eigenvalue $\lambda = -1$.

(16) $\begin{bmatrix} -1 & 1 \\ 0 & -1 \end{bmatrix}$ **(17)** $\begin{bmatrix} -1 & 1 & 0 \\ 0 & -1 & 1 \\ 0 & 0 & 1 \end{bmatrix}$ **(18)** $\begin{bmatrix} 0 & 4 & 2 \\ -1 & 4 & 1 \\ -1 & -7 & -4 \end{bmatrix}$

(19) $\begin{bmatrix} 3 & -2 & 2 \\ 2 & -2 & 1 \\ -9 & 9 & -4 \end{bmatrix}$ **(20)** $\begin{bmatrix} 2 & 0 & 3 \\ 2 & -1 & 1 \\ -1 & 0 & -2 \end{bmatrix}$

(21) Find a generalized eigenvector of type 3 corresponding to $\lambda = 3$ and a generalized eigenvector of type 2 corresponding to $\lambda = 4$ for

$$\mathbf{A} = \begin{bmatrix} 4 & 1 & 0 & 0 & 1 \\ 0 & 4 & 0 & 0 & 0 \\ 0 & 0 & 3 & 1 & 0 \\ 0 & 0 & 0 & 3 & 2 \\ 0 & 0 & 0 & 0 & 3 \end{bmatrix}$$

(22) The vector $[1 \quad 1 \quad 1 \quad 0]^T$ is known to be a generalized eigenvector of type 3 corresponding to the eigenvalue 2 for

$$\mathbf{A} = \begin{bmatrix} 2 & 2 & 1 & 1 \\ 0 & 2 & -1 & 0 \\ 0 & 0 & 2 & 0 \\ 0 & 0 & 0 & 1 \end{bmatrix}$$

Construct a chain from this vector.

(23) Redo Problem 22 for the generalized eigenvector $[0 \quad 0 \quad 1 \quad 0]^T$, which is also of type 3 corresponding to the same eigenvalue and matrix.

(24) The vector $[0 \quad 0 \quad 0 \quad 0 \quad 1]^T$ is known to be a generalized eigenvector of type 4 corresponding to the eigenvalue 1 for

$$\mathbf{A} = \begin{bmatrix} 1 & 0 & 1 & 0 & -1 \\ 0 & 1 & 0 & 0 & 0 \\ 0 & 0 & 1 & -1 & 2 \\ 0 & 0 & 0 & 1 & 1 \\ 0 & 0 & 0 & 0 & 1 \end{bmatrix}$$

Construct a chain from this vector.

(25) Redo Problem 24 for the generalized eigenvector $[0 \quad 0 \quad 0 \quad 1 \quad 0]^T$, which is of type 3 corresponding to the same eigenvalue and matrix.

(26) The vector $[1 \quad 0 \quad 0 \quad 0 \quad -1]^T$ is known to be a generalized eigenvector of type 3 corresponding to the eigenvalue 3 for

$$\mathbf{A} = \begin{bmatrix} 4 & 1 & 0 & 0 & 1 \\ 0 & 4 & 0 & 0 & 0 \\ 0 & 0 & 3 & 1 & 0 \\ 0 & 0 & 0 & 3 & 2 \\ 0 & 0 & 0 & 0 & 3 \end{bmatrix}$$

Construct a chain from this vector.

(27) Redo Problem 26 for the generalized eigenvector $[0 \quad 1 \quad 0 \quad 0 \quad 0]^T$, which is of type 2 corresponding to the eigenvalue 4 for the same matrix.

(28) Find a generalized eigenvector of type 2 corresponding to the eigenvalue -1 for

$$\mathbf{A} = \begin{bmatrix} -1 & 1 \\ 0 & -1 \end{bmatrix}$$

and construct a chain from this vector.

(29) Find a generalized eigenvector of type 2 corresponding to the eigenvalue -1 for

$$\mathbf{A} = \begin{bmatrix} -1 & 1 & 0 \\ 0 & -1 & 1 \\ 0 & 0 & 1 \end{bmatrix}$$

and construct a chain from this vector.

(30) Find a generalized eigenvector of type 2 corresponding to the eigenvalue -1 for

$$\mathbf{A} = \begin{bmatrix} 0 & 4 & 2 \\ -1 & 4 & 1 \\ -1 & -7 & -4 \end{bmatrix}$$

and construct a chain from this vector.

(31) Find a generalized eigenvector of type 4 corresponding to the eigenvalue 2 for

$$\mathbf{A} = \begin{bmatrix} 2 & 1 & 3 & -1 \\ 0 & 2 & -1 & 4 \\ 0 & 0 & 2 & 1 \\ 0 & 0 & 0 & 2 \end{bmatrix}$$

and construct a chain from this vector.

(32) Find a generalized eigenvector of type 3 corresponding to the eigenvalue 3 for

$$\mathbf{A} = \begin{bmatrix} 4 & 1 & 1 & 2 & 2 \\ -1 & 2 & 1 & 3 & 0 \\ 0 & 0 & 3 & 0 & 0 \\ 0 & 0 & 0 & 2 & 1 \\ 0 & 0 & 0 & 1 & 2 \end{bmatrix}$$

and construct a chain from this vector.

(33) Prove that a generalized eigenvector of type 1 is an eigenvector.

(34) Prove that a generalized eigenvector of any type cannot be a zero vector.

(35) Let $T: \mathbb{V} \to \mathbb{V}$ be a linear transformation. Prove that the following sets are invariant subspaces under T.

(a) $\{\mathbf{0}\}$,

(b) $\mathbb{V}$,

(c) $span\{\mathbf{v}_1, \mathbf{v}_2, \ldots, \mathbf{v}_k\}$ where each vector is an eigenvector of T (not necessarily corresponding to the same eigenvalue).

(36) Let $\mathbb{V}$ be a finite-dimensional vector space. Prove that $\mathbb{V}$ is the direct sum of two subspaces $\mathbb{U}$ and $\mathbb{W}$ if and only if (i) each vector in $\mathbb{V}$ can be written as the sum of a vector in $\mathbb{U}$ with a vector in $\mathbb{W}$, and (ii) the only vector common to both $\mathbb{U}$ and $\mathbb{W}$ is the zero vector.

(37) Let $\mathbb{B}$ be a basis of k-vectors for $\mathbb{U}$, an invariant subspace of the linear transformation $T: \mathbb{V} \to \mathbb{V}$, and let $\mathbb{C}$ be a basis for $\mathbb{W}$, another subspace (but not invariant) with $\mathbb{V} = \mathbb{U} \oplus \mathbb{W}$. Show that the matrix representation of T with respect to the basis $\mathbb{B} \cup \mathbb{C}$ has the partitioned form

$$\mathbf{A} = \begin{bmatrix} \mathbf{A}_1 & \mathbf{A}_2 \\ 0 & \mathbf{A}_3 \end{bmatrix}$$

with $\mathbf{A}_1$ having order $k \times k$.

(38) Determine the length of the chains in a canonical basis if each chain is associated with the same eigenvalue λ and if a full set of index numbers is given by each of the following.

(a) $\rho_3 = \rho_2 = \rho_1 = 1$, (b) $\rho_3 = \rho_2 = \rho_1 = 2$,
(c) $\rho_3 = 1,\ \rho_2 = \rho_1 = 2$, (d) $\rho_3 = 1,\ \rho_2 = 2,\ \rho_1 = 3$,
(e) $\rho_3 = \rho_2 = 1,\ \rho_1 = 3$, (f) $\rho_3 = 3,\ \rho_2 = 4,\ \rho_1 = 3$,
(g) $\rho_2 = 2,\ \rho_1 = 4$, (h) $\rho_2 = 4,\ \rho_1 = 2$,
(i) $\rho_2 = 2,\ \rho_1 = 3$, (j) $\rho_2 = \rho_1 = 2$.

In Problems 39 through 45, find a canonical basis for the given matrices.

(39) $\begin{bmatrix} 3 & 1 \\ -1 & 1 \end{bmatrix}$.

(40) $\begin{bmatrix} 7 & 3 & 3 \\ 0 & 1 & 0 \\ -3 & -3 & 1 \end{bmatrix}$.

(41) $\begin{bmatrix} 5 & 1 & -1 \\ 0 & 5 & 2 \\ 0 & 0 & 5 \end{bmatrix}$.

(42) $\begin{bmatrix} 5 & 1 & 2 \\ 0 & 3 & 0 \\ 2 & 1 & 5 \end{bmatrix}$.

(43) $\begin{bmatrix} 2 & 1 & 0 & -1 \\ 0 & 2 & 1 & 1 \\ 0 & 0 & 2 & 0 \\ 0 & 0 & 0 & 2 \end{bmatrix}$.

(44) $\begin{bmatrix} 3 & 1 & 0 & -1 \\ 0 & 3 & 1 & 0 \\ 0 & 0 & 4 & 1 \\ 0 & 0 & 0 & 4 \end{bmatrix}$.

(45) $\begin{bmatrix} 4 & 1 & 1 & 0 & 0 & -1 \\ 0 & 4 & 2 & 0 & 0 & 1 \\ 0 & 0 & 4 & 1 & 0 & 1 \\ 0 & 0 & 0 & 5 & 1 & 0 \\ 0 & 0 & 0 & 0 & 5 & 2 \\ 0 & 0 & 0 & 0 & 0 & 4 \end{bmatrix}$

In Problems 46 through 50, a full set of index numbers are specified for the eigenvalue 2 of multiplicity 5 for a 5×5 matrix **A**. In each case, find a matrix in Jordan canonical form that is similar to **A**. Assume that a canonical basis is ordered so that chains of length 1 appear before chains of length 2, which appear before chains of length 3, and so on.

(46) $\rho_3 = \rho_2 = 1,\ \rho_1 = 3$.

(47) $\rho_3 = 1,\ \rho_2 = \rho_1 = 2$.

(48) $\rho_{2,} = 2,\ \rho_1 = 3$.

(49) $\rho_4 = \rho_3 = \rho_2 = 1,\ \rho_1 = 2$.

(50) $\rho_5 = \rho_4 = \rho_3 = \rho_2 = \rho_1 = 1$.

In Problems 51 through 56, a full set of index numbers are specified for the eigenvalue 3 of multiplicity 6 for a 6×6 matrix **A**. In each case, find a matrix in Jordan canonical form that is similar to **A**. Assume that a canonical basis is ordered so that chains of length 1 appear before chains of length 2, which appear before chains of length 3, and so on.

(51) $\rho_3 = \rho_2 = \rho_1 = 2$.

(52) $\rho_3 = 1,\ \rho_2 = 2,\ \rho_1 = 3$.

(53) $\rho_3 = \rho_2 = 1,\ \rho_1 = 4$.

(54) $\rho_2 = \rho_1 = 3$.

(55) $\rho_2 = 2,\ \rho_1 = 4$.

(56) $\rho_2 = 1,\ \rho_1 = 5$.

(57) A canonical basis for a linear transformation $T\colon \mathbb{R}^4 \to \mathbb{R}^4$ contains three chains corresponding to the eigenvalue 2: two chains $\mathbf{x}_1$ and $\mathbf{y}_1$, each of length 1, and one chain $\mathbf{w}_1$, $\mathbf{w}_2$ of length 2. Find the matrix representation of T with respect to this canonical basis, ordered as follows.

(a) $\{\mathbf{x}_1, \mathbf{y}_1, \mathbf{w}_1, \mathbf{w}_2\}$,

(b) $\{\mathbf{y}_1, \mathbf{w}_1, \mathbf{w}_2, \mathbf{x}_1\}$,

(c) $\{\mathbf{w}_1, \mathbf{w}_2, \mathbf{x}_1, \mathbf{y}_1\}$,

(d) $\{\mathbf{w}_1, \mathbf{w}_2, \mathbf{y}_1, \mathbf{x}_1\}$,

(58) A canonical basis for a linear transformation $T\colon \mathbb{R}^6 \to \mathbb{R}^6$ contains two chains corresponding to the eigenvalue 3: one chain $\mathbf{x}_1$ of length 1 and one chain $\mathbf{y}_1$, $\mathbf{y}_2$ of length 2, and two chains corresponding to the eigenvalue 5: one chain $\mathbf{u}_1$ of length 1 and one chain of $\mathbf{v}_1$, $\mathbf{v}_2$ of length 2. Find the matrix representation of T with respect to this canonical basis, ordered as follows.

(a) $\{\mathbf{x}_1, \mathbf{y}_1, \mathbf{y}_2, \mathbf{u}_1, \mathbf{v}_1, \mathbf{v}_2\}$,

(b) $\{\mathbf{y}_1, \mathbf{y}_2, \mathbf{x}_1, \mathbf{u}_1, \mathbf{v}_1, \mathbf{v}_2\}$,

(c) $\{\mathbf{x}_1, \mathbf{u}_1, \mathbf{v}_1, \mathbf{v}_2, \mathbf{y}_1, \mathbf{y}_2\}$,

(d) $\{\mathbf{y}_1, \mathbf{y}_2, \mathbf{v}_1, \mathbf{v}_2, \mathbf{x}_1, \mathbf{u}_1\}$,

(e) $\{\mathbf{x}_1, \mathbf{u}_1, \mathbf{y}_1, \mathbf{y}_2, \mathbf{v}_1, \mathbf{v}_2\}$,

(f) $\{\mathbf{v}_1, \mathbf{v}_2, \mathbf{u}_1, \mathbf{x}_1, \mathbf{y}_1, \mathbf{y}_2\}$.

In Problems 59 through 73, find a matrix representation in Jordan canonical form for the given linear transformation.

(59) $T\begin{bmatrix} a \\ b \end{bmatrix} = \begin{bmatrix} 2a - 3b \\ a - 2b \end{bmatrix}$.

(60) $T\begin{bmatrix} a \\ b \end{bmatrix} = \begin{bmatrix} 3a + b \\ -a + 5b \end{bmatrix}$.

(61) $T\begin{bmatrix} a \\ b \end{bmatrix} = \begin{bmatrix} 2a - b \\ a + 4b \end{bmatrix}$.

(62) $T\begin{bmatrix} a \\ b \end{bmatrix} = \begin{bmatrix} a + 2b \\ -a + 4b \end{bmatrix}$.

(63) $T\begin{bmatrix} a \\ b \end{bmatrix} = \begin{bmatrix} 2a + b \\ 2a + 3b \end{bmatrix}$.

(64) $T\begin{bmatrix} a \\ b \end{bmatrix} = \begin{bmatrix} 2a - 5b \\ a - 2b \end{bmatrix}$.

(65) $T\begin{bmatrix} a \\ b \\ c \end{bmatrix} = \begin{bmatrix} 9a + 3b + 3c \\ 3b \\ -3a - 3b + 3c \end{bmatrix}$.

(66) $T\begin{bmatrix} a \\ b \\ c \end{bmatrix} = \begin{bmatrix} 2a + 2b - 2c \\ 2b + c \\ 2c \end{bmatrix}$.

(67) $T\begin{bmatrix} a \\ b \\ c \end{bmatrix} = \begin{bmatrix} b + 2c \\ -2b \\ 2a + b \end{bmatrix}$.

(68) $T\begin{bmatrix} a \\ b \\ c \end{bmatrix} = \begin{bmatrix} 2a - c \\ 2a + b - 2c \\ -a + 2c \end{bmatrix}$.

(69) $T\begin{bmatrix} a \\ b \\ c \end{bmatrix} = \begin{bmatrix} a + b + c \\ 0 \\ a + 2b + 2c \end{bmatrix}$.

(70) $T\begin{bmatrix} a \\ b \\ c \end{bmatrix} = \begin{bmatrix} a + 2b + 3c \\ 2a + 4b + 6c \\ 3a + 6b + 9c \end{bmatrix}$.

(71) $T\begin{bmatrix} a \\ b \\ c \\ c \end{bmatrix} = \begin{bmatrix} 3a + b - d \\ 3b + c + d \\ 3c \\ 3d \end{bmatrix}$.

(72) $T\begin{bmatrix} a \\ b \\ c \\ c \end{bmatrix} = \begin{bmatrix} a + b - d \\ b + c \\ 2c + d \\ 2d \end{bmatrix}$

(73) $T\begin{bmatrix} a \\ b \\ c \\ d \\ e \\ f \\ g \end{bmatrix} = \begin{bmatrix} -a-c+d+e+3f \\ b \\ 2a+b+2c-d-d-6f \\ -2a-c+2d+e+3f \\ e \\ f \\ -a-b+d+2e+4f+g \end{bmatrix}$, with $\lambda = 1$ as the only eigenvalue.

(74) The *generalized null space of an* $n \times n$ *matrix* $\mathbf{A}$ *and eigenvalue* λ, denoted by $\mathbb{N}_\lambda(\mathbf{A})$, is the set of all vectors $\mathbf{x} \in \mathbb{R}^n$ such that $(\mathbf{A} - \lambda\mathbf{I})^k \mathbf{x} = \mathbf{0}$ for some non-negative integer k. Show that if $\mathbf{x}$ is a generalized eigenvector of any type corresponding to λ, then $\mathbf{x} \in \mathbb{N}_\lambda(\mathbf{A})$.

(75) Prove that $\mathbb{N}_\lambda(\mathbf{A})$, as defined in Problem 74, is a subspace of $\mathbb{R}^n$.

(76) Prove that every square matrix $\mathbf{A}$ commutes with $(\mathbf{A} - \lambda\mathbf{I})^n$ for every positive integer n and every scalar λ.

(77) Prove that $\mathbb{N}_\lambda(\mathbf{A})$ is an invariant subspace of $\mathbb{R}^n$ under $\mathbf{A}$.

(78) Prove that if $\mathbf{A}$ has order $n \times n$ and $\mathbf{x} \in \mathbb{N}_\lambda(\mathbf{A})$, then $(\mathbf{A} - \lambda\mathbf{I})^n \mathbf{x} = \mathbf{0}$.

APPENDIX B
Markov Chains

Eigenvalues and eigenvectors arise naturally in the study of matrix representations of linear transformations, but that is far from their only use. In this Appendix, we present an application to those probabilistic systems known as Markov chains.

An elementary understanding of Markov chains requires only a little knowledge of probabilities; in particular, that probabilities describe the likelihoods of different events occurring, that probabilities are numbers between 0 and 1, and that if the set of all possible events is limited to a finite number that are mutually exclusive then the sum of the probabilities of each event occurring is 1. Significantly more probability theory is needed to prove the relevant theorems about Markov chains, so we limit ourselves in this section to simply understanding the application.

▶DEFINITION 1

A finite *Markov chain* is a set of objects (perhaps people), a set of consecutive time periods (perhaps five-year intervals), and a finite set of different states (perhaps employed and unemployed) such that

(i) during any given time period, each object is in only one state (although different objects can be in different states) and
(ii) the probability that an object will move from one state to another state (or remain in the same state) over a time period depends only on the beginning and ending states. ◀

We denote the states as state 1, state 2, state 3, through state N, and let p_{ij} designate the probability of moving in one time period into state i from state j $(i, j = 1, 2, \ldots, N)$. The matrix $\mathbf{P} = [p_{ij}]$ is called a *transition matrix*.

A transition matrix for an N-state Markov chain is an $N \times N$ matrix with nonnegative entries; the sum of the entries in each column is 1.

Example 1 Construct a transition matrix for the following Markov chain. A traffic control administrator in the Midwest classifies each day as either clear or cloudy. Historical data show that the probability of a clear day following a cloudy day is 0.6, whereas the probability of a clear day following a clear day is 0.9.

Solution: Although one can conceive of many other classifications such as rainy, very cloudy, partly sunny, and so on, this particular administrator opted for only two, so we have just two states: clear and cloudy, and each day must fall into one and only one of these two states. Arbitrarily, we take clear to be state 1 and cloudy to be state 2. The natural time unit is 1 day. We are given that $p_{12}=0.6$, so it must follow that $p_{22}=0.4$, because after a cloudy, day the next day must be either clear or cloudy and the probability that one or the other of these two events occurring is 1. Similarly, we are given that $p_{11}=0.9$, so it also follows that $p_{21}=0.1$. The transition matrix is

$$\mathbf{P} = \begin{bmatrix} \text{clear} & \text{cloudy} \\ 0.9 & 0.6 \\ 0.1 & 0.4 \end{bmatrix} \begin{matrix} \text{clear} \\ \text{cloudy} \end{matrix}$$

Example 2 Construct a transition matrix for the following Markov chain. A medical survey lists individuals as thin, normal, or obese. A review of yearly checkups from doctors' records showed that 80% of all thin people remained thin 1 year later while the other 20% gained enough weight to be reclassified as normal. For individuals of normal weight, 10% became thin, 60% remained normal, and 30% became obese the following year. Of all obese people, 90% remained obese 1 year later while the other 10% lost sufficient weight to fall into the normal range. Although some thin people became obese a year later, and vice versa, their numbers were insignificant when rounded to two decimals.

Solution: We take state 1 to be thin, state 2 to be normal, and state 3 to be obese. One time period equals 1 year. Converting each percent to its decimal representation so that it may also represent a probability, we have $p_{21}=0.2$, the probability of an individual having normal weight after being thin the previous year, $p_{32}=0.3$, the probability of an individual becoming obese 1 year after having a normal weight, and, in general,

$$\mathbf{P} = \begin{bmatrix} \text{thin} & \text{normal} & \text{obese} \\ 0.8 & 0.1 & 0 \\ 0.2 & 0.6 & 0.1 \\ 0 & 0.3 & 0.9 \end{bmatrix} \begin{matrix} \text{thin} \\ \text{normal} \\ \text{obese} \end{matrix}$$

Powers of a transition matrix have the same properties of a transition matrix: all elements are between 0 and 1, and every column sum equals 1 (see Problem 20). Furthermore,

▶THEOREM 1

If **P** *is a transition matrix for a finite Markov chain, and if $p_{ij}^{(k)}$ denotes the i-j element of* $\mathbf{P}^k$, *the kth power of* **P**, *then $p_{ij}^{(k)}$ is the probability of moving to state i from state j in k time periods.*◀

For the transition matrix created in Example 2, we calculate the second and third powers as

$$\mathbf{P}^2 = \begin{bmatrix} \text{thin} & \text{normal} & \text{obese} \\ 0.66 & 0.14 & 0.01 \\ 0.28 & 0.41 & 0.15 \\ 0.06 & 0.45 & 0.84 \end{bmatrix} \begin{matrix} \text{thin} \\ \text{normal} \\ \text{obese} \end{matrix}$$

and

$$\mathbf{P}^3 = \begin{bmatrix} \text{thin} & \text{normal} & \text{obese} \\ 0.556 & 0.153 & 0.023 \\ 0.306 & 0.319 & 0.176 \\ 0.138 & 0.528 & 0.801 \end{bmatrix} \begin{matrix} \text{thin} \\ \text{normal} \\ \text{obese} \end{matrix}$$

Here, $p_{11}^{(2)}$ $5=0.66$ is the probability of a thin person remaining thin 2 years later, $p_{32}^{(2)}$ $6=0.45$ is the probability of a normal person becoming fat 2 years later, while $p_{13}^{(2)}$ $7=0.023$ is the probability of a fat person becoming thin 3 years later.

For the transition matrix created in Example 1, we calculate the second power to be

$$\mathbf{P}^2 = \begin{bmatrix} \text{clear} & \text{cloudy} \\ 0.87 & 0.78 \\ 0.13 & 0.22 \end{bmatrix} \begin{matrix} \text{clear} \\ \text{cloudy} \end{matrix}$$

Consequently, $p_{12}^{(2)}$ $9=0.78$ is the probability of a cloudy day being followed by a clear day 2 days later, while $p_{22}^{(2)}$ $10=0.22$ is the probability of a cloudy day being followed by a cloudy day 2 days later. Calculating the 10th power of this same transition matrix and rounding all entries to four decimal places for presentation purposes, we have

$$\mathbf{P}^{10} = \begin{bmatrix} \text{clear} & \text{cloudy} \\ 0.8571 & 0.8571 \\ 0.1429 & 0.1429 \end{bmatrix} \begin{matrix} \text{clear} \\ \text{cloudy} \end{matrix} \tag{B.1}$$

Since $p_{11}^{(10)}$ $12=p_{12}^{(10)}$ $13=0.8571$, it follows that the probability of having a clear day 10 days after a cloudy day is the same as the probability of having a clear day 10 days after a clear day.

An object in a Markov chain must be in one and only one state at any time, but that state is not always known with certainty. Often, probabilities are provided to describe the likelihood of an object being in any one of the states at any given time. These probabilities can be combined into an n-tuple. A *distribution vector* **d** for an N-state Markov chain at a given time is an N-dimensional column matrix

A distribution vector for an N-state Markov chain at a given time is a column matrix whose ith component is the probability that an object is in the ith state at that given time.

having as its components, one for each state, the probabilities that an object in the system is in each of the respective states at that time.

Example 3 Find the distribution vector for the Markov chain described in Example 1 if the current day is known to be cloudy.

Solution: The objects in the system are days, which are classified as either clear, state 1, or cloudy, state 2. We are told with certainty that the current day is cloudy, so the probability that the day is cloudy is 1 and the probability that the day is clear is 0. Therefore,

$$\mathbf{d} = \begin{bmatrix} 0 \\ 1 \end{bmatrix}$$

Example 4 Find the distribution vector for the Markov chain described in Example 2 if it is known that currently 7% of the population is thin, 31% of population is of normal weight, and 62% of the population is obese.

Solution: The objects in the system are people. Converting the stated percentages into their decimal representations, we have

$$\mathbf{d} = \begin{bmatrix} 0.07 \\ 0.31 \\ 0.62 \end{bmatrix}$$

Different time periods can have different distribution vectors, so we let $\mathbf{d}^{(k)}$ denote a distribution vector *after* k time periods. In particular, $\mathbf{d}^{(1)}$ is a distribution vector after 1 time period, $\mathbf{d}^{(2)}$ is a distribution vector after 2 time periods, and $\mathbf{d}^{(10)}$ is a distribution vector after 10 time periods. An initial distribution vector for the beginning of a Markov chain is designated by $\mathbf{d}^{(0)}$. The distribution vectors for various time periods are related.

▶THEOREM 2

If **P** *is a transition matrix for a Markov chain, then*

$$\mathbf{d}^{(k)} = \mathbf{P}^k\mathbf{d}^{(10)} = \mathbf{P}\mathbf{d}^{(k-1)},$$

where $\mathbf{P}^k$ *denotes the* k*th power of* **P**.◀

For the distribution vector and transition matrix created in Examples 1 and 3, we calculate

$$\mathbf{d}^{(1)} = \mathbf{P}\mathbf{d}^{(0)} = \begin{bmatrix} 0.9 & 0.6 \\ 0.1 & 0.4 \end{bmatrix} \begin{bmatrix} 0 \\ 1 \end{bmatrix} = \begin{bmatrix} 0.6 \\ 0.4 \end{bmatrix}$$

$$\mathbf{d}^{(2)} = \mathbf{P}^2\mathbf{d}^{(0)} = \begin{bmatrix} 0.87 & 0.78 \\ 0.13 & 0.22 \end{bmatrix} \begin{bmatrix} 0 \\ 1 \end{bmatrix} = \begin{bmatrix} 0.78 \\ 0.22 \end{bmatrix} \tag{B.2}$$

$$\mathbf{d}^{(10)} = \mathbf{P}^{10}\mathbf{d}^{(0)} = \begin{bmatrix} 0.8571 & 0.8571 \\ 0.1429 & 0.1429 \end{bmatrix} \begin{bmatrix} 0 \\ 1 \end{bmatrix} = \begin{bmatrix} 0.8571 \\ 0.1429 \end{bmatrix}$$

The probabilities of following a cloudy day with a cloudy day after 1 time period, 2 time periods, and 10 time periods, respectively, are 0.4, 0.22, and 0.1429.

For the distribution vector and transition matrix created in Examples 2 and 4, we calculate

$$\mathbf{d}^{(3)} = \mathbf{P}^3\mathbf{d}^{(0)} = \begin{bmatrix} 0.556 & 0.153 & 0.023 \\ 0.306 & 0.319 & 0.176 \\ 0.138 & 0.528 & 0.801 \end{bmatrix} \begin{bmatrix} 0.07 \\ 0.31 \\ 0.62 \end{bmatrix} = \begin{bmatrix} 0.10061 \\ 0.22943 \\ 0.66996 \end{bmatrix}$$

Rounding to three decimal places, we have that the probabilities of an arbitrarily chosen individual being thin, normal weight, or obese after three time periods (years) are, respectively, 0.101, 0.229, and 0.700.

The 10th power of the transition matrix created in Example 1 is given by Eq. (B.1) as

$$\mathbf{P}^{10} = \begin{bmatrix} 0.8571 & 0.8571 \\ 0.1429 & 0.1429 \end{bmatrix}$$

Continuing to calculate successively higher powers of **P**, we find that each is identical to $\mathbf{P}^{10}$ when we round all entries to four decimal places. Convergence is a bit slower for the transition matrix associated with Example 3, but it also occurs. As we calculate successively higher powers of that matrix, we find that

$$\mathbf{P}^{10} = \begin{bmatrix} 0.2283 & 0.1287 & 0.0857 \\ 0.2575 & 0.2280 & 0.2144 \\ 0.5142 & 0.6433 & 0.6999 \end{bmatrix}$$

$$\mathbf{P}^{20} = \begin{bmatrix} 0.1294 & 0.1139 & 0.1072 \\ 0.2277 & 0.2230 & 0.2210 \\ 0.6429 & 0.6631 & 0.6718 \end{bmatrix} \tag{B.3}$$

and

$$\lim_{n\to\infty} \mathbf{P}^n = \begin{bmatrix} 0.1111 & 0.1111 & 0.1111 \\ 0.2222 & 0.2222 & 0.2222 \\ 0.6667 & 0.6667 & 0.6667 \end{bmatrix}$$

where all entries have been rounded to four decimal places for presentation purposes.

A transition matrix is regular if one of its powers has only positive elements.

Not all transition matrices have powers that converge to a limiting matrix **L**, but many do. A transition matrix for a finite Markov chain is *regular* if it or one of its powers contains only positive elements. Powers of a regular matrix always converge to a limiting matrix **L**.

The transition matrix created in Example 1 is regular because all of its elements are positive. The transition matrix **P** created in Example 2 is also regular because all elements of $\mathbf{P}^2$, its second power, are positive. In contrast, the transition matrix

$$\mathbf{P} = \begin{bmatrix} 0 & 1 \\ 1 & 0 \end{bmatrix}$$

is not regular because each of its powers is either itself or the 2×2 identity matrix, both of which contain zero entries.

By definition, some power of a regular matrix **P**, say the mth, contains only positive elements. Since the elements of **P** are nonnegative, it follows from matrix multiplication that every power of **P** greater than m must also have all positive components. Furthermore, if $\mathbf{L} = \lim_{k\to\infty} \mathbf{P}^k$, then it is also true that $\mathbf{L} = \lim_{k\to\infty} \mathbf{P}^{k-1}$. Therefore,

$$\mathbf{L} = \lim_{k\to\infty} \mathbf{P}^k = \lim_{k\to\infty} \left(\mathbf{P}\mathbf{P}^{k-1}\right) = \mathbf{P}\left(\lim_{k\to\infty} \mathbf{P}^{k-1}\right) = \mathbf{PL} \tag{B.4}$$

Denote the columns of **L** as $\mathbf{x}_1, \mathbf{x}_2, \ldots, \mathbf{x}_N$, respectively, so that $\mathbf{L} = [\mathbf{x}_1\ \mathbf{x}_2, \ldots \mathbf{x}_N]$. Then equation (C.4) becomes

$$[\mathbf{x}_1, \mathbf{x}_2, \ldots, \mathbf{x}_N] = \mathbf{P}[\mathbf{x}_1, \mathbf{x}_2, \ldots, \mathbf{x}_N]$$

where $\mathbf{x}_j = \mathbf{P}\mathbf{x}_j$, $(j = 1, 2, \ldots, N)$, or $\mathbf{P}\mathbf{x}_j = (1)\mathbf{x}_j$. Thus, each column of **L** is an eigenvector of **P** corresponding to the eigenvalue 1. We have proved part of the following important result.

▶THEOREM 3

*If an $N \times N$ transition matrix **P** is regular, then successive integral powers of **P** converge to a limiting matrix **L** whose columns are eigenvectors of **P** associated with eigenvalue $\lambda = 1$. The components of this eigenvector are positive and sum to unity.*◀

Even more is true. If $\mathbf{P}$ is regular, then its eigenvalue $\lambda = 1$ has multiplicity 1, and there is only one linearly independent eigenvector associated with that eigenvalue. This eigenvector will be in terms of one arbitrary constant, which is uniquely determined by the requirement that the sum of the components is 1. Thus, each column of $\mathbf{L}$ is the *same* eigenvector.

We define the *limiting state distribution vector* for an N-state Markov chain as an N-dimensional column vector $\mathbf{d}^{(\infty)}$ having as its components the limiting probabilities that an object in the system is in each of the respective states after a large number of time periods. That is,

$$\mathbf{d}^{(\infty)} = \lim_{n\to\infty} \mathbf{d}^{(n)}$$

Consequently,

$$\mathbf{d}^{(\infty)} = \lim_{n\to\infty} \mathbf{d}^{(n)} = \lim_{n\to\infty} \left(\mathbf{P}^n \mathbf{d}^{(0)}\right) = \left(\lim_{n\to\infty} \mathbf{P}^n\right)\mathbf{d}^{(0)} = \mathbf{L}\mathbf{d}^{(0)}$$

Each column of $\mathbf{L}$ is identical to every other column, so each row of $\mathbf{L}$ contains a single number repeated N times. Combining this with the fact that $\mathbf{d}^{(0)}$ has components that sum to 1, it follows that the product $\mathbf{L}\mathbf{d}^{(0)}$ is equal to each of the identical columns of $\mathbf{L}$. That is, $\mathbf{d}^{(\infty)}$ is the eigenvector of $\mathbf{P}$ corresponding to $\lambda = 1$, having the sum of its components equal to 1.

Example 5 Find the limiting state distribution vector for the Markov chain described in Example 1.

The limiting state distribution vector for a transition matrix $\mathbf{P}$ is the unique eigenvector of $\mathbf{P}$ corresponding to $\lambda = 1$, *having the sum of its components equal to 1.*

Solution: The transition matrix is

$$\mathbf{P} = \begin{bmatrix} 0.9 & 0.6 \\ 0.1 & 0.4 \end{bmatrix}$$

which is regular. Eigenvectors for this matrix have the form

$$x = \begin{bmatrix} x \\ y \end{bmatrix}$$

Eigenvectors corresponding to $\lambda = 1$ satisfy the matrix equation $(\mathbf{P} - 1\mathbf{I})\mathbf{x} = \mathbf{0}$, or equivalently, the set of equations

$$\begin{aligned} -0.1x + 0.9y &= 0 \\ 0.1x - 0.6y &= 0 \end{aligned}$$

Solving by Gaussian elimination, we find $x = 6y$ with y arbitrary. Thus,

$$x = \begin{bmatrix} 6x \\ y \end{bmatrix}$$

If we choose y so that the sum of the components of $\mathbf{x}$ sum to 1, we have $7y=1$, or $y=1/7$. The resulting eigenvector is the limiting state distribution vector, namely,

$$\mathbf{d}^{(\infty)} = \begin{bmatrix} 6/7 \\ 1/7 \end{bmatrix}$$

Furthermore,

$$\mathbf{L} = \begin{bmatrix} 6/7 & 6/7 \\ 1/7 & 1/7 \end{bmatrix}$$

Over the long run, 6 out of 7 days will be clear and 1 out of 7 days will be cloudy. We see from Eqs. (B.1) and (B.2) that convergence to four decimal places for the limiting state distribution and $\mathbf{L}$ is achieved after 10 time periods.

Example 6 Find the limiting state distribution vector for the Markov chain described in Example 2.

Solution: The transition matrix is

$$\mathbf{P} = \begin{bmatrix} 0.8 & 0.1 & 0 \\ 0.2 & 0.6 & 0.1 \\ 0 & 0.3 & 0.9 \end{bmatrix}$$

$\mathbf{P}^2$ has only positive elements, so $\mathbf{P}$ is regular. Eigenvectors for this matrix have the form

$$\mathbf{x} = \begin{bmatrix} x \\ y \\ x \end{bmatrix}$$

Eigenvectors corresponding to $\lambda=1$ satisfy the matrix equation $(\mathbf{P}-1\mathbf{I})\mathbf{x}=\mathbf{0}$, or equivalently, the set of equations

$$\begin{aligned} -0.2x + 0.1y &= 0 \\ 0.2x - 0.4y + 0.1z &= 0 \\ 0.3y - 0.1z &= 0 \end{aligned}$$

Solving by Gaussian elimination, we find $x=(1/6)z$, $y=(1/3)z$, with z arbitrary. Thus,

$$x = \begin{bmatrix} z/6 \\ z/3 \\ z \end{bmatrix}$$

We choose z so that the sum of the components of $\mathbf{x}$ sum to 1, hence $(1/6)z+(1/3)z+z=1$, or $z=2/3$. The resulting eigenvector is the limiting state distribution vector, namely,

$$\mathbf{d}^{(\infty)} = \begin{bmatrix} 1/9 \\ 2/9 \\ 6/9 \end{bmatrix}$$

Furthermore,

$$\mathbf{L} = \begin{bmatrix} 1/9 & 1/9 & 1/9 \\ 2/9 & 2/9 & 2/9 \\ 6/9 & 6/9 & 6/9 \end{bmatrix}$$

Compare $\mathbf{L}$ with Eq. (B.3). The components of $\mathbf{d}^{(\infty)}$ imply that, over the long run, one out of nine people will be thin, two out of nine people will be of normal weight, and six out of nine people will be obese.

PROBLEMS APPENDIX B

(1) Determine which of the following matrices cannot be transition matrices and explain why:

(a) $\begin{bmatrix} 0.15 & 0.57 \\ 0.85 & 0.43 \end{bmatrix}$, (b) $\begin{bmatrix} 0.27 & 0.74 \\ 0.63 & 0.16 \end{bmatrix}$,

(c) $\begin{bmatrix} 0.45 & 0.53 \\ 0.65 & 0.57 \end{bmatrix}$, (d) $\begin{bmatrix} 1.27 & 0.23 \\ -0.27 & 0.77 \end{bmatrix}$,

(e) $\begin{bmatrix} 1 & 1/2 & 0 \\ 0 & 1/3 & 0 \\ 0 & 1/6 & 0 \end{bmatrix}$, (f) $\begin{bmatrix} 1/2 & 1/2 & 1/3 \\ 1/4 & 1/3 & 1/4 \\ 1/4 & 1/6 & 7/12 \end{bmatrix}$,

(g) $\begin{bmatrix} 0.34 & 0.18 & 0.53 \\ 0.38 & 0.42 & 0.21 \\ 0.35 & 0.47 & 0.19 \end{bmatrix}$, (h) $\begin{bmatrix} 0.34 & 0.32 & -0.17 \\ 0.78 & 0.65 & 0.80 \\ -0.12 & 0.03 & 0.37 \end{bmatrix}$.

(2) Construct a transition matrix for the following Markov chain: Census figures show a population shift away from a large midwestern metropolitan city to its suburbs. Each year, 5% of all families living in the city move to the suburbs, while during the same time period, only 1% of those living in the suburbs move into the city. *Hint:* Take state 1 to represent families living in the city, state 2 to represent families living in the suburbs, and 1 year as one time period.

(3) Construct a transition matrix for the following Markov chain: Every 4 years, voters in a New England town elect a new mayor because a town ordinance prohibits mayors from succeeding themselves. Past data indicate that a Democratic mayor is succeeded by another Democrat 30% of the time and by a Republican 70% of the time. A Republican mayor, however, is succeeded by another Republican 60% of the time and by a Democrat 40% of the time. *Hint:* Take state 1 to represent a Republican mayor in office, state 2 to represent a Democratic mayor in office, and 4 years as one time period.

(4) Construct a transition matrix for the following Markov chain: The apple harvest in New York orchards is classified as poor, average, or good. Historical data indicate that if the harvest is poor 1 year then there is a 40% chance of having a good harvest the next year, a 50% chance of having an average harvest, and a 10% chance of having another poor harvest. If a harvest is average 1 year, the chance of a poor, average, or good harvest the next year is 20%, 60%, and 20%, respectively. If a harvest is good, then the chance of a poor, average, or good harvest the next year is 25%, 65%, and 10%, respectively. *Hint:* Take state 1 to be a poor harvest, state 2 to be an average harvest, state 3 to be a good harvest, and 1 year as one time period.

(5) Construct a transition matrix for the following Markov chain: Brand X and brand Y control the majority of the soap powder market in a particular region, and each has promoted its own product extensively. As a result of past advertising campaigns, it is known that over a two-year period of time, 10% of brand Y customers change to brand X and 25% of all other customers change to brand X. Furthermore, 15% of brand X customers change to brand Y and 30% of all other customers change to brand Y. The major brands also lose customers to smaller competitors, with 5% of brand X customers switching to a minor brand during a two-year time period and 2% of brand Y customers doing likewise. All other customers remain loyal to their past brand of soap powder. *Hint:* Take state 1 to be a brand X customer, state 2 a brand Y customer, state 3 another brand's customer, and 2 years as one time period.

(6) (a) Calculate $\mathbf{P}^2$ and $\mathbf{P}^3$ for the two-state transition matrix:

$$\mathbf{P} = \begin{bmatrix} 0.1 & 0.4 \\ 0.9 & 0.6 \end{bmatrix}$$

(b) Determine the probability of an object beginning in state 1 and remaining in state 1 after two time periods.

(c) Determine the probability of an object beginning in state 1 and ending in state 2 after two time periods.

(d) Determine the probability of an object beginning in state 1 and ending in state 2 after three time periods.

(e) Determine the probability of an object beginning in state 2 and remaining in state 2 after three time periods.

(7) Consider a two-state Markov chain. List the number of ways an object in state 1 can end in state 1 after three time periods.

(8) Consider the Markov chain described in Problem 2. Determine (a) the probability a family living in the city will find themselves in the suburbs after 2 years, and (b) the probability a family living in the suburbs will find themselves living in the city after 2 years.

(9) Consider the Markov chain described in Problem 3. Determine (a) the probability there will be a Republican mayor 8 years after a Republican mayor serves, and (b) the probability there will be a Republican mayor 12 years after a Republican mayor serves.

(10) Consider the Markov chain described in Problem 4. It is known that this year that the apple harvest was poor. Determine (a) the probability next year's harvest will be poor, and (b) the probability that the harvest in 2 years will be poor.

(11) Consider the Markov chain described in Problem 5. Determine (a) the probability that a brand X customer will remain a brand X customer after 4 years, (b) after 6 years, and (c) the probability that a brand X customer will become a brand Y customer after 4 years.

(12) Consider the Markov chain described in Problem 2. (a) Explain the significance of each component of $\mathbf{d}^{(0)} = [0.6 \quad 0.4]^{\mathrm{T}}$. (b) Use this vector to find $\mathbf{d}^{(1)}$ and $\mathbf{d}^{(2)}$.

(13) Consider the Markov chain described in Problem 5. (a) Explain the significance of each component of $\mathbf{d}^{(0)} = [0.4 \quad 0.5 \quad 0.1]^{\mathrm{T}}$. (b) Use this vector to find $\mathbf{d}^{(1)}$ and $\mathbf{d}^{(2)}$.

(14) Consider the Markov chain described in Problem 3. (a) Determine an initial distribution vector if the town currently has a Democratic mayor, and (b) show that the components of $\mathbf{d}^{(1)}$ are the probabilities that the next mayor will be a Republican and a Democrat, respectively.

(15) Consider the Markov chain described in Problem 4. (a) Determine an initial distribution vector if this year's crop is known to be poor, (b) Calculate $\mathbf{d}^{(2)}$ and use it to determine the probability that the harvest will be good in 3 years.

(16) Find the limiting distribution vector for the Markov chain described in Problem 2, and use it to determine the probability that a family eventually will reside in the city.

(17) Find the limiting distribution vector for the Markov chain described in Problem 3, and use it to determine the probability of having a Republican mayor over the long run.

(18) Find the limiting distribution vector for the Markov chain described in Problem 4, and use it to determine the probability of having a good harvest over the long run.

(19) Find the limiting distribution vector for the Markov chain described in Problem 5, and use it to determine the probability that a person will become a Brand Y customer over the long run.

(20) Use mathematical induction to prove that if **P** is a transition matrix for an n-state Markov chain, then any integral power of **P** has the properties that (a) all elements are nonnegative numbers between 0 and 1, and (b) the sum of the elements in each column is 1.

(21) A nonzero row vector **y** is a *left eigenvector* for a matrix **A** if there exists a scalar λ such that $\mathbf{yA}=\lambda\mathbf{y}$. Prove that if **x** and λ are a corresponding pair of eigenvectors and eigenvalues for a matrix **B**, then $\mathbf{x}^{\mathrm{T}}$ and λ are a corresponding pair of left eigenvectors and eigenvalues for $\mathbf{B}^{\mathrm{T}}$.

(22) Show directly that the n-dimensional row vector $\mathbf{y}=[1\ 1\ 1\ \ldots\ 1]$ is a left eigenvector for any $N\times N$ transition matrix **P**. Then, using the results of Problem 20, deduce that $\lambda=1$ is an eigenvalue for any transition matrix.

(23) Prove that every eigenvalue λ of a transition matrix **P** satisfies the inequality $|\lambda|\leq 1$. *Hint:* Let $\mathbf{x}=[x_1\quad x_2\ \ldots\ x_{\mathrm{N}}]^{\mathrm{T}}$ be an eigenvector of **P** corresponding to the eigenvalue λ, and let $x_{\mathrm{i}}=\max\ \{x_1, x_2, \ldots, x_{\mathrm{N}}\}$. Consider the ith component of the vector equation $\mathbf{Px}=\lambda\mathbf{x}$, and show that $|\lambda|\ |x_{\mathrm{i}}|\leq|x_{\mathrm{i}}|$.

(24) A state in a Markov chain is *absorbing* if no objects in the system can leave the state after they enter it. Describe the ith column of a transition matrix for a Markov chain in which the ith state is absorbing.

(25) Prove that a transition matrix for a Markov chain with one or more absorbing states cannot be regular.

APPENDIX C

More on Spanning Trees of Graphs

As we explored in Chapter 5, Sections 5–7, some networks can be modeled by graphs, these graphs can be represented by some special matrices, and the eigenvalues and characteristic polynomials in turn tell us some important information about the graphs (and hence, the networks they represent).

A graph is *bipartite* if its vertex set can be decomposed into two disjoint sets such that no two graph vertices within the same set are adjacent. In Figure C.1, we observe that vertices 2, 4, and 6 do not have any edges between them, but each has an edge between it and vertices 1, 3, and 5. Vertices 1, 3, and 5 have no edges between them as well. Thus, the graph is bipartite.

We first state four theorems about adjacency matrices of bipartite graphs.

▶THEOREM 1

A graph G is bipartite if and only if $-r$ is an adjacency matrix eigenvalue of G (where r denotes the largest positive eigenvalue of G). ◀

Proof: We will prove this in the forward direction only, i.e., if G is bipartite, then $-r$ is an eigenvalue of G. Assume G is bipartite, then $A(G)$ can be written in the form $\begin{pmatrix} 0 & B \\ B^t & 0 \end{pmatrix}$ where B is a $p \times q$ matrix. Let r be the largest eigenvalue of G and let $\begin{bmatrix} x \\ y \end{bmatrix}$ denote its corresponding eigenvector, where x is a $p \times 1$ column vector and y is a $q \times 1$ column vector. It follows that $B \cdot y = r \cdot x$ and that $(B^t) \cdot x = r \cdot y$. Also, $B \cdot (-y) = (-r) \cdot x$ and $(B^t) \cdot x = (-r) \cdot (-y)$. Therefore, $(-r)$ is an eigenvalue and $\begin{bmatrix} x \\ -y \end{bmatrix}$ is the corresponding eigenvector.

We present another theorem about adjacency matrix eigenvalues for bipartite graphs.

▶THEOREM 2

The adjacency matrix eigenvalues of a graph G are paired if and only if G is a bipartite graph. ◀

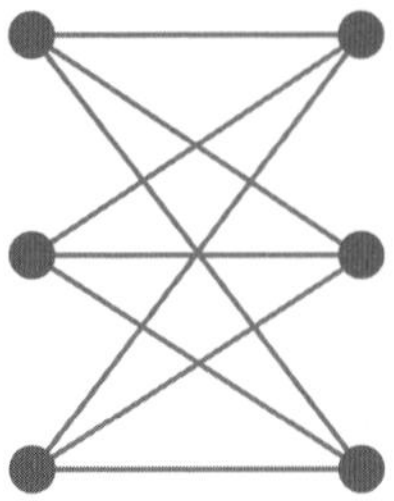

FIGURE C.1

Proof: *Necessity*: If G is bipartite, then the eigenvalues of G are paired. Assume G is bipartite, then $A(G)$ can be written in the form $\begin{pmatrix} 0 & B \\ B^t & 0 \end{pmatrix}$, where B is a $p \times q$ matrix. Let λ be an eigenvalue of G and let its corresponding eigenvector be $\begin{bmatrix} x \\ y \end{bmatrix}$, where x is a $p \times 1$ vector and y is a $q \times 1$ column matrix. It follows that $B \cdot y = \lambda \cdot x$ and that $(B^t) \cdot x = \lambda \cdot y$. Also, $B \cdot (-y) = (-\lambda) \cdot x$ and $(B^t) \cdot x = (-\lambda) \cdot (-y)$. Therefore, $(-\lambda)$ is an eigenvalue and $\begin{bmatrix} x \\ -y \end{bmatrix}$ is the corresponding eigenvector, and hence the eigenvalues of G are paired. *Sufficiency*. If the eigenvalues of G are paired, then G is bipartite. From Theorem 1, we know that if $(-r)$ is an eigenvalue of G, then G is bipartite. Since the eigenvalues of the graph are paired, then it follows that the largest eigenvalue, denoted by r, is paired. Hence, $-r$ is also an eigenvalue. Therefore, G is bipartite.

A *complete bipartite graph* is a set of graph vertices decomposed into 2 disjoint sets such that no two graph vertices within the same set are adjacent but **every** pair of graph vertices in the 2 disjoint sets are adjacent. It is represented by $K_{p,q}$, where p and q represent the number of nodes in the partite sets. The graph in Figure C.1 is a representative of this special type of bipartite graph, namely, $K_{3,3}$.

►THEOREM 3

The adjacency matrix eigenvalues of the complete bipartite graph $k_{p,q}$ are zero with multiplicity $(p+q-2)$ and $\pm\sqrt{pq}$. ◄

Proof: Let A be the adjacency matrix of a complete bipartite graph $K_{p,q}$. Then A can be written in the form $\begin{pmatrix} 0 & B \\ B^t & 0 \end{pmatrix}$, where B denotes the $p \times q$ matrix consisting entirely of ones (since complete). By looking at the matrix $A - \lambda I_{p+q}$, we can use basic row reduction techniques to see that $p+q-2$ of the eigenvalues are zero. We also know from Theorem 2 that if a graph is bipartite, its eigenvalues are paired. Therefore, the remaining 2 eigenvalues are paired, call them k and $-k$. From Theorem 3, part (iii), of Chapter 5, section 6, we know that the sum of the squares of the eigenvalues of G is equal to twice the number of edges in G, so $(0_1^2+0_2^2+\cdots+0_{p+q-2}^2)+k^2+(-k)^2=2e$, and for any complete bipartite graph, the number of edges is equal to $p.q$. Thus, we have $2k^2=2(p.q)$, and so $k=\sqrt{pq}$ and $-k=-\sqrt{pq}$.

Example 1 The graph in Figure C.1 has adjacency matrix $A=\begin{bmatrix}0&1&0&1&0&1\\1&0&1&0&1&0\\0&1&0&1&0&1\\1&0&1&0&1&0\\0&1&0&1&0&1\\1&0&1&0&1&0\end{bmatrix}$. This matrix has characteristic polynomial $\lambda^6-9\lambda^4=(\lambda-0)(\lambda-0)(\lambda-0)(\lambda-0)(\lambda-3)(\lambda+3)$, so its eigenvalues are 3, 0, 0, 0, 0, −3. Since p and q are both 3, $(p+q-2)=(3+3-2)=4$, confirming that 0 occurs with multiplicity 4, and $\pm\sqrt{pq}=\pm\sqrt{3.3}=\pm3$.

Our final Theorem for bipartite graphs refers to Theorem 3, part (iii), of Chapter 5, section 6, on the adjacency matrix raised to a power.

►THEOREM 4

A graph G is bipartite if and only if no power of its adjacency matrix A consists entirely of strictly positive entries (i.e., every power of A must contain zeros). ◄

Proof: We will sketch the proof of the sufficiency only, i.e., if a graph G is bipartite, then no power of its adjacency matrix A consists entirely of strictly positive entries. Again, since bipartite, A can be written in the form $\begin{pmatrix}0&B\\B^t&0\end{pmatrix}$.

$$A^2=\begin{pmatrix}0&B\\B^t&0\end{pmatrix}*\begin{pmatrix}0&B\\B^t&0\end{pmatrix}=\begin{pmatrix}BB^t&0\\0&B^tB\end{pmatrix}$$

$$A^3=\begin{pmatrix}BB^t&0\\0&B^tB\end{pmatrix}*\begin{pmatrix}0&B\\B^t&0\end{pmatrix}=\begin{pmatrix}0&BB^tB\\B^tBB^t&0\end{pmatrix}$$

$$A^4=\begin{pmatrix}0&BB^tB\\B^tBB^t&0\end{pmatrix}*\begin{pmatrix}0&B\\B^t&0\end{pmatrix}=\begin{pmatrix}BB^tBB^t&0\\0&B^tBB^tB\end{pmatrix}$$

And so on So for all k, this demonstrates that A^k will have entries that are zero. Since G is bipartite, the vertices can be split into two disjoint sets $\{\mathbf{u}_1,\mathbf{u}_2,\ldots\mathbf{u}_n\}$ and $\{\mathbf{v}_1,\mathbf{v}_2,\ldots,\mathbf{v}_n\}$. There is no path of even length from any $\mathbf{u}_x$ to $\mathbf{v}_y$. Therefore, following Theorem 3, part (iii), of Chapter 5, section 6, there will be zeros in every even power of the adjacency matrix. In the same manner, there is no path of odd length from any $\mathbf{u}_x$ to $\mathbf{u}_y$ or $\mathbf{v}_x$ to $\mathbf{v}_y$ and thus, there will be zeros in every odd power of the adjacency matrix. Hence, no power of A consists entirely of strictly positive entries.

We defined a *regular graph* to be one in which all n of its vertices have the same degree, r. The adjacency matrix eigenvalues for these graphs can help us to determine their number of spanning trees, and we know from Theorem 4 of Chapter 5 that the maximum eigenvalue of such graphs is r itself.

►THEOREM 5

The number of spanning trees for a graph G on n vertices that is regular of degree r and having adjacency matrix eigenvalues $a_1 \le a_2 \le \cdots \le a_n = r$ *is* $\frac{1}{n}\prod_{i-1}^{n-1}(r - a_i)$. ◄

Proof: Recall that the Laplacian matrix $L = diag(d_1, d_2, \ldots, d_n) - A$, where A is the adjacency matrix of the graph G and $d_1, d_2, \ldots, d_n$ are the degrees of the vertices. Thus, if G is a regular graph, each vertex having degree equal to r, it follows that $\lambda_i = r - a_i$ for $I = 1, 2, \ldots, n-1$.

Example 2 Besides being complete bipartite, the graph in Figure C.1 is regular of degree 3. As seen in Example 1, its adjacency matrix has entries of each row after the first one is formed by shifting the entries (and wrapping around the last entry back to the beginning). This is called a *circulant matrix.* When a graph's vertices can be ordered in such a way that its adjacency matrix is a circulant matrix, the graph is called a *circulant graph*. As stated earlier, this matrix has eigenvalues 3, 0, 0, 0, 0, −3. We eliminate the largest eigenvalue (namely, $r = 3$) and produce the product $\frac{1}{6}(3-0)^4(3-(-3))^1 = 81$, meaning the graph in Figure C.1 has 81 spanning trees.

In Chapter 5, section 7, we discussed the calculation of the number of spanning trees using Laplacian eigenvalues. Consider a graph on n vertices. If we then consider a complete graph on n vertices, K_n, and "subtract" the edges of G from K_n, then the resulting n vertex graph is called *the complement of* G, denoted $\overline{G}$. We present the following theorem specifying the relationship between the Laplacian eigenvalues of G and those of $\overline{G}$.

►THEOREM 6

If a graph G has Laplacian eigenvalues $0 \le \lambda_1 \le \lambda_2 \le \cdots \le \lambda_n$, *then* $\overline{G}$ *has Laplacian eigenvalues* $\overline{\lambda_k} = n - \lambda_{n-k}$, $k = 2, \ldots n$. ◄

Example 3 The graph in Figure C.2 depicts the complement of the graph in Figure C.1. Note that the complement of $K_{3,3}$ is two disjoint K_3 graphs. (In fact, the complement of any complete bipartite graph $K_{p,q}$ is K_p and K_q. In question 16 of Chapter 5, section 7, the Laplacian eigenvalues for the complete graph K_n are shown to be n with multiplicity $n-1$ and 0. Thus, the Laplacian eigenvalues for a graph comprised of two disjoint K_3 graphs are 0, 0, 3, 3, 3, 3, and then the Laplacian eigenvalues for $K_{3,3}$ are 6-0, 6-3, 6-3, 6-3, 6-3, and the obligatory zero eigenvalue, i.e., 0, 3, 3, 3, 3, 6. We will use this technique to determine the number of spanning trees in complete bipartite graphs in exercises 7-12 in this appendix.

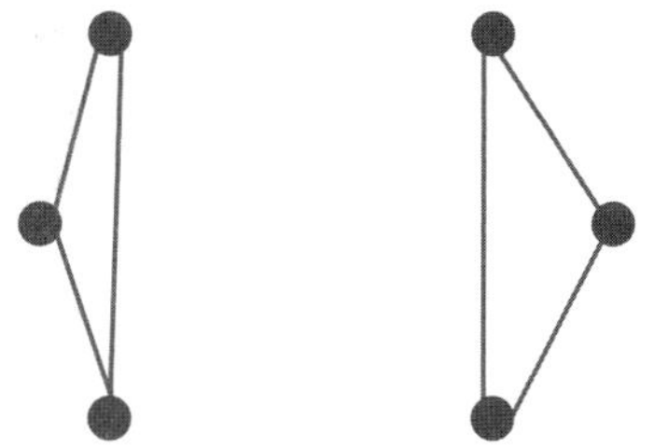

FIGURE C.2

PROBLEMS APPENDIX C

We recommend the use of computer software to assist in the computation of characteristic polynomials and eigenvalues in this appendix, particularly problem 6.

Complete graphs are also regular graphs. In fact, every K_n is regular of degree $n-1$. Exercises 1-5 deal with complete graphs and conjecture a general formula for their number of spanning trees using the adjacency matrix.

(1) (a) Draw K_3.
(b) Find the adjacency matrix for K_3.
(c) Find the eigenvalues for K_3.

(2) (a) Draw K_4.
(b) Find the adjacency matrix for K_4.
(c) Find the eigenvalues for K_4.

(3) (a) Draw K_5.
(b) Find the adjacency matrix for K_5.
(c) Find the eigenvalues for K_5.

(4) What pattern do you see in your responses to questions 1c, 2c, and 3c? Can you formulate a conjecture about the adjacency eigenvalues for any complete graph K_n?

(5) Using your response to question (4), and using Theorem 5 of this appendix, can you generalize a formula for the number of spanning trees for any graph K_n? How does this compare with the one that you found in Chapter 5, section 7?

Figure C.3 depicts a very famous and interesting graph, the Petersen Graph. It is regular of degree 3.

(6) (a) Find the adjacency matrix for the Petersen Graph.
(b) Find the adjacency eigenvalues for the Petersen graph.
(c) Does Theorem 1 of this appendix apply to the Petersen graph? What, if anything, can be concluded?
(d) Does Theorem 2 of this appendix apply to the Petersen graph? What, if anything, can be concluded?
(e) Use Theorem 5 to determine the number of spanning trees in the Petersen graph.
(f) Find the Laplacian eigenvalues for the Petersen graph.

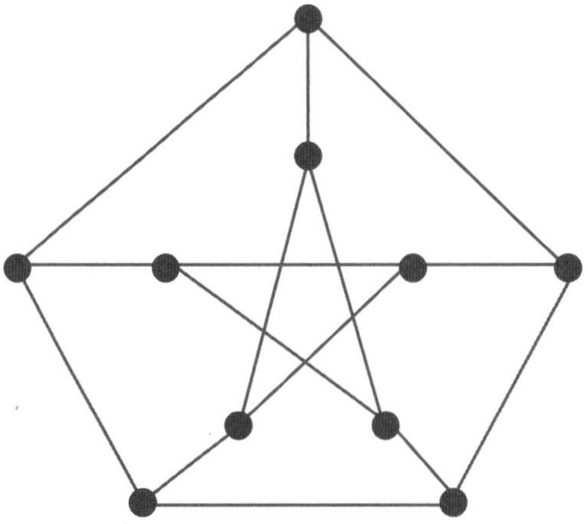

FIGURE C.3

(g) Use Theorem 6 of Section 5.7 to calculate the number of spanning trees in the Petersen graph. Does your answer agree with the one you found in part (e)?

Exercises 7-12 deal with complete bipartite graphs and conjecture a general formula for their number of spanning trees using the Laplacian matrix.

(7) (a) Draw $\boldsymbol{K}_{4,4}$.
(b) Find $\overline{K_{4,4}}$.
(c) Use Theorem 6 and Example 3 of this section to compute the Laplacian eigenvalues for $\boldsymbol{K}_{4,4}$.

(8) (a) Draw $\boldsymbol{K}_{4,5}$.
(b) Find $\overline{K_{4,5}}$.
(c) Use Theorem 6 and Example 3 of this section to compute the Laplacian eigenvalues for $\boldsymbol{K}_{4,5}$.

(9) (a) Draw $\boldsymbol{K}_{4,6}$.
(b) Find $\overline{K_{4,6}}$.
(c) Use Theorem 6 and Example 3 of this section to compute the Laplacian eigenvalues for $\boldsymbol{K}_{4,6}$.

(10) (a) Draw $\boldsymbol{K}_{5,5}$.
(b) Find $\overline{K_{5,5}}$.
(c) Use Theorem 6 and Example 3 of this section to compute the Laplacian eigenvalues for $\boldsymbol{K}_{5,5}$.

(11) (a) Draw $\boldsymbol{K}_{5,7}$.
(b) Find $\overline{K_{5,7}}$.
(c) Use Theorem 6 and Example 3 of this section to compute the Laplacian eigenvalues for $\boldsymbol{K}_{5,7}$.

(12) (a) Based on your answers to part (c) of questions 7-11, conjecture a formula for the Laplacian eigenvalues for a complete bipartite graph.
(b) Based on your answer to part (a) of this question, conjecture a formula for the number of spanning trees for a complete bipartite graph.

A graph is *complete multipartite* on k parts if its vertex set can be decomposed into k disjoint sets such that no vertex is adjacent to any of the other vertices in its part, but each vertex is adjacent to all the vertices in the $K-1$ other parts. A complete multipartite graph having k parts of order $r_1, r_2, \ldots, r_k$, respectively, on $n = r_1 + r_2 + \ldots + r_k$ nodes is denoted $\mathbf{K}_{r_1, r_2, \ldots, r_k}$. In problems 13-17, we will formulate a conjecture about the number of spanning trees of such graphs, but consider only multipartite sets with equal part sizes, i.e., $\mathbf{K}_{r_1, r_2, \ldots, r_k}$ with $r_1 = r_2 = \ldots = r_k$. We will also confine our study to complete multipartite graphs having exactly three parts. We will refer to the graphs under consideration as *regular complete tripartite graphs*.

(13) (a) Draw $K_{3,3,3}$.
(b) Determine $\overline{K_{3,3,3}}$.
(c) Use Theorem 6 and Example 3 of this section to compute the Laplacian eigenvalues for $K_{3,3,3}$.

(14) For $K_{4,4,4}$.
(a) Determine $\overline{K_{4,4,4}}$.
(b) Use Theorem 6 and Example 3 of this section to compute the Laplacian eigenvalues for $K_{4,4,4}$.

(15) For $K_{5,5,5}$.
(a) Determine $\overline{K_{5,5,5}}$.
(b) Use Theorem 6 and Example 3 of this section to compute the Laplacian eigenvalues for $K_{5,5,5}$.

(16) For $K_{6,6,6}$.
(a) Determine $\overline{K_{6,6,6}}$.
(b) Use Theorem 6 and Example 3 of this section to compute the Laplacian eigenvalues for $K_{6,6,6}$.

(17) (a) Based on your answers to part (c) of question 13 and part (b) of questions 14-16, conjecture a formula for the Laplacian eigenvalues for a regular complete tripartite graphs.
(b) Based on your answer to part (a) of this question, conjecture a formula for the number of spanning trees for regular complete tripartite graphs.

APPENDIX D

Technology

As we have demonstrated in this text, linear algebra is a very powerful tool. It can be applied to such diverse areas as differential equations (see Chapter 5) and to least-squares techniques (see Chapter 6). Yet the actual calculations needed to arrive at solutions can be very tedious. The computation of higher-order determinants (see Chapter 1) and the application of the **QR** algorithm (see Section 6.3) can likewise require much time.

The field of numerical analysis can assist with calculations and, if appropriate, with approximations. But even when numerical techniques are uses, one almost always needs computational assistance in the form of technology.

One of the most useful tools is **MATLAB**® (http://www.mathworks.com/prod ucts/matlab/). This software is employed by educators and is very useful with respect to many topics in linear algebra.

Another software package is **MATHEMATICA**® (http://www.wolfram.com/). To illustrate this computer algebra system, the reader is asked to refer to the modeling problem of Section 5.4.

The syntax for the system of differential equations is given by:

$$\begin{aligned} \mathit{DSolve}[\{&\mathrm{S1[t]} = 10 - \mathrm{S1[t]}/20, \mathrm{S2'[t]} = \mathrm{S1[t]}/50 + \mathrm{S3[t]}/25 - (6/50)^{*}\mathrm{S2[t]}, \mathrm{S3'[t]} \\ &= \mathrm{S3[t]}/25 + \mathrm{S2[t]}/50, \mathrm{S1[0]} = 0, \mathrm{S2[0]} = 0, \mathrm{S3[0]} = 0\}, \\ &\{\mathrm{S1}[t], \mathrm{S2}[t], \mathrm{S3}[t]\}, t] \end{aligned}$$

The solution, obtained by hitting the "Shift" and "Enter" keys simultaneously, is as follows:

$$[\mathrm{t}] \to 200\,\mathrm{e}^{-t/20}\left(-1+\mathrm{e}^{t/20}\right),\ \mathrm{S2}[\mathrm{t}] \to \frac{1}{1491}$$

$$\left(50\,\mathrm{e}^{-\frac{t}{20}-\frac{1}{50}(-2+3\sqrt{2})t}\left(-1512\,\mathrm{e}^{\frac{1}{50}(-2+3\sqrt{2})t} - 330\,\mathrm{e}^{\frac{t}{20}+\frac{3\sqrt{2}t}{35}+\frac{1}{50}(-2-3\sqrt{2})t}\right.\right.$$

$$-205\sqrt{2}\,\mathrm{e}^{\frac{t}{20}+\frac{3\sqrt{2}t}{25}+\frac{1}{50}(-2-3\sqrt{2})t} + 1182\,\mathrm{e}^{\frac{t}{20}+\frac{1}{50}(-2+3\sqrt{2})t}$$

$$+205\sqrt{2}\,\mathrm{e}^{\frac{t}{20}+\frac{1}{50}(-2+3\sqrt{2})t} + 330\,\mathrm{e}^{\frac{t}{20}+\frac{1}{50}(-2-3\sqrt{2})t+\frac{1}{50}(-2+3\sqrt{2})t}$$

$$+205\sqrt{2}\,\mathrm{e}^{\frac{t}{20}+\frac{1}{50}(-2-3\sqrt{2})t+\frac{1}{50}(-2+3\sqrt{2})t} - 756\,\mathrm{e}^{\frac{t}{10}+\frac{1}{50}(-2-3\sqrt{2})t+\frac{1}{50}(-2+3\sqrt{2})t}$$

$$-560\sqrt{2}\,\mathrm{e}^{\frac{t}{10}+\frac{1}{50}(-2-3\sqrt{2})t+\frac{1}{50}(-2+3\sqrt{2})t} + 330\,\mathrm{e}^{\frac{t}{20}+\frac{1}{25}(-2+3\sqrt{2})t}$$

$$-205\sqrt{2}\,\mathrm{e}^{\frac{t}{20}+\frac{1}{25}(-2+3\sqrt{2})t} - 756\,\mathrm{e}^{\frac{t}{10}+\frac{1}{25}(-2+3\sqrt{2})t} + 560\sqrt{2}\,\mathrm{e}^{\frac{t}{10}+\frac{1}{25}(-2+3\sqrt{2})t}$$

$$+756\,\mathrm{e}^{\frac{t}{20}+\frac{1}{50}(-2-3\sqrt{2})t+\frac{1}{100}(1+6\sqrt{2})t} + 560\sqrt{2}\,\mathrm{e}^{\frac{t}{20}+\frac{1}{50}(-2-3\sqrt{2})t+\frac{1}{100}(1+6\sqrt{2})t}$$

$$\left.\left.+756\,\mathrm{e}^{\frac{t}{20}+\frac{1}{50}(-2+3\sqrt{2})t+\frac{1}{100}(1+6\sqrt{2})t} - 560\sqrt{2}\,\mathrm{e}^{\frac{t}{20}+\frac{1}{50}(-2+3\sqrt{2})t+\frac{1}{100}(1+6\sqrt{2})t}\right)\right),$$

$$\mathrm{S3}[\mathrm{t}] \to -\frac{1}{1491}\left(50\,\mathrm{e}^{-\frac{t}{20}-\frac{1}{50}(-2+3\sqrt{2})t}\left(-336\,\mathrm{e}^{\frac{1}{50}(-2+3\sqrt{2})t}\right.\right.$$

$$+45\,\mathrm{e}^{\frac{t}{20}+\frac{3\sqrt{2}t}{25}+\frac{1}{50}(-2-3\sqrt{2})t} - 85\sqrt{2}\,\mathrm{e}^{\frac{t}{20}+\frac{3\sqrt{2}t}{25}+\frac{1}{50}(-2-3\sqrt{2})t} + 381\,\mathrm{e}^{\frac{t}{20}+\frac{1}{50}(-2+3\sqrt{2})t}$$

$$+85\sqrt{2}\,\mathrm{e}^{\frac{t}{20}+\frac{1}{50}(-2+3\sqrt{2})t} - 45\,\mathrm{e}^{\frac{t}{20}+\frac{1}{50}(-2-3\sqrt{2})t+\frac{1}{50}(-2+3\sqrt{2})t}$$

$$+85\sqrt{2}\,\mathrm{e}^{\frac{t}{20}+\frac{1}{50}(-2-3\sqrt{2})t+\frac{1}{50}(-2+3\sqrt{2})t} - 168\,\mathrm{e}^{\frac{t}{10}+\frac{1}{50}(-2-3\sqrt{2})t+\frac{1}{50}(-2+3\sqrt{2})t}$$

$$-14\sqrt{2}\,\mathrm{e}^{\frac{t}{10}+\frac{1}{50}(-2-3\sqrt{2})t+\frac{1}{50}(-2+3\sqrt{2})t} - 45\,\mathrm{e}^{\frac{t}{20}+\frac{1}{25}(-2+3\sqrt{2})t}$$

$$-85\sqrt{2}\,\mathrm{e}^{\frac{t}{20}+\frac{1}{25}(-2+3\sqrt{2})t} - 168\,\mathrm{e}^{\frac{t}{10}+\frac{1}{25}(-2+3\sqrt{2})t} + 14\sqrt{2}\,\mathrm{e}^{\frac{t}{10}+\frac{1}{25}(-2+3\sqrt{2})t}$$

$$+168\,\mathrm{e}^{\frac{t}{20}+\frac{1}{50}(-2-3\sqrt{2})t+\frac{1}{100}(1+6\sqrt{2})t} + 14\sqrt{2}\,\mathrm{e}^{\frac{t}{10}+\frac{1}{50}(-2-3\sqrt{2})t+\frac{1}{100}(1+6\sqrt{2})t}$$

$$\left.\left.+168\,\mathrm{e}^{\frac{t}{20}+\frac{1}{50}(-2+3\sqrt{2})t+\frac{1}{100}(1+6\sqrt{2})t} - 14\sqrt{2}\,\mathrm{e}^{\frac{t}{20}+\frac{1}{50}(-2+3\sqrt{2})t+\frac{1}{100}(1+6\sqrt{2})t}\right)\right)$$

One readily sees why this problem would be difficult to solve without technology.

APPENDIX E

Mathematical Induction

In Problems 1 through 10, prove the given propositions using mathematical induction. First show proposition is true for $k=1$. Then show proposition is true for $k=n+1$ assuming proposition is true for $k=n$.

If a proposition is true for $n=1$ and also if the proposition is true for $n=k$ whenever it is assumed true for $n=k-1$, then the proposition is true for all natural numbers $n=1, 2, 3, \ldots$

(1) $1+2+\cdots+n+n(n+1)/2.$

(2) $1+3+5+\cdots+(2n-1)=n^2.$

(3) $1^2+2^2+\cdots+n^2+n(n+1)(2n+1)/6.$

(4) $1^3+2^3+\cdots+n^3+n^2(n+1)^2/4.$

(5) $1^2+3^2+5^2+\cdots+(2n-1)^2=n(4n^2-1)/3.$

(6) $\sum_{k=1}^{n}\left(3k^2-k\right)=n^2(n+1).$

(7) $\sum_{k=1}^{n}\frac{1}{k(k+1)}=\frac{n}{(n+1)}.$

(8) $\sum_{k=1}^{n}2^{k-1}=2^n-1.$

(9) For any real number $x\neq 1$, $\sum_{k=1}^{n}x^{k-1}=\frac{x^n-1}{x-1}.$

(10) 7^n+2 is a multiple of 3.

Answers and Hints to Selected Problems

CHAPTER 1

Section 1.1

(1) **A** is 2×2, **B** is 2×2, **C** is 2×2, **D** is 4×2, **E** is 4×2,
F is 4×2, **G** is 2×3, **H** is 3×3, **J** is 1×5.

(2) $a_{12}=2$, a_{31} does not exist;
$b_{12}=6$, b_{31} does not exist;
$c_{12}=0$, c_{31} does not exist;
$d_{12}=1$, $d_{31}=3$;
$e_{12}=2$, $e_{31}=5$;
$f_{12}=1$, $f_{31}=0$;
$g_{12}=1/3$, e_{31} does not exist;
$h_{12}=\sqrt{3}$, $h_{31}=\sqrt{5}$;
$j_{12}=0$, j_{31} does not exist.

(3) $a_{11}=1$, $a_{21}=3$, b_{32} does not exist, $d_{32}=-2$,
d_{23} does not exist, $e_{22}=-2$, $g_{23}=-5/6$,
$h_{33}=\sqrt{3}$, j_{21} does not exist.

(4) A, B, C, and H.

(5) J is a row matrix.

(6) $\begin{bmatrix} 1 \\ 2 \\ 3 \\ 4 \end{bmatrix}$.

(7) $[1 \quad 4 \quad 9 \quad 16 \quad 25]$.

(8) $A = \begin{bmatrix} 1 & -1 \\ -1 & 1 \end{bmatrix}$.

(9) $A = \begin{bmatrix} 1 & 1/2 & 1/3 \\ 2 & 1 & 2/3 \\ 3 & 3/2 & 1 \end{bmatrix}$.

(10) $B = \begin{bmatrix} 1 & 0 & -1 \\ 0 & -1 & -2 \\ -1 & -2 & -3 \end{bmatrix}$.

(11) $C = \begin{bmatrix} 1 & 1 & 1 & 1 \\ 1 & 2 & 3 & 4 \end{bmatrix}$.

(12) $\mathbf{D} = \begin{bmatrix} 1 & -1 & -2 & -3 \\ 3 & 0 & -1 & -2 \\ 4 & 5 & 0 & -1 \end{bmatrix}$.

(13) $\begin{bmatrix} 2 & 4 \\ 6 & 8 \end{bmatrix}$.

(14) $\begin{bmatrix} -5 & -10 \\ -15 & -20 \end{bmatrix}$.

(15) $\begin{bmatrix} 9 & 3 \\ -3 & 6 \\ 9 & -6 \\ 6 & 18 \end{bmatrix}$.

(16) $\begin{bmatrix} -20 & 20 \\ 0 & -20 \\ 50 & -30 \\ 50 & 10 \end{bmatrix}$.

(17) $\begin{bmatrix} 0 & -1 \\ 1 & 0 \\ 0 & 0 \\ -2 & -2 \end{bmatrix}$.

(18) $\begin{bmatrix} 6 & 8 \\ 10 & 12 \end{bmatrix}$.

(19) $\begin{bmatrix} 0 & 2 \\ 6 & 1 \end{bmatrix}$.

(20) $\begin{bmatrix} 1 & 3 \\ -1 & 0 \\ 8 & -5 \\ 7 & 7 \end{bmatrix}$.

(21) $\begin{bmatrix} 3 & 2 \\ -2 & 2 \\ 3 & -2 \\ 4 & 8 \end{bmatrix}$.

(22) Does not exist.

(23) $\begin{bmatrix} -4 & -4 \\ -4 & -4 \end{bmatrix}$.

(24) $\begin{bmatrix} -2 & -2 \\ 0 & -7 \end{bmatrix}$.

(25) $\begin{bmatrix} 5 & -1 \\ -1 & 4 \\ -2 & 1 \\ -3 & 5 \end{bmatrix}$.

(26) $\begin{bmatrix} 3 & 0 \\ 0 & 2 \\ 3 & -2 \\ 0 & 4 \end{bmatrix}$.

(27) $\begin{bmatrix} 17 & 22 \\ 27 & 32 \end{bmatrix}$.

(28) $\begin{bmatrix} 5 & 6 \\ 3 & 18 \end{bmatrix}$.

(29) $\begin{bmatrix} -0.1 & 0.2 \\ 0.9 & -0.2 \end{bmatrix}$.

(30) $\begin{bmatrix} 4 & -3 \\ -1 & 4 \\ -10 & 6 \\ -8 & 0 \end{bmatrix}$.

(31) $\mathbf{X} = \begin{bmatrix} 4 & 4 \\ 4 & 4 \end{bmatrix}$.

(32) $\mathbf{Y} = \begin{bmatrix} -11 & -12 \\ -11 & -19 \end{bmatrix}$.

(33) $\mathbf{X} = \begin{bmatrix} 11 & 1 \\ -3 & 8 \\ 4 & -3 \\ 1 & 17 \end{bmatrix}$.

(34) $\mathbf{Y} = \begin{bmatrix} -1.0 & 0.5 \\ 0.5 & -1.0 \\ 2.5 & -1.5 \\ 1.5 & -0.5 \end{bmatrix}$.

(35) $\mathbf{R} = \begin{bmatrix} -2.8 & -1.6 \\ 3.6 & -9.2 \end{bmatrix}$.

(36) $\mathbf{S} = \begin{bmatrix} -1.5 & 1.0 \\ -1.0 & -1.0 \\ -1.5 & 1.0 \\ 2.0 & 0 \end{bmatrix}$.

(37) $\begin{bmatrix} -\theta^3 + 6\theta^2 + \theta & 6\theta - 6 \\ 21 & -\theta^3 - 2\theta^2 - \theta + 6/\theta \end{bmatrix}$.

(38) $[a_{ij}] + [b_{ij}] = [a_{ij} + b_{ij}] = [b_{ij} + a_{ij}] = [b_{ij}] + [a_{ij}]$.

(39) $[a_{ij}] + [0_{ij}] = [a_{ij} + 0_{ij}] = [a_{ij} + 0] = [a_{ij}]$.

(40) $(\lambda_1 + \lambda_2)[a_{ij}] = [(\lambda_1 + \lambda_2)a_{ij}] = [\lambda_1 a_{ij} + \lambda_2 a_{ij}] = [\lambda_1 a_{ij}] + [\lambda_2 a_{ij}] = \lambda_1 [a_{ij}] + \lambda_2 [a_{ij}]$.

(41) $(\lambda_1 \lambda_2)[a_{ij}] = [(\lambda_1 \lambda_2)a_{ij}] = [\lambda_1(\lambda_2 a_{ij})] = \lambda_1[\lambda_2 a_{ij}] = \lambda_1(\lambda_2[a_{ij}])$.

(42)
$$\begin{array}{c} \\ \\ \\ \end{array}\begin{array}{cccc} \textit{Refrigerators} & \textit{Stoves} & \textit{Washing machines} & \textit{Dryers} \end{array}$$
$$\begin{bmatrix} 3 & 5 & 3 & 4 \\ 0 & 2 & 9 & 5 \\ 4 & 2 & 0 & 0 \end{bmatrix}\begin{array}{l} \textit{store 1} \\ \textit{store 2} \\ \textit{store 3} \end{array}$$

(43) $\begin{bmatrix} 72 & 12 & 16 \\ 45 & 32 & 16 \\ 81 & 10 & 35 \end{bmatrix}$.

(44)
$$\begin{array}{cc} \textit{Purchase price} & \textit{Interest rate} \end{array}$$
$$\begin{bmatrix} 1,000 & 0.07 \\ 2,000 & 0.075 \\ 3,000 & 0.0725 \end{bmatrix}\begin{array}{l} \textit{first cerificate} \\ \textit{second cerificate} \\ \textit{third cerificate} \end{array}$$

(45) (a) $[200 \quad 150]$, (b) $[600 \quad 450]$, (c) $[550 \quad 350]$.

(46) (b) $[11 \quad 2 \quad 6 \quad 3]$, (c) $[9 \quad 4 \quad 10 \quad 8]$.

(47) (d) $[10,500 \quad 6,000 \quad 4,500]$, (e) $[35,500 \quad 14,500 \quad 3,300]$.

Section 1.2

(1) (a) 2×2, (b) 4×4, (c) 2×1,

(d) Not defined, (e) 4×2, (f) 2×4,

(g) 4×2, (h) Not defined, (i) Not defined,

(j) 1×4, (k) 4×4, (l) 4×2.

(2) $\begin{bmatrix} 19 & 22 \\ 43 & 50 \end{bmatrix}$. **(3)** $\begin{bmatrix} 23 & 24 \\ 31 & 46 \end{bmatrix}$. **(4)** $\begin{bmatrix} 5 & -4 & 3 \\ 9 & -8 & 7 \end{bmatrix}$.

(5) $\mathbf{A} = \begin{bmatrix} 13 & -12 & 11 \\ 17 & -16 & 15 \end{bmatrix}$.

(6) Not defined. **(7)** $[-5 \quad -6]$. **(8)** $[-9 \quad -10]$.

(9) $[-7 \quad 4 \quad -1]$. **(10)** Not defined. **(11)** $\begin{bmatrix} 1 & -3 \\ 7 & -3 \end{bmatrix}$.

(12) $\begin{bmatrix} 2 & -2 & 2 \\ 7 & -4 & 1 \\ -8 & 4 & 0 \end{bmatrix}.$

(13) $[1 \quad 3]$.

(14) Not defined.

(15) Not defined.

(16) Not defined.

(17) $\begin{bmatrix} -1 & -2 & -1 \\ 1 & 0 & -3 \\ 1 & 3 & 5 \end{bmatrix}.$

(18) $\begin{bmatrix} 2 & -2 & 1 \\ 2 & 0 & 0 \\ 1 & -2 & 2 \end{bmatrix}.$

(19) $[-1 \quad 1 \quad 5]$.

(21) $\mathbf{AB} = \mathbf{0}$.

(22) $\mathbf{AB} = \mathbf{AC} = \begin{bmatrix} 8 & 6 \\ 4 & 3 \end{bmatrix}.$

(23) $\mathbf{AB} = \mathbf{CB} = \begin{bmatrix} 8 & 16 \\ 2 & 4 \end{bmatrix}.$

(24) $\begin{bmatrix} x + 2y \\ 3x + 4y \end{bmatrix}.$

(25) $\begin{bmatrix} x - z \\ 3x + y + z \\ x + 3y \end{bmatrix}.$

(26) $\begin{bmatrix} a_{11}x + a_{12}y \\ a_{21}x + a_{22}y \end{bmatrix}.$

(27) $\begin{bmatrix} b_{11}x + b_{12}y + b_{13}z \\ b_{21}x + b_{22}y + b_{23}z \end{bmatrix}.$

(28) $\begin{bmatrix} 0 & 0 \\ 0 & 0 \end{bmatrix}.$

(29) $\begin{bmatrix} 0 & 40 \\ -16 & 8 \end{bmatrix}.$

(30) $\begin{bmatrix} 0 & 0 & 0 \\ 0 & 0 & 0 \\ 0 & 0 & 0 \end{bmatrix}.$

(33) Let the ith row of an $m \times n$ matrix $\mathbf{A}$ be 0. If $\mathbf{C} = \mathbf{AB}$, then for $j = 1,2,\ldots,n$,

$$c_{ij} = \sum_{k=1}^{n} a_{ik}b_{kj} = \sum_{k=1}^{n} (0)b_{kj} = 0$$

(34)

$$\begin{bmatrix} 1 & 2 \\ 1 & 4 \end{bmatrix} \begin{bmatrix} 1 & 1 \\ 0 & 0 \end{bmatrix} = \begin{bmatrix} 1 & 1 \\ 3 & 3 \end{bmatrix}.$$

(35) Let they jth column of an $m \times n$ matrix $\mathbf{B}$ be 0. If $\mathbf{C} = \mathbf{AB}$, then for $i = 1,2,\ldots,m$,

$$c_{ij} = \sum_{k=1}^{n} a_{ik}b_{kj} = \sum_{k=1}^{n} a_{ik}(0) = 0$$

(36) $\begin{bmatrix} 1 & 0 \\ 1 & 0 \end{bmatrix} \begin{bmatrix} 1 & 2 \\ 3 & 4 \end{bmatrix} = \begin{bmatrix} 1 & 2 \\ 1 & 2 \end{bmatrix}.$

(37) $$[a_{ij}]([b_{ij}][c_{ij}]) = [a_{ij}]\left[\sum_{k=1}^{n} b_{ik}c_{kj}\right] = \left[\sum_{p=1}^{m} a_{ip}\left(\sum_{k=1}^{n} b_{pk}c_{kj}\right)\right]$$
$$= \left[\sum_{p=1}^{m}\sum_{k=1}^{n} a_{ip}b_{pk}c_{kj}\right] = \left[\sum_{k=1}^{n}\sum_{p=1}^{m} a_{ip}b_{pk}c_{kj}\right]$$
$$= \left[\sum_{k=1}^{n}\left(\sum_{p=1}^{m} a_{ip}b_{pk}\right)c_{kj}\right]$$
$$= \left[\sum_{p=1}^{m} a_{ip}b_{pj}\right] [c_{ij}] = ([a_{ij}][b_{ij}])[c_{ij}]$$

(39) $\begin{bmatrix} 2 & 3 \\ 4 & -5 \end{bmatrix}\begin{bmatrix} x \\ y \end{bmatrix} = \begin{bmatrix} 10 \\ 11 \end{bmatrix}.$ **(40)** $\begin{bmatrix} 5 & 20 \\ -1 & 4 \end{bmatrix}\begin{bmatrix} x \\ y \end{bmatrix} = \begin{bmatrix} 80 \\ -64 \end{bmatrix}.$

(41) $\begin{bmatrix} 3 & 3 \\ 6 & -8 \\ -1 & 2 \end{bmatrix}\begin{bmatrix} x \\ y \end{bmatrix} = \begin{bmatrix} 100 \\ 300 \\ 500 \end{bmatrix}.$ **(42)** $\begin{bmatrix} 1 & 3 \\ 2 & -1 \\ -2 & -6 \\ 4 & -9 \\ -6 & 3 \end{bmatrix}\begin{bmatrix} x \\ y \end{bmatrix} = \begin{bmatrix} 4 \\ 1 \\ -8 \\ -5 \\ -3 \end{bmatrix}.$

(43) $\begin{bmatrix} 1 & 1 & -1 \\ 3 & 2 & 4 \end{bmatrix}\begin{bmatrix} x \\ y \\ z \end{bmatrix} = \begin{bmatrix} 0 \\ 0 \end{bmatrix}.$ **(44)** $\begin{bmatrix} 2 & -1 & 0 \\ 0 & -4 & -1 \end{bmatrix}\begin{bmatrix} x \\ y \\ z \end{bmatrix} = \begin{bmatrix} 12 \\ 15 \end{bmatrix}.$

(45) $\begin{bmatrix} 1 & 2 & -2 \\ 2 & 1 & 1 \\ -1 & 1 & -1 \end{bmatrix}\begin{bmatrix} x \\ y \\ z \end{bmatrix} = \begin{bmatrix} -1 \\ 5 \\ -2 \end{bmatrix}.$ **(46)** $\begin{bmatrix} 2 & 1 & -1 \\ 1 & 2 & 1 \\ 3 & -1 & 2 \end{bmatrix}\begin{bmatrix} x \\ y \\ z \end{bmatrix} = \begin{bmatrix} 0 \\ 0 \\ 0 \end{bmatrix}.$

(47) $\begin{bmatrix} 1 & 1 & 1 \\ 2 & 1 & 3 \\ 1 & 3 & 0 \end{bmatrix}\begin{bmatrix} x \\ y \\ z \end{bmatrix} = \begin{bmatrix} 2 \\ 4 \\ 1 \end{bmatrix}.$ **(48)** $\begin{bmatrix} 1 & 2 & -1 \\ 2 & -1 & 2 \\ 2 & 2 & -1 \\ 1 & 2 & 1 \end{bmatrix}\begin{bmatrix} x \\ y \\ z \end{bmatrix} = \begin{bmatrix} 5 \\ 1 \\ 7 \\ 3 \end{bmatrix}.$

(49) $\begin{bmatrix} 5 & 3 & 2 & 4 \\ 1 & 1 & 0 & 1 \\ 3 & 2 & 2 & 0 \\ 1 & 1 & 2 & 3 \end{bmatrix}\begin{bmatrix} x \\ y \\ z \\ w \end{bmatrix} = \begin{bmatrix} 5 \\ 0 \\ -3 \\ 4 \end{bmatrix}.$

(50) $\begin{bmatrix} 2 & -1 & 1 & -1 \\ 1 & 2 & -1 & 2 \\ 1 & -3 & 2 & -3 \end{bmatrix}\begin{bmatrix} x \\ y \\ z \\ w \end{bmatrix} = \begin{bmatrix} 1 \\ -1 \\ 2 \end{bmatrix}.$

(51) (a) $\mathbf{pn} = [38{,}000]$, which is the total revenue for the flight.

(b) $\mathbf{np} = \begin{bmatrix} 26{,}000 & 45{,}500 & 65{,}000 \\ 4{,}000 & 7{,}000 & 10{,}000 \\ 2{,}000 & 3{,}500 & 5{,}000 \end{bmatrix}$, which is of no significance.

(52) (a) $\mathbf{hP} = [9{,}625\ 9{,}762.50\ 9{,}887.50\ 10{,}100\ 9{,}887.50]$, which tabulates the value of the portfolio each day.
(b) $\mathbf{Ph}$ does not exist.

(53) $\mathbf{Tw} = [14.00\ 65.625\ 66.50]^{\mathrm{T}}$, which tabulates the cost of producing each product.

(54) $\mathbf{qTw} = [33{,}862.50]$, which is the cost of producing all items on order.

(55) $\mathbf{FC} = \begin{bmatrix} 613 & 625 \\ 887 & 960 \\ 1870 & 1915 \end{bmatrix}$, which tabulates the number of each gender in each state of sickness.

Section 1.3

(1) (a) $(\mathbf{AB})^{\mathrm{T}} = \mathbf{B}^{\mathrm{T}}\mathbf{A}^{\mathrm{T}} = \begin{bmatrix} -3 & -1 \\ 6 & -7 \\ 3 & 4 \end{bmatrix}$, $\mathbf{A}^{\mathrm{T}}\mathbf{B}^{\mathrm{T}}$ is not defined.

(b) $(\mathbf{AB})^{\mathrm{T}} = \mathbf{B}^{\mathrm{T}}\mathbf{A}^{\mathrm{T}} = \begin{bmatrix} 18 & 40 \\ 24 & 52 \end{bmatrix}$, $\mathbf{A}^{\mathrm{T}}\mathbf{B}^{\mathrm{T}}$ $\begin{bmatrix} 8 & 18 & 28 \\ 10 & 22 & 34 \\ 12 & 26 & 40 \end{bmatrix}$.

(c) $(\mathbf{AB})^{\mathrm{T}} = \mathbf{B}^{\mathrm{T}}\mathbf{A}^{\mathrm{T}} = \begin{bmatrix} 27 & 11 & 22 \\ 8 & -19 & 56 \\ -4 & 11 & -23 \end{bmatrix}$, $\mathbf{A}^{\mathrm{T}}\mathbf{B}^{\mathrm{T}}$ $\begin{bmatrix} 8 & 2 & -15 \\ 54 & 3 & 2 \\ -27 & 6 & -36 \end{bmatrix}$.

(2) $\begin{bmatrix} 7 & 4 & -1 \\ 6 & 1 & 0 \\ 2 & 2 & -6 \end{bmatrix}$.

(3) $\mathbf{x}^{\mathrm{T}}\mathbf{x} = [29]$ $\mathbf{xx}^{\mathrm{T}} = \begin{bmatrix} 4 & 6 & 8 \\ 6 & 9 & 12 \\ 8 & 12 & 16 \end{bmatrix}$.

(4) (a) $\mathbf{BA}^{\mathrm{T}}$, (b) $2\mathbf{A}^{\mathrm{T}} + \mathbf{B}$, (c) $(\mathbf{B}^{\mathrm{T}} + \mathbf{C})\mathbf{A}$, (d) $\mathbf{AB} + \mathbf{C}^{\mathrm{T}}$,
(e) $\mathbf{A}^{\mathrm{T}}\mathbf{A}^{\mathrm{T}} + \mathbf{A}^{\mathrm{T}}\mathbf{A} - \mathbf{AA}^{\mathrm{T}} - \mathbf{AA}$.

(5) (a), (b), and (d).

(6) $\begin{bmatrix} a & b \\ c & d \end{bmatrix}$, $[a\ \ b]$, $[c\ \ d]$, $\begin{bmatrix} b \\ c \end{bmatrix}$, $\begin{bmatrix} d \\ c \end{bmatrix}$, $[a]$, $[b]$, $[c]$, $[d]$.

(7) Partition $\mathbf{A}$ into four 2×2 submatrices. Then $\mathbf{A}^2 = \left[\begin{array}{cc|cc} 18 & 6 & 0 & 0 \\ 12 & 6 & 0 & 0 \\ \hline 0 & 0 & 1 & 0 \\ 0 & 0 & 3 & 4 \end{array}\right]$.

(8) Partition **B** into four 2×2 submatrices. Then $\mathbf{B}^2 = \left[\begin{array}{cc|cc} 7 & 8 & 0 & 0 \\ -4 & -1 & 0 & 0 \\ \hline 0 & 0 & 5 & 1 \\ 0 & 0 & 1 & 2 \end{array}\right]$.

(9) $\mathbf{AB} = \left[\begin{array}{cc|cc} 11 & 9 & 0 & 0 \\ 4 & 6 & 0 & 0 \\ \hline 0 & 0 & 2 & 1 \\ 0 & 0 & 4 & -1 \end{array}\right]$.

(10) $\mathbf{A}^2 = \left[\begin{array}{cc|cccc} 1 & 0 & 0 & 0 & 0 & 0 \\ 0 & 4 & 0 & 0 & 0 & 0 \\ \hline 0 & 0 & 0 & 0 & 1 & 0 \\ 0 & 0 & 0 & 0 & 0 & 1 \\ 0 & 0 & 0 & 0 & 0 & 0 \\ 0 & 0 & 0 & 0 & 0 & 0 \end{array}\right]$, $\mathbf{A}^3 = \left[\begin{array}{cc|cccc} 1 & 0 & 0 & 0 & 0 & 0 \\ 0 & 8 & 0 & 0 & 0 & 0 \\ \hline 0 & 0 & 0 & 0 & 0 & 1 \\ 0 & 0 & 0 & 0 & 0 & 0 \\ 0 & 0 & 0 & 0 & 0 & 0 \\ 0 & 0 & 0 & 0 & 0 & 0 \end{array}\right]$,

$$\mathbf{A}^n = \begin{bmatrix} 1 & 0 & 0 & 0 & 0 & 0 \\ 0 & 2^n & 0 & 0 & 0 & 0 \\ 0 & 0 & 0 & 0 & 0 & 0 \\ 0 & 0 & 0 & 0 & 0 & 0 \\ 0 & 0 & 0 & 0 & 0 & 0 \\ 0 & 0 & 0 & 0 & 0 & 0 \end{bmatrix}, \qquad n = 4,5,6,\ldots$$

(11) **A**, **B**, **F**, **M**, **N**, **R**, and **T**.

(12) **E**, **F**, **H**, **K**, **L**, **M**, **N**, **R**, and **T**.

(13) Yes.

(14) No, see **H** and **L** in Problem 11.

(15) Yes, see **L** in Problem 11.

(16) $\mathbf{AB} = \mathbf{BA} = \begin{bmatrix} -5 & 0 & 0 \\ 0 & 9 & 0 \\ 0 & 0 & 2 \end{bmatrix}$.

(18) No.

(19) If $\mathbf{D} = [d_{ij}]$ is a diagonal matrix, then they jth column of **AD** is they jth column of **A** multiplied by d_{jj}.

(20) If $\mathbf{D} = [d_{ij}]$ is a diagonal matrix, then the ith row of **DA** is the ith row of **A** multiplied by d_{ii}.

(21) Let $\mathbf{A} = [a_{ij}]$. Then $(\mathbf{A}^{\mathrm{T}})^{\mathrm{T}} = [a_{ji}]^{\mathrm{T}} = [a_{ij}] = \mathbf{A}$.

(22) Let $\mathbf{A} = [a_{ij}]$. Then $(\lambda\mathbf{A})^{\mathrm{T}} = [\lambda a_{ji}] = \lambda[a_{ji}] = \lambda\mathbf{A}^{\mathrm{T}}$.

(23) $(\mathbf{A}+\mathbf{B})^{\mathrm{T}} = ([a_{ij}]+[b_{ij}])^{\mathrm{T}} = [a_{ij}+b_{ij}]^{\mathrm{T}} = [a_{ji}+\mathrm{bji}] = [a_{ji}]+[\mathrm{bji}] = \mathbf{A}^{\mathrm{T}}+\mathbf{B}^{\mathrm{T}}$.

(24) $(\mathrm{ABC})^{\mathrm{T}}=[(\mathrm{AB})\mathrm{C}]^{\mathrm{T}}=\mathrm{C}^{\mathrm{T}}(\mathrm{AB})^{\mathrm{T}}=\mathrm{C}^{\mathrm{T}}(\mathrm{B}^{\mathrm{T}}\mathrm{A}^{\mathrm{T}})$.

(25) $\mathbf{B}^{\mathrm{T}}=\left[(\mathrm{A}+\mathrm{A}^{\mathrm{T}})/2\right]^{\mathrm{T}}=\frac{1}{2}(\mathrm{A}+\mathrm{A}^{\mathrm{T}})^{\mathrm{T}}=\frac{1}{2}\left[\mathrm{A}^{\mathrm{T}}+(\mathrm{A}^{\mathrm{T}})^{\mathrm{T}}\right]=\frac{1}{2}(\mathrm{A}^{\mathrm{T}}+\mathrm{A})=\mathbf{B}$.

(26)
$$\begin{aligned}\mathrm{C}^{\mathrm{T}}&=\left[(\mathrm{A}-\mathrm{A}^{\mathrm{T}})/2\right]^{\mathrm{T}}=\frac{1}{2}(\mathrm{A}-\mathrm{A}^{\mathrm{T}})^{\mathrm{T}}=\frac{1}{2}\left[\mathrm{A}^{\mathrm{T}}-(\mathrm{A}^{\mathrm{T}})^{\mathrm{T}}\right]\\&=\frac{1}{2}(\mathrm{A}^{\mathrm{T}}-\mathrm{A})=-\frac{1}{2}(\mathrm{A}-\mathrm{A}^{\mathrm{T}})=-\mathrm{C}.\end{aligned}$$

(27) $\mathrm{A}=\frac{1}{2}(\mathrm{A}+\mathrm{A}^{\mathrm{T}})+\frac{1}{2}(\mathrm{A}-\mathrm{A}^{\mathrm{T}})$.

(28) $$\begin{bmatrix}1&7/2&-1/2\\7/2&1&5\\-1/2&5&-8\end{bmatrix}+\begin{bmatrix}0&3/2&-1/2\\-3/2&0&-2\\1/2&2&0\end{bmatrix}.$$

(29) $$\begin{bmatrix}6&3/2&1\\3/2&0&-4\\1&-4&2\end{bmatrix}+\begin{bmatrix}0&-1/2&2\\1/2&0&3\\-2&-3&0\end{bmatrix}.$$

(30) $(\mathrm{AA}^{\mathrm{T}})^{\mathrm{T}}=(\mathrm{A}^{\mathrm{T}})^{\mathrm{T}}\mathrm{A}^{\mathrm{T}}=\mathrm{AA}^{\mathrm{T}}$

(31) Each diagonal element must equal its own negative and, therefore, must be zero.

(33) For any $n\times n$ matrix $\mathbf{A}$, consider sequentially the equations $\mathbf{AD}_{\mathrm{i}}=\mathbf{D}_{\mathrm{i}}\mathbf{A}$, where all the elements in $\mathbf{D}_i$ $(i=1,2,\ldots,n)$ are zero except for a single 1 in the i-i position.

Section 1.4

(1) (a) No. (b) Yes.

(2) (a) Yes. (b) No. (c) Yes.

(3) $k=1$.

(4) $k=1/12$.

(5) k is arbitrary; any value will work.

(6) $$\begin{aligned}x+2y&=5\\y&=8\end{aligned}$$
Solution: $x=-11,\ y=8$

(7) $$\begin{aligned}x-2y+3z&=10\\y-5z&=-3\\z&=4\end{aligned}$$
Solution: $x=32,\ y=17, z=4$

(8) $$\begin{aligned}x_1-3x_2+12x_3&=40\\x_2-5x_3&=-200\\x_3&=25\end{aligned}$$
Solution: $x_1=-410$, $x_2=-50$, $x_3=25$

(9) $$\begin{aligned}x+3y&=-8\\y+4z&=2\\0&=0\end{aligned}$$
Solution: $x=-14+12z$, $y=2-4z$

(10) $x_1 - 7x_2 + 2x_3 = 0$
$x_2 - x_3 = 0$
$0 = 0$
Solution: $x_1 = 5x_3$,
$x_2 = x_3$,
x_3 is arbitrary

(11) $x_1 - x_2 = 1$
$x_2 - 2x_3 = 2$
$x_3 = -3$
$0 = 1$
No solution

(12) $x=51, y=23$.

(13) $x=-103, y=18$.

(14) $x=18.5, y=-6$.

(15) $x=y=0$.

(16) $x=3y$, y is arbitrary.

(17) $x=-3/29, y=-2/29, z=41/29$.

(18) $x=3/23, y=28/23, z=-32/23$.

(19) $x=48/35, y=-10/35, z=-9/35$.

(20) No solution.

(21) $x=2y-z$, y and z are arbitrary.

(22) $x=y=z=0$.

(23) $x_1=-x_3$, $x_2=0$, x_3 is arbitrary.

(24) $x_1=x_2-2x_3$, x_2 and x_3 are arbitrary.

(25) $x_1=1, x_2=-2$.

(26) $x_1 = \frac{5}{7} - x_3$, $x_2 = -\frac{6}{7}$, x_3 is arbitrary.

(27) $x_1=-3, x_2=4$.

(28) $x_1=13/3, x_2=x_3=-5/3$.

(29) No solution.

(30) Each equation graphs as a plane. If the planes do not intersect, the equations have no solutions. If the planes do intersect, their intersection is either a line or a plane, each yielding infinitely many solutions.

(31) $\mathbf{Au}=\mathbf{A}(\mathbf{y}+\mathbf{z})=\mathbf{Ay}+\mathbf{Az}=\mathbf{b}+\mathbf{0}=\mathbf{b}$.

(32) (a) α can be any real number.
(b) $\alpha=1$.

(33) $50r + 60s = 70,000$
$30r + 40s = 45,000$
Solution: $r = 500, s = 750$

(34) $5d + 0.25b = 200$
$10d + b = 500$
Solution: $d = 30, b = 200$

(35) $8,000A + 3,0005B + 1,000C = 70,000$
$5,000A + 12,0005B + 10,000C = 181,000$
$1,000A + 3,0005B + 2,000C = 41,000$
Solution: $A = 5, \quad B = 8, \quad C = 6$

(36) $b + 0.05c + 0.05s = 20,000$
$c = 8,000$
$0.03c + s = 12,000$
Solution: $b = \$19,012$

(37) (a) $C = 800,000 + 30B$
$S = 40B$
(b) Add the additional equation $S=C$. Then $B=80{,}000$.

(38) $-0.60p_1 + 030p_2 + 0.50p_3 = 0$
$0.40p_1 - 0.75p_2 + 0.350p_3 = 0$
$0.20p_1 + 0.45p_2 + 0.85p_3 = 0$
Solution: $p_1 = (48/33)p_3, p_2 = (41/33)p_3, p_3$is arbitrary.

(39) $(-1/2)p_1 + (1/3)p_2 + (1/6)p_3 = 0$
$(1/4)p_1 - (2/3)p_2 + (1/3)p_3 = 0$
$(1/4)p_1 + (1/3)p_2 - (1/2)p_3 = 0$
Solution: $p_1 = (8/9)p_3, p_2 = (5/6)p_3, p_3$is arbitrary.

$-0.85p_1 + 0.10p_2 + 0.15p_4 = 0$
$0.20p_1 - 0.60p_2 + \frac{1}{3}p_3 + 0.40p_4 = 0$

(40) $0.30p_1 + 0.15p_2 - \frac{2}{3}p_3 + 0.45p_4 = 0$
$0.35p_1 + 0.35p_2 + \frac{1}{3}p_3 - p_4 = 0$
Solution: $p_1 \approx 0.3435p_4, p_2 \approx 1.4195p_4, p_3 \approx 1.1489p_4, p_4$is arbitrary.

(41) 4. **(42)** 5. **(43)** 4. **(44)** 9. **(45)** 3. **(46)** 4.

Section 1.5

(1) –2. **(2)** 38. **(3)** 38. **(4)** –2.
(5) 82. **(6)** –82. **(7)** 9. **(8)** –20.
(9) 21. **(10)** –6. **(11)** 22. **(12)** 0.
(13) –9. **(14)** –33. **(15)** 15. **(16)** –5.
(17) –10. **(18)** 0. **(19)** 0. **(20)** 0.
(21) 119. **(22)** –8. **(23)** 22. **(24)** –7.
(25) –40. **(26)** 52. **(27)** 25. **(28)** 0.

(29) 0. **(30)** –11. **(31)** 0.

(32) 0 and 2. **(33)** –1 and 4. **(34)** 2 and 3.

(35) $\pm\sqrt{6}$. **(36)** $\lambda^2 - 9\lambda - 2$.

(37) $\lambda^2 - 9\lambda + 38$. **(38)** $\lambda^2 - 13\lambda - 2$.

(39) $\lambda^2 - 8\lambda + 9$. **(40)** $\lambda^3 + 7\lambda + 22$.

(41) $\lambda^3 + 4\lambda^2 - 17\lambda$. **(42)** $-\lambda^3 + 6\lambda - 9$.

(43) $-\lambda^3 + 10\lambda^2 - 22\lambda - 33$.

(44) $|\mathbf{A}| = 11,\ \ |\mathbf{B}| = 5,\ \ |\mathbf{AB}| = 55$.

(45) 3. **(46)** 24. **(47)** 28.

(48) –1. **(49)** 0. **(50)** –311.

(51) –10. **(52)** 0. **(53)** –5.

(54) 0. **(55)** 0. **(56)** 119.

(57) –9. **(58)** –33. **(59)** 15.

(60) 2187. **(61)** 52. **(62)** 25.

(63) 0. **(64)** 0. **(65)** 152.

(66) Multiply the first row by 2, the second row by –1, and the second column by 2.

(67) Apply the third elementary row operation with the third row to make the first two rows identical.

(68) Multiply the first column by 1/2, the second column by 1/3, to obtain identical columns.

(69) Interchange the second and third rows, and then transpose.

(70) Use the third column to simplify both the first and second columns.

(71) Factor the numbers –1, 2, 2, and 3 from the third row, second row, first column, and second column, respectively.

(72) Factor a 5 from the third row. Then use this new third row to simplify the second row and the new second row to simplify the first row.

(74) $\det\left\{3\begin{bmatrix}1 & 3\\ -3 & 4\end{bmatrix}\right\} = \begin{vmatrix}3 & 9\\ -9 & 12\end{vmatrix} = 117 = 9(13) = (3)^2\begin{vmatrix}1 & 3\\ -3 & 4\end{vmatrix}.$

(75) $\det\left\{-2\begin{bmatrix}2 & 3\\ -3 & -2\end{bmatrix}\right\} = \begin{vmatrix}-4 & -6\\ 6 & 4\end{vmatrix} = 20 = 4(5) = (-2)^2\begin{vmatrix}2 & 3\\ -3 & -2\end{vmatrix}.$

(76) $\det\left\{-1\begin{bmatrix}1 & 2 & -2\\ 1 & 3 & 3\\ 2 & 5 & 0\end{bmatrix}\right\} = \begin{vmatrix}-1 & -2 & 2\\ -1 & -3 & -3\\ -2 & -5 & 0\end{vmatrix} = 1 = (-1)(-1) = (-1)^3\begin{vmatrix}1 & 2 & -2\\ 1 & 3 & 3\\ 2 & 5 & 0\end{vmatrix}.$

Section 1.6

(1) (c)

(2) None.

(3) $\begin{bmatrix} 0 & 1 \\ 1 & 0 \end{bmatrix}$.

(4) $\begin{bmatrix} 3 & 0 \\ 0 & 1 \end{bmatrix}$.

(5) $\begin{bmatrix} 3 & 0 \\ 0 & -5 \end{bmatrix}$.

(6) $\begin{bmatrix} 1 & 0 & 0 \\ 0 & -5 & 0 \\ 0 & 0 & 1 \end{bmatrix}$.

(7) $\begin{bmatrix} 1 & 0 \\ 3 & 1 \end{bmatrix}$.

(8) $\begin{bmatrix} 1 & 3 \\ 0 & 1 \end{bmatrix}$.

(9) $\begin{bmatrix} 1 & 0 & 0 \\ 0 & 1 & 3 \\ 0 & 0 & 1 \end{bmatrix}$.

(10) $\begin{bmatrix} 1 & 0 & 0 \\ 0 & 1 & 0 \\ 5 & 0 & 1 \end{bmatrix}$.

(11) $\begin{bmatrix} 1 & 0 & 0 & 0 & 0 & 0 \\ 0 & 0 & 0 & 1 & 0 & 0 \\ 0 & 0 & 1 & 0 & 0 & 0 \\ 0 & 1 & 0 & 0 & 0 & 0 \\ 0 & 0 & 0 & 0 & 1 & 0 \\ 0 & 0 & 0 & 0 & 0 & 1 \end{bmatrix}$.

(12) $\begin{bmatrix} 1 & 0 \\ 0 & 7 \end{bmatrix}$.

(13) $\begin{bmatrix} 1/2 & 0 \\ 0 & 1 \end{bmatrix}$.

(14) $\begin{bmatrix} 1 & -2 \\ 0 & 1 \end{bmatrix}$.

(15) $\begin{bmatrix} 1 & 0 \\ 3 & 1 \end{bmatrix}$.

(16) $\begin{bmatrix} 1 & 0 \\ -1 & 1 \end{bmatrix}$.

(17) $\begin{bmatrix} 1 & 0 & 0 \\ 0 & 1/2 & 0 \\ 0 & 0 & 1 \end{bmatrix}$.

(18) $\begin{bmatrix} 0 & 1 & 0 \\ 1 & 0 & 0 \\ 0 & 0 & 1 \end{bmatrix}$.

(19) $\begin{bmatrix} 1 & 0 & -3 \\ 0 & 1 & 0 \\ 0 & 0 & 1 \end{bmatrix}$.

(20) $\begin{bmatrix} 1 & 0 & 0 \\ 0 & 1 & 2 \\ 0 & 0 & 1 \end{bmatrix}$.

(21) $\begin{bmatrix} 1 & 0 & 0 & 0 \\ 0 & 1 & 0 & 0 \\ 0 & 0 & 0 & 1 \\ 0 & 0 & 1 & 0 \end{bmatrix}$.

(22) $\begin{bmatrix} 1 & 0 & 0 & 0 \\ 0 & 1 & 0 & 0 \\ 3 & 0 & 1 & 0 \\ 0 & 0 & 0 & 1 \end{bmatrix}$.

(23) $\begin{bmatrix} 4 & -1 \\ -3 & 1 \end{bmatrix}$.

(24) $\frac{1}{3}\begin{bmatrix} 2 & -1 \\ -1 & 2 \end{bmatrix}$.

(25) Does not exist.

(26) $\frac{1}{2}\begin{bmatrix} 1 & 1 & -1 \\ 1 & -1 & 1 \\ -1 & 1 & 1 \end{bmatrix}$.

(27) $\begin{bmatrix} 0 & 1 & 0 \\ 0 & 0 & 1 \\ 1 & 0 & 0 \end{bmatrix}$.

(28) $\begin{bmatrix} -1 & -1 & 1 \\ 6 & 5 & -4 \\ -3 & -2 & 2 \end{bmatrix}$.

(29) Does not exist.

(30) $\frac{1}{2}\begin{bmatrix} 1 & 0 & 0 \\ -5 & 2 & 0 \\ 1 & -2 & 2 \end{bmatrix}$.

(31) $\frac{1}{6}\begin{bmatrix} 3 & -1 & -8 \\ 0 & 2 & 1 \\ 0 & 0 & 3 \end{bmatrix}$.

(32) $\begin{bmatrix} 9 & -5 & -2 \\ 5 & -3 & -1 \\ -36 & 21 & 8 \end{bmatrix}$.

(33) $\frac{1}{17}\begin{bmatrix} 1 & 7 & -2 \\ 7 & -2 & 3 \\ -2 & 3 & 4 \end{bmatrix}$.

(34) $\frac{1}{17}\begin{bmatrix} 14 & 5 & -6 \\ -5 & -3 & 7 \\ 13 & 1 & -8 \end{bmatrix}$.

(35) Does not exist.

(36) $\frac{1}{33}\begin{bmatrix} 5 & 3 & 1 \\ -6 & 3 & 12 \\ -8 & 15 & 5 \end{bmatrix}$.

(37) $\frac{1}{4}\begin{bmatrix} 0 & -4 & 4 \\ 1 & 5 & -4 \\ 3 & 7 & -8 \end{bmatrix}$.

(38) $\frac{1}{4}\begin{bmatrix} 4 & -4 & -4 & -4 \\ 0 & 4 & 2 & 5 \\ 0 & 0 & 2 & 3 \\ 0 & 0 & 0 & -2 \end{bmatrix}$.

(39) $\begin{bmatrix} 1 & 0 & 0 & 0 \\ 2 & -1 & 0 & 0 \\ -8 & 3 & 1/2 & 0 \\ -25 & 10 & 2 & -1 \end{bmatrix}$.

(41) (a) $\begin{bmatrix} 4 & -1 \\ -3 & 1 \end{bmatrix}$. (b) $\begin{bmatrix} 4 & -6 \\ -6 & 12 \end{bmatrix}$.

(42) $x=1$, $y=-2$.

(43) $a=-3$, $b=4$.

(44) $x=5/4$, $y=1/2$.

(45) $l=1$, $p=3$.

(46) Not possible; the coefficient matrix is singular.

(47) $x=-8$, $y=5$, $z=3$.

(48) $x=y=z=1$.

(49) $l=1$, $m=-2$, $n=0$.

(50) Not possible; the coefficient matrix is singular.

(51) $x=y=1$, $z=2$.

(52) (a) $x=70$, $y=-40$.

(53) (a) $x=13/3$, $y=-5/3$, $z=-5/3$.

(b) $x=113/30$, $y=-34/30$, $z=-31/30$.

(c) $x=79/15$, $y=-32/15$, $z=-38/15$.

(d) $x=41/10$, $y=-18/10$, $z=-17/10$.

(54) (a) $\mathbf{A}^{-2}=\begin{bmatrix} 11 & -4 \\ -8 & 3 \end{bmatrix}$, $\mathbf{A}^{-3}=\begin{bmatrix} 41 & -15 \\ -30 & 11 \end{bmatrix}$.

(b) $A^{-2} = \begin{bmatrix} 9 & -20 \\ -4 & 9 \end{bmatrix}, \quad A^{-3} = \begin{bmatrix} -38 & 85 \\ 17 & -38 \end{bmatrix}.$

(c) $A^{-2} = \begin{bmatrix} 19 & -5 \\ -15 & 4 \end{bmatrix}, \quad A^{-3} = \begin{bmatrix} 91 & -24 \\ -72 & 19 \end{bmatrix}.$

(d) $A^{-2} = \begin{bmatrix} 1 & -2 & 1 \\ 0 & 1 & -2 \\ 0 & 0 & 1 \end{bmatrix}, \quad A^{-3} = \begin{bmatrix} 1 & -3 & 3 \\ 0 & 1 & -3 \\ 0 & 0 & 1 \end{bmatrix}.$

(e) $A^{-2} = \begin{bmatrix} 1 & -4 & -4 \\ 0 & 1 & 2 \\ 0 & 0 & 1 \end{bmatrix}, \quad A^{-3} = \begin{bmatrix} 1 & -6 & -9 \\ 0 & 1 & 3 \\ 0 & 0 & 1 \end{bmatrix}.$

(56) Use the result of Problem 19 or Problem 20 of Section 1.3.

(58) $(BA^{-1})^T(A^{-1}B^T)^{-1} = \left[(A^{-1})^T B^T\right)\left[(B^T)^{-1}(A^{-1})^{-1}\right] = \left[(A^T)^{-1}B^T\right)\right]$

$$\left[(B^T)^{-1}A\right] = A^{-1}\left[B^T(B^T)^{-1}\right]A = A^{-1}IA = A^{-1}A = I$$

(60) $[(1/\lambda)A^{-1}][\lambda A] = (1/\lambda)(\lambda)A^{-1}A = 1I = I.$

(61) $(ABC)^{-1} = [(AB)C]^{-1} = C^{-1}\,(AB)^{-1} = C^{-1}\,(B^{-1}A^{-1})$

Section 1.7

(1) $\begin{bmatrix} 1 & 0 \\ 3 & 1 \end{bmatrix}\begin{bmatrix} 1 & 1 \\ 0 & 1 \end{bmatrix}, \quad \mathbf{x} = \begin{bmatrix} 10 \\ -9 \end{bmatrix}.$

(2) $\begin{bmatrix} 1 & 0 \\ 0.5 & 1 \end{bmatrix}\begin{bmatrix} 2 & 1 \\ 0 & 1.5 \end{bmatrix}, \quad \mathbf{x} = \begin{bmatrix} 8 \\ -5 \end{bmatrix}.$

(3) $\begin{bmatrix} 1 & 0 \\ 0.625 & 1 \end{bmatrix}\begin{bmatrix} 8 & 3 \\ 0 & 0.125 \end{bmatrix}, \quad \mathbf{x} = \begin{bmatrix} -400 \\ 1275 \end{bmatrix}.$

(4) $\begin{bmatrix} 1 & 0 & 0 \\ 1 & 1 & 0 \\ 0 & -1 & 1 \end{bmatrix}\begin{bmatrix} 1 & 1 & 0 \\ 0 & -1 & 1 \\ 0 & 0 & 2 \end{bmatrix}, \quad \mathbf{x} = \begin{bmatrix} 3 \\ 1 \\ -2 \end{bmatrix}.$

(5) $\begin{bmatrix} 1 & 0 & 0 \\ -1 & 1 & 0 \\ -2 & -2 & 1 \end{bmatrix}\begin{bmatrix} -1 & 2 & 0 \\ 0 & -1 & 1 \\ 0 & 0 & 5 \end{bmatrix}, \quad \mathbf{x} = \begin{bmatrix} 5 \\ 2 \\ -1 \end{bmatrix}.$

(6) $\begin{bmatrix} 1 & 0 & 0 \\ 2 & 1 & 0 \\ -1 & 0 & 1 \end{bmatrix}\begin{bmatrix} 2 & 1 & 3 \\ 0 & -1 & -6 \\ 0 & 0 & 1 \end{bmatrix}, \quad \mathbf{x} = \begin{bmatrix} -10 \\ 0 \\ 10 \end{bmatrix}.$

(7) $\begin{bmatrix} 1 & 0 & 0 \\ \frac{4}{3} & 1 & 0 \\ 1 & \frac{-21}{8} & 1 \end{bmatrix}\begin{bmatrix} 3 & 2 & 1 \\ 0 & -\frac{8}{3} & -\frac{8}{3} \\ 0 & 0 & \frac{1}{8} \end{bmatrix}, \quad \mathbf{x} = \begin{bmatrix} 10 \\ -10 \\ 40 \end{bmatrix}.$

(8) $\begin{bmatrix} 1 & 0 & 0 \\ 2 & 1 & 0 \\ -1 & -0.75 & 1 \end{bmatrix}\begin{bmatrix} 1 & 2 & -1 \\ 0 & -4 & 3 \\ 0 & 0 & 4.25 \end{bmatrix}$, $\mathbf{x} = \begin{bmatrix} 79 \\ 1 \\ 1 \end{bmatrix}$.

(9) $\begin{bmatrix} 1 & 0 & 0 \\ 0 & 1 & 0 \\ 0 & 0 & 1 \end{bmatrix}\begin{bmatrix} 1 & 2 & -1 \\ 0 & 2 & 1 \\ 0 & 0 & 1 \end{bmatrix}$, $\mathbf{x} = \begin{bmatrix} 19 \\ -3 \\ 5 \end{bmatrix}$.

(10) $\begin{bmatrix} 1 & 0 & 0 \\ 3 & 1 & 0 \\ 1 & \frac{1}{2} & 1 \end{bmatrix}\begin{bmatrix} 1 & 0 & 0 \\ 0 & 2 & 0 \\ 0 & 0 & 2 \end{bmatrix}$, $\mathbf{x} = \begin{bmatrix} 2 \\ -1 \\ 1/2 \end{bmatrix}$.

(11) $\begin{bmatrix} 1 & 0 & 0 & 0 \\ 1 & 1 & 0 & 0 \\ 1 & 1 & 1 & 0 \\ 0 & 1 & 2 & 1 \end{bmatrix}\begin{bmatrix} 1 & 0 & 1 & 1 \\ 0 & 1 & -1 & 0 \\ 0 & 0 & 1 & -1 \\ 0 & 0 & 0 & 3 \end{bmatrix}$, $\mathbf{x} = \begin{bmatrix} 1 \\ -5 \\ 2 \\ 1 \end{bmatrix}$.

(12) $\begin{bmatrix} 1 & 0 & 0 & 0 \\ \frac{1}{2} & 1 & 0 & 0 \\ 0 & 0 & 1 & 0 \\ 0 & \frac{2}{7} & \frac{5}{7} & 1 \end{bmatrix}\begin{bmatrix} 2 & 1 & -1 & 3 \\ 0 & \frac{7}{2} & \frac{5}{2} & -\frac{1}{2} \\ 0 & 0 & -1 & 1 \\ 0 & 0 & 0 & \frac{3}{7} \end{bmatrix}$, $\mathbf{x} \approx \begin{bmatrix} 266.67 \\ -166.67 \\ 166.67 \\ 266.67 \end{bmatrix}$.

(13) $\begin{bmatrix} 1 & 0 & 0 & 0 \\ 1 & 1 & 0 & 0 \\ 1 & 1 & 1 & 0 \\ 0 & -1 & -2 & 1 \end{bmatrix}\begin{bmatrix} 1 & 2 & 1 & 1 \\ 0 & -1 & 1 & 0 \\ 0 & 0 & -1 & 1 \\ 0 & 0 & 0 & 3 \end{bmatrix}$, $\mathbf{x} = \begin{bmatrix} 10 \\ 10 \\ 10 \\ -10 \end{bmatrix}$.

(14) $\begin{bmatrix} 1 & 0 & 0 & 0 \\ 1 & 1 & 0 & 0 \\ -2 & 1.5 & 1 & 0 \\ 0.5 & 0 & 0.25 & 1 \end{bmatrix}\begin{bmatrix} 2 & 0 & 2 & 0 \\ 0 & 2 & -2 & 6 \\ 0 & 0 & 8 & -8 \\ 0 & 0 & 0 & 3 \end{bmatrix}$, $\mathbf{x} = \begin{bmatrix} -2.5 \\ -1.5 \\ 1.5 \\ 2.0 \end{bmatrix}$.

(15) (a) $x=5$, $y=-2$; (b) $x=-5/7$, $y=1/7$.

(16) (a) $x=1$, $y=0$, $z=2$; (b) $x=140$, $y=-50$, $z=-20$.

(17) (a) $\begin{bmatrix} 8 \\ -3 \\ -1 \end{bmatrix}$, (b) $\begin{bmatrix} 2 \\ 0 \\ 0 \end{bmatrix}$, (c) $\begin{bmatrix} 35 \\ 5 \\ 15 \end{bmatrix}$, (d) $\begin{bmatrix} -0.5 \\ 1.5 \\ 1.5 \end{bmatrix}$.

(18) (a) $\begin{bmatrix} -1 \\ -1 \\ 1 \\ 1 \end{bmatrix}$, (b) $\begin{bmatrix} 0 \\ 0 \\ 0 \\ 0 \end{bmatrix}$, (c) $\begin{bmatrix} 80 \\ 50 \\ -10 \\ 20 \end{bmatrix}$, (d) $\begin{bmatrix} -1/3 \\ 1/3 \\ 1/3 \\ 1/3 \end{bmatrix}$.

(21) (d) $\mathbf{A}$ is singular.

CHAPTER 2

Section 2.1

(1)

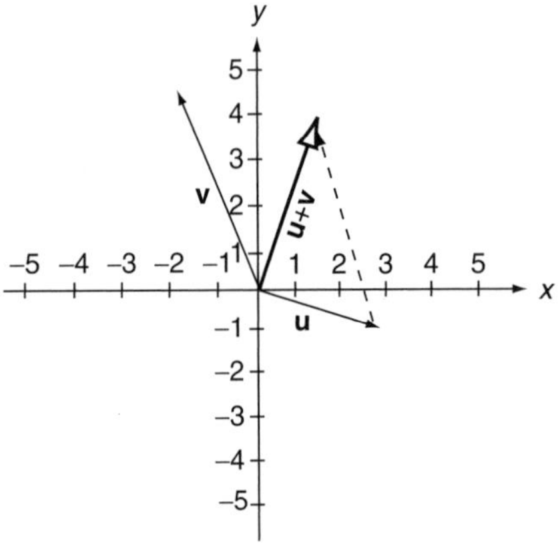

(4)

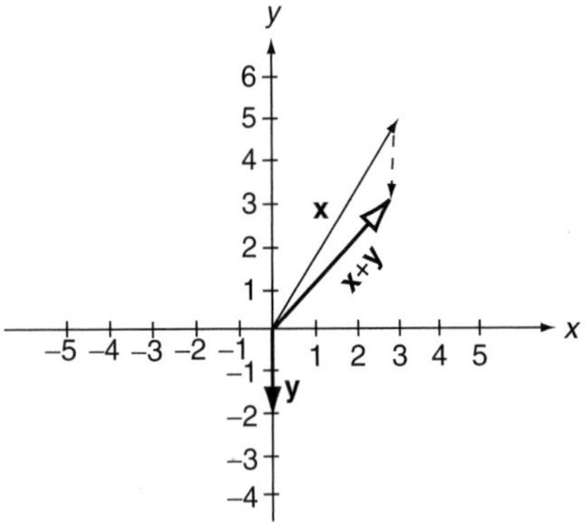

(6)

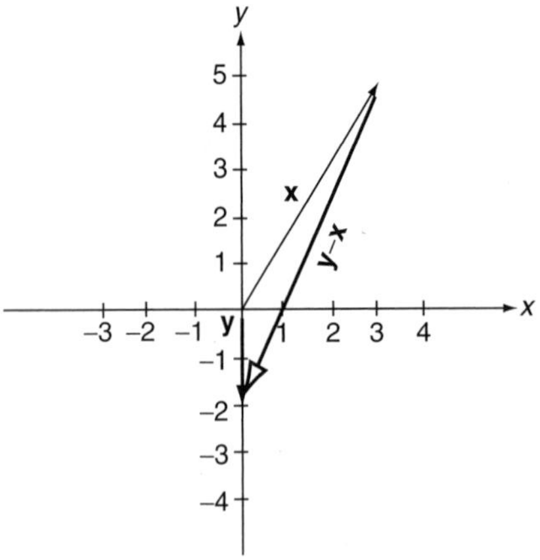

(7)

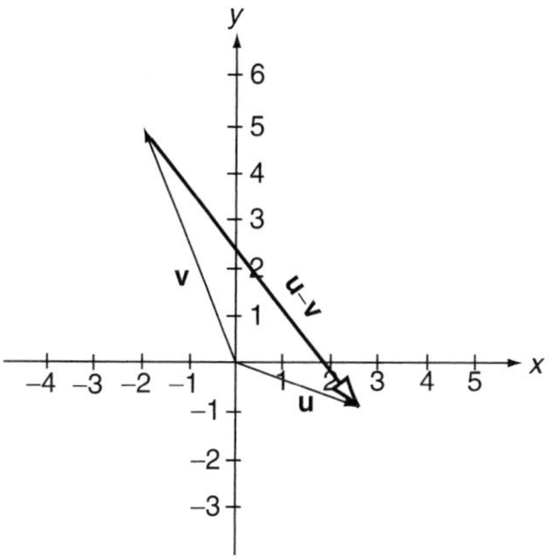
y
6
5
4
3
2
1
v
u–v
x
–4 –3 –2 –1 1 2 3 4 5
u
–1
–2
–3

(10)

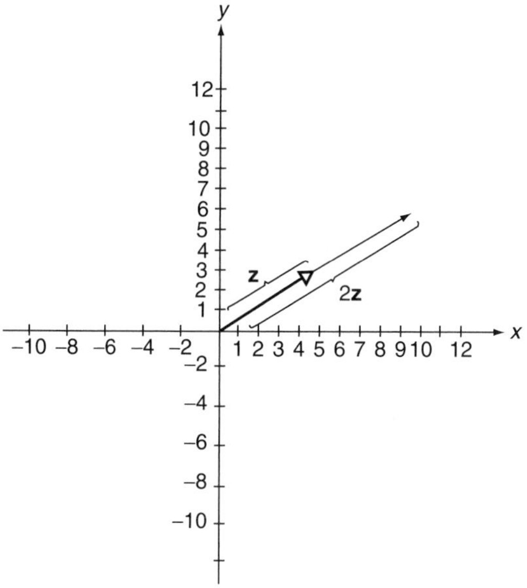
y
12
10
9
8
7
6
5
4
3
2
1
z
2z
x
–10 –8 –6 –4 –2 1 2 3 4 5 6 7 8 9 10 12
–2
–4
–6
–8
–10

(12)

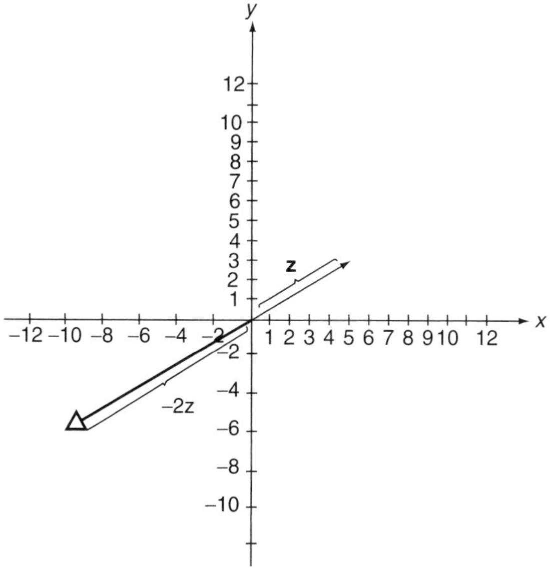
y
12
10
9
8
7
6
5
4
3
2
1
z
x
–12 –10 –8 –6 –4 –2 1 2 3 4 5 6 7 8 9 10 12
–2
–4
–2z
–6
–8
–10

(16)

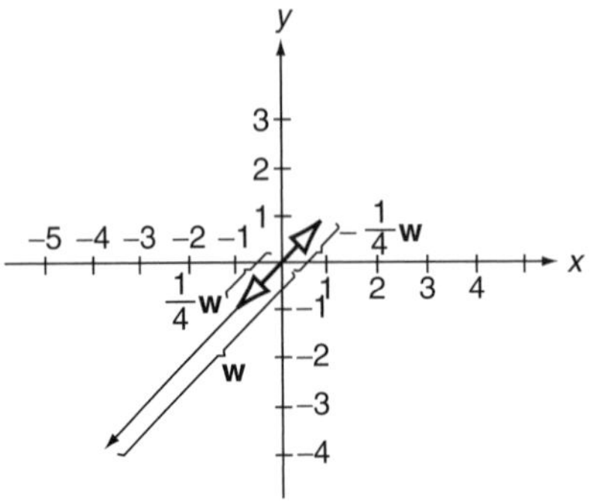

(17) (a) 341.57°, (b) 111.80°, (c) 225°, (d) 59.04°, (e) 270°.

(19) $\sqrt{2}$. **(20)** 5. **(21)** $\sqrt{5}$. **(22)** $\sqrt{3}$.

(23) $\sqrt{3/4}$. **(24)** $\sqrt{3}$. **(25)** $\sqrt{15}$. **(26)** 2.

(27) $\sqrt{2}$. **(28)** $\sqrt{39}$. **(29)** $\sqrt{5}$.

(30)

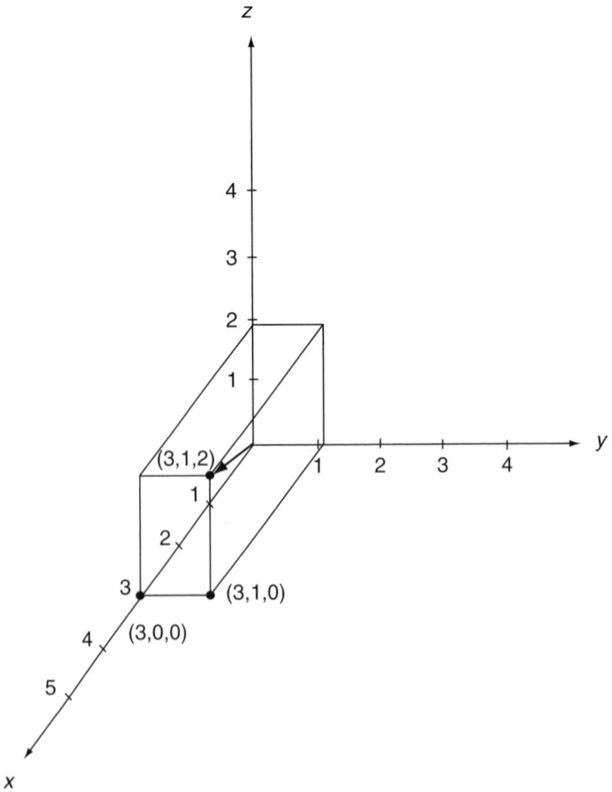

(33)

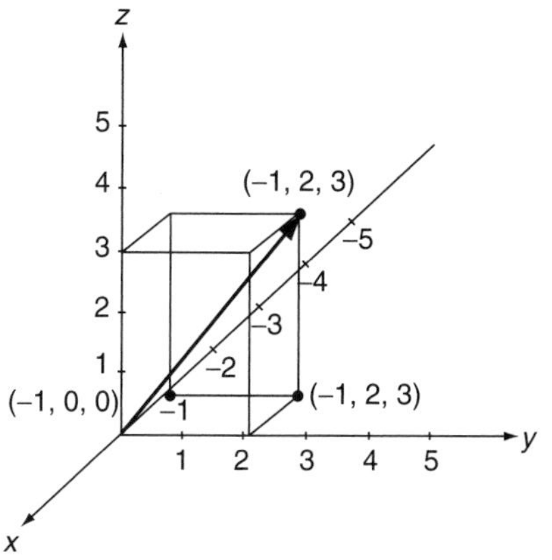
z
5
4
3
2
1
(−1, 2, 3)
(−1, 0, 0)
(−1, 2, 3)
−1
−2
−3
−4
−5
1 2 3 4 5
y
x

(38)

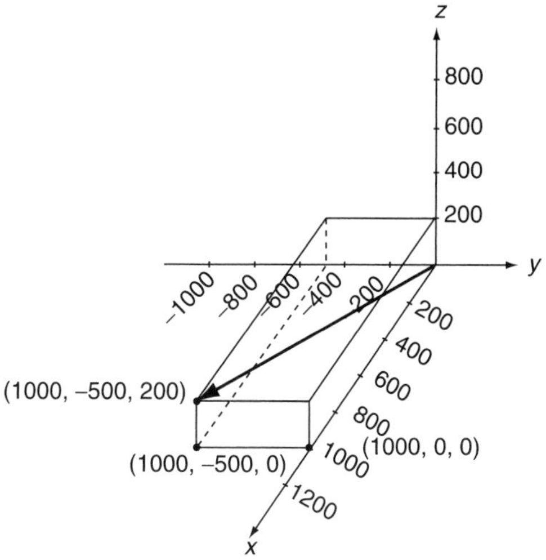
z
800
600
400
200
y
−1000 −800 −600 −400 −200
200 400 600 800 1000 1200
(1000, −500, 200)
(1000, −500, 0)
(1000, 0, 0)
x

(39)

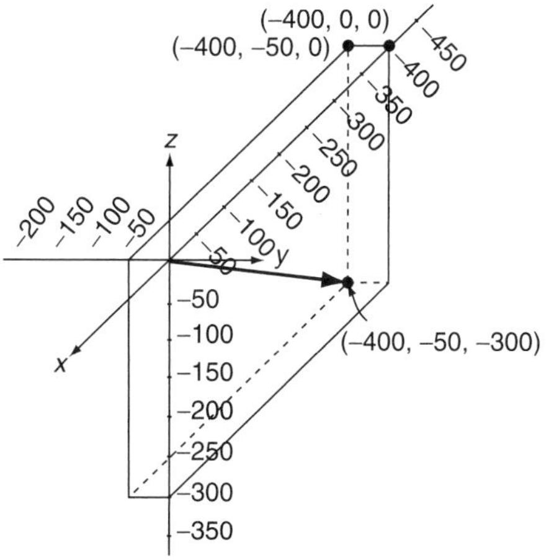
(−400, 0, 0)
(−400, −50, 0)
−450
−400
−350
−300
−250
−200
−150
−100
−50
y
z
−200 −150 −100 −50
−50
−100
−150
−200
−250
−300
−350
(−400, −50, −300)
x

(40) Not normalized. **(41)** Not normalized. **(42)** Normalized.

(43) Normalized. **(44)** Not normalized. **(45)** Normalized.

(46) Normalized. **(47)** Normalized. **(48)** Normalized.

Section 2.2

(1) Vector space. **(2)** Violates (A1). **(3)** Vector space.

(4) Vector space. **(5)** Violates (A1). **(6)** Violates (A1).

(7) Vector space. **(8)** Vector space. **(9)** Violates (S3).

(10) Violates (S3). **(11)** Violates (S3). **(12)** Violates (S3).

(13) Violates (A4). **(14)** Violates (A5). **(15)** Violates (S4).

(16) Violates (S5). **(17)** Violates (S5). **(18)** Violates (S3).

(19) Vector space. **(20)** Violates (A1). **(21)** Vector space.

(22) Violates (S4,S5). **(23)** Violates (S3). **(24)** Vector space.

(26) Let $\mathbf{0}_1$ and $\mathbf{0}_2$ be two zero vectors. Then $\mathbf{0}_1 = \mathbf{0}_1 + \mathbf{0}_2 = \mathbf{0}_2$.

(27) $\mathbf{v} \oplus (\mathbf{u} - \mathbf{v}) = \mathbf{v} \oplus (\mathbf{u} \oplus -\mathbf{v}) = \mathbf{v} \oplus (-\mathbf{v} \oplus \mathbf{u}) = (\mathbf{v} \oplus -\mathbf{v}) \oplus \mathbf{u} = 0 \oplus \mathbf{u} = \mathbf{u}$.

(28)
$$\begin{aligned} \mathbf{v} &= 0 \oplus \mathbf{v} = (-\mathbf{u} \oplus \mathbf{u}) \oplus \mathbf{v} = -\mathbf{u} \oplus (\mathbf{u} \oplus \mathbf{v}) \\ &= -\mathbf{u} \oplus (\mathbf{u} \oplus \mathbf{w}) = (-\mathbf{u} \oplus \mathbf{u}) \oplus \mathbf{w} = 0 \oplus \mathbf{w} = 0. \end{aligned}$$

(29) $\mathbf{u} \oplus \mathbf{u} = 1 \odot \mathbf{u} \oplus 1 \odot \mathbf{u} = (1+1) \odot \mathbf{u} = 2 \odot \mathbf{u}$.

(30) Given $2 \odot \mathbf{u} = 2 \odot \mathbf{v}$. Then,

$$\mathbf{u} = 1 \odot \mathbf{u} = (\tfrac{1}{2} \odot 2) \odot \mathbf{u} = \tfrac{1}{2} \odot (2 \odot \mathbf{u}) = \tfrac{1}{2} \odot (2 \odot \mathbf{v}) = (\odot 2)\tfrac{1}{2} \odot \mathbf{v} = 1 \odot \mathbf{v} = \mathbf{v}.$$

(31) First show that $-\beta \odot \mathbf{u} = (-\beta) \odot \mathbf{u}$. Then

$\mathbf{0} = (\alpha \odot \mathbf{u}) \oplus (-\beta \odot \mathbf{u}) = \alpha \odot \mathbf{u} \oplus (-\beta) \odot \mathbf{u} = [\alpha + (-\beta)] \odot \mathbf{u} = (\alpha - \beta) \odot \mathbf{u}$ and the result follows from Theorem 7.

(32) $\mathbf{0} \oplus \mathbf{0} = \mathbf{0}$. Thus, $\mathbf{0}$ is an additive inverse of $\mathbf{0}$, and the additive inverse is unique.

Section 2.3

Problems 4, 5, 9, 11, 13, 16, 20, 21, and 22 are *not* subspaces; all the others are subspaces.

(24) (a) and (c). **(25)** (a) and (c).

(26) (a) and (c). **(27)** (a) and (c).

(28) All except (f). **(29)** (b), (c), (d), and (e).

(30) (b), (c), (d), and (e).

(31) (a), (b), and (c).

(32) $\left\{ \begin{bmatrix} x \\ y \end{bmatrix} \in \mathbb{R}^2 \middle| y = 2x \right\}$.

(33) $\left\{ \begin{bmatrix} a & b \\ c & d \end{bmatrix} \in \mathbb{M}_{2\times 2} \middle| d = 0 \right\}$.

(34) $\{a_3t^3 + a_2t^2 + a_tt + a_0e\,\mathbb{P}^3 | a_0 = 0\}$.

(35) The straight line through the origin defined by the equation $y=5x$ in an $x-y$ coordinate system.

(36) Yes **(37)** No **(38)** Yes **(39)** Yes

(41) Given that $\mathbf{u} = \sum_{i=1}^{n} c_i\mathbf{v}_i$ and $\mathbf{v}_i = \sum_{j=1}^{m} a_{ij}\mathbf{w}_j$. Then,

$$\mathbf{u} = \sum_{i=1}^{n} c_i \left(\sum_{j=1}^{m} a_{ij}\mathbf{w}_j \right) = \sum_{i=1}^{n}\sum_{j=1}^{m} c_i a_{ij}\mathbf{w}_j = \sum_{j=1}^{m} \left(\sum_{i=1}^{n} c_i a_{ij} \right) \mathbf{w}_j.$$

Define $d_j = \sum_{i=1}^{n} c_i a_{ij}$.

(42) Denote the columns of $\mathbf{A}$ as $\mathbf{A}_1, \mathbf{A}_2, \ldots, \mathbf{A}_n$ and $\mathbf{x} = [x_1x_2\ldots x_n]^{\mathrm{T}}$. Then $y = x_1\mathbf{A}_1 + x_2\mathbf{A}_2 + \ldots + x_n\mathbf{A}_n$.

(43) Let $\mathbf{Ay} = \mathbf{Az} = 0$. Then $\mathbf{A}(\alpha\mathbf{y} + \beta\mathbf{z}) = \alpha(\mathbf{Ay}) + \beta(\mathbf{Az}) = \alpha(0) + \beta(\mathbf{0}) = \mathbf{0}$.

(44) $\mathbf{A}(2\mathbf{x}) = 2(\mathbf{Ax}) = 2\mathbf{b} \neq \mathbf{b}$.

(45) $\mathbf{u}+\mathbf{v}$ and $\mathbf{u}-\mathbf{v}$ belong to $span\{\mathbf{u},\mathbf{v}\}$. Also, $\mathbf{u} = \frac{1}{2}(\mathbf{u}+\mathbf{v}) + \frac{1}{2}(\mathbf{u}-\mathbf{v})$ and $\mathbf{v} = \frac{1}{2}(\mathbf{u}+\mathbf{v}) - \frac{1}{2}(\mathbf{u}-\mathbf{v})$, so u and v belong to $span\{\mathbf{u}+\mathbf{v}, \mathbf{u}-\mathbf{v}\}$.

(46) $\mathbf{u}+\mathbf{v}$, $\mathbf{v}+\mathbf{w}$, and $\mathbf{u}+\mathbf{w}$ belong to $span\{\mathbf{u},\mathbf{v},\mathbf{w}\}$. Also,

$$\mathbf{u} = \tfrac{1}{2}(\mathbf{u}+\mathbf{v}) - \tfrac{1}{2}(\mathbf{v}+\mathbf{w}) + \tfrac{1}{2}(\mathbf{u}+\mathbf{w})$$
$$\mathbf{v} = \tfrac{1}{2}(\mathbf{u}+\mathbf{v}) + \tfrac{1}{2}(\mathbf{v}+\mathbf{w}) - \tfrac{1}{2}(\mathbf{u}+\mathbf{w})$$
$$\mathbf{w} = -\tfrac{1}{2}(\mathbf{u}+\mathbf{v}) + \tfrac{1}{2}(\mathbf{v}+\mathbf{w}) + \tfrac{1}{2}(\mathbf{u}+\mathbf{w})$$

so $\mathbf{u}$, $\mathbf{v}$, and $\mathbf{w}$ belong to $span\{\mathbf{u}+\mathbf{v}, \mathbf{v}+\mathbf{w}, \mathbf{u}+\mathbf{w}\}$.

(48) $\mathbb{W}$ contains all linear combinations of vectors in $\mathbb{S}$, hence it contains all vectors in the $span(\mathbb{S})$.

Section 2.4

(1) Independent. **(2)** Independent. **(3)** Dependent.

(4) Dependent. **(5)** Independent. **(6)** Dependent.

(7) Independent. **(8)** Dependent. **(9)** Dependent.

(10) Dependent.
(11) Independent.
(12) Dependent.
(13) Independent.
(14) Independent.
(15) Dependent.
(16) Independent.
(17) Dependent.
(18) Dependent.
(19) Dependent.
(20) Independent.
(21) Dependent.
(22) Independent.
(23) Independent.
(24) Dependent.
(25) Independent.
(26) Independent.
(27) Dependent.
(28) Independent.
(29) Dependent.
(30) Dependent.

(31) One vector is a scalar multiple of the other.

(32) $\mathbf{v}_2$ is not a scalar multiple of $\mathbf{v}_1$, and the result follows from Theorem 1.

(34) $\mathbf{0} = c_1(\mathbf{u}+\mathbf{v}) + c_2(\mathbf{u} - \mathbf{v}) = (c_1+c_2)\mathbf{u} + (c_1 - c_2)\mathbf{v}$. Then $(c_1+c_2)=0$ and $(c_1 - c_2)=0$, whereupon $c_1 = c_2 = 0$.

(35) $\mathbf{0} = c_1(\mathbf{v}_1 - \mathbf{v}_2) + c_2(\mathbf{v}_1 + \mathbf{v}_3) + c_3(\mathbf{v}_2 - \mathbf{v}_3)$
$= (c_1+c_2)\mathbf{v}_1 + (-c_1+c_3)\mathbf{v}_2 + (c_2 - c_3)\mathbf{v}_3$.

Then $(c_1+c_2)=0$, $(-c_1+c_3)=0$, and $(c_2 - c_3)=0$, whereupon $c_1=c_2=c_3=0$.

(36) $\mathbf{0} = c_1(\mathbf{v}_1+\mathbf{v}_2+\mathbf{v}_3) + c_2(\mathbf{v}_2+\mathbf{v}_3) + c_3(\mathbf{v}_3)$
$= (c_1)\mathbf{v}_1 + (c_1+c_2)\mathbf{v}_2 + (c_1+c_2+c_3)\mathbf{v}_3$.

Then $(c_1)=0$, $(c_1+c_2)=0$, and $(c_1+c_2+c_3)=0$, whereupon $c_1=c_2=c_3=0$.

(38) Let $\mathbf{R}_1, \mathbf{R}_2, \ldots, \mathbf{R}_p$ be the nonzero rows, and form the equation

$$c_1\mathbf{R}_1 + c_2\mathbf{R}_2 + \ldots + c_p\mathbf{R}_p = 0$$

Let k be the column containing the first nonzero element in $\mathbf{R}_1$. Since no other row has an element in this column, it follows that the kth component on the left side of the equation, after it is summed, is just c_1. Thus, $c_1=0$. Now repeat this argument for the second row, using $c_1=0$ and conclude that $c_2=0$.

(39) Consider $c_1\mathbf{x}_1 + c_2\mathbf{x}_2 + \ldots + c_k\mathbf{x}_k = \mathbf{0}$. Then $c_1\mathbf{A}\mathbf{x}_1 + c_2\mathbf{A}\mathbf{x}_2 + \ldots + c_k\mathbf{A}\mathbf{x}_k = \mathbf{A}\mathbf{0} = \mathbf{0}$ and $c_1\mathbf{y}_1 + c_2\mathbf{y}_2 + \ldots + c_k\mathbf{y}_k = 0$, whereupon $c_1 = c_2 = \ldots = c_k = 0$.

(40) Nothing.

(41) Nothing.

(43) If $\{\mathbf{v}_1, \mathbf{v}_2, \ldots, \mathbf{v}_k\}$ is linearly dependent, then there exists a set of scalars $c_1, c_2, \ldots, c_k$, not all zero such that $c_1\mathbf{v}_1 + c_2\mathbf{v}_2 + \ldots + c_k\mathbf{v}_k = \mathbf{0}$. For the set $\{\mathbf{v}_1, \mathbf{v}_2, \ldots, \mathbf{v}_k, \mathbf{w}_1, \mathbf{w}_2, \ldots, w_r\}$, we have $c_1\mathbf{v}_1 + c_2\mathbf{v}_2 + \ldots + c_k\mathbf{v}_k + 0\mathbf{w}_1 + 0\mathbf{w}_2 + \ldots + 0\mathbf{w}_r = \mathbf{0}$.

(44) (a) 1, (b) 0, (c) 0.

(45) If the columns are linearly independent, then the parallelpiped generated by the three vectors collapses into either a parallelogram, a line segment, or the origin (Theorem 4 of Section 2.4), all of which have zero volume.

(46) It is the product of the diagonal elements.

(47) It must be zero.

Section 2.5

(1) (a), (b), (c), (d), and (f).

(2) (a), (c), (e), and (f).

(3) (a), (b), (c), (e), and (f).

(4) (e), (f), and (g).

(5) (a), (b), (c), and (d).

(6) (a), (b), and (d).

(7) (c), (d), and (e).

(8) (b), (c), (d), and (f).

(9) (a) $[-2 \quad 3]^T$, (b) $[0 \quad 1]^T$.

(10) (a) $[0 \quad 2]^T$, (b) $[4 \quad -2]^T$.

(11) (a) $[2 \quad -1]^T$, (b) $[-2 \quad 1]^T$.

(12) (a) $[0 \quad 1]^T$, (b) $[-0.7 \quad 0.4]^T$.

(13) (a) $[-50 \quad 30]^T$, (b) $[-10 \quad 6]^T$.

(14) (a) $[1 \quad 1 \quad 0]^T$, (b) $[1 \quad 0 \quad 0]^T$, (c) $[0 \quad 1 \quad 0]^T$.

(15) (a) $[2 \quad -1]^T$, (b) $[1 \quad 1]^T$.

(16) (a) $[0 \quad 1 \quad -1 \quad 0]^T$, (b) $[0 \quad 1 \quad -1 \quad 1]^T$

(17) Denote the spanning set as $\{\mathbf{x}_1, \mathbf{x}_2, \ldots, \mathbf{x}_n \mathbf{v}\}$ with $\mathbf{v} = \sum_{k=1}^{n} d_k x_k$. If $\mathbf{y} \in \mathbb{V}$, then

$$\mathbf{y} = \sum_{k=1}^{n} c_k \mathbf{x}_k + c_{n+1} v = \sum_{k=1}^{n} c_k \mathbf{x}_k + c_{n+1} \sum_{k=1}^{n} d_k x_x = \sum_{k=1}^{n} (c_k + c_{n+1} d_k) \mathbf{x}_k.$$

(18) Delete any zero vectors from the set. Order the remaining vectors in the spanning set, and then consider them one at a time in the order selected. Determine whether each vector is a linear combination of the ones preceding it. If it is, delete it and use Problem 17 to conclude that the remaining set spans $\mathbb{V}$. After all vectors have been considered and, perhaps, deleted, the set remaining has the property that no vector is a linear combination of the vectors preceding it.

(19) First four matrices.

(20) $\{[1 \quad 1], [1 \quad 2]\}$.

(21) $\{t^2 + t,\ t + 1,\ t^2 + 1\}$.

(23) $\{t^2+2t-3,\ t^2+5t,\ 2t^2-4,\ t^3\}$.

(24) $\{[1\ \ 2\ \ 1]^T, [1\ \ 2\ \ 0]^T, [1\ \ 0\ \ 0]^T\}$.

(25) $\{t^3+t^2+t,\ t^2+t+1,\ t+1,\ t^3\}$.

(26) If it did, then the process described in Problem 18 would yield a basis having less vectors than the dimension of the vector space.

(27) If the second vector is not a scalar multiple of the first vector, then the second vector is not a linear combination of the first, and the two vectors are linearly independent.

(28) Choose a basis for $\mathbb{W}$, then use the results of Problem 22.

(29) Let $\{\mathbf{w}_1, \mathbf{w}_2, \ldots, \mathbf{w}_m\}$ be a basis for $\mathbb{W}$ and extend it into a basis for $\mathbb{V}$.

(30) Use Problem 26.

(31) Use Problem 18.

Section 2.6

(1) $[1\ \ 1\ \ 2]^T, [0\ \ 1\ \ 4/3]^T$

(2) $[1\ \ 1\ \ 2]^T, [0\ \ 1\ \ 4]^T, [0\ \ 0\ \ 1]^T$.

(3) $\begin{bmatrix} 1 & \frac{1}{2} & 1 \end{bmatrix}^T$.

(4) $[1\ \ 0\ \ 2]^T, [0\ \ 1\ \ 1]^T$.

(5) First two vectors.

(6) $[1\ \ 0\ \ 0\ \ 1], [0\ \ 1\ \ 0\ \ 0], [0\ \ 0\ \ 1\ \ 1]$.

(7) $[1\ \ 0\ \ -1\ \ 1], [0\ \ 1\ \ 3\ \ -2], [0\ \ 0\ \ 1\ \ 0]$.

(8) $[1\ \ 0\ \ 2\ \ 1], [0\ \ 1\ \ 2\ \ 0], \left[0\ \ 0\ \ 1\ \ \tfrac{1}{2}\right] [0\ \ 0\ \ 0\ \ 1]$.

(9) $[1\ \ 2\ \ 4\ \ 0], [0\ \ 1\ \ 4/3\ \ -1/3]$.

(10) $t^2+t+1,\ t+1$.

(11) $t^2+1,\ t+1,\ \frac{1}{4}$.

(12) $t,\ 1$.

(13) First two vectors.

(14) First two vectors.

(15) $t^3+t^2-t,\ t^2+t+1$.

(16) $t^3+\frac{1}{2}t^2+\frac{1}{2},\ t^2+t,\ t+1$.

(17) $t^3+3t^2,\ t^2+1,\ t+1$.

(18) First two vectors.

(19) $\begin{bmatrix} 1 & 1 \\ 1 & 0 \end{bmatrix}, \begin{bmatrix} 0 & 1 \\ 1 & 0 \end{bmatrix}, \begin{bmatrix} 0 & 0 \\ 1 & 0 \end{bmatrix}$.

(20) First two vectors.

(21) $\begin{bmatrix} 1 & 3 \\ 1 & 2 \end{bmatrix}, \begin{bmatrix} 0 & 1 \\ 0 & 1 \end{bmatrix}$.

(22) Independent.

(23) Independent.

(24) Dependent.

(25) Independent. **(26)** Dependent. **(27)** Dependent.

(28) Dependent. **(29)** Independent. **(30)** Independent.

(31) Dependent. **(32)** Independent. **(33)** Dependent.

(34) Dependent **(35)** Independent. **(36)** Independent.

(37) Dependent. **(38)** Independent. **(39)** Independent.

(40) Dependent. **(41)** Independent. **(42)** Dependent.

(43) Dependent. **(44)** No

(45) $k=$ row rank $\leq$ number of rows $= m$. Also, each row, considered as a row matrix, is an n-tuple and, therefore, an element in an n-dimensional vector space. Every subset of such vectors contains most n-linearly independent vectors (Corollary 1 of Section 2.5), thus $k \leq n$.

Section 2.7

(1) 2. **(2)** 1. **(3)** 2. **(4)** 1. **(5)** 1.

(6) 2. **(7)** 3. **(8)** 2. **(9)** 3. **(10)** 4.

(11) Row rank ≤ 3. **(12)** Column rank ≤ 2. **(13)** 0.

(14) (a) No, (b) Yes.

(15) (a) Yes, (b) Yes, (c) No.

(16) (a) Yes, (b) No, (c) Yes.

(17) Consistent with no arbitrary unknowns: $x=2/3$, $y=1/3$

(18) Inconsistent.

(19) Consistent with one arbitrary unknown: $x=(1/2)\,(3-2z)$, $y=-1/2$

(20) Consistent with two arbitrary unknowns: $x=(1/7)\,(11-5z-2w)$, $y=(1/7)\,(1-3z+3w)$.

(21) Consistent with no arbitrary unknowns: $x=y=1$, $z=-1$.

(22) Consistent with no arbitrary unknowns: $x=y=0$.

(23) Consistent with no arbitrary unknowns: $x=y=z=0$.

(24) Consistent with one arbitrary unknown: $x=-z$, $y=z$.

(25) Consistent with two arbitrary unknowns: $x=z-7w$, $y=2z-2w$.

(26) That row can be transformed into a zero row using elementary row operations.

(27) Transform the matrix to row-reduced form by elementary row operations; at least one row will be zero.

(28) Use Theorem 1 and Theorem 10 of section 1.5.

CHAPTER 3

Section 3.1

(1) Function; image $= \{1, 2, 3, 4, 5\}$.

(2) Not a function.

(3) Not a function.

(4) Function; image $= \{2, 4\}$.

(5) Not a function.

(6) Function; image $= \{10, 30, 40, 50\}$.

(7) Not a function.

(8) Function; image $= \{6\}$.

(9) Function; image $= \{a, c, d, f\}$.

(10) Function; image $= \{a, b, c, d, f\}$.

(11) Not a function.

(12) Function; image $= \{2, 4, 6, 8, 10\}$.

(13) Not a function.

(14) Function; image $= \{\text{blue, yellow}\}$.

(15) Function; image $= \{10.3, 18.6, 22.7\}$.

(16) Function.

(17) Function.

(18) Function.

(19) Function.

(20) Not a function.

(21) Not a function.

(22) Function.

(23) Not a function.

(24) Function.

(25) Function.

(26) Not a function.

(27) A function when the domain is restricted to be all real numbers excluding $-3 < x < 3$.

(28) Not a function.

(29) A function when the domain is limited to $-4 \leq x \leq 4$.

(30) (a) No, (b) Yes, (c) No, (d) Yes.

(31) (a) 2, (b) 0, (c) 6, (d) $4x^2 - 6x + 2$.

(32) (a) 1, (b) 3, (c) $8x^2 - 2x$, (d) $2a^2 + 4ab + 2b^2 - a - b$.

(33) (a) -9, (b) -1, (c) $8z^3 - 1$, (d) $a^3 - 3a^2b + 3ab^2 - b^3 - 1$.

(34) Neither is onto.

(35) 1, 12, and 15.

(36) Figure 3.1 is one-to-one; Figure 3.2 is not.

(37) 1, 6, and 10.

Section 3.2

(1) (a) [4 9], (b) [−2 15], (c) [−16 600], (d) [0 −21].

(2) (a) [4 1], (b) [1 3], (c) [−6 198], (d) [2 −9].

(3) (a) [3 3], (b) [1 −3], (c) [0 0], (d) [5 3].

(4) (a) [−2 1], (b) [1 −6], (c) [2 2], (d) [−2 −1].

(5) (a) [−4 0], (b) [13 0], (c) [−2 0], (d) [0 0].

(6) (a) [2 3 3], (b) [−1 1 −1], (c) [3 6 2], (d) [0 0 2].

(7) (a) [2 1 0], (b) [−1 −6 5], (c) [3 0 3], (d) [0 0 0].

(8) (a) $\begin{bmatrix} 3 & 1 \\ 4 & 2 \end{bmatrix}$, (b) $\begin{bmatrix} 3 & 1 \\ 3 & -1 \end{bmatrix}$, (c) $\begin{bmatrix} -5 & 10 \\ 0 & 20 \end{bmatrix}$, (d) $\begin{bmatrix} 13 & 28 \\ 44 & -32 \end{bmatrix}$.

(9) (a) $\begin{bmatrix} 3 & 0 \\ 0 & -1 \end{bmatrix}$, (b) $\begin{bmatrix} 0 & 0 \\ 0 & 0 \end{bmatrix}$, (c) $\begin{bmatrix} 30 & 0 \\ 0 & -5 \end{bmatrix}$, (d) $\begin{bmatrix} -4 & 0 \\ 0 & -31 \end{bmatrix}$.

(10) (a) $5t^2 - 7t$, (b) $-t^2 + 2t$,

(c) $-3t^2 + 3t$, (d) $-3t^2 + 3t$.

(11) Linear. **(12)** Not linear. **(13)** Not linear. **(14)** Not linear.

(15) Linear. **(16)** Linear. **(17)** Linear. **(18)** Not linear.

(19) Linear. **(20)** Linear. **(21)** Not linear. **(22)** Linear.

(23) Linear. **(24)** Linear. **(25)** Not linear. **(26)** Linear.

(27) Not linear. **(28)** Linear. **(29)** Linear. **(30)** Linear.

(31) Not linear. **(32)** Not linear. **(33)** Linear. **(34)** Linear.

(35) Linear. **(36)** Linear. **(37)** Linear. **(38)** Linear.

(39) Linear. **(40)** Not linear. **(41)** Linear. **(42)** Not linear.

(43) $I(\alpha\mathbf{u} + \beta\mathbf{v}) = \alpha\mathbf{u} + \beta\mathbf{v} = \alpha I(\mathbf{u}) + \beta I(\mathbf{v})$.

(44) If $\mathbf{v} \in \mathbb{V}$, then $\mathbf{v} = \sum_{i=1}^{n} c_i\mathbf{v}_i$ and

$$\mathbf{L}(\mathbf{v}) = L\left(\sum_{i=1}^{n} c_i\mathbf{v}_i\right) = \sum_{i=1}^{n} c_i L(\mathbf{v}_i) = \sum_{i=1}^{n} c_i\mathbf{v}_i = \mathbf{v}.$$

(45) $0(\alpha\mathbf{u} + \beta\mathbf{v}) = \mathbf{0} = \alpha\mathbf{0} + \beta\mathbf{0} = \alpha 0(\mathbf{u}) + \beta 0(\mathbf{v})$.

(46) If $\mathbf{v} \in \mathbb{V}$, then $\mathbf{v} = \sum_{i=1}^{n} c_i\mathbf{v}_i$ and

$$\mathbf{L}(\mathbf{v}) = L\left(\sum_{i=1}^{n} c_i\mathbf{v}_i\right) = \sum_{i=1}^{n} c_i L(\mathbf{v}_i) = \sum_{i=1}^{n} c_i 0 = 0.$$

(47) $T(\alpha\mathbf{u} + \beta\mathbf{v}) = T(\alpha\mathbf{u}) + T(\beta\mathbf{v})$ from equation (3.2)

$= \alpha T(\mathbf{u}) + \beta T(\mathbf{v})$ from equation (3.3)

(48) Not Linear.

(49) -2.

(50) $3\mathbf{u} - 4\mathbf{v}$.

(51) $2\mathbf{v}$.

(52) $L(\mathbf{v}_1 + \mathbf{v}_2 + \mathbf{v}_3) = L[\mathbf{v}_1 + (\mathbf{v}_2 + \mathbf{v}_3)] = L(\mathbf{v}_1) + L(\mathbf{v}_2 + \mathbf{v}_3) = L(\mathbf{v}_1) + L(\mathbf{v}_2) + L(\mathbf{v}_3)$.

(53) $(S+T)(\alpha\mathbf{u}+\beta\mathbf{v}) = S(\alpha\mathbf{u}+\beta\mathbf{v}) + T(\alpha\mathbf{u}+\beta\mathbf{v})$
$= [\alpha S(\mathbf{u}) + \beta S(\mathbf{v})] + [\alpha T(\mathbf{u}) + \beta T(\mathbf{v})]$
$= \alpha[S(\mathbf{u}) + T(\mathbf{u})] + \beta[S(\mathbf{v}) + T(\mathbf{v})]$
$= \alpha(S+T)(\mathbf{u}) + \beta(S+T)(\mathbf{v})$.

(54) $(kT)\,(\alpha\mathbf{u}+\beta\mathbf{v}) = k[T(\alpha\mathbf{u}+\beta\mathbf{v})] = k[\alpha T(\mathbf{u}) + \beta T(\mathbf{v})]$
$= \alpha[kT(\mathbf{u})] + \beta[kT(\mathbf{v})]$
$= \alpha(kT)(\mathbf{u}) + \beta(kT)(\mathbf{v})$.

(55) $(ST)\,(\alpha\mathbf{u}+\beta\mathbf{v}) = S[T(\alpha\mathbf{u}+\beta\mathbf{v})] = S[\alpha T(\mathbf{u}) + \beta T(\mathbf{v})]$
$= \alpha S[T(\mathbf{u})] + \beta S[T(\mathbf{v})] = \alpha(ST)(\mathbf{u}) + \beta(ST)(\mathbf{v})$.

(56) (a) $[3 \quad 6]$, (b) $[-2 \quad 0]$, (c) $[7 \quad 9]$,
(d) $[3 \quad 3]$. (e) $[-2 \quad -6]$, (f) $[-8 \quad -9]$.

(57) (a) $[3 \quad -4]$, (b) $[6 \quad -4]$, (c) $[-3 \quad -1]$,
(d) $[-3 \quad 1]$. (e) $[-6 \quad 6]$, (f) $[6 \quad -1]$.

(58) (a) $[10 \quad -2]$, (b) $[0 \quad 0]$, (c) $[15 \quad -3]$,
(d) $[5 \quad -1]$. (e) $[-10 \quad 2]$, (f) $[-15 \quad 3]$.

(59) (a) $[-2 \quad -3]$, (b) $[4 \quad 6]$, (c) $[-8 \quad -12]$,
(d) $[-4 \quad -6]$. (e) $[0 \quad 0]$, (f) $[10 \quad 15]$.

(60) (a) $[5 \quad -1]$, (b) $[2 \quad 6]$, (c) $[5 \quad -9]$,
(d) $[1 \quad -5]$. (e) $[-6 \quad -2]$, (f) $[-4 \quad 12]$.

(61) $L^2\,[a \quad b] = L(L[a \quad b]) = L([a \quad 0]) = [a \quad 0] = L[a \quad b]$.

(62) $(LM)[a \quad b] = L(M[a \quad b]) = L[0 \quad b\,] = [0 \quad 0] = 0$.

Section 3.3

(1) $\begin{bmatrix} 1 & 2 \\ 1 & 0 \\ 0 & 2 \end{bmatrix}$. **(2)** $\begin{bmatrix} 2 & 3 \\ 0 & -1 \\ 2 & 4 \end{bmatrix}$.

(3) $\begin{bmatrix} 1 & 0 \\ 0 & 2 \\ 0 & 0 \end{bmatrix}$. **(4)** $\begin{bmatrix} 0 & -1 \\ 2 & 4 \\ 0 & 0 \end{bmatrix}$.

(5) $\begin{bmatrix} 2 & 5 & 4 \\ -2 & -1 & 5 \end{bmatrix}$. **(6)** $\begin{bmatrix} -1 & 3 & 0 \\ -3 & -8 & 9 \end{bmatrix}$.

(7) $\begin{bmatrix} 1 & 9/2 & 13/2 \\ 1 & 1/2 & -5/2 \end{bmatrix}$.

(8) $\begin{bmatrix} -5/2 & -1 & 9/2 \\ 3/2 & 4 & -9/2 \end{bmatrix}$.

(9) $\begin{bmatrix} 1 & 3 & 0 \\ 8 & -2 & -6 \end{bmatrix}$.

(10) $\begin{bmatrix} 25 & 30 \\ -45 & 50 \end{bmatrix}$.

(11) $\begin{bmatrix} 550 & 150 \\ 50 & 250 \end{bmatrix}$.

(12) $\begin{bmatrix} 5/2 & 3 \\ -14 & 4 \end{bmatrix}$.

(13) $\begin{bmatrix} 55 & 15 \\ -100 & 20 \end{bmatrix}$.

(14) $\begin{bmatrix} -185/3 & 25/3 \\ 85/3 & 115/3 \end{bmatrix}$.

(15) $\begin{bmatrix} 2 & 0 \\ -1 & 3 \end{bmatrix}$.

(16) $\begin{bmatrix} -10/3 & 8/3 \\ 8/3 & -1/3 \end{bmatrix}$.

(17) $\begin{bmatrix} 1 & 0 & 0 \\ 0 & 1 & 0 \\ 0 & 0 & 1 \\ 0 & 0 & 0 \end{bmatrix}$.

(18) $\begin{bmatrix} 1 & 1 & 0 \\ 1/2 & 0 & 1/2 \\ 1/2 & 0 & 1/2 \\ 0 & 1 & 1 \end{bmatrix}$.

(19) $\begin{bmatrix} 3/2 & 1 & 1 & -1/2 \\ 3/2 & -1 & -1 & 1/2 \\ -3/2 & 1 & 1 & 1/2 \end{bmatrix}$.

(20) $\begin{bmatrix} 5/2 & 2 & -3/2 \\ -1/2 & 0 & 5/2 \end{bmatrix}$.

(21) $\begin{bmatrix} -2 & -7 & 3 \\ 4 & 16 & -6 \\ 0 & -7 & 6 \end{bmatrix}$.

(22) $\begin{bmatrix} 5 & 6 & 1 \\ -3 & -4 & 0 \\ 2 & 1 & 3 \\ 2 & 8 & -3 \end{bmatrix}$.

(23) $\begin{bmatrix} 5 & 6 & 1 \\ -3 & 3 & -6 \\ -4 & -16 & 6 \\ 4 & 9 & 0 \end{bmatrix}$.

(24) $\begin{bmatrix} 1 & 2 & 0 & 0 \\ 1 & -1 & 0 & 0 \\ 2 & -1 & 4 & 4 \\ 4 & -1 & 3 & 5 \end{bmatrix}$.

(25) $\begin{bmatrix} -1/3 & 0 & -11/3 & 3 \\ 2/3 & 1 & 10/3 & 0 \end{bmatrix}$.

(26) $\begin{bmatrix} 4 \\ -2 \\ 6 \end{bmatrix}$.

(27) $\begin{bmatrix} 1 \\ 3 \\ -2 \end{bmatrix}$.

(28) $\begin{bmatrix} -2 \\ -8 \\ -6 \end{bmatrix}$.

(29) $\begin{bmatrix} 5 \\ 23 \end{bmatrix}$.

(30) $\begin{bmatrix} 8 \\ 18 \end{bmatrix}$.

(31) $\begin{bmatrix} 2 \\ -9 \end{bmatrix}$.

(32) $\begin{bmatrix} -65 \\ -240 \end{bmatrix}$.

(33) $6t - 2$.

(34) $6t - 2$.

(35) $\begin{bmatrix} 0 \\ 10 \end{bmatrix}$.

(36) $\begin{bmatrix} -4 \\ 9 \\ -19 \end{bmatrix}$.

(37) $\begin{bmatrix} 0 \\ -20 \\ 28 \end{bmatrix}$.

(38) $(5a - b)t - 2a$

(39) $(2a + 2b)t + (4a + 3b)$.

(40) $(4a - b + 3c)t^2 + (3a - 2b + 2c)t + (2a - 2b + c)$.

(41) $\begin{bmatrix} 3a + c & 2a + 2c - 2d \\ 2a - b + 2c - d & a \end{bmatrix}$.

Section 3.4

(1) $\begin{bmatrix} -1 & 0 \\ 1 & 1 \end{bmatrix}$.

(2) $\begin{bmatrix} 2 & 1 \\ -1 & 0 \end{bmatrix}$.

(3) $\begin{bmatrix} -1 & 1 \\ 1 & 0 \end{bmatrix}$.

(4) $\begin{bmatrix} 0 & 1 \\ 1 & 1 \end{bmatrix}$.

(5) $\begin{bmatrix} -1 & -1 \\ 3 & 4 \end{bmatrix}$.

(6) $\begin{bmatrix} -4 & -1 \\ 3 & 1 \end{bmatrix}$.

(7) $\begin{bmatrix} -10 & -10 \\ 30 & -10 \end{bmatrix}$.

(8) $\frac{1}{2}\begin{bmatrix} 1 & 1 & -1 \\ -1 & 1 & 1 \\ 1 & -1 & 1 \end{bmatrix}$.

(9) $\begin{bmatrix} 1 & -1 & 0 \\ 0 & 1 & -1 \\ 0 & 0 & 1 \end{bmatrix}$.

(10) $\begin{bmatrix} 1 & 1 & 1 \\ 0 & 1 & 1 \\ 0 & 0 & 1 \end{bmatrix}$.

(11) $\frac{1}{2}\begin{bmatrix} 1 & 2 & 1 \\ -1 & 0 & 1 \\ 1 & 0 & 1 \end{bmatrix}$.

(12) $\begin{bmatrix} 1 & 0 & 3/2 \\ 0 & 1 & 3/2 \\ 0 & 0 & -1/2 \end{bmatrix}$.

(13) $\begin{bmatrix} 0 & -1 & -2 \\ 1 & 0 & 2 \\ 0 & 1 & 1 \end{bmatrix}$.

(14) (a) $\begin{bmatrix} 1 & 5 \\ 1 & -2 \end{bmatrix}$, (b) $\begin{bmatrix} -4 & -5 \\ 1 & 3 \end{bmatrix}$.

(15) (a) $\begin{bmatrix} 17 & 23 \\ -13 & -18 \end{bmatrix}$, (b) $\begin{bmatrix} 1 & -1 \\ -5 & -2 \end{bmatrix}$.

(16) (a) $\begin{bmatrix} 2 & 0 \\ 6 & 5 \end{bmatrix}$, (b) $\begin{bmatrix} 5 & 0 \\ 0 & 2 \end{bmatrix}$.

(17) (a) $\begin{bmatrix} -1 & 0 \\ 3 & 2 \end{bmatrix}$, (b) $\begin{bmatrix} -1 & 0 \\ 0 & 2 \end{bmatrix}$.

(18) (a) $\begin{bmatrix} 3 & 0 \\ 0 & -1 \end{bmatrix}$, (b) $\begin{bmatrix} 15 & 4 \\ -48 & -13 \end{bmatrix}$.

(19) (a) $\begin{bmatrix} -13 & -8 \\ 24 & 15 \end{bmatrix}$, (b) $\begin{bmatrix} -1 & 0 \\ 8 & 3 \end{bmatrix}$.

(20) (a) $\begin{bmatrix} 1 & 0 \\ 0 & 1 \end{bmatrix}$, (b) $\begin{bmatrix} 1 & 0 \\ 0 & 1 \end{bmatrix}$.

(21) (a) $\begin{bmatrix} 0 & 0 \\ 0 & 0 \end{bmatrix}$, (b) $\begin{bmatrix} 0 & 0 \\ 0 & 0 \end{bmatrix}$.

(22) (a) $\begin{bmatrix} 3 & -1 & 1 \\ 2 & 0 & -2 \\ 3 & -3 & 1 \end{bmatrix}$, (b) $\begin{bmatrix} 2 & 0 & 0 \\ 0 & -2 & 0 \\ 0 & 0 & 4 \end{bmatrix}$.

(23) (a) $\begin{bmatrix} 3 & -1 & 1 \\ 2 & 0 & -2 \\ 3 & -3 & 1 \end{bmatrix}$, (b) $\begin{bmatrix} 1 & 0 & 3 \\ -1 & 2 & -1 \\ 3 & 0 & 1 \end{bmatrix}$.

(24) (a) $\begin{bmatrix} 1 & -1 & 0 \\ 0 & 2 & 0 \\ 1 & 0 & 3 \end{bmatrix}$, (b) $\begin{bmatrix} 1/2 & -1 & -3/2 \\ 3/2 & 3 & 3/2 \\ -1/2 & 0 & 5/2 \end{bmatrix}$.

(25) (a) $\begin{bmatrix} 1 & 0 & 0 \\ 0 & 2 & 0 \\ 0 & 0 & -3 \end{bmatrix}$, (b) $\begin{bmatrix} 1 & -1 & -1 \\ 0 & 2 & 5 \\ 0 & 0 & -3 \end{bmatrix}$.

(26) If $\mathbf{PA}=\mathbf{BP}$, then $\mathbf{P} = \begin{bmatrix} a & b \\ 0 & 0 \end{bmatrix}$, which is singular.

(27) If $\mathbf{PA}=\mathbf{BP}$, then $\mathbf{P} = \frac{d}{3}\begin{bmatrix} -2 & 1 \\ 0 & 3 \end{bmatrix}$ with d arbitrary. Choose $d \neq 0$ to make $\mathbf{P}$ invertible.

(29) Given that transition matrices $\mathbf{P}_1$ and $\mathbf{P}_2$ exist such that $A=P_1^{-1}BP_1$ and $B=P_2^{-1}CP_2$.

Then

$$\mathbf{A} = \mathbf{P}_1^{-1}(\mathbf{P}_2^{-1}\mathbf{C}\mathbf{P}_2)\mathbf{P}_1 = \mathbf{A} = (\mathbf{P}_1^{-1}\mathbf{P}_2^{-1})\mathbf{C}(\mathbf{P}_2\mathbf{P}_1) = \mathbf{A} = (\mathbf{P}_2\mathbf{P}_1)^{-1}\mathbf{C}(\mathbf{P}_2\mathbf{P}_1).$$

Take $\mathbf{P}=\mathbf{P}_2\mathbf{P}_1$.

(30) If $\mathbf{A}=\mathbf{P}^{-1}\mathbf{BP}$, then

$$\mathbf{A}^2=\mathbf{AA}=(\mathbf{P}^{-1}\mathbf{BP})(\mathbf{P}^{-1}\mathbf{BP})=(\mathbf{P}^{-1}\mathbf{B})(\mathbf{PP}^{-1})(\mathbf{BP})=(\mathbf{P}^{-1}\mathbf{B})\mathbf{I}(\mathbf{BP})=\mathbf{P}^{-1}\mathbf{B}^2\mathbf{P}.$$

(32) If $\mathbf{A}=\mathbf{P}^{-1}\mathbf{BP}$, then $\mathbf{A}^{\mathrm{T}}=(\mathbf{P}^{-1}\mathbf{BP})^{\mathrm{T}}=\mathbf{P}^{\mathrm{T}}\mathbf{BT}(\mathbf{P}^{-1})^{\mathrm{T}}=\mathbf{P}^{\mathrm{T}}\mathbf{B}^{\mathrm{T}}(\mathbf{P}^{\mathrm{T}})^{-1}$. Take the new transition matrix to be $\mathbf{P}^{\mathrm{T}}$.

(33) Take $\mathbf{P}=\mathbf{I}$.

(35) If $\mathbf{A}=\mathbf{P}^{-1}\mathbf{BP}$, then $\mathbf{B}=\mathbf{PAP}^{-1}$. First show that $\mathbf{PA}^{-1}\mathbf{P}^{-1}$ is the inverse of $\mathbf{B}$. Next, $\mathbf{A}^{-1}=(\mathbf{P}^{-1}\mathbf{BP})^{-1}=\mathbf{P}^{-1}\mathbf{B}^{-1}(\mathbf{P}^{-1})^{-1}=\mathbf{P}^{-1}\mathbf{B}^{-1}\mathbf{P}$.

(36) $\mathbf{P}$ can be any invertible 2×2 matrix.

(37) $\mathbf{P}$ can be any invertible 2×2 matrix.

(38) $\mathbf{P}=\begin{bmatrix} 0 & 1 & 0 & 0 & \dots & 0 \\ 0 & 0 & 1 & 0 & \dots & 0 \\ 0 & 0 & 0 & 1 & \dots & 0 \\ \vdots & \vdots & \vdots & \vdots & & \vdots \\ 0 & 0 & 0 & 0 & \dots & 1 \\ 1 & 0 & 0 & 0 & \dots & 0 \end{bmatrix}$.

Section 3.5

(1) (b) and (c).

(2) (b) and (c).

(3) (a), (b), (c), and (d).

(4) (b) and (d).

(5) (d).

(6) (a) $[1\quad 0\quad 1]$, (b) $[1\quad 0\quad -1]$, (c) $[2\quad 0\quad 0]$, (d) $[1\quad 0\quad 2]$.

(7) (a) Not in the range. (b) $[1\quad 0\quad 0]$, (c) $[2\quad 0\quad 0]$, (d) Not in the range.

(8) (a) Not in the range, (b) $\begin{bmatrix} 1 & 0 \\ 0 & 0 \end{bmatrix}$, (c) Not in the range, (d) $\begin{bmatrix} 3 & 0 \\ 0 & 5 \end{bmatrix}$.

(9) (a) $\begin{bmatrix} 1 & 1 \\ 1 & 1 \end{bmatrix}$, (b) $\begin{bmatrix} 0 & 0 \\ 1 & 0 \end{bmatrix}$, (c) $\begin{bmatrix} 1 & 0 \\ 0 & 0 \end{bmatrix}$, (d) $\begin{bmatrix} 0 & -5 \\ 3 & 0 \end{bmatrix}$.

(10) (a) Not in the range. (b) t^2-2, (c) -3, (d) Not in the range.

(11) Nullity is 0, rank is 2, one-to-one and onto.

(12) Nullity is 0, rank is 2, one-to-one and onto.

(13) Nullity is 1, rank is 1, neither one-to-one nor onto.

(14) Nullity is 1, rank is 2, not one-to-one but onto.

(15) Nullity is 1, rank is 2, not one-to-one but onto.

(16) Nullity is 2, rank is 1, neither one-to-one nor onto.

(17) Nullity is 0, rank is 2, one-to-one but not onto.

(18) Nullity is 0, rank is 2, one-to-one but not onto.

(19) Nullity is 0, rank is 2, one-to-one but not onto.

(20) Nullity is 1, rank is 1, neither one-to-one nor onto.

(21) Nullity is 1, rank is 1, neither one-to-one nor onto.

(22) Nullity is 2, rank is 1, not one-to-one but onto.

(23) Nullity is 3, rank is 0, neither one-to-one nor onto.

(24) Nullity is 0, rank is 4, one-to-one and onto.

(25) Nullity is 2, rank is 2, neither one-to-one nor onto.

(26) Nullity is 3, rank is 1, neither one-to-one nor onto.

(27) Nullity is 3, rank is 1, not one-to-one but onto.

(28) Nullity is 2, rank is 1, neither one-to-one nor onto.

(29) Nullity is 1, rank is 2, neither one-to-one nor onto.

(30) Nullity is 3, rank is 0, neither one-to-one nor onto.

(31) (b) and (d).

(32) (a) and (d).

(33) (b) and (c).

(34) $\left\{\begin{bmatrix}-2\\1\end{bmatrix}\right\}$ for the kernel; $\left\{\begin{bmatrix}1\\2\end{bmatrix}\right\}$ for the range.

(35) The kernel contains only the zero vector; the range is $\mathbb{R}^2$.

(36) $\left\{\begin{bmatrix}1\\1\\0\end{bmatrix}, \begin{bmatrix}0\\0\\1\end{bmatrix}\right\}$ for the kernel; $\left\{\begin{bmatrix}1\\-1\end{bmatrix}\right\}$ for the range.

(37) $\left\{\begin{bmatrix}0\\1\\0\end{bmatrix}\right\}$ for the kernel; the range is $\mathbb{R}^2$.

(38) $\left\{\begin{bmatrix}-1\\-1\\0\end{bmatrix}\right\}$ for the kernel; $\left\{\begin{bmatrix}1\\2\\3\end{bmatrix}, \begin{bmatrix}0\\1\\1\end{bmatrix}\right\}$ for the range.

(39) $\left\{\begin{bmatrix}-1\\1\\0\end{bmatrix}, \begin{bmatrix}-1\\0\\1\end{bmatrix}\right\}$ for the kernel; $\left\{\begin{bmatrix}1\\1\\1\end{bmatrix}\right\}$ for the range.

(40) The kernel contains only the zero vector; the range is $\mathbb{R}^3$.

(41) The kernel contains only the zero vector; the range is $\mathbb{R}^1$.

(42) $\left\{ \begin{bmatrix} -1 \\ 1 \\ 0 \\ 0 \end{bmatrix}, \begin{bmatrix} -2 \\ 0 \\ 1 \\ 0 \end{bmatrix}, \begin{bmatrix} -2 \\ 0 \\ 0 \\ 1 \end{bmatrix} \right\}$ for the kernel; the range is $\mathbb{R}^1$.

(43) They are the same.

(44) Rank of T = dimension of $\mathbb{W}$. Use Corollary 1 and the fact that the nullity of T is nonnegative.

(45) $dim(\mathbb{M}_{2\times 2}) = 4 > 3 = dim(\mathbb{P}^2)$.

(46) $dim(\mathbb{R}^2) = 3 > 2 = dim(\mathbb{R}^2)$.

(47) If $\sum_{i=1}^{k} c_i \mathbf{v}_i = \mathbf{0}$, then

$$0 = T(0) = T\left(\sum_{i=1}^{k} c_i v_i\right) = \sum_{i=1}^{k} c_i T\mathbf{v}_i = \sum_{i=1}^{k} c_i \mathbf{w}_i, \quad \text{and } c_1 = c_2 = \ldots c_k = 0.$$

(48) If $dim\ (\mathbb{W}) < dim\ (\mathbb{V})$, then the nullity of $\mathbb{V}$ is greater than zero and many vectors map into the zero vector.

(49) $dim(\mathbb{R}^3) = 3 < 4 = dim(\mathbb{R}^4)$.

(50) $dim(\mathbb{R}^3) = 3 < 4 = dim(\mathbb{M}_{2\times 2})$.

(51) If $\mathbf{w} \in Im(T)$, then there exists a vector $\mathbf{v} \in \mathbb{V}$ such that $T(\mathbf{v}) = \mathbf{w}$. Since $\mathbf{v} = \sum_{i=1}^{p} c_i \mathbf{v}_i$, it follows that

$$\mathbf{w} = T(\mathbf{v}) = T\left(\sum_{i=1}^{p} c_i \mathbf{v}_i\right) = \sum_{i=1}^{p} c_i T(\mathbf{v}_i).$$

(52) $0 = \sum_{i=1}^{k} c_i T(\mathbf{v}_i)$ implies that $0 = \mathrm{T}\left(\sum_{i=1}^{k} c_i \mathbf{v}_i\right)$. Then $\sum_{i=1}^{k} c_i \mathbf{v}_i = 0$ if T is one-to-one and $c_1 = c_2 = \ldots = c_n = 0$ if $\{\mathbf{v}_1, \mathbf{v}_2 \ldots, \mathbf{v}_n\}$ is linearly independent. Conversely, let $\{\mathbf{v}_1, \mathbf{v}_2 \ldots, \mathbf{v}_n\}$ be a basis for $\mathbb{V}$. This set is linearly independent, and by hypothesis so is $\{T(\mathbf{v}_1), T(\mathbf{v}_2) \ldots, T(\mathbf{v}_n)\}$.

If $T(\mathbf{u}) = T(\mathbf{v})$, with $\mathbf{u} = \sum_{i=1}^{n} c_i \mathbf{v}_i$ and $v = \sum_{i=1}^{n} d_i \mathbf{v}_i$, then

$$\sum_{i=1}^{n} c_i T(\mathbf{v}_i) = T(\mathbf{u}) = T(\mathbf{v}) = \sum_{i=1}^{n} d_i T(\mathbf{v}_i) \text{ and } \sum_{i=1}^{n} (c_i - d_i) T(\mathbf{v}_i) = 0,$$

whereupon $c_i - d_i = 0 (i = 1, 2, \ldots, n)$, and $\mathbf{u} = \mathbf{v}$.

(53) Let $\{\mathbf{v}_1, \mathbf{v}_2 \ldots, \mathbf{v}_n\}$ be a basis for $\mathbb{V}$. We are given that $\{T(\mathbf{v}_1), T(\mathbf{v}_2) \ldots, T(\mathbf{v}_n)\}$. is a basis for $\mathbb{W}$. If $T(\mathbf{u}) = T(\mathbf{v})$, with $\mathbf{u} = \sum_{i=1}^{n} c_i \mathbf{v}_i$ and $v = \sum_{i=1}^{n} d_i \mathbf{v}_i$, then $\sum_{i=1}^{n} c_i T(\mathbf{v}_i) = T(\mathbf{u}) = T(\mathbf{v}) = \sum_{i=1}^{n} d_i T(\mathbf{v}_i)$, and it follows from Theorem 5 of Section 2.5 that $\mathbf{u} = \mathbf{v}$.

(54) Let the dimension of $\mathbb{V} = n$. T is one-to-one if and only if $\nu(T) = 0$ (Theorem 5) if and only if the rank of T equals n (Corollary 1) if and only if an $n \times n$ matrix representation of T has rank n if and only if the matrix has an inverse (Theorem 5 of Section 2.7).

(55) T is onto if and only if T is one-to-one (Theorem 6). Then use the results of Problem 54.

CHAPTER 4

Section 4.1

(1) (a), (d), (e), (g), and (i).

(2) $\lambda = 3$ for (a), (e), and (g); $\lambda = 5$ for (d) and (i).

(3) (b), (c), (d), (e), and (g).

(4) $\lambda = 2$ for (b); $\lambda = 1$ for (c) and (d); $\lambda = 3$ for (e) and (g).

(5) (a), (b), and (d).

(6) $\lambda = -2$ for (a); $\lambda = -1$ for (b); $\lambda = 2$ for (d).

(7) (a), (c), and (d).

(8) $\lambda = -2$ for (b) and (c); $\lambda = 1$ for (d).

(9) $\left\{\begin{bmatrix} 2 \\ 1 \end{bmatrix}\right\}$ for $\lambda = 2$; $\left\{\begin{bmatrix} 1 \\ 1 \end{bmatrix}\right\}$ for $\lambda = 3$.

(10) $\left\{\begin{bmatrix} 1 \\ -1 \end{bmatrix}\right\}$ for $\lambda = 1$; $\left\{\begin{bmatrix} 1 \\ 2 \end{bmatrix}\right\}$ for $\lambda = 4$.

(11) $\left\{\begin{bmatrix} 3 \\ -2 \end{bmatrix}\right\}$ for $\lambda = 1$; $\left\{\begin{bmatrix} 1 \\ 2 \end{bmatrix}\right\}$ for $\lambda = 4$.

(12) $\left\{\begin{bmatrix} 1 \\ -1 \end{bmatrix}\right\}$ for $\lambda = -3$; $\left\{\begin{bmatrix} 2 \\ 3 \end{bmatrix}\right\}$ for $\lambda = 12$.

(13) $\left\{\begin{bmatrix} 1 \\ 1 \end{bmatrix}\right\}$ for $\lambda = 3$; $\left\{\begin{bmatrix} 1 \\ -2 \end{bmatrix}\right\}$ for $\lambda = -3$.

(14) No real eigenvalues.

(15) $\left\{\begin{bmatrix}1\\0\end{bmatrix}\right\}$ for $\lambda = 3$; with multiplicity 2.

(16) $\left\{\begin{bmatrix}1\\0\end{bmatrix}, \begin{bmatrix}0\\1\end{bmatrix}\right\}$ for $\lambda = 3$; with multiplicity 2.

(17) $\left\{\begin{bmatrix}1\\1\end{bmatrix}\right\}$ for $\lambda = t$; $\left\{\begin{bmatrix}-1\\2\end{bmatrix}\right\}$ for $\lambda = -2t$.

(18) $\left\{\begin{bmatrix}1\\1\end{bmatrix}\right\}$ for $\lambda = 2\theta$; $\left\{\begin{bmatrix}-2\\1\end{bmatrix}\right\}$ for $\lambda = 3\theta$.

(19) $\left\{\begin{bmatrix}0\\1\\0\end{bmatrix}\right\}$ for $\lambda = 2$; $\left\{\begin{bmatrix}1\\1\\1\end{bmatrix}\right\}$ for $\lambda = 4$; $\left\{\begin{bmatrix}-1\\0\\1\end{bmatrix}\right\}$ for $\lambda = -2$.

(20) $\left\{\begin{bmatrix}1\\-4\\1\end{bmatrix}\right\}$ for $\lambda = 1$; $\left\{\begin{bmatrix}0\\1\\0\end{bmatrix}\right\}$ for $\lambda = 2$; $\left\{\begin{bmatrix}-1\\0\\1\end{bmatrix}\right\}$ for $\lambda = 3$.

(21) $\left\{\begin{bmatrix}1\\-4\\1\end{bmatrix}\right\}$ for $\lambda = 2$; $\left\{\begin{bmatrix}0\\1\\0\end{bmatrix}\right\}$ for $\lambda = 3$; $\left\{\begin{bmatrix}-1\\0\\1\end{bmatrix}\right\}$ for $\lambda = 4$.

(22) $\left\{\begin{bmatrix}1\\0\\-1\end{bmatrix}, \begin{bmatrix}1\\-1\\-0\end{bmatrix}\right\}$ for $\lambda = 1$; with multiplicity 2; $\left\{\begin{bmatrix}1\\0\\1\end{bmatrix}\right\}$ for $\lambda = 3$.

(23) $\left\{\begin{bmatrix}1\\0\\-1\end{bmatrix}\right\}$ for $\lambda = 1$; with multiplicity 2; $\left\{\begin{bmatrix}1\\0\\1\end{bmatrix}\right\}$ for $\lambda = 3$.

(24) $\left\{\begin{bmatrix}3\\0\\-1\end{bmatrix}, \begin{bmatrix}-1\\5\\-3\end{bmatrix}\right\}$ for $\lambda = 0$; with multiplicity 2 $\left\{\begin{bmatrix}1\\2\\3\end{bmatrix}\right\}$ for $\lambda = 14$.

(25) $\left\{\begin{bmatrix}1\\3\\9\end{bmatrix}\right\}$ for $\lambda = 3$; with multiplicity 3.

(26) $\left\{\begin{bmatrix}-1\\0\\1\end{bmatrix}, \begin{bmatrix}-2\\1\\0\end{bmatrix}\right\}$ for $\lambda = 3$; with multiplicity 2 $\left\{\begin{bmatrix}1\\2\\1\end{bmatrix}\right\}$ for $\lambda = 9$.

(27) $\left\{\begin{bmatrix}0\\1\\1\end{bmatrix}\right\}$ for $\lambda = 1$; $\left\{\begin{bmatrix}-1\\0\\1\end{bmatrix}\right\}$ for $\lambda = -2$ $\left\{\begin{bmatrix}1\\1\\1\end{bmatrix}\right\}$ for $\lambda = 5$.

(28) $\left\{\begin{bmatrix}-1\\1\\0\end{bmatrix}\right\}$ for $\lambda = 2$; $\left\{\begin{bmatrix}1\\1\\1\end{bmatrix}\right\}$ for $\lambda = 3$; $\left\{\begin{bmatrix}-1\\-1\\2\end{bmatrix}\right\}$ for $\lambda = 6$.

(29) $\left\{\begin{bmatrix}1\\1\\1\\1\end{bmatrix}\right\}$ for $\lambda = 1$; with multiplicity 4.

(30) $\left\{\begin{bmatrix}1\\0\\0\\0\end{bmatrix}, \begin{bmatrix}0\\1\\1\\1\end{bmatrix}\right\}$ for $\lambda = 1$; with multiplicity 4.

(31) $\left\{\begin{bmatrix}-1\\0\\0\\1\end{bmatrix}, \begin{bmatrix}-1\\0\\1\\0\end{bmatrix}, \begin{bmatrix}-1\\0\\0\\1\end{bmatrix}\right\}$ for $\lambda = 1$ with multiplicity 3; $\left\{\begin{bmatrix}0\\1\\1\\1\end{bmatrix}\right\}$ for $\lambda = 4$.

(32) $\left\{\begin{bmatrix}0\\-1\\1\\0\end{bmatrix}, \begin{bmatrix}-1\\-1\\0\\1\end{bmatrix}\right\}$ for $\lambda = 2$ with multiplicity 2; $\left\{\begin{bmatrix}1\\0\\0\\0\end{bmatrix}\right\}$ for $\lambda = 3$ with multiplicity 2.

(33) $\left\{\begin{bmatrix}2/\sqrt{5}\\1/\sqrt{5}\end{bmatrix}\right\}$ for $\lambda = 2$; $\left\{\begin{bmatrix}1/\sqrt{2}\\1/\sqrt{2}\end{bmatrix}\right\}$ for $\lambda = 3$.

(34) $\left\{\begin{bmatrix}1/\sqrt{2}\\-1/\sqrt{2}\end{bmatrix}\right\}$ for $\lambda = 1$; $\left\{\begin{bmatrix}1/\sqrt{5}\\2/\sqrt{5}\end{bmatrix}\right\}$ for $\lambda = 4$.

(35) $\left\{\begin{bmatrix}3/\sqrt{13}\\-2/\sqrt{13}\end{bmatrix}\right\}$ for $\lambda = 0$; $\left\{\begin{bmatrix}1/\sqrt{5}\\2/\sqrt{5}\end{bmatrix}\right\}$ for $\lambda = 8$.

(36) $\left\{\begin{bmatrix}0\\1\\0\end{bmatrix}\right\}$ for $\lambda = 2$; $\left\{\begin{bmatrix}1/\sqrt{3}\\1/\sqrt{3}\\1/\sqrt{3}\end{bmatrix}\right\}$ for $\lambda = 4$; $\left\{\begin{bmatrix}-1/\sqrt{2}\\0\\1/\sqrt{2}\end{bmatrix}\right\}$ for $\lambda = -2$.

(37) $\left\{\begin{bmatrix} 1/\sqrt{18} \\ -4/\sqrt{18} \\ 1/\sqrt{18} \end{bmatrix}\right\}$ for $\lambda = 1$; $\left\{\begin{bmatrix} 0 \\ 1 \\ 0 \end{bmatrix}\right\}$ for $\lambda = 2$; $\left\{\begin{bmatrix} -1/\sqrt{2} \\ 0 \\ 1/\sqrt{2} \end{bmatrix}\right\}$ for $\lambda = 3$.

(38) $\{-5t+(3-\sqrt{34})\}$ for $\lambda = \sqrt{34}$; $\{-5t+(3+\sqrt{34})\}$ for $\lambda = -\sqrt{34}$

(39) $\{-5t+2\}$ for $\lambda = 1\{t-1\}$ for $\lambda = -2$

(40) $\{t^2+1, t\}$ for $\lambda = 1\{-t^2-2t+1\}$ for $\lambda = 3$

(41) $\left\{\begin{bmatrix} 1 \\ -1 \end{bmatrix}\right\}$ for $\lambda = 3$ with multiplicity 2.

(42) $\left\{\begin{bmatrix} 5 \\ 3 \end{bmatrix}\right\}$ for $\lambda = 10$; $\left\{\begin{bmatrix} 2 \\ -3 \end{bmatrix}\right\}$ for $\lambda = -11$;

(43) $\left\{\begin{bmatrix} 5 \\ -4 \\ 1 \end{bmatrix}\right\}$ for $\lambda = 0$ $\left\{\begin{bmatrix} -1 \\ 0 \\ 1 \end{bmatrix}\right\}$ for $\lambda = 2$ with multiplicity 2.

(44) $\left\{\begin{bmatrix} 1 \\ 0 \\ -1 \end{bmatrix}, \begin{bmatrix} 1 \\ 2 \\ 1 \end{bmatrix}\right\}$ for $\lambda = 2$ with multiplicity 2; $\left\{\begin{bmatrix} 1 \\ -1 \\ 1 \end{bmatrix}\right\}$ for $\lambda = 5$.

(45) $\left\{\begin{bmatrix} 1 & 1 \\ 0 & 1 \end{bmatrix}\right\}$ for $\lambda = 1$ with multiplicity 3.

(46) $\{1\}$ for $\lambda=0$ of multiplicity 2.

(47) $\{1\}$ for $\lambda=0$ of multiplicity 3.

(48) $\{t, 1\}$ for $\lambda=0$ of multiplicity 3.

(49) $\{e\}\{e^{3t}\}$ for $\lambda=3$; $\{e^{-3t}\}$ for $\lambda=-3$.

(50) $\{e^{3t}, e^{-3t}\}$ for $\lambda=9$ of multiplicity 2.

(51) No real eigenvalues.

(52) $\{\sin t, \cos t\}$ for $\lambda=1$ of multiplicity 2.

(53) $\{\sin 2t, \cos 2t\}$ for $\lambda=4$ of multiplicity 2.

(54) Expanding by the first column,

$$\begin{vmatrix} -\lambda & 1 & 0 & \cdots & 0 & 0 \\ 0 & -\lambda & 1 & \cdots & 0 & 0 \\ \vdots & \vdots & \vdots & \ddots & \vdots & \vdots \\ 0 & 0 & 0 & \cdots & 1 & 0 \\ 0 & 0 & 0 & \cdots & -\lambda & 1 \\ -a_0 & -a_1 & -a_2 & \cdots & -a_{n-1} & -a_{n-\lambda} \end{vmatrix}$$

$$= -\lambda \begin{vmatrix} -\lambda & 1 & 0 & \cdots & 0 & 0 \\ 0 & -\lambda & 0 & \cdots & 0 & 0 \\ \vdots & \vdots & \vdots & \ddots & \vdots & \vdots \\ 0 & 0 & 0 & \cdots & -\lambda & 1 \\ -a_1 & -a_2 & -a_3 & \cdots & -a_{n-1} & -a_n - \lambda \end{vmatrix}$$

$$+ (-1)^n a_0 \begin{vmatrix} 1 & 0 & \cdots & 0 & 0 \\ -\lambda & 1 & \cdots & 0 & 0 \\ \vdots & \vdots & \ddots & \vdots & \vdots \\ 0 & 0 & \cdots & 1 & 0 \\ 0 & 0 & \cdots & -\lambda & 1 \end{vmatrix}$$

Use the induction hypothesis on the first determinant and note that the second determinant is the determinant of a lower triangular matrix.

Section 4.2

(1) 9.

(2) 9.2426.

(3) $5+8+\lambda=-4$, $\lambda=-17$.

(4) $(5)(8)\lambda=-4$, $\lambda=-0.1$

(5) Their product is -24.

(6) (a) $-6, 8$; (b) $-15, 20$; (c) $-6, 1$; (d) $1, 8$;

(7) (a) $4, 4, 16$; (b) $-8, 8, 64$; (c) $6, -6, -12$; (d) $1, 5, 7$.

(8) (a) $2\mathbf{A}$, (b) $5\mathbf{A}$, (c) $\mathbf{A}^2$, (d) $\mathbf{A}+3\mathbf{I}$.

(9) (a) $2\mathbf{A}$, (b) $\mathbf{A}^2$, (c) $\mathbf{A}^3$, (d) $\mathbf{A} - 2\mathbf{I}$.

(10) $8 + 2 = 10 = (5 + \sqrt{18}) + (5 - \sqrt{18}) = \lambda_1 + \lambda_2$;

$\det(\mathbf{A}) = 7 = (5 + \sqrt{18})\,(5 - \sqrt{18}) = \lambda_1\lambda_2$.

(11) $1+2+7=10=0+(5 + \sqrt{10}) + (5 - \sqrt{10}) = \lambda_1 + \lambda_2 + \lambda_3$;

$\det(\mathbf{A})=0=(0)(5 + \sqrt{10})\,(5 - \sqrt{10}) = \lambda_1\lambda_2\lambda_3$.

(12) $\dfrac{1}{5+\sqrt{18}} = \dfrac{1}{7}(5-\sqrt{18})$ and $\dfrac{1}{5-\sqrt{18}} = \dfrac{1}{7}\left(5+\sqrt{18}\right)$ for Problem 10.

The matrix in Problem 11 has a zero eigenvalue and no inverse.

(13) $\mathbf{A} = \begin{bmatrix} 1 & 2 \\ 4 & 3 \end{bmatrix}$ has eigenvalues –1 and 5; $\mathbf{A} = \begin{bmatrix} 2 & -1 \\ 3 & -2 \end{bmatrix}$ has eigenvalues 1 and –1; $\mathbf{A}+\mathbf{B} = \begin{bmatrix} 3 & 1 \\ 7 & 1 \end{bmatrix}$ has eigenvalues $2 \pm 2\sqrt{2}$.

(14) Use **A** and **B** from the solution to Problem 13. Then $\mathbf{AB} = \begin{bmatrix} 8 & -5 \\ 14 & -10 \end{bmatrix}$ has eigenvalues $-1 \pm \sqrt{11}$.

(15) $\mathbf{x} = \begin{bmatrix} -1 \\ 1 \end{bmatrix}$ is an eigenvector of $\mathbf{A} = \begin{bmatrix} 1 & 2 \\ 4 & 3 \end{bmatrix}$, but $\mathbf{A}^T\mathbf{x} = \begin{bmatrix} 1 & 4 \\ 2 & 3 \end{bmatrix}\begin{bmatrix} -1 \\ 1 \end{bmatrix} = \begin{bmatrix} 3 \\ 1 \end{bmatrix} \neq \lambda\mathbf{x}$ for any real constant λ.

(16) $(\mathbf{A} - c\mathbf{I})\mathbf{x} = \mathbf{Ax} - c\mathbf{x} = \lambda\mathbf{x} - c\mathbf{x} = (\lambda - c)\mathbf{x}$.

(17) $\det(\mathbf{A}^T - \lambda\mathbf{I}) = \det(\mathbf{A} - \lambda\mathbf{I})^T = (\det(\mathbf{A} - \lambda\mathbf{I})$.

(23) (a) $\mathbf{A}^2 - 4\mathbf{A} - 5\mathbf{I} = \begin{bmatrix} 1 & 2 \\ 4 & 3 \end{bmatrix}\begin{bmatrix} 1 & 2 \\ 4 & 3 \end{bmatrix} - 4\begin{bmatrix} 1 & 2 \\ 4 & 3 \end{bmatrix} - 5\begin{bmatrix} 1 & 0 \\ 0 & 1 \end{bmatrix} = \begin{bmatrix} 0 & 0 \\ 0 & 0 \end{bmatrix}$.

(b) $\mathbf{A}^2 - 5\mathbf{A} = \begin{bmatrix} 1 & 2 \\ 2 & 4 \end{bmatrix}\begin{bmatrix} 1 & 2 \\ 2 & 4 \end{bmatrix} - 4\begin{bmatrix} 1 & 2 \\ 2 & 4 \end{bmatrix} - 5\begin{bmatrix} 1 & 0 \\ 0 & 1 \end{bmatrix} = \begin{bmatrix} 0 & 0 \\ 0 & 0 \end{bmatrix}$.

(24) Use the results of Problem 18 and Theorem 7 of Appendix A.

(25)
$$\mathbf{A}^n + a_{n-1}\mathbf{A}^{n-1} + \ldots a_1\mathbf{A} + a_0\mathbf{I} = 0$$
$$\mathbf{A}[\mathbf{A}^{n-1} + a_{n-1}\mathbf{A}^{n-2} + \ldots + a_1\mathbf{I}] = -a_0\mathbf{I}$$
or
$$\mathbf{A}\left[-\frac{1}{a_0}(\mathbf{A}^{n-1} + a_{n-1}\mathbf{A}^{n-2} + \ldots + a_1\mathbf{I})\right] = \mathbf{I}$$

Thus, $(-1/a_0)(\mathbf{A}^{n-1} + a_{n-1}\mathbf{A}^{n-2} + \ldots + a_1\mathbf{I})$ is the inverse of **A**.

(26) (a) $\mathbf{A}^{-1} = \begin{bmatrix} -2 & 1 \\ 3/2 & -1/2 \end{bmatrix}$, (b) since $a_0 = 0$, the inverse does not exist,

(c) since $a_0 = 0$, the inverse does not exist,

(d) $\mathbf{A}^{-1} = \begin{bmatrix} -1/3 & -1/3 & 2/3 \\ -1/3 & 1/6 & 1/6 \\ 1/2 & 1/4 & -1/4 \end{bmatrix}$, (e) $\mathbf{A}^{-1} = \begin{bmatrix} 1 & 0 & 0 & 0 \\ 0 & -1 & 0 & 0 \\ 0 & 0 & -1 & 0 \\ 0 & 0 & 0 & 1 \end{bmatrix}$.

Section 4.3

(1) Yes; $\mathbf{M} = \begin{bmatrix} 3 & 1 \\ 1 & 1 \end{bmatrix}$, $\mathbf{D} = \begin{bmatrix} 1 & 0 \\ 0 & -1 \end{bmatrix}$.

(2) No, if the vector space is the set of real two-tuples.

(3) No.

(4) No.

(5) Yes; $\mathbf{M} = \begin{bmatrix} -10 & 0 & 0 \\ 1 & 1 & -3 \\ 8 & 2 & 1 \end{bmatrix}$, $\mathbf{D} = \begin{bmatrix} 1 & 0 & 0 \\ 0 & 3 & 0 \\ 0 & 0 & -4 \end{bmatrix}$.

(6) Yes; $\mathbf{M} = \begin{bmatrix} 1 & 0 & 1 \\ -2 & -2 & 0 \\ 0 & 1 & 1 \end{bmatrix}$, $\mathbf{D} = \begin{bmatrix} 3 & 0 & 0 \\ 0 & 3 & 0 \\ 0 & 0 & 7 \end{bmatrix}$.

(7) Yes; $\mathbf{M} = \begin{bmatrix} 3 & -2 & 1 \\ 0 & 1 & 2 \\ -1 & 0 & 3 \end{bmatrix}$, $\mathbf{D} = \begin{bmatrix} 0 & 0 & 0 \\ 0 & 0 & 0 \\ 0 & 0 & 14 \end{bmatrix}$.

(8) Yes; $\mathbf{M} = \begin{bmatrix} 1 & 0 & 1 \\ 0 & 1 & -1 \\ -1 & 1 & 1 \end{bmatrix}$, $\mathbf{D} = \begin{bmatrix} 2 & 0 & 0 \\ 0 & 2 & 0 \\ 0 & 0 & 5 \end{bmatrix}$.

(9) No.

(10) No.

(11) No.

(12) Yes; $\{3t+1, t+1\}$.

(13) Yes; $\{3t+1, -t+3\}$.

(14) Yes; $\{-10t^2+t+8, t+2, -3t+1\}$.

(15) Yes; $\{t^2-2t, -2t+1, t^2+1\}$.

(16) No.

(17) Yes, $\left\{ \begin{bmatrix} 3 & 0 \\ 0 & -1 \end{bmatrix}, \begin{bmatrix} -1 & 5 \\ 0 & -3 \end{bmatrix}, \begin{bmatrix} 1 & 2 \\ 0 & 3 \end{bmatrix} \right\}$.

(18) No.

(19) Yes, $\left\{ \begin{bmatrix} 1 & 0 \\ 0 & -1 \end{bmatrix}, \begin{bmatrix} 1 & 0 \\ 2 & 1 \end{bmatrix}, \begin{bmatrix} 1 & 0 \\ -1 & 1 \end{bmatrix} \right\}$.

(20) No.

(21) No.

Section 4.4

(1) $\begin{bmatrix} e^{-1} & 0 \\ 0 & e^{0} \end{bmatrix}$.

(2) $\begin{bmatrix} e^{2} & 0 \\ 0 & e^{3} \end{bmatrix}$.

(3) $\begin{bmatrix} e^{-7} & 0 \\ 0 & e^{-7} \end{bmatrix}$.

(4) $\begin{bmatrix} 1 & 0 \\ 0 & 1 \end{bmatrix}$.

(5) $\begin{bmatrix} e^{-7} & e^{-7} \\ 0 & e^{-7} \end{bmatrix}$.

(6) $\begin{bmatrix} e^{2} & e^{2} \\ 0 & e^{2} \end{bmatrix}$.

(7) $\begin{bmatrix} e^{3} & e^{3} \\ 0 & e^{3} \end{bmatrix}$.

(8) $\begin{bmatrix} 1 & 1 \\ 0 & 1 \end{bmatrix}$.

(9) $\begin{bmatrix} e^{2} & 0 & 0 \\ 0 & e^{3} & 0 \\ 0 & 0 & e^{4} \end{bmatrix}$.

(10) $\begin{bmatrix} e^{1} & 0 & 0 \\ 0 & e^{-5} & 0 \\ 0 & 0 & e^{-1} \end{bmatrix}$.

(11) $\begin{bmatrix} e^{2} & 0 & 0 \\ 0 & e^{2} & 0 \\ 0 & 0 & e^{2} \end{bmatrix}$.

(12) $\begin{bmatrix} e^{2} & e^{2} & \frac{1}{2}e^{2} \\ 0 & e^{2} & e^{2} \\ 0 & 0 & e^{2} \end{bmatrix}$.

(13) $\begin{bmatrix} e^{-1} & e^{-1} & \frac{1}{2}e^{-1} \\ 0 & e^{-1} & e^{-1} \\ 0 & 0 & e^{-1} \end{bmatrix}$.

(14) $\begin{bmatrix} 1 & 0 & 1/2 \\ 0 & 1 & 1 \\ 0 & 0 & 1 \end{bmatrix}$.

(15) $\begin{bmatrix} e^{-1} & 0 & 0 \\ 0 & e^{-1} & e^{-1} \\ 0 & 0 & e^{-1} \end{bmatrix}$.

(16) $\begin{bmatrix} e^{2} & 0 & 0 \\ 0 & e^{2} & e^{2} \\ 0 & 0 & e^{2} \end{bmatrix}$.

(17) $\begin{bmatrix} e^{1} & 0 & 0 & 0 \\ 0 & e^{5} & 0 & 0 \\ 0 & 0 & e^{-5} & 0 \\ 0 & 0 & 0 & e^{3} \end{bmatrix}$.

(18) $\begin{bmatrix} e^{-5} & 0 & 0 & 0 \\ 0 & e^{-5} & 0 & 0 \\ 0 & 0 & e^{-5} & 0 \\ 0 & 0 & 0 & e^{-5} \end{bmatrix}$.

(19) $\begin{bmatrix} e^{-5} & 0 & 0 & 0 \\ 0 & e^{-5} & 0 & 0 \\ 0 & 0 & e^{-5} & e^{-5} \\ 0 & 0 & 0 & e^{-5} \end{bmatrix}$.

(20) $\begin{bmatrix} e^{-5} & 0 & 0 & 0 \\ 0 & e^{-5} & e^{-5} & \frac{1}{2}e^{-5} \\ 0 & 0 & e^{-5} & e^{-5} \\ 0 & 0 & 0 & e^{-5} \end{bmatrix}$.

(21) $\begin{bmatrix} e^{-5} & 0 & 0 & 0 \\ 0 & e^{-5} & 0 & 0 \\ 0 & 0 & e^{-5} & e^{-5} \\ 0 & 0 & 0 & e^{-5} \end{bmatrix}$.

(22) $\begin{bmatrix} e^{-5} & e^{-5} & \frac{1}{2}e^{-5} & \frac{1}{6}e^{-5} \\ 0 & e^{-5} & e^{-5} & \frac{1}{2}e^{-5} \\ 0 & 0 & e^{-5} & e^{-5} \\ 0 & 0 & 0 & e^{-5} \end{bmatrix}$.

(23) $\begin{bmatrix} e^2 & 0 & e^2 \\ 0 & e^2 & 0 \\ 0 & 0 & e^2 \end{bmatrix}$.

(24) $\frac{1}{7}\begin{bmatrix} 3e^5+4e^{-2} & 3e^5-3e^{-2} \\ 4e^5-4e^{-2} & 4e^5+3e^{-2} \end{bmatrix}$.

(25) $e^3\begin{bmatrix} 2 & -1 \\ 1 & 0 \end{bmatrix}$.

(26) $e^2\begin{bmatrix} 0 & 1 & 3 \\ -1 & 2 & 5 \\ 0 & 0 & 1 \end{bmatrix}$.

(27) $e^{\pi}\begin{bmatrix} 1 & \pi/3 & \pi^2/12-\pi \\ 0 & 1 & \pi/2 \\ 0 & 0 & 1 \end{bmatrix}$.

(28) $e^2\begin{bmatrix} 1 & 1 & 1 \\ 0 & 1 & 2 \\ 0 & 0 & 1 \end{bmatrix}$.

(29) $\begin{bmatrix} e & 0 & 0 \\ -e+2e^2 & 2e^2 & -e^2 \\ e^2 & e^2 & 0 \end{bmatrix}$.

(30) $e^{\mathbf{A}} = \begin{bmatrix} e & 3e \\ 0 & e \end{bmatrix}$ and $e^{-\mathbf{A}} = \begin{bmatrix} e^{-1} & -3e^{-1} \\ 0 & e^{-1} \end{bmatrix}$.

(31) $e^{\mathbf{A}} = \begin{bmatrix} \frac{1}{2}(e^{8i}+e^{-8i}) & \frac{-i}{16}(e^{8i}-e^{-8i}) \\ 4i(e^{8i}-e^{-8i}) & \frac{1}{2}(e^{8i}+e^{-8i}) \end{bmatrix}$ and $e^{-\mathbf{A}} = \begin{bmatrix} \frac{1}{2}(e^{8i}+e^{-8i}) & -\frac{i}{16}(e^{8i}-e^{-8i}) \\ -4i(e^{8i}-e^{-8i}) & \frac{1}{2}(e^{8i}+e^{-8i}) \end{bmatrix}$.

(32) $e^{\mathbf{A}} = \begin{bmatrix} 1 & 1 & 1/2 \\ 0 & 1 & 1 \\ 0 & 0 & 1 \end{bmatrix}$ and $e^{-\mathbf{A}} = \begin{bmatrix} 1 & -1 & 1/2 \\ 0 & 1 & -1 \\ 0 & 0 & 1 \end{bmatrix}$. **A** has no inverse.

(33) $e^{\mathbf{A}} = \begin{bmatrix} e & e-1 \\ 0 & 1 \end{bmatrix}$, $e^{\mathbf{B}} = \begin{bmatrix} 1 & e-1 \\ 0 & 1 \end{bmatrix}$, $e^{\mathbf{A}}e^{\mathbf{B}} = \begin{bmatrix} e & 2e^2-2e \\ 0 & e \end{bmatrix}$,
$e^{\mathbf{B}}e^{\mathbf{A}} = \begin{bmatrix} e & 2e-2 \\ 0 & e \end{bmatrix}$, $e^{\mathbf{A}+\mathbf{B}} = \begin{bmatrix} e & 2e \\ 0 & e \end{bmatrix}$.

(34) $\mathbf{A} = \begin{bmatrix} 1 & 0 \\ 0 & 2 \end{bmatrix}$, $\mathbf{B} = \begin{bmatrix} 3 & 0 \\ 0 & 4 \end{bmatrix}$.

(36) $1/7\begin{bmatrix} 3e^{8t}+4e^{t} & 4e^{8t}-4e^{t} \\ 3e^{8t}-3e^{t} & 4e^{8t}+3e^{t} \end{bmatrix}$.

(37) $\begin{bmatrix} (2/\sqrt{3})\sinh\sqrt{3}t + \cosh\sqrt{3}t & (1/\sqrt{3})\sinh\sqrt{3}t \\ (-1/\sqrt{3})\sinh\sqrt{3}t & (-2/\sqrt{3})\sinh\sqrt{3}t + \cosh\sqrt{3}t \end{bmatrix}.$

Note: $\sinh\sqrt{3}t = \frac{e^{\sqrt{3}t} - e^{-\sqrt{3}t}}{2}$ and $\cosh\sqrt{3}t = \frac{e^{\sqrt{3}t} + e^{-\sqrt{3}t}}{2}$.

(38) $e^{3t}\begin{bmatrix} 1+t & t \\ -t & 1-t \end{bmatrix}.$

(39) $\begin{bmatrix} 1.4e^{2t} - 0.4e^{7t} & 0.2e^{2t} - 0.2e^{7t} \\ 2.8e^{2t} - 2.8e^{7t} & -0.4e^{-2t} + 1.4e^{-7t} \end{bmatrix}.$

(40) $-\begin{bmatrix} 0.8e^{-2t} + 0.2e^{-7t} & 0.4e^{-2t} - 0.4e^{-7t} \\ 0.4e^{-2t} - 0.4e^{-7t} & 0.2e^{-2t} + 0.8e^{-7t} \end{bmatrix}.$

(41) $\begin{bmatrix} 0.5e^{-4t} + 0.5e^{-16t} & 0.5e^{-4t} - 0.5e^{-16t} \\ 0.5e^{-4t} - 0.5e^{-16t} & 0.5e^{-4t} + 0.5e^{-16t} \end{bmatrix}.$

(42) $e^{2t}\begin{bmatrix} 1 & t \\ 0 & 1 \end{bmatrix}.$ **(43)** $\begin{bmatrix} 1 & t & t^2/2 \\ 0 & 1 & t \\ 0 & 0 & 1 \end{bmatrix}.$

(44) $\frac{1}{12}\begin{bmatrix} 12e^{t} & 0 & 0 \\ -9e^{t} + 14e^{3t} - 5e^{-3t} & 8e^{3t} + 4e^{-3t} & 4e^{3t} - 4e^{-3t} \\ -24e^{t} + 14e^{3t} + 10e^{-3t} & 8e^{3t} - 8e^{-3t} & 4e^{3t} + 8e^{-3t} \end{bmatrix}$

(45) $e^{-t}\begin{bmatrix} 1 & t & t^2/2 \\ 0 & 1 & t \\ 0 & 0 & 1 \end{bmatrix}.$ **(46)** $e^{4t}\begin{bmatrix} 1 & t & 0 \\ 0 & 1 & 0 \\ 0 & 0 & 1 \end{bmatrix}.$

(47) $\begin{bmatrix} e^{2t} & te^{2t} & 0 \\ 0 & e^{2t} & 0 \\ 0 & 0 & e^{-t} \end{bmatrix}.$ **(48)** $(1/2)\begin{bmatrix} -e^{-t} + 3e^{t} & -3e^{-t} + 3e^{t} & 0 \\ e^{-t} - e^{t} & 3e^{-t} - e^{t} & 0 \\ 2te^{t} & 2te^{t} & 2e^{t} \end{bmatrix}.$

(49) $e^{2t}\begin{bmatrix} 1+t & t & 0 \\ -t & 1-t & 0 \\ t - \frac{1}{2}t^2 & 2t - \frac{1}{2}t^2 & 1 \end{bmatrix}.$

(50) $e^{t}\begin{bmatrix} t^2 + 4t - 7/2 & 2t^2 - 2t & 2t \\ 2t - \frac{t^2}{2} & t^2 - t + 1 & t \\ \frac{3t^2}{2} - 7t & -3t^2 + 5t & -3t + 1 \end{bmatrix}.$

(51) $\begin{bmatrix} \cos(8t) & \frac{1}{8}\sin(8t) \\ -8\sin(8t) & \cos(8t) \end{bmatrix}$.

(52) $\begin{bmatrix} 2\sin(t)+\cos(t) & 5\sin(t) \\ -\sin(t) & -2\sin(t)+\cos(t) \end{bmatrix}$.

(53) $\dfrac{1}{3}e^{-4t}\begin{bmatrix} 4\sin(3t)+3\cos(3t) & \sin(3t) \\ -25\sin(3t) & -4\sin(3t)+3\cos(3t) \end{bmatrix}$.

(54) $e^{4t}\begin{bmatrix} -\sin t+\cos t & \sin t \\ -2\sin t & \sin t+\cos t \end{bmatrix}$.

(55) $\dfrac{1}{3}\begin{bmatrix} e^{-t}+8e^{2t} & 0 & 4e^{-t}+8e^{2t} \\ 3e^{2t}+6te^{2t} & 6e^{2t} & 3e^{2t}+6te^{2t} \\ -e^{-t}-2e^{2t} & 0 & -4e^{-t}-2e^{2t} \end{bmatrix}$

$$= \frac{1}{3}\begin{bmatrix} -e^{-t}+4e^{2t} & 0 & -4e^{-t}+4e^{2t} \\ 3te^{2t} & 3e^{2t} & 3te^{2t} \\ e^{-t}-e^{2t} & 0 & 4e^{-t}-e^{2t} \end{bmatrix}\begin{bmatrix} 3 & 0 & 4 \\ 1 & 2 & 1 \\ -1 & 0 & -2 \end{bmatrix}.$$

(56) $\begin{bmatrix} -\sin t & \cos t \\ -\cos t & -\sin t \end{bmatrix} = \begin{bmatrix} \cos t & \sin t \\ -\sin t & \cos t \end{bmatrix}\begin{bmatrix} 0 & 1 \\ -1 & 0 \end{bmatrix}$.

(57) $d\mathbf{A}^2(t)/dt = \begin{bmatrix} 2t+40t^4 & 6t^2+4te^t+2t^3e^t \\ 16t^3+12t^2e^t+4t^3e^t & 40t^2+2e^{2t} \end{bmatrix}$,

$$2\mathbf{A}(t)d\mathbf{A}(t)/dt = \begin{bmatrix} 2t+48t^4 & 8t^2+4t^2e^t \\ 8t^3+24t^2e^t & 32t^4+2e^{2t} \end{bmatrix}.$$

Section 4.5

(1)

TABLE

Iteration	Eigenvector Components		Eigenvalue
0	1.0000	1.0000	
1	0.6000	1.0000	5.0000
2	0.5238	1.0000	4.2000
3	0.5059	1.0000	4.0476
4	0.5015	1.0000	4.0118
5	0.5004	1.0000	4.0029

(2)

TABLE

Iteration	Eigenvector Components		Eigenvalue
0	1.0000	1.0000	
1	0.5000	1.0000	10.0000
2	0.5000	1.0000	8.0000
3	0.5000	1.0000	8.0000

(3) **TABLE**

Iteration	Eigenvector Components		Eigenvalue
0	1.0000	1.0000	
1	0.6000	1.0000	15.0000
2	0.6842	1.0000	11.4000
3	0.6623	1.0000	12.1579
4	0.6678	1.0000	11.9610
5	0.6664	1.0000	12.0098

(4) **TABLE**

Iteration	Eigenvector Components		Eigenvalue
0	1.0000	1.0000	
1	0.5000	1.0000	2.0000
2	0.2500	1.0000	4.0000
3	0.2000	1.0000	5.0000
4	0.1923	1.0000	5.2000
5	0.1912	1.0000	5.2308

(5) **TABLE**

Iteration	Eigenvector Components		Eigenvalue
0	1.0000	1.0000	
1	1.0000	0.6000	10.0000
2	1.0000	0.5217	9.2000
3	1.0000	0.5048	9.0435
4	1.0000	0.5011	9.0096
5	1.0000	0.5002	9.0021

(6) **TABLE**

Iteration	Eigenvector Components		Eigenvalue
0	1.0000	1.0000	
1	1.0000	0.4545	11.0000
2	1.0000	0.4175	9.3636
3	1.0000	0.4145	9.2524
4	1.0000	0.4142	9.2434
5	1.0000	0.4142	9.2427

(7)

TABLE

Iteration	Eigenvector Components			Eigenvalue
0	1.0000	1.0000	1.0000	
1	0.2500	1.0000	0.8333	12.0000
2	0.0763	1.0000	0.7797	9.8333
3	0.0247	1.0000	0.7605	9.2712
4	0.0081	1.0000	0.7537	9.0914
5	0.0027	1.0000	0.7513	9.0310

(8)

TABLE

Iteration	Eigenvector Components			Eigenvalue
0	1.0000	1.0000	1.0000	
1	0.6923	0.6923	1.0000	13.0000
2	0.5586	0.7241	1.0000	11.1538
3	0.4723	0.6912	1.0000	11.3448
4	0.4206	0.6850	1.0000	11.1471
5	0.3883	0.6774	1.0000	11.1101

(9)

TABLE

Iteration	Eigenvector Components			Eigenvalue
0	1.0000	1.0000	1.0000	
1	0.4000	0.7000	1.0000	20.0000
2	0.3415	0.6707	1.0000	16.4000
3	0.3343	0.6672	1.0000	16.0488
4	0.3335	0.6667	1.0000	16.0061
5	0.3333	0.6667	1.0000	16.0008

(10)

TABLE

Iteration	Eigenvector Components			Eigenvalue
0	1.0000	1.0000	1.0000	
1	0.4000	1.0000	0.3000	–20.0000
2	1.0000	0.7447	0.0284	–14.1000
3	0.5244	1.0000	–0.3683	–19.9504
4	1.0000	0.7168	–0.5303	–18.5293
5	0.6814	1.0000	–0.7423	–20.3976

(11) $\begin{bmatrix}1\\1\\1\end{bmatrix}$ is a linear combination of $\begin{bmatrix}1\\-4\\1\end{bmatrix}$ and $\begin{bmatrix}0\\1\\0\end{bmatrix}$, which are eigenvectors corresponding to $\lambda = 1$ and $\lambda = 2$, not $\lambda = 3$. Thus, the power method converges to $\lambda = 2$.

(12) There is no single dominant eigenvalue. Here, $|\lambda_1| = |\lambda_2| = \sqrt{34}$.

(13) If we shift by $\lambda = 4$, the power method on $\mathbf{A} = \begin{bmatrix} -2 & 1 \\ 2 & -1 \end{bmatrix}$ will not work because one eigenvalue is zero. We will not be able to obtain a dominant value for this matrix regardless of the initial value used (see answer to problem 25).

(14) Shift by $\lambda = 16$. The power method on $\mathbf{A} = \begin{bmatrix} -13 & 2 & 3 \\ 2 & -10 & 6 \\ 3 & 6 & -5 \end{bmatrix}$ converges after three iterations to $\mu = -14$. $\lambda + \mu = 2$.

(15)

TABLE

Iteration	Eigenvector Components		Eigenvalue
0	1.0000	1.0000	
1	–0.3333	1.0000	0.6000
2	1.0000	–0.7778	0.6000
3	–0.9535	1.0000	0.9556
4	1.0000	–0.9904	0.9721
5	–0.9981	1.0000	0.9981

(16)

TABLE

Iteration	Eigenvector Components		Eigenvalue
0	1.0000	–0.5000	
1	–0.8571	1.0000	0.2917
2	1.0000	–0.9615	0.3095
3	–0.9903	1.0000	0.3301
4	1.0000	–0.9976	0.3317
5	–0.9994	1.0000	0.3331

(17)

TABLE

Iteration	Eigenvector Components		Eigenvalue
0	1.0000	1.0000	
1	0.2000	1.0000	0.2778
2	–0.1892	1.0000	0.4111
3	–0.2997	1.0000	0.4760
4	–0.3258	1.0000	0.4944
5	–0.3316	1.0000	0.4987

(18)

TABLE

Iteration	Eigenvector Components		Eigenvalue
0	1.0000	1.0000	
1	−0.2000	1.0000	0.7143
2	−0.3953	1.0000	1.2286
3	−0.4127	1.0000	1.3123
4	−0.4141	1.0000	1.3197
5	−0.4142	1.0000	1.3203

(19)

TABLE

Iteration	Eigenvector Components			Eigenvalue
0	1.0000	1.0000	1.0000	
1	1.0000	0.4000	−0.2000	0.3125
2	1.0000	0.2703	−0.4595	0.4625
3	1.0000	0.2526	−0.4949	0.4949
4	1.0000	0.2503	−0.4994	0.4994
5	1.0000	0.2500	−0.4999	0.4999

(20)

TABLE

Iteration	Eigenvector Components			Eigenvalue
0	1.0000	1.0000	1.0000	
1	0.3846	1.0000	0.9487	−0.1043
2	0.5004	0.7042	1.0000	−0.0969
3	0.3296	0.7720	1.0000	−0.0916
4	0.3857	0.6633	1.0000	−0.0940
5	0.3244	0.7002	1.0000	−0.0907

(21)

TABLE

Iteration	Eigenvector Components			Eigenvalue
0	1.0000	1.0000	1.0000	
1	−0.6667	1.0000	−0.6667	−1.5000
2	−0.3636	1.0000	−0.3636	1.8333
3	−0.2963	1.0000	−0.2963	1.2273
4	−0.2712	1.0000	−0.2712	1.0926
5	−0.2602	1.0000	−0.2602	1.0424

(22) We cannot construct an **LU** decomposition. Shift as explained in Problem 13.

(23) We cannot solve for $\mathbf{Lx}_1 = \mathbf{y}$ uniquely for $\mathbf{x}_1$, because one eigenvalue is zero. Since the rows are linearly dependent, we have at least one "free variable". In this case, we could try to eliminate a variable to obtain a system which is both "smaller" and "determined".

(24) Yes, on occasion.

(25) Inverse power method applied to $\mathbf{A} = \begin{bmatrix} -7 & 2 & 3 \\ 2 & -4 & 6 \\ 3 & 6 & 1 \end{bmatrix}$ converges to $\mu = 1/6$. $\lambda + 1/\mu = 10 + 6 = 16$.

(26) Inverse power method applied to $\mathbf{A} = \begin{bmatrix} 27 & -17 & 7 \\ -17 & 21 & 1 \\ 7 & 1 & 11 \end{bmatrix}$ converges to $\mu = 1/3$. $\lambda + 1/\mu = -25 + 3 = -22$.

CHAPTER 5

Section 5.2

(1) $\mathbf{x}(t) = \begin{bmatrix} x(t) \\ y(t) \end{bmatrix}, \mathbf{A} = \begin{bmatrix} 2 & 3 \\ 4 & 5 \end{bmatrix}, \mathbf{f}(t) = \begin{bmatrix} 0 \\ 0 \end{bmatrix}, \mathbf{c} = \begin{bmatrix} 6 \\ 7 \end{bmatrix}, t_0 = 0.$

(2) $\mathbf{x}(t) = \begin{bmatrix} y(t) \\ z(t) \end{bmatrix}, \mathbf{A} = \begin{bmatrix} 3 & 2 \\ 4 & 1 \end{bmatrix}, \mathbf{f}(t) = \begin{bmatrix} 0 \\ 0 \end{bmatrix}, \mathbf{c} = \begin{bmatrix} 1 \\ 1 \end{bmatrix}, t_0 = 0.$

(3) $\mathbf{x}(t) = \begin{bmatrix} x(t) \\ y(t) \end{bmatrix}, \mathbf{A} = \begin{bmatrix} -3 & 3 \\ 4 & -4 \end{bmatrix}, \mathbf{f}(t) = \begin{bmatrix} 1 \\ -1 \end{bmatrix}, \mathbf{c} = \begin{bmatrix} 0 \\ 0 \end{bmatrix}, t_0 = 0.$

(4) $\mathbf{x}(t) = \begin{bmatrix} x(t) \\ y(t) \end{bmatrix}, \mathbf{A} = \begin{bmatrix} 3 & 0 \\ 2 & 0 \end{bmatrix}, \mathbf{f}(t) = \begin{bmatrix} t \\ t+1 \end{bmatrix}, \mathbf{c} = \begin{bmatrix} 1 \\ -1 \end{bmatrix}, t_0 = 0.$

(5) $\mathbf{x}(t) = \begin{bmatrix} x(t) \\ y(t) \end{bmatrix}, \mathbf{A} = \begin{bmatrix} 3 & 7 \\ 1 & 1 \end{bmatrix}, \mathbf{f}(t) = \begin{bmatrix} 2 \\ 2t \end{bmatrix}, \mathbf{c} = \begin{bmatrix} 2 \\ -3 \end{bmatrix}, t_0 = 1.$

(6) $\mathbf{x}(t) = \begin{bmatrix} u(t) \\ v(t) \\ w(t) \end{bmatrix}, \mathbf{A} = \begin{bmatrix} 1 & 1 & 1 \\ 1 & -3 & 1 \\ 0 & 1 & 1 \end{bmatrix}, \mathbf{f}(t) = \begin{bmatrix} 0 \\ 0 \\ 0 \end{bmatrix}, \mathbf{c} = \begin{bmatrix} 0 \\ 1 \\ -1 \end{bmatrix}, t_0 = 4.$

(7) $\mathbf{x}(t) = \begin{bmatrix} x(t) \\ y(t) \\ z(t) \end{bmatrix}, \mathbf{A} = \begin{bmatrix} 0 & 6 & 1 \\ 1 & 0 & -3 \\ 0 & -2 & 0 \end{bmatrix}, \mathbf{f}(t) = \begin{bmatrix} 0 \\ 0 \\ 0 \end{bmatrix}, \mathbf{c} = \begin{bmatrix} 10 \\ 10 \\ 20 \end{bmatrix}, t_0 = 0.$

(8) $\mathbf{x}(t) = \begin{bmatrix} r(t) \\ s(t) \\ u(t) \end{bmatrix}, \mathbf{A} = \begin{bmatrix} 1 & -3 & 1 \\ 1 & -1 & 0 \\ 2 & 1 & -1 \end{bmatrix}, \mathbf{f}(t) = \begin{bmatrix} \sin t \\ t^2+1 \\ \cos t \end{bmatrix}, \mathbf{c} = \begin{bmatrix} 4 \\ -2 \\ 5 \end{bmatrix}, t_0 = 1.$

(9) $\mathbf{x}(t) = \begin{bmatrix} x_1(t) \\ x_2(t) \end{bmatrix}, \mathbf{A} = \begin{bmatrix} 0 & 1 \\ 3 & 2 \end{bmatrix}, \mathbf{f}(t) = \begin{bmatrix} 0 \\ 0 \end{bmatrix}, \mathbf{c} = \begin{bmatrix} 4 \\ 5 \end{bmatrix}, t_0 = 0.$

(10) $\mathbf{x}(t) = \begin{bmatrix} x_1(t) \\ x_2(t) \end{bmatrix}, \mathbf{A} = \begin{bmatrix} 0 & 1 \\ 1 & -1 \end{bmatrix}, \mathbf{f}(t) = \begin{bmatrix} 0 \\ 0 \end{bmatrix}, \mathbf{c} = \begin{bmatrix} 2 \\ 0 \end{bmatrix}, t_0 = 1.$

(11) $\mathbf{x}(t) = \begin{bmatrix} x_1(t) \\ x_2(t) \end{bmatrix}, \mathbf{A} = \begin{bmatrix} 0 & 1 \\ 1 & 0 \end{bmatrix}, \mathbf{f}(t) = \begin{bmatrix} 0 \\ t^2 \end{bmatrix}, \mathbf{c} = \begin{bmatrix} -3 \\ 3 \end{bmatrix}, t_0 = 0.$

(12) $\mathbf{x}(t) = \begin{bmatrix} x_1(t) \\ x_2(t) \end{bmatrix}, \mathbf{A} = \begin{bmatrix} 0 & 1 \\ 3 & 2 \end{bmatrix}, \mathbf{f}(t) = \begin{bmatrix} 0 \\ 2 \end{bmatrix}, \mathbf{c} = \begin{bmatrix} 0 \\ 0 \end{bmatrix}, t_0 = 0.$

(13) $\mathbf{x}(t) = \begin{bmatrix} x_1(t) \\ x_2(t) \end{bmatrix}, \mathbf{A} = \begin{bmatrix} 0 & 1 \\ -2 & 3 \end{bmatrix}, \mathbf{f}(t) = \begin{bmatrix} 0 \\ e^{-t} \end{bmatrix}, \mathbf{c} = \begin{bmatrix} 2 \\ 2 \end{bmatrix}, t_0 = 1.$

(14) $\mathbf{x}(t) = \begin{bmatrix} x_1(t) \\ x_2(t) \\ x_3(t) \end{bmatrix}, \mathbf{A} = \begin{bmatrix} 0 & 1 & 0 \\ 0 & 0 & 1 \\ 1 & 0 & -1 \end{bmatrix}, \mathbf{f}(t) = \begin{bmatrix} 0 \\ 0 \\ 0 \end{bmatrix}, \mathbf{c} = \begin{bmatrix} 2 \\ 1 \\ -205 \end{bmatrix}, t_0 = -1.$

(15) $\mathbf{x}(t) = \begin{bmatrix} x_1(t) \\ x_2(t) \\ x_3(t) \\ x_4(t) \end{bmatrix}, \mathbf{A} = \begin{bmatrix} 0 & 1 & 0 & 0 \\ 0 & 0 & 1 & 0 \\ 0 & 0 & 0 & 1 \\ 0 & 1 & -1 & 0 \end{bmatrix}, \mathbf{f}(t) = \begin{bmatrix} 0 \\ 0 \\ 0 \\ 1 \end{bmatrix}, \mathbf{c} = \begin{bmatrix} 1 \\ 2 \\ \pi \\ e^3 \end{bmatrix}, t_0 = 0.$

(16) $\mathbf{x}(t) = \begin{bmatrix} x_1(t) \\ x_2(t) \\ x_3(t) \\ x_4(t) \\ x_5(t) \\ x_6(t) \end{bmatrix}, \mathbf{A} = \begin{bmatrix} 0 & 1 & 0 & 0 & 0 & 0 \\ 0 & 0 & 1 & 0 & 0 & 0 \\ 0 & 0 & 0 & 1 & 0 & 0 \\ 0 & 0 & 0 & 0 & 1 & 0 \\ 0 & 0 & 0 & 0 & 0 & 1 \\ 0 & 0 & 0 & 0 & -4 & 0 \end{bmatrix}, \mathbf{f}(t) = \begin{bmatrix} 0 \\ 0 \\ 0 \\ 0 \\ 0 \\ t^2 - t \end{bmatrix}, \mathbf{c} = \begin{bmatrix} 2 \\ 1 \\ 0 \\ 2 \\ 1 \\ 0 \end{bmatrix}, t_0 = \pi.$

Section 5.3

(3) (a) $e^{-3t} \begin{bmatrix} 1 & -t & t^2/t \\ 0 & 1 & -t \\ 0 & 0 & 1 \end{bmatrix},$ (b) $e^{3(t-2)} \begin{bmatrix} 1 & (t-2) & (t-2)^2/2 \\ 0 & 1 & (t-2) \\ 0 & 0 & 1 \end{bmatrix},$

(c) $e^{3(t-s)} \begin{bmatrix} 1 & (t-s) & (t-s)^2/2 \\ 0 & 1 & (t-s) \\ 0 & 0 & 1 \end{bmatrix},$

(d) $e^{-3(t-2)}\begin{bmatrix} 1 & -(t-2) & (t-2)^2/2 \\ 0 & 1 & -(t-2) \\ 0 & 0 & 1 \end{bmatrix}.$

(4) (a) $\frac{1}{6}\begin{bmatrix} 2e^{-5t}+4e^{t} & 2e^{-5t}-2e^{t} \\ 4e^{-5t}-4e^{t} & 4e^{-5t}+2e^{t} \end{bmatrix},$

(b) $\frac{1}{6}\begin{bmatrix} 2e^{-5s}+4e^{s} & 2e^{-5s}-2e^{s} \\ 4e^{-5s}-4e^{s} & 4e^{-5s}+2e^{s} \end{bmatrix},$

(c) $\frac{1}{6}\begin{bmatrix} 2e^{5(t-3)}+4e^{-(t-3)} & 2e^{5(t-3)}-2e^{-(t-3)} \\ 4e^{5(t-3)}-4e^{-(t-3)} & 4e^{5(t-3)}+2e^{-(t-3)} \end{bmatrix}.$

(5) (a) $\frac{1}{3}\begin{bmatrix} \sin 3t+3\cos 3t & -5\sin 3t \\ 2\sin 3t & -\sin 3t+3\cos 3t \end{bmatrix},$

(b) $\frac{1}{3}\begin{bmatrix} \sin 3s+3\cos 3s & -5\sin 3s \\ 2\sin 3s & -\sin 3s+3\cos 3s \end{bmatrix},$

(c) $\frac{1}{3}\begin{bmatrix} \sin 3(t-s)+3\cos 3(t-s) & -5\sin 3(t-s) \\ 2\sin 3(t-s) & -\sin 3(t-s)+3\cos 3(t-s) \end{bmatrix}.$

(6) Only (c). **(7)** Only (c). **(8)** Only (b).

(9) $x(t)=5e^{(t-2)}-3e^{-(t-2)}$, $y(t)=5e^{(t-2)}-e^{-(t-2)}$.

(10) $x(t)=2e^{(t-1)}-1$, $y(t)=2e^{(t-1)}-1$.

(11) $x(t)=k_3e^{t}+3k4e^{-t}$, $y(t)=k_3e^{t}+k_4e^{-t}$.

(12) $x(t)=k_3e^{t}+3k_4e^{-t}-1$, $y(t)=k_3e^{t}+k_4e^{-t}-1$.

(13) $x(t)=\cos 2t-(1/6)\sin 2t+(1/3)\sin t$.

(14) $x(t)=\frac{t^4}{24}+\frac{5t^2}{4}-\frac{2t}{3}-\frac{5}{8}$.

(15) $x(t)=(2/9)e^{2t}+(5/9)e^{-t}-(11/3)te^{-t}$.

(16) $x(t)=-8\cos t-6\sin t+8+6t$, $y(t)=4\cos t-2\sin t-3$.

Section 5.4

Note that the units are *kg of sugar* for problems (1) through (4) and *tons of pollution* for problems (5) through (7).

(1) $x(t)=500e^{-7t/100}\left(-1+e^{7t/100}\right)$
$y(t)=5e^{-7t/100}\left(-100-7t+100e^{7t/100}\right)$

(2) $x(t)\to 500$, $y(t)\to 437.5$ (kg of sugar)

(3) $x(t) \to 500$, $y(t) \to 583.333$ (kg of sugar)

(4) $x(t) = -495e^{-\frac{7t}{100}} + 500$

$$y(t) = -\frac{3475}{100}te^{-\frac{7t}{100}} - 488e^{-\frac{7t}{100}} + 500$$

(5) $x(t) = \frac{1}{14}e^{-6t}\left(-1 - 7e^{6t} + 8e^{7t}\right)$

$$y(t) = \frac{2e^{-6t}}{21}\left(6e^{7t} - 7e^{6t} + 1\right)$$

$x(t) \to \infty$, $y(t) \to \infty$ as $t \to \infty$

(6) $x(t) = \frac{1}{14}e^{-6t}\left(-1 - 7e^{6t} + 8e^{7t}\right)$

$$y(t) = \frac{-2}{21}e^{-6t}(-1 + e^{t})^2\left(1 + 2e^{t} + 3e^{2t} + 4e^{3t} + 5e^{4t} + 6e^{5t}\right)$$

$x(t) \to \infty$, $y(t) \to -\infty$ as $t \to \infty$ (Note that this is not a realistic model, due to the behavior of $y(t)$.)

(7) $x(t) \to -\infty$, $y(t) \to \infty$ as $t \to \infty$.

Section 5.6

(1) $\begin{bmatrix} 0 & 1 & 0 & 0 \\ 1 & 0 & 1 & 0 \\ 0 & 1 & 0 & 1 \\ 0 & 0 & 1 & 0 \end{bmatrix}$ **(2)** $\lambda^4 - 3\lambda^2 + 1$ **(3)** 3 edges, 0 triangles

(4) (a) $\begin{bmatrix} 0 & 2 & 0 & 1 \\ 2 & 0 & 3 & 0 \\ 0 & 3 & 0 & 2 \\ 1 & 0 & 2 & 0 \end{bmatrix}$, (b) entry (1,1)=0, (1,2)=2, (1,3)=0, (1,4)=

1

(5) $\frac{-1 \pm \sqrt{5}}{2}, \frac{1 \pm \sqrt{5}}{2}$ **(8)** $\begin{bmatrix} 0 & 1 & 0 & 0 \\ 1 & 0 & 1 & 1 \\ 0 & 1 & 0 & 1 \\ 0 & 1 & 1 & 0 \end{bmatrix}$

(9) $\lambda^4 - 4\lambda^2 - 2\lambda + 1$ **(10)** 4 edges, 1 triangle

(11) (a) $\begin{bmatrix} 0 & 3 & 1 & 1 \\ 3 & 2 & 4 & 4 \\ 1 & 4 & 2 & 3 \\ 1 & 4 & 3 & 2 \end{bmatrix}$, (b) entry (2,1)=3, (2,2)=2, (2,3)=4, (2,4)=4

(12) approx.. -1.48, -1, 0.31, 2.17

(14) $\begin{bmatrix} 0 & 1 & 0 & 0 \\ 0 & 0 & 1 & 0 \\ 1 & 0 & 0 & 0 \\ 0 & 1 & 0 & 0 \end{bmatrix}$ **(15)** $\lambda^4 - 2\lambda^2 + 1$ **(16)** -1,-1,1,1

(18) $\begin{bmatrix} 0 & 1 & 0 & 0 \\ 1 & 0 & 0 & 1 \\ 1 & 0 & 0 & 1 \\ 0 & 1 & 1 & 0 \end{bmatrix}$ **(19)** $\lambda^4 - 4\lambda^2$

(20) 0,0,-2,2 (23) hint: use definition of eigenvalue and let eigenvector $x = [1,1,\ldots,1]$

Section 5.7

(2) $L = D - A = \begin{bmatrix} 0 & -1 & 0 & 0 \\ -1 & 2 & -1 & 0 \\ 0 & -1 & 2 & -1 \\ 0 & 0 & -1 & 1 \end{bmatrix}$

(3) $\lambda^4 - 6\lambda^3 + 10\lambda^2 - 4\lambda$

(4) 0,2, $2 \pm \sqrt{2}$

(5) 1 spanning tree (i.e., the graph itself)

(8) $L = D - A = \begin{bmatrix} 1 & -1 & 0 & 0 \\ -1 & 3 & -1 & -1 \\ 0 & -1 & 2 & -1 \\ 0 & -1 & -1 & 2 \end{bmatrix}$

(9) $\lambda^4 - 8\lambda^3 + 19\lambda^2 - 12\lambda$

(10) 0,1,3,4

(11) 3 spanning trees

(13) (b) $\begin{bmatrix} 2 & -1 & -1 \\ -1 & 2 & -1 \\ -1 & -1 & 2 \end{bmatrix}$ (c) 0,3,3

(14) (b) $\begin{bmatrix} 3 & -1 & -1 & -1 \\ -1 & 3 & -1 & -1 \\ -1 & -1 & 3 & -1 \\ -1 & -1 & -1 & 3 \end{bmatrix}$ (c) 0,4,4,4

(15) (b) $\begin{bmatrix} 4 & -1 & -1 & -1 & -1 \\ -1 & 4 & -1 & -1 & -1 \\ -1 & -1 & 4 & -1 & -1 \\ -1 & -1 & -1 & 4 & -1 \\ -1 & -1 & -1 & -1 & 4 \end{bmatrix}$ (c) 0,5,5,5,5

(16) For K_n, 0 with multiplicity 1 and n with multiplicity n -1

(17) $t(G) = \frac{1}{n} n^{n-1} = n^{n-2}$, which is Cayley's formula

CHAPTER 6

Section 6.1

(1) (a) 11, (b) $\sqrt{5}$, (c) no.
(2) (a) 0, (b) $\sqrt{2}$, (c) yes.
(3) (a) −50, (b) $\sqrt{74}$, (c) no.
(4) (a) 0, (b) $\sqrt{68}$, (c) yes.
(5) (a) 0, (b) 5, (c) yes.
(6) (a) 6, (b) $\sqrt{5}$, (c) no.
(7) (a) 26, (b) $\sqrt{24}$, (c) no.
(8) (a) −30, (b) $\sqrt{38}$, (c) no.
(9) (a) 0, (b) $\sqrt{1400}$, (c) yes.
(10) (a) 7/24, (b) $\dfrac{\sqrt{21}}{8}$, (c) no.
(11) (a) 2, (b) $\sqrt{3}$, (c) no.
(12) (a) 0, (b) $\sqrt{3}$, (c) yes.
(13) (a) 0, (b) $\sqrt{2}$, (c) yes.
(14) (a) 1, (b) 1, (c) no.
(15) (a) 1/12, (b) $\frac{\sqrt{3}}{2}$, (c) no.
(16) (a) −13, (b) $\sqrt{55}$, (c) no.
(17) Inner product undefined.
(18) (a) $[3/5 \quad 4/5]^{\mathrm{T}}$,

(b) $\left[20/\sqrt{425} \quad 5/\sqrt{425}\right]^T$,

(c) $\left[1/\sqrt{21} \quad 2/\sqrt{21} \quad 4\sqrt{21}\right]^T$,

(d) $\left[-4/\sqrt{34} \quad 3/\sqrt{34} \quad -3\sqrt{34}\right]^T$,

(e) $\left[\sqrt{3}/3 \quad \sqrt{3}/3 \quad \sqrt{3}/3\right]^T$,

(f) $\left[1/\sqrt{55} \quad 2/\sqrt{55} \quad 3\sqrt{55} \quad 4\sqrt{55} \quad 5\sqrt{55}\right]^T$.

(19) 36.9°. **(20)** 14.0°. **(21)** 78.7°.
(22) 90°. **(23)** 118.5°. **(24)** 50.8°.
(25) 19.5°. **(26)** 17.7°. **(27)** violates 1c
(28) With $\mathbf{x} = [x_1 \quad x_2 \quad x_3 \ldots x_n]^{\mathrm{T}}$ and $\mathbf{y} = [y_1 \quad y_2 \quad y_3 \ldots y_n]^{\mathrm{T}}$,

$$<\lambda\mathbf{x},\mathbf{y}> = (\lambda x_1)(y_1) + (\lambda x_2)(y_2) + \ldots + (\lambda x_n)(y_n)$$
$$= \lambda[x_1y_1 + x_2y_2 + x_3y_3 + \ldots + x_ny_n] = \lambda <\mathbf{x},\mathbf{y}>.$$

(29)
$$\left\langle \begin{bmatrix} x_1 \\ x_2 \\ \vdots \\ x_n \end{bmatrix} + \begin{bmatrix} z_1 \\ z_2 \\ \vdots \\ z_n \end{bmatrix}, \begin{bmatrix} y_1 \\ y_2 \\ \vdots \\ y_n \end{bmatrix} \right\rangle = \left\langle \begin{bmatrix} x_1 + z_1 \\ x_2 + z_2 \\ \vdots \\ x_n + z_n \end{bmatrix}, \begin{bmatrix} y_1 \\ y_2 \\ \vdots \\ y_n \end{bmatrix} \right\rangle$$
$$= (x_1 + z_1)y_1 + (x_2 + z_2)y_2 + \ldots (x_n + z_n)y_n$$
$$= (x_1y_1 + x_2z_2 + \ldots + x_ny_n) + (z_1y_1 + z_2y_2 + \ldots + z_ny_n)$$
$$= \langle \mathbf{x}, \mathbf{y} \rangle + \langle \mathbf{z}, \mathbf{y} \rangle$$

(30) $\langle \mathbf{0}, \mathbf{y} \rangle = \langle 0\mathbf{0}, \mathbf{y} \rangle = 0\langle \mathbf{0}, \mathbf{y} \rangle = 0.$

(31) $\|\mathbf{x}+\mathbf{y}\|^2 = \langle \mathbf{x}+\mathbf{y}, \mathbf{x}+\mathbf{y} \rangle = \langle \mathbf{x}, \mathbf{x} \rangle + \langle \mathbf{x}, \mathbf{y} \rangle + \langle \mathbf{y}, \mathbf{x} \rangle + \langle \mathbf{y}, \mathbf{y} \rangle$
$= \langle \mathbf{x}, \mathbf{x} \rangle + 0 + 0 + \langle \mathbf{y}, \mathbf{y} \rangle = \|\mathbf{x}\|^2 + \|\mathbf{y}\|^2.$

(32) Note: $\langle \mathbf{x}, \mathbf{y} \rangle = 0$ when **x** and **y** are orthogonal.

$\|\mathbf{x}+\mathbf{y}\| = \|\mathbf{x}-\mathbf{y}\| \Leftrightarrow \|\mathbf{x}+\mathbf{y}\|^2 = \|\mathbf{x}-\mathbf{y}\|^2 \Leftrightarrow$
$\|\mathbf{x}\|^2 + 2\langle \mathbf{x}, \mathbf{y} \rangle + \|\mathbf{y}\|^2 = \|\mathbf{x}\|^2 - 2\langle \mathbf{x}, \mathbf{y} \rangle + \|\mathbf{y}\|^2 \Leftrightarrow 2\langle \mathbf{x}, \mathbf{y} \rangle = -2\langle \mathbf{x}, \mathbf{y} \rangle \Leftrightarrow$
$4\langle \mathbf{x}, \mathbf{y} \rangle = 0 \Leftrightarrow \langle \mathbf{x}, \mathbf{y} \rangle = 0.$

(33) $\|\mathbf{x}+\mathbf{y}\|^2 + \|\mathbf{x}-\mathbf{y}\|^2 = \langle \mathbf{x}+\mathbf{y}, \mathbf{x}+\mathbf{y} \rangle + \langle \mathbf{x}-\mathbf{y}, \mathbf{x}-\mathbf{y} \rangle$
$= [\langle \mathbf{x}, \mathbf{x} \rangle + \langle \mathbf{x}, \mathbf{y} \rangle + \langle \mathbf{y}, \mathbf{x} \rangle + \langle \mathbf{y}, \mathbf{y} \rangle] + \langle \mathbf{x}, \mathbf{x} \rangle - [\langle \mathbf{x}, \mathbf{y} \rangle - \langle \mathbf{y}, \mathbf{x} \rangle + \langle \mathbf{y}, \mathbf{y} \rangle]$
$2\langle \mathbf{x}, \mathbf{x} \rangle + 2\langle \mathbf{y}, \mathbf{x} \rangle = 2\|\mathbf{x}\|^2 + 2\|\mathbf{y}\|^2$

(31) $\|\mathbf{x}+\mathbf{y}\|^2 - \|\mathbf{x}-\mathbf{y}\|^2 = \langle x+\mathbf{y}, x+\mathbf{y} \rangle - \langle \mathbf{x}-\mathbf{y}, \mathbf{x}-\mathbf{y} \rangle$
$= [\langle \mathbf{x}, \mathbf{x} \rangle + \langle \mathbf{x}, \mathbf{y} \rangle + \langle \mathbf{y}, \mathbf{x} \rangle + \langle \mathbf{y}, \mathbf{y} \rangle] - \langle \mathbf{x}, \mathbf{x} \rangle - [\langle \mathbf{x}, \mathbf{y} \rangle - \langle \mathbf{y}, \mathbf{x} \rangle + \langle \mathbf{y}, \mathbf{y} \rangle]$
$= 2\langle \mathbf{x}, \mathbf{y} \rangle + 2\langle \mathbf{y}, \mathbf{x} \rangle = 4\langle \mathbf{x}.\mathbf{y} \rangle$

(35) $\langle \mathbf{x}, \alpha\mathbf{y} + \beta\mathbf{z} \rangle = \langle \alpha\mathbf{y} + \beta\mathbf{z}, \mathbf{x} \rangle = \langle \alpha\mathbf{y}, \mathbf{x} \rangle + \langle \beta\mathbf{z}, \mathbf{x} \rangle = \alpha\langle \mathbf{y}, \mathbf{x} \rangle + \beta\langle \mathbf{z}, \mathbf{x} \rangle$
$= \alpha(0) + \beta(0)\ 0$

(36) $0 \le \|\lambda\mathbf{x} + \mathbf{y}\|^2 = \langle \lambda\mathbf{x} + \mathbf{y}, \lambda\mathbf{x} + \mathbf{y} \rangle$
(a) $= \langle \lambda\mathbf{x}, \lambda\mathbf{x} \rangle - \langle \lambda\mathbf{x}, \mathbf{y} \rangle - \langle \mathbf{y}, \lambda\mathbf{x} \rangle + \langle \mathbf{y}, \mathbf{y} \rangle$
$= \lambda^2\langle \mathbf{x}, \mathbf{x} \rangle - \lambda\langle \mathbf{x}, \mathbf{y} \rangle - \lambda\langle \mathbf{y}, \mathbf{x} \rangle + \langle \mathbf{y}, \mathbf{y} \rangle$

(37) From Problem 35, $0 = \|\lambda\mathbf{x} - \mathbf{y}\|^2$ if and only if $\lambda\mathbf{x} - \mathbf{y} = 0$ if and only if $\mathbf{y} = \lambda\mathbf{x}$.

(38) (a) 48.2°, (b) 121.4°, (c) 180°, (d) 32.6°, (e) 26.0°.

(40) (a) 44, (b) −23, (c) −17, (d) −16, (e) −11, (f) 53.

(41) (a) 22, (b) −11.5, (c) −8.5.

(42) 145. **(43)** 27. **(44)** 32.

(47) (a) −8, (b) −5, (c) 7, (d) 3, (e) 2, (f) 1.

(48) (a) −22, (b) −184, (c) 22, (d) −21.

(50) (a) –763/60, (b) -325/6, (c) 107/30, (d) 113/6, (e) 303/70, (f) 2.

Section 6.2

(1) (a) $\begin{bmatrix} 1.6 \\ 0.8 \end{bmatrix}$, (b) $\begin{bmatrix} -0.6 \\ 1.2 \end{bmatrix}$.

(2) (a) $\begin{bmatrix} 0.7059 \\ 1.1765 \end{bmatrix}$, (b) $\begin{bmatrix} 0.2941 \\ -0.1765 \end{bmatrix}$.

(3) (a) $\begin{bmatrix} 0.5 \\ 0.5 \end{bmatrix}$, (b) $\begin{bmatrix} 2.5 \\ -2.5 \end{bmatrix}$.

(4) (a) $\begin{bmatrix} 0 \\ 0 \end{bmatrix}$, (b) $\begin{bmatrix} 4 \\ -1 \end{bmatrix}$.

(5) (a) $\begin{bmatrix} -0.7529 \\ -3.3882 \end{bmatrix}$, (b) $\begin{bmatrix} -6.2471 \\ 1.3882 \end{bmatrix}$.

(6) (a) $\begin{bmatrix} 1 \\ 0 \\ 1 \end{bmatrix}$, (b) $\begin{bmatrix} 1 \\ 1 \\ -1 \end{bmatrix}$.

(7) (a) $\begin{bmatrix} 8/9 \\ 8/9 \\ 4/9 \end{bmatrix}$, (b) $\begin{bmatrix} 1/9 \\ 1/9 \\ -4/9 \end{bmatrix}$.

(8) (a) $\begin{bmatrix} 1.2963 \\ 3.2407 \\ 3.2407 \end{bmatrix}$, (b) $\begin{bmatrix} -1.2963 \\ -0.2407 \\ 0.7593 \end{bmatrix}$.

(9) (a) $\begin{bmatrix} 2/3 \\ 2/3 \\ 2/3 \\ 0 \end{bmatrix}$, (b) $\begin{bmatrix} -2/3 \\ 1/3 \\ 1/3 \\ 1 \end{bmatrix}$.

(10) (a) $\begin{bmatrix} -7/6 \\ 7/3 \\ 0 \\ 7/6 \end{bmatrix}$, (b) $\begin{bmatrix} 13/6 \\ -1/3 \\ 3 \\ 17/6 \end{bmatrix}$.

(11) $\begin{bmatrix} 3 \\ 5 \end{bmatrix} = \frac{29}{5}\begin{bmatrix} 3/5 \\ 4/5 \end{bmatrix} - \frac{3}{5}\begin{bmatrix} 4/5 \\ -3/5 \end{bmatrix}$.

(12) $\begin{bmatrix} 3 \\ 5 \end{bmatrix} = \frac{8}{\sqrt{2}}\begin{bmatrix} 1/\sqrt{2} \\ 1/\sqrt{2} \end{bmatrix} - \frac{2}{\sqrt{2}}\begin{bmatrix} 1/\sqrt{2} \\ -1/\sqrt{2} \end{bmatrix}$.

(13) $\begin{bmatrix} 2 \\ -3 \end{bmatrix} = \frac{-4}{\sqrt{5}}\begin{bmatrix} 1/\sqrt{2} \\ 1/\sqrt{2} \end{bmatrix} - \frac{7}{\sqrt{5}}\begin{bmatrix} -2/\sqrt{5} \\ 1/\sqrt{5} \end{bmatrix}$.

(14) $$\begin{bmatrix}1\\2\\3\end{bmatrix}=\frac{11}{5}\begin{bmatrix}3/5\\4/5\\0\end{bmatrix}-\frac{2}{5}\begin{bmatrix}4/5\\-3/5\\0\end{bmatrix}+3\begin{bmatrix}0\\0\\1\end{bmatrix}.$$

(15) $$\begin{bmatrix}10\\0\\-20\end{bmatrix}=6\begin{bmatrix}3/5\\4/5\\0\end{bmatrix}+8\begin{bmatrix}4/5\\-3/5\\0\end{bmatrix}-20\begin{bmatrix}0\\0\\1\end{bmatrix}.$$

(16) $$\begin{bmatrix}10\\0\\-20\end{bmatrix}=\frac{10}{\sqrt{2}}\begin{bmatrix}1/\sqrt{2}\\1/\sqrt{2}\\0\end{bmatrix}-\frac{50}{\sqrt{6}}\begin{bmatrix}-1/\sqrt{6}\\1/\sqrt{6}\\2/\sqrt{6}\end{bmatrix}-\frac{10}{\sqrt{3}}\begin{bmatrix}1/\sqrt{3}\\-1/\sqrt{3}\\1/\sqrt{3}\end{bmatrix}.$$

(17) $$\begin{bmatrix}10\\0\\-20\end{bmatrix}=-\frac{10}{\sqrt{2}}\begin{bmatrix}-1/\sqrt{2}\\1/\sqrt{2}\\0\end{bmatrix}+\frac{50}{\sqrt{6}}\begin{bmatrix}1/\sqrt{6}\\1/\sqrt{6}\\-2/\sqrt{6}\end{bmatrix}+\frac{10}{\sqrt{3}}\begin{bmatrix}-1/\sqrt{3}\\-1/\sqrt{3}\\-1/\sqrt{3}\end{bmatrix}.$$

(18) $2t+1=0.4(0.6t-0.8)+2.2(0.8t+0.6).$

(19) $t^2+2t+3=-1.8(0.6t^2-0.8)+2.6(0.8t^2+0.6)+2(t).$

(20) $t^2-1=1.4(0.6t^2-0.8)+0.2(0.8t^2+0.6)+0(t).$

(21) $$\begin{bmatrix}1&1\\-1&2\end{bmatrix}=\frac{3}{\sqrt{3}}\begin{bmatrix}1/\sqrt{3}&1/\sqrt{3}\\-1/\sqrt{3}&0\end{bmatrix}+\frac{2}{\sqrt{3}}\begin{bmatrix}0&-1/\sqrt{3}\\-1/\sqrt{3}&1/\sqrt{3}\end{bmatrix}.$$
$$+\frac{2}{\sqrt{3}}\begin{bmatrix}1/\sqrt{3}&0\\1/\sqrt{3}&1/\sqrt{3}\end{bmatrix}+\frac{2}{\sqrt{3}}\begin{bmatrix}-1/\sqrt{3}&1/\sqrt{3}\\0&1/\sqrt{3}\end{bmatrix}.$$

(22) $$\begin{bmatrix}1&2\\3&4\end{bmatrix}=\frac{11}{5}\begin{bmatrix}3/5&4/5\\0&0\end{bmatrix}-\frac{11}{5}\begin{bmatrix}4/5&-3/5\\0&0\end{bmatrix}$$
$$-\frac{7}{5}\begin{bmatrix}0&0\\3/5&-4/5\end{bmatrix}-\frac{24}{5}\begin{bmatrix}0&0\\-4/5&-3/5\end{bmatrix}.$$

(23) $$\begin{bmatrix}4&5\\-6&7\end{bmatrix}=\frac{9-6\sqrt{2}}{2}\begin{bmatrix}1/2&1/2\\1/\sqrt{2}&0\end{bmatrix}-\frac{9+6\sqrt{2}}{2}\begin{bmatrix}-1/2&-1/2\\1/\sqrt{2}&0\end{bmatrix}$$
$$+\frac{1+7\sqrt{2}}{2}\begin{bmatrix}-1/2&1/2\\0&1/\sqrt{2}\end{bmatrix}-\frac{1-7\sqrt{2}}{2}\begin{bmatrix}1/2&-1/2\\0&1/\sqrt{2}\end{bmatrix}.$$

(24) $\begin{bmatrix}1/\sqrt{5}\\2/\sqrt{5}\end{bmatrix},\begin{bmatrix}1/\sqrt{5}\\-1/\sqrt{5}\end{bmatrix}.$

(25) $\begin{bmatrix}1/\sqrt{2}\\1/\sqrt{2}\end{bmatrix},\begin{bmatrix}-1/\sqrt{2}\\1/\sqrt{2}\end{bmatrix}.$

(26) $\begin{bmatrix} 2/\sqrt{13} \\ -2/\sqrt{13} \end{bmatrix}, \begin{bmatrix} 2/\sqrt{13} \\ 3/\sqrt{13} \end{bmatrix}.$

(27) $\begin{bmatrix} 1/\sqrt{6} \\ 2/\sqrt{6} \\ 1/\sqrt{6} \end{bmatrix}, \begin{bmatrix} 1/\sqrt{3} \\ -1/\sqrt{3} \\ 1/\sqrt{3} \end{bmatrix}, \begin{bmatrix} -1/\sqrt{2} \\ 0 \\ 1/\sqrt{2} \end{bmatrix}.$

(28) $\begin{bmatrix} 2/\sqrt{5} \\ 1/\sqrt{5} \\ 0 \end{bmatrix}, \begin{bmatrix} -2/\sqrt{45} \\ 4/\sqrt{45} \\ 5/\sqrt{45} \end{bmatrix}, \begin{bmatrix} 1/3 \\ -2/3 \\ 2/3 \end{bmatrix}.$

(29) $\begin{bmatrix} 1/\sqrt{2} \\ 1/\sqrt{2} \\ 0 \end{bmatrix}, \begin{bmatrix} 1/\sqrt{3} \\ -1/\sqrt{3} \\ 1/\sqrt{3} \end{bmatrix}, \begin{bmatrix} -1/\sqrt{6} \\ 1/\sqrt{6} \\ 2/\sqrt{6} \end{bmatrix}.$

(30) $\begin{bmatrix} 0 \\ 3/5 \\ 4/5 \end{bmatrix}, \begin{bmatrix} 3/5 \\ 16/25 \\ -12/25 \end{bmatrix}, \begin{bmatrix} 4/5 \\ -12/25 \\ 9/25 \end{bmatrix}.$

(31) $\begin{bmatrix} 0 \\ 1/\sqrt{3} \\ 1/\sqrt{3} \\ 1/\sqrt{3} \end{bmatrix}, \begin{bmatrix} 3/\sqrt{15} \\ -2/\sqrt{15} \\ 1/\sqrt{15} \\ 1/\sqrt{15} \end{bmatrix}, \begin{bmatrix} 3/\sqrt{35} \\ 3/\sqrt{35} \\ -4/\sqrt{35} \\ 1/\sqrt{35} \end{bmatrix}, \begin{bmatrix} 1/\sqrt{7} \\ 1/\sqrt{7} \\ 1/\sqrt{7} \\ -2/\sqrt{7} \end{bmatrix}.$

(32) $\begin{bmatrix} 1/\sqrt{2} \\ 1/\sqrt{2} \\ 0 \\ 0 \end{bmatrix}, \begin{bmatrix} -1/\sqrt{6} \\ 1/\sqrt{6} \\ -2/\sqrt{6} \\ 0 \end{bmatrix}, \begin{bmatrix} 1/\sqrt{3} \\ -1/\sqrt{3} \\ -1/\sqrt{3} \\ 0 \end{bmatrix}, \begin{bmatrix} 0 \\ 0 \\ 0 \\ -1 \end{bmatrix}.$

(33) One of the **q** vectors becomes zero.

(34) $\left\langle \mathbf{a}, \mathbf{x} - \frac{\langle \mathbf{a}, \mathbf{x} \rangle}{\langle \mathbf{a}, \mathbf{a} \rangle}\mathbf{a} \right\rangle = \langle \mathbf{a}, \mathbf{x} \rangle - \left\langle \mathbf{a}, \frac{\langle \mathbf{a}, \mathbf{x} \rangle}{\langle \mathbf{a}, \mathbf{a} \rangle}\mathbf{a} \right\rangle = \langle \mathbf{a}, \mathbf{x} \rangle - \frac{\langle \mathbf{a}, \mathbf{x} \rangle}{\langle \mathbf{a}, \mathbf{a} \rangle}\langle \mathbf{a}, \mathbf{a} \rangle = 0.$

(35) $\|s\mathbf{x} + t\mathbf{y}\|^2 = \langle s\mathbf{x} + t\mathbf{y}, s\mathbf{x} + t\mathbf{y} \rangle = s^2\langle \mathbf{x}, \mathbf{x} \rangle + 2st\langle \mathbf{x}, \mathbf{y} \rangle + t^2\langle \mathbf{y}, \mathbf{y} \rangle$
$= s^2(1) + st(0) + t^2(1)$

(36) An identity matrix.

(37) Set $\mathbf{y}=\mathbf{x}$ and use part *(a)* of Theorem 1 of Section 6.1.

(38) Denote the columns of **A** as $\mathbf{A}_1, \mathbf{A}_2, \ldots, \mathbf{A}_n$, and the elements of **y** as $y_1, y_2, \ldots y_n$, respectively. Then, $\mathbf{Ay}$ $\mathbf{A}_1 y_1 + \mathbf{A}_2 y_2 + \ldots + \mathbf{A}_n y_n$ and $\langle \mathbf{Ay}, \mathbf{p} \rangle = y_1 \langle \mathbf{A}_1, \mathbf{p} \rangle + y_2 \langle \mathbf{A}_2, \mathbf{p} \rangle + \ldots + y_n \langle \mathbf{A}n, \mathbf{p} \rangle$.

(39) Use Theorem 1.

Section 6.3

(1) $\begin{bmatrix} \sqrt{5} & \frac{2\sqrt{5}}{5} \\ 0 & \frac{6\sqrt{5}}{5} \\ 0 & 0 \end{bmatrix}$

(2) $\begin{bmatrix} 0.4472 & 0.8944 \\ 0.8944 & -0.4472 \end{bmatrix} \begin{bmatrix} 2.2361 & 1.7889 \\ 0.0000 & 1.3416 \end{bmatrix}.$

(3) $\begin{bmatrix} 0.7071 & -0.7071 \\ 0.7071 & 0.7071 \end{bmatrix} \begin{bmatrix} 1.4142 & 5.6569 \\ 0.0000 & 1.4142 \end{bmatrix}.$

(4) $\begin{bmatrix} 0.8321 & 0.5547 \\ -0.5547 & 0.8321 \end{bmatrix} \begin{bmatrix} 3.6056 & 0.8321 \\ 0.0000 & 4.1603 \end{bmatrix}.$

(5) $\begin{bmatrix} 0.3333 & 0.8085 \\ 0.6667 & 0.1617 \\ 0.6667 & -0.5659 \end{bmatrix} \begin{bmatrix} 3.0000 & 2.6667 \\ 0.0000 & 1.3744 \end{bmatrix}.$

(6) $\begin{bmatrix} 0.3015 & -0.2752 \\ 0.3015 & -0.8808 \\ 0.9045 & 0.3853 \end{bmatrix} \begin{bmatrix} 3.3166 & 4.8242 \\ 0.0000 & 1.6514 \end{bmatrix}.$

(7) $\begin{bmatrix} 0.7746 & 0.4034 \\ -0.5164 & 0.5714 \\ 0.2582 & 0.4706 \\ -0.2582 & 0.5378 \end{bmatrix} \begin{bmatrix} 3.8730 & 0.2582 \\ 0.0000 & 1.9833 \end{bmatrix}.$

(8) $\begin{bmatrix} 0.8944 & -0.2981 & 0.3333 \\ 0.4472 & 0.5963 & -0.6667 \\ 0.0000 & 0.7454 & 0.6667 \end{bmatrix} \begin{bmatrix} 2.2361 & 0.4472 & 1.7889 \\ 0.0000 & 1.3416 & 0.8944 \\ 0.0000 & 0.0000 & 2.0000 \end{bmatrix}.$

(9) $\begin{bmatrix} 0.7071 & 0.5774 & -0.4082 \\ 0.7071 & -0.5774 & 0.4082 \\ 0.0000 & 0.5774 & 0.8165 \end{bmatrix} \begin{bmatrix} 1.4142 & 1.4142 & 2.8284 \\ 0.0000 & 1.7321 & 0.5774 \\ 0.0000 & 0.0000 & 0.8165 \end{bmatrix}.$

(10) $\begin{bmatrix} 0.00 & 0.60 & 0.80 \\ 0.60 & 0.64 & -0.48 \\ 0.80 & -0.48 & 0.36 \end{bmatrix} \begin{bmatrix} 5 & 3 & 7 \\ 0 & 5 & 2 \\ 0 & 0 & 1 \end{bmatrix}.$

(11) $\begin{bmatrix} 0.0000 & 0.7746 & 0.5071 \\ 0.5774 & -0.5164 & 0.5071 \\ 0.5774 & 0.2582 & -0.6761 \\ 0.5774 & 0.2582 & 0.1690 \end{bmatrix} \begin{bmatrix} 1.7321 & 1.1547 & 1.1547 \\ 0.0000 & 1.2970 & 0.5164 \\ 0.0000 & 0.0000 & 1.1832 \end{bmatrix}.$

(12) $\begin{bmatrix} 0.7071 & -0.4082 & 0.5774 \\ 0.7071 & 0.4082 & -0.5774 \\ 0.0000 & -0.8165 & -0.5774 \\ 0.0000 & 0.0000 & 0.0000 \end{bmatrix} \begin{bmatrix} 1.4142 & 0.7071 & 0.7071 \\ 0.0000 & 1.2247 & 0.4082 \\ 0.0000 & 0.0000 & 1.1547 \end{bmatrix}.$

(13) $\mathbf{A}_1 = \mathbf{R}_0\mathbf{Q}_0 + 7\mathbf{I}$

$$= \begin{bmatrix} 19.3132 & -1.2945 & 0.0000 \\ 0.0000 & 7.0231 & -0.9967 \\ 0.0000 & 0.0000 & 0.0811 \end{bmatrix} \begin{bmatrix} -0.3624 & 0.0756 & 0.9289 \\ 0.0000 & -0.9967 & 0.0811 \\ 0.9320 & 0.0294 & 0.3613 \end{bmatrix}$$

$$+7 \begin{bmatrix} 1 & 0 & 0 \\ 0 & 1 & 0 \\ 0 & 0 & 1 \end{bmatrix} = \begin{bmatrix} 0.0000 & 2.7499 & 17.8357 \\ -0.9289 & -0.0293 & 0.2095 \\ 0.0756 & 0.0024 & 70.293 \end{bmatrix}$$

(14) $\mathbf{A}_1 = \mathbf{R}_0\mathbf{Q}_0 - 14\mathbf{I}$

$$= \begin{bmatrix} 24.3721 & -17.8483 & 3.8979 \\ 0.0000 & 8.4522 & -4.6650 \\ 0.0000 & 0.0000 & 3.6117 \end{bmatrix} \begin{bmatrix} 0.6565 & -0.6250 & 0.4223 \\ -0.6975 & -0.2898 & 0.6553 \\ 0.2872 & 0.7248 & 0.6262 \end{bmatrix}$$

$$-14 \begin{bmatrix} 1 & 0 & 0 \\ 0 & 1 & 0 \\ 0 & 0 & 1 \end{bmatrix} = \begin{bmatrix} 15.5690 & -7.2354 & 1.0373 \\ -7.2354 & -19.8307 & 2.6178 \\ 1.0373 & 2.6178 & -11.7383 \end{bmatrix}$$

(15) Shifr by 4.

$$\mathbf{R}_0 = \begin{bmatrix} 4.1231 & -0.9701 & 0.0000 & 13.5820 \\ 0.0000 & 4.0073 & -09982 & -4.1982 \\ 0.0000 & 0.0000 & 4.0005 & 12.9509 \\ 0.0000 & 0.0000 & 0.0000 & 3.3435 \end{bmatrix}$$

$$\mathbf{Q}_0 = \begin{bmatrix} -0.9701 & -0.2349 & -0.0586 & -0.0151 \\ 0.225 & -0.9395 & -0.2344 & -0.0605 \\ 0.0000 & 0.2495 & -0.9376 & -0.2421 \\ 0.0000 & 0.0000 & 0.2500 & -0.9683 \end{bmatrix}$$

$$\mathbf{A}_1 = \mathbf{R}_0\mathbf{Q}_0 + 4\mathbf{I} = \begin{bmatrix} -0.2353 & -0.0570 & 3.3809 & -13.1545 \\ 0.9719 & -0.0138 & -1.0529 & 4.0640 \\ 0.0000 & 0.9983 & 3.1864 & -13.5081 \\ 0.0000 & 0.0000 & 0.8358 & 0.7626 \end{bmatrix}$$

(16) 7.2077, − 0.1039 ± 1.5769i. **(17)** −11, − 22, 17. **(18)** 2, 3, 9.

(19) Method fails. $\mathbf{A}_0 - 7\mathbf{I}$ does not have linearly independent columns, so no **QR** decomposition is possible.

(20) 2, 2, 16. **(21)** 2, $3 \pm i$. **(22)** $\pm i$, $2 \pm 3i$

(23) $3.1265 \pm 1.2638i$, $-2.6265 \pm 0.7590i$.

(24) 0.0102, 0.8431, 3.8581, 30.2887.

(25) Each diagonal element of the upper triangular matrix **R** is the magnitude of a nonzero vector (see Theorem 3 of Section 6.2) and is, therefore, nonzero. Use Theorems 4 and 10 of Section 1.5.

(26) $\mathbf{Q}^T\mathbf{Q} = \mathbf{I}$. Thus, $\mathbf{Q}^T = \mathbf{Q}^{-1}$.

(27) $\mathbf{A}_k\mathbf{R}_{k-1} = \mathbf{R}_{k-1}\mathbf{Q}_{k-1}\mathbf{R}_{k-1} = \mathbf{R}_{k-1}\mathbf{A}_{k-1}$.

Set $\mathbf{P} = (\mathbf{R}_{k-1})^{-1}$ and use Theorem 1 of Section 4.1.

Section 6.4

(1) $x \approx 2.225$, $y \approx 1.464$.

(2) $x \approx 3$, $y \approx 3$.

(3) $x \approx 9.879$, $y \approx 18.398$.

(4) $x \approx -1.174$, $y \approx 8.105$.

(5) $x \approx 1.512$, $y \approx 0.639$, $z \approx 0.945$.

(6) $x \approx 7.845$, $y \approx 1.548$, $z \approx 5.190$.

(7) $x \approx 81.003$, $y \approx 50.870$, $z \approx 38.801$.

(8) $x \approx 2.818$, $y \approx -0.364$, $z \approx -1.364$.

(9) (b) $y = 2.3x + 8.1$ (c) 21.9.

(10) (a) $\mathrm{y} = 1.2\mathrm{x} + 19.4$,

(b) $\mathrm{y} = 1.2(60) + 19.4 = 91.4$, so total mlb attendance in 2020 projects to 91.4 million

(11) (a) $y = 0.27x + 10.24$, (b) 12.4.

(12) $$m = \frac{N\sum_{i=1}^{N} x_i y_i - \sum_{i=1}^{N} x_i \sum_{i=1}^{N} y_i}{N\sum_{i=1}^{N} x_i^2 - \left(\sum_{i=1}^{N} x_i\right)^2}, \quad c = \frac{\sum_{i=1}^{N} y_i \sum_{i=1}^{N} x_i^2 - \sum_{i=1}^{N} x_i \sum_{i=1}^{N} x_i y_i}{N\sum_{i=1}^{N} x_i^2 - \left(\sum_{i=1}^{N} x_i\right)^2}.$$

If $N\sum_{i=1}^{N} x_i^2$ is near $\left(\sum_{i=1}^{N} x_i\right)^2$, then the denominator is near 0.

(13) $\sum_{i=1}^{N} x_i' = 0$ so the denominators for m and c found in Problem 13 reduce to $\sum_{i=1}^{N} (x_i')^2$.

(14) $y = 2.3x' + 15$.

(15) (a) $y = -0.198x' + 21.18$ (b) year 2000 is coded as $x' = 30$; $y(30) = 15.24$.

(21) $\mathbf{E} = \begin{bmatrix} 0.842 \\ 0.211 \\ -2.311 \end{bmatrix}$.

(22) $\mathbf{E} = \begin{bmatrix} 0.161 \\ 0.069 \\ -0.042 \\ -0.172 \end{bmatrix}$.

Section 6.5

(1) (a) *span* $\left\{ \begin{bmatrix} 1 \\ 0 \\ 0 \end{bmatrix}, \begin{bmatrix} 0 \\ -1 \\ 1 \end{bmatrix} \right\}$, (b) $\begin{bmatrix} 0 \\ 1/2 \\ 1/2 \end{bmatrix}$.

(2) (a) *span* $\left\{ \begin{bmatrix} -1 \\ 1 \\ 0 \end{bmatrix}, \begin{bmatrix} -1 \\ 0 \\ 1 \end{bmatrix} \right\}$, (b) $\begin{bmatrix} 2/3 \\ 2/3 \\ 2/3 \end{bmatrix}$.

(3) (a) *span* $\left\{ \begin{bmatrix} -1 \\ 2 \\ 0 \end{bmatrix}, \begin{bmatrix} -1 \\ 0 \\ 2 \end{bmatrix} \right\}$, (b) $\begin{bmatrix} 1 \\ 1/2 \\ 1/2 \end{bmatrix}$.

(4) (a) *span* $\left\{ \begin{bmatrix} 1 \\ -2 \\ 1 \end{bmatrix} \right\}$, (b) $\begin{bmatrix} 7/6 \\ 4/6 \\ 1/6 \end{bmatrix}$.

(5) (a) *span* $\left\{ \begin{bmatrix} 1 \\ -4 \\ 2 \end{bmatrix} \right\}$, (b) $\begin{bmatrix} 8/7 \\ 3/7 \\ 2/7 \end{bmatrix}$.

(6) (a) *span* $\left\{ \begin{bmatrix} 1 \\ 0 \\ 1 \end{bmatrix} \right\}$, (b) $\begin{bmatrix} 1 \\ 1 \\ 0 \end{bmatrix}$.

(7) (a) *span* $\left\{ \begin{bmatrix} -1 \\ 1 \\ 0 \end{bmatrix} \right\}$, (b) $\begin{bmatrix} 1 \\ 1 \\ 0 \end{bmatrix}$.

(8) Same as Problem 2

(9) (a) *span* $\left\{ \begin{bmatrix} 0 \\ -1 \\ 1 \end{bmatrix} \right\}$, (b) $\begin{bmatrix} 1 \\ 1/2 \\ 1/2 \end{bmatrix}$.

(10) (a) $\{\mathbf{0}\}$, (b) $\begin{bmatrix} 1 \\ 1 \\ 0 \end{bmatrix}$.

(11) (a) *span* $\left\{\begin{bmatrix}1\\0\\0\\0\end{bmatrix}, \begin{bmatrix}0\\1\\0\\0\end{bmatrix}, \begin{bmatrix}0\\0\\-1\\1\end{bmatrix}\right\}$, (b) $\begin{bmatrix}0\\0\\1/2\\1/2\end{bmatrix}$.

(12) (a) *span* $\left\{\begin{bmatrix}1\\0\\0\\0\end{bmatrix}, \begin{bmatrix}0\\-1\\1\\0\end{bmatrix}, \begin{bmatrix}0\\-1\\0\\1\end{bmatrix}\right\}$, (b) $\begin{bmatrix}0\\1/3\\1/3\\1/3\end{bmatrix}$.

(13) (a) *span* $\left\{\begin{bmatrix}1\\0\\0\\0\end{bmatrix}, \begin{bmatrix}0\\0\\-1\\1\end{bmatrix}\right\}$, (b) $\begin{bmatrix}0\\0\\1/2\\1/2\end{bmatrix}$.

(14) (a) *span* $\left\{\begin{bmatrix}1\\0\\0\\0\end{bmatrix}, \begin{bmatrix}0\\0\\1\\0\end{bmatrix}\right\}$, (b) $\begin{bmatrix}0\\0\\0\\0\end{bmatrix}$.

(15) (a) *span* $\left\{\begin{bmatrix}-1\\1\\0\\0\end{bmatrix}, \begin{bmatrix}-1\\0\\1\\1\end{bmatrix}\right\}$, (b) $\begin{bmatrix}3/5\\3/5\\4/5\\-1/5\end{bmatrix}$.

(16) (a) *span* $\left\{\begin{bmatrix}-2\\1\\1\\1\end{bmatrix}\right\}$, (b) $\begin{bmatrix}5/7\\1/7\\8/7\\1/7\end{bmatrix}$.

(17) (a) *span* $\left\{\begin{bmatrix}-1\\1\\0\\0\end{bmatrix}, \begin{bmatrix}-1\\0\\1\\0\end{bmatrix}\right\}$, (b) $\begin{bmatrix}2/3\\2/3\\2/3\\0\end{bmatrix}$.

(18) (a) *span* $\left\{\begin{bmatrix}1\\-1\\-1\\1\end{bmatrix}\right\}$, (b) $\begin{bmatrix}1\\0\\1\\0\end{bmatrix}$.

(19) (a) $\{\mathbf{0}\}$, (b) $\begin{bmatrix}1\\0\\1\\0\end{bmatrix}$.

(20) (a) $span\left\{\begin{bmatrix}0\\-1\\1\\1\end{bmatrix}\right\}$, (b) $\begin{bmatrix}1\\1/3\\2/3\\-1/3\end{bmatrix}$.

(21) No.

(22) $\|\mathbf{x}\|^2 = \langle \mathbf{u}+\mathbf{u}^{\perp}, \mathbf{u}+\mathbf{u}^{\perp}\rangle = \langle \mathbf{u},\mathbf{u}\rangle + \langle \mathbf{u},\mathbf{u}^{\perp}\rangle + \langle \mathbf{u}^{\perp},\mathbf{u}\rangle + \langle \mathbf{u}^{\perp},,\mathbf{u}^{\perp}\rangle$
$= \langle \mathbf{u},\mathbf{u}\rangle + 0 + 0 + \langle \mathbf{u}^{\perp},,\mathbf{u}^{\perp}\rangle = \|\mathbf{u}\|^2 + \|\mathbf{u}^{\perp}\|^2.$

(23) Let $\mathbb{B} = \{\mathbf{u}_1,\ \mathbf{u}_2\ldots,\ \mathbf{u}_r\}$ **and** $\mathbb{C} = \{\mathbf{w}_1, \mathbf{w}_2\ldots, \mathbf{w}_s\}$. If $v \in \mathbb{V}$, then there exists a $\mathbf{u} \in \mathbb{U}$ and $\mathbf{w} \in \mathbb{W}$ such that $\mathrm{v}=\mathbf{u}+\mathbf{w}$. But $\mathbf{u} = \sum_{i=1}^{r} c_i\mathbf{u}_i$ and $\mathbf{w} = \sum_{j=1}^{s} d_j\mathbf{w}_i$ for scalars $c_1,\ldots,c_r,$ and $d_1,\ldots,d_s$. Then, $\mathbf{v} = \sum_{i=1}^{r} c_i\mathbf{u}_i + \sum_{j=1}^{s} d_j\mathbf{w}_j$ and $\mathbb{B}\cup\mathbb{C}$ is a spanning set for $\mathbb{V}$. Consider the equation $\sum_{i=1}^{r} c_i\mathbf{u}_i + \sum_{j=1}^{s} d_j\mathbf{w}_j = 0$. Since $\sum_{i=1}^{r}(0)\mathbf{u}_i + \sum_{j=1}^{s}(0)\mathbf{w}_j = 0$, it follows from uniqueness that $c_i=0(i=1,2,\ldots,r)$ and $d_j=0(j=1,2,\ldots,s)$. Thus, $\mathbb{B}\cup\mathbb{C}$ is linearly independent.

(24) Let $\mathbf{v}\in\mathbb{U}$ with basic $\mathbb{B} = \{\mathbf{u}_1,\mathbf{u}_2,\ldots,\mathbf{u}_r\}$. Then $\mathbf{v} = \sum_{i=1}^{r} c_i\mathbf{u}_i$ for scalars $c_1,\ \ldots,\ c_r$.

Let $\mathbf{v}\in\mathbb{W}$ with basic $\mathbb{C} = \{\mathbf{w}_1,\mathbf{w}_2,\ldots,\mathbf{w}_s\}$. Then $\mathbf{v} = \sum_{j=1}^{s} d_j w_i$ for scalars $d_1,\ \ldots,\ d_s$.

$0 = \mathbf{v}-\mathbf{v} = \sum_{i=1}^{r} c_i\mathbf{u}_i - \sum_{j=1}^{s} d_j\mathbf{w}_j$. But $0 = \sum_{i=1}^{r}(0)\mathbf{u}_i - \sum_{j=1}^{s}(0)\mathbf{w}_j$, so it follows from uniqueness that $c_i=0(i=1,2,\ldots,r)$ and $d_j=0(j=1,\ 2,\ \ldots,\ s)$. Thus, $\mathbf{v} = \sum_{i=1}^{r} c_i\mathbf{u}_i = \sum_{j=1}^{s}(0)\mathbf{u}_i = 0$.

(25) Use the results of Problem 23.

APPENDICES

Appendix A

(1) (a) Yes, (b) No, (c) No, (d) Yes,
(e) Yes, (f) Yes.

(2) (a) Yes, (b) Yes, (c) No, (d) Yes,
(e) No, (f) Yes.

(3) (a) Yes, (b) No, (c) Yes, (d) No,
(e) Yes, (f) Yes.

(4) (a) No, (b) Yes, (c) No, (d) Yes,
(e) Yes, (f) Yes.

(5) (a) No, (b) No, (c) No, (d) Yes,
(e) No, (f) No.

(6) (a) Yes, (b) Yes, (c) No, (d) No,
(e) Yes, (f) Yes.

(7) (a) No, (b) No, (c) Yes, (d) No.

(8) (a) No, (b) Yes, (c) No, (d) Yes.

(9) (a) Yes, (b) Yes, (c) Yes, (d) No.

(10) (a) Yes, (b) No, (c) Yes, (d) No.

(11) $\left\{ \begin{bmatrix} a \\ b \\ c \end{bmatrix} \in \mathbf{R}^3 \middle| b = 0 \right\}$.

(12) $\begin{bmatrix} 7/2 & -1/2 & 0 & 0 \\ 1/2 & 5/2 & 0 & 0 \\ 0 & 0 & 4 & 1 \\ 0 & 0 & 0 & 4 \end{bmatrix}$.

(13) $\begin{bmatrix} 2 & 0 & 0 & 0 \\ 0 & 5/2 & -1/2 & 0 \\ 0 & 1/2 & 3/2 & -1 \\ 0 & 0 & 0 & 2 \end{bmatrix}$.

(14) $\begin{bmatrix} 2 & 0 & 0 & 0 \\ 0 & 2 & 0 & 0 \\ 0 & 0 & 3 & 1 \\ 0 & 0 & 0 & 3 \end{bmatrix}$.

(15) (a) Yes, (b) No, (c) Yes, (d) Yes, (e) No, (f) No.

(16) $\begin{bmatrix} 0 \\ 1 \end{bmatrix}$. **(17)** $\begin{bmatrix} 0 \\ 1 \\ 0 \end{bmatrix}$. **(18)** $\begin{bmatrix} 0 \\ 0 \\ 1 \end{bmatrix}$. **(19)** $\begin{bmatrix} 0 \\ 0 \\ 1 \end{bmatrix}$. **(20)** $\begin{bmatrix} 1 \\ 0 \\ -1 \end{bmatrix}$.

(21) For $\lambda = 3, x_3 = \begin{bmatrix} 1 \\ 0 \\ 0 \\ 0 \\ -1 \end{bmatrix}$, and for $\lambda = 4, x_2 = \begin{bmatrix} 0 \\ 1 \\ 0 \\ 0 \\ 0 \end{bmatrix}$.

(22) $x_3 = \begin{bmatrix} 1 \\ 1 \\ 1 \\ 0 \end{bmatrix}, x_2 = \begin{bmatrix} 3 \\ -1 \\ 0 \\ 0 \end{bmatrix}, x_1 = \begin{bmatrix} -2 \\ 0 \\ 0 \\ 0 \end{bmatrix}$.

(23) $x_3 = \begin{bmatrix} 0 \\ 0 \\ 1 \\ 0 \end{bmatrix}, x_2 = \begin{bmatrix} 1 \\ -1 \\ 0 \\ 0 \end{bmatrix}, x_1 = \begin{bmatrix} -2 \\ 0 \\ 0 \\ 0 \end{bmatrix}$.

(24) $x_4 = \begin{bmatrix} 0 \\ 0 \\ 0 \\ 0 \\ 1 \end{bmatrix}, x_3 = \begin{bmatrix} -1 \\ 0 \\ 2 \\ 1 \\ 0 \end{bmatrix}, x_2 = \begin{bmatrix} 2 \\ 0 \\ -1 \\ 0 \\ 0 \end{bmatrix}, x_1 = \begin{bmatrix} -1 \\ 0 \\ 0 \\ 0 \\ 0 \end{bmatrix}$.

(25) $x_3 = \begin{bmatrix} 0 \\ 0 \\ 0 \\ 1 \\ 0 \end{bmatrix}, x_2 = \begin{bmatrix} 0 \\ 0 \\ -1 \\ 0 \\ 0 \end{bmatrix}, x_1 = \begin{bmatrix} -1 \\ 0 \\ 0 \\ 0 \\ 0 \end{bmatrix}$.

(26) $x_3 = \begin{bmatrix} 1 \\ 0 \\ 0 \\ 0 \\ -1 \end{bmatrix}, x_2 = \begin{bmatrix} 0 \\ 0 \\ 0 \\ -2 \\ 0 \end{bmatrix}, x_1 = \begin{bmatrix} 0 \\ 0 \\ -2 \\ 0 \\ 0 \end{bmatrix}$.

(27) $x_2 = \begin{bmatrix} 0 \\ 1 \\ 0 \\ 0 \\ 0 \end{bmatrix}, x_1 = \begin{bmatrix} 1 \\ 0 \\ 0 \\ 0 \\ 0 \end{bmatrix}$. **(28)** $x_2 = \begin{bmatrix} 0 \\ 1 \end{bmatrix}, x_1 = \begin{bmatrix} 1 \\ 0 \end{bmatrix}$.

(29) $x_2 = \begin{bmatrix} 0 \\ 1 \\ 0 \end{bmatrix}, x_1 = \begin{bmatrix} 1 \\ 0 \\ 0 \end{bmatrix}$. **(30)** $x_2 = \begin{bmatrix} 0 \\ 0 \\ 1 \end{bmatrix}, x_1 = \begin{bmatrix} 2 \\ 1 \\ -3 \end{bmatrix}$.

(31) $x_4 = \begin{bmatrix} 0 \\ 0 \\ 0 \\ 1 \end{bmatrix}, x_3 = \begin{bmatrix} -1 \\ 4 \\ 1 \\ 0 \end{bmatrix}, x_2 = \begin{bmatrix} 7 \\ -1 \\ 0 \\ 0 \end{bmatrix}, x_1 = \begin{bmatrix} -1 \\ 0 \\ 0 \\ 0 \end{bmatrix}$.

(32) $x_3 = \begin{bmatrix} 0 \\ 0 \\ 1 \\ 0 \\ 0 \end{bmatrix}, x_2 = \begin{bmatrix} 1 \\ 1 \\ 0 \\ 0 \\ 0 \end{bmatrix}, x_1 = \begin{bmatrix} 2 \\ -2 \\ 0 \\ 0 \\ 0 \end{bmatrix}.$

(33) $\mathbf{x}$ is a generalized eigenvector of type 1 corresponding to the eigenvalue λ if $(\mathbf{A} - \lambda\mathbf{I})^1\ \mathbf{x}=\mathbf{0}$ and $(\mathbf{A} - \lambda\mathbf{I})^0\ \mathbf{x}\neq 0$. That is, if $\mathbf{Ax}=\lambda\mathbf{x}$ and $\mathbf{x}\neq\mathbf{0}$.

(34) If $\mathbf{x}=0$, then $(\mathbf{A} - \lambda\mathbf{I})^n\mathbf{x}=(\mathbf{A} - \lambda\mathbf{I})^n\ \mathbf{0}=\mathbf{0}$ for every positive integer n.

(35) (a) Use Theorem 1 of Section 3.5.

(b) By the definition of T, $T(\mathbf{v}) \in \mathbb{V}$ for each $\mathbf{v} \in \mathbb{V}$.

(c) Let $T(\mathbf{v}_i)=\lambda_i\mathbf{v}_i$. If $\mathbf{v} \in span\{\mathbf{v}_1, \mathbf{v}_2, \ldots, \mathbf{v}_k\}$, then there exist scalars $c_1, c_2, \ldots, c_k$ such that $\mathbf{v} = \sum_{i=1}^{k} c_i\mathbf{v}_i$. Consequently, $T(v) = T\left(\sum_{i=1}^{k} c_i v_i\right) = \sum_{i=1}^{k} c_i T(v_i) = \sum_{i=1}^{k} c_i(\lambda_i v_i) = \sum_{i=1}^{k} (c_i\lambda_i)v_i$, which also belongs to $span\{\mathbf{v}_1, \mathbf{v}_2, \ldots, \mathbf{v}_k\}$.

(36) If $\mathbb{V} = \mathbb{U} \oplus \mathbb{W}$, then (i) and (ii) follow from the definition of a direct sum and Problem 24 of Section 6.5. To show the converse, assume that $\mathbf{v}=\mathbf{u}_1+\mathbf{w}_1$ and also $\mathbf{v}=\mathbf{u}_2+\mathbf{w}_2$, where $\mathbf{u}_1$ and $\mathbf{u}_2$ are vectors in $\mathbb{U}$, and $\mathbf{w}_1$ and $\mathbf{w}_2$ are vectors in $\mathbb{W}$. Then $\mathbf{0}=\mathbf{v} - \mathbf{v}=(\mathbf{u}_1+\mathbf{w}_1)=(\mathbf{u}_2+\mathbf{w}_2)=(\mathbf{u}_1 - \mathbf{u}_2)+(\mathbf{w}_1 - \mathbf{w}_2)$, or $(\mathbf{u}_1 - \mathbf{u}_2)=(\mathbf{w}_2 - \mathbf{w}_1)$. The left-side of this last equation is in $\mathbb{U}$, and the right side is in $\mathbb{W}$. Both sides are equal, so both sides are in $\mathbb{U}$ and $\mathbb{W}$. It follows from (*ii*) that $(\mathbf{u}_1 - \mathbf{u}_2)=\mathbf{0}$ and $(\mathbf{w}_2 - \mathbf{w}_1)=\mathbf{0}$. Thus, $\mathbf{u}_1=\mathbf{u}_2$ and $\mathbf{w}_1=\mathbf{w}_2$.

(38) (a) One chain of length 3;

(b) two chains of length 3;

(c) one chain of length 3, and one chain of length 2;

(d) one chain of length 3, one chain of length 2, and one chain of length 1;

(e) one chain of length 3 and two chains of length 1;

(f) cannot be done, the numbers as given are not compatible;

(g) two chains of length 2, and two chains of length 1;

(h) cannot be done, the numbers as given are not compatible;

(i) two chains of length 2 and one chain of length 1;

(j) two chains of length 2.

(39) $\mathbf{x}_2 = \begin{bmatrix} 0 \\ 1 \end{bmatrix}, x_1 = \begin{bmatrix} 1 \\ -1 \end{bmatrix}.$

(40) $\mathbf{x}_1 = \begin{bmatrix} -1 \\ 1 \\ 1 \end{bmatrix}$ corresponds to $\lambda=1$ and $\mathbf{y}_2 = \begin{bmatrix} 0 \\ 0 \\ 1 \end{bmatrix}, \mathbf{y}_1 = \begin{bmatrix} 3 \\ 0 \\ -3 \end{bmatrix}$ correspond to $\lambda=4$.

(41) $x_3 = \begin{bmatrix} 0 \\ 0 \\ 1 \end{bmatrix}, x_2 = \begin{bmatrix} -1 \\ 2 \\ 0 \end{bmatrix}, x_1 = \begin{bmatrix} 2 \\ 0 \\ 0 \end{bmatrix}.$

(42) $x_1 = \begin{bmatrix} 1 \\ -2 \\ 0 \end{bmatrix}, y_1 = \begin{bmatrix} 0 \\ -2 \\ 1 \end{bmatrix}$ both correspond to $\lambda=3$ and $\mathbf{z}_1 = \begin{bmatrix} 1 \\ 0 \\ 1 \end{bmatrix}$. corresponds to $\lambda=7$

(43) $x_3 = \begin{bmatrix} 0 \\ 0 \\ 0 \\ 1 \end{bmatrix}, x_2 = \begin{bmatrix} -1 \\ 1 \\ 0 \\ 0 \end{bmatrix}, x_1 = \begin{bmatrix} 1 \\ 0 \\ 0 \\ 0 \end{bmatrix}, y_1 = \begin{bmatrix} 0 \\ -1 \\ 1 \\ -1 \end{bmatrix}.$

(44) $x_2 = \begin{bmatrix} 0 \\ 1 \\ 0 \\ 0 \end{bmatrix}, x_1 = \begin{bmatrix} 1 \\ 0 \\ 0 \\ 0 \end{bmatrix}$ correspond to $\lambda=3$ and $y_2 = \begin{bmatrix} 3 \\ 1 \\ 0 \\ -1 \end{bmatrix}, y_1 = \begin{bmatrix} -1 \\ -1 \\ -1 \\ 0 \end{bmatrix}$ correspond to $\lambda=4$.

(45) $x_4 = \begin{bmatrix} 0 \\ 0 \\ 0 \\ 2 \\ -2 \\ 1 \end{bmatrix}, x_3 = \begin{bmatrix} -1 \\ 1 \\ 2 \\ 0 \\ 0 \\ 0 \end{bmatrix}, x_2 = \begin{bmatrix} 3 \\ 4 \\ 0 \\ 0 \\ 0 \\ 0 \end{bmatrix}, x_1 = \begin{bmatrix} 4 \\ 0 \\ 0 \\ 0 \\ 0 \\ 0 \end{bmatrix}$. correspond to $\lambda=4$, and

$y_2 = \begin{bmatrix} -5 \\ -2 \\ 0 \\ 1 \\ 1 \\ 0 \end{bmatrix}, y_1 = \begin{bmatrix} 3 \\ 2 \\ 1 \\ 1 \\ 0 \\ 0 \end{bmatrix}$ correspond to $\lambda=5$.

(46) $\begin{bmatrix} 2 & 0 & 0 & 0 & 0 \\ 0 & 2 & 0 & 0 & 0 \\ 0 & 0 & 2 & 1 & 0 \\ 0 & 0 & 0 & 2 & 1 \\ 0 & 0 & 0 & 0 & 2 \end{bmatrix}.$ **(47)** $\begin{bmatrix} 2 & 1 & 0 & 0 & 0 \\ 0 & 2 & 0 & 0 & 0 \\ 0 & 0 & 2 & 1 & 0 \\ 0 & 0 & 0 & 2 & 1 \\ 0 & 0 & 0 & 0 & 2 \end{bmatrix}.$

(48) $$\begin{bmatrix} 2 & 0 & 0 & 0 & 0 \\ 0 & 2 & 1 & 0 & 0 \\ 0 & 0 & 2 & 0 & 0 \\ 0 & 0 & 0 & 2 & 1 \\ 0 & 0 & 0 & 0 & 2 \end{bmatrix}.$$

(49) $$\begin{bmatrix} 2 & 0 & 0 & 0 & 0 \\ 0 & 2 & 1 & 0 & 0 \\ 0 & 0 & 2 & 1 & 0 \\ 0 & 0 & 0 & 2 & 1 \\ 0 & 0 & 0 & 0 & 2 \end{bmatrix}.$$

(50) $$\begin{bmatrix} 2 & 1 & 0 & 0 & 0 \\ 0 & 2 & 1 & 0 & 0 \\ 0 & 0 & 2 & 1 & 0 \\ 0 & 0 & 0 & 2 & 1 \\ 0 & 0 & 0 & 0 & 2 \end{bmatrix}.$$

(51) $$\begin{bmatrix} 3 & 1 & 0 & 0 & 0 & 0 \\ 0 & 3 & 1 & 0 & 0 & 0 \\ 0 & 0 & 3 & 0 & 0 & 0 \\ 0 & 0 & 0 & 3 & 1 & 0 \\ 0 & 0 & 0 & 0 & 3 & 1 \\ 0 & 0 & 0 & 0 & 0 & 3 \end{bmatrix}.$$

(52) $$\begin{bmatrix} 3 & 0 & 0 & 0 & 0 & 0 \\ 0 & 3 & 1 & 0 & 0 & 0 \\ 0 & 0 & 3 & 0 & 0 & 0 \\ 0 & 0 & 0 & 3 & 1 & 0 \\ 0 & 0 & 0 & 0 & 3 & 1 \\ 0 & 0 & 0 & 0 & 0 & 3 \end{bmatrix}.$$

(53) $$\begin{bmatrix} 3 & 0 & 0 & 0 & 0 & 0 \\ 0 & 3 & 0 & 0 & 0 & 0 \\ 0 & 0 & 3 & 0 & 0 & 0 \\ 0 & 0 & 0 & 3 & 1 & 0 \\ 0 & 0 & 0 & 0 & 3 & 1 \\ 0 & 0 & 0 & 0 & 0 & 3 \end{bmatrix}.$$

(54) $$\begin{bmatrix} 3 & 1 & 0 & 0 & 0 & 0 \\ 0 & 3 & 0 & 0 & 0 & 0 \\ 0 & 0 & 3 & 1 & 0 & 0 \\ 0 & 0 & 0 & 3 & 0 & 0 \\ 0 & 0 & 0 & 0 & 3 & 1 \\ 0 & 0 & 0 & 0 & 0 & 3 \end{bmatrix}.$$

(55) $$\begin{bmatrix} 3 & 0 & 0 & 0 & 0 & 0 \\ 0 & 3 & 0 & 0 & 0 & 0 \\ 0 & 0 & 3 & 1 & 0 & 0 \\ 0 & 0 & 0 & 3 & 0 & 0 \\ 0 & 0 & 0 & 0 & 3 & 1 \\ 0 & 0 & 0 & 0 & 0 & 3 \end{bmatrix}.$$

(56) $$\begin{bmatrix} 3 & 0 & 0 & 0 & 0 & 0 \\ 0 & 3 & 0 & 0 & 0 & 0 \\ 0 & 0 & 3 & 0 & 0 & 0 \\ 0 & 0 & 0 & 3 & 0 & 0 \\ 0 & 0 & 0 & 0 & 3 & 1 \\ 0 & 0 & 0 & 0 & 0 & 3 \end{bmatrix}.$$

(57) (a) $$\begin{bmatrix} 2 & 0 & 0 & 0 \\ 0 & 2 & 0 & 0 \\ 0 & 0 & 2 & 1 \\ 0 & 0 & 0 & 2 \end{bmatrix},$$ (b) $$\begin{bmatrix} 2 & 0 & 0 & 0 \\ 0 & 2 & 1 & 0 \\ 0 & 0 & 2 & 0 \\ 0 & 0 & 0 & 2 \end{bmatrix},$$

(c) $$\begin{bmatrix} 2 & 1 & 0 & 0 \\ 0 & 2 & 0 & 0 \\ 0 & 0 & 2 & 0 \\ 0 & 0 & 0 & 2 \end{bmatrix},$$ (d) $$\begin{bmatrix} 2 & 1 & 0 & 0 \\ 0 & 2 & 0 & 0 \\ 0 & 0 & 2 & 0 \\ 0 & 0 & 0 & 2 \end{bmatrix}.$$

(58) (a) $\begin{bmatrix} 3 & 0 & 0 & 0 & 0 & 0 \\ 0 & 3 & 1 & 0 & 0 & 0 \\ 0 & 0 & 3 & 0 & 0 & 0 \\ 0 & 0 & 0 & 5 & 0 & 0 \\ 0 & 0 & 0 & 0 & 5 & 1 \\ 0 & 0 & 0 & 0 & 0 & 5 \end{bmatrix}$, (b) $\begin{bmatrix} 3 & 1 & 0 & 0 & 0 & 0 \\ 0 & 3 & 0 & 0 & 0 & 0 \\ 0 & 0 & 3 & 0 & 0 & 0 \\ 0 & 0 & 0 & 5 & 0 & 0 \\ 0 & 0 & 0 & 0 & 5 & 1 \\ 0 & 0 & 0 & 0 & 0 & 5 \end{bmatrix}$,

(c) $\begin{bmatrix} 3 & 0 & 0 & 0 & 0 & 0 \\ 0 & 5 & 0 & 0 & 0 & 0 \\ 0 & 0 & 5 & 1 & 0 & 0 \\ 0 & 0 & 0 & 5 & 0 & 0 \\ 0 & 0 & 0 & 0 & 3 & 1 \\ 0 & 0 & 0 & 0 & 0 & 3 \end{bmatrix}$, (d) $\begin{bmatrix} 3 & 1 & 0 & 0 & 0 & 0 \\ 0 & 3 & 0 & 0 & 0 & 0 \\ 0 & 0 & 5 & 1 & 0 & 0 \\ 0 & 0 & 0 & 5 & 0 & 0 \\ 0 & 0 & 0 & 0 & 3 & 0 \\ 0 & 0 & 0 & 0 & 0 & 5 \end{bmatrix}$,

(e) $\begin{bmatrix} 3 & 0 & 0 & 0 & 0 & 0 \\ 0 & 5 & 0 & 0 & 0 & 0 \\ 0 & 0 & 3 & 1 & 0 & 0 \\ 0 & 0 & 0 & 3 & 0 & 0 \\ 0 & 0 & 0 & 0 & 5 & 1 \\ 0 & 0 & 0 & 0 & 0 & 5 \end{bmatrix}$, (f) $\begin{bmatrix} 5 & 1 & 0 & 0 & 0 & 0 \\ 0 & 5 & 0 & 0 & 0 & 0 \\ 0 & 0 & 5 & 0 & 0 & 0 \\ 0 & 0 & 0 & 3 & 0 & 0 \\ 0 & 0 & 0 & 0 & 3 & 1 \\ 0 & 0 & 0 & 0 & 0 & 3 \end{bmatrix}$.

(59) $\begin{bmatrix} 1 & 0 \\ 0 & -1 \end{bmatrix}$ with basis $\left\{ \begin{bmatrix} 3 \\ 1 \end{bmatrix}, \begin{bmatrix} 1 \\ 1 \end{bmatrix} \right\}$.

(60) $\begin{bmatrix} 4 & 1 \\ 0 & 4 \end{bmatrix}$ with basis $\left\{ \begin{bmatrix} 1 \\ 1 \end{bmatrix}, \begin{bmatrix} 0 \\ 1 \end{bmatrix} \right\}$.

(61) $\begin{bmatrix} 3 & 1 \\ 0 & 3 \end{bmatrix}$ with basis $\left\{ \begin{bmatrix} -1 \\ 1 \end{bmatrix}, \begin{bmatrix} 1 \\ 0 \end{bmatrix} \right\}$.

(62) $\begin{bmatrix} 2 & 0 \\ 0 & 3 \end{bmatrix}$ with basis $\left\{ \begin{bmatrix} 2 \\ 1 \end{bmatrix}, \begin{bmatrix} 1 \\ 1 \end{bmatrix} \right\}$.

(63) $\begin{bmatrix} 4 & 0 \\ 0 & 1 \end{bmatrix}$ with basis $\left\{ \begin{bmatrix} 1 \\ 2 \end{bmatrix}, \begin{bmatrix} 1 \\ -1 \end{bmatrix} \right\}$.

(64) Not similar to a real matrix in Jordan canonical form. If matrices are allowed to be complex, then $\begin{bmatrix} i & 0 \\ 0 & -i \end{bmatrix}$ with basis $\left\{ \begin{bmatrix} 2+i \\ 1 \end{bmatrix}, \begin{bmatrix} 2-i \\ 1 \end{bmatrix} \right\}$.

(65) $\begin{bmatrix} 3 & 0 & 0 \\ 0 & 6 & 1 \\ 0 & 0 & 6 \end{bmatrix}$ with baiss $\left\{ \begin{bmatrix} -1 \\ 1 \\ 1 \end{bmatrix}, \begin{bmatrix} 3 \\ 0 \\ -3 \end{bmatrix}, \begin{bmatrix} 0 \\ 0 \\ 1 \end{bmatrix} \right\}$.

(66) $\begin{bmatrix} 2 & 1 & 0 \\ 0 & 2 & 1 \\ 0 & 0 & 2 \end{bmatrix}$ with baiss $\left\{ \begin{bmatrix} 2 \\ 0 \\ 0 \end{bmatrix}, \begin{bmatrix} -2 \\ 1 \\ 0 \end{bmatrix}, \begin{bmatrix} 0 \\ 0 \\ 1 \end{bmatrix} \right\}$.

(67) $\begin{bmatrix} -2 & 0 & 0 \\ 0 & -2 & 0 \\ 0 & 0 & 2 \end{bmatrix}$ with baiss $\left\{ \begin{bmatrix} 1 \\ -2 \\ 0 \end{bmatrix}, \begin{bmatrix} 0 \\ -2 \\ 1 \end{bmatrix}, \begin{bmatrix} 1 \\ 0 \\ 1 \end{bmatrix} \right\}$.

(68) $\begin{bmatrix} 1 & 0 & 0 \\ 0 & 1 & 0 \\ 0 & 0 & 3 \end{bmatrix}$ with baiss $\left\{ \begin{bmatrix} 0 \\ 1 \\ 0 \end{bmatrix}, \begin{bmatrix} 1 \\ 0 \\ 1 \end{bmatrix}, \begin{bmatrix} 1 \\ 2 \\ -1 \end{bmatrix} \right\}$.

(69) $\begin{bmatrix} 0 & 0 & 0 \\ 0 & 2 & 1 \\ 0 & 0 & 2 \end{bmatrix}$ with basis $\left\{ \begin{bmatrix} 5 \\ -4 \\ 1 \end{bmatrix}, \begin{bmatrix} -1 \\ 0 \\ 1 \end{bmatrix}, \begin{bmatrix} 1 \\ 0 \\ 0 \end{bmatrix} \right\}$.

(70) $\begin{bmatrix} 0 & 0 & 0 \\ 0 & 0 & 0 \\ 0 & 0 & 14 \end{bmatrix}$ with basis $\left\{ \begin{bmatrix} 3 \\ 0 \\ -1 \end{bmatrix}, \begin{bmatrix} -1 \\ 5 \\ -3 \end{bmatrix}, \begin{bmatrix} 1 \\ 2 \\ 3 \end{bmatrix} \right\}$.

(71) $\begin{bmatrix} 3 & 0 & 0 & 0 \\ 0 & 3 & 1 & 0 \\ 0 & 0 & 3 & 1 \\ 0 & 0 & 0 & 3 \end{bmatrix}$ with basis $\left\{ \begin{bmatrix} 0 \\ -1 \\ 1 \\ -1 \end{bmatrix}, \begin{bmatrix} 1 \\ 0 \\ 0 \\ 0 \end{bmatrix}, \begin{bmatrix} -1 \\ 1 \\ 0 \\ 0 \end{bmatrix}, \begin{bmatrix} 0 \\ 0 \\ 0 \\ 1 \end{bmatrix} \right\}$.

(72) $\begin{bmatrix} 1 & 1 & 0 & 0 \\ 0 & 1 & 0 & 0 \\ 0 & 0 & 2 & 1 \\ 0 & 0 & 0 & 2 \end{bmatrix}$ with basis $\left\{ \begin{bmatrix} 1 \\ 0 \\ 0 \\ 0 \end{bmatrix}, \begin{bmatrix} 0 \\ 1 \\ 0 \\ 0 \end{bmatrix}, \begin{bmatrix} -1 \\ -1 \\ -1 \\ 0 \end{bmatrix}, \begin{bmatrix} 3 \\ 1 \\ 0 \\ -1 \end{bmatrix} \right\}$.

(73) $\begin{bmatrix} 1 & 0 & 0 & 0 & 0 & 0 & 0 \\ 0 & 1 & 0 & 0 & 0 & 0 & 0 \\ 0 & 0 & 1 & 1 & 0 & 0 & 0 \\ 0 & 0 & 0 & 1 & 1 & 0 & 0 \\ 0 & 0 & 0 & 0 & 1 & 0 & 0 \\ 0 & 0 & 0 & 0 & 0 & 1 & 1 \\ 0 & 0 & 0 & 0 & 0 & 0 & 1 \end{bmatrix}$ with basis

$$\left\{ \begin{bmatrix} 0 \\ 0 \\ -1 \\ -2 \\ 1 \\ 0 \\ 0 \end{bmatrix}, \begin{bmatrix} 1 \\ 3 \\ 1 \\ 0 \\ 0 \\ 1 \\ 0 \end{bmatrix}, \begin{bmatrix} -1 \\ 0 \\ 1 \\ -1 \\ 0 \\ 0 \\ 0 \end{bmatrix}, \begin{bmatrix} 0 \\ 0 \\ 1 \\ 0 \\ 0 \\ 0 \\ -1 \end{bmatrix}, \begin{bmatrix} 0 \\ 1 \\ 0 \\ 0 \\ 0 \\ 0 \\ 0 \end{bmatrix}, \begin{bmatrix} -2 \\ 0 \\ 2 \\ -2 \\ 0 \\ 0 \\ -1 \end{bmatrix}, \begin{bmatrix} 1 \\ 0 \\ 0 \\ 0 \\ 0 \\ 0 \\ 0 \end{bmatrix} \right\}.$$

(74) If $\mathbf{x}$ is a generalized eigenvector of type m corresponding to the eigenvalue λ, then $(\mathbf{A} - \lambda\mathbf{I})^m\mathbf{x} = \mathbf{0}$.

(75) Let $\mathbf{u}$ and $\mathbf{v}$ belong to $N\lambda(\mathbf{A})$. Then there exist nonnegative integers m and n such that $(\mathbf{A}-\lambda\mathbf{I})^m\mathbf{u}=\mathbf{0}$ and $(\mathbf{A}-\lambda\mathbf{I})^n\mathbf{v}=\mathbf{0}$. If $n \geq m$, then $(\mathbf{A}-\lambda\mathbf{I})^n\mathbf{u}=(\mathbf{A}-\lambda\mathbf{I})^{n-m}(\mathbf{A}-\lambda\mathbf{I})^m\mathbf{u}=(\mathbf{A}-\lambda\mathbf{I})^{n-m}\mathbf{0}=\mathbf{0}$. For any scalars α

and β, $(\mathbf{A}-\lambda\mathbf{I})^n(\alpha\mathbf{u}+\beta\mathbf{v})=\alpha[(\mathbf{A}-\lambda\mathbf{I})^n\mathbf{u}]+\beta[(\mathbf{A}-\lambda\mathbf{I})^n\mathbf{v})]=\alpha\mathbf{0}+\beta\mathbf{0}=\mathbf{0}$. The reasoning is similar if $m>n$.

(76) $(\mathbf{A}-\lambda\mathbf{I})_n$ is an nth degree polynomial in **A**, and **A** commutes with every polynomial in **A**.

(77) If $(\mathbf{A}-\lambda\mathbf{I})^k\mathbf{x}=0$, then $(\mathbf{A}-\lambda\mathbf{I})^k(\mathbf{A}\mathbf{x})=A[(\mathbf{A}-\boldsymbol{\lambda}\mathbf{I})^k\mathbf{x}]=\mathbf{A}\mathbf{0}=\mathbf{0}$.

(78) If this was not so, then there exists a vector $\mathbf{x}\in\mathbb{R}^n$ such that $(\mathbf{A}-\boldsymbol{\lambda}\mathbf{I})^k=\mathbf{0}$ and $(\mathbf{A}-\boldsymbol{\lambda}\mathbf{I})^{k-1}\neq\mathbf{0}$ with $k>n$. Therefore, **x** is a generalized eigenvector of type k with $k>n$. The chain propagated by **x** is a linearly independent set of k vectors in $\mathbb{R}^n$ with $k>n$. This contradicts Theorem 3 of Section 2.5.

Appendix B

(1) Matrix (a) can be a transition matrix. The other matrices are not transition matrices because: (b) Second column sum is less than unity. (c) Both column sums are greater than unity. (d) Matrix contains a negative element. (e) Third column sum is less than unity. (f) Third column sum is greater than unity. (g) None of the column sums is unity. (h) Matrix contains negative elements.

(2) $\begin{bmatrix}0.95 & 0.01\\ 0.05 & 0.99\end{bmatrix}$.

(3) $\begin{bmatrix}0.6 & 0.7\\ 0.4 & 0.3\end{bmatrix}$.

(4) $\begin{bmatrix}0.10 & 0.20 & 0.25\\ 0.50 & 0.60 & 0.65\\ 0.40 & 0.20 & 0.10\end{bmatrix}$.

(5) $\begin{bmatrix}0.80 & 0.10 & 0.25\\ 0.15 & 0.88 & 0.30\\ 0.05 & 0.02 & 0.45\end{bmatrix}$.

(6) (a) $\mathbf{P}^2=\begin{bmatrix}0.37 & 0.63\\ 0.28 & 0.72\end{bmatrix}$ and $\mathbf{P}^3=\begin{bmatrix}0.289 & 0.316\\ 0.711 & 0.684\end{bmatrix}$,
(b) 0.37,
(c) 0.63,
(d) 0.711,
(e) 0.684.

(7) $1\to1\to1\to1$, $1\to1\to2\to1$, $1\to2\to1\to1$, $1\to2\to2\to1$.

(8) (a) 0.097,
(b) 0.0194.

(9) (a) 0.64,
(b) 0.636.

(10) (a) 0.1,
(b) 0.21.

(11) (a) 0.6675,
(b) 0.577075,
(c) 0.267.

(12) (a) There is a 0.6 probability that an individual chosen at random initially will live in the city; thus, 60% of the population initially lives in the city, while 40% lives in the suburbs.
(b) $\mathbf{d}^{(1)} = [0.574 \ 0.426]^T$,
(c) $\mathbf{d}^{(2)} = [0.54956 \ 0.45044]^T$.

(13) (a) 40% of customers now use brand X, 50% use brand Y, and 10% use other brands.
(b) $\mathbf{d}^{(1)} = [0.395 \ 0.490 \ 0.075]^T$,
(c) $\mathbf{d}^{(2)} = [0.3875 \ 0.47343 \ 0.06330]^T$.

(14) (a) $\mathbf{d}^{(0)} = [0 \ 1]^T$,
(b) $\mathbf{d}^{(1)} = [0.7 \ 0.3]^T$.

(15) (a) $\mathbf{d}^{(0)} = [0 \ 1 \ 0]^T$,
(b) $\mathbf{d}^{(3)} = [0.192 \ 0.592 \ 0.216]^T$. There is a probability of 0.216 that the harvest will be good in three years.

(16) (a) $[1/6 \ 5/6]^T$,
(b) 1/6.

(17) $[7/11 \quad 4/11]^T$; probability of having a Republican is $7/11 \approx 0.636$.

(18) $[23/120 \quad 71/120 \quad 26/120]^T$; probability of a good harvest is $26/120 \approx 0.217$.

(19) $[40/111 \quad 65/111 \quad 6/111]^T$; probability of a person using brand Y is $65/111 = 0.586$.

Appendix C

(1) (b) $\begin{bmatrix} 0 & 1 & 1 \\ 1 & 0 & 1 \\ 1 & 1 & 0 \end{bmatrix}$. (c) $-1, -1, 2$

(2) (b) $\begin{bmatrix} 0 & 1 & 1 & 1 \\ 1 & 0 & 1 & 1 \\ 1 & 1 & 0 & 1 \\ 1 & 1 & 1 & 0 \end{bmatrix}$. (c) $-1, -1, -1, 3$

(3) (b) $\begin{bmatrix} 0 & 1 & 1 & 1 & 1 \\ 1 & 0 & 1 & 1 & 1 \\ 1 & 1 & 0 & 1 & 1 \\ 1 & 1 & 1 & 0 & 1 \\ 1 & 1 & 1 & 1 & 0 \end{bmatrix}$. (c) $-1, -1, -1, -1, 4$

(4) The adjacency eigenvalues of K_n are $n-1$ copies of -1 and one copy of $n-1$.

(5) $t(G) = \frac{1}{n}(n-1-(-1))^{n-1} = n^{n-2}$, , which is Cayley's Theorem.

(6) (b) 5 copies of 1, 4 copies of -2, 1 copy of 3,
(c) an eigenvalue of 3 but none of -3 so not bipartite,
(d) not paired
(e) 2000 spanning trees
(f) 0,2,2,2,2,2,5,5,5,5
(g) 2000

(7) (b) complement is two disjoint K_4 graphs,
(c) 8 with multiplicity 1, 4 with multiplicity 6, 0 with multiplicity 1

(8) (b) complement is disjoint K_5 and K_4 graphs
(c) 9 with multiplicity 1, 5 with multiplicity 3, 4 with multiplicity 4, 0 with multiplicity 1

(9) (b) complement is disjoint K_6 and K_4 graphs,
(c) 10 with multiplicity 1, 6 with multiplicity 3, 4 with multiplicity 5, 0 with multiplicity 1

(10) (b) complement is two disjoint K_5 graphs,
(c) 10 with multiplicity 1, 5 with multiplicity 8, 0 with multiplicity 1

(11) (b) complement is disjoint K_5 and K_7 graphs,
(c) 12 with multiplicity 1, 7 with multiplicity 4, 5 with multiplicity 6, 0 with multiplicity 1

(12) (a) For K_{pq} has eigenvalues 0 with multiplicity 1, q with multiplicity $p-1$, p with multiplicity $q-1$, $p+q$ with multiplicity 1,
(b) $t(G) = \frac{1}{p+q}(q)^{p-1}(p)^{q-1}(p+q) = (q)^{p-1}(p)^{q-1}$

(13) (b) complement is three disjoint K_3 graphs,
(c) 9 with multiplicity 2, 6 with multiplicity 6, 0 with multiplicity 1

(14) (a) complement is three disjoint K_4 graphs,
(b) 12 with multiplicity 2, 8 with multiplicity 9, 0 with multiplicity 1

(15) (a) complement is three disjoint K_5 graphs,
(b) 15 with multiplicity 2, 10 with multiplicity 12, 0 with multiplicity 1

(16) (a) complement is three disjoint K_6 graphs,
(b) 18 with multiplicity 2, 12 with multiplicity 15, 0 with multiplicity 1,

(17) (a) complement is three disjoint K_r graphs, graph has eigenvalues $3r$ with multiplicity 2, $2r$ with multiplicity $3(r\text{-}1)$, 0 with multiplicity 1,
(b) $t(G) = \frac{1}{3r}(3r)^2(2r)^{3(r-1)}$

Appendix D

(1) $x = 30$ x *model* bicycles; $y = 20$ y *model* bicycles; $P = \$410$.

(2) $x = 35$ x *model* bicycles; $y = 0$ y *model* bicycles; $P = \$3500$.

(3) $x = 120$ x *model* bicycles; $y = 120$ y *model* bicycles; $P = \$2640$.

Appendix E

(1) $(1+2+\cdots+n)+(n+1) = n(n+1)/2+(n+1) = (n+1)(n+2)/2$.

(2) $[1+3+5+\ldots+(2\mathbf{n}-1)]+(2\mathbf{n}+1) = \mathbf{n}^2+(2\mathbf{n}+1) = (\mathbf{n}+1)^2$.

(3)
$$\begin{aligned}&(1^2+2^2+\ldots+n^2)+(n+1)^2\\ &\quad = n(n+1)(2n+1)/6+(n+1)^2\\ &\quad = (n+1)[n(2n+1)/6+(n+1)]\\ &\quad = (n+1)[2n^2+7n+6]/6\\ &\quad = (n+1)(n+2)(2n+3)/6\end{aligned}$$

(4)
$$\begin{aligned}&(1^3+2^3+\ldots+n^3)+(n+1)^3\\ &\quad = n^2(n+1)^2/4+(n+1)^3\\ &\quad = (n+1)^2[n^2/4+(n+1)]\\ &\quad = (n+1)^2(n+2)^2/4.\end{aligned}$$

(5)
$$\begin{aligned}&\left[1^2+3^2+5^2+\ldots+(2n-1)^2\right]+(2n+1)^2\\ &\quad = n(4n^2-1)/3+(2n+1)^2\\ &\quad = n(2n-1)(2n+1)/3+(2n+1)^2\\ &\quad = (2n+1)[n(2n-1)/3+(2n+1)]\\ &\quad = (2n+1)(2n+3)(n+1)/3\\ &\quad = [2(n+1)-1][2(n+1)+1](n+1)/3\\ &\quad = \left[4(n+1)^2-1\right](n+1)/3\end{aligned}$$

(6)
$$\begin{aligned}\sum_{k=1}^{n+1}(3k^2-k) &= \sum_{k=1}^{n}(3k^2-k)+\left[3(n+1)^2-(n+1)\right]\\ &= n^2(n+1)+\left[3(n+1)^2+(n+1)\right]\\ &= (n+1)[n^2+3(n+1)+1]\\ &= (n+1)(n+2)(n+1)\\ &= (n+1)^2(n+2)\end{aligned}$$

(7) $$\sum_{k=1}^{n+1}\frac{1}{k(k+1)}$$
$$=\sum_{k=1}^{n}\frac{1}{k(k+1)}+\frac{1}{(n+1)(n+2)}$$
$$=\frac{n}{n+1}+\frac{1}{(n+1)(n+2)}$$
$$=\frac{n^2+2n+1}{(n+1)(n+2)}$$
$$=\frac{n+1}{n+2}$$

(8) $$\sum_{k=1}^{n+1}2^{k-1}=\sum_{k=1}^{n}2^{k-1}+2^n=[2^n-1]+2^n=2(2^n)-1=2^{n+1}-1.$$

(9) $$\sum_{k=1}^{n+1}x^{k-1}=\sum_{k=1}^{n}x^{k-1}+x^n=\frac{x^n-1}{x-1}+x^n$$
$$=\frac{x^n-1+x^n(x-1)}{x-1}=\frac{x^{n+1}-1}{x-1}.$$

(10) $7^{n+1}+2=7^n(6+1)+2=6(7^n)+(7^n+1).6(7^n)$ is a multiple of 3 because 6 is, and (7^n+1) is a multiple of 3 by the induction hypothesis.

Index

Note: Page numbers followed by *f* indicate figures and *t* indicate tables.